Acta Numerica 2006

Acta Numerica

Volume 15   2006

CAMBRIDGE UNIVERSITY PRESS

CAMBRIDGE UNIVERSITY PRESS
Cambridge, New York, Melbourne, Madrid, Cape Town,
Singapore, São Paulo, Delhi, Tokyo, Mexico City

Cambridge University Press
The Edinburgh Building, Cambridge CB2 8RU, UK

Published in the United States of America by Cambridge University Press, New York

www.cambridge.org
Information on this title: www.cambridge.org/9780521174336

First published 2006
First paperback edition 2011

A catalogue record for this publication is available from the British Library

ISBN 978-0-521-86815-0 Hardback
ISBN 978-0-521-17433-6 Paperback

Contents

Acta Numerica (2006), pp. 1–155
doi: 10.1017/S0962492906210018

Finite element exterior calculus, homological techniques, and applications

Douglas N. Arnold[*]
*Institute for Mathematics and its Applications
and School of Mathematics,
University of Minnesota, Minneapolis, MN 55455, USA*
E-mail: `arnold@ima.umn.edu`

Richard S. Falk[†]
*Department of Mathematics,
Rutgers University, Piscataway, NJ 08854, USA*
E-mail: `falk@math.rutgers.edu`

Ragnar Winther[‡]
*Centre of Mathematics for Applications
and Department of Informatics,
University of Oslo, PO Box 1053, 0316 Oslo, Norway*
E-mail: `ragnar.winther@cma.uio.no`

Dedicated to Carme, Rena, and Rita

Finite element exterior calculus is an approach to the design and understanding of finite element discretizations for a wide variety of systems of partial differential equations. This approach brings to bear tools from differential geometry, algebraic topology, and homological algebra to develop discretizations which are compatible with the geometric, topological, and algebraic structures which underlie well-posedness of the PDE problem being solved. In the finite element exterior calculus, many finite element spaces are revealed as spaces of piecewise polynomial differential forms. These connect to each other in discrete subcomplexes of elliptic differential complexes, and are also related to the continuous elliptic complex through projections which commute with the complex differential. Applications are made to the finite element discretization of a variety of problems, including the Hodge Laplacian, Maxwell's equations, the equations of elasticity, and elliptic eigenvalue problems, and also to preconditioners.

[*] Supported in part by NSF grant DMS-0411388.
[†] Supported in part by NSF grant DMS03-08347.
[‡] Supported by the Norwegian Research Council.

CONTENTS

1. Introduction

The finite element method is one of the greatest advances in numerical computing of the past century. It has become an indispensable tool for simulation of a wide variety of phenomena arising in science and engineering. A tremendous asset of finite elements is that they not only provide a methodology to develop numerical algorithms for simulation, but also a theoretical framework in which to assess the accuracy of the computed solutions. In this paper we survey and develop the finite element exterior calculus, a new theoretical approach to the design and understanding of finite element discretizations for a wide variety of systems of partial differential equations. This approach brings to bear tools from differential geometry, algebraic topology, and homological algebra to develop discretizations which are compatible with the geometric, topological, and algebraic structures which underlie well-posedness of the PDE problem being solved. Applications treated here include the finite element discretization of the Hodge Laplacian (which includes many problems as particular cases), Maxwell's equations of electromagnetism, the equations of elasticity, and elliptic eigenvalue problems, and the construction of preconditioners.

To design a finite element method to solve a problem in partial differential equations, the problem is first given a weak or variational formulation which characterizes the solution among all elements of a given space of functions on the domain of interest. A finite element method for this problem proceeds with the construction of a finite-dimensional subspace of the given function space where the solution is sought, and then the specification of a unique element of this subspace as the solution of an appropriate set of equations on this finite-dimensional space. In finite element methods, the subspace is constructed from a triangulation or simplicial decomposition of the given domain, using spaces of polynomials on each simplex, pieced together by a certain assembly process. Because the finite element space so constructed is a subspace of the space where the exact solution is sought, one can consider the difference between the exact and finite element solution and measure it via appropriate norms, seminorms, or functionals. Generally speaking, error bounds can be obtained in terms of three quantities: the approximation error, which measures the error in the best approximation of the exact solution possible within the finite element space, the consistency error, which measures the extent to which the equations used to select the finite element solution from the finite element space reflect the continuous problem, and the stability constant, which measures the well-posedness of the finite-dimensional problem. The approximation properties of finite element spaces are well understood, and the consistency of finite element methods is usually easy to control (in fact, for all the methods considered in this paper there is no consistency error in the sense that the exact solution will satisfy the natural extension of the finite element equations to solution space). In marked contrast, the stability of finite element procedures can be very subtle. For many important problems, the development of stable finite element methods remains extremely challenging or even out of reach, and in other cases it is difficult to assess the stability of methods of interest. Lack of stable methods not only puts some important problems beyond the reach of simulation, but has also led to spectacular and costly failures of numerical methods.

It should not be surprising that stability is a subtle matter. Establishing stability means proving the well-posedness of the discrete equations, uniformly in the discretization parameters. Proving the well-posedness of PDE problems is, of course, the central problem of the theory of PDEs. While there are PDE problems for which this is a simple matter, for many important problems it lies deep, and a great deal of mathematics, including analysis, geometry, topology, and algebra, has been developed to establish the well-posedness of various PDE problems. So it is to be expected that a great deal of mathematics bears as well on the stability of numerical methods for PDE. An important but insufficiently appreciated point is that *approximability and consistency together with well-posedness of the continuous problem do not imply stability.* For example, one may consider a PDE problem whose solution is characterized by a variational principle, *i.e.*, as the unique critical point of some functional on some function space, and define a finite element method by seeking a critical point of the same functional (so there is no consistency error) on a highly accurate finite element space (based on small elements and/or high-order polynomials, so there is arbitrary low approximability error), and yet such a method will very often be unstable and therefore not convergent. Analogously, in the finite difference methodology, one may start with a PDE problem stated in strong form and replace the derivatives in the equation by consistent finite differences, and yet obtain a finite difference method which is unstable.

As mentioned, the well-posedness of many PDE problems reflects geometrical, algebraic, topological, and homological structures underlying the problem, formalized by exterior calculus, Hodge theory, and homological algebra. In recent years there has been a growing realization that stability of numerical methods can be obtained by designing methods which are *compatible* with these structures in the sense that they reproduce or mimic them (not just approximate them) on the discrete level. See, for example, Arnold (2002), and the volume edited by Arnold, Bochev, Lehoucq, Nicolaides and Shashkov (2006a). In the present paper, the compatibility is mostly related to elliptic complexes which are associated with the PDEs under consideration, mostly the de Rham complex and its variants and another complex associated with the equations of linear elasticity. Our finite element spaces will arise as the spaces in finite-dimensional subcomplexes which inherit the cohomology and other features of the exact complexes. The inheritance will generally be established by cochain projections: projection operators from the infinite-dimensional spaces in the original elliptic complex which map onto the finite element subspaces and commute with the differential operators of the complex. Thus the main theme of the paper is the development of finite element subcomplexes of certain elliptic differential complexes and cochain projections onto them, and their implications and applications in numerical PDEs. We refer to this theme and the mathematical framework we construct to carry it out, as *finite element exterior calculus*.

We mention some of the computational challenges which motivated the development of finite element exterior calculus and which it has helped to address successfully. These are challenges both of understanding the poor behaviour of seemingly reasonable numerical methods, and of developing effective methods. In each case, the finite element exterior calculus provides an explanation for the difficulties experienced with naive methods, and also points to a practical finite element solution.

- The system $\sigma = \operatorname{grad} u$, $\operatorname{div} u = f$ (the Poisson equation written in first-order form) is among the simplest and most basic PDEs. But even in one dimension, the stability of finite element discretizations for it are hard to predict. For example, the use of classical continuous piecewise linear elements for both σ and u is unstable. See Arnold, Falk and Winther (2006b) for a discussion and numerical examples.

This is one of many problems handled in a unified manner by the finite element exterior calculus approach to the Hodge Laplacian given in Section 7.

- A standard variational form for the vector Poisson problem is to minimize the energy $\int[(|\operatorname{div} u|^2 + |\operatorname{curl} u|^2)/2 - fu]dx$ over vector fields satisfying the boundary condition, say, $u \cdot n = 0$ on the boundary. This problem is well-posed, but a standard finite element method, which proceeds by minimizing the energy over a piecewise polynomial subspace of the energy space, will inevitably *converge to the wrong solution* on some domains, *e.g.*, nonconvex polygons or polyhedra. See the end of Section 2.2 for more on this.

- With naive choices of finite element spaces, the computed spectra of some elliptic eigenvalue problems related to the vector Laplacian or Maxwell's equation bear no relation to the true spectrum, or include spurious eigenvalues that converge to a point outside the true spectrum as the mesh is refined. See, *e.g.*, Boffi (2006) for more on this and numerical examples. Application of the finite element exterior calculus makes the computation and numerical analysis of such eigenvalue problems straightforward, as explained in Section 8.

- The equations of elasticity are classically among the most important applications of the finite element method. A natural variational formulation for them is the Hellinger–Reissner principle. Despite four decades of searching, stable finite elements for this variational principle have proved elusive. The first such using polynomial shape functions were presented in Arnold and Winther (2002) for plane elasticity, using the ideas of finite element exterior calculus and spurring further development. New families of stable elements for elasticity in arbitrary dimensions are presented in Section 11.

- Multigrid methods based on the additive Schwartz smoothers used for standard finite element methods do not work when applied to some other problems involving the $H(\operatorname{curl})$-inner product. But a different choice of the additive Schwartz smoother does lead to effective multigrid solvers. See Section 10.

The finite element exterior calculus is not the only compatible discretization approach in active development. It certainly has motivations and features in common with mimetic finite difference methods (Bochev and Hyman 2006), covolume methods (Nicolaides and Trapp 2006), and the discrete exterior calculus (Desbrun, Hirani, Leok and Marsden 2005). Moreover, there are coincidences of methods in simple cases. However, there are also major differences. In mimetic finite differences and the discrete exterior calculus, the

fundamental object is a simplicial cochain, a function assigning a number to
each simplex of dimension k of the simplicial complex. There is, of course,
a close relation between differential forms and simplicial cochains. A differential form of degree k can be integrated over a simplex of dimension k
to give a number, and de Rham's theorem states that this map from the
de Rham complex to the simplicial cochain complex induces an isomorphism
on cohomology. It is less obvious how to go from a cochain to a differential
form. An answer was provided by Whitney, who constructed a one-sided
inverse of the de Rham map by associating to any k-cochain a piecewise linear differential k-form. Bossavit (1988) recognized that the Whitney forms
coincided with low-order finite element spaces that had been developed for
electromagnetism. In view of all this, there is a very close relationship between mimetic finite differences and discrete exterior calculus, on the one
hand, and the Whitney form complex of the finite element exterior calculus. But the finite element exterior calculus described in this paper involves
much more than the Whitney forms, namely two families of finite element
spaces of differential forms, $\mathcal{P}_r\Lambda^k(\mathcal{T}_h)$ and $\mathcal{P}_r^-\Lambda^k(\mathcal{T}_h)$, with a rich structure
of interconnections between them. Of these, the spaces $\mathcal{P}_1^-\Lambda^k(\mathcal{T}_h)$ are the
Whitney forms and isomorphic to simplicial cochains, while the others do
not naturally fit in the simplicial cochain formalism. As pointed out at the
start of this Introduction, a great strength of finite element methods is that
the discrete solution belongs to the same function space as the exact solution and so comparison is natural. For this reason we view the realization
of finite elements as differential forms, rather than as discrete objects which
mimic differential forms, as highly desirable.

Much of the foundation of finite element exterior calculus was developed
by many people over a long period of time. Besides the work of Whitney
and Bossavit already mentioned, we signal the contributions of Hiptmair
(1999a, 2001, 2002), especially to the aspects of the theory that are relevant
for electromagnetic problems. Interest in the subject grew with the 2002
presentation at the International Congress of Mathematicians (Arnold 2002)
where the strong connection between numerical stability and cochain projections of elliptic complexes was first emphasized. Many other references will
be made throughout the paper, but the literature is too large to reference
anywhere near all the relevant work.

This paper is more than a survey of an existing theory. We also present a
number of results here which either are new, or appeared only recently in our
other work. For example, we emphasize the relevance of the Koszul complex
in constructing finite element subspaces of spaces of differential forms and
determining their properties. The Koszul differential leads to the family of
finite element spaces $\mathcal{P}_r^-\Lambda^k(\mathcal{T}_h)$, including the Whitney forms ($r = 1$).

The two families of spaces $\mathcal{P}_r\Lambda^k(\mathcal{T}_h)$ and $\mathcal{P}_r^-\Lambda^k(\mathcal{T}_h)$ include among them,
for special values of r and k, the Lagrange finite elements and most of the

stable finite element spaces that have been used to solve mixed formulations of the Poisson or Maxwell's equations. A new aspect of this paper is the development of bases and degrees of freedom for all these spaces in a unified fashion that is not possible when the spaces are studied in isolation from each other. In particular, the degrees of freedom we present are in some cases more natural than the ones that have appeared previously in the literature.

As alluded to, a key feature of our approach is the arrangement of these spaces of finite element differential forms into finite-dimensional subcomplexes of the de Rham complex. In fact, we show that for each polynomial degree there are 2^{n-1} such subcomplexes in n dimensions which reproduce the same cohomology. Some of these complexes had been studied before, especially the celebrated complex of Whitney forms and its higher degree generalizations, but many are new. Some of these new complexes turn out to be essential for the discretization of the elasticity system.

Another direction which is new to this paper is the use of group representation theory to characterize the spaces of polynomial differential forms $\mathcal{P}_r \Lambda^k(\mathbb{R}^n)$ and $\mathcal{P}_r^- \Lambda^k(\mathbb{R}^n)$ which are used to construct the finite element differential forms. In Section 3.4, we show that these spaces are nearly the only ones with a certain affine invariance property.

The development of finite element methods with weakly imposed symmetry for the elasticity equations as presented here appeared only recently in our work. A key tool, not previously used in numerical analysis, is the BGG resolution. The presentation here is more general and simpler than the previous versions, and the treatment of traction boundary conditions first appears here.

A great strength of the exterior calculus approach is the way it unifies seemingly different problems in a common framework. A familiar example is the unification of the three main first-order differential operators of vector calculus, grad, div, and curl, as a single operator, the exterior derivative, d. In a similar way, the Hodge Laplacian, $d\delta + \delta d$, where δ is the formal adjoint of d, unifies many important second-order differential operators. By studying the numerical solution of source and eigenvalue properties and preconditioners for the Hodge Laplacian in generality, we simultaneously treat many different problems, including the standard finite element methods with Lagrange elements for the scalar Laplacian, mixed finite elements for the Laplacian, different mixed finite element formulations for the vector Laplacian, curl curl problems arising in electromagnetics in different formulations, and div-curl systems.

We also mention some more technical aspects of the current presentation that distinguish it from others that appear in the literature. We treat domains of full topological generality, that is, with arbitrary Betti numbers. Thus our goal is not to show that the finite element de Rham subcomplexes we construct are exact, but rather that they reproduce the de Rham cohomology of the domain. When we consider boundary value problems for the Hodge Laplacian, we must take into account the harmonic forms, and so well-posedness depends on a finite number of compatibility conditions of the data and uniqueness requires a finite number of side conditions, and we reproduce this situation on the discrete level.

Another technical aspect of the presentation is that we use not only the canonical commuting projection operators into the finite element spaces, but also smoothed variants of these, which still commute with the exterior derivative and which have better boundedness properties. These allow simpler and cleaner analysis in many places. The construction of such bounded cochain projections is a recent and active subject of research. We present an approach here inspired by Schöberl (2005) and Christiansen (2005), while recognizing that it is very possibly not the last word in this matter.

The contents of the paper are as follows. In the first part of the paper, we develop the finite element exterior calculus, starting with a review of the necessary exterior algebra, exterior calculus, and Hodge theory. We then turn to polynomial differential forms which are intimately related to the Koszul complex. Here we develop the polynomial spaces $\mathcal{P}_r\Lambda^k(\mathbb{R}^n)$ and $\mathcal{P}_r^-\Lambda^k(\mathbb{R}^n)$ and find bases and degrees of freedom for them, and also characterize them through affine invariance. We then assemble these spaces into finite element differential forms, which are the main objects of interest, and derive the finite element de Rham subcomplexes and cochain projections which are the key tools of finite element exterior calculus. With these we easily present the basic finite element discrete Hodge theory.

In the second part of the paper, we apply these tools to concrete problems: discretization of the Hodge Laplacian, eigenvalue problems, Maxwell's equations, and preconditioning. The final application, and the most substantial, is to mixed discretizations of elasticity. Here we use some additional tools: vector-valued differential forms and the BGG resolution.

PART ONE

Exterior calculus, finite elements, and homology

2. Exterior algebra and exterior calculus

In this section we recall the basic objects and results of exterior algebra and exterior calculus. This material can be found, in varying degrees of generality and in varying notation, in many sources, including Arnold (1978), Bott and Tu (1982), Federer (1969), Jänich (2001), Lang (1995), and Taylor (1996).

2.1. Exterior algebra

Alternating algebraic forms on a vector space
Let V be a real vector space of dimension n. For each positive integer k, we denote by $\mathrm{Alt}^k V$ the space of alternating k-linear maps $V \times \cdots \times V \to \mathbb{R}$. We refer to such maps as *alternating algebraic k-forms on V* or simply algebraic k-forms. Thus, an algebraic k-form on V assigns to a k-tuple $(v_1, \ldots, v_k)$ of elements of V a real number $\omega(v_1, \ldots, v_k)$, with the mapping linear in each argument, and reversing sign when any two arguments are interchanged. It is natural to set $\mathrm{Alt}^0 V = \mathbb{R}$. Note that $\mathrm{Alt}^1 V$ is just the dual space V^* of V, the space of covectors.

Exterior product
Given $\omega \in \mathrm{Alt}^j V$ and $\eta \in \mathrm{Alt}^k V$, we define their *exterior product* or *wedge product* $\omega \wedge \eta \in \mathrm{Alt}^{j+k} V$ by

$$(\omega \wedge \eta)(v_1, \ldots, v_{j+k})$$
$$= \sum_\sigma (\mathrm{sign}\,\sigma)\omega(v_{\sigma(1)}, \ldots, v_{\sigma(j)})\eta(v_{\sigma(j+1)}, \ldots, v_{\sigma(j+k)}), \quad v_i \in V, \quad (2.1)$$

where the sum is over all permutations σ of $\{1, \ldots, j+k\}$, for which $\sigma(1) < \sigma(2) < \cdots \sigma(j)$ and $\sigma(j+1) < \sigma(j+2) < \cdots \sigma(j+k)$. The exterior product is bilinear and associative, and satisfies the anti-commutativity condition

$$\eta \wedge \omega = (-1)^{jk}\omega \wedge \eta, \quad \omega \in \mathrm{Alt}^j V, \ \eta \in \mathrm{Alt}^k V.$$

This can be summarized by the statement that the direct sum $\mathrm{Alt}\,V := \bigoplus_k \mathrm{Alt}^k V$ is a *anti-commutative graded algebra* called the Grassmann algebra or exterior algebra of V^*. (In the context of graded algebras, anti-commutativity is often referred to simply as commutativity.)

Pullback

A linear transformation of vector spaces $L : V \to W$ induces a linear transformation $L^* : \operatorname{Alt} W \to \operatorname{Alt} V$ called the pullback, given by

$$L^*\omega(v_1,\ldots,v_k) = \omega(Lv_1,\ldots,Lv_k), \quad \omega \in \operatorname{Alt}^k W, \quad v_1,\ldots,v_k \in V.$$

The pullback acts contravariantly: if $U \xrightarrow{K} V \xrightarrow{L} W$, then $\operatorname{Alt} W \xrightarrow{L^*} \operatorname{Alt} V \xrightarrow{K^*} \operatorname{Alt} U$, and $K^* \circ L^* = (L \circ K)^*$. The pullback also respects the algebra structure in the sense that $L^*(\omega \wedge \eta) = L^*\omega \wedge L^*\eta$.

A particular case is when L is the inclusion i_V of a subspace V into W. Then the pullback defines a surjection i_V^* of $\operatorname{Alt} W$ onto $\operatorname{Alt} V$. If in addition W has an inner product, so that the orthogonal projection $\pi_V : W \to V$ is defined, then its pullback defines an injection of $\operatorname{Alt} V$ into $\operatorname{Alt} W$ and the pullback of the composition $W \xrightarrow{\pi_V} V \xrightarrow{i} W$ associates to each $\omega \in \operatorname{Alt}^k W$ its *tangential part* with respect to V, given by

$$(\pi_V^* i_V^* \omega)(v_1,\ldots,v_k) = \omega(\pi_V v_1,\ldots,\pi_V v_k).$$

Thus the tangential part of ω vanishes if and only the image of ω in $\operatorname{Alt}^k V$ vanishes. We may also speak of the *normal part* of ω with respect to V, defined to be $\omega - \pi_V^* i_V^* \omega$. (For $k > 1$, this is not generally the same as the tangential part with respect to the orthogonal complement of V.)

Bases

Let $v_1,\ldots,v_n$ be some basis of V. Then an algebraic k-form ω is uniquely determined by its values $\omega(v_{\sigma(1)},\ldots,v_{\sigma(k)})$ for each increasing map $\sigma : \{1,\ldots,k\} \to \{1,\ldots,n\}$, and these values may be assigned arbitrarily. Thus

$$\dim \operatorname{Alt}^k V = \binom{n}{k},$$

and, in particular, $\operatorname{Alt}^k V = 0$ for $k > n$.

Associated to the given basis is the dual basis $\mu_1,\ldots,\mu_n$ for V^*, defined by $\mu_i(v_j) = \delta_{ij}$. For $\sigma,\rho : \{1,\ldots,k\} \to \{1,\ldots,n\}$ increasing, we have

$$\mu_{\sigma(1)} \wedge \cdots \wedge \mu_{\sigma(k)}(v_{\rho(1)},\ldots,v_{\rho(k)}) = \begin{cases} 1, & \text{if } \sigma = \rho, \\ 0, & \text{otherwise,} \end{cases}$$

so the $\binom{n}{k}$ algebraic k-forms $\mu_{\sigma(1)} \wedge \cdots \wedge \mu_{\sigma(k)}$ form a basis for $\operatorname{Alt}^k V$ naturally associated to the given basis of V.

Interior product

To an algebraic k-form ω and a vector $v \in V$, we may associate an algebraic $(k-1)$-form $\omega \lrcorner v$ called the interior product, or the contraction of ω by v, defined by

$$\omega \lrcorner v(v_1,\ldots,v_{k-1}) = \omega(v,v_1,\ldots,v_{k-1}).$$

(We take $\omega \lrcorner v$ to be 0 if ω is an algebraic 0-form.) Since the forms are alternating, repeated contraction vanishes:

$$(\omega \lrcorner v) \lrcorner v = 0, \quad \omega \in \mathrm{Alt}^k V.$$

Moreover, it is easy to check that

$$(\omega \wedge \eta) \lrcorner v = (\omega \lrcorner v) \wedge \eta + (-1)^k \omega \wedge (\eta \lrcorner v), \quad \omega \in \mathrm{Alt}^k V, \ \eta \in \mathrm{Alt}^l V.$$

Inner product, orientation, and volume form

If the vector space V is endowed with an inner product, then $\mathrm{Alt}^k V$ is naturally endowed with an inner product by the formula

$$\langle \omega, \eta \rangle = \sum_\sigma \omega(v_{\sigma(1)}, \ldots, v_{\sigma(k)}) \eta(v_{\sigma(1)}, \ldots, v_{\sigma(k)}), \quad \omega, \eta \in \mathrm{Alt}^k V, \quad (2.2)$$

where the sum is over increasing sequences $\sigma : \{1, \ldots, k\} \to \{1, \ldots, n\}$ and $v_1, \ldots, v_n$ is any *orthonormal* basis (the right-hand side being independent of the choice of orthonormal basis). The space $\mathrm{Alt}^n V$, $n = \dim V$, has dimension 1, and we uniquely determine an element in it by giving its value on a single ordered basis. Its value on any list of n vectors is then the same value multiplied by the determinant of the matrix expressing the vectors in terms of the specified basis. In particular, we determine an algebraic n-form, unique up to sign, by requiring it to take the value 1 on some orthonormal basis. It will then take the values ± 1 on all orthonormal bases. We fix the sign by *orienting* the vector space, *i.e.*, by designating one ordered basis as positively oriented (and then all bases will be positively or negatively oriented according to the sign of the determinant of the change-of-basis matrix). The resulting uniquely determined algebraic n-form is called the *volume form* on the oriented vector space V.

Bivectors

When we treat the equations of elasticity, we will make use of the space $V \wedge V$ of bivectors, which we now define. (There is an entire algebra of multivectors, in analogy to the exterior algebra of multi-covectors, but we shall not have need of this.)

For v and w elements of a vector space V, define $v \wedge w = v \otimes w - w \otimes v \in V \otimes V$, and let $V \wedge V$ denote the subspace of $V \otimes V$ spanned by elements of the form $v \wedge w$. If $v_1, \ldots, v_n$ denotes a basis of V, then the vectors $v_i \wedge v_j$, $i < j$, form a basis for $V \wedge V$, and so $\dim V \wedge V = \binom{n}{2}$. The mapping $v \otimes w \mapsto (v \wedge w)/2$ defines a linear operator $\mathrm{skw} : V \otimes V \to V \wedge V$.

The space $V \otimes V$ is identified with $\mathcal{L}(V^*, V)$, the space of linear operators from V^* to V, by $(v \otimes w)(f) = f(w)v$. If we assume that V has an inner product, then V^* is identified with V, and so $V \otimes V$ with $\mathcal{L}(V, V)$. The subspace $V \wedge V$ is then the subspace of linear operators which are skew-symmetric with respect to the inner product and skw is the map which takes a linear operator to its skew-symmetric part.

The Hodge star operation

Let V be an oriented inner product space with volume form vol. Given $\omega \in \mathrm{Alt}^k V$, we obtain a linear map $L_\omega : \mathrm{Alt}^{n-k} V \to \mathbb{R}$, by composing the map $\mu \mapsto \omega \wedge \mu$ with the canonical isomorphism of $\Lambda^n V$ onto $\mathbb{R}$ (given by $c\,\mathrm{vol} \mapsto c$). By the Riesz representation theorem, there exists an element $\star\omega \in \mathrm{Alt}^{n-k} V$ such that $L_\omega(\mu) = \langle \star\omega, \mu \rangle$. In other words,

$$\omega \wedge \mu = \langle \star\omega, \mu \rangle \mathrm{vol}, \quad \omega \in \mathrm{Alt}^k V, \ \mu \in \mathrm{Alt}^{n-k} V.$$

The linear map $\omega \mapsto \star\omega$ taking $\mathrm{Alt}^k V$ into $\mathrm{Alt}^{n-k} V$ is the Hodge star operator; there is one for each k with $0 \le k \le n$. From the definition we find that

$$\omega(e_{\sigma(1)}, \ldots, e_{\sigma(k)}) = (\mathrm{sign}\,\sigma)\star\omega(e_{\sigma(k+1)}, \ldots, e_{\sigma(n)}),$$

for any positively oriented orthonormal basis $e_1, \ldots, e_k$ and any permutation σ.

Applying this, we see that

$$\star(\star\omega) = (-1)^{k(n-k)}\omega, \quad \omega \in \mathrm{Alt}^k V,$$

and, consequently, that the Hodge star operator is an isometry.

Let W be a subspace of V, and let $\omega = \omega_\| + \omega_\perp$ denote the decomposition of an algebraic form into its tangential and normal parts with respect to W. By taking an orthonormal basis of W and extending it to an orthonormal basis of V, we find that

$$(\star\omega)_\| = \star(\omega_\perp), \quad (\star\omega)_\perp = \star(\omega_\|).$$

In particular, the image of $\star\omega$ in $\mathrm{Alt}^k W$ (under the pullback of the inclusion) vanishes if and only if the normal component of ω vanishes.

There is also a Hodge star operation for multivectors. In two dimensions it takes bivectors to scalars (by $e_1 \wedge e_2 \mapsto 1$) and in three dimensions it takes bivectors to vectors (via $e_1 \wedge e_2 \mapsto e_3$, etc.)

The case $V = \mathbb{R}^n$

Finally, we consider the case $V = \mathbb{R}^n$, and note that we have a natural identification of $\mathrm{Alt}^0 \mathbb{R}^n$ and $\mathrm{Alt}^n \mathbb{R}^n$ with $\mathbb{R}$. In fact, $\mathrm{Alt}^0 \mathbb{R}^n$ is defined to equal $\mathbb{R}$, while the general element $\omega \in \mathrm{Alt}^n \mathbb{R}^n$ can be written

$$\omega(v_1, \ldots, v_n) = c \det[v_1 | \cdots | v_n],$$

for some $c \in \mathbb{R}$, so $\omega \leftrightarrow c$ is the desired identification. For $n > 1$, we also have natural identifications of $\mathrm{Alt}^1 \mathbb{R}^n$ and $\mathrm{Alt}^{n-1} \mathbb{R}^n$ with $\mathbb{R}^n$. The identification of $\mathrm{Alt}^1 \mathbb{R}^n$ with $\mathbb{R}^n$ is the usual Riesz identification of V^* with

V based on an inner product on V (for $\mathbb{R}^n$ we use the Euclidean inner product, of course). The identification of $\mathrm{Alt}^{n-1}\mathbb{R}^n$ with $\mathbb{R}^n$ identifies a vector $v \in \mathbb{R}^n$ with the algebraic $(n-1)$-form

$$(v_1, \ldots, v_{n-1}) \mapsto \det[v|v_1|\cdots|v_{n-1}].$$

The canonical basis $e_1, \ldots, e_n$ of $\mathbb{R}^n$ gives rise to a dual basis of $(\mathbb{R}^n)^*$ which will often be denoted $dx_1, \ldots, dx_n$ (the reason for this notation will be made clear in Section 2.3). Thus $dx_i(e_j) = \delta_{ij}$ and $dx_i(v) = e_i \cdot v = v_i$.

For $n = 3$, each of the spaces $\mathrm{Alt}^k \mathbb{R}^3$ may be identified either with $\mathbb{R}$ ($k = 0, 3$) or with $\mathbb{R}^3$ ($k = 1, 2$). Thus all the operations on exterior forms correspond to operations on scalars and/or vectors. The correspondences are listed in Table 2.1 (overleaf).

A similar set of correspondences exist for $n = 2$, although in this case, for operations involving $\mathrm{Alt}^1 \mathbb{R}^2$ there are two possibilities since there are two *different* identifications of $\mathrm{Alt}^1 \mathbb{R}^2$ with $\mathbb{R}^2$ (as 1-forms or as $(n-1)$-forms).

2.2. Exterior calculus on manifolds

Manifolds

The natural setting for exterior calculus is a (sufficiently) smooth manifold Ω of finite dimension n, with or without boundary. At each point $x \in \Omega$, the tangent space $T_x\Omega$ is a vector space of dimension n. The *tangent bundle* consists of all pairs (x, v) with $x \in \Omega$, $v \in T_x\Omega$. It is a smooth manifold of dimension $2n$. The sections of this bundle, that is, the maps $x \in \Omega \mapsto v(x) \in T_x\Omega$, are the *vector fields* on Ω.

If $\phi : \Omega \to \Omega'$ is a smooth map of manifolds, and $x \in \Omega$, then the tangent map $D\phi_x$ is a linear map from $T_x\Omega$ to $T_{\phi(x)}\Omega'$. In case $\Omega' = \mathbb{R}$, so ϕ is a smooth scalar-valued function on Ω, we write $\partial_v \phi(x)$ for $D\phi_x(v)$, $x \in \Omega$, $v \in T_x\Omega$, the directional derivative of ϕ at x in the direction given by v. In case $\Omega \subset \mathbb{R}$, so ϕ is a curve in Ω', we write $d\phi(t)/dt = D\phi_t(1)$.

Differential forms

Applying the exterior algebra construction to the tangent spaces, we obtain the exterior forms bundle, a smooth manifold whose elements are pairs (x, μ) with $x \in \Omega$, $\mu \in \mathrm{Alt}^k T_x\Omega$. A *differential k-form* is a section of this bundle, i.e., a map ω which associates to each $x \in \Omega$ an element $\omega_x \in \mathrm{Alt}^k T_x\Omega$. Thus, if ω is a differential k-form on Ω, $x \in \Omega$, and $v_1, \ldots, v_k \in T_x\Omega$, then $\omega_x(v_1, \ldots, v_k) \in \mathbb{R}$. If the map

$$x \in \Omega \mapsto \omega_x\big(v_1(x), \ldots, v_k(x)\big) \in \mathbb{R}, \qquad (2.3)$$

is smooth (infinitely differentiable) whenever the v_i are smooth vector fields, then we say that ω is a smooth differential k-form. We denote by $\Lambda^k(\Omega)$ the space of all smooth differential k-forms on Ω. Note that $\Lambda^0(\Omega) = C^\infty(\Omega)$ and $\Lambda^1(\Omega)$ is the space of smooth covector fields.

Table 2.1. Correspondence between alternating algebraic forms on $\mathbb{R}^3$ and scalars/vectors.

correspondence	
$\mathrm{Alt}^0 \mathbb{R}^3 = \mathbb{R}$	$c \leftrightarrow c$
$\mathrm{Alt}^1 \mathbb{R}^3 \xrightarrow{\cong} \mathbb{R}^3$	$u_1\,dx_1 + u_2\,dx_2 + u_3\,dx_3 \leftrightarrow u$
$\mathrm{Alt}^2 \mathbb{R}^3 \xrightarrow{\cong} \mathbb{R}^3$	$u_3\,dx_1 \wedge dx_2 - u_2\,dx_1 \wedge dx_3$
	$\qquad\qquad +u_1\,dx_2 \wedge dx_3 \leftrightarrow u$
$\mathrm{Alt}^3 \mathbb{R}^3 \xrightarrow{\cong} \mathbb{R}$	$c \leftrightarrow c\,dx_1 \wedge dx_2 \wedge dx_3$

exterior product	
$\wedge : \mathrm{Alt}^1 \mathbb{R}^3 \times \mathrm{Alt}^1 \mathbb{R}^3 \to \mathrm{Alt}^2 \mathbb{R}^3$	$\times : \mathbb{R}^3 \times \mathbb{R}^3 \to \mathbb{R}^3$
$\wedge : \mathrm{Alt}^1 \mathbb{R}^3 \times \mathrm{Alt}^2 \mathbb{R}^3 \to \mathrm{Alt}^3 \mathbb{R}^3$	$\cdot : \mathbb{R}^3 \times \mathbb{R}^3 \to \mathbb{R}$

pullback by a linear map $L : \mathbb{R}^3 \to \mathbb{R}^3$	
$L^* : \mathrm{Alt}^0 \mathbb{R}^3 \to \mathrm{Alt}^0 \mathbb{R}^3$	$\mathrm{id} : \mathbb{R} \to \mathbb{R}$
$L^* : \mathrm{Alt}^1 \mathbb{R}^3 \to \mathrm{Alt}^1 \mathbb{R}^3$	$L^T : \mathbb{R}^3 \to \mathbb{R}^3$
$L^* : \mathrm{Alt}^2 \mathbb{R}^3 \to \mathrm{Alt}^2 \mathbb{R}^3$	$(\det L)L^{-1} : \mathbb{R}^3 \to \mathbb{R}^3$
$L^* : \mathrm{Alt}^3 \mathbb{R}^3 \to \mathrm{Alt}^3 \mathbb{R}^3$	$(\det L) : \mathbb{R} \to \mathbb{R} \quad (c \mapsto c \det L)$

interior product with a vector $v \in \mathbb{R}^3$	
$\lrcorner v : \mathrm{Alt}^1 \mathbb{R}^3 \to \mathrm{Alt}^0 \mathbb{R}^3$	$v \cdot : \mathbb{R}^3 \to \mathbb{R}$
$\lrcorner v : \mathrm{Alt}^2 \mathbb{R}^3 \to \mathrm{Alt}^1 \mathbb{R}^3$	$v \times : \mathbb{R}^3 \to \mathbb{R}^3$
$\lrcorner v : \mathrm{Alt}^3 \mathbb{R}^3 \to \mathrm{Alt}^2 \mathbb{R}^3$	$v : \mathbb{R} \to \mathbb{R}^3 \quad (c \mapsto cv)$

inner product and volume form	
inner product on $\mathrm{Alt}^k \mathbb{R}^3$ induced	dot product on $\mathbb{R}$ and $\mathbb{R}^3$
$\quad$ by dot product on $\mathbb{R}^3$	
$\mathrm{vol} = dx_1 \wedge dx_2 \wedge dx_3$	$(v_1, v_2, v_3) \mapsto \det[v_1|v_2|v_3]$

Hodge star	
$\star : \mathrm{Alt}^0 \mathbb{R}^3 \to \mathrm{Alt}^3 \mathbb{R}^3$	$\mathrm{id} : \mathbb{R} \to \mathbb{R}$
$\star : \mathrm{Alt}^1 \mathbb{R}^3 \to \mathrm{Alt}^2 \mathbb{R}^3$	$\mathrm{id} : \mathbb{R}^3 \to \mathbb{R}^3$

Exterior product
For each k, $\Lambda^k(\Omega)$ is not only an (infinite-dimensional) vector space, but also a module with respect to the ring $C^\infty(\Omega)$ of smooth functions on Ω: if $\omega \in \Lambda^k(\Omega)$ and $f \in C^\infty(\Omega)$, the product $f\omega$ belongs to $\Lambda^k(\Omega)$. The exterior product of algebraic forms may be applied pointwise to define the exterior product of differential forms:

$$(\omega \wedge \eta)_x = \omega_x \wedge \eta_x.$$

In this way we obtain the anti-commutative graded algebra

$$\Lambda(\Omega) = \bigoplus_k \Lambda^k(\Omega).$$

C^m spaces of differentiable differential forms
We may also consider spaces of differential forms with less smoothness. If the map (2.3) merely belongs to $C^m(\Omega)$ for some $m \geq 0$ whenever the v_i are smooth vector fields, then we say that ω is a C^m differential k-form, the space of all such we denote by $C^m\Lambda^k(\Omega)$.

Integration of differential forms
Differential forms can be integrated and differentiated without recourse to any additional structure, such as a measure or a metric, on the manifold Ω. If f is an oriented, piecewise smooth k-dimensional submanifold of the manifold Ω, and ω is a continuous k-form, then the integral $\int_f \omega$ is well-defined. Thus, for example, 0-forms can be evaluated at points, 1-forms can be integrated over directed curves, and 2-forms can be integrated over oriented surfaces.

Exterior differentiation
The *exterior derivative* d : $\Lambda(\Omega) \to \Lambda(\Omega)$ is a graded linear operator of degree $+1$, i.e., d maps $\Lambda^k(\Omega)$ into $\Lambda^{k+1}(\Omega)$ for each $k \geq 0$. We give a formula for Ω a domain in $\mathbb{R}^n$. In this case, we may identify each tangent space $T_x\Omega$ with $\mathbb{R}^n$, and hence for given $\omega \in \Lambda^k(\Omega)$ and vectors $v_1, \ldots, v_k$, obtain a smooth mapping $\Omega \to \mathbb{R}$ given by

$$x \mapsto \omega_x(v_1, \ldots, v_k).$$

We then define

$$d\omega_x(v_1, \ldots, v_{k+1}) = \sum_{j=1}^{k+1} (-1)^{j+1} \partial_{v_j} \omega_x(v_1, \ldots, \hat{v}_j, \ldots, v_{k+1}),$$

$$\omega \in \Lambda^k, v_1, \ldots, v_{k+1} \in V,$$

where the hat is used to indicate a suppressed argument. For Ω a general manifold, a similar but more involved expression can be used to define exterior differentiation. In this case the vectors v_i must be replaced by smoothly varying vector fields, and additional terms arise due to the noncommutativity of the vector fields. See Lang (1995, Chapter V, Section 3).

We recall two key properties of exterior differentiation. It is a differential: $d \circ d = 0$; and it satisfies a Leibniz rule with respect to the wedge product:

$$d(\omega \wedge \eta) = d\omega \wedge \eta + (-1)^j \omega \wedge d\eta, \quad \omega \in \Lambda^j(\Omega), \quad \eta \in \Lambda^k(\Omega).$$

Pullback

A smooth map $\phi : \Omega \to \Omega'$ between manifolds provides a pullback of differential forms from Ω' to Ω. Namely, if ω is a differential k-form on Ω', we define the *pullback* $\phi^*\omega \in \Lambda^k(\Omega)$ by

$$(\phi^*\omega)_x = (D\phi_x)^* \omega_{\phi(x)}$$

i.e.,

$$(\phi^*\omega)_x(v_1, \ldots, v_k) = \omega_{\phi(x)}(D\phi_x(v_1), \ldots, D\phi_x(v_k)).$$

The pullback respects exterior products and differentiation:

$$\phi^*(\omega \wedge \eta) = \phi^*\omega \wedge \phi^*\eta, \quad \phi^*(d\omega) = d(\phi^*\omega), \quad \omega, \eta \in \Lambda(\Omega').$$

If ϕ is an orientation-preserving diffeomorphism, then we also have

$$\int_\Omega \phi^*\omega = \int_{\Omega'} \omega, \quad \omega \in \Lambda^n(\Omega').$$

If Ω' is a submanifold of Ω, then the pullback of the inclusion $\Omega' \hookrightarrow \Omega$ is the *trace* map $\mathrm{Tr}_{\Omega,\Omega'} : \Lambda(\Omega) \to \Lambda(\Omega')$. If the domain Ω is clear from context, we may write $\mathrm{Tr}_{\Omega'}$ instead of $\mathrm{Tr}_{\Omega,\Omega'}$, and we usually abbreviate $\mathrm{Tr}_{\Omega,\partial\Omega}$ as simply Tr. Note that, if Ω' is a submanifold of positive codimension and $k > 0$, then the vanishing of $\mathrm{Tr}_{\Omega,\Omega'} \omega$ on Ω' for $\omega \in \Lambda^k(\Omega)$ does *not* imply that $\omega_x \in \mathrm{Alt}^k T_x\Omega$ vanishes for $x \in \Omega'$, only that it vanishes when applied to k-tuples of vectors tangent to Ω', or, in other words, that the tangential part of ω_x with respect to $T_x\Omega'$ vanishes.

Stokes' theorem and integration by parts

Integration of differential forms and exterior differentiation are related via Stokes' theorem. If Ω is an oriented n-manifold with boundary $\partial\Omega$ (endowed with the induced orientation), then

$$\int_\Omega d\omega = \int_{\partial\Omega} \mathrm{Tr}\,\omega, \quad \omega \in \Lambda^{n-1}(\Omega).$$

Combining with the Leibniz rule, we get the integration by parts formula

$$\int_\Omega d\omega \wedge \eta = (-1)^{k-1} \int_\Omega \omega \wedge d\eta + \int_{\partial\Omega} \mathrm{Tr}\,\omega \wedge \mathrm{Tr}\,\eta, \qquad (2.4)$$

if $\omega \in \Lambda^k(\Omega), \eta \in \Lambda^{n-k-1}(\Omega)$.

Interior product
Clearly we may form the interior product of a differential k-form ω with a vector field v, to obtain a $(k-1)$-form: $(\omega \lrcorner v)_x = \omega_x \lrcorner v_x$.

Inner product
Suppose Ω is an oriented Riemannian manifold, so that each tangent space is endowed with an inner product, and so also are the spaces $\mathrm{Alt}^k\, T_x\Omega$. Moreover, there is a unique *volume form*, vol in $\Lambda^n(\Omega)$ such that at each $x \in \Omega$, vol_x is the volume form associated with the oriented inner product space $T_x\Omega$. Therefore we can define the integral of any function $f \in \Lambda^0(\Omega)$ simply as $\int_\Omega f \,\mathsf{vol}$. In particular, we can define the L^2-inner product of any two differential k-forms on Ω as the integral of their pointwise inner product:

$$\langle \omega, \eta \rangle_{L^2 \Lambda^k} = \int_\Omega \langle \omega_x, \eta_x \rangle \,\mathsf{vol} = \int \omega \wedge \star \eta. \tag{2.5}$$

The completion of $\Lambda^k(\Omega)$ in the corresponding norm defines the Hilbert space $L^2 \Lambda^k(\Omega)$.

Sobolev spaces of differential forms
On an oriented Riemannian manifold, we may also define the Sobolev spaces $H^s(\Omega)$ and $W_p^s(\Omega)$ of functions with $s \geq 0$ derivatives in $L^2(\Omega)$ and $L^p(\Omega)$, respectively. We may then define the spaces $H^s \Lambda^k(\Omega)$ consisting of differential forms for which the quantities in (2.3) belong to $H^s(\Omega)$. These Sobolev spaces are Hilbert spaces.

For a differential form to belong to $H^s \Lambda^k(\Omega)$, all its partial derivatives of order at most s (in some coordinate system) must be square integrable. A different notion is obtained by only considering the exterior derivatives. We define another Hilbert space:

$$H\Lambda^k(\Omega) = \{\, \omega \in L^2 \Lambda^k(\Omega) \mid \mathrm{d}\omega \in L^2 \Lambda^{k+1}(\Omega) \,\}.$$

The norm is defined by

$$\|\omega\|_{H\Lambda^k}^2 = \|\omega\|_{H\Lambda}^2 := \|\omega\|_{L^2 \Lambda^k}^2 + \|\mathrm{d}\omega\|_{L^2 \Lambda^{k+1}}^2.$$

The space $H\Lambda^0(\Omega)$ coincides with $H^1 \Lambda^0(\Omega)$ (or simply $H^1(\Omega)$), while the space $H\Lambda^n(\Omega)$ coincides with $L^2 \Lambda^n(\Omega)$. For $0 < k < n$, $H\Lambda^k(\Omega)$ is contained strictly between $H^1 \Lambda^k(\Omega)$ and $L^2 \Lambda^k(\Omega)$.

The de Rham complex
The de Rham complex is the sequence of spaces and mappings

$$0 \to \Lambda^0(\Omega) \xrightarrow{\mathrm{d}} \Lambda^1(\Omega) \xrightarrow{\mathrm{d}} \cdots \xrightarrow{\mathrm{d}} \Lambda^n(\Omega) \to 0.$$

Since $\mathrm{d} \circ \mathrm{d} = 0$, we have

$$\mathcal{R}\big(\mathrm{d}: \Lambda^{k-1}(\Omega) \to \Lambda^k(\Omega)\big) \subset \mathcal{N}\big(\mathrm{d}: \Lambda^k(\Omega) \to \Lambda^{k+1}(\Omega)\big)),$$

for $k = 0, 1, \ldots, n$, which is to say that this sequence is a *cochain complex*. The de Rham cohomology spaces are the quotient spaces. They are finite-dimensional vector spaces whose dimensions are the Betti numbers of the manifold Ω. For a bounded connected region in $\mathbb{R}^3$, the zeroth Betti number is 1, the first Betti number is the genus (number of handles), the second Betti number is one less than the number of connected components of the boundary (number of holes), and the third Betti number is 0.

For an oriented Riemannian manifold, we obtain the same cohomology spaces from the L^2 de Rham complex[1]:

$$0 \to H\Lambda^0(\Omega) \xrightarrow{\mathrm{d}} H\Lambda^1(\Omega) \xrightarrow{\mathrm{d}} \cdots \xrightarrow{\mathrm{d}} H\Lambda^n(\Omega) \to 0. \qquad (2.6)$$

The kth cohomology space is isomorphic to the space of $\mathfrak{H}^k(\Omega)$ of *harmonic k-forms* on Ω:

$$\mathfrak{H}^k(\Omega) = \{\, \omega \in H\Lambda^k(\Omega) \mid \mathrm{d}\omega = 0,\ \langle \omega, \mathrm{d}\eta \rangle_{L^2\Lambda^k} = 0\ \forall \eta \in H\Lambda^{k-1} \,\}.$$

The isomorphism simply associates to a harmonic form ω the cohomology class it represents.

In the case of a contractible domain Ω, all the cohomology spaces vanish, except the one of lowest order, which is the constants. In other words, the extended de Rham complex

$$0 \to \mathbb{R} \xrightarrow{\subset} \Lambda^0(\Omega) \xrightarrow{\mathrm{d}} \Lambda^1(\Omega) \xrightarrow{\mathrm{d}} \cdots \xrightarrow{\mathrm{d}} \Lambda^n(\Omega) \to 0$$

is exact.

The coderivative operator

The coderivative operator $\delta : \Lambda^k(\Omega) \to \Lambda^{k-1}(\Omega)$ is defined

$$\star\delta\omega = (-1)^k \, \mathrm{d} \star \omega, \quad \omega \in \Lambda^k(\Omega). \qquad (2.7)$$

It follows directly from (2.4), (2.5), and (2.7), that

$$\langle \mathrm{d}\omega, \eta \rangle = \langle \omega, \delta\eta \rangle + \int_{\partial\Omega} \mathrm{Tr}\,\omega \wedge \mathrm{Tr}\star\eta, \quad \omega \in \Lambda^k(\Omega), \eta \in \Lambda^{k+1}(\Omega). \qquad (2.8)$$

Thus $\delta : \Lambda^{k+1}(\Omega) \to \Lambda^k(\Omega)$ is the formal adjoint of $\mathrm{d} : \Lambda^k(\Omega) \to \Lambda^{k+1}(\Omega)$ with respect to the L^2-inner product: we have

$$\langle \mathrm{d}\omega, \eta \rangle = \langle \omega, \delta\eta \rangle$$

if ω or η vanish near the boundary.

In analogy with $H\Lambda^k(\Omega)$, we define the space

$$H^*\Lambda^k(\Omega) = \{\, \omega \in L^2\Lambda^k(\Omega) \mid \delta\omega \in L^2\Lambda^{k-1}(\Omega) \,\}.$$

---

[1] The L^2 de Rham complex is often written $0 \to L^2\Lambda^0(\Omega) \xrightarrow{\mathrm{d}} \cdots \xrightarrow{\mathrm{d}} L^2\Lambda^n(\Omega) \to 0$ where the d are taken as *unbounded* operators with the $H\Lambda^k(\Omega)$ spaces as domains. This is an equivalent notion.

Clearly $H^*\Lambda^k(\Omega) = \star H\Lambda^{k-n}(\Omega)$, and so the dual complex

$$0 \leftarrow H^*\Lambda^0(\Omega) \overset{\delta}{\leftarrow} H^*\Lambda^n(\Omega) \overset{\delta}{\leftarrow} \cdots \overset{\delta}{\leftarrow} H^*\Lambda^n(\Omega) \leftarrow 0 \qquad (2.9)$$

contains the same information as the de Rham complex.

Boundary traces

Using the theory of traces in Sobolev space, we find that the trace operator $\mathrm{Tr} : \Lambda^k(\Omega) \to \Lambda^k(\partial\Omega)$ extends by continuity to a mapping of $H^1\Lambda^k(\Omega)$ onto the Sobolev space $H^{1/2}\Lambda^k(\partial\Omega)$. Of course, the trace cannot be extended to all of $L^2\Lambda(\Omega)$. However, we can give a meaning to the trace of $\omega \in H\Lambda(\Omega)$ as follows. Given $\rho \in H^{1/2}\Lambda^k(\partial\Omega)$, let $\bar{\star}\rho \in H^{1/2}\Lambda^{n-k-1}(\partial\Omega)$ denote the Hodge star of ρ with respect to the boundary. Then we can find $\eta \in H^1\Lambda^{n-k-1}(\Omega)$ with $\mathrm{Tr}\,\eta = \bar{\star}\rho$ and

$$\|\eta\|_{H^1} \le c\|\bar{\star}\rho\|_{H^{1/2}} \le c\|\rho\|_{H^{1/2}}.$$

Now for $\omega \in \Lambda^k(\Omega)$, (2.4) gives

$$\langle \mathrm{Tr}\,\omega, \rho \rangle = \int_{\partial\Omega} (\mathrm{Tr}\,\omega) \wedge \bar{\star}\rho = \int_{\partial\Omega} \mathrm{Tr}\,\omega \wedge \mathrm{Tr}\,\eta$$

$$= \int_{\Omega} [d\omega \wedge \eta + (-1)^k \omega \wedge \delta\eta] \le c\|\omega\|_{H\Lambda}\|\eta\|_{H^1}$$

$$\le c\|\omega\|_{H\Lambda}\|\rho\|_{H^{1/2}}.$$

It follows that we can extend Tr to a bounded operator on $H\Lambda^k(\Omega)$ with values in $H^{-1/2}\Lambda^k(\partial\Omega)$, the dual of $H^{1/2}\Lambda^k(\partial\Omega)$. We may then define

$$\mathring{H}\Lambda^k(\Omega) = \{\, \omega \in H\Lambda^k(\Omega) \mid \mathrm{Tr}\,\omega = 0 \,\}.$$

If $\omega \in H^*\Lambda^k(\Omega)$, then $\star\omega \in H\Lambda^{n-k}(\Omega)$, so $\mathrm{Tr}(\star\omega)$ is well-defined. Clearly,

$$\mathring{H}^*\Lambda^k(\Omega) := \star\mathring{H}\Lambda^{n-k}(\Omega) = \{\, \omega \in H^*\Lambda^k(\Omega) \mid \mathrm{Tr}(\star\omega) = 0 \,\}.$$

We recall that for ω smooth, $\mathrm{Tr}\,\omega$ vanishes at some $x \in \partial\Omega$ if and only if the tangential part of ω_x vanishes, and $\mathrm{Tr}(\star\omega)$ vanishes if and only if the normal part vanishes.

We can use the coderivative operator and the trace operator to characterize the orthogonal complement of the range of d and δ. From the adjoint equation (2.8), we see that

$$\{\, \omega \in L^2\Lambda^k(\Omega) \mid \langle \omega, d\eta \rangle_{L^2\Lambda^k} = 0 \,\, \forall \eta \in H\Lambda^{k-1}(\Omega) \,\}$$

$$= \{\, \omega \in \mathring{H}^*\Lambda^k(\Omega) \mid \delta\omega = 0 \,\} \quad (2.10)$$

and

$$\{\, \omega \in L^2\Lambda^k(\Omega) \mid \langle \omega, \delta\eta \rangle_{L^2\Lambda^k} = 0 \,\, \forall \eta \in H\Lambda^{k+1}(\Omega) \,\}$$

$$= \{\, \omega \in \mathring{H}\Lambda^k(\Omega) \mid d\omega = 0 \,\}. \quad (2.11)$$

The first then gives us an expression for the harmonic forms:

$$\mathfrak{H}^k = \{\, \omega \in H\Lambda^k(\Omega) \cap \mathring{H}^*\Lambda^k(\Omega) \mid d\omega = 0, \delta\omega = 0 \,\}. \qquad (2.12)$$

In words, the harmonic forms are determined by the differential equations $d\omega = 0$ and $\delta\omega = 0$ together with the boundary conditions $\text{Tr} \star\omega = 0$.

Cohomology with boundary conditions

Let $\mathring{\Lambda}^k(\Omega)$ denote the subspace of $\Lambda^k(\Omega)$ consisting of smooth k-forms with compact support. Since pullbacks commute with exterior differentiation, $\text{Tr}\, d\omega = d\,\text{Tr}\,\omega$, and so $d\mathring{\Lambda}^k(\Omega) \subset \mathring{\Lambda}^{k+1}(\Omega)$. The de Rham complex with compact support is then

$$0 \to \mathring{\Lambda}^0(\Omega) \xrightarrow{\text{d}} \mathring{\Lambda}^1(\Omega) \xrightarrow{\text{d}} \cdots \xrightarrow{\text{d}} \mathring{\Lambda}^n(\Omega) \to 0. \qquad (2.13)$$

Since the closure of $\mathring{\Lambda}^k(\Omega)$ in $H\Lambda^k(\Omega)$ is $\mathring{H}\Lambda^k(\Omega)$, the L^2 version of the complex (2.13), with the same cohomology, is

$$0 \to \mathring{H}\Lambda^0(\Omega) \xrightarrow{\text{d}} \mathring{H}\Lambda^1(\Omega) \xrightarrow{\text{d}} \cdots \xrightarrow{\text{d}} \mathring{H}\Lambda^n(\Omega) \to 0. \qquad (2.14)$$

The cohomology space is again isomorphic to a space of harmonic forms, in this case,

$$\mathring{\mathfrak{H}}^k(\Omega) = \{\, \omega \in \mathring{H}\Lambda^k(\Omega) \mid d\omega = 0,\ \langle \omega, d\eta\rangle_{L^2\Lambda^k} = 0 \ \forall \eta \in \mathring{H}\Lambda^{k-1}(\Omega) \,\}.$$

In analogy to (2.10), (2.11), and (2.12), we have

$$\{\, \omega \in L^2\Lambda^k(\Omega) \mid \langle \omega, d\eta\rangle_{L^2\Lambda^k} = 0\ \forall \eta \in \mathring{H}\Lambda^{k-1}(\Omega) \,\}$$
$$= \{\, \omega \in H^*\Lambda^k(\Omega) \mid \delta\omega = 0 \,\},$$

$$\{\, \omega \in L^2\Lambda^k(\Omega) \mid \langle \omega, \delta\eta\rangle_{L^2\Lambda^k} = 0\ \forall \eta \in \mathring{H}^*\Lambda^{k+1}(\Omega) \,\}$$
$$= \{\, \omega \in H\Lambda^k(\Omega) \mid d\omega = 0 \,\},$$
$$\mathring{\mathfrak{H}}^k = \{\, \omega \in \mathring{H}\Lambda^k(\Omega) \cap H^*\Lambda^k(\Omega) \mid d\omega = 0, \delta\omega = 0 \,\}.$$

It is a simple untangling of the definitions to see that $\star\mathring{\mathfrak{H}}^k(\Omega) = \mathfrak{H}^{n-k}(\Omega)$. Thus there is an isomorphism between the kth de Rham cohomology space and the $(n - k)$th cohomology space with boundary conditions. This is called *Poincaré duality*.

For a contractible domain Ω, the only nonvanishing cohomology space is now that of highest order, and the extended de Rham complex

$$0 \to \mathring{\Lambda}^0(\Omega) \xrightarrow{\text{d}} \mathring{\Lambda}^1(\Omega) \xrightarrow{\text{d}} \cdots \xrightarrow{\text{d}} \mathring{\Lambda}^n(\Omega) \xrightarrow{\int} \mathbb{R} \to 0$$

is exact.

Homological algebra

The language of homological algebra was invented to clarify the common algebraic structures behind a variety of constructions in different branches of mathematics, for example the de Rham cohomology of differential geometry and simplicial homology in algebraic topology. Here we introduce some of the basic definitions of homological algebra which will be useful below, including cochain maps, cochain projections, and cochain homotopies. More details can be found in many places, *e.g.*, Hilton and Stammbach (1997).

A *cochain complex* is a sequence of real vector spaces (or more generally modules or abelian groups, but we will only use vector spaces) and maps:

$$\cdots \to V_{k-1} \xrightarrow{\mathrm{d}_{k-1}} V_k \xrightarrow{\mathrm{d}_k} V_{k+1} \to \cdots$$

with $\mathrm{d}_{k+1} \circ \mathrm{d}_k = 0$. Equivalently, we may think of a cochain complex as the graded algebra $V = \bigoplus V_k$ equipped with a graded linear operator $\mathrm{d} : V \to V$ of degree $+1$ satisfying $\mathrm{d} \circ \mathrm{d} = 0$. A *chain complex* is the same thing except that the indices decrease, and all the definitions below apply *mutatis mutandis* to chain complexes. The de Rham complex $(\Lambda(\Omega), \mathrm{d})$ and its variants (2.6), (2.13), and (2.14) are examples of cochain complexes, while the dual complex (2.9) and its variants are chain complexes. All the complexes we consider are nonnegative in that $V_k = 0$ for $k < 0$, and bounded in that $V_n = 0$ for n sufficiently large.

Given a cochain complex V, the elements of $\mathcal{N}(\mathrm{d}_k)$ are called the k-cocycles and the elements of $\mathcal{R}(\mathrm{d}_k)$ the k-coboundaries. The quotient space $H^k(V) := \mathcal{N}(\mathrm{d}_k)/\mathcal{R}(\mathrm{d}_k)$ is the kth cohomology space.

Given two cochain complexes V and V', a set of maps $f_k : V_k \to V'_k$ satisfying $\mathrm{d}'_k f_k = f_{k+1} \mathrm{d}_k$ (*i.e.*, is a graded linear map $f : V \to V'$ of degree 0 satisfying $\mathrm{d}' f = f \mathrm{d}$) is called a *cochain map*. When f is a cochain map, f_k maps k-cochains to k-cochains and k-coboundaries to k-coboundaries, and hence induces a map $H^k(f) : H^k(V) \to H^k(V')$.

If V is a cochain complex and V' a subcomplex (*i.e.*, $V'_k \subset V_k$ and $\mathrm{d}V'_k \subset V'_{k+1}$), then the inclusion $i : V' \to V$ is a cochain map and so induces a map of cohomology $H^k(V') \to H^k(V)$. If there exists a *cochain projection* of V onto V', *i.e.*, a cochain map π such that $\pi_k : V_k \to V'_k$ restricts to the identity on V'_k, then $\pi \circ i = \mathrm{id}_{V'}$, so $H^k(\pi) \circ H^k(i) = \mathrm{id}_{H^k(V')}$. We conclude that in this case $H^k(i)$ is injective and $H^k(\pi)$ is surjective. In particular, if one of the cohomology spaces $H^k(V)$ vanishes, then so does $H^k(V')$. We shall use this property frequently.

Given a cochain map $f : V \to V'$, a graded linear map $\kappa : V \to V'$ of degree -1 (*i.e.*, a sequence of maps $\kappa_k : V_k \to V'_{k-1}$) is called a *contracting cochain homotopy* for f, if

$$f_k = \mathrm{d}'_{k-1} \kappa_k + \kappa_{k+1} \mathrm{d}_k : V_k \to V_{k+1}.$$

If there exists a contracting homotopy for f, then f induces the zero map on cohomology (since if z is a k-cocycle, then $fz = \mathrm{d}'\kappa z + \kappa \mathrm{d} z = \mathrm{d}'\kappa z$ is a k-coboundary). In particular, if the identity cochain map on V admits a contracting homotopy, then the cohomology of V vanishes. We shall see an example of this when we discuss the Koszul complex (Theorem 3.1).

Cycles and boundaries of the de Rham complex

We have four variants of the L^2 de Rham complex: (2.6), the dual version (2.9), and versions of each of these incorporating boundary conditions, namely (2.14) and the dual analogue. For each of these we have the corresponding spaces of cycles and boundaries (for brevity we use the term cycles and boundaries to refer as well to cocycles and coboundaries). We denote these by

$$\mathfrak{Z}^k = \{\,\omega \in H\Lambda^k(\Omega) \mid d\omega = 0\,\}, \quad \mathfrak{Z}^{*k} = \{\,\omega \in H^*\Lambda^k(\Omega) \mid \delta\omega = 0\,\},$$

$$\mathring{\mathfrak{Z}}^k = \{\,\omega \in \mathring{H}\Lambda^k(\Omega) \mid d\omega = 0\,\}, \quad \mathring{\mathfrak{Z}}^{*k} = \{\,\omega \in \mathring{H}^*\Lambda^k(\Omega) \mid \delta\omega = 0\,\},$$

and

$$\mathfrak{B}^k = dH\Lambda^{k-1}(\Omega), \quad \mathfrak{B}^{*k} = \delta\Lambda^{k+1}(\Omega),$$

$$\mathring{\mathfrak{B}}^k = d\mathring{H}\Lambda^{k-1}(\Omega), \quad \mathring{\mathfrak{B}}^{*k} = \delta\mathring{\Lambda}^{k+1}(\Omega).$$

Each of the spaces of cycles is obviously closed in the space $H\Lambda^k(\Omega)$ or $H^*\Lambda^k(\Omega)$, as appropriate. Each is closed in $L^2\Lambda^k(\Omega)$ as well. For example, we show that $\mathring{\mathfrak{Z}}^k$ is L^2 closed. Suppose that $\omega_n \in \mathring{\mathfrak{Z}}^k$ and ω_n converges to some ω in $L^2\Lambda^k(\Omega)$. We must show that $\omega \in \mathring{H}\Lambda^k(\Omega)$ and $d\omega = 0$. Now convergence of ω_n to ω in L^2 implies convergence of $d\omega_n$ to $d\omega$ in H^{-1}. Since $d\omega_n = 0$, we conclude that $d\omega = 0$, so $\omega \in H\Lambda^k(\Omega)$. Since we have convergence of both ω_n and $d\omega_n$, we have that ω_n converges to ω in $H\Lambda^k(\Omega)$, and so ω must belong to $\mathring{\mathfrak{Z}}^k$.

The spaces of boundaries are all closed in $L^2\Lambda^k(\Omega)$ as well. This will follow from Poincaré's inequality, in the next subsubsection.

For any of these subspaces of $L^2\Lambda^k(\Omega)$, we use the superscript $\perp$ to denote its orthogonal complement in that space. Note that the orthogonal complement of $\mathfrak{Z}^k$ in $H\Lambda^k(\Omega)$ is just $H\Lambda^k(\Omega) \cap \mathfrak{Z}^{k\perp}$, and a similar relationship holds for all the other spaces. Equation (2.10) can be rewritten as $\mathfrak{B}^{k\perp} = \mathfrak{Z}^{*k}$. In fact, it is easy to verify using (2.8) that

$$\mathfrak{Z}^{k\perp} \subset \mathfrak{B}^{k\perp} = \mathring{\mathfrak{Z}}^{*k}, \quad \mathfrak{Z}^{*k\perp} \subset \mathfrak{B}^{*k\perp} = \mathring{\mathfrak{Z}}^{k}, \tag{2.15}$$

$$\mathring{\mathfrak{Z}}^{k\perp} \subset \mathring{\mathfrak{B}}^{k\perp} = \mathfrak{Z}^{*k}, \quad \mathring{\mathfrak{Z}}^{*k\perp} \subset \mathring{\mathfrak{B}}^{*k\perp} = \mathfrak{Z}^{k}. \tag{2.16}$$

Compactness and Poincaré's inequality

If Ω is a smoothly bounded oriented Riemannian manifold with boundary, then the space $H\Lambda^k(\Omega) \cap \mathring{H}^*\Lambda^k(\Omega)$ is a subspace of $H^1\Lambda^k(\Omega)$, and there holds the estimate (Gaffney 1951)

$$\|\omega\|_{H^1\Lambda^k} \le c(\|d\omega\| + \|\delta\omega\| + \|\omega\|).$$

(A similar result holds for $\mathring{H}\Lambda^k(\Omega) \cap H^*\Lambda^k(\Omega)$.) We can then apply Rellich's lemma to conclude that $H\Lambda^k(\Omega) \cap \mathring{H}^*\Lambda^k(\Omega)$ is *compactly* embedded in $L^2\Lambda^k(\Omega)$.

For polyhedral Ω, the space $H\Lambda^k(\Omega) \cap \mathring{H}^*\Lambda^k(\Omega)$ may not be embedded in $H^1\Lambda^k(\Omega)$. However the following theorem, which is proved in Picard (1984) for manifolds with Lipschitz boundary, states that the compact embedding into L^2 still holds.

Theorem 2.1. The embeddings of $H\Lambda^k(\Omega) \cap \mathring{H}^*\Lambda^k(\Omega)$ and $\mathring{H}\Lambda^k(\Omega) \cap H^*\Lambda^k(\Omega)$ into $L^2\Lambda^k(\Omega)$ are compact.

From this we obtain Poincaré's inequality by a standard compactness argument.

Theorem 2.2. There exists a constant c such that

$$\|\omega\| \leq c(\|d\omega\| + \|\delta\omega\|)$$

for $\omega \in H\Lambda^k(\Omega) \cap \mathring{H}^*\Lambda^k(\Omega) \cap \mathfrak{H}^{k\perp}$ or $\mathring{H}\Lambda^k(\Omega) \cap H^*\Lambda^k(\Omega) \cap \mathring{\mathfrak{H}}^{k\perp}$.

Proof. We will give the proof for $H\Lambda^k(\Omega) \cap \mathring{H}^*\Lambda^k(\Omega) \cap \mathfrak{H}^{k\perp}$, the other case being similar. If the result were not true, we could find a sequence of $\omega_n \in H\Lambda^k(\Omega) \cap \mathring{H}^*\Lambda^k(\Omega) \cap \mathfrak{H}^{k\perp}$ such that $\|\omega_n\| = 1$ while $d\omega_n \to 0$ and $\delta\omega_n \to 0$ in L^2. The sequence is certainly bounded in $H\Lambda^k(\Omega) \cap \mathring{H}^*\Lambda^k(\Omega)$, and, so precompact in $L^2\Lambda^k(\Omega)$ by Theorem 2.1. Passing to a subsequence, we have ω_n converges in $L^2\Lambda^k(\Omega)$ to some $\omega \in L^2\Lambda^k(\Omega)$. Clearly $d\omega = 0$, $\delta\omega = 0$, and the convergence holds in $H\Lambda^k(\Omega)$. Thus $\omega \in H\Lambda^k(\Omega) \cap \mathring{H}^*\Lambda^k(\Omega)$ and so $\omega \in \mathfrak{H}^k$ by (2.12), but also in $\mathfrak{H}^{k\perp}$, and so $\omega = 0$. But $\|\omega\| = \lim \|\omega_n\| = 1$, which is a contradiction. $\square$

A special, but very useful, case is if $\omega \in \mathfrak{Z}^{k\perp}$. Then $\omega \in H\Lambda^k(\Omega) \cap \mathring{H}^*\Lambda^k(\Omega) \cap \mathfrak{H}^{k\perp}$, so

$$\|\omega\| \leq c\|d\omega\|, \quad \omega \in \mathfrak{Z}^{k\perp}. \tag{2.17}$$

Of course the analogous result for the coderivative holds as well.

Theorem 2.3. The spaces of boundaries, $\mathfrak{B}^k$, $\mathfrak{B}^{*k}$, $\mathring{\mathfrak{B}}^k$, and $\mathring{\mathfrak{B}}^{*k}$, are closed in $L^2\Lambda^k(\Omega)$.

Proof. We prove that $\mathring{\mathfrak{B}}^k$ is closed, the other cases being similar. Suppose $\omega_n \in \mathring{\mathfrak{B}}^k$ and $\omega_n \to \omega$ in $L^2\Lambda^k(\Omega)$. There exist $\eta_n \in \mathring{\mathfrak{Z}}^{(k-1)\perp}$ with $\omega_n = d\eta_n$. From the fact that the sequence $d\eta_n$ is convergent in $L^2\Lambda^k(\Omega)$ and Poincaré's inequality, we find that η_n is Cauchy with respect to the L^2-norm and so η_n converges to some $\eta \in L^2\Lambda^{k-1}(\Omega)$. Necessarily $d\eta = \omega$, so η_n converges to η in $H\Lambda^k(\Omega)$, so $\eta \in \mathring{H}\Lambda^k(\Omega)$ and $\omega = d\eta \in \mathring{\mathfrak{B}}^k$. $\square$

Hodge decomposition

At this point, the Hodge decomposition (or really two Hodge decompositions, with different boundary conditions), is just a matter of gathering results.

From the first equality in (2.15), we have an orthogonal decomposition

$$L^2\Lambda^k(\Omega) = \mathfrak{B}^k \oplus \mathring{\mathfrak{Z}}^{*k}.$$

But $\mathring{\mathfrak{B}}^{*k}$ is a closed subspace of $\mathring{\mathfrak{Z}}^{*k}$ whose orthogonal complement is, by the second equality in (2.16),

$$\mathring{\mathfrak{Z}}^{*k} \cap \mathfrak{Z}^k = \mathfrak{H}^k.$$

Thus

$$L^2\Lambda^k(\Omega) = \mathfrak{B}^k \oplus \mathfrak{H}^k \oplus \mathring{\mathfrak{B}}^{*k},$$

which is the first Hodge decomposition. The second follows analogously. The following equation summarizes both, together with the relations of the various spaces discussed:

$$
\overbrace{\phantom{\mathfrak{B}^k \oplus \mathfrak{H}^k}}^{\mathfrak{Z}^k = \mathring{\mathfrak{B}}^{*k\perp}}
\overbrace{\phantom{\mathfrak{B}^k}}^{\mathfrak{Z}^{k\perp}}
\overbrace{\phantom{\mathfrak{B}^k \oplus \mathfrak{H}^k}}^{\mathring{\mathfrak{Z}}^k = \mathfrak{B}^{*k\perp}}
\overbrace{\phantom{\mathfrak{B}^k}}^{\mathring{\mathfrak{Z}}^{k\perp}}
$$
$$L^2\Lambda^k(\Omega) = \underbrace{\mathfrak{B}^k \oplus \underbrace{\mathfrak{H}^k \oplus \mathring{\mathfrak{B}}^{*k}}_{\mathring{\mathfrak{Z}}^{*k} = \mathfrak{B}^{k\perp}}}_{\mathring{\mathfrak{Z}}^{*k\perp}} = \underbrace{\underbrace{\mathring{\mathfrak{B}}^k \oplus \mathfrak{H}^k}_{\mathfrak{Z}^{*k\perp}} \oplus \mathfrak{B}^{*k}}_{\mathfrak{Z}^{*k} = \mathring{\mathfrak{B}}^{k\perp}} \qquad (2.18)$$

The Hodge Laplacian

The differential operator $\delta d + d\delta$ maps k-forms to k-forms. It is called the Hodge Laplacian. Here we briefly consider boundary value problems for the Hodge Laplacian, taking a variational approach.

Consider first the problem of minimizing the quadratic functional

$$\mathcal{J}(u) = \frac{1}{2}\langle du, du \rangle + \frac{1}{2}\langle \delta u, \delta u \rangle - \langle f, u \rangle$$

over $u \in H\Lambda^k(\Omega) \cap \mathring{H}^*\Lambda^k(\Omega)$ where $f \in L^2\Lambda^k(\Omega)$ is given. Here we run into a problem if the space of harmonic forms $\mathfrak{H}^k$ does not vanish. If the L^2 projection $P_{\mathfrak{H}^k}f$ is not zero, then $\mathcal{J}$ admits no minimum, since we could take $u = cP_{\mathfrak{H}^k}f$ for c arbitrarily large. If we insist that f be orthogonal to $\mathfrak{H}^k$, then there exists a minimizer, but it is not unique, because the addition of a harmonic function to u will not change $\mathcal{J}(u)$. To avoid this difficulty, we can work in the orthogonal complement of $\mathfrak{H}^k$. That is, we assume that $f \in \mathfrak{H}^{k\perp}$, and seek $u \in X := H\Lambda^k(\Omega) \cap \mathring{H}^*\Lambda^k(\Omega) \cap \mathfrak{H}^{k\perp}$ minimizing $\mathcal{J}(u)$. Now X is a Hilbert space with norm

$$\|u\|_X = \|u\|_{H\Lambda} + \|u\|_{H^*\Lambda}.$$

By Poincaré's inequality, Theorem 2.2, the square root of the quadratic part of $J(u)$ defines an equivalent norm:

$$\|u\|_X \approx \|du\| + \|\delta u\|, \quad u \in X.$$

Therefore it is easy to prove that there exists a unique minimizer $u \in X$ which satisfies the Euler–Lagrange equations

$$\langle du, dv \rangle + \langle \delta u, \delta v \rangle = \langle f, v \rangle, \quad v \in X.$$

In fact, these equations are satisfied for all $v \in H\Lambda^k(\Omega) \cap \mathring{H}^*\Lambda^k(\Omega)$, since they are trivially satisfied for $v \in \mathfrak{H}^k$ (using the fact that $f \perp \mathfrak{H}^k$). Thus, in view of (2.8), the solution u satisfies the differential equation $(\delta d + d\delta)u = f$ and also the boundary condition $\mathrm{Tr}(\star du) = 0$ in a weak sense. We have also imposed the boundary condition $\mathrm{Tr}(\star u) = 0$ and orthogonality to $\mathfrak{H}^k$, i.e., we have well-posedness, in an appropriate sense, of the following boundary value problem: given $f \in L^2\Lambda^k(\Omega)$ orthogonal to $\mathfrak{H}^k$, find u such that

$$(\delta d + d\delta)u = f \text{ in } \Omega, \quad \mathrm{Tr}(\star u) = 0, \ \mathrm{Tr}(\star du) = 0 \text{ on } \partial\Omega, \quad u \perp \mathfrak{H}^k.$$

A slightly different way to handle the harmonic functions, which we shall follow when we discuss mixed formulations in Section 7, is to impose the orthogonality condition via a Lagrange multiplier $p \in \mathfrak{H}^k$. Then the weak problem is to find $u \in H\Lambda^k(\Omega) \cap \mathring{H}^*\Lambda^k(\Omega)$ and $p \in \mathfrak{H}^k$ such that

$$\langle du, dv \rangle + \langle \delta u, \delta v \rangle + \langle p, v \rangle = \langle f, v \rangle, \quad v \in H\Lambda^k(\Omega) \cap \mathring{H}^*\Lambda^k(\Omega),$$
$$\langle u, q \rangle = 0, \quad q \in \mathfrak{H}^k.$$

The differential equation is now $(\delta d + d\delta)u + p = f$, which has a solution for any f. We get $p = P_{\mathfrak{H}^k}f$ and so p vanishes if $f \perp \mathfrak{H}^k$.

Returning to the first approach, we could have also chosen the space X to be $\mathring{H}\Lambda^k(\Omega) \cap H^*\Lambda^k(\Omega) \cap \mathring{\mathfrak{H}}^{k\perp}$. In this case we obtain the same differential equation, but the weakly imposed boundary condition $\mathrm{Tr}(\delta u) = 0$ and the strongly imposed boundary condition $\mathrm{Tr}(u) = 0$, and the orthogonality assumed on f and imposed on u is to $\mathring{\mathfrak{H}}^k$ rather than to $\mathfrak{H}^k$.

The finite element solution of this problem might seem straightforward, at least when the harmonic functions vanish, but for some domains it most definitely is not. Take, for instance, the case where Ω is a contractible but nonconvex polyhedron in three dimensions and the form degree $k = 1$ or 2. It is shown in Costabel (1991) that the norm on the space $X = H\Lambda^k(\Omega) \cap \mathring{H}^*\Lambda^k(\Omega)$ is equivalent to the $H^1\Lambda^k(\Omega)$-norm when restricted to $X^1 := H^1\Lambda^k(\Omega) \cap \mathring{H}^*\Lambda^k(\Omega)$, but that X^1 is a closed infinite-codimensional subspace of X, and that except for very nongeneric data, the solution u to the Hodge Laplacian problem will belong to X but not to X^1.

Now if we triangulate the domain and use piecewise smooth (*e.g.*, finite element) forms in X to approximate the solution, we will not converge to the solution. For a piecewise smooth form which belongs to $H\Lambda^k(\Omega) \cap H^*\Lambda^k(\Omega)$ always belongs to H^1, and so if it belongs to X, it belongs to X^1. Thus our approximate solutions will remain in a closed subspace which does not contain the exact solution, and so cannot converge to it.

The mixed formulation we present in Section 7 will not suffer from this (very serious) defect.

2.3. Exterior calculus on $\mathbb{R}^n$

Global coordinates

Suppose that Ω is an open subset of $\mathbb{R}^n$. Then we have the global coordinate functions x_i, $i = 1, \ldots, n$. Moreover, each of the tangent spaces $T_x\Omega$ may be identified naturally with $\mathbb{R}^n$, and so with each other. This simplifies matters greatly, especially from the computational point of view. Note that the exterior derivative of the coordinate function x_i is the functional that takes a vector $v \in \mathbb{R}^n$ to its ith component $v_i \in \mathbb{R}$. This explains the notation $\mathrm{d}x_i$ introduced for this functional in Section 2.1.

A general element of $\Lambda^k(\Omega)$ may be written

$$\omega_x = \sum_{1 \le \sigma(1) < \cdots < \sigma(k) \le n} a_\sigma \mathrm{d}x_{\sigma(1)} \wedge \cdots \wedge \mathrm{d}x_{\sigma(k)},$$

where the $a_\sigma \in C^\infty(\Omega)$. If we allow instead $a_\sigma \in C^p(\Omega)$, $a_\sigma \in L^2(\Omega)$, $a_\sigma \in H^s(\Omega)$, etc., we obtain the spaces $C^p\Lambda^k(\Omega)$, $L^2\Lambda^k(\Omega)$, $H^s\Lambda^k(\Omega)$, etc.

The volume form is simply $\mathrm{d}x_1 \wedge \cdots \wedge \mathrm{d}x_n$.

In terms of the global coordinates, the exterior derivative also has a simple expression:

$$\mathrm{d}\sum a_\sigma \mathrm{d}x_{\sigma(1)} \wedge \cdots \wedge \mathrm{d}x_{\sigma(k)} = \sum_\sigma \sum_{i=1}^n \frac{\partial a_\sigma}{\partial x_i} \mathrm{d}x_i \wedge \mathrm{d}x_{\sigma(1)} \wedge \cdots \wedge \mathrm{d}x_{\sigma(k)}.$$

Proxy fields

Based on the identification of $\mathrm{Alt}^0 \mathbb{R}^n$ and $\mathrm{Alt}^n \mathbb{R}^n$ with $\mathbb{R}$ and of $\mathrm{Alt}^1 \mathbb{R}^n$ and $\mathrm{Alt}^{n-1} \mathbb{R}^n$ with $\mathbb{R}^n$, we may identify each 0-form and n-form with a scalar-valued function and, for $n > 1$, each 1-form and $(n-1)$-form with a vector field. The associated fields are called *proxy fields* for the forms. For $n = 2$ we have two *different* identifications of $\Lambda^1(\Omega)$ with $C^\infty(\Omega; \mathbb{R}^2)$, *i.e.*, two ways to associate a proxy field to a 1-form.

Interpreted in terms of the proxy fields, the exterior derivative operators $\mathrm{d} : \Lambda^0(\Omega) \to \Lambda^1(\Omega)$ and $\mathrm{d} : \Lambda^{n-1}(\Omega) \to \Lambda^n(\Omega)$ become $\mathrm{grad} : C^\infty(\Omega) \to C^\infty(\Omega; \mathbb{R}^n)$ and $\mathrm{div} : C^\infty(\Omega; \mathbb{R}^n) \to C^\infty(\Omega)$, respectively.

Table 2.2. Correspondences between a differential forms ω on $\Omega \subset \mathbb{R}^3$ and scalar/vector fields w on Ω. In the integrals, f denotes a submanifold of dimension k, and $\mathcal{H}_k$ denotes the k-dimensional Hausdorff measure (Lebesgue measure for $k = 3$). The unit tangent t for $k = 1$ and unit normal n for $k = 2$ are determined by the orientation of f.

k	$\Lambda^k(\Omega)$	$H\Lambda^k(\Omega)$	$d\omega$	$\int_f \omega$	$\kappa\omega$
0	$C^\infty(\Omega)$	$H^1(\Omega)$	$\operatorname{grad} w$	$w(f)$	0
1	$C^\infty(\Omega;\mathbb{R}^3)$	$H(\operatorname{curl},\Omega;\mathbb{R}^3)$	$\operatorname{curl} w$	$\int_f w \cdot t\, d\mathcal{H}_1$	$x \mapsto x \cdot w(x)$
2	$C^\infty(\Omega;\mathbb{R}^3)$	$H(\operatorname{div},\Omega;\mathbb{R}^3)$	$\operatorname{div} w$	$\int_f w \cdot n\, d\mathcal{H}_2$	$x \mapsto x \times w(x)$
3	$C^\infty(\Omega)$	$L^2(\Omega)$	0	$\int_f w\, d\mathcal{H}_3$	$x \to x\, w(x)$

Table 2.2 summarizes correspondences between differential forms and their proxy fields in the case $\Omega \subset \mathbb{R}^3$. (The last column refers to the Koszul differential, introduced in the next section.)

For $\Omega \subset \mathbb{R}^3$, the de Rham complex becomes

$$0 \to C^\infty(\Omega) \xrightarrow{\operatorname{grad}} C^\infty(\Omega;\mathbb{R}^3) \xrightarrow{\operatorname{curl}} C^\infty(\Omega;\mathbb{R}^3) \xrightarrow{\operatorname{div}} C^\infty(\Omega) \to 0,$$

and the L^2 de Rham complex

$$0 \to H^1(\Omega) \xrightarrow{\operatorname{grad}} H(\operatorname{curl},\Omega;\mathbb{R}^3) \xrightarrow{\operatorname{curl}} H(\operatorname{div},\Omega;\mathbb{R}^3) \xrightarrow{\operatorname{div}} L^2(\Omega) \to 0.$$

For $\Omega \subset \mathbb{R}^2$, the de Rham complex becomes

$$0 \to C^\infty(\Omega) \xrightarrow{\operatorname{grad}} C^\infty(\Omega;\mathbb{R}^2) \xrightarrow{\operatorname{rot}} C^\infty(\Omega) \to 0,$$

or

$$0 \to C^\infty(\Omega) \xrightarrow{\operatorname{curl}} C^\infty(\Omega;\mathbb{R}^2) \xrightarrow{\operatorname{div}} C^\infty(\Omega) \to 0,$$

depending on which of the two identification we choose for $\Lambda^1(\mathbb{R}^2)$.

For Ω any bounded domain in $\mathbb{R}^n$ with Lipschitz boundary, it is known that the divergence operator maps $H^1(\Omega;\mathbb{R}^n)$ onto $L^2(\Omega)$ and $\mathring{H}^1(\Omega;\mathbb{R}^n)$ onto the orthogonal complement of the constants in $L^2(\Omega)$ (see Girault and Raviart (1986, Corollary 2.4)). Since we will need this result in the final section of the paper, we state it here in the language of differential forms.

Theorem 2.4. Let Ω be a bounded domain in $\mathbb{R}^n$ with a Lipschitz boundary. Then, for all $\mu \in L^2\Lambda^n(\Omega)$ there exists $\eta \in H^1\Lambda^{n-1}(\Omega)$ satisfying $d\eta = \mu$. If, in addition, $\int_\Omega \mu = 0$, then we can choose $\eta \in \mathring{H}^1\Lambda^{n-1}(\Omega)$.

3. Polynomial differential forms and the Koszul complex

In this section we consider spaces of polynomial differential forms, which lead to a variety of subcomplexes of the de Rham complex. These will be used in later sections to construct finite element spaces of differential forms. A key tool will be the Koszul differential and the associated Koszul complex. The material in the first two subsections can be extracted from the literature, to which some references are given, but the goals, context, and level of generality are often quite different from ours, so we intend the presentation here to be self-contained. In Section 3.3 we introduce the spaces $\mathcal{P}_r^- \Lambda^k$, which will be of great importance later. (We introduced these spaces under the name $\mathcal{P}_{r-1}^+ \Lambda^k$ in Arnold *et al.* (2006*b*), but have changed the indexing in order to have the graded multiplication property (3.16).) In Section 3.4 we determine all finite-dimensional spaces of polynomial differential forms which are invariant under affine transformations. To the best of our knowledge, this result is new. In the following subsection, we exhibit a wide variety of polynomial subcomplexes of the de Rham complex (essentially 2^{n-1} of them associated to each polynomial degree). These will lead to finite element de Rham subcomplexes in the following sections. Some of these have appeared in the literature previously, but the systematic derivation of all of them first appeared in Arnold *et al.* (2006*b*).

3.1. Polynomial differential forms

Let $\mathcal{P}_r(\mathbb{R}^n)$ and $\mathcal{H}_r(\mathbb{R}^n)$ denote the spaces of polynomials in n variables of degree at most r and of homogeneous polynomial functions of degree r, respectively. We interpret these spaces to be the zero space if $r < 0$. The space of all polynomial functions is $\mathcal{P}(\mathbb{R}^n) = \bigoplus_{r=0}^{\infty} \mathcal{H}_r(\mathbb{R}^n)$, a commutative graded algebra. We can then define spaces of polynomial differential forms, $\mathcal{P}_r \Lambda^k(\mathbb{R}^n)$, $\mathcal{H}_r \Lambda^k(\mathbb{R}^n)$, *etc.* For brevity, we will at times suppress $\mathbb{R}^n$ from the notation and write simply $\mathcal{P}_r$, $\mathcal{H}_r$, $\mathcal{P}_r \Lambda^k$, *etc.*

We note for future reference that

$$\dim \mathcal{P}_r \Lambda^k(\mathbb{R}^n) = \dim \mathcal{P}_r(\mathbb{R}^n) \cdot \dim \mathrm{Alt}^k \, \mathbb{R}^n$$
$$= \binom{n+r}{n}\binom{n}{k} = \binom{r+k}{r}\binom{n+r}{n-k}, \qquad (3.1)$$

and $\dim \mathcal{H}_r \Lambda^k(\mathbb{R}^n) = \dim \mathcal{P}_r \Lambda^k(\mathbb{R}^{n-1})$.

The space of polynomial differential forms

$$\mathcal{P}\Lambda = \bigoplus_{r=0}^{\infty} \bigoplus_{k=0}^{n} \mathcal{H}_r \Lambda^k$$

is called the *Koszul algebra* (Guillemin and Sternberg 1999, Chapter 3.1).

For each polynomial degree $r \geq 0$ we get a homogeneous polynomial subcomplex of the de Rham complex:

$$0 \to \mathcal{H}_r \Lambda^0 \xrightarrow{\mathrm{d}} \mathcal{H}_{r-1} \Lambda^1 \xrightarrow{\mathrm{d}} \cdots \xrightarrow{\mathrm{d}} \mathcal{H}_{r-n} \Lambda^n \to 0. \tag{3.2}$$

We shall verify below the exactness of this sequence. More precisely, the cohomology vanishes if $r > 0$ and also for $r = 0$ except in the lowest degree, where the cohomology space is $\mathbb{R}$ (reflecting the fact that the constants are killed by the gradient).

Taking the direct sum of the homogeneous polynomial de Rham complexes over all polynomial degrees gives the polynomial de Rham complex:

$$0 \to \mathcal{P}_r \Lambda^0 \xrightarrow{\mathrm{d}} \mathcal{P}_{r-1} \Lambda^1 \xrightarrow{\mathrm{d}} \cdots \xrightarrow{\mathrm{d}} \mathcal{P}_{r-n} \Lambda^n \to 0 \tag{3.3}$$

for which the cohomology space is $\mathbb{R}$ in the lowest degree, and vanishes otherwise.

It is easy to see that if $\phi : \mathbb{R}^n \to \mathbb{R}^n$ is a *linear* map, then

$$\phi^*(\mathcal{H}_r \Lambda^k) \subset \mathcal{H}_r \Lambda^k, \quad \phi^*(\mathcal{P}_r \Lambda^k) \subset \mathcal{P}_r \Lambda^k,$$

and if $\phi : \mathbb{R}^n \to \mathbb{R}^n$ is an *affine* map, then

$$\phi^*(\mathcal{P}_r \Lambda^k) \subset \mathcal{P}_r \Lambda^k. \tag{3.4}$$

3.2. *The Koszul complex*

Let $x \in \mathbb{R}^n$. Since there is a natural identification of $\mathbb{R}^n$ with the tangent space $T_0 \mathbb{R}^n$ at the origin, there is a vector in $T_0 \mathbb{R}^n$ corresponding to x. (The origin is chosen for convenience here, but we could use any other point instead.) Then the translation map $y \mapsto y + x$ induces an isomorphism from $T_0 \mathbb{R}^n$ to $T_x \mathbb{R}^n$, and so there is an element $X(x) \in T_x \mathbb{R}^n$ corresponding to x. (Essentially $X(x)$ is the vector based at x which points opposite to the origin, and whose length is $|x|$.) The interior product with the vector field X, $\kappa := \lrcorner X$, maps $\Lambda^k(\mathbb{R}^n)$ to $\Lambda^{k-1}(\mathbb{R}^n)$ by the formula

$$(\kappa \omega)_x(v_1, \ldots, v_{k-1}) = \omega_x(X(x), v_1, \ldots, v_{k-1}).$$

From the similar properties for the interior product of algebraic forms, we have that

$$\kappa \circ \kappa = 0 \tag{3.5}$$

and

$$\kappa(\omega \wedge \eta) = (\kappa \omega) \wedge \eta + (-1)^k \omega \wedge (\kappa \eta), \quad \omega \in \Lambda^k, \ \eta \in \Lambda^l. \tag{3.6}$$

In terms of coordinates, if $\omega_x = \sum_\sigma a_\sigma(x) \mathrm{d}x_{\sigma(1)} \wedge \cdots \wedge \mathrm{d}x_{\sigma(k)}$, then

$$(\kappa \omega)_x = \sum_\sigma \sum_{i=1}^k (-1)^{i+1} a_\sigma(x) x_{\sigma(i)} \mathrm{d}x_{\sigma(1)} \wedge \cdots \wedge \widehat{\mathrm{d}x_{\sigma(i)}} \wedge \cdots \wedge \mathrm{d}x_{\sigma(k)}.$$

Note that κ maps $\mathcal{H}_r\Lambda^k$ to $\mathcal{H}_{r+1}\Lambda^{k-1}$, i.e., κ increases polynomial degree and decreases form degree, the exact opposite of the exterior derivative d.

Another useful formula is the pullback of $\kappa\omega$ under a linear or affine map. First suppose that $\phi : \mathbb{R}^n \to \mathbb{R}^n$ is linear. Then if ω is a k-form on $\mathbb{R}^n$, we have

$$\phi^*\kappa\omega = \kappa\phi^*\omega.$$

Indeed,

$$(\phi^*\kappa\omega)_x(v_1,\ldots,v_{k-1}) = (\kappa\omega)_{\phi x}(\phi v_1,\ldots,\phi v_{k-1}) = \omega_{\phi x}(\phi x, \phi v_1,\ldots,\phi v_{k-1})$$
$$= (\phi^*\omega)_x(x, v_1,\ldots,v_{k-1}) = (\kappa\phi^*\omega)_x(v_1,\ldots,v_{k-1}).$$

For the case of an affine mapping $\phi x = \psi x + b$, with ψ linear and $b \in \mathbb{R}^n$, a similar computation gives

$$(\phi^*\kappa\omega)_x(v_1,\ldots,v_{k-1}) = (\kappa\omega)_{\phi x}(\psi v_1,\ldots,\psi v_{k-1})$$
$$= \omega_{\phi x}(\psi x + b, \psi x, \psi v_1,\ldots,\psi v_{k-1})$$
$$= \omega_{\phi x}(\psi x, \psi v_1,\ldots,\psi v_{k-1}) + \omega_{\phi x}(b, \psi v_1,\ldots,\psi v_{k-1})$$
$$= (\kappa\phi^*\omega)_x(v_1,\ldots,v_{k-1}) + \mu_x(v_1,\ldots,v_{k-1}),$$

where $\mu_x(v_1,\ldots,v_{k-1}) = \omega_{\phi x}(b, \psi v_1,\ldots,\psi v_{k-1})$, so $\mu \in \mathcal{P}_r\Lambda^{k-1}$. Thus

$$\phi^*\kappa\omega - \kappa\phi^*\omega \in \mathcal{P}_r\Lambda^{k-1}, \quad \omega \in \mathcal{H}_r\Lambda^k. \tag{3.7}$$

The operator κ maps the Koszul algebra $\mathcal{P}\Lambda$ to itself. There it is called the *Koszul operator* (Guillemin and Sternberg 1999, Chapter 3.1), and gives rise to the homogeneous *Koszul complex* (Loday 1992, Chapter 3.4.6),

$$0 \to \mathcal{H}_{r-n}\Lambda^n \xrightarrow{\kappa} \mathcal{H}_{r-n+1}\Lambda^{n-1} \xrightarrow{\kappa} \cdots \xrightarrow{\kappa} \mathcal{H}_r\Lambda^0 \to 0. \tag{3.8}$$

We show below that this complex is exact for $r > 0$. Adding over polynomial degrees, we obtain the Koszul complex (for any $r \geq 0$),

$$0 \to \mathcal{P}_{r-n}\Lambda^n \xrightarrow{\kappa} \mathcal{P}_{r-n+1}\Lambda^{n-1} \xrightarrow{\kappa} \cdots \xrightarrow{\kappa} \mathcal{P}_r\Lambda^0 \to 0,$$

for which all the cohomology spaces vanish, except the rightmost, which is equal to $\mathbb{R}$.

To prove the exactness of the homogeneous polynomial de Rham and Koszul complexes, we establish a key connection between the exterior derivative and the Koszul differential. In the language of homological algebra, this says that the Koszul operator is a contracting homotopy for the homogeneous polynomial de Rham complex.

Theorem 3.1.

$$(\mathrm{d}\kappa + \kappa\mathrm{d})\omega = (r + k)\omega, \quad \omega \in \mathcal{H}_r\Lambda^k. \tag{3.9}$$

Proof. It suffices to prove the result for $\omega = f \, dx_{\sigma(1)} \wedge \cdots \wedge dx_{\sigma(k)}$ where $1 \leq \sigma(1) < \cdots < \sigma(k) \leq n$, and $f \in \mathcal{H}_r$. To simplify notation, we may as well assume that $\sigma(i) = i$, so $\omega = f \, dx_1 \wedge \cdots \wedge dx_k$. Now

$$\kappa d\omega = \kappa \left(\sum_{i=1}^{n} \frac{\partial f}{\partial x_i} \, dx_i \wedge dx_1 \wedge \cdots \wedge dx_k \right)$$

$$= \sum_{i=1}^{n} \kappa \left(\frac{\partial f}{\partial x_i} \, dx_i \wedge dx_1 \wedge \cdots \wedge dx_k \right)$$

$$= \sum_{i=1}^{n} \frac{\partial f}{\partial x_i} \left[x_i \, dx_1 \wedge \cdots \wedge dx_k \right.$$

$$\left. + \sum_{j=1}^{k} (-1)^j x_j \, dx_i \wedge dx_1 \wedge \cdots \wedge \widehat{dx_j} \wedge \cdots \wedge dx_k \right]$$

$$= r\omega + \sum_{i=1}^{n} \sum_{j=1}^{k} (-1)^j \frac{\partial f}{\partial x_i} x_j \, dx_i \wedge dx_1 \wedge \cdots \wedge \widehat{dx_j} \wedge \cdots \wedge dx_k.$$

In the last step we have used Euler's identity $\sum_i x_i \partial f / \partial x_i = rf$ for $f \in \mathcal{H}_r$. On the other hand,

$$d\kappa\omega = d \left[\sum_{j=1}^{k} (-1)^{j-1} f x_j \, dx_1 \wedge \cdots \wedge \widehat{dx_j} \wedge \cdots \wedge dx_k \right]$$

$$= \sum_{j=1}^{k} \sum_{i=1}^{n} (-1)^j \, {}^1 \frac{\partial f x_j}{\partial x_i} \, dx_i \wedge dx_1 \wedge \cdots \wedge \widehat{dx_j} \wedge \cdots \wedge dx_k$$

$$= \sum_{j=1}^{k} \sum_{i=1}^{n} (-1)^{j-1} \left(f \delta_{ij} + \frac{\partial f}{\partial x_i} x_j \right) dx_i \wedge dx_1 \wedge \cdots \wedge \widehat{dx_j} \wedge \cdots \wedge dx_k$$

$$= k\omega + \sum_{j=1}^{k} \sum_{i=1}^{n} (-1)^{j-1} \frac{\partial f}{\partial x_i} x_j \, dx_i \wedge dx_1 \wedge \cdots \wedge \widehat{dx_j} \wedge \cdots \wedge dx_k.$$

Adding these two expressions gives the desired result. □

Remark. An alternative proof of the theorem is based on the *homotopy formula* of differential geometry (see Lang (1995, Chapter V, Proposition 5.3) or Taylor (1996, Chapter 1, Proposition 13.1)), which states that for any vector field v on a manifold

$$d(\omega \lrcorner v) + (d\omega) \lrcorner v = \mathcal{L}_v \omega.$$

Here $\mathcal{L}_v \omega$ denotes the *Lie derivative* of ω with respect to v, defined by

$$\mathcal{L}_v \omega = \frac{d}{dt} (\alpha_t^* \omega) \Big|_{t=0},$$

where $x \mapsto \alpha_t(x) \in \Omega$ is the *flow* defined for (x,t) in a neighbourhood of $\Omega \times \{0\}$ in $\Omega \times \mathbb{R}$ by $d\alpha_t(x)/dt = v(\alpha_t(x))$ and $\alpha_0(x) = x$. For the vector field $v = X$, it is easy to check that the flow is simply $\alpha_t(x) = e^t x$, so

$$(\alpha_t^* \omega)_x(v_1, \ldots, v_k) = \omega_{e^t x}(e^t v_1, \ldots, e^t v_k) = e^{(r+k)t} \omega_x(v_1, \ldots, v_k),$$

for $\omega \in \mathcal{H}_r \Lambda^k$. Differentiating with respect to t and setting $t = 0$, we obtain the desired result.

As a simple consequence of Theorem 3.9, we prove the injectivity of d on the range of κ and *vice versa*.

Theorem 3.2. If $d\kappa\omega = 0$ for some $\omega \in \mathcal{P}\Lambda$, then $\kappa\omega = 0$. If $\kappa d\omega = 0$ for some $\omega \in \mathcal{P}\Lambda$, then $d\omega = 0$.

Proof. We may assume that $\omega \in \mathcal{H}_r \Lambda^k$ for some $r, k \geq 0$. If $r = k = 0$, the result is trivial, so we may assume that $r + k > 0$. Then

$$(r + k)\kappa\omega = \kappa(d\kappa + \kappa d)\omega = 0,$$

if $d\kappa\omega = 0$, so $\kappa\omega = 0$ in this case. Similarly,

$$(r + k)d\omega = d(d\kappa + \kappa d)\omega = 0,$$

if $\kappa d\omega = 0$. $\qquad\qquad\qquad\qquad\qquad\qquad\qquad\qquad\qquad\qquad\qquad\square$

Another easy application of (3.9) is to establish the claimed cohomology of the Koszul complex and polynomial de Rham complex. Suppose that $\omega \in \mathcal{H}_r \Lambda^k$ for some $r, k \geq 0$ with $r + k > 0$, and that $\kappa\omega = 0$. From (3.9), we see that $\omega = \kappa\eta$ with $\eta = d\omega/(r + k) \in \mathcal{H}_{r-1} \Lambda^{k+1}$. This establishes the exactness of the homogeneous Koszul complex (3.8) (except when $r = 0$ and the sequence reduces to $0 \to \mathbb{R} \to 0$). A similar argument establishes the exactness of (3.2).

Another immediate but important consequence of (3.9) is a direct sum decomposition of $\mathcal{H}_r \Lambda^k$ for $r, k \geq 0$ with $r + k > 0$:

$$\mathcal{H}_r \Lambda^k = \kappa \mathcal{H}_{r-1} \Lambda^{k+1} \oplus d\mathcal{H}_{r+1} \Lambda^{k-1}. \qquad (3.10)$$

Indeed, if $\omega \in \mathcal{H}_r \Lambda^k$, then $\eta = d\omega/(r+k) \in \mathcal{H}_{r-1} \Lambda^{k+1}$ and $\mu = \kappa\omega/(r+k) \in \mathcal{H}_{r+1} \Lambda^{k-1}$ and $\omega = \kappa\eta + d\mu$, so $\mathcal{H}_r \Lambda^k = \kappa \mathcal{H}_{r-1} \Lambda^{k+1} + d\mathcal{H}_{r+1} \Lambda^{k-1}$. Also, if $\omega \in \kappa \mathcal{H}_{r-1} \Lambda^{k+1} \cap d\mathcal{H}_{r+1} \Lambda^{k-1}$, then $d\omega = \kappa\omega = 0$ (since $d \circ d = \kappa \circ \kappa = 0$), and so, by (3.9), $\omega = 0$. This shows that the sum is direct. Since $\mathcal{P}_r \Lambda^k = \bigoplus_{j=0}^r \mathcal{H}_j \Lambda^k$, we also have

$$\mathcal{P}_r \Lambda^k = \kappa \mathcal{P}_{r-1} \Lambda^{k+1} \oplus d\mathcal{P}_{r+1} \Lambda^{k-1}. \qquad (3.11)$$

We now use the exactness of the Koszul complex to compute the dimension of the summands in (3.10).

Theorem 3.3. Let $r \geq 0$, $1 \leq k \leq n$, for integers r, k, and n. Then

$$\dim \kappa \mathcal{H}_r \Lambda^k(\mathbb{R}^n) = \dim \mathrm{d}\mathcal{H}_{r+1}\Lambda^{k-1}(\mathbb{R}^n) = \binom{n+r}{n-k}\binom{r+k-1}{k-1}. \quad (3.12)$$

Proof. Applying κ to both sides of (3.10), we have

$$\kappa \mathcal{H}_r \Lambda^k(\mathbb{R}^n) = \kappa \mathrm{d}\mathcal{H}_{r+1}\Lambda^{k-1}(\mathbb{R}^n).$$

But κ is injective on the range of d by Theorem 3.2. Thus the first equality of (3.12) holds.

We turn to the proof of the dimension formula

$$\dim \kappa \mathcal{H}_r \Lambda^k(\mathbb{R}^n) = \binom{n+r}{n-k}\binom{r+k-1}{k-1}. \quad (3.13)$$

From the exactness properties of the Koszul complex, we know that the Koszul operator is injective on $\mathcal{H}_0\Lambda^k(\mathbb{R}^n)$ for all $k \geq 1$ and on $\mathcal{H}_r\Lambda^n(\mathbb{R}^n)$ for all $r \geq 0$, so the formula is trivially verified in these cases. Now the range of κ acting on $\mathcal{H}_r\Lambda^k(\mathbb{R}^n)$ is equal to the dimension of $\mathcal{H}_r\Lambda^k(\mathbb{R}^n)$ minus the dimension of the null space of κ on that space. By the exactness of the Koszul complex, the null space is $\kappa \mathcal{H}_{r-1}\Lambda^{k+1}(\mathbb{R}^n)$. By (3.1)

$$\dim \mathcal{H}_r\Lambda^k(\mathbb{R}^n) = \dim \mathcal{P}_r(\mathbb{R}^n)\Lambda^k(\mathbb{R}^{n-1}) = \binom{n+r-1}{n-1}\binom{n}{k}.$$

Thus

$$\dim \kappa \mathcal{H}_r\Lambda^k(\mathbb{R}^n) = \binom{n+r-1}{n-1}\binom{n}{k} - \dim \kappa \mathcal{H}_{r-1}\Lambda^{k+1}(\mathbb{R}^n). \quad (3.14)$$

The dimension formula (3.13) follows from this equation and a backward induction on k, the case $k = n$ being known. Indeed suppose that (3.12) holds for all r with k replaced by $k+1$. Substituting this (with r replaced by $r-1$) into (3.14) and using the binomial identity

$$\binom{n+r-1}{n-1}\binom{n}{k} - \binom{n+r-1}{n-k-1}\binom{r+k-1}{k} = \binom{n+r}{n-k}\binom{r+k-1}{k-1},$$

we obtain the result. $\qquad \square$

3.3. The space $\mathcal{P}_r^-\Lambda^k$

Let $r \geq 1$. Obviously, $\mathcal{P}_r\Lambda^k = \mathcal{P}_{r-1}\Lambda^k + \mathcal{H}_r\Lambda^k$. In view of (3.10), we may define a space of k-forms intermediate between $\mathcal{P}_{r-1}\Lambda^k$ and $\mathcal{P}_r\Lambda^k$ by

$$\mathcal{P}_r^-\Lambda^k = \mathcal{P}_{r-1}\Lambda^k + \kappa \mathcal{H}_{r-1}\Lambda^{k+1} = \mathcal{P}_{r-1}\Lambda^k + \kappa \mathcal{P}_{r-1}\Lambda^{k+1}.$$

Note that the first sum is direct, while the second need not be. An equivalent definition is

$$\mathcal{P}_r^-\Lambda^k = \{\, \omega \in \mathcal{P}_r\Lambda^k \mid \kappa\omega \in \mathcal{P}_r\Lambda^{k-1} \,\}.$$

Note that $\mathcal{P}_r^- \Lambda^0 = \mathcal{P}_r \Lambda^0$ and $\mathcal{P}_r^- \Lambda^n = \mathcal{P}_{r-1}\Lambda^n$, but for $0 < k < n$, $\mathcal{P}_r^- \Lambda^k$ is contained strictly between $\mathcal{P}_{r-1}\Lambda^k$ and $\mathcal{P}_r \Lambda^k$. For $r \leq 0$, we set $\mathcal{P}_r^- \Lambda^k = 0$.

From (3.12), we have

$$
\begin{aligned}
\dim \mathcal{P}_r^- \Lambda^k(\mathbb{R}^n) &= \dim \mathcal{P}_{r-1}\Lambda^k + \dim \kappa \mathcal{H}_{r-1}\Lambda^{k+1} \\
&= \binom{n+r-1}{n}\binom{n}{k} + \binom{n+r-1}{n-k-1}\binom{r+k-1}{k} \\
&= \binom{r+k-1}{k}\binom{n+r}{n-k},
\end{aligned}
\tag{3.15}
$$

where the last step is a simple identity.

Analogous to the obvious closure relation

$$
\mathcal{P}_r \Lambda^k \wedge \mathcal{P}_s \Lambda^l \subset \mathcal{P}_{r+s}\Lambda^{k+l},
$$

the $\mathcal{P}_r^- \Lambda^k$ spaces satisfy

$$
\mathcal{P}_r^- \Lambda^k \wedge \mathcal{P}_s^- \Lambda^l \subset \mathcal{P}_{r+s}^- \Lambda^{k+l},
\tag{3.16}
$$

first proved in Christiansen (2005). To prove (3.16), it suffices to show that

$$
\kappa \mathcal{H}_{r-1}\Lambda^{k+1} \wedge \kappa \mathcal{H}_{s-1}\Lambda^{l+1} \subset \kappa \mathcal{H}_{r+s-1}\Lambda^{k+l-1}.
$$

By the exactness of the Koszul complex, it is enough to show that $\kappa(\kappa\omega \wedge \kappa\mu) = 0$. But this follows immediately from (3.5) and (3.6). Taking $l = 0$ and noting that $\mathcal{P}_s^- \Lambda^l = \mathcal{P}_s$, we get

$$
p \in \mathcal{P}_s, \ \omega \in \mathcal{P}_r^- \Lambda^k \implies p\omega \in \mathcal{P}_{r+s}^- \Lambda^k.
\tag{3.17}
$$

We close by noting a simple consequence of Lemma 3.2.

Theorem 3.4. If $\omega \in \mathcal{P}_r^- \Lambda^k$ and $d\omega = 0$, then $\omega \in \mathcal{P}_{r-1}\Lambda^k$.

Proof. Write $\omega = \omega_1 + \kappa\omega_2$ with $\omega_1 \in \mathcal{P}_{r-1}\Lambda^k$ and $\omega_2 \in \mathcal{P}_{r-1}\Lambda^{k+1}$. Then

$$
d\omega = 0 \implies d\kappa\omega_2 = 0 \implies \kappa\omega_2 = 0 \implies \omega \in \mathcal{P}_{r-1}\Lambda^k. \qquad \square
$$

3.4. Invariant spaces of polynomial differential forms

We have already noted in (3.4) that the spaces $\mathcal{P}_r \Lambda^k$ of polynomial differential forms are affine-invariant, *i.e.*, mapped into themselves by the pullback of affine transformations of $\mathbb{R}^n$. This is a stronger property than linear invariance (invariance under the pullback of linear transformations). For example, $\mathcal{H}_r \Lambda^k$ is linear-invariant, but not affine-invariant. Let us explain the significance of affine invariance for finite element spaces of differential forms. In the next section we define the space $\mathcal{P}_r \Lambda^k(T)$ for an n-simplex T to be the space of restrictions of polynomials in $\mathcal{P}_r \Lambda^k$ to the simplex. In the following section we define the finite element space $\mathcal{P}_r \Lambda_-^k(\mathcal{T}_h)$ for a simplicial complex $\mathcal{T}_h$ consisting of piecewise polynomial differential forms which restrict to $\mathcal{P}_r \Lambda^k(T)$ on each $T \in \mathcal{T}_h$. Another possible construction would be to select a single reference simplex $\hat{T}$, and define $\mathcal{P}_r \Lambda^k(\hat{T})$, and then to define $\mathcal{P}_r \Lambda^k(T) = \Phi^*(\mathcal{P}_r \Lambda^k(\hat{T}))$ for any other simplex T, where $\Phi : \hat{T} \to T$ is an affine isomorphism. Affine invariance shows that these two definitions of $\mathcal{P}_r \Lambda^k(T)$ are the same (and so the space does not depend on the choice of affine isomorphism Φ of $\hat{T}$ on T).

It is a relatively easy matter to see that the only finite-dimensional affine-invariant spaces of polynomial 0-forms (i.e., ordinary polynomial functions) are the spaces $\mathcal{P}_r \Lambda^0$ for $r = 0, 1, 2, \ldots$ (and this will follow from the techniques below). Similarly, the only finite-dimensional affine-invariant spaces of polynomial n-forms are the spaces $\mathcal{P}_r \Lambda^n$. However, for $0 < k < n$, there are other affine-invariant spaces of polynomial k-forms. Specifically, the spaces $\mathcal{P}_r^- \Lambda^k$, $r = 1, 2, \ldots$, are affine-invariant. In this subsection, we shall determine *all* the affine-invariant subspaces of polynomial k-forms.

First we note that the decomposition of $\mathcal{H}_r \Lambda^k$ given in (3.10) is a decomposition into subspaces which are linear-invariant (but not affine-invariant). Indeed, if $\phi : \mathbb{R}^n \to \mathbb{R}^n$ is linear, then from the relations $\phi^* d = d\phi^*$ (which holds for any transformation ϕ), and $\phi^* \kappa = \kappa \phi^*$ (which holds for ϕ linear), and the invariance under linear transformations of the spaces of the homogeneous forms, we have

$$\phi^* d\mathcal{H}_{r+1}\Lambda^{k-1} = d\phi^*\mathcal{H}_{r+1}\Lambda^{k-1} \subset d\mathcal{H}_{r+1}\Lambda^{k-1},$$
$$\phi^* \kappa\mathcal{H}_{r-1}\Lambda^{k+1} = \kappa\phi^*\mathcal{H}_{r-1}\Lambda^{k+1} \subset \kappa\mathcal{H}_{r-1}\Lambda^{k+1}.$$

This establishes the invariance of the summands.

The same argument shows that the space $d\mathcal{P}_{r+1}\Lambda^{k-1}$ is invariant under affine transformations:

$$\phi^* d\mathcal{P}_{r+1}\Lambda^{k-1} = d\phi^*\mathcal{P}_{r+1}\Lambda^{k-1} \subset d\mathcal{P}_{r+1}\Lambda^{k-1}. \tag{3.18}$$

For the range of κ, we can only get a weaker result using (3.7), namely

$$\phi^* \kappa\mathcal{P}_{r-1}\Lambda^{k+1} \subset \kappa\phi^*\mathcal{P}_{r-1}\Lambda^{k+1} + \mathcal{P}_{r-1}\Lambda^k \subset \kappa\mathcal{P}_{r-1}\Lambda^{k+1} + \mathcal{P}_{r-1}\Lambda^k. \tag{3.19}$$

We now combine these results to find several affine-invariant subspaces of $\mathcal{P}_r \Lambda^k$. Below we shall show that these are the only such subspaces.

Theorem 3.5. Let $0 < k < n$, $r \geq 0$, $-1 \leq s \leq r$, for integers k, n, r, and s. Then the space

$$X(r, s, k, n) := d\mathcal{P}_{r+1}\Lambda^{k-1}(\mathbb{R}^n) + \kappa\mathcal{P}_s\Lambda^{k+1}(\mathbb{R}^n)$$

is an affine-invariant subspace of $\mathcal{P}\Lambda^k(\mathbb{R}^n)$. Furthermore:

- (case $s = r$) $X(r, r, k, n) = \mathcal{P}_{r+1}^-\Lambda^k(\mathbb{R}^n)$;
- (case $s = r - 1$) $X(r, r - 1, k, n) = \mathcal{P}_r\Lambda^k(\mathbb{R}^n)$;
- (case $s < r - 1$) if $-1 \le s < r - 1$, then

$$X(r, s, k, n) = \mathcal{P}_{s+1}\Lambda^k(\mathbb{R}^n) + d\mathcal{P}_{r+1}\Lambda^{k-1}(\mathbb{R}^n)$$
$$= \{\, \omega \in \mathcal{P}_r\Lambda^k(\mathbb{R}^n) \mid d\omega \in \mathcal{P}_s\Lambda^k(\mathbb{R}^n) \,\}.$$

This space is contained strictly between $\mathcal{P}_{s+1}\Lambda^k$ and $\mathcal{P}_r\Lambda^k$, but does not contain $\mathcal{P}_{s+2}\Lambda^k$.

Proof.   From (3.18), (3.19), (3.11), and the fact that $s \le r$, we have

$$\phi^* X(r, s, k, n) \subset d\mathcal{P}_{r+1}\Lambda^{k-1} + \kappa\mathcal{P}_s\Lambda^{k+1} + \mathcal{P}_s\Lambda^k$$
$$= d\mathcal{P}_{r+1}\Lambda^{k-1} + \kappa\mathcal{P}_s\Lambda^{k+1} + d\mathcal{P}_{s+1}\Lambda^{k-1} + \kappa\mathcal{P}_{s-1}\Lambda^{k-1}$$
$$= d\mathcal{P}_{r+1}\Lambda^{k-1} + \kappa\mathcal{P}_s\Lambda^{k+1} = X(r, s, k, n),$$

which is the claimed invariance. The bulleted points are then simple observations.   □

Thus, for $0 < k < n$, there are three distinct types of finite-dimensional affine-invariant spaces of polynomial differential k-forms (and, as we shall soon show, these are the only ones):

- the spaces $\mathcal{P}_r\Lambda^k$ of all polynomial k-forms up to a given degree, r;
- the reduced spaces $\mathcal{P}_r^-\Lambda^k$; and
- the spaces consisting of all $\omega \in \mathcal{P}_r\Lambda^k$ for which the exterior derivative $d\omega$ is constrained to belong to $\mathcal{P}_s\Lambda^k$ for some $-1 \le s < r - 1$ (with $s = -1$ corresponding to the constraint $d\omega = 0$).

In this paper we will investigate spaces of piecewise polynomial differential forms for which the pieces belong to one of the spaces $\mathcal{P}_r\Lambda^k$ or $\mathcal{P}_r^-\Lambda^k$, *i.e.*, to spaces of the first or second kind listed. The third class of spaces will not be considered. Up until now, these spaces have not been widely used as mixed finite element spaces, and it is not clear that there is a motivation to do so. However, a vector-valued analogue of these spaces played a major role in the development of stable mixed finite elements for elasticity in Arnold and Winther (2002).

In the remainder of this subsection, we show that the spaces $X(r, s, k, n)$ given in Theorem 3.5 are the *only* finite-dimensional affine-invariant subspaces of $\mathcal{P}\Lambda^k$. This result will not be needed later, and so the reader uninterested in the proof may safely skip ahead to Section 3.5.

Theorem 3.6.   Let $0 < k < n$, and suppose that $X \subset \mathcal{P}\Lambda^k$ is a non-zero finite-dimensional subspace satisfying $\phi^* X \subset X$ for all affine maps

$\phi : \mathbb{R}^n \to \mathbb{R}^n$. Then $X = X(r, s, k, n)$ for some integers r, s, with $r \geq 0$, $-1 \leq s \leq r$.

The proof will be based on the representation theory of the general linear group, for which we will first summarize the main results needed. These results may be gleaned from Fulton and Harris (1991), especially Section 6.1. Via the pullback, the group $\mathrm{GL}(\mathbb{R}^n)$ acts on $\mathcal{P}\Lambda^k$ and on its subspace $\mathcal{H}_r\Lambda^k$. From equation (6.9) of Fulton and Harris (1991) in the case $\lambda = (r)$ and $m = k$, and the accompanying discussion, we find that $\mathcal{H}_r\Lambda^k$ (which is $\mathrm{Sym}^r V \otimes \bigwedge^k V$ in the notation of Fulton and Harris (1991)) has precisely two nonzero proper invariant subspaces under this action. Since we have already established that $d\mathcal{H}_{r+1}\Lambda^{k-1}$ and $\kappa\mathcal{H}_{r-1}\Lambda^{k+1}$ are such subspaces, the decomposition (3.10) is the decomposition of $\mathcal{H}_r\Lambda^k$ into *irreducible* linear-invariant subspaces. Moreover, all the nonzero spaces $d\mathcal{H}_{r+1}\Lambda^{k-1}$, $\kappa\mathcal{H}_{r-1}\Lambda^{k+1}$ (for varying r and k) are inequivalent as representations, because, as explained in Fulton and Harris (1991), they are the images of projections associated with different partitions (or different Young diagrams). This means that there does not exist a linear isomorphism between any two of them which commutes with the pullback action. Consequently, we may write down the decomposition of $\mathcal{P}\Lambda^k = \bigoplus_{r=0} \mathcal{H}_r\Lambda^k$ into irreducible linear-invariant subspaces,

$$\mathcal{P}\Lambda^k = \bigoplus_{j=1}^{\infty} d\mathcal{H}_j\Lambda^{k-1} \oplus \bigoplus_{i=0}^{\infty} \kappa\mathcal{H}_i\Lambda^{k+1}, \tag{3.20}$$

and from this decomposition we can read off all the finite-dimensional linear-invariant subspaces of $\mathcal{P}_r\Lambda^k$: they are just the sums of some finite number of the summands appearing in (3.20).

The next step is to determine which of the linear-invariant subspaces is actually affine-invariant. We shall do this by considering the effect of the pullback by the translation operation. First we introduce some notation. Let

$$\pi_r^{\mathrm{d}} : \mathcal{P}\Lambda^k \to d\mathcal{H}_{r+1}\Lambda^{k-1}, \qquad \pi_s^{\kappa} : \mathcal{P}\Lambda^k \to \kappa\mathcal{H}_{s-1}\Lambda^{k+1}$$

denote the projections determined by the decomposition (3.20). Denote by $\tau : \mathbb{R}^n \to \mathbb{R}^n$ the unit translation in the x_1 direction:

$$\tau(x) = (x_1 + 1, x_2, \ldots, x_n).$$

The proof of Theorem 3.6 will follow easily from the next lemma.

Lemma 3.7. Let $0 < k < n$.

(1) For any $r \geq 1$, there exists $\omega \in d\mathcal{H}_{r+1}\Lambda^{k-1}$ such that $\pi_{r-1}^{\mathrm{d}}(\tau^*\omega) \neq 0$.

(2) For any $s \geq 2$, there exists $\omega \in \kappa\mathcal{H}_{s-1}\Lambda^{k+1}$ such that $\pi_{s-1}^{\kappa}(\tau^*\omega) \neq 0$.

(3) For any $s \geq 1$, there exists $\omega \in \kappa\mathcal{H}_{s-1}\Lambda^{k+1}$ such that $\pi_{s-1}^{\mathrm{d}}(\tau^*\omega) \neq 0$.

Proof. For the proof we will exhibit such forms ω explicitly, and verify the result by direct computation.

(1) Let

$$\omega = (r+1)^{-1}\, d(x_1^{r+1}\, dx_2 \wedge \cdots \wedge dx_k) = x_1^r\, dx_1 \wedge \cdots \wedge dx_k \in d\mathcal{H}_{r+1}\Lambda^{k-1}.$$

Then

$$\tau^*\omega = (x_1+1)^r\, dx_1 \wedge \cdots \wedge dx_k$$
$$= x_1^r\, dx_1 \wedge \cdots \wedge dx_k + rx_1^{r-1}\, dx_1 \wedge \cdots \wedge dx_k + \cdots,$$

where we have expanded by polynomial degree. The term of degree $r-1$ is

$$rx_1^{r-1}\, dx_1 \wedge \cdots \wedge dx_k = d(x_1^r\, dx_2 \wedge \cdots \wedge dx_k) \in d\mathcal{H}_r\Lambda^{k-1}.$$

Therefore $\pi_{r-1}^d(\tau^*\omega) = rx_1^{r-1}\, dx_1 \wedge \cdots \wedge dx_k \neq 0$.

(2) Let

$$\omega = \kappa(x_1^{s-1}\, dx_1 \wedge \cdots \wedge dx_{k+1})$$

$$= x_1^{s-1} \sum_{j=1}^{k+1}(-1)^{j+1}x_j\, dx_1 \wedge \cdots \wedge \widehat{dx_j} \wedge \cdots \wedge dx_{k+1} \in \kappa\mathcal{H}_{s-1}\Lambda^{k+1}.$$

Then

$$\tau^*\omega = (x_1+1)^s\, dx_2 \wedge \cdots \wedge dx_{k+1}$$

$$+ (x_1+1)^{s-1} \sum_{j=2}^{k+1}(-1)^{j+1}x_j\, dx_1 \wedge \cdots \wedge \widehat{dx_j} \wedge \cdots \wedge dx_{k+1}.$$

Letting $\mu = \pi_{s-1}(\tau^*\omega)$, where $\pi_{s-1} : \mathcal{P}\Lambda^k \to \mathcal{H}_{s-1}\Lambda^k$ is the projection onto homogeneous polynomial forms of degree $s-1$, we get

$$\mu = sx_1^{s-1}\, dx_2 \wedge \cdots \wedge dx_{k+1}$$

$$+ (s-1)x_1^{s-2} \sum_{j=2}^{k+1}(-1)^{j+1}x_j\, dx_1 \wedge \cdots \wedge \widehat{dx_j} \wedge \cdots \wedge dx_{k+1}.$$

Then

$$d\mu = s(s-1)x_1^{s-1}\, dx_1 \wedge \cdots \wedge dx_{k+1}$$

$$+ (s-1)x_1^{s-2} \sum_{j=2}^{k+1}(-1)^{j+1}\, dx_j \wedge \cdots \wedge \widehat{dx_j} \wedge \cdots \wedge dx_{k+1}$$

$$= (s^2-1)x_1^{s-1}\, dx_1 \wedge \cdots \wedge dx_{k+1} \neq 0.$$

Thus $\mu \in \mathcal{H}_{s-1}\Lambda^k$, but $\mu \notin d\mathcal{H}_s\Lambda^{k+1}$. Therefore $\pi_{s-1}^d(\tau^*\omega) = \pi_{s-1}^d\mu \neq 0$.

(3) Let

$$\omega = \kappa(x_n^s \, dx_1 \wedge \cdots \wedge dx_{k+1}) = x_n^s \sum_{j=1}^{k+1} (-1)^{j+1} x_j \, dx_1 \wedge \cdots \wedge \widehat{dx_j} \wedge \cdots \wedge dx_{k+1}.$$

Then $\tau^* \omega = \omega + \mu$ where $\mu = x_n^s \, dx_2 \wedge \cdots \wedge dx_{k+1}$. Now

$$\kappa\mu = x_n^s \sum_{j=2}^{k+1} (-1)^j x_j \, dx_2 \wedge \cdots \wedge \widehat{dx_j} \wedge \cdots \wedge dx_{k+1} \neq 0.$$

Thus $\mu \in \mathcal{H}_s \Lambda^{k-1}$ but $\mu \notin \kappa \mathcal{H}_{s-1} \Lambda^k$, so $\pi_{s+1}^d \tau^* \omega = \pi_{s+1}^d \tau^* \mu \neq 0$. $\qquad \square$

Proof of Theorem 3.6. Finally, we complete the proof of Theorem 3.6. If X is an affine-invariant subspace of $\mathcal{P}\Lambda^k$, it is *a fortiori* linear-invariant, and so a sum of finitely many summands from (3.20). Also $\tau^* X \subset X$. Suppose that $d\mathcal{H}_{r+1}\Lambda^{k-1}$ is the highest degree summand in the range of d. It follows from the first statement of the lemma that $d\mathcal{H}_r \Lambda^{k-1}$ must then be a summand as well, and so, by induction, $d\mathcal{P}_{r+1}\Lambda^{k-1} \subset X$. Similarly, if $\kappa \mathcal{H}_s \Lambda^{k+1}$ is the highest degree summand in the range of κ, the second statement of the lemma ensures that $\kappa \mathcal{P}_s \Lambda^{k+1} \subset X$. Finally, the third statement of the lemma ensures that $r \geq s$. Thus $X = d\mathcal{P}_{r+1}\Lambda^{k-1} + \kappa \mathcal{P}_s \Lambda^{k+1} = X(r, s, k, n)$. This completes the proof. $\qquad \square$

3.5. Exact sequences of polynomial differential forms

We have seen the polynomial de Rham complex (3.3) is a subcomplex of the de Rham complex on $\mathbb{R}^n$ for which cohomology vanishes except for the constants at the lowest order. In other words, the sequence

$$\mathbb{R} \hookrightarrow \mathcal{P}_r \Lambda^0 \xrightarrow{d} \mathcal{P}_{r-1} \Lambda^1 \xrightarrow{d} \cdots \xrightarrow{d} \mathcal{P}_{r-n} \Lambda^n \to 0 \qquad (3.21)$$

is exact for any $r \geq 0$ (some of the spaces vanish if $r < n$). Such a complex, namely one which begins with the inclusion of $\mathbb{R}$, has vanishing cohomology, and terminates at 0, is called a *resolution* of $\mathbb{R}$.

As we shall soon verify, the complex

$$\mathbb{R} \hookrightarrow \mathcal{P}_r^- \Lambda^0 \xrightarrow{d} \mathcal{P}_r^- \Lambda^1 \xrightarrow{d} \cdots \xrightarrow{d} \mathcal{P}_r^- \Lambda^n \to 0 \qquad (3.22)$$

is another resolution of $\mathbb{R}$, for any $r > 0$. Note that in this complex, involving the $\mathcal{P}_r^- \Lambda^k$ spaces, the polynomial degree r is held fixed, while in (3.21), the polynomial degree decreases as the form order increases. Recall that the 0th order spaces in these complexes, $\mathcal{P}_r \Lambda^0$ and $\mathcal{P}_r^- \Lambda^0$, coincide. In fact, the complex (3.21) is a subcomplex of (3.22), and these two are the extreme cases of a set of 2^{n-1} different resolutions of $\mathbb{R}$, each a subcomplex of the next, and all of which have the space $\mathcal{P}_r \Lambda^0$ in the 0th order.

To prove all this, we first prove a simple lemma.

Lemma 3.8.

(1) For $r \geq 1$, $d\mathcal{P}_r^- \Lambda^k \subset d\mathcal{P}_r \Lambda^k \subset \mathcal{P}_{r-1} \Lambda^{k+1} \subset \mathcal{P}_r^- \Lambda^{k+1}$.

(2) The following four restrictions of d each have the same kernel:

$$d : \mathcal{P}_r \Lambda^k \to \mathcal{P}_{r-1} \Lambda^{k+1}, \quad d : \mathcal{P}_r \Lambda^k \to \mathcal{P}_r^- \Lambda^{k+1},$$
$$d : \mathcal{P}_{r+1}^- \Lambda^k \to \mathcal{P}_r \Lambda^{k+1}, \quad d : \mathcal{P}_{r+1}^- \Lambda^k \to \mathcal{P}_{r+1}^- \Lambda^{k+1}.$$

(3) The following four restrictions of d each have the same image:

$$d : \mathcal{P}_r \Lambda^k \to \mathcal{P}_{r-1} \Lambda^{k+1}, \quad d : \mathcal{P}_r \Lambda^k \to \mathcal{P}_r^- \Lambda^{k+1},$$
$$d : \mathcal{P}_r^- \Lambda^k \to \mathcal{P}_{r-1} \Lambda^{k+1}, \quad d : \mathcal{P}_r^- \Lambda^k \to \mathcal{P}_r^- \Lambda^{k+1}.$$

Proof. The first statement is clear. To prove the second, we need to show that if $\omega \in \mathcal{P}_{r+1}^- \Lambda^k$ with $d\omega = 0$, then $\omega \in \mathcal{P}_r \Lambda^k$. This follows from Theorem 3.4. Finally, to prove the third statement, it suffices to note that $\mathcal{P}_r \Lambda^k = \mathcal{P}_r^- \Lambda^k + d\mathcal{P}_{r+1} \Lambda^{k-1}$, so $d\mathcal{P}_r \Lambda^k = d\mathcal{P}_r^- \Lambda^k$. □

We now exhibit 2^{n-1} resolutions of $\mathbb{R}$, subcomplexes of the de Rham complex, beginning $\mathbb{R} \hookrightarrow \mathcal{P}_r \Lambda^0$. In view of the lemma, we may continue the complex with the map $d : \mathcal{P}_r \Lambda^0 \to \mathcal{P}_{r-1} \Lambda^1$ or $d : \mathcal{P}_r \Lambda^0 \to \mathcal{P}_r^- \Lambda^1$. The former is a subcomplex of the latter. With either choice, the cohomology vanishes at the first position. (In the former case we use the cohomology of (3.21), and in the latter we get the same result thanks to the lemma.)

Next, if we made the first choice, we can continue the complex with either $d : \mathcal{P}_{r-1} \Lambda^1 \to \mathcal{P}_{r-2} \Lambda^2$ or $d : \mathcal{P}_{r-1} \Lambda^1 \to \mathcal{P}_{r-1}^- \Lambda^2$. Or, if we made the second choice, we can continue with either $d : \mathcal{P}_r^- \Lambda^1 \to \mathcal{P}_{r-1} \Lambda^2$ or $d : \mathcal{P}_r^- \Lambda^1 \to \mathcal{P}_r^- \Lambda^2$. In any case, we may use the lemma and the exactness of (3.21) to see that the second cohomology space vanishes. Continuing in this way at each order, $k = 1, \ldots, n-1$, we have two choices for the space of k-forms (but only one choice for $k = n$, since $\mathcal{P}_{r-1} \Lambda^n$ coincides with $\mathcal{P}_r^- \Lambda^n$), and so we obtain 2^{n-1} complexes. These form a totally ordered set with respect to subcomplexes. For $r \geq n$ these are all distinct (but for small r some coincide because the later spaces vanish).

In the case $n = 3$, the four complexes so obtained are:

$$\mathbb{R} \hookrightarrow \mathcal{P}_r \Lambda^0 \xrightarrow{d} \mathcal{P}_{r-1} \Lambda^1 \xrightarrow{d} \mathcal{P}_{r-2} \Lambda^2 \xrightarrow{d} \mathcal{P}_{r-3} \Lambda^3 \to 0,$$

$$\mathbb{R} \hookrightarrow \mathcal{P}_r \Lambda^0 \xrightarrow{d} \mathcal{P}_{r-1} \Lambda^1 \xrightarrow{d} \mathcal{P}_{r-1}^- \Lambda^2 \xrightarrow{d} \mathcal{P}_{r-2} \Lambda^3 \to 0,$$

$$\mathbb{R} \hookrightarrow \mathcal{P}_r \Lambda^0 \xrightarrow{d} \mathcal{P}_r^- \Lambda^1 \xrightarrow{d} \mathcal{P}_{r-1} \Lambda^2 \xrightarrow{d} \mathcal{P}_{r-2} \Lambda^3 \to 0,$$

$$\mathbb{R} \hookrightarrow \mathcal{P}_r \Lambda^0 \xrightarrow{d} \mathcal{P}_r^- \Lambda^1 \xrightarrow{d} \mathcal{P}_r^- \Lambda^2 \xrightarrow{d} \mathcal{P}_{r-1} \Lambda^3 \to 0.$$

3.6. Change of origin

We defined the Koszul differential as $\kappa = \lrcorner X$, where $X(x)$ is the translation to x of the vector pointing from the origin in $\mathbb{R}^n$ to x. The choice of the origin as a base point is arbitrary – any point in $\mathbb{R}^n$ could be used. That is, if $y \in \mathbb{R}^n$, we can define a vector field X_y by assigning to each point x the translation to x of the vector pointing from y to x, and then define a Koszul operator $\kappa_y = \lrcorner X_y$. It is easy to check that for $\omega \in \mathcal{P}_{r-1}\Lambda^{k+1}$ and any two points $y, y' \in \mathbb{R}^n$, the difference $\kappa_y \omega - \kappa_{y'} \omega \in \mathcal{P}_{r-1}\Lambda^k$. Hence the space

$$\mathcal{P}_r^- \Lambda^k = \mathcal{P}_{r-1}\Lambda^k + \kappa_y \mathcal{P}_{r-1}\Lambda^{k+1}$$

does not depend on the particular choice of the point y. This observation is important, because it allows us to define $\mathcal{P}_r^-\Lambda^k(V)$ for any affine subspace V of $\mathbb{R}^n$. We simply set

$$\mathcal{P}_r^- \Lambda^k(V) = \mathcal{P}_{r-1}\Lambda^k(V) + \kappa_y \mathcal{P}_{r-1}\Lambda^{k+1}(V),$$

where y is any point of V. Note that if $\omega \in \mathcal{P}_r^-\Lambda^k(\mathbb{R}^n)$, then the trace of ω on V belongs to $\mathcal{P}_r^-\Lambda^k(V)$.

4. Polynomial differential forms on a simplex

Having introduced the spaces of polynomial differential forms $\mathcal{P}_r\Lambda^k(\mathbb{R}^n)$ and $\mathcal{P}_r^-\Lambda^k(\mathbb{R}^n)$, we now wish to create finite element spaces of differential forms. These will be obtained using a triangulation of the domain and assembling spaces of polynomial differential forms on each of the simplices in the triangulation. First, for each simplex T of the triangulation, we specify a space of *shape functions*. This will be either $\mathcal{P}_r\Lambda^k(T)$ or $\mathcal{P}_r^-\Lambda^k(T)$, where these denote the spaces of forms obtained by restricting the forms in $\mathcal{P}_r\Lambda^k(\mathbb{R}^n)$ and $\mathcal{P}_r^-\Lambda^k(\mathbb{R}^n)$, respectively, to T. It is also necessary to specify how these pieces are assembled to obtain a global space – in other words to specify the degree of interelement continuity. To this end, we need to specify a set of *degrees of freedom* for the shape spaces associated to a simplex – that is, a basis for the dual space – in which each degree of freedom is associated with a particular subsimplex. When a subsimplex is shared by more than one simplex in the triangulation, we will insist that the degrees of freedom associated with that subsimplex be single-valued, and this will determine the interelement continuity. This assembly process is an important part of the architecture of finite element codes, and the specification of a geometrically structured set of degrees of freedom distinguishes a finite element space from an arbitrary piecewise polynomial space. The association of the degrees of freedom to subsimplices gives a decomposition of the dual space of the shape functions on T into a direct sum of subspaces indexed by all the subsimplices of T. It is really this geometric decomposition of the dual space that determines the interelement continuity rather than the particular choice of degrees of freedom, since we may choose any convenient basis for each space in the decomposition and obtain the same assembled finite element space.

For computation with finite elements we also need a basis, not only for the dual space, but also for the space of shape functions itself, which similarly decomposes the spaces into subspaces indexed by the subsimplices. This way the necessary discretization matrices can be computed simplex by simplex and assembled into a global matrix. In this section we define such bases and decompositions for $\mathcal{P}_r\Lambda^k(T)$ and $\mathcal{P}_r^-\Lambda^k(T)$. There have been several papers which have discussed bases of these spaces for use in computation in particular cases, e.g., Webb (1999), Hiptmair (2001), Ainsworth and Coyle (2003), Gopalakrishnan, García-Castillo and Demkowicz (2005). In our presentation, we emphasize the fact, first noted in Arnold et al. (2006b), that the construction of the degrees of freedom and basis for the $\mathcal{P}_r\Lambda^k$ spaces requires the use of $\mathcal{P}_r^-\Lambda^k$ spaces on the subsimplices, and vice versa. Therefore, the two families of spaces must be studied together to get optimal results.

4.1. Simplices and barycentric coordinates

For $1 \le k \le n$, let $\Sigma(k,n)$ denote the set of increasing maps $\{1,\ldots,k\} \to \{1,\ldots,n\}$ and for $0 \le k \le n$, let $\Sigma_0(k,n)$ denote the set of increasing maps $\{0,\ldots,k\} \to \{0,\ldots,n\}$. These sets have cardinality $\binom{n}{k}$ and $\binom{n+1}{k+1}$ respectively. For $\sigma \in \Sigma(k,n)$ with $k < n$ we define $\sigma^* \in \Sigma(n-k,n)$ such that $\mathcal{R}(\sigma) \cup \mathcal{R}(\sigma^*) = \{1,\ldots,n\}$ and, similarly, for $\sigma \in \Sigma_0(k,n)$ with $k < n$, we define $\sigma^* \in \Sigma_0(n-k-1,n)$ such that $\mathcal{R}(\sigma) \cup \mathcal{R}(\sigma^*) = \{0,\ldots,n\}$. For $\sigma \in \Sigma(k,n)$ we define $(0,\sigma) \in \Sigma_0(k,n)$ by $(0,\sigma)(0) = 0$, $(0,\sigma)(j) = \sigma(j)$, $j = 1,\ldots,k$.

Let $x_0, x_1, \ldots, x_n$ be $n+1$ points in general position in $\mathbb{R}^n$ ordered so that the vectors $x_1 - x_0, \ldots, x_n - x_0$ give a positively oriented frame. Then the closed convex hull of these points, which we denote by $[x_0, \ldots, x_n]$, is the n-simplex with the vertices x_i. Call this simplex T. For each $\sigma \in \Sigma_0(k,n)$, the set $f_\sigma = [x_{\sigma(0)}, \ldots, x_{\sigma(k)}]$ is a subsimplex of dimension k. There are $\binom{n+1}{k+1}$ subsimplices of dimension k, with the vertices being the subsimplices of dimension 0, and T itself being the only subsimplex of dimension n. For $k < n$, f_{σ^*} is the $(n-k-1)$-dimensional subsimplex of T opposite to the k-subsimplex f_σ. We denote the set of subsimplices of dimension k of T by $\Delta_k(T)$, and the set of all subsimplices of T by $\Delta(T)$.

We denote by $\lambda_0, \ldots, \lambda_n$ the barycentric coordinate functions, so $\lambda_i \in \mathcal{P}_1(\mathbb{R}^n)$ is determined by the equations $\lambda_i(x_j) = \delta_{ij}$, $0 \le i, j \le n$. The λ_i form a basis for $\mathcal{P}_1(\mathbb{R}^n)$ and satisfy $\sum_i \lambda_i \equiv 1$. We have

$$T = \{\, x \in \mathbb{R}^n \mid \lambda_i(x) \ge 0, \quad i = 0, \ldots, n \,\},$$

and for the subsimplices

$$f_\sigma = \{ x \in T \mid \lambda_i(x) = 0, \ i \in \mathcal{R}(\sigma^*) \}.$$

In particular, the subsimplices of codimension one, or faces, of T are

$$F_i := [x_0, \ldots, \hat{x}_i, \ldots, x_n] = \{ x \in T \mid \lambda_i(x) = 0 \}, \quad i = 0, \ldots, n.$$

For a subsimplex $f = f_\sigma$, the functions $\lambda_{\sigma(0)}, \ldots, \lambda_{\sigma(k)}$ are the barycentric coordinates of f. Note that they are defined on all of $\mathbb{R}^n$. Their restrictions to f depends only on f, but their values off f depend on all the vertices of T. There is an isomorphism between the space $\mathcal{P}_r(f)$ of polynomial functions on f of degree at most r, i.e., the restrictions of functions in $\mathcal{P}_r(\mathbb{R}^n)$ to f, and the space $\mathcal{H}_r(\mathbb{R}^{k+1})$ of homogeneous polynomials of degree r in $k+1$ variables. Namely each $p \in \mathcal{P}_r(f)$ may be expressed as

$$p(x) = q\big(\lambda_{\sigma(0)}(x), \ldots, \lambda_{\sigma(k)}(x)\big), \quad x \in f,$$

for a unique $q \in \mathcal{H}_r(\mathbb{R}^{k+1})$. The right-hand side is defined for all $x \in \mathbb{R}^n$, and so provides a way to extend functions from $\mathcal{P}_r(f)$ to $\mathcal{P}_r(\mathbb{R}^n)$. We shall call this extension $E_{f,T}(p)$. Note that $E_{f,T}(p)$ vanishes on all subsimplices of T disjoint from f.

The bubble function associated to a subsimplex $f = f_\sigma$ of T is given by the product

$$b_f = \lambda_\sigma := \lambda_{\sigma(0)} \lambda_{\sigma(1)} \cdots \lambda_{\sigma(k)} \in \mathcal{P}_k(\mathbb{R}^n).$$

It vanishes on any subsimplex of T which does not contain f, and, in particular, on all the subsimplices of dimension less than k. Thus its restriction to f belongs to $\mathring{\mathcal{P}}_{k+1}(f)$, the subspace of $\mathcal{P}_{k+1}(f)$ consisting of polynomials functions on f which vanish on the boundary of f.

The vectors $t_i := x_i - x_0$, $i = 1, \ldots, n$, form a basis for $\mathbb{R}^n$. The dual basis functions are the 1-forms $d\lambda_i \in \mathrm{Alt}^1 \mathbb{R}^n$. Note that $d\lambda_0 = -\sum_{i=1}^n d\lambda_i$ is not included in the basis. Similarly, for any face $f = f_\sigma$, the restrictions of $d\lambda_{\sigma(1)}, \ldots, d\lambda_{\sigma(k)}$ to the tangent space V of f at any point of f (V is independent of the point), give a basis for $\mathrm{Alt}^1 V$.

The algebraic k-forms $(d\lambda)_\sigma := d\lambda_{\sigma(1)} \wedge \cdots \wedge d\lambda_{\sigma(k)}$, $\sigma \in \Sigma(k, n)$, form a basis for Alt^k. Therefore, a differential k-form ω can be expressed in the form

$$\omega = \sum_{\sigma \in \Sigma(k,n)} a_\sigma (d\lambda)_\sigma, \tag{4.1}$$

for some coefficient functions $a_\sigma : T \to \mathbb{R}$. The coefficient functions in this expansion are uniquely determined and can be recovered using the dual basis:

$$a_\sigma(x) = \omega_x(t_{\sigma(1)}, \ldots, t_{\sigma(k)}).$$

For any simplex f, let $|f| = \int_f \mathrm{vol}_f$ denote its k-dimensional volume. Then $\mathrm{vol}_T(t_1,\ldots,t_n) = n!\,|T|$, while $\mathrm{d}\lambda_1 \wedge \cdots \wedge \mathrm{d}\lambda_n(t_1,\ldots,t_n) = 1$. Thus $\mathrm{d}\lambda_1 \wedge \cdots \wedge \mathrm{d}\lambda_n = 1/(n!|T|)\,\mathrm{vol}_T$. Similarly, if f is the k-simplex with vertices $x_{\sigma(0)},\ldots,x_{\sigma(k)}$, then

$$\mathrm{d}\lambda_{\sigma(1)} \wedge \cdots \wedge \mathrm{d}\lambda_{\sigma(k)} = \pm\frac{1}{k!\,|f|}\,\mathrm{vol}_f.$$

It follows that

$$\int_f \lambda_{\sigma(j)}\mathrm{d}\lambda_{\sigma(1)}\wedge\cdots\wedge\mathrm{d}\lambda_{\sigma(k)} = \pm\frac{1}{k!\,|f|}\int_f \lambda_{\sigma(j)}\,\mathrm{vol}_f = \frac{1}{(k+1)!}, \quad j = 0,\ldots,k. \tag{4.2}$$

4.2. Degrees of freedom and basis for $\mathcal{P}_r(T)$

Before proceeding to the case of general $\mathcal{P}_r\Lambda^k(T)$ and $\mathcal{P}_r^-\Lambda^k(T)$, we consider some simple cases. First we consider the familiar case of $\mathcal{P}_r(T) = \mathcal{P}_r\Lambda^0(T) = \mathcal{P}_r^-\Lambda^0(T)$, which will correspond to the Lagrange finite element spaces.

It is well known (and is easily shown, and will follow from the more general result below) that an element of $\mathcal{P}_r(T)$ vanishes if it vanishes at the vertices, its moments of degree at most $r-2$ vanish on each edge, its moments of degree at most $r-3$ vanish on each 2-subsimplex, *etc.* Let us associate to each $f \in \Delta(T)$, a subspace of the dual space of $\mathcal{P}_r(T)^*$ by

$$W(f) := \left\{ \phi \in \mathcal{P}_r(T)^* \,\middle|\, \phi(p) = \int_f pq\,\mathrm{vol}_f \text{ for some } q \in \mathcal{P}_{r-\dim f-1}(f) \right\}.$$

In case $\dim f = 0$, *i.e.*, $f = \{x_i\}$ for some vertex x_i, then $\mathcal{P}_r(f) = \mathbb{R}$ and $\int_f p\,\mathrm{vol}_f = p(x_i)$, so $W(f)$ is the span of this evaluation functional at the vertex. Thus if $\phi(p) = 0$ for all $p \in \sum_f W(f)$, then p vanishes. Therefore, the $W(f)$ span $\mathcal{P}_r(T)^*$, *i.e.*, $\sum_{f\in\Delta(T)} W(f) = \mathcal{P}_r(T)^*$. Obviously $\mathcal{P}_{r-\dim f-1}(f)$ maps onto $W(f)$ for each f. But it is easy to check that

$$\sum_{f\in\Delta(T)} \dim \mathcal{P}_{r-\dim f-1}(f) = \dim \mathcal{P}_r(T)^*,$$

from which we conclude that $\mathcal{P}_{r-\dim f-1}(f) \cong W(f)$ and

$$\mathcal{P}_r(T)^* = \bigoplus_{f\in\Delta(T)} W(f).$$

This is the desired geometrical decomposition of the dual space of $\mathcal{P}_r(T)$. As mentioned, it is then easy to generate a set of degrees of freedom by choosing a convenient basis for each of the spaces $\mathcal{P}_{r-\dim f-1}(f)$, for example the monomials of degree $r - \dim f - 1$ in the barycentric coordinates for f.

Next we give the geometrical decomposition of $\mathcal{P}_r(T)$ itself. We start with the monomial basis in barycentric coordinates (sometimes called the

Bernstein basis): the basis functions are the polynomials $\lambda^I = \lambda_0^{i_0} \cdots \lambda_n^{i_n}$ where $I = (i_0, \ldots, i_n) \in \mathbb{N}^{n+1}$ is a multi-index for which the i_j sum to r. We then associate the monomial λ^I to the simplex f whose vertices are the x_j for which $i_j > 0$. Thus λ_0^r is associated to the vertex x_0, the $r - 1$ monomials $\lambda_0^{r-1}\lambda_1$, $\lambda_0^{r-2}\lambda_1^2$, $\ldots$, $\lambda_0\lambda_1^{r-1}$ are associated to the edge $[x_0, x_1]$, etc.

Define $V(f)$ to be the span of the monomials so associated with f. Obviously

$$\mathcal{P}_r(T) = \bigoplus_{f \in \Delta(T)} V(f),$$

and it is easy to see that $\mathcal{P}_{r-\dim f - 1}(f) \cong V(f) = E_{f,T}\mathring{\mathcal{P}}_r(f)$ with the isomorphism given by $p \mapsto b_f E_{f,T}(p)$.

4.3. Degrees of freedom and basis for the Whitney forms

A differential k-form on T can be integrated over a k-simplex $f \in \Delta_k(T)$ and thus associates to each such f a real number. (In the language of algebraic topology, a differential form determines a simplicial k-cochain.) Given any simplicial k-cochain, $i.e.$, any choice of real numbers, one for each $f \in \Delta_k(T)$, Whitney showed how to define a differential form corresponding to that cochain. Namely, he associated to $f = f_\sigma \in \Delta_k(T)$ a differential k-form on T, which we shall call the Whitney form associated to the subsimplex f, given by

$$\phi_\sigma := \sum_{i=0}^{k} (-1)^i \lambda_{\sigma(i)} \, d\lambda_{\sigma(0)} \wedge \cdots \wedge \widehat{d\lambda_{\sigma(i)}} \wedge \cdots \wedge d\lambda_{\sigma(k)} \qquad (4.3)$$

(Whitney 1957, equation (12), p. 139). Now if $f' = f_\rho$ is a k-subsimplex different from f, then for some i, $\sigma(i) \notin \mathcal{R}(\rho)$, and so the trace of $\lambda_{\sigma(i)}$ and $d\lambda_{\sigma(i)}$ both vanish on f'. Thus $\text{Tr}_{f'} \phi_\sigma = 0$. On the other hand, it follows from (4.2) that $\int_f \phi_\sigma = \pm 1/k!$. Thus the Whitney k-forms corresponding to the k-subsimplices of T span a subspace of $\mathcal{P}_1 \Lambda^k(T)$ which is isomorphic to the space of k-dimensional simplicial cochains.

The next result asserts that the space spanned by the Whitney k-forms is precisely the space $\mathcal{P}_1^- \Lambda^k(T)$.

Theorem 4.1. The Whitney k-forms ϕ_σ corresponding to $f_\sigma \in \Delta_k(T)$ form a basis for $\mathcal{P}_1^- \Lambda^k(T)$.

Proof. It is enough to show that $\phi_\sigma \in \mathcal{P}_1^- \Lambda^k(T)$, since we have already seen that the ϕ_σ are linearly independent, and that their number equals $\binom{n+1}{k+1} = \dim \mathcal{P}_1^- \Lambda^k(T)$; see (3.15). Now for each i, $\kappa d\lambda_i = \lambda_i - \lambda_i(0)$, so $\kappa d\lambda_i$ differs from λ_i by a constant. Combining this with the Leibniz rule for κ (3.6), we conclude that $\kappa(d\lambda_{\sigma(0)} \wedge \cdots \wedge d\lambda_{\sigma(k)})$ differs from ϕ_σ by a constant k-form. Thus $\phi_\sigma \in \mathcal{P}_0 \Lambda^k(T) + \kappa \mathcal{P}_0 \Lambda^{k+1}(T) = \mathcal{P}_1^- \Lambda^k(T)$. $\qquad \square$

The trace of a function in $\mathcal{P}_1^- \Lambda^k(T)$ on a k-dimensional face f belongs to $\mathcal{P}_1^- \Lambda^k(f) = \mathcal{P}_0 \Lambda^k(f)$, a 1-dimensional space. Thus, for $f = f_\sigma$, $\mathrm{Tr}_f \, \phi_\sigma$ is a nonzero constant multiple of vol_f, while on the other k-dimensional subsimplices the trace of ϕ_σ vanishes.

Based on the Whitney forms, we again have geometrical decompositions

$$\mathcal{P}_1^- \Lambda^k(T) = \bigoplus_{f \in \Delta(T)} V(f), \quad \mathcal{P}_1^- \Lambda^k(T)^* = \bigoplus_{f \in \Delta(T)} W(f),$$

where now $V(f)$ and $W(f)$ vanish unless $\dim f = k$ and for $f = f_\sigma \in \Delta_k(T)$, $V(f)$ and $W(f)$ are 1-dimensional:

$$V(f) = \mathbb{R}\phi_\sigma, \quad W(f) = \mathbb{R}[\int_f \mathrm{Tr}_f(\cdot)].$$

4.4. A basis for $\mathcal{P}_r^- \Lambda^k(T)$

In this subsection we display a first basis for $\mathcal{P}_r^- \Lambda^k(T)$, analogous to the basis (4.1) for $\mathcal{P}_r \Lambda^k(T)$. This basis is *not* well adapted for numerical computation, because it does not admit an appropriate geometric decomposition. Such a basis, which is more difficult to obtain, will be constructed in Section 4.7. We begin with two lemmas.

Lemma 4.2. Let x be a vertex of T. Then the Whitney forms corresponding to the k-subsimplices that contain x are linearly independent over the ring of polynomials $\mathcal{P}(T)$.

Proof. Without loss of generality, we may assume that the vertex $x = x_0$, so we must prove that if

$$\omega = \sum_{\substack{\sigma \in \Sigma_0(k,n) \\ \sigma(0)=0}} p_\sigma \phi_\sigma$$

vanishes on T for some polynomials $p_\sigma \in \mathcal{P}(T)$, then all the $p_\sigma = 0$. From the definition (4.3), we see that if $\sigma(0) = 0$, then

$$\phi_\sigma(x_0) = d\lambda_{\sigma(1)} \wedge \cdots \wedge d\lambda_{\sigma(k)}.$$

Thus the values of the Whitney forms ϕ_σ at the vertex x_0 form a basis for $\mathrm{Alt}^1 \mathbb{R}^n$. By continuity, we conclude that there is a neighbourhood N of x_0 such that for $x \in N$, the algebraic k-forms $\phi_\sigma(x)$, $\sigma \in \Sigma_0(k,n)$, $\sigma(0) = 0$, are linearly independent. Since

$$\sum_{\substack{\sigma \in \Sigma_0(k,n) \\ \sigma(0)=0}} p_\sigma(x)\phi_\sigma(x) = \omega(x) = 0,$$

we conclude that for each σ, $p_\sigma(x)$ vanishes for all $x \in N$, whence $p_\sigma \equiv 0$. $\qquad \square$

Lemma 4.3. Suppose that

$$\omega = \sum_{\sigma \in \Sigma_0(k,n)} p_\sigma \phi_\sigma$$

where

$$p_\sigma(x) = a_\sigma\left(\lambda_{\sigma(0)}(x), \lambda_{\sigma(0)+1}(x), \ldots, \lambda_n(x)\right),$$

for some polynomial a_σ in $n - \sigma(0) + 1$ variables. If ω vanishes, then each of the a_σ vanishes.

Proof. If one of the $a_\sigma \neq 0$, choose $\rho \in \Sigma_0(k,n)$ such that $a_\rho \neq 0$ but $a_\sigma = 0$ if $\sigma(0) > j := \rho(0)$. Let $f = [x_j, x_{j+1}, \ldots, x_n]$. Then $\mathrm{Tr}_f \phi_\sigma = 0$ if $\sigma(0) < j$ and $a_\sigma = 0$ if $\sigma(0) > j$, so

$$\mathrm{Tr}_f \omega = \sum_{\substack{\sigma \in \Sigma_0(k,n) \\ \sigma(0)=j}} \mathrm{Tr}_f(p_\sigma) \, \mathrm{Tr}_f \phi_\sigma.$$

Now $\mathrm{Tr}_f \phi_\sigma$ is the Whitney form on f associated to the subsimplex f_σ of f, so the preceding lemma (applied to f in place of T) implies that $\mathrm{Tr}_f(p_\sigma) = 0$ for all σ with $\sigma(0) = j$, and, in particular, for $\sigma = \rho$. Since $p_\rho = a_\rho(\lambda_j, \ldots, \lambda_n)$, this implies that $a_\rho = 0$, a contradiction. $\square$

Theorem 4.4. For each $\sigma \in \Sigma_0(k,n)$ let $a_\sigma \in \mathcal{H}_{r-1}(\mathbb{R}^{n-\sigma(0)+1})$. Then the k-form

$$\omega = \sum_{\sigma \in \Sigma_0(k,n)} a_\sigma(\lambda_{\sigma(0)}, \lambda_{\sigma(0)+1}, \ldots, \lambda_n)\phi_\sigma \tag{4.4}$$

belongs to $\mathcal{P}_r^- \Lambda^k(T)$. Moreover, each $\omega \in \mathcal{P}_r^- \Lambda^k(T)$ can be written in the form (4.4) for a unique choice of polynomials $a_\sigma \in \mathcal{H}_{r-1}(\mathbb{R}^{n-\sigma(0)+1})$.

Proof. For $a_\sigma \in \mathcal{H}_{r-1}(\mathbb{R}^{n-\sigma(0)+1})$, $a_\sigma(\lambda_{\sigma(0)}, \lambda_{\sigma(0)+1}, \ldots, \lambda_n) \in \mathcal{P}_{r-1}(T)$, so $\omega \in \mathcal{P}_r^- \Lambda^k(T)$ by (3.17).

Thus (4.4) defines a linear mapping

$$\bigoplus_{\sigma \in \Sigma_0(k,n)} \mathcal{H}_{r-1}(\mathbb{R}^{n-\sigma(0)+1}) \to \mathcal{P}_r^- \Lambda^k(T).$$

By the preceding lemma, we know that this map is injective, so to complete the proof of the theorem it suffices to check that

$$\sum_{\sigma \in \Sigma_0(k,n)} \dim \mathcal{H}_{r-1}(\mathbb{R}^{n-\sigma(0)+1}) = \dim \mathcal{P}_r^- \Lambda^k(T) = \binom{r+k-1}{k}\binom{n+r}{r+k}.$$

To see this, observe that for a fixed j, $0 \leq j \leq n - k$, we have by (3.1)

$$\sum_{\substack{\sigma \in \Sigma_0(k,n) \\ \sigma(0)=j}} \dim \mathcal{H}_{r-1}(\mathbb{R}^{n-\sigma(0)+1})$$

$$= \dim \mathcal{P}_{r-1}\Lambda^k(\mathbb{R}^{n-j}) = \binom{r+k-1}{k}\binom{n-j+r-1}{r+k-1}.$$

Therefore, we have

$$\sum_{\sigma \in \Sigma_0(k,n)} \dim \mathcal{H}_{r-1}(\mathbb{R}^{n-\sigma(0)+1})$$

$$= \binom{r+k-1}{k}\sum_{j=0}^{n-k}\binom{n-j+r-1}{r+k-1} = \binom{r+k-1}{k}\binom{n+r}{r+k}.$$

Here, we have used the identity

$$\sum_{j=0}^{s}\binom{m+j}{m} = \binom{m+s+1}{m+1},$$

which is easily established by induction on s. $\qquad\square$

Of course the theorem implies that we obtain a basis for $\mathcal{P}_r^-\Lambda^k(T)$, by choosing any convenient basis, e.g., the monomial basis, for each of the spaces $\mathcal{H}_{r-1}(\mathbb{R}^{n-\sigma(0)+1})$.

4.5. Geometrical decomposition of $\mathcal{P}_r\Lambda^k(T)^*$

We now turn to the general case of $\mathcal{P}_r\Lambda^k(T)$ and $\mathcal{P}_r^-\Lambda^k(T)$, providing a geometrical decomposition of each space and its dual, as was done for $\mathcal{P}_r\Lambda^0(T)$ and $\mathcal{P}_1^-\Lambda^k(T)$ in Sections 4.2 and 4.3. In this subsection, we construct the decomposition of $\mathcal{P}_r\Lambda^k(T)^*$ in which the summand associated to a subsimplex is isomorphic to a space of polynomial differential forms (of the $\mathcal{P}_s^-$ type) on the subsimplex. As a consequence, we obtain (via choice of bases for the summands), a set of degrees of freedom for $\mathcal{P}_r\Lambda^k(T)$, with each degree of freedom associated to a subsimplex of T. This will allow the construction of finite element differential forms based on the $\mathcal{P}_r\Lambda^k$ spaces in the next section.

The decomposition, which is given in Theorem 4.10, will be built up in a sequence of results.

Lemma 4.5. An element $\omega \in \mathcal{P}_r\Lambda^k(T)$ has vanishing trace on the faces $F_1, \ldots, F_n$ (but not necessarily on F_0) if and only if it can be written in the form

$$\omega = \sum_{\sigma \in \Sigma(k,n)} p_\sigma \lambda_{\sigma^*}(\mathrm{d}\lambda)_\sigma, \qquad (4.5)$$

for some $p_\sigma \in \mathcal{P}_{r-n+k}(T)$.

Proof. Both λ_i and the trace of $d\lambda_i$ vanish on the face F_i, so all the forms $\lambda_{\sigma^*}(d\lambda)_\sigma$ vanish on each F_i, $i = 1, \ldots, n$. Thus if ω has the form (4.5), then its traces vanish as claimed.

On the other hand, suppose the $\omega \in \mathcal{P}_r\Lambda^k(T)$ has such vanishing traces. Write ω as in (4.1) with the $a_\sigma \in \mathcal{P}_r(T)$. We must show that λ_{σ^*} is a divisor of a_σ. For $1 \leq i \leq k$, and $1 \leq j \leq n - k$, the vertices x_0, and $x_{\sigma(i)}$ belong to the face $F_{\sigma^*(j)}$. Therefore the vector $t_{\sigma(i)}$ is tangent to the face, and so

$$p_\sigma(x) = \omega_x(t_{\sigma(1)}, \ldots, t_{\sigma(k)})$$

must vanish on $F_{\sigma^*(j)}$. Thus $\lambda_{\sigma^*(j)}$ divides $p_\sigma(x)$ for each $1 \leq j \leq n - k$, i.e., λ_{σ^*} divides p_σ as claimed. $\square$

Lemma 4.6. Let $\omega \in \mathcal{P}_r\Lambda^k(T)$. Suppose that $\operatorname{Tr}_{F_i} \omega = 0$ for $i = 1, \ldots, n$ and that

$$\int \omega \wedge \eta = 0, \quad \eta \in \mathcal{P}_{r-n+k}\Lambda^{n-k}(T). \tag{4.6}$$

Then $\omega = 0$.

Proof. Write ω as in (4.5), and set

$$\eta = \sum_{\sigma \in \Sigma(k,n)} (\pm)_\sigma p_\sigma (d\lambda)_{\sigma^*} \in \mathcal{P}_{r-n+k}\Lambda^{n-k}(T),$$

where $(\pm)_\sigma$ is the sign of the permutation

$$(\sigma(1), \ldots, \sigma(k), \sigma^*(1), \ldots, \sigma^*(n - k)).$$

Then

$$0 = \int \omega \wedge \mu = \left(\sum_\sigma \int p_\sigma^2 \lambda_{\sigma^*}\right) d\lambda_1 \wedge \cdots \wedge d\lambda_n.$$

Since $\lambda_{\sigma^*} > 0$ on the interior of T, we conclude that all the $p_\sigma \equiv 0$ and so $\omega = 0$. $\square$

Lemma 4.7. Let $\omega \in \mathring{\mathcal{P}}_r\Lambda^k(T)$. Suppose that

$$\int \omega \wedge \eta = 0, \quad \eta \in \mathcal{P}_{r-n+k}^-\Lambda^{n-k}(T). \tag{4.7}$$

Then $\omega = 0$.

Proof. If $\omega \in \mathring{\mathcal{P}}_r\Lambda^k(T)$, then $d\omega \in \mathring{\mathcal{P}}_{r-1}\Lambda^{k+1}(T)$, and

$$\int d\omega \wedge \mu = \pm \int \omega \wedge d\mu, \quad \mu \in \Lambda^{n-k-1}. \tag{4.8}$$

Now if $\mu \in \mathcal{P}_{r-n+k}\Lambda^{n-k-1}(T)$, then

$$d\mu \in \mathcal{P}_{r-n+k-1}\Lambda^{n-k}(T) \subset \mathcal{P}_{r-n+k}^-\Lambda^{n-k}(T),$$

so

$$\int d\omega \wedge \mu = 0, \quad \eta \in \mathcal{P}_{r-n+k}\Lambda^{n-k-1}(T).$$

Applying Lemma 4.6, we conclude that $d\omega = 0$, and hence, by (4.8), that

$$\int \omega \wedge \eta = 0, \quad \eta \in \Lambda^{n-k-1}(T).$$

Together with the hypothesis (4.7), we find that the hypothesis (4.6) of the previous lemma is fulfilled, and so ω vanishes.   □

Now for some $0 \le k \le n$, $r \ge 1$, let $\omega \in \mathcal{P}_r\Lambda^k(T)$ and $f \in \Delta_k(T)$. The trace of ω on ∂f certainly vanishes (since it is a k-form on a manifold of dimension $k-1$), so if $\int_f \operatorname{Tr}_f \omega \wedge \eta = 0$ for all $\eta \in \mathcal{P}_r^-\Lambda^0(f)$, then, by the previous lemma applied to f, $\operatorname{Tr}_f \omega = 0$. Therefore, if we assume that

$$\int_f \operatorname{Tr}_f \omega \wedge \eta = 0, \quad \eta \in \mathcal{P}_r^-\Lambda^0(f), \quad f \in \Delta_k(T),$$

we conclude that $\operatorname{Tr}_f \omega = 0$ for all $f \in \Delta_k(T)$. If we then assume also that

$$\int_f \operatorname{Tr}_f \omega \wedge \eta = 0, \quad \eta \in \mathcal{P}_{r-1}^-\Lambda^1(f), \quad f \in \Delta_{k+1}(T),$$

we may apply the lemma again to conclude that $\operatorname{Tr}_f \omega = 0$ for all $f \in \Delta_{k+1}(T)$. Continuing in this way, we obtain the following theorem.

Theorem 4.8. Let $0 \le k \le n$, $r \ge 1$. Suppose that $\omega \in \mathcal{P}_r\Lambda^k(T)$ satisfies

$$\int_f \operatorname{Tr}_f \omega \wedge \eta = 0, \quad \eta \in \mathcal{P}_{r+k-\dim f}^-\Lambda^{\dim f-k}(f), \quad f \in \Delta(T). \qquad (4.9)$$

Then $\omega = 0$.

Note that since $\mathcal{P}_s^-\Lambda^k$ vanishes, if $s \le 1$ or $k < 0$, the only subsimplices that contribute in (4.9) have $k \le \dim f \le \min(n, r+k-1)$.

The association of η to the linear functional $\omega \mapsto \int_f \operatorname{Tr}_f \omega \wedge \eta$, gives a surjection of $\mathcal{P}_{r+k-\dim f}^-\Lambda^{\dim f-k}(f)$ onto

$$W(f) := \left\{ \phi \in \mathcal{P}_r\Lambda^k(T)^* \mid \right. \qquad (4.10)$$

$$\left. \phi(\omega) = \int_f \operatorname{Tr}_f \omega \wedge \eta \text{ for some } \eta \in \mathcal{P}_{r+k-\dim f}^-\Lambda^{\dim f-k}(f) \right\}.$$

The following theorem, a simple dimension count, will imply that all these surjections are isomorphisms, and that the $W(f)$ span $\mathcal{P}_r\Lambda^k(T)^*$.

Theorem 4.9. Let $0 \le k \le n$, $r \ge 1$. Then

$$\sum_{f \in \Delta(T)} \dim \mathcal{P}^-_{r+k-\dim f} \Lambda^{\dim f - k}(f) = \dim \mathcal{P}_r \Lambda^k(T). \qquad (4.11)$$

Proof. Using the convention that $\binom{m}{l} = 0$ for $l < 0$ or $l > m$, the fact that the number of simplices of dimension j is $\binom{n+1}{j+1}$, and the dimension formula (3.15), we see that the left-hand side equals

$$\sum_j \binom{n+1}{j+1} \binom{r-1}{j-k} \binom{r+k}{k}$$

$$= \binom{r+k}{k} \sum_{j=0}^{n-k} \binom{n+1}{j+k+1} \binom{r-1}{j}$$

$$= \binom{r+k}{k} \sum_{j=0}^{n-k} \binom{n+1}{n-k-j} \binom{r-1}{j}$$

$$= \binom{r+k}{k} \binom{n+r}{n-k} = \dim \mathcal{P}_r \Lambda^k(T).$$

In the last step we have used the binomial identity

$$\sum_j \binom{l}{p-j} \binom{m}{j} = \binom{l+m}{p},$$

which can be deduced by matching coefficients of x^p after binomial expansion of the equation $(x+1)^l (x+1)^m = (x+1)^{l+m}$. $\qquad\square$

Combining Theorems 4.8 and 4.11, we have the main result of this subsection.

Theorem 4.10. Let $0 \le k \le n$, $r \ge 1$. For each $f \in \Delta(T)$, define $W(f)$ by (4.10). (Note that $W(f) = 0$ unless $k \le \dim f \le \min(n, r+k-1)$.) Then $W(f) \cong \mathcal{P}^-_{r+k-\dim f} \Lambda^{\dim f - k}(f)$ and

$$\mathcal{P}_r \Lambda^k(T)^* = \bigoplus_{f \in \Delta(T)} W(f).$$

We also have, as a corollary of the above, an isomorphism of $\mathring{\mathcal{P}}_r \Lambda^k(T)^*$ with $\mathcal{P}^-_{r+k-n} \Lambda^{n-k}(T)$; see Lemma 4.7. Thus

$$\dim \mathring{\mathcal{P}}_r \Lambda^k(T)^* = \binom{r-1}{n-k} \binom{r+k}{k}.$$

4.6. Geometrical decomposition of $\mathcal{P}^-_r \Lambda^k(T)^*$

In a very similar manner, we obtain a geometrical decomposition of the dual of $\mathcal{P}^-_r \Lambda^k(T)$. First we prove an analogue of Lemma 4.7.

Lemma 4.11. Let $\omega \in \mathring{\mathcal{P}}_r^- \Lambda^k(T)$. Suppose that

$$\int \omega \wedge \eta = 0, \quad \eta \in \mathcal{P}_{r-n+k-1} \Lambda^{n-k}(T).$$

Then $\omega = 0$.

Proof. Since $\omega \in \mathring{\mathcal{P}}_r^- \Lambda^k(T)$, $d\omega \in \mathring{\mathcal{P}}_{r-1} \Lambda^{k+1}(T)$. Therefore

$$\int d\omega \wedge \mu = \pm \int \omega \wedge d\mu, \quad \mu \in \Lambda^{n-k-1}.$$

Now if $\mu \in \mathcal{P}_{r-n+k} \Lambda^{n-k-1}(T)$, then $d\mu$ belongs to $\mathcal{P}_{r-n+k-1} \Lambda^{n-k}(T)$ which is contained in $\mathcal{P}_{r-n+k}^- \Lambda^{n-k}(T)$, so $\int d\omega \wedge \mu = 0$. Thus $\int d\omega \wedge \mu = 0$ for $\mu \in \mathcal{P}_{r-n+k} \Lambda^{n-k-1}(T)$. Applying Lemma 4.6 with r replaced by $r-1$ and k replaced by $k+1$, we conclude that $d\omega = 0$. Theorem 3.4 then implies that $\omega \in \mathcal{P}_{r-1} \Lambda^k(T)$, and we can apply Lemma 4.6 again, this time with r replaced by $r-1$, to conclude that ω vanishes.   □

Just as for Theorem 4.8, a finite induction based on this result gives the desired decomposition.

Theorem 4.12. Let $0 \le k \le n$, $r \ge 1$. Suppose that $\omega \in \mathcal{P}_r^- \Lambda^k(T)$ satisfies

$$\int_f \mathrm{Tr}_f \, \omega \wedge \eta = 0, \quad \eta \in \mathcal{P}_{r+k-\dim f-1} \Lambda^{\dim f - k}(f), \quad f \in \Delta(T).$$

Then $\omega = 0$.

Note that since $\mathcal{P}_s^- \Lambda^k$ vanishes if $s \le 1$ or $k < 0$, again, the only subsimplices that contribute have $k \le \dim f \le \min(n, r+k-1)$.

The dimension count is a simple calculation like that of Theorem 4.9.

Theorem 4.13. Let $0 \le k \le n$, $r \ge 1$. Then

$$\sum_{f \in \Delta(T)} \dim \mathcal{P}_{r+k-\dim f-1} \Lambda^{\dim f - k}(f) = \dim \mathcal{P}_r^- \Lambda^k(T). \qquad (4.12)$$

Combining Theorems 4.12 and 4.12, we get the desired decomposition.

Theorem 4.14. Let $0 \le k \le n$, $r \ge 1$. For each $f \in \Delta(T)$, define

$$W(f) := \left\{ \phi \in \mathcal{P}_r^- \Lambda^k(T)^* \, \middle| \right.$$

$$\left. \phi(\omega) = \int_f \mathrm{Tr}_f \, \omega \wedge \eta \text{ for some } \eta \in \mathcal{P}_{r+k-\dim f-1} \Lambda^{\dim f - k}(f) \right\}.$$

(Note that $W(f) = 0$ unless $k \le \dim f \le \min(n, r+k-1)$.) Then $W(f) \cong \mathcal{P}_{r+k-\dim f-1} \Lambda^{\dim f - k}(f)$ and

$$\mathcal{P}_r^- \Lambda^k(T)^* = \bigoplus_{f \in \Delta(T)} W(f).$$

Moreover, $\mathring{\mathcal{P}}_r^- \Lambda^k(T)^* \cong \mathcal{P}_{r+k-n-1}\Lambda^{n-k}(T)$; see Lemma 4.11. Therefore,

$$\dim \mathring{\mathcal{P}}_r^- \Lambda^k(T) = \binom{n}{k}\binom{r+k-1}{n}.$$

4.7. Geometrical decomposition of $\mathcal{P}_r^- \Lambda^k$

Finally we give geometrical decompositions of the spaces themselves. From these we can easily obtain bases for use in computation. In this subsection we consider the spaces $\mathcal{P}_r^- \Lambda^k(T)$, and in the next, the spaces $\mathcal{P}_r\Lambda^k(T)$. Since we have already treated the case $k = 0$, we assume that $1 \le k \le n$.

Our main result will be the following decomposition.

Theorem 4.15. Let $k, r \ge 1$. Then

$$\mathcal{P}_r^- \Lambda^k(T) = \bigoplus_{f \in \Delta(T)} V(f),$$

where $V(f) = 0$ if $\dim f < k$ or $\dim f \ge r + k$ and

$$V(f) \cong \mathring{\mathcal{P}}_{r+k-\dim f - 1}\Lambda^{\dim f - k}(f)$$

otherwise.

We will build up to the proof with a number of preliminary results. First we obtain a representation of forms in $\mathcal{P}_r^- \Lambda^k(T)$ with vanishing trace. In it we use the notation $(0, \rho)$ to denote the extension of a sequence $\rho : \{1, \ldots, k\} \to \{1, \ldots, n\}$ to the sequence $(0, \rho) : \{0, \ldots, k\} \to \{0, \ldots, n\}$ determined by $(0, \rho)(0) = 0$. Note that for any $\rho \in \Sigma(k, n)$, the elements $\lambda_{\rho^*}\phi_{(0,\rho)}$ of $\mathcal{P}_{n-k+1}^- \Lambda^k(T)$ have vanishing trace.

Theorem 4.16. For $1 \le k \le n$, $r \ge n + 1 - k$, the map

$$\sum_{\rho \in \Sigma(k,n)} a_\rho (d\lambda)_{\rho^*} \mapsto \sum_{\rho \in \Sigma(k,n)} a_\rho \lambda_{\rho^*}\phi_{(0,\rho)},$$

where the $a_\rho \in \mathcal{P}_{r+k-n-1}(T)$, defines an isomorphism of $\mathcal{P}_{r+k-n-1}\Lambda^{n-k}(T)$ onto $\mathring{\mathcal{P}}_r^- \Lambda^k(T)$.

Proof. The map is an injection according to Lemma 4.2, and Theorem 4.12 implies that $\dim \mathring{\mathcal{P}}_r^- \Lambda^k(T) \le \dim \mathcal{P}_{r+k-n-1}\Lambda^{n-k}(T)$. The result follows. $\square$

Corollary 4.17. Let $f = f_\sigma \in \Delta(T)$ with $k \le \dim f \le r + k - 1$. Then the map

$$\sum_{\rho \in \Sigma(k,\dim f)} a_\rho (d\lambda)_{\sigma \circ \rho^*} \mapsto \sum_{\rho \in \Sigma(k,\dim f)} a_\rho \lambda_{\sigma \circ \rho^*}\phi_{\sigma \circ (0,\rho)},$$

where the $a_\rho \in \mathcal{P}_{r+k-\dim f - 1}(f)$, defines an isomorphism

$$\mathcal{P}_{r+k-\dim f - 1}\Lambda^{\dim f - k}(f) \cong \mathring{\mathcal{P}}_r^- \Lambda^k(f).$$

Proof. This is just the theorem applied to f rather than T. ☐

From the unique representation

$$\omega = \sum_{\rho \in \Sigma(k, \dim f)} a_\rho \lambda_{\sigma \circ \rho^*} \phi_{\sigma \circ (0,\rho)},$$

we obtain an extension operator

$$E^k_{f,T} : \mathring{\mathcal{P}}^-_r \Lambda^k(f) \to \mathcal{P}^-_r \Lambda^k(T).$$

Namely, we write the coefficients a_ρ in terms of the barycentric coordinates on f, so the entire expression is in terms of these barycentric coordinates, and so extends to the whole of $\mathbb{R}^n$. We then obviously have

$$\mathrm{Tr}_f(E^k_{f,T}\omega) = \omega, \quad \omega \in \mathring{\mathcal{P}}^-_r \Lambda^k(f).$$

Proposition 4.18. If $\omega \in \mathring{\mathcal{P}}^-_r \Lambda^k(f)$, then $\mathrm{Tr}_F(E^k_{f,T}\omega)$ vanishes for each face F opposite a vertex on f.

Proof. Let x_i be a vertex of f and F the face opposite (on which $\lambda_i = 0$). For each $\rho \in \Sigma(k, \dim f)$, i is either in $\mathcal{R}(\sigma \circ \rho^*)$ or in $\mathcal{R}(\sigma \circ (0,\rho))$. In the first case, $\mathrm{Tr}_F(\lambda_{\sigma \circ \rho^*})$ vanishes and in the latter, $\mathrm{Tr}_F(\phi_{\sigma \circ (0,\rho)})$ vanishes. Thus each term in the expression for $E^k_{f,T}\omega$ has vanishing trace on F. ☐

Corollary 4.19. Let $f, g \in \Delta(T)$ with dimensions at least k. If $f \not\subseteq g$, then

$$\mathrm{Tr}_g(E^k_{f,T}\omega) = 0, \quad \omega \in \mathring{\mathcal{P}}^-_r \Lambda^k(f).$$

Proof. There exists a vertex of f which does not belong to g, and so g is contained in the face opposite this vertex. ☐

Proposition 4.20. For each $f \in \Delta(T)$ with $\dim f \geq k$, let $\omega_f \in \mathring{\mathcal{P}}^-_r \Lambda^k(f)$. If

$$\sum_{\substack{f \in \Delta(T) \\ \dim f \geq k}} E^k_{f,T}\omega_f = 0,$$

then each $\omega_f = 0$.

Proof. If some ω_f does not vanish, there is a face g of minimal dimension such that $\omega_g \neq 0$, so that g does not contain any of the other faces f for which $\omega_f \neq 0$. Taking the trace of the sum on g and invoking the corollary, we conclude that $\omega_g = 0$, a contradiction. ☐

We can now prove Theorem 4.15. We define

$$V(f) = E^k_{f,T}\mathring{\mathcal{P}}^-_r \Lambda^k(f),$$

for f with $k \leq \dim f < r + k$. By the previous proposition, the sum of the $V(f)$ is direct, and we need only prove that it equals all of $\mathcal{P}^-_r \Lambda^k(T)$. But we have already shown, in Theorem 4.14, that the dimensions of $V(f)$ sum to the dimension of $\mathcal{P}_r^- \Lambda^k(T)$, and this completes the proof.

4.8. Geometrical decomposition of $\mathcal{P}_r \Lambda^k$

In complete analogy with Theorem 4.15 above, we will establish a corresponding geometrical decomposition for the spaces $\mathcal{P}_r \Lambda^k(T)$.

Theorem 4.21. Let $k, r \geq 1$. Then

$$\mathcal{P}_r \Lambda^k(T) = \bigoplus_{f \in \Delta(T)} V(f),$$

where $V(f) = 0$ if $\dim f < k$ or $\dim f \geq r + k$ and

$$V(f) \cong \mathcal{P}_{r+k-\dim f}^- \Lambda^{\dim f - k}(f)$$

otherwise.

As above, we start the construction by defining an isomorphism onto the subspace of forms in $\mathcal{P}_r \Lambda^k(T)$ with vanishing trace, $\overset{\circ}{\mathcal{P}}_r \Lambda^k(T)$. Recall that for $\rho \in \Sigma_0(n - k, n)$, $\rho^* \in \Sigma_0(k - 1, n)$ is defined such that $\lambda_\rho(d\lambda)_{\rho^*} \in \overset{\circ}{\mathcal{P}}_{n-k+1} \Lambda^k(T)$. In the definition of the isomorphism below, we will also use the representation (4.4) of any element $\mathcal{P}_r^- \Lambda^k(T)$.

Theorem 4.22. For $1 \leq k \leq n$, $r \geq n - k + 1$, the map

$$\sum_{\rho \in \Sigma_0(n-k,n)} a_\rho \phi_\rho \mapsto \sum_{\rho \in \Sigma_0(n-k,n)} a_\rho \lambda_\rho (d\lambda)_{\rho^*},$$

where $a_\rho = a_\rho(\lambda_{\rho(0)}, \lambda_{\rho(0)+1}, \ldots, \lambda_n) \in \mathcal{P}_{r+k-n-1}(T)$, defines an isomorphism of $\mathcal{P}_{r+k-n}^- \Lambda^{n-k}(T)$ onto $\overset{\circ}{\mathcal{P}}_r \Lambda^k(T)$.

Proof. As Theorem 4.8 implies that $\dim \overset{\circ}{\mathcal{P}}_r \Lambda^k(T) \leq \dim \mathcal{P}_{r+k-n}^- \Lambda^{n-k}(T)$, it is enough to show that the map is injective. However, if

$$\sum_{\rho \in \Sigma_0(n-k,n)} a_\rho \lambda_\rho (d\lambda)_{\rho^*} \equiv 0,$$

and the coefficients a_ρ are of the form given above, then we must have

$$\sum_{\substack{\rho \in \Sigma_0(n-k,n) \\ \rho(0)=0}} a_\rho \lambda_\rho (d\lambda)_{\rho^*} \equiv 0,$$

since none of the other terms will have λ_0 as factor. Furthermore, the set $\{(d\lambda)_{\rho^*}\}$, where ρ is taken to be in $\Sigma_0(n - k, n)$ with $\rho(0) = 0$, is a basis for $\text{Alt}^k \mathbb{R}^n$. Therefore, all a_ρ, with $\rho(0) = 0$, are identically equal to zero. A simple inductive argument on $\rho(0)$ now implies that all a_ρ are identically equal to zero. This completes the proof. $\square$

If we apply this result to a subsimplex f instead of T, we immediately obtain the following.

Corollary 4.23. Let $f = f_\sigma \in \Delta(T)$ with $k \leq \dim f \leq r + k - 1$. Then the map

$$\sum_{\rho \in \Sigma_0(\dim f - k, \dim f)} a_\rho \phi_{\sigma \circ \rho} \mapsto \sum_{\rho \in \Sigma_0(\dim f - k, \dim f)} a_\rho \lambda_{\sigma \circ \rho} (d\lambda)_{\sigma \circ \rho^*},$$

where $a_\rho = a_\rho(\lambda_{\sigma \circ \rho(0)}, \lambda_{\sigma(\rho(0)+1)}, \ldots, \lambda_{\sigma(\dim f)}) \in \mathcal{P}_{r+k-\dim f-1}(f)$, defines an isomorphism of $\mathcal{P}_{r+k-\dim f}^- \Lambda^{\dim f - k}(f)$ onto $\mathring{\mathcal{P}}_r \Lambda^k(f)$.

From the unique representation

$$\sum_{\rho \in \Sigma_0(\dim f - k, \dim f)} a_\rho \lambda_{\sigma \circ \rho} (d\lambda)_{\sigma \circ \rho^*},$$

where $a_\rho = a_\rho(\lambda_{\sigma \circ \rho(0)}, \lambda_{\sigma(\rho(0)+1)}, \ldots, \lambda_{\sigma(\dim f)}) \in \mathcal{P}_{r+k-\dim f-1}(f)$, we obtain an extension operator

$$E_{f,T}^k : \mathring{\mathcal{P}}_r \Lambda^k(f) \to \mathcal{P}_r \Lambda^k(T),$$

which is analogous, but not identical, to the extension operator introduced in Section 4.7. Furthermore, as above we can show that if $f, g \in \Delta(T)$, with dimensions at least k and $f \not\subseteq g$, then

$$\mathrm{Tr}_g(E_{f,T}^k \omega) = 0, \quad \omega \in \mathring{\mathcal{P}}_r \Lambda^k(f).$$

We can therefore again conclude that the sum of

$$V(f) = E_{f,T}^k \mathring{\mathcal{P}}_r \Lambda^k(f), \quad f \in \Delta(T), \dim f \geq k$$

is direct. Finally, it follows from Theorem 4.10 that

$$\dim \mathcal{P}_r \Lambda^k(T) = \sum_{\substack{f \in \Delta(T) \\ \dim f \geq k}} \dim V(f).$$

Hence, Theorem 4.21 is established.

4.9. The canonical projection operators

In this subsection, we shall define projection operators Π from the spaces $C^0 \Lambda^k(T)$ of continuous differential forms on T onto the spaces $\mathcal{P}_r \Lambda^k(T)$ and $\mathcal{P}_r^- \Lambda^k(T)$, and show the commutativity property $d\Pi = \Pi d$. By Theorems 4.8 and 4.12, for $0 \leq k \leq n$, $r \geq 1$, $\omega \in \mathcal{P}_r \Lambda^k(T)$ is uniquely determined by the quantities

$$\int_f \mathrm{Tr}_f \, \omega \wedge \eta, \quad \eta \in \mathcal{P}_{r+k-\dim f}^- \Lambda^{\dim f - k}(f), \quad f \in \Delta(T),$$

and $\omega \in \mathcal{P}_r^- \Lambda^k(T)$ is uniquely determined by the quantities

$$\int_f \mathrm{Tr}_f \omega \wedge \eta, \quad \eta \in \mathcal{P}_{r+k-\dim f-1}\Lambda^{\dim f-k}(f), \quad f \in \Delta(T).$$

Hence, for $\omega \in \Lambda^k(T)$, we define the projection operators Π_r^k mapping $\Lambda^k(T)$ to $\mathcal{P}_r\Lambda^k(T)$ and Π_{r-}^k mapping $\Lambda^k(T)$ to $\mathcal{P}_r^-\Lambda^k(T)$ by the relations

$$\int_f \mathrm{Tr}_f(\omega - \Pi_r^k \omega) \wedge \eta = 0, \quad \eta \in \mathcal{P}_{r+k-\dim f}^-\Lambda^{\dim f-k}(f), \quad f \in \Delta(T),$$

and

$$\int_f \mathrm{Tr}_f(\omega - \Pi_{r-}^k \omega) \wedge \eta = 0, \quad \eta \in \mathcal{P}_{r+k-\dim f-1}\Lambda^{\dim f-k}(f), \quad f \in \Delta(T).$$

An easy consequence of the definitions is that the projection operators commute with affine transformations. That is, if $\Phi : T' \to T$ is an affine map, then the following diagrams commute:

$$
\begin{array}{ccc}
\Lambda^k(T) & \overset{\Phi^*}{\longrightarrow} & \Lambda^k(T') \\
\Pi \downarrow & & \Pi \downarrow \\
\mathcal{P}_r\Lambda^k(T) & \overset{\Phi^*}{\longrightarrow} & \mathcal{P}_r\Lambda^k(T')
\end{array}
\qquad
\begin{array}{ccc}
\Lambda^k(T) & \overset{\Phi^*}{\longrightarrow} & \Lambda^k(T') \\
\Pi \downarrow & & \Pi \downarrow \\
\mathcal{P}_r^-\Lambda^k(T) & \overset{\Phi^*}{\longrightarrow} & \mathcal{P}_r^-\Lambda^k(T').
\end{array}
$$

This follows from the facts that the pullback by an affine isomorphism respects the trace operation, wedge product, and integral, and that the $\mathcal{P}_r\Lambda^k$ and $\mathcal{P}_r^-\Lambda^k$ spaces are affine-invariant.

We now show that these projection operators commute with d. More specifically, we establish the following lemma.

Lemma 4.24. The following four diagrams commute:

$$
\begin{array}{ccc}
\Lambda^k(T) & \overset{d}{\longrightarrow} & \Lambda^{k+1}(T) \\
\Pi \downarrow & & \Pi \downarrow \\
\mathcal{P}_r\Lambda^k(T) & \overset{d}{\longrightarrow} & \mathcal{P}_{r-1}\Lambda^{k+1}(T)
\end{array}
\qquad
\begin{array}{ccc}
\Lambda^k(T) & \overset{d}{\longrightarrow} & \Lambda^{k+1}(T) \\
\Pi \downarrow & & \Pi \downarrow \\
\mathcal{P}_r\Lambda^k(T) & \overset{d}{\longrightarrow} & \mathcal{P}_r^-\Lambda^{k+1}(T)
\end{array}
$$

$$
\begin{array}{ccc}
\Lambda^k(T) & \overset{d}{\longrightarrow} & \Lambda^{k+1}(T) \\
\Pi \downarrow & & \Pi \downarrow \\
\mathcal{P}_r^-\Lambda^k(T) & \overset{d}{\longrightarrow} & \mathcal{P}_r^-\Lambda^{k+1}(T)
\end{array}
\qquad
\begin{array}{ccc}
\Lambda^k(T) & \overset{d}{\longrightarrow} & \Lambda^{k+1}(T) \\
\Pi \downarrow & & \Pi \downarrow \\
\mathcal{P}_r^-\Lambda^k(T) & \overset{d}{\longrightarrow} & \mathcal{P}_{r-1}\Lambda^{k+1}(T).
\end{array}
$$

Proof. Let Π_r^k denote the projection onto $\mathcal{P}_r\Lambda^k(T)$. Then the first diagram asserts that $\mu := d\Pi_r^k\omega - \Pi_{r-1}^{k+1}d\omega = 0$. From Lemma 3.8, we know that $d\Pi_r^k\omega - \Pi_{r-1}^{k+1}d\omega \in \mathcal{P}_{r-1}\Lambda^{k+1}(T)$. Thus $\mu = 0$ if

$$\int_f \mathrm{Tr}_f\, \mu \wedge \eta = 0, \quad \eta \in \mathcal{P}_{r+k-\dim f}^-\Lambda^{\dim f-k-1}(f), \quad f \in \Delta(T).$$

But, by Stokes' theorem and the fact that $\mathrm{Tr}_f\, d = d_f\, \mathrm{Tr}_f$, we get

$$\int_f \mathrm{Tr}_f\, d\Pi_r^k\omega \wedge \eta = \int_f d_f\, \mathrm{Tr}_f\, \Pi_r^k\omega \wedge \eta$$

$$= (-1)^{k-1} \int_f \mathrm{Tr}_f\, \Pi_r^k\omega \wedge d_f\eta + \int_{\partial f} \mathrm{Tr}\, \Pi_r^k\omega \wedge \mathrm{Tr}\, \eta$$

$$= (-1)^{k-1} \int_f \mathrm{Tr}_f\, \omega \wedge d_f\eta + \int_{\partial f} \mathrm{Tr}\, \omega \wedge \mathrm{Tr}\, \eta$$

$$= \int_f d_f\, \mathrm{Tr}_f\, \omega \wedge \eta = \int_f \mathrm{Tr}_f\, \Pi_{r-1}^{k+1}d\omega \wedge \eta.$$

A similar argument is used to establish the other three identities.   □

5. Finite element differential forms and their cohomology

Throughout this section, we assume that Ω is a bounded polyhedral domain in $\mathbb{R}^n$ which is partitioned into a finite set of n-simplices $\mathcal{T}$. These n-simplices determine a simplicial decomposition of Ω. That is, their union is the closure of Ω, and the intersection of any two is either empty or a common subsimplex of each. Adopting the terminology of the two-dimensional case, we will refer to $\mathcal{T}$ as a *triangulation* of Ω. In this section we will define two families of spaces of finite element differential forms with respect to the triangulation $\mathcal{T}$, denoted $\mathcal{P}_r\Lambda^k(\mathcal{T})$ and $\mathcal{P}_r^-\Lambda^k(\mathcal{T})$, which are subspaces of the corresponding Sobolev space $H\Lambda^k(\Omega)$. In particular, we will show how these finite element spaces lead to a number of finite element subcomplexes of the de Rham complex.

5.1. Finite element differential forms

Interelement continuity
Recall that $\Delta_j(T)$ denotes the set of all j-dimensional subsimplices of the simplex T. Furthermore, we let $\Delta_j(\mathcal{T})$ be the set of all j-dimensional subsimplices generated by $\mathcal{T}$, i.e.,

$$\Delta_j(\mathcal{T}) = \bigcup_{T \in \mathcal{T}} \Delta_j(T).$$

It follows from Stokes' theorem that if $\omega \in L^2\Lambda^k(\Omega)$ is piecewise smooth with respect to the triangulation $\mathcal{T}$, then $\omega \in H\Lambda^k(\Omega)$ if and only if $\mathrm{Tr}\,\omega$ is single-valued at all $f \in \Delta_{n-1}(\mathcal{T})$. In other words, if $T_1, T_2 \in \mathcal{T}$ have the common face $f \in \Delta_{n-1}(\mathcal{T})$, then

$$\mathrm{Tr}_{T_1,f}\,\omega = \mathrm{Tr}_{T_2,f}\,\omega,$$

where $\mathrm{Tr}_{T,f}$ denotes the trace on f derived from $\omega|_T$. However, if $\mathrm{Tr}_{T_1,f}\,\omega = \mathrm{Tr}_{T_2,f}\,\omega$, then clearly $\mathrm{Tr}_{T_1,g}\,\omega = \mathrm{Tr}_{T_2,g}\,\omega$ for all $g \in \Delta_j(f)$, $k \le j \le n-1$. Hence, we have the following characterization of piecewise smooth functions in $H\Lambda^k(\Omega)$.

Lemma 5.1. Let $\omega \in L^2\Lambda^k(\Omega)$ be piecewise smooth with respect to the triangulation $\mathcal{T}$. The following statements are equivalent:

(1) $\omega \in H\Lambda^k(\Omega)$,

(2) $\mathrm{Tr}\,\omega$ is single-valued for all $f \in \Delta_{n-1}(\mathcal{T})$,

(3) $\mathrm{Tr}\,\omega$ is single-valued for all $f \in \Delta_j(\mathcal{T})$, $k \le j \le n-1$.

Shape functions and degrees of freedom
For $r \ge 1$, the spaces $\mathcal{P}_r\Lambda^k(\mathcal{T})$ and $\mathcal{P}_r^-\Lambda^k(\mathcal{T})$ are defined using the polynomial spaces $\mathcal{P}_r\Lambda^k(T)$ and $\mathcal{P}_r^-\Lambda^k(T)$, respectively, as shape functions, and enforcing just enough interelement continuity to ensure inclusion in $H\Lambda^k(\Omega)$. That is, we define

$$\mathcal{P}_r\Lambda^k(\mathcal{T}) = \{\omega \in H\Lambda^k(\Omega) \mid \omega|_T \in \mathcal{P}_r\Lambda^k, \; T \in \mathcal{T}\},$$

$$\mathcal{P}_r^-\Lambda^k(\mathcal{T}) = \{\omega \in H\Lambda^k(\Omega) \mid \omega|_T \in \mathcal{P}_r^-\Lambda^k, \; T \in \mathcal{T}\}.$$

Hence, all elements of these spaces have to satisfy the continuity requirements specified in Lemma 5.1.

Degrees of freedom for these spaces are easily determined from the case of a single simplex treated in Section 4. For the space $\mathcal{P}_r\Lambda^k(\mathcal{T})$ we use

$$\int_f \mathrm{Tr}_f\,\omega \wedge \nu, \quad \nu \in \mathcal{P}_{r-j+k}^-\Lambda^{j-k}(f), \quad f \in \Delta_j(\mathcal{T}), \tag{5.1}$$

for $k \le j \le \min(n, r + k - 1)$. By Lemma 5.1, the quantities in (5.1) are well-defined for $\omega \in \mathcal{P}_r\Lambda^k(\mathcal{T})$. In view of Section 4.8, the quantities corresponding to $f \subset T$ determine $\omega|_T \in \mathcal{P}_r\Lambda^k(T)$. Therefore, if we choose any basis for the test spaces $\mathcal{P}_{r-j+k}^-\Lambda^{j-k}(f)$ and assign arbitrary values to $\int_f \mathrm{Tr}_f\,\omega\wedge\nu$ for basis elements ν and for all $f \in \Delta_j(\mathcal{T})$, we determine a unique piecewise polynomial ω. Moreover, if $F \in \Delta_{n-1}(\mathcal{T})$, then $\mathrm{Tr}_F\,\omega \in \mathcal{P}_r\Lambda^k(F)$ and so is determined by the quantities in (5.1) corresponding to $f \subset F$.

Table 5.1. Correspondences between finite element differential forms and the classical finite element spaces for $n = 2$.

k	$\Lambda_h^k(\Omega)$	Classical finite element space
0	$\mathcal{P}_r\Lambda^0(\mathcal{T})$	Lagrange elements of degree $\leq r$
1	$\mathcal{P}_r\Lambda^1(\mathcal{T})$	Brezzi–Douglas–Marini $H(\mathrm{div})$ elements of degree $\leq r$
2	$\mathcal{P}_r\Lambda^2(\mathcal{T})$	discontinuous elements of degree $\leq r$
0	$\mathcal{P}_r^-\Lambda^0(\mathcal{T})$	Lagrange elements of degree $\leq r$
1	$\mathcal{P}_r^-\Lambda^1(\mathcal{T})$	Raviart–Thomas $H(\mathrm{div})$ elements of order $r - 1$
2	$\mathcal{P}_r^-\Lambda^2(\mathcal{T})$	discontinuous elements of degree $\leq r - 1$

Table 5.2. Correspondences between finite element differential forms and the classical finite element spaces for $n = 3$.

k	$\Lambda_h^k(\Omega)$	Classical finite element space
0	$\mathcal{P}_r\Lambda^0(\mathcal{T})$	Lagrange elements of degree $\leq r$
1	$\mathcal{P}_r\Lambda^1(\mathcal{T})$	Nédélec 2nd-kind $H(\mathrm{curl})$ elements of degree $\leq r$
2	$\mathcal{P}_r\Lambda^2(\mathcal{T})$	Nédélec 2nd-kind $H(\mathrm{div})$ elements of degree $\leq r$
3	$\mathcal{P}_r\Lambda^3(\mathcal{T})$	discontinuous elements of degree $\leq r$
0	$\mathcal{P}_r^-\Lambda^0(\mathcal{T})$	Lagrange elements of degree $\leq r$
1	$\mathcal{P}_r^-\Lambda^1(\mathcal{T})$	Nédélec 1st-kind $H(\mathrm{curl})$ elements of order $r - 1$
2	$\mathcal{P}_r^-\Lambda^2(\mathcal{T})$	Nédélec 1st-kind $H(\mathrm{div})$ elements of order $r - 1$
3	$\mathcal{P}_r^-\Lambda^3(\mathcal{T})$	discontinuous elements of degree $\leq r - 1$

Thus $\mathrm{Tr}_F\,\omega$ is single-valued and so $\omega \in H\Lambda^k(\Omega)$. This establishes that we have a set of degrees of freedom $\mathcal{P}_r\Lambda^k(\mathcal{T})$ (and so this is truly a finite element space).

Analogously, the degrees of freedom for the space $\mathcal{P}_r^-\Lambda^k(\mathcal{T})$ are given by

$$\int_f \mathrm{Tr}_f\,\omega \wedge \nu, \quad \nu \in \mathcal{P}_{r-j+k-1}\Lambda^{j-k}(f), \quad f \in \Delta_j(\mathcal{T}), \qquad (5.2)$$

for $k \leq j \leq \min(n, r + k - 1)$.

In two and three dimensions we may use proxy fields to identify these spaces of finite element differential forms with finite element spaces of scalar and vector functions. In Tables 5.1 and 5.2, we summarize the correspondences between spaces of finite element differential forms and classical finite element spaces: the Lagrange elements (Ciarlet 1978); the Raviart–Thomas elements introduced in two dimensions by Raviart and Thomas (1977) and generalized to three dimensions by Nédélec (1980); the Brezzi–Douglas–Marini elements introduced by Brezzi, Douglas and Marini (1985) and generalized to three dimensions by Nédélec (1986) and Brezzi, Douglas, Durán and Fortin (1987); and the spaces of discontinuous elements of degree r, i.e., all piecewise polynomials of degree no greater than r.

5.2. The canonical projections

In Section 4.9, we defined for each simplex T a projection $\Pi_T : C^0 \Lambda^k(T) \to \mathcal{P}_r \Lambda^k(T)$. We can then define $\Pi = \Pi_T : C^0 \Lambda^k(\Omega) \to \mathcal{P}_r \Lambda^k(T)$ by

$$(\Pi_T \omega)|_T = \Pi_T(\omega|_T).$$

(Note that the degrees of freedom used to determine Π_T ensure that the traces of $\Pi_T \omega$ on faces are single-valued.) Equivalently, $\Pi_T \omega$ is determined by the equations

$$\int_f \mathrm{Tr}_f \Pi_T \omega \wedge \nu = \int_f \mathrm{Tr}_f \omega \wedge \nu, \quad \nu \in \mathcal{P}_{r-j+k}^- \Lambda^{j-k}(f), \quad f \in \Delta_j(T),$$

for $k \leq j \leq \min(n, r + k - 1)$. Of course, an analogous operator $\Pi = \Pi_T : C^0 \Lambda^k(\Omega) \to \mathcal{P}_r^- \Lambda^k(T)$ is defined similarly. From Lemma 4.24, we get commutativity of the projections with the exterior derivative.

Theorem 5.2. The following four diagrams commute:

$$
\begin{array}{ccc}
\Lambda^k(\Omega) & \xrightarrow{\ d\ } & \Lambda^{k+1}(\Omega) \\
\Pi \downarrow & & \Pi \downarrow \\
\mathcal{P}_r \Lambda^k(T) & \xrightarrow{\ d\ } & \mathcal{P}_{r-1} \Lambda^{k+1}(T)
\end{array}
\qquad
\begin{array}{ccc}
\Lambda^k(\Omega) & \xrightarrow{\ d\ } & \Lambda^{k+1}(\Omega) \\
\Pi \downarrow & & \Pi \downarrow \\
\mathcal{P}_r \Lambda^k(T) & \xrightarrow{\ d\ } & \mathcal{P}_r^- \Lambda^{k+1}(T)
\end{array}
$$

$$
\begin{array}{ccc}
\Lambda^k(\Omega) & \xrightarrow{\ d\ } & \Lambda^{k+1}(\Omega) \\
\Pi \downarrow & & \Pi \downarrow \\
\mathcal{P}_r^- \Lambda^k(T) & \xrightarrow{\ d\ } & \mathcal{P}_r^- \Lambda^{k+1}(T)
\end{array}
\qquad
\begin{array}{ccc}
\Lambda^k(\Omega) & \xrightarrow{\ d\ } & \Lambda^{k+1}(\Omega) \\
\Pi \downarrow & & \Pi \downarrow \\
\mathcal{P}_r^- \Lambda^k(T) & \xrightarrow{\ d\ } & \mathcal{P}_{r-1} \Lambda^{k+1}(T)
\end{array}
$$

5.3. Error estimates

Now we consider error bounds for $\|\omega - \Pi\omega\|$. To this end we need to consider not just a single triangulation $\mathcal{T}$ but a family of triangulations $\mathcal{T}_h$ of Ω indexed by the discretization parameter

$$h = \max_{T \subset \mathcal{T}_h} \operatorname{diam} T.$$

We assume that the discretization parameter runs over a set of positive values bounded by some $h_{\max}$ and accumulating at zero.

In order to obtain error estimates, we will assume that there exists a constant $C_{\mathrm{mesh}} > 0$, called the *mesh regularity constant*, for which

$$|h|^n \leq C_{\mathrm{mesh}}|T|, \quad T \in \mathcal{T}_h, \tag{5.3}$$

where $|T|$ denotes the volume of the simplex T. This assumption has two consequences. First, it enforces the shape regularity of the triangulation family, meaning that each simplex is Lipschitz diffeomorphic to a ball with uniform bounds on the Lipschitz constants of the diffeomorphism and its inverse. Second it implies that the family $\{\mathcal{T}_h\}$ is quasi-uniform in the sense that ratio of diameters of elements in $\mathcal{T}_h$ are bounded uniformly in h. The quasi-uniformity property is too restrictive for many applications, and often it can be avoided by more sophisticated analysis, using, instead of (5.3), the weaker assumption $|\operatorname{diam} T|^n \leq C|T|$. However, here we shall assume quasi-uniformity in order to avoid some technical difficulties, and so use (5.3).

We now state bounds for the error in the canonical projection of a differential form, in terms of the $W_p^s \Lambda^k(\Omega)$ seminorm of the form and a power of the mesh parameter h.

Theorem 5.3. Denote by Π_h the canonical projection of $\Lambda^k(\Omega)$ onto either $\mathcal{P}_r \Lambda^k(\mathcal{T}_h)$ or $\mathcal{P}_{r+1}^- \Lambda^k(\mathcal{T}_h)$. Let $1 \leq p \leq \infty$ and $(n-k)/p < s \leq r+1$. Then Π_h extends boundedly to $W_p^s \Lambda^k(\Omega)$, and there exists a constant c independent of h, such that

$$\|\omega - \Pi_h \omega\|_{L^p \Lambda^k(\Omega)} \leq Ch^s |\omega|_{W_p^s \Lambda^k(\Omega)}, \quad \omega \in W_p^s \Lambda^k(\Omega).$$

Proof. We shall show the result element by element. That is, we shall show that Π_T extends boundedly to $W_p^s \Lambda^k(T)$ and

$$\|\omega - \Pi_T \omega\|_{L^p \Lambda^k(T)} \leq Ch^s |\omega|_{W_p^s \Lambda^k(T)}, \quad \omega \in \Lambda^k(T), \tag{5.4}$$

for each element $T \in \mathcal{T}_h$, with the constant independent of T and h.

Since $s > (n-k)/p$, the Sobolev embedding theorem implies that $\omega \in W_p^s(T)$ admits traces on subsimplices of dimension k and so the operator Π_T is bounded on $W_p^s(T)$ with the bound

$$\|(I - \Pi_T)\omega\|_{L^p \Lambda^k(T)} \leq c\|\omega\|_{W_p^s \Lambda^k(T)},$$

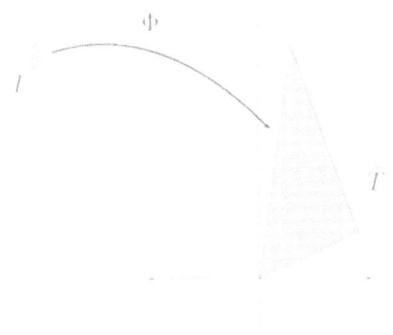

Figure 5.1. Scaling of a simplex
by dilation and translation.

for some constant c (which may depend on T). But $I - \Pi_T$ annihilates polynomial forms of degree r or less, and $s \leq r + 1$, and so the Bramble–Hilbert lemma (Brenner and Scott 2002, Lemma 4.3.8) implies that

$$\|(I - \Pi_T)\omega\|_{L^p\Lambda^k(T)} \leq c(T)|\omega|_{W_p^s\Lambda^k(T)}, \tag{5.5}$$

where now the Sobolev seminorm appears on the right-hand side. We choose $c(T)$ to be the least constant c for which (5.5) holds. Then $c(T)$ is a continuous function of T, or, more precisely, of its vertices, which belong to the open set

$$\{ (x_0, \ldots, x_n) \in (\mathbb{R}^n)^{n+1} \mid \det[x_1 - x_0, \ldots, x_n - x_0] \neq 0 \}.$$

To get from (5.5) to (5.4) we use scaling. Let x_0 denote the first vertex of T, and define $\Phi(x) = (x - x_0)/h$, an invertible affine map depending on T and h which maps T to $\hat{T} := \Phi(T)$ (see Figure 5.1). For $\omega \in W_p^s\Lambda^k(T)$, define $\hat{\omega} = \Phi^{-1*}\omega \in W_p^s\Lambda^k(\hat{T})$. It is easy to check that $\Pi_{\hat{T}}\hat{\omega} = \Phi^{-1*}(\Pi_T\omega)$ and so that

$$\|(I - \Pi_{\hat{T}})\hat{\omega}\|_{L^p\Lambda^k(\hat{T})} = h^{k-n/p}\|(I - \Pi_T)\omega\|_{L^p\Lambda^k(T)},$$

$$|\hat{\omega}|_{W_p^s\Lambda^k(\hat{T})} = h^{s+k-n/p}|\omega|_{W_p^s\Lambda^k(T)}.$$

Therefore

$$\|(I - \Pi_T)\omega\|_{L^p\Lambda^k(T)} = h^{n/p-k}\|(I - \Pi_{\hat{T}})\hat{\omega}\|_{L^p\Lambda^k(\hat{T})}$$

$$\leq c(\hat{T})h^{n/p-k}|\hat{\omega}|_{W_p^s\Lambda^k(\hat{T})}$$

$$= c(\hat{T})h^s\|(I - \Pi_T)\omega\|_{L^p\Lambda^k(T)},$$

whence

$$c(T) = c(\hat{T})h^s. \tag{5.6}$$

But the mesh regularity assumption implies that there exist positive constants K_1 and K_2 depending only on C_{mesh} such that

$$\operatorname{diam} \hat{T} \le K_1, \quad |\hat{T}| \ge K_2.$$

Together with the fact that $\hat{T}$ has a vertex at the origin, these conditions confine the vertices to a compact set, and so there is a constant C, independent of $T \in \mathcal{T}_h$ and h so that $c(\hat{T}) \le C$. Combining with (5.5) and (5.6), we obtain (5.4) and hence the theorem. □

Remark. This proof uses the standard elements of localization, Sobolev embedding, the Bramble–Hilbert lemma, and scaling, but is a little unusual in that we scale by dilation and translation only, and thus we get a family of scaled elements $\hat{T}$, which is, in an appropriate sense, compact. The more standard proof uses affine scaling to a fixed reference element and so avoids the compactness argument. We have chosen to use dilation and compactness here, because we will use the same technique again when bounding smoothed projections.

5.4. Projections and smoothing

The estimates just obtained for the canonical projections are not sufficient for our needs. For example, they do not furnish bounds for the projection into $\mathcal{P}_r \Lambda^1(\mathcal{T}_h)$ in terms of the H^1-norm in three dimensions ($s = 1$, $p = 2$, $k = 1$, $n = 3$ is not allowed). They provide no estimate in terms of any $p = 2$ based Sobolev norm for the projection into $\mathcal{P}_1 \Lambda^0(\mathcal{T}_h)$ (*i.e.*, the nodal interpolant) in more than three dimensions.

The difficulty with the canonical projection is that, owing to its dependence on traces on subsimplices, it is not bounded on spaces with insufficient smoothness. This same problem shows up in that it is not easy to use the canonical projections in a cochain projection from a version of the de Rham sequence onto a discrete subcomplex. If the spaces in the de Rham sequence are smooth enough that the projection operators are bounded on them, they are generally too smooth to include the finite element spaces.

In finite element theory, the Clément interpolant (Clément 1975) is often invoked to overcome problems of this sort. In our situation, the Clément interpolant would be defined by assigning to a given form in $\omega \in L^2 \Lambda^k(\Omega)$ a finite element differential form $\breve{\Pi}_h \omega$ specified by the degrees of freedom in the finite element space. For a degree of freedom associated with a subsimplex f, the Clément interpolant takes the value of this degree of freedom not from ω, but rather from *the L^2-projection of ω into the space of polynomials forms of degree r on the union of the simplices in $\mathcal{T}_h$ containing f.* This construction yields an operator which is bounded on L^2 and satisfies optimal order error bounds:

$$\|\omega - \check{\Pi}_h\omega\|_{L^p\Lambda^k(\Omega)} \leq ch^s\|\omega\|_{W_p^s\Lambda^k(\Omega)}, \quad 0 \leq s \leq r+1.$$

However, $\check{\Pi}_h$ is not a projection (it does not leave the finite element subspace fixed), and it does not commute with the exterior derivative. And so it, too, is not sufficient for our purposes.

To overcome these difficulties, in this subsection we will construct an alternative set of projection operators $\bar{\Pi}_h$ which are bounded from $L^2\Lambda^k(\Omega)$ to the finite element space. The approach we take is highly influenced by recent work by Schöberl (2005) and Christiansen (2005), partially unpublished. Briefly, to project a form ω we will first extend it to a slightly larger domain and then regularize. Regularization commutes with exterior differentiation, and the extension is chosen so that it commutes with exterior differentiation as well. Next we use the canonical projection to project the regularized form into the subspace. This procedure gives an operator which is bounded on L^2 and which commutes with the exterior derivative, but which is not a projection. We remedy this by multiplying by the inverse of the operator restricted to the subspace (which can be shown to exist). The resulting operator still commutes with exterior differentiation, is still bounded on L^2, and is a projection. From the last two properties, we easily obtain optimal error estimates.

The extension operator

Since we will make use of a smoothing operator defined by convolution with a mollifier function, we will need to extend functions in $H\Lambda^k(\Omega)$ to a fixed larger domain $\tilde{\Omega}$ where the closure of Ω, $\bar{\Omega}$ is contained in $\tilde{\Omega}$. Let the extended domain be of the form $\tilde{\Omega} = \bar{\Omega} \cup \Omega_o$, where $\bar{\Omega} \cap \Omega_o = \emptyset$. Following Schöberl (2005), we utilize the compactness of the boundary $\partial\Omega$ to construct the outer neighbourhood Ω_o, a corresponding interior neighbourhood of $\partial\Omega$, $\Omega_i \subset \Omega$, and a Lipschitz continuous bijection $\Psi : \Omega_o \cup \partial\Omega \to \Omega_i \cup \partial\Omega$, with the additional properties that $\Psi(x) = x$ on $\partial\Omega$.

Using the mapping Ψ, we define an extension operator $E : H\Lambda^k(\Omega) \to H\Lambda^k(\tilde{\Omega})$ by

$$(E\omega)_x = (\Psi^*\omega)_x, \quad x \in \Omega_o.$$

This operator clearly maps $L^2\Lambda^k(\Omega)$ boundedly into $L^2\Lambda^k(\tilde{\Omega})$, and since $d \circ \Psi^* = \Psi^* \circ d$, we obtain that $E \in \mathcal{L}(H\Lambda^k(\Omega), H\Lambda^k(\tilde{\Omega}))$.

The smoothing operator

As in Christiansen (2005), we will perform a construction where we combine the canonical projections introduced above with a standard smoothing operator defined by convolution with an approximate Dirac delta function.

For parameters $\epsilon > 0$, we will employ smoothing operators $R_\epsilon : L^2\Lambda^k(\Omega) \rightarrow \Lambda^k(\Omega)$ of the form

$$R_\epsilon\omega = (\rho_\epsilon * E\omega)|_\Omega.$$

Here $*$ denotes the convolution product, and the mollifier function ρ_ϵ is of the form $\rho_\epsilon(x) = \epsilon^{-n}\rho(x/\epsilon)$, where $\rho : \mathbb{R}^n \rightarrow \mathbb{R}$ is C^∞, is nonnegative with compact support, and with integral equal to 1. Note that for ϵ sufficiently small, the operator $R_\epsilon : L^2\Lambda^k(\Omega) \rightarrow \Lambda^k(\Omega)$ is well-defined, and we have the commutativity property $\mathrm{d}R_\epsilon = R_\epsilon\mathrm{d}$.

The smoothing parameter has to be related to the triangulation $\mathcal{T}_h$. In fact, for more general triangulations than we allow here, it seems that an x-dependent smoothing parameter is required. We refer to Christiansen (2005) for such a discussion. However, since we assume that the triangulation is quasi-uniform (see (5.3)), we can choose the smoothing parameter proportional to the mesh parameter h. Hence, our construction will use the smoothing operator $R_{\epsilon h}$, where $\epsilon > 0$ will be chosen independently of the triangulation $\mathcal{T}_h$. We will take ϵ sufficiently small that:

- for $x \in \bar{\Omega}$, the ball of radius $\epsilon h_{\max}$ about x is contained in $\Omega \cup \Omega_o$;

- for each h, if $x \in T$ for some $T \in \mathcal{T}_h$, then the ball of radius ϵh about x is contained in the union of the simplices in $\mathcal{T}_h$ which intersect T.

Uniform bounds

Now we let Λ_h^k denote either $\mathcal{P}_r\Lambda^k(\mathcal{T}_h)$ or $\mathcal{P}_r^-\Lambda^k(\mathcal{T}_h)$ and denote by Π_h the canonical interpolation operator onto Λ_h^k.

Since we have restricted to ϵ sufficiently small, $R_{\epsilon h}$ maps $L^2\Lambda^k(\Omega)$ into $C\Lambda^k(\Omega)$ for all h. Then the map $\Pi_h R_{\epsilon h} : L^2\Lambda^k(\Omega) \rightarrow \Lambda_h^k$ is certainly bounded. The following lemma, as we shall prove below via a scaling argument, states that the bound is uniform in h (but not in ϵ).

Lemma 5.4. For ϵ sufficiently small as above, there exists a constant $c(\epsilon)$ such that the operator norm satisfies

$$\|\Pi_h R_{\epsilon h}\|_{\mathcal{L}(L^2\Lambda^k(\Omega), L^2\Lambda^k(\Omega))} \leq c(\epsilon),$$

for all h.

The restriction $\Pi_h R_{\epsilon h}|_{\Lambda_h^k}$ maps Λ_h^k into itself. The following lemma, which we shall also prove below using scaling, states that it converges to the identity as $\epsilon \rightarrow 0$, uniformly in h. Lemma 5.5 uses the notation $\|\cdot\|_{\mathcal{L}(\Lambda_h^k, \Lambda_h^k)}$ to denote the L^2-operator norm of an operator $\Lambda_h^k \rightarrow \Lambda_h^k$.

Lemma 5.5. There exists a constant c, independent of h and ϵ, such that

$$\|I - \Pi_h R_{\epsilon h}|_{\Lambda_h^k}\|_{\mathcal{L}(\Lambda_h^k, \Lambda_h^k)} \leq c\epsilon.$$

In view of this lemma, we can choose ϵ sufficiently small that

$$\|I - \Pi_h R_{\epsilon h}|_{\Lambda_h^k}\|_{\mathcal{L}(\Lambda_h^k, \Lambda_h^k)} \leq 1/2 \tag{5.7}$$

for all h. It follows that $\Pi_h R_{\epsilon h}|_{\Lambda_h^k}$ is invertible and that its inverse J_h^ϵ : $\Lambda_h^k \to \Lambda_h^k$ satisfies

$$\|J_h^\epsilon\|_{\mathcal{L}(\Lambda_h^k, \Lambda_h^k)} \leq 2. \tag{5.8}$$

The smoothed projections

Combining these results, we can easily complete the construction of the smoothed projections. We fix ϵ sufficiently small as above and also so that (5.7) holds. Then, for this fixed ϵ, we set

$$\tilde{\Pi}_h = J_h^\epsilon \Pi_h R_{\epsilon h}.$$

This operator has all the properties we need.

Theorem 5.6. The smoothed projection $\tilde{\Pi}_h$ is a projection of $L^2\Lambda^k(\Omega)$ onto Λ_h^k which commutes with the exterior derivative and satisfies

$$\|\omega - \tilde{\Pi}_h\omega\| \leq ch^s\|\omega\|_s, \quad \omega \in H^s\Lambda^k(\Omega), \ 0 \leq s \leq r+1.$$

Moreover, for all $\omega \in L^2\Lambda^k(\Omega)$, $\tilde{\Pi}_h\omega \to \omega$ in L^2 as $h \to 0$.

Proof. By construction, $\tilde{\Pi}_h$ is a bounded linear operator from $L^2\Lambda^k(\Omega)$ to Λ_h^k which commutes with d, so we need only establish the error bounds. Using Lemma 5.4 and (5.8), we have that $\tilde{\Pi}_h$ is uniformly bounded in $\mathcal{L}(L^2\Lambda^k(\Omega), L^2\Lambda^k(\Omega))$. Thus from the projection property we have

$$\|\omega - \tilde{\Pi}_h\omega\| = \inf_{\mu \in \Lambda_h^k} \|(I - \tilde{\Pi}_h)(\omega - \mu)\| \leq c \inf_{\mu \in \Lambda_h^k} \|\omega - \mu\| \leq ch^s\|\omega\|_s,$$

where the last inequality can be proved, *e.g.*, by using the Clément interpolant. The final pointwise estimate follows from the estimate for $s = 1$ and the fact that $H^1\Lambda^k(\Omega)$ is dense in $L^2\Lambda^k(\Omega)$. $\qquad\square$

Proofs of the uniform bounds

We now turn to the proofs of Lemmas 5.4 and 5.5. As in the proof of Theorem 5.3, we use localization, scaling and compactness, but the situation is complicated by the fact that the smoothing operator is not entirely local: the restriction of $R_{\epsilon h}\omega$ to a simplex T is not determined by $\omega|_T$. However, since we have assumed that ϵ is sufficiently small, $(R_{\epsilon h}\omega)|_T$ is determined by $\omega|_{T^\star}$, where $T^\star = \bigcup \mathcal{T}_h(T)$ and

$$\mathcal{T}_h(T) := \{ T' \in \mathcal{T}_h \mid T' \cap T \neq \emptyset \}$$

Figure 5.2. Macro-element of an element T:
the triangulation is $\mathcal{T}_h(T)$, the triangulated
region is $T^\star$, and the shaded simplex is T.

is the macro-element in $\mathcal{T}_h$ determined by T (see Figure 5.2). We shall write
$\Lambda_h^k(T)$ and $\Lambda_h^k(T^\star)$ for the space of restrictions of elements of Λ_h^k to T or
$T^\star$. The former is just the polynomial space $\mathcal{P}_r\Lambda^k(T)$ or $\mathcal{P}_r^-\Lambda^k(T)$, while
the latter is equal to either $\mathcal{P}_r\Lambda^k(\mathcal{T}_h(T))$ or $\mathcal{P}_r^-\Lambda^k(\mathcal{T}_h(T))$. Now the shape
regularity property implies bounded overlap of the $T^\star$, so

$$\sum_{T \in \mathcal{T}_h} \|\omega\|_{H^s\Lambda^k(T^\star)} \le c\|\omega\|_{H^s\Lambda^k(\Omega)}.$$

Therefore, to prove Lemma 5.4, it suffices to show that

$$\|\Pi_T R_{\epsilon h}\|_{\mathcal{L}(L^2\Lambda^k(T^\star), L^2\Lambda^k(T))} \le c(\epsilon)$$

with $c(\epsilon)$ uniform over $T \in \mathcal{T}_h$ and over h. (There are some small modifications needed for the simplices T intersecting $\partial\Omega$.) Similarly, for Lemma 5.5
it is sufficient to show that

$$\|I - \Pi_T R_{\epsilon h}\|_{\mathcal{L}(\Lambda_h^k(T^\star), \Lambda_h^k(T))} \le c\epsilon,$$

uniformly over $T \in \mathcal{T}_h$ and over h. To prove these we shall employ scaling.
    Let $\Phi(x) = (x - x_0)/h$, where x_0 is the first vertex of T. Thus Φ maps T
onto a simplex $\hat{T}$ with a vertex at the origin and diameter bounded above
and below by positive constants depending only on C_{mesh}. It also maps $T^\star$
onto $\hat{T}^\star := \Phi_T(T^\star)$ (see Figure 5.3). Then $\Phi^{*-1}\Pi_h\Phi^* : \Lambda^k(\hat{T}) \to \Lambda_h^k(\hat{T})$
is just the canonical projection $\Pi_{\hat{T}}$ onto the polynomial space $\Lambda_h^k(\hat{T})$, and
$\Phi^{*-1}R_{\epsilon h}\Phi^* : L^2\Lambda^k(\hat{T}^\star) \to \Lambda_h^k(\hat{T})$ is just the smoothing operator R_ϵ (because
of our choice of regularization parameter ϵh, proportional to h). Thus we
find that

$$\|\Pi_{\hat{T}} R_\epsilon\|_{\mathcal{L}(L^2\Lambda^k(\hat{T}^\star), L^2\Lambda^k(\hat{T}))} = \|\Pi_T R_{\epsilon h}\|_{\mathcal{L}(L^2\Lambda^k(T^\star), L^2\Lambda^k(T))}$$

and

$$\|I - \hat{\Pi}_{\hat{T}} R_\epsilon\|_{\mathcal{L}(\Lambda_h^k(\hat{T}^\star), \Lambda_h^k(\hat{T}))} = \|I - \Pi_T R_{\epsilon h}\|_{\mathcal{L}(\Lambda_h^k(T^\star), \Lambda_h^k(T))}.$$

Thus we must show that, for fixed ϵ,

$$\|\Pi_{\hat{T}} R_\epsilon\|_{\mathcal{L}(L^2\Lambda^k(\hat{T}^\star), L^2\Lambda^k(\hat{T}))} \le c(\epsilon), \qquad (5.9)$$

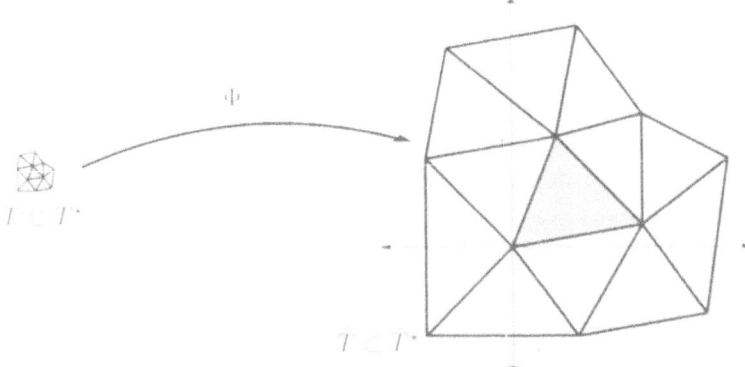

Figure 5.3. Scaling of a simplex
and its macro-element.

uniformly over $T \in \mathcal{T}_h$ and over h, and that

$$\|I - \Pi_{\hat{T}} R_\epsilon\|_{\mathcal{L}(\Lambda_h^k(\hat{T}^\star), \Lambda_h^k(\hat{T}))} \le c\epsilon, \tag{5.10}$$

uniformly over $T \in \mathcal{T}_h$, h, and ϵ.

For such T and h, the configuration of simplices in $\hat{T}^\star$ varies over a compact set, and hence it is sufficient to show (5.9) for any single simplex T with a macro-element neighbourhood $T^\star$. But this is evident: R_ϵ is bounded $L^2\Lambda^k(\hat{T}^\star) \to C\Lambda^k(\hat{T})$ (though not uniformly in ϵ) and $\Pi_{\hat{T}}$ is bounded $C\Lambda^k(\hat{T}) \to \Lambda_h^k(\hat{T})$. This proves (5.9) and hence Lemma 5.4.

Finally, to prove (5.10), and hence Lemma 5.5, we will derive (5.10) from a more general bound. We will show that there is a constant c, independent of ϵ, such that

$$\|\Pi_{\hat{T}}(I - R_\epsilon)\omega\|_{L^2\Lambda^k(\hat{T})} \le c\epsilon \sum_{T' \in \mathcal{T}(\hat{T})} \|\omega\|_{W_\infty^1(T')} \tag{5.11}$$

for all $\omega \in H\Lambda^k(\hat{T}^\star)$ such that $\omega|_{T'} \in H^1\Lambda^k(T')$ for $T' \in \mathcal{T}(\hat{T})$. Here $\mathcal{T}(\hat{T})$ denotes the set of n simplices which defines $\hat{T}^\star$. In fact, (5.11) will immediately imply (5.10) since $\Pi_{\hat{T}}\omega = \omega$ and

$$\sum_{T' \in \mathcal{T}(\hat{T})} \|\omega\|_{W_\infty^1(T')} \le c\|\omega\|_{L^2\Lambda^k(\hat{T}^\star)}$$

for any $\omega \in \Lambda_h^k(\hat{T}^\star)$.

In order to establish (5.11), recall that the space $\Lambda_h^k(\hat{T})$ is either of the form $\mathcal{P}_r\Lambda^k(\hat{T})$ or $\mathcal{P}_r^-\Lambda^k(\hat{T})$, for a suitable $r \ge 1$. As a consequence of the degrees of freedom for these spaces, given by Theorems 4.10 and 4.14, to prove (5.11) it is enough to show that for a given $f \in \Delta(\hat{T})$, with $\dim f \geq k$, and $\eta \in \Lambda^{\dim f - k}(f)$ we have

$$\left| \int_f (I - R_\epsilon) \omega \wedge \eta \right| \leq c\epsilon \sum_{T' \in T(\hat{T})} \|\omega\|_{W^1_\infty(T')} \tag{5.12}$$

for all $\omega \in H\Lambda^k(\hat{T}^\star)$ such that $\omega|_{T'} \in W^1_\infty\Lambda^k(T')$ for $T' \in T(\hat{T})$. Here the constant c is independent of ϵ and ω, but it is allowed to depend on the test function η. Recall that the integration operator $\int_f$ is to be interpreted as the evaluation operator at the point f when $\dim f = 0$.

To show this bound assume first that $\dim f > 0$ and that $\omega \in H\Lambda^k(\hat{T}^\star) \cap W^1_\infty\Lambda^k(T')$, $T' \in T(\hat{T})$. We will decompose the f into f_ϵ and $f \setminus f_\epsilon$, where

$$f_\epsilon = \{x \in f \mid \operatorname{dist}(x, \partial f) \geq C\epsilon\}.$$

Here the constant $C > 0$ is chosen such that, for any point $x \in f_\epsilon$, the ball of radius ϵ with centre at x will only intersect the elements of $T(\hat{T})$ which have f as a subsimplex. A consequence of this construction is that, if $x \in f_\epsilon$ and $v_1, v_2, \ldots, v_k$ are unit tangent vectors to f, then $\omega_y(v_1, \ldots, v_k)$ is continuous for $|x - y| \leq \epsilon$, and

$$|\omega_x(v_1, \ldots, v_k) - \omega_y(v_1, \ldots, v_k)| \leq \epsilon \sum_{T' \in T(\hat{T})} \|\omega\|_{W^1_\infty(T')}.$$

However, this implies that

$$\left| \int_{f_\epsilon} (I - R_\epsilon) \omega \wedge \eta \right| \leq c\epsilon \sum_{T' \in T(\hat{T})} \|\omega\|_{W^1_\infty(T')},$$

where the constant c is independent of ϵ and ω. Finally, it is straightforward to see that

$$\left| \int_{f \setminus f_\epsilon} (I - R_\epsilon) \omega \wedge \eta \right| \leq \left| \int_{f \setminus f_\epsilon} \omega \wedge \eta \right| + \left| \int_{f \setminus f_\epsilon} R_\epsilon \omega \wedge \eta \right| \leq c\epsilon \|\omega\|_{L^\infty \Lambda^k(\hat{T}^\star)}.$$

Hence we have verified the bound (5.12) when $\dim f > 0$. If $\dim f = 0$, the bound (5.12) still holds by a simple modification of the proof.

5.5. Discrete de Rham complexes

As in Section 3.5, the spaces $\mathcal{P}_r\Lambda^k(\mathcal{T}_h)$ and $\mathcal{P}_r^-\Lambda^k(\mathcal{T}_h)$ lead to a collection of discrete de Rham complexes, essentially 2^{n-1} for each polynomial degree. To this end, we observe that

$$d\mathcal{P}_r\Lambda^k(\mathcal{T}_h) \subset \mathcal{P}_{r-1}\Lambda^{k+1}(\mathcal{T}_h) \subset \mathcal{P}_r^-\Lambda^{k+1}(\mathcal{T}_h),$$

and

$$d\mathcal{P}_r^-\Lambda^k(\mathcal{T}_h) \subset \mathcal{P}_{r-1}\Lambda^{k+1}(\mathcal{T}_h) \subset \mathcal{P}_r^-\Lambda^{k+1}(\mathcal{T}_h).$$

This leads to discrete complexes of the form

$$0 \to \Lambda_h^0 \xrightarrow{d} \Lambda_h^1 \xrightarrow{d} \cdots \xrightarrow{d} \Lambda_h^n \to 0, \tag{5.13}$$

where for each map of the form $\Lambda_h^k \xrightarrow{d} \Lambda_h^{k+1}$, we can substitute one of the four choices

$$\mathcal{P}_r\Lambda^k(\mathcal{T}_h) \xrightarrow{d} \mathcal{P}_{r-1}\Lambda^{k+1}(\mathcal{T}_h), \qquad \mathcal{P}_r\Lambda^k(\mathcal{T}_h) \xrightarrow{d} \mathcal{P}_r^-\Lambda^{k+1}(\mathcal{T}_h),$$

$$\mathcal{P}_r^-\Lambda^k(\mathcal{T}_h) \xrightarrow{d} \mathcal{P}_r^-\Lambda^{k+1}(\mathcal{T}_h), \qquad \mathcal{P}_r^-\Lambda^k(\mathcal{T}_h) \xrightarrow{d} \mathcal{P}_{r-1}\Lambda^{k+1}(\mathcal{T}_h).$$

Each complex so obtained is a subcomplex of the L^2 de Rham complex (2.6). Making use of the smoothed projections, we obtain in each case a commuting diagram:

$$
\begin{array}{ccccccc}
0 \to H\Lambda^0(\Omega) & \xrightarrow{d} & H\Lambda^1(\Omega) & \xrightarrow{d} & \cdots & \xrightarrow{d} & H\Lambda^n(\Omega) \to 0 \\
\downarrow{\tilde{\Pi}^0} & & \downarrow{\tilde{\Pi}^1} & & & & \downarrow{\tilde{\Pi}^n} \\
0 \to \quad \Lambda_h^0 & \xrightarrow{d} & \Lambda_h^1 & \xrightarrow{d} & \cdots & \xrightarrow{d} & \Lambda_h^n \quad \to 0.
\end{array}
$$

Thus the projections give a cochain projection from the de Rham sequence to the discrete de Rham sequence, and so induce a surjection on cohomology. In fact as we shall now prove, following Christiansen (2005), this is in each case an isomorphism on cohomology.

The simplest finite element de Rham complex is the complex of Whitney forms,

$$0 \to \mathcal{P}_1^-\Lambda^0(\mathcal{T}_h) \xrightarrow{d} \mathcal{P}_1^-\Lambda^1(\mathcal{T}_h) \xrightarrow{d} \cdots \xrightarrow{d} \mathcal{P}_1^-\Lambda^n(\mathcal{T}_h) \to 0. \tag{5.14}$$

That the cohomology of this complex is isomorphic with the de Rham cohomology is a known, but deep result. It follows from de Rham's theorem, since the cohomology of the complex of Whitney forms is equal to the simplicial cohomology associated with the triangulation $\mathcal{T}_h$, and de Rham's theorem states that this simplicial cohomology is isomorphic to the de Rham cohomology.

Considering the Whitney forms complex (5.14) as a subcomplex of (5.13), the canonical projections Π_h define cochain projections. Note that the Π_h are defined on the finite element spaces Λ_h^k, because all of the trace moments they require are single-valued on Λ_h^k. From the commuting diagram

$$
\begin{array}{ccccccc}
0 \to \quad \Lambda_h^0 & \xrightarrow{d} & \Lambda_h^1 & \xrightarrow{d} & \cdots & \xrightarrow{d} & \Lambda_h^n \quad \to 0 \\
\downarrow{\Pi_h} & & \downarrow{\Pi_h} & & & & \downarrow{\Pi_h} \\
0 \to \mathcal{P}_1^-\Lambda^0(\mathcal{T}_h) & \xrightarrow{d} & \mathcal{P}_1^-\Lambda^1(\mathcal{T}_h) & \xrightarrow{d} & \cdots & \xrightarrow{d} & \mathcal{P}_1^-\Lambda^n(\mathcal{T}_h) \to 0,
\end{array}
$$

we conclude that the cohomology of the top row, which we have already seen to be an image of the de Rham cohomology, maps onto the cohomology of the bottom row, which is isomorphic to the de Rham cohomology. Hence the dimension of all the corresponding cohomology groups are equal and both cochain projections induce an isomorphism on cohomology.

5.6. Discrete Hodge decompositions and discrete Poincaré inequality

We close this section with the discrete analogue of the Hodge decomposition and of Poincaré's inequality.

Discrete Hodge decomposition and discrete harmonic functions

Let

$$\mathfrak{Z}_h^k = \{\,\omega \in \Lambda_h^k \mid d\omega = 0\,\}, \quad \mathfrak{B}_h^k = d\Lambda_h^{k-1},$$

denote the spaces of finite element cycles and boundaries. We have $\mathfrak{B}_h^k \subset \mathfrak{Z}_h^k$ so $\mathfrak{Z}_h^{k\perp} \subset \mathfrak{B}_h^{k\perp}$, where the orthogonal complements are taken within the space Λ_h^k with respect to the $L^2\Lambda^k$-norm (or the $H\Lambda^k$-norm, which gives the same result). The orthogonal complement of $\mathfrak{Z}_h^{k\perp}$ inside $\mathfrak{B}_h^{k\perp}$ is

$$\mathfrak{H}_h^k := \mathfrak{B}_h^{k\perp} \cap \mathfrak{Z}_h^k = \{\,\omega \in \Lambda_h^k \mid d\omega = 0, \ \langle \omega, d\tau \rangle = 0 \ \forall \tau \in \Lambda_h^{k-1}\,\},$$

the space of discrete harmonic forms. We have seen above that this space has the dimension of the kth de Rham cohomology space, which is the kth Betti number of the domain. We note that $\mathfrak{B}_h^k \subset \mathfrak{B}^k$ and $\mathfrak{Z}_h^k \subset \mathfrak{Z}^k$, but $\mathfrak{Z}_h^{k\perp}$ is not generally contained in $\mathfrak{Z}^{k\perp}$ and $\mathfrak{H}_h^k$ is not generally contained in $\mathfrak{H}^k$.

The discrete Hodge decomposition is a simple consequence of the definitions:

$$\Lambda_h^k = \mathfrak{B}_h^k \oplus \mathfrak{B}_h^{k\perp} = \mathfrak{B}_h^k \oplus \mathfrak{H}_h^k \oplus \mathfrak{Z}_h^{k\perp}.$$

We now make some important observations about discrete harmonic forms. The following theorem shows that they can be computed as the elements of the null space of a finite element matrix.

Theorem 5.7. Consider the homogeneous linear system: find $(\sigma_h, u_h) \in \Lambda_h^{k-1} \times \Lambda_h^k$ such that

$$\langle \sigma_h, \tau \rangle = \langle d\tau, u_h \rangle, \quad \tau \in \Lambda_h^{k-1},$$
$$\langle d\sigma_h, v \rangle + \langle du_h, dv \rangle = 0, \quad v \in \Lambda_h^k.$$

Then (σ_h, u_h) is a solution if and only if $\sigma_h = 0$ and $u_h \in \mathfrak{H}_h^k$.

Proof. Clearly $(0, u_h)$ is a solution if $u_h \in \mathfrak{H}_h^k$. On the other hand, if (σ_h, u_h) is a solution, by taking $\tau = \sigma_h$, $v = u_h$, and combining the two equations, we find that $\|\sigma_h\|^2 + \|du_h\|^2 = 0$, so that $\sigma_h = 0$ and $du_h = 0$. Then the first equation implies that $\langle d\tau, u_h \rangle = 0$ for all $\tau \in \Lambda_h^{k-1}$, so indeed $u_h \in \mathfrak{H}_h^k$. $\qquad\square$

We know that the space of discrete harmonic k-forms has the same dimension as the space of harmonic k-forms. In the next theorem we show that the discrete harmonic forms also provide good approximation of the harmonic forms. We shall use the smoothed projection operator $\tilde{\Pi}_h$, but, as can be seen from the proof, any projection operator which commutes with d could be used instead.

Theorem 5.8. Let $\sigma \in \mathfrak{H}^k$. Then there exists $\sigma_h \in \mathfrak{H}_h^k$ such that

$$\|\sigma - \sigma_h\| \leq \|\sigma - \tilde{\Pi}_h\sigma\|. \tag{5.15}$$

Proof. First we show that there exists a unique $(\sigma_h, u_h) \in \Lambda_h^k \times \mathfrak{B}_h^{k+1}$ such that

$$\langle \sigma_h, \tau \rangle - \langle d\tau, u_h \rangle = \langle \sigma, \tau \rangle, \quad \tau \in \Lambda_h^k,$$
$$\langle d\sigma_h, v \rangle = 0, \quad v \in \mathfrak{B}_h^{k+1}. \tag{5.16}$$

This is a finite-dimensional linear system, so we just need to show that if (σ_h, u_h) is a solution when $\sigma = 0$, then $\sigma_h = 0$ and $u_h = 0$. Choosing $\tau = \sigma_h$ and $v = u_h$, we get that $\sigma_h = 0$, and then that $\langle d\tau, u_h \rangle = 0$ for all $\tau \in \Lambda_h^k$, i.e., $u_h \in \mathfrak{B}_h^{(k+1)\perp}$. Thus $u_h = 0$.

Next we show that for the solution of (5.16), $\sigma_h \in \mathfrak{H}_h^k$. The second equation immediately gives $d\sigma_h = 0$, i.e., $\sigma_h \in \mathfrak{Z}_h^k$. Taking $\tau = d\rho$, $\rho \in \Lambda_h^{k-1}$, we get

$$\langle \sigma_h, d\rho \rangle = \langle \sigma, d\rho \rangle = 0, \quad \rho \in \Lambda_h^{k-1},$$

where the last equation holds because σ is harmonic. Thus $\sigma_h \in \mathfrak{B}_h^{k\perp}$, i.e., $\sigma_h \in \mathfrak{H}_h^k$.

Finally, we have from the first equation in (5.16) that

$$\langle \sigma - \sigma_h, \tilde{\Pi}_h\sigma - \sigma_h \rangle = \langle d(\sigma_h - \tilde{\Pi}_h\sigma), u_h \rangle = 0,$$

since $d\sigma_h = 0$ and $d\tilde{\Pi}_h\sigma = \tilde{\Pi}_h d\sigma = 0$. The inequality (5.15) follows immediately. $\qquad\square$

It is also the case, as we show below, that a discrete harmonic k-form can be approximated well by a harmonic k-form.

Lemma 5.9. Let $p \in \mathfrak{H}_h^k$ be a discrete harmonic k-form. There exists a $r \in \mathfrak{H}^k$ such that $\|r\| \leq \|p\|$ and

$$\|p - r\| \leq \|(I - \tilde{\Pi}_h)r\|.$$

Proof. We define $r = p - d\sigma$, where $d\sigma \in \mathfrak{B}^k$ is the L^2-projection of p onto $\mathfrak{B}^k$, i.e.,

$$\langle d\sigma, d\tau \rangle = \langle p, d\tau \rangle, \quad \tau \in H\Lambda^{k-1}(\Omega).$$

Note that $r \in \mathfrak{Z}^k$ since both p and $d\sigma$ belong to this space. On the other hand, the definition of σ implies that

$$\langle r, d\tau \rangle = 0, \quad \tau \in H\Lambda^{k-1}(\Omega).$$

Hence, $r \in \mathfrak{Z}^k \cap \mathfrak{B}^{k\perp} = \mathfrak{H}^k$. Furthermore, the bound $\|r\| = \|p - d\sigma\| \le \|p\|$ is a consequence of the fact that $d\sigma$ is the L^2-projection of p. Finally, to derive the error bound, observe that, since $p \in \mathfrak{H}_h^k$, we have

$$\langle r - p, d\tau \rangle = -\langle p, d\tau \rangle = 0, \quad \tau \in \Lambda_h^{k-1}.$$

Therefore,

$$
\begin{aligned}
\|r - p\|^2 &= \langle r - p, d\sigma \rangle = \langle r - p, d(I - \tilde{\Pi}_h)\sigma \rangle \\
&\le \|r - p\| \|(I - \tilde{\Pi}_h)d\sigma\| = \|r - p\| \|(I - \tilde{\Pi}_h)r\|,
\end{aligned}
$$

which gives the desired bound for $\|r - p\|$.   □

The following approximation result, relating forms in $\mathfrak{Z}_h^{k\perp}$ to forms in $\mathfrak{Z}^{k\perp}$, will also prove useful.

Lemma 5.10. If $u \in \mathfrak{Z}_h^{k\perp}$ and $w \in \mathfrak{Z}^{k\perp}$ satisfies $dw = du$, then

$$\|u - w\| \le \|w - \tilde{\Pi}_h w\|.$$

Proof. Since $w \in \mathfrak{Z}^{k\perp}$, there exists $z \in \mathring{\mathfrak{B}}^{k+1}$ such that $w = \delta z$, i.e.,

$$\langle w, v \rangle = \langle dv, z \rangle, \quad v \in H\Lambda^k(\Omega).$$

Similarly, since $u \in \mathfrak{Z}_h^{k\perp}$, there exists $z_h \in \mathfrak{B}_h^{k+1}$ such that

$$\langle u, v \rangle = \langle dv, z_h \rangle, \quad v \in \Lambda_h^k.$$

Hence,

$$\langle w - u, v \rangle = \langle dv, z - z_h \rangle, \quad v \in \Lambda_h^k.$$

Choosing $v = \tilde{\Pi}_h w - u$ and noting that $dv = 0$, we get

$$\|w - u\|^2 = \langle w - u, w - \tilde{\Pi}_h w \rangle \le \|w - u\| \|w - \tilde{\Pi}_h w\|.$$

The result follows immediately.   □

A discrete Poincaré inequality
Using Lemma 5.10, we prove the analogue of (2.17).

Theorem 5.11. There is a positive constant c, independent of h, such that

$$\|w\| \le c\|dw\|, \quad w \in \mathfrak{Z}_h^{k\perp}.$$

Proof. Define $\eta \in \mathfrak{Z}^{k\perp} \subset H\Lambda^k(\Omega)$ by $d\eta = dw$ (so η is the L^2-projection of w into $\mathfrak{Z}^{k\perp}$). By (2.17),

$$\|\eta\| \le c\|dw\|.$$

Hence, it is enough to show that $\|\omega\| \leq c\|\eta\|$. But this follows immediately from Lemma 5.10 and the boundedness of $\bar{\Pi}_h$. $\square$

6. Differential forms with values in a vector space

In Section 11 at the end of this paper, in which we study discretizations of the equations of elasticity, we will need to use differential forms with values in a vector space. We introduce the necessary ideas here, which are straightforward extensions of the material in Section 2. Federer (1969) is one reference for this material. At the end of this section, we consider a particular operator acting on vector- and bivector-valued algebraic forms, and establish some properties which will be needed later.

Let V and W be finite-dimensional vector spaces. We then define the space $\text{Alt}^k(V; W)$ of alternating k-linear forms on V with values in W. This is a vector space of dimension $\binom{\dim V}{k} \dim W$. There is a natural identification of $(\text{Alt}^k V) \otimes W$ with $\text{Alt}^k(V; W)$ given by

$$(\omega \otimes w)(v_1, \ldots, v_k) = \omega(v_1, \ldots, v_k)w, \quad \omega \in \text{Alt}^k V, \ w \in W, \ v_1, \ldots, v_k \in V.$$

For $k = 0$ and 1, we have $\text{Alt}^0(V; W) = W$ and $\text{Alt}^1(V; W) = \mathcal{L}(V, W)$, the space of linear operators from $V \to W$. Most of the definitions of Section 2.1 carry over without difficulty. For some we require an inner product on W, which we shall assume is given (and denoted with a dot). For example, the exterior product maps $\text{Alt}^j(V; W) \times \text{Alt}^k(V; W) \to \text{Alt}^{j+k} V$ (the range space is scalar-valued). It is defined by the analogue of (2.1)

$$(\omega \wedge \eta)(v_1, \ldots, v_{j+k})$$
$$= \sum_{\sigma}(\text{sign } \sigma)\omega(v_{\sigma(1)}, \ldots, v_{\sigma(j)}) \cdot \eta(v_{\sigma(j+1)}, \ldots, v_{\sigma(j+k)}), \quad v_i \in V,$$

where the sum is again over all permutations σ of $\{1, \ldots, j + k\}$, for which $\sigma(1) < \sigma(2) < \cdots \sigma(j)$ and $\sigma(j + 1) < \sigma(j + 2) < \cdots \sigma(j + k)$. Assuming that V also has an inner product, we get an inner product on $\text{Alt}^k(V; W)$ in analogy with (2.2):

$$\langle \omega, \eta \rangle = \sum_{\sigma} \omega(v_{\sigma(1)}, \ldots, v_{\sigma(k)}) \cdot \eta(v_{\sigma(1)}, \ldots, v_{\sigma(k)}), \quad \omega, \eta \in \text{Alt}^k(V; W),$$

where the sum is over increasing sequences $\sigma : \{1, \ldots, k\} \to \{1, \ldots, n\}$ and $v_1, \ldots, v_n$ is any orthonormal basis. Assuming also an orientation on V, the Hodge star operation is again defined by

$$\omega \wedge \mu = \langle \star\omega, \mu \rangle \text{vol}, \quad \omega \in \text{Alt}^k(V; W), \ \mu \in \text{Alt}^{n-k}(V; W).$$

Both sides of this equation are elements of the 1-dimensional space $\text{Alt}^n V$ of real-valued n-forms on V.

For a manifold Ω, we define the space $\Lambda^k(\Omega; W)$ of differential k-forms with values in W in the obvious way, i.e., as forms ω on Ω, such that at each point $x \in \Omega$, $\omega_x \in \text{Alt}^k(T_x\Omega; W)$. By taking the tensor product of the de Rham complex with W, we get the *vector-valued de Rham complex*

$$0 \to \Lambda^0(\Omega; W) \xrightarrow{\text{d}} \Lambda^1(\Omega; W) \xrightarrow{\text{d}} \cdots \xrightarrow{\text{d}} \Lambda^n(\Omega; W) \to 0. \qquad (6.1)$$

Here d represents the W-valued exterior derivative $\text{d} \otimes \text{id}_W$ where d is the ordinary exterior derivative. The cohomology is just the tensor product of the ordinary de Rham cohomology with W.

When Ω is an open subset of $\mathbb{R}^n$ (the only case we require), we can write an arbitrary element of $\Lambda^k(\Omega; W)$ as $\sum a_\sigma \text{d}x_{\sigma(1)} \wedge \cdots \wedge \text{d}x_{\sigma(k)}$ with the a_σ functions from $\Omega \to W$. In Section 11 we will use two different vector spaces W, namely $V = \mathbb{R}^n$ (to be thought of as the tangent space to Ω at any point; we use the linear structure and the Euclidean inner product on V but our approach is basis independent) and $V \wedge V$, the space of bivectors, defined in Section 2.1 and identifiable with the space of skew-symmetric linear operators on V.

In treating the equations of elasticity on a domain $\Omega \subset \mathbb{R}^n$, we shall represent the stress as an element $\sigma \in \text{Alt}^{n-1}(\Omega; V)$. This is natural, because the stress is a quantity that, when integrated over surfaces (submanifolds of dimension $n - 1$), gives the force vector (or covector – in view of the inner product, we will not draw this distinction). Let us relate this to the usual definition of the stress, a second-order tensor (or matrix) defined at each point $x \in \Omega$, which, when multiplied by the normal vector to a surface passing through the point, yields the surface force density acting on the surface. The tensor of course represents an element of $\mathcal{L}(V, V) = \text{Alt}^1(V; V)$. That operator is simply $\star\sigma_x$.

To close this section, we consider the operator $S = S_k : \text{Alt}^k(V; V) \to \text{Alt}^{k+1}(V; V \wedge V)$, defined by

$$(S\omega)(v_1, \ldots, v_{k+1}) = \sum_{j=1}^{k+1}(-1)^{j+1}v_j \wedge \omega(v_1, \ldots, \hat{v}_j, \ldots, v_{k+1}),$$

$$v_1, \ldots, v_{k+1} \in V, \quad (6.2)$$

for V an inner product space of dimension n. Of particular importance for our work in elasticity are the cases $k = n - 1$ and $k = n - 2$. The operator $S_{n-1} : \Lambda^{n-1}(V; V) \to \Lambda^n(V; V \wedge V)$, in particular, is a familiar operator in disguise. This is revealed by composing with the Hodge star isomorphism on both sides.

Proposition 6.1. The composition

$$\text{Alt}^1(V; V) \xrightarrow{\star} \text{Alt}^{n-1}(V; V) \xrightarrow{S} \text{Alt}^n(V; V \wedge V) \xrightarrow{\star} \text{Alt}^0(V; V \wedge V)$$

is equal to $(-1)^n 2\,\text{skw}$.

Proof. Let $v, w \in V$, and view $v \otimes w$ as an element of $\mathrm{Alt}^1(V; V)$. We shall show that $\star S(\star(v \otimes w)) = (-1)^n v \wedge w$. Since such elements span $\mathrm{Alt}^1(V; V)$, this gives the result. The calculation is straightforward. Let $e_1, \ldots, e_n$ be a positively oriented orthonormal basis for V. Then

$$\star S(\star(v \otimes w)) = S(\star(v \otimes w))(e_1, \ldots, e_n)$$

$$= \sum_{j=1}^n (-1)^{j+1} e_j \wedge (\star(v \otimes w))(e_1, \ldots, \hat{e}_j, \ldots, e_n)$$

$$= \sum_{j=1}^n (-1)^{j+1} e_j \wedge (-1)^{n-j} (v \otimes w)(e_j)$$

$$= (-1)^{n+1} \sum_{j=0}^n e_j \wedge v(w \cdot e_j) = (-1)^n \sum_{j=1}^n v \wedge w,$$

where we have substituted w for $\sum(w_j \cdot e_j)e_j$ in the last step. $\qquad\square$

Finally we consider the operator S_k for $k = n - 2$. In this case the dimensions of the domain and range coincide:

$$\dim \mathrm{Alt}^{n-2}(V; V) = \binom{n}{2} n = \dim \mathrm{Alt}^{n-1}(V; V \wedge V).$$

In fact, the operator is an isomorphism. To prove this, we first establish two lemmas.

Lemma 6.2. Let $e_1, \ldots, e_n$ be a positively oriented orthonormal basis of V, and let $\omega \in \mathrm{Alt}^{n-2}(V; V)$. Then

$$(S_{n-2}\omega)(e_1, \ldots, \hat{e}_i, \ldots, e_n) = (-1)^{i+1} \sum_{j=1}^n e_j \wedge (\star\omega)(e_i, e_j), \quad i = 1, \ldots, n.$$

Proof. It suffices to prove the case $i = n$, since we may always reorder the basis elements (possibly changing orientation). Then, from the definition of S_{n-2},

$$(S_{n-2}\omega)(e_1, \ldots, e_{n-1}) = \sum_{j=1}^{n-1} (-1)^{j+1} e_j \wedge \omega(e_1, \ldots, \hat{e}_j, \ldots, e_{n-1})$$

$$= \sum_{j=1}^{n-1} (-1)^{j+1} (-1)^{n-j-1} e_j \wedge \star\omega(e_j, e_n)$$

$$= (-1)^{n+1} \sum_{j=1}^n e_j \wedge \star\omega(e_n, e_j),$$

where we have used the fact that $\star\omega(e_n, e_n) = 0$. $\qquad\square$

Lemma 6.3. Suppose $\mu \in \mathrm{Alt}^2(V; V)$ satisfies

$$\sum_{j=1}^{n} e_j \wedge \mu(e_i, e_j) = 0, \quad i = 1, \ldots, n,$$

for some orthonormal basis $e_1, \ldots, e_n$ of V. Then $\mu = 0$.

Proof. We may expand $\mu(e_i, e_j) = \sum_{k=1}^{n} \mu_{ijk} e_k$, for some coefficients $\mu_{ijk} \in \mathbb{R}$ satisfying

$$\mu_{ijk} = -\mu_{jik}. \tag{6.3}$$

Now

$$0 = \sum_{j,k=1}^{n} \mu_{ijk} e_j \wedge e_k = \sum_{j<k} (\mu_{ijk} - \mu_{ikj}) e_j \wedge e_k,$$

whence we conclude

$$\mu_{ijk} = \mu_{ikj}. \tag{6.4}$$

But (6.3) and (6.4) imply that μ vanishes:

$$\mu_{ijk} = -\mu_{jik} = -\mu_{jki} = \mu_{kji} = \mu_{kij} = -\mu_{ikj} = -\mu_{ijk}. \qquad \square$$

Theorem 6.4. The operator $S_{n-2} : \mathrm{Alt}^{n-2}(V, V) \to \mathrm{Alt}^{n-1}(V, V \wedge V)$ is an isomorphism.

Proof. If $S_{n-2}\omega = 0$, then Lemmas 6.2 and 6.3 imply that $\star\omega$ vanishes, so ω vanishes. Then S_{n-2} is injective, and since its domain and range have equal dimension, it is an isomorphism. $\qquad \square$

PART TWO

Applications to discretization
of differential equations

7. The Hodge Laplacian

In this section, we consider the discretization of boundary value problems associated to the Hodge Laplacian, $d\delta+\delta d$, by mixed finite element methods. After first obtaining a mixed variational formulation of these boundary value problems, we then translate to the language of partial differential equations in the case when $n = 3$. The aim here is to show that these formulations in this general setting include many of the problems important in applications. The first main result of the section is to establish the well-posedness of the mixed formulation. We then turn to finite element discretization using the finite element spaces developed in Part 1 of the paper. Using the tools developed for these spaces, we are easily able to establish stability of the mixed finite element approximation. By standard finite element theory, this gives a quasi-optimal error estimate for the variables being approximated. It is well known, however, that since this estimate couples together all the variables being approximated, it does not always give the best result for the approximation of each variable separately, and these more refined results are needed in some applications, and in particular for the approximation of the eigenvalue problem associated to the Hodge Laplacian. Hence, we end the section with a detailed error analysis of these mixed finite element methods.

7.1. Mixed formulation of the Hodge Laplacian

Let Ω be a domain in $\mathbb{R}^n$ and $0 \le k \le n$ an integer. Given $f \in L^2\Lambda^k(\Omega)$, define $\mathcal{J} : H\Lambda^{k-1}(\Omega) \times H\Lambda^k(\Omega) \times \mathfrak{H}^k \to \mathbb{R}$ by

$$\mathcal{J}(\tau, v, q) = \frac{1}{2}\langle\tau, \tau\rangle - \langle d\tau, v\rangle - \frac{1}{2}\langle dv, dv\rangle - \langle v, q\rangle + \langle f, v\rangle.$$

Then a critical point $(\sigma, u, p) \in H\Lambda^{k-1}(\Omega) \times H\Lambda^k(\Omega) \times \mathfrak{H}^k$ of $\mathcal{J}$ is determined by the equations

$$\langle\sigma, \tau\rangle = \langle d\tau, u\rangle, \quad \tau \in H\Lambda^{k-1}(\Omega),$$
$$\langle d\sigma, v\rangle + \langle du, dv\rangle + \langle v, p\rangle = \langle f, v\rangle, \quad v \in H\Lambda^k(\Omega), \qquad (7.1)$$
$$\langle u, q\rangle = 0, \quad q \in \mathfrak{H}^k.$$

In this formulation, p is a Lagrange multiplier corresponding to the constraint given by the third equation of (7.1). However, even if p is eliminated by incorporating this constraint into the space $H\Lambda^k(\Omega)$, the critical point would still be a saddle point – a minimizer with respect to σ and a maximizer with respect to u – and could not generally be obtained from a constrained minimization problem for σ via introduction of an additional Lagrange multiplier.

Letting $P_{\mathfrak{H}^k}$ denote the L^2-projection into $\mathfrak{H}^k$, equations (7.1) are weak formulations of the equations

$$\sigma = \delta u, \quad \mathrm{d}\sigma + \delta \mathrm{d}u + p = f, \quad P_{\mathfrak{H}^k} u = 0, \tag{7.2}$$

respectively, and, since $p = P_{\mathfrak{H}^k} f$, together give the Hodge–Laplace problem $(\mathrm{d}\delta + \delta \mathrm{d})u = f - P_{\mathfrak{H}^k} f$, where δ is the Hodge star operator defined previously. Also implied are the natural boundary conditions that the trace of $\star u$ and the trace of $\star \mathrm{d}u$ on $\partial\Omega$ both must vanish.

If, instead, we seek a critical point $(\sigma, u, p) \in \mathring{H}\Lambda^{k-1}(\Omega) \times \mathring{H}\Lambda^k(\Omega) \times \mathring{\mathfrak{H}}^k$, then we obtain the essential boundary conditions that the trace of σ as a $(k-1)$-form on $\partial\Omega$ and the trace of u as a k-form on $\partial\Omega$ both must vanish.

7.2. Splitting of the mixed formulation

By using the Hodge decomposition (2.18), we can split the problem (7.1) into three simpler problems. First, we write $f = f_{\mathrm{d}} + f_{\mathfrak{H}} + f_{\delta}$, where $f_{\mathrm{d}} \in \mathfrak{B}^k = \mathrm{d}(H\Lambda^{k-1}(\Omega))$, $f_{\mathfrak{H}} \in \mathfrak{H}^k$, and $f_{\delta} \in \mathring{\mathfrak{B}}^{*k} = \delta \mathring{H}^* \Lambda^{k+1}(\Omega)$.

Now let (σ, u, p) be a solution of (7.1). From the second equation in (7.1), it follows immediately that $p_{\mathfrak{H}} = f_{\mathfrak{H}}$, and from the third equation it follows that $u \in \mathfrak{H}^{k\perp}$, so $u = u_{\mathrm{d}} + u_{\delta}$ with $u_{\mathrm{d}} \in \mathfrak{B}^k$ and $u_{\delta} \in \mathring{\mathfrak{B}}^{*k}$.

Taking $v \in \mathring{\mathfrak{B}}^{*k}$ we find that $u_{\delta} \in \mathring{\mathfrak{B}}^{*k}$ satisfies

$$\langle \mathrm{d}u_{\delta}, \mathrm{d}v \rangle = \langle f_{\delta}, v \rangle, \quad v \in \mathring{\mathfrak{B}}^{*k}. \tag{7.3}$$

Taking $v \in \mathfrak{B}^k$ we find that $(\sigma, u_{\mathrm{d}}) \in H\Lambda^{k-1}(\Omega) \times \mathfrak{B}^k$ satisfies

$$\langle \sigma, \tau \rangle = \langle \mathrm{d}\tau, u_{\mathrm{d}} \rangle, \ \tau \in H\Lambda^{k-1}(\Omega), \qquad \langle \mathrm{d}\sigma, v \rangle = \langle f_{\mathrm{d}}, v \rangle, \ v \in \mathfrak{B}^k. \tag{7.4}$$

The converse reasoning is also straightforward, and so we have the following theorem.

Theorem 7.1. Suppose that $(\sigma, u, p) \in H\Lambda^{k-1}(\Omega) \times H\Lambda^k(\Omega) \times \mathfrak{H}^k$ solves (7.1) and that f has the Hodge decomposition $f_{\mathrm{d}} + f_{\mathfrak{H}} + f_{\delta}$, with $f_{\mathrm{d}} \in \mathfrak{B}^k$, $f_{\mathfrak{H}} \in \mathfrak{H}^k$, $f_{\delta} \in \mathring{\mathfrak{B}}^{*k}$. Then $p = f_{\mathfrak{H}}$ and u has the Hodge decomposition $u_{\mathrm{d}} + u_{\delta}$ with $u_{\mathrm{d}} \in \mathfrak{B}^k$ and $u_{\delta} \in \mathring{\mathfrak{B}}^{*k}$, where u_{δ} solves (7.3) and (σ, u_{d}) solves (7.4). Conversely, if $p = f_{\mathfrak{H}}$, $u_{\delta} \in \mathring{\mathfrak{B}}^{*k}$ solves (7.3), and $(\sigma, u_{\mathrm{d}}) \in H\Lambda^{k-1}(\Omega) \times \mathfrak{B}^k$ solves (7.4), then, setting $u = u_{\mathrm{d}} + u_{\delta}$, $(\sigma, u, p) \in H\Lambda^{k-1}(\Omega) \times H\Lambda^k(\Omega) \times \mathfrak{H}^k$ solves (7.1).

In this section we consider the solution to the Hodge Laplacian problem (7.1), but our results also apply to the solutions of (7.4) and (7.3) since these are just the special cases when $f = f_{\mathrm{d}} \in \mathfrak{B}^k$ or $f = f_{\delta} \in \mathring{\mathfrak{B}}^{*k}$.

Note that (7.4) is a weak formulation of the equations

$$\sigma_d = \delta u_d, \quad d\sigma_d = f_d, \quad du_d = 0,$$

together with the natural boundary condition that the trace of $\star u_d$ on $\partial\Omega$ vanishes and the side condition that $u_d \perp \mathfrak{H}^k$. Eliminating σ_d from the system, it becomes

$$d\delta u_d = f_d, \quad du_d = 0,$$

with the indicated boundary condition and side condition. Analogously, (7.3) is a weak formulation of the equations

$$\delta d u_\delta = f_\delta, \quad \delta u_\delta = 0,$$

together with the essential boundary condition that the trace of $\star u_\delta$ on $\partial\Omega$ vanishes and the same side condition.

7.3. Variable coefficients

We have considered the mixed formulation (7.2) without introducing coefficients. But we may easily generalize to allow coefficients. Let $A : L^2\Lambda^{k-1}(\Omega) \to L^2\Lambda^{k-1}(\Omega)$ and $B : L^2\Lambda^{k+1}(\Omega) \to L^2\Lambda^{k+1}(\Omega)$ be bounded, symmetric, positive definite operators with respect to the standard inner products in $L^2\Lambda^{k-1}(\Omega)$ and $L^2\Lambda^{k+1}(\Omega)$. Then we may define equivalent inner products:

$$\langle \sigma, \tau \rangle_A := \langle A\sigma, \tau \rangle, \quad \langle \omega, \mu \rangle_B := \langle B\omega, \mu \rangle,$$

for $\sigma, \tau \in L^2\Lambda^{k-1}(\Omega)$, $\omega, \mu \in L^2\Lambda^{k+1}(\Omega)$. We may then consider, as a generalization of (7.1), the problem of finding $(\sigma, u, p) \in H\Lambda^{k-1}(\Omega) \times H\Lambda^k(\Omega) \times \mathfrak{H}^k$ determined by the equations

$$\langle \sigma, \tau \rangle_A = \langle d\tau, u \rangle, \quad \tau \in H\Lambda^{k-1}(\Omega),$$
$$\langle d\sigma, v \rangle + \langle du, dv \rangle_B + \langle v, p \rangle = \langle f, v \rangle, \quad v \in H\Lambda^k(\Omega),$$
$$\langle u, q \rangle = 0, \quad q \in \mathfrak{H}^k.$$

This is a weak formulation of the differential equations and side condition

$$A\sigma = \delta u, \quad d\sigma + \delta(Bdu) + p = f, \quad P_{\mathfrak{H}^k} u = 0,$$

and the boundary conditions $\mathrm{Tr}(\star u) = 0$, $\mathrm{Tr}[\star(Bdu)] = 0$ on $\partial\Omega$.

We may split the problem as in the previous subsection, and obtain the two reduced problems, namely

$$A\sigma_d = \delta u_d, \quad d\sigma_d = f_d, \quad du_d = 0,$$

and

$$\delta(Bdu_\delta) = f_\delta, \quad \delta u_\delta = 0.$$

Although these more general problems are important for applications, their treatment is no more complicated, except notationally, than the simple case where A and B are the identity, and so we shall continue to consider that case only.

7.4. Translation to the language of partial differential equations

Let us consider more concretely the situation in $n = 3$ dimensions, identifying the spaces $H\Lambda^k(\Omega)$ with function spaces as described in Section 2.3. For $k = 3$, (7.1) becomes: find $\sigma \in H(\text{div}, \Omega; \mathbb{R}^3)$, $u \in L^2(\Omega)$ such that

$$\int_\Omega \sigma \cdot \tau \, dx = \int_\Omega \text{div} \, \tau u \, dx, \quad \tau \in H(\text{div}, \Omega; \mathbb{R}^3),$$

$$\int_\Omega \text{div} \, \sigma v \, dx = \int_\Omega (f - p) v \, dx, \quad v \in L^2(\Omega), \qquad \int_\Omega u q \, dx = 0, \quad q \in \mathfrak{H}^k.$$

This is the standard mixed formulation for the Dirichlet problem for the Poisson equation. The first equation is equivalent to the differential equation $\sigma = -\text{grad} \, u$ and the boundary condition $u = 0$, while the second equation is equivalent to $\text{div} \, \sigma = f$. In this case, $\mathfrak{H}^k = 0$, so $p = 0$ and the last equation is not needed. If, instead, we seek $\sigma \in H_0(\text{div}, \Omega; \mathbb{R}^3)$, then the boundary condition $u = 0$ is replaced by the boundary condition $\sigma \nu = 0$. Then $\mathfrak{H}^k = \mathbb{R}$, and so $p = \int_\Omega f \, dx / \text{meas}(\Omega)$ and $\int_\Omega u \, dx = 0$. These are the only boundary value problems when $k = 3$. Since $du = 0$, this problem is already of the form (7.4).

For $k = 2$, the unknowns $\sigma \in H(\text{curl}, \Omega; \mathbb{R}^3)$ and $u \in H(\text{div}, \Omega; \mathbb{R}^3)$ satisfy the differential equations

$$\sigma = \text{curl} \, u, \qquad \text{curl} \, \sigma - \text{grad} \, \text{div} \, u = f - p,$$

the auxiliary condition $P_{\mathfrak{H}}^k u = 0$, and the boundary conditions $u \times \nu = 0$, $\text{div} \, u = 0$ on $\partial\Omega$, so this is a mixed formulation for the vectorial Poisson equation

$$(\text{curl} \, \text{curl} - \text{grad} \, \text{div}) u = f - p, \tag{7.5}$$

with the auxiliary variable $\sigma = \text{curl} \, u$. If, instead, we seek $\sigma \in H_0(\text{curl}, \Omega; \mathbb{R}^3)$ and $u \in H_0(\text{div}, \Omega; \mathbb{R}^3)$, then we obtain the boundary conditions $\sigma \times \nu = 0$ and $u \cdot \nu = 0$. When $k = 2$, (7.4) becomes

$$\sigma_d = \text{curl} \, u_d, \qquad \text{curl} \, \sigma_d = f_d, \qquad \text{div} \, u_d = 0,$$

while problem (7.3) becomes

$$-\text{grad} \, \text{div} \, u_\delta = f_\delta, \qquad \text{curl} \, u_\delta = 0.$$

In fact, since $f_\delta = \text{grad} \, F$ for some F, this problem has the equivalent form

$$-\text{div} \, u_\delta = F, \qquad \text{curl} \, u_\delta = 0.$$

For $k = 1$, (7.1) is a different mixed formulation of the vectorial Poisson equation (7.5). Now $\sigma \in H^1(\Omega)$ and $u \in H(\text{curl}, \Omega; \mathbb{R}^3)$ satisfy the differential equations

$$\sigma = -\operatorname{div} u, \qquad \operatorname{grad} \sigma + \operatorname{curl} \operatorname{curl} u = f - p,$$

the auxiliary condition $P_{\mathfrak{H}^k} u = 0$, and the boundary conditions $u \cdot \nu = 0$, $(\operatorname{curl} u) \times \nu = 0$. If, instead, we seek $\sigma \in H_0^1(\Omega)$ and $u \in H_0(\text{curl}, \Omega; \mathbb{R}^3)$, then we obtain the boundary conditions $\sigma = 0$ and $u \times \nu = 0$. When $k = 1$, (7.4) becomes

$$\sigma_{\mathrm{d}} = -\operatorname{div} u_{\mathrm{d}}, \qquad \operatorname{grad} \sigma_{\mathrm{d}} = f_{\mathrm{d}}, \qquad \operatorname{curl} u_{\mathrm{d}} = 0,$$

while problem (7.3) becomes

$$\operatorname{curl} \operatorname{curl} u_\delta = f_\delta, \qquad \operatorname{div} u_\delta = 0.$$

Finally, we interpret the case $k = 0$. In this case $H\Lambda^{-1}(\Omega) = 0$, so $\sigma = 0$ and we can ignore the first equation of (7.1). Then $u \in H^1(\Omega)$ and $p \in \mathfrak{H}^0 = \mathbb{R}$ satisfy

$$\int_\Omega \operatorname{grad} u \cdot \operatorname{grad} v \, \mathrm{d}x = \int_\Omega (f - p)v \, \mathrm{d}x, \; v \in H^1(\Omega), \qquad \int_\Omega uq \, \mathrm{d}x = 0, \; q \in \mathbb{R}.$$

Thus, $p = \int_\Omega f \, \mathrm{d}x / \operatorname{meas}(\Omega)$, and we just have the usual weak formulation of the Neumann problem for the Poisson equation $-\Delta u = f - p$. If, instead, we seek $u \in H_0^1(\Omega)$, then $p = 0$ and we obtain the usual weak formulation of the Dirichlet problem for Poisson's equation. For $k = 0$, problem (7.4) is vacuous while problem (7.3) becomes $-\Delta u_\delta = f_\delta$.

7.5. Well-posedness of the mixed formulation

To discuss the well-posedness of the system (7.1), we let $B : [H\Lambda^{k-1}(\Omega) \times H\Lambda^k(\Omega) \times \mathfrak{H}^k] \times [H\Lambda^{k-1}(\Omega) \times H\Lambda^k(\Omega) \times \mathfrak{H}^k] \to \mathbb{R}$ denote the bounded bilinear form

$$B(\sigma, u, p; \tau, v, q) = \langle \sigma, \tau \rangle - \langle \mathrm{d}\tau, u \rangle + \langle \mathrm{d}\sigma, v \rangle + \langle \mathrm{d}u, \mathrm{d}v \rangle + \langle v, p \rangle - \langle u, q \rangle.$$

Well-posedness of the system (7.1) is equivalent to the inf-sup condition for B (Babuška and Aziz 1972), i.e., we must establish the following result.

Theorem 7.2. There exist constants $\gamma > 0$, $C < \infty$ such that, for any $(\sigma, u, p) \in H\Lambda^{k-1}(\Omega) \times H\Lambda^k(\Omega) \times \mathfrak{H}^k$, there exists $(\tau, v, q) \in H\Lambda^{k-1}(\Omega) \times H\Lambda^k(\Omega) \times \mathfrak{H}^k$ with

$$B(\sigma, u, p; \tau, v, q) \geq \gamma(\|\sigma\|_{H\Lambda}^2 + \|u\|_{H\Lambda}^2 + \|p\|^2), \tag{7.6}$$
$$\|\tau\|_{H\Lambda} + \|v\|_{H\Lambda} + \|q\| \leq C(\|\sigma\|_{H\Lambda} + \|u\|_{H\Lambda} + \|p\|). \tag{7.7}$$

Proof. By the Hodge decomposition, given $u \in H\Lambda^k(\Omega)$, there exist forms $u_{\mathrm{d}} \in \mathfrak{B}^k$, $u_{\mathfrak{H}} \in \mathfrak{H}^k$, and $u_\delta \in \mathfrak{B}^{*k}$, such that

$$u = u_{\mathrm{d}} + u_{\mathfrak{H}} + u_\delta, \qquad \|u\|^2 = \|u_{\mathrm{d}}\|^2 + \|u_{\mathfrak{H}}\|^2 + \|u_\delta\|^2. \tag{7.8}$$

Since $u_{\mathrm{d}} \in \mathfrak{B}^k$, $u_{\mathrm{d}} = d\rho$, for some $\rho \in \mathfrak{Z}^{k-1\perp}$. Since $\mathfrak{B}^{*k} = \mathfrak{Z}^{k\perp}$ and $du_\delta = du$, we get using the Poincaré inequality (2.17) that

$$\|\rho\| \le K'\|u_{\mathrm{d}}\|, \qquad \|u_\delta\| \le K\|du\|, \tag{7.9}$$

where K and K' are constants independent of ρ and u_δ. Let $\tau = \sigma - t\rho \in H\Lambda^{k-1}(\Omega)$, $v = u + d\sigma + p \in H\Lambda^k(\Omega)$, and $q = p - u_{\mathfrak{H}} \in \mathfrak{H}^k$, with $t = 1/(K')^2$. Using (7.8) and (7.9), and a simple computation, we get

$$B(\sigma, u, p; \tau, v, q)$$

$$= \|\sigma\|^2 + \|d\sigma\|^2 + \|du\|^2 + \|p\|^2 + t\|u_{\mathrm{d}}\|^2 + \|u_{\mathfrak{H}}\|^2 - t\langle\sigma, \rho\rangle$$

$$\ge \frac{1}{2}\|\sigma\|^2 + \|d\sigma\|^2 + \|du\|^2 + \|p\|^2 + t\|u_{\mathrm{d}}\|^2 + \|u_{\mathfrak{H}}\|^2 - \frac{t^2}{2}\|\rho\|^2$$

$$\ge \frac{1}{2}\|\sigma\|^2 + \|d\sigma\|^2 + \|du\|^2 + \|p\|^2 + \|u_{\mathfrak{H}}\|^2 + \|u_{\mathrm{d}}\|^2(t - t^2(K')^2/2)$$

$$\ge \frac{1}{2}\|\sigma\|^2 + \|d\sigma\|^2 + \frac{1}{2}\|du\|^2 + \|p\|^2 + \|u_{\mathfrak{H}}\|^2 + \frac{1}{2(K')^2}\|u_{\mathrm{d}}\|^2 + \frac{1}{2K^2}\|u_\delta\|^2$$

$$\ge \frac{1}{2}\|\sigma\|^2 + \|d\sigma\|^2 + \frac{1}{2}\|du\|^2 + \frac{1}{2(K'')^2}\|u\|^2 + \|p\|^2,$$

where $K'' = \max(K', K, 1/\sqrt{2})$. Hence, we obtain (7.6) with $\gamma > 0$ depending only on K and K'. The upper bound (7.7) follows easily from (7.8) and (7.9). □

Remark. If, instead, we consider the form B over the space $[\mathring{H}\Lambda^{k-1}(\Omega) \times \mathring{H}\Lambda^k(\Omega) \times \mathring{\mathfrak{H}}^k] \times [\mathring{H}\Lambda^{k-1}(\Omega) \times \mathring{H}\Lambda^k(\Omega) \times \mathring{\mathfrak{H}}^k]$ then the stability result is still valid. The proof must be modified to use the Hodge decomposition $u = u_{\mathrm{d}} + u_{\mathfrak{H}} + u_\delta$, where now $u_{\mathrm{d}} \in \mathring{\mathfrak{B}}^k$, $u_{\mathfrak{H}} \in \mathring{\mathfrak{H}}^k$, and $u_\delta \in \mathfrak{B}^{*k}$.

7.6. Well-posedness of discretizations of the mixed formulation

We next consider discrete versions of these results. Suppose we are given a triangulation, and let

$$0 \to \Lambda_h^0 \xrightarrow{\mathrm{d}} \Lambda_h^1 \xrightarrow{\mathrm{d}} \cdots \xrightarrow{\mathrm{d}} \Lambda_h^n \to 0 \tag{7.10}$$

denote any of the 2^{n-1} finite element de Rham complexes (for each value of the degree) discussed previously. Recall we have a commuting diagram of the form

$$0 \to H\Lambda^0(\Omega) \xrightarrow{\;d\;} H\Lambda^1(\Omega) \xrightarrow{\;d\;} \cdots \xrightarrow{\;d\;} H\Lambda^n(\Omega) \to 0$$

$$\tilde{\Pi}_h^0 \Big\downarrow \qquad\qquad \tilde{\Pi}_h^1 \Big\downarrow \qquad\qquad\qquad \tilde{\Pi}_h^n \Big\downarrow \qquad\qquad (7.11)$$

$$0 \to \quad \Lambda_h^0 \quad \xrightarrow{\;d\;} \quad \Lambda_h^1 \quad \xrightarrow{\;d\;} \cdots \xrightarrow{\;d\;} \quad \Lambda_h^n \quad \to 0,$$

where the $\tilde{\Pi}_h^k$ are bounded projections, $i.e.$,

$$\|\tilde{\Pi}_h^k \omega\|_{H\Lambda^k} \le C \|\omega\|_{H\Lambda^k}, \qquad \omega \in H\Lambda^k(\Omega), \qquad (7.12)$$

with the constant C independent of ω and h. We note that the canonical interpolation operators associated to the standard finite element spaces do not satisfy these conditions, since their definition requires more regularity. However, the new projection operators discussed in Section 5 do satisfy these conditions, so we can assume we have such projection operators. Of course, we would also like the results presented below to apply to problems with essential boundary conditions. In that case, we would need projection operators $\tilde{\Pi}_h^k$ mapping $\mathring{H}\Lambda^k$ to $\mathring{\Lambda}_h^k$ that again satisfy (7.12). Although our construction in Section 5 did not include this case, we believe that such a construction is also possible, and we shall assume that we have such projection operators in this case too.

Under these conditions, we shall next demonstrate stability of the finite element method: find $\sigma_h \in \Lambda_h^{k-1}$, $u_h \in \Lambda_h^k$, $p_h \in \mathfrak{H}_h^k$ such that

$$\langle \sigma_h, \tau \rangle = \langle d\tau, u_h \rangle, \qquad \tau \in \Lambda_h^{k-1},$$

$$\langle d\sigma_h, v \rangle + \langle du_h, dv \rangle + \langle v, p_h \rangle = \langle f, v \rangle, \qquad v \in \Lambda_h^k, \qquad (7.13)$$

$$\langle u, q \rangle = 0, \qquad q \in \mathfrak{H}_h^k.$$

In view of the discrete de Rham complexes obtained in Section 5.5, this result can be applied to prove the stability of four different families of mixed methods for the Hodge Laplacian problem, using any of the four choices of spaces

$$\mathcal{P}_r^- \Lambda^{k-1}(\mathcal{T}_h) \times \mathcal{P}_r^- \Lambda^k(\mathcal{T}_h), \qquad \mathcal{P}_r \Lambda^{k-1}(\mathcal{T}_h) \times \mathcal{P}_r^- \Lambda^k(\mathcal{T}_h),$$
$$\qquad\qquad (7.14)$$
$$\mathcal{P}_{r+1}^- \Lambda^{k-1}(\mathcal{T}_h) \times \mathcal{P}_r \Lambda^k(\mathcal{T}_h), \qquad \mathcal{P}_{r+1} \Lambda^{k-1}(\mathcal{T}_h) \times \mathcal{P}_r \Lambda^k(\mathcal{T}_h),$$

to discretize the $(k-1)$-forms and the k-forms, respectively. Since the $(k-1)$-forms disappear for $k=0$, and since we have $\mathcal{P}_r^- \Lambda^0(\mathcal{T}_h) = \mathcal{P}_r \Lambda^0(\mathcal{T}_h)$ and $\mathcal{P}_r^- \Lambda^n(\mathcal{T}_h) = \mathcal{P}_{r-1} \Lambda^n(\mathcal{T}_h)$, these reduce to a single family of methods for $k = 0$ (namely, the use of the standard Lagrange elements for the standard Laplacian problem), and to two families of methods for $k=1$ or $k=n$.

Stability of the method (7.13) is equivalent to the inf-sup condition for B restricted to the finite element spaces (Babuška and Aziz 1972), $i.e.$, we must establish the following result.

Theorem 7.3. There exist constants $\gamma > 0$, $C < \infty$ independent of h such that, for any $(\sigma, u, p) \in \Lambda_h^{k-1} \times \Lambda_h^k \times \mathfrak{H}_h^k$, there exists $(\tau, v, q) \in \Lambda_h^{k-1} \times \Lambda_h^k \times \mathfrak{H}_h^k$ with

$$B(\sigma, u, p; \tau, v, q) \geq \gamma(\|\sigma\|_{H\Lambda}^2 + \|u\|_{H\Lambda}^2 + \|p\|^2),$$
$$\|\tau\|_{H\Lambda} + \|v\|_{H\Lambda} + \|q\| \leq C(\|\sigma\|_{H\Lambda} + \|u\|_{H\Lambda} + \|p\|).$$

Proof. The proof in the discrete case closely follows the proof given above for the continuous case. By the discrete Hodge decomposition, given $u \in \Lambda_h^k$, there exist forms $u_{\mathrm{d}} \in \mathfrak{B}_h^k$, $u_{\mathfrak{H}} \in \mathfrak{H}_h^k$, and $u_\delta \in \mathfrak{Z}_h^{k\perp}$, such that

$$u = u_{\mathrm{d}} + u_{\mathfrak{H}} + u_\delta, \qquad \|u\|^2 = \|u_{\mathrm{d}}\|^2 + \|u_{\mathfrak{H}}\|^2 + \|u_\delta\|^2. \tag{7.15}$$

Since $u_{\mathrm{d}} \in \mathfrak{B}_h^k$, $u_{\mathrm{d}} = \mathrm{d}\rho$, for some $\rho \in \mathfrak{Z}_h^{(k-1)\perp}$. Since $\mathrm{d}u_\delta = \mathrm{d}u$, we get using the discrete Poincaré inequality, Theorem 5.11, that

$$\|\rho\| \leq K'\|u_{\mathrm{d}}\|, \qquad \|u_\delta\| \leq K\|\mathrm{d}u\|, \tag{7.16}$$

where K and K' are constants independent of ρ, u_δ, and h. The result now follows by applying the same proof as in the continuous case, where we use (7.15) and (7.16) in place of (7.8) and (7.9). To handle other boundary conditions, the discrete Hodge decomposition must be modified as in the continuous case. $\qquad\square$

From this stability result, we then obtain the following quasi-optimal error estimates.

Theorem 7.4. Let $(\sigma, u, p) \in H\Lambda^{k-1}(\Omega) \times H\Lambda^k(\Omega) \times \mathfrak{H}^k$ be the solution of problem (7.1) and let $(\sigma_h, u_h, p_h) \in \Lambda_h^{k-1} \times \Lambda_h^k \times \mathfrak{H}_h^k$ be the solution of problem (7.13). Then

$$\|\sigma - \sigma_h\|_{H\Lambda} + \|u - u_h\|_{H\Lambda} + \|p - p_h\| \tag{7.17}$$
$$\leq C\left(\inf_{\tau \in \Lambda_h^{k-1}} \|\sigma - \tau\|_{H\Lambda} + \inf_{v \in \Lambda_h^k} \|u - v\|_{H\Lambda} + \inf_{q \in \mathfrak{H}_h^k} \|p - q\| + \|P_{\mathfrak{H}_h^k} u\| \right),$$

where $P_{\mathfrak{H}_h^k} u$ denotes the L^2-projection of u into $\mathfrak{H}_h^k$. Moreover,

$$\|P_{\mathfrak{H}_h^k} u\| \leq \inf_{r \in \mathfrak{H}^k} \|q - r\| \inf_{v_{\mathrm{d}} \in \mathfrak{B}_h^k} \|u_{\mathrm{d}} - v_{\mathrm{d}}\| \leq \varepsilon_h \inf_{v_{\mathrm{d}} \in \mathfrak{B}_h^k} \|u_{\mathrm{d}} - v_{\mathrm{d}}\|, \tag{7.18}$$

where u_{d} is the L^2-projection of u into $\mathfrak{B}^k$, $q = P_{\mathfrak{H}_h^k} u / \|P_{\mathfrak{H}_h^k} u\|$ and

$$\varepsilon_h = \sup_{\substack{r \in \mathfrak{H}^k \\ \|r\|=1}} \|(I - \tilde{\Pi}_h)r\|.$$

Proof. First observe that (σ, u, p) satisfies

$$B(\sigma, u, p; \tau_h, v_h, q_h) = \langle f, v_h \rangle - \langle u, q_h \rangle, \quad (\tau_h, v_h, q_h) \in \Lambda_h^{k-1} \times \Lambda_h^k \times \mathfrak{H}_h^k.$$

Let $\tau \in \Lambda_h^{k-1}$, $u \in \Lambda_h^k$, $q \in \mathfrak{H}_h^k$. Then, for any $(\tau_h, v_h, q_h) \in \Lambda_h^{k-1} \times \Lambda_h^k \times \mathfrak{H}_h^k$, we have

$$B(\sigma_h - \tau, u_h - v, p_h - q; \tau_h, v_h, q_h)$$
$$= B(\sigma - \tau, u - v, p - q; \tau_h, v_h, q_h) + \langle u, q_h \rangle$$
$$= B(\sigma - \tau, u - v, p - q; \tau_h, v_h, q_h) + \langle P_{\mathfrak{H}_h^k} u, q_h \rangle$$
$$\leq C(\|\sigma - \tau\|_{H\Lambda} + \|u - v\|_{H\Lambda} + \|p - q\| + \|P_{\mathfrak{H}_h^k} u\|)$$
$$\times (\|\tau_h\|_{H\Lambda} + \|v_h\|_{H\Lambda} + \|q_h\|).$$

Theorem 7.3 then gives

$$\|\sigma_h - \tau\|_{H\Lambda} + \|u_h - v\|_{H\Lambda} + \|p_h - q\|$$
$$\leq C(\|\sigma - \tau\|_{H\Lambda} + \|u - v\|_{H\Lambda} + \|p - q\| + \|P_{\mathfrak{H}_h^k} u\|),$$

from which (7.17) follows by the triangle inequality.

Now $u \perp \mathfrak{H}^k$, so $u = u_{\mathrm{d}} + u_\delta$, with $u_{\mathrm{d}} \in \mathfrak{B}^k$ and $u_\delta \in \mathfrak{Z}^{k\perp}$. Since $\mathfrak{H}_h^k \subset \mathfrak{Z}^k$, $P_{\mathfrak{H}_h^k} u_\delta = 0$, while, by the discrete Hodge decomposition, $P_{\mathfrak{H}_h^k} v_{\mathrm{d}} = 0$ for all $v_{\mathrm{d}} \in \mathfrak{B}_h^k$. Let $q = P_{\mathfrak{H}_h^k} u / \|P_{\mathfrak{H}_h^k} u\| \in \mathfrak{H}_h^k$. For any $v_{\mathrm{d}} \in \mathfrak{B}_h^k$ we have

$$\|P_{\mathfrak{H}_h^k} u\| = \langle u_{\mathrm{d}} - v_{\mathrm{d}}, q \rangle = \inf_{r \in \mathfrak{H}^k} \langle u_{\mathrm{d}} - v_{\mathrm{d}}, q - r \rangle$$
$$\leq \|u_{\mathrm{d}} - v_{\mathrm{d}}\| \inf_{r \in \mathfrak{H}^k} \|q - r\|.$$

Furthermore, by Lemma 5.9, we can find $r \in \mathfrak{H}^k$ with $\|r\| \leq 1$ and $\|q - r\| \leq \|(I - \tilde{\Pi}_h)r\| \leq \varepsilon_h$, and hence (7.18) follows.   □

Remark. Let u be as in the theorem above. Since $u \perp \mathfrak{H}^k$, it follows that if $\mathfrak{H}_h^k \subset \mathfrak{H}^k$, then $P_{\mathfrak{H}_h^k} u = 0$. On the other hand, if $\Lambda_h^{k-1} \times \Lambda_h^k$ is one of the choices given in (7.14) and u_d and all elements of $\mathfrak{H}^k$ are sufficiently smooth, then

$$\|P_{\mathfrak{H}_h^k} u\| \leq \varepsilon_h \|(I - \tilde{\Pi}_h)u_d\| = \mathcal{O}(h^{2r}).$$

If the solution (σ, u, p) is sufficiently smooth, we then obtain the following order of convergence estimates.

Corollary 7.5. If $\Lambda_h^{k-1} \times \Lambda_h^k$ is one of the choices given in (7.14) and $\|P_{\mathfrak{H}_h^k} u\| = \mathcal{O}(h^r)$, then

$$\|\sigma - \sigma_h\|_{H\Lambda} + \|u - u_h\|_{H\Lambda} + \|p - p_h\| = \mathcal{O}(h^r).$$

Proof. This result follows from the previous theorem by using the approximation properties of the subspaces given in Theorems 5.6 and 5.8.   □

Remark. As noted earlier, problems (7.3) and (7.4) are special cases of problem (7.1) when $f = f_\delta \in \mathring{\mathfrak{B}}^{*k}$ or $f = f_d \in \mathfrak{B}^k$. Although these reduced problems have a simpler form, they are not so easy to approximate directly by finite element methods, since that would involve finding a basis for finite element subspaces of $\mathfrak{B}^k$ or $\mathring{\mathfrak{B}}^{*k}$. However, since they are equivalent to problem (7.1), one can use the discretization of problem (7.1) with standard finite element spaces to find good approximations to problems of this type.

Remark. We also note that an early use of a discrete Hodge decomposition and discrete Poincaré inequality to establish stability of mixed finite element methods appears in the work of Fix, Gunzburger and Nicolaides (1981), in connection with the grid decomposition principle. See also Bochev and Gunzburger (2005) for a more recent exposition.

7.7. Regularity properties

To obtain order of convergence estimates below, we will need to make some assumptions about the domain Ω that will ensure that the solution of problem (7.1) has some regularity beyond merely belonging to the space in which we seek the solution.

We shall say that the domain Ω is *s-regular* if, for $w \in H\Lambda^k(\Omega) \cap \mathring{H}^* \Lambda^k(\Omega)$ or $\mathring{H}\Lambda^k(\Omega) \cap H^* \Lambda^k(\Omega)$, $w \in H^s \Lambda^k(\Omega)$ and

$$\|w\|_{H^s\Lambda} \leq C(\|w\| + \|\mathrm{d}w\| + \|\delta w\|), \qquad (7.19)$$

for some $0 < s \leq 1$. A smoothly bounded domain is 1-regular (Gaffney 1951) and a Lipschitz domain is 1/2-regular (Mitrea, Mitrea and Taylor 2001, Theorem 11.2). For Ω convex, (7.19) holds for $s = 1$ and the term $\|w\|$ may be omitted (Mitrea 2001, Corollary 5.2). A Lipschitz polyhedron in $\mathbb{R}^3$ is *s*-regular for some $1/2 < s \leq 1$ (see Amrouche, Bernardi, Dauge and Girault (1998) and Costabel (1991)). Also note that by Poincaré's inequality (Theorem 2.2), for $w \in H\Lambda^k(\Omega) \cap \mathring{H}^* \Lambda^k(\Omega) \cap \mathfrak{H}^{k\perp}$ or $\mathring{H}\Lambda^k(\Omega) \cap H^* \Lambda^k(\Omega) \cap \mathring{\mathfrak{H}}^{k\perp}$, we may also omit the term $\|w\|$.

7.8. Improved error estimates: basic bounds

As is well known from the theory of mixed finite element methods (Falk and Osborn 1980, Douglas and Roberts 1985), it is sometimes possible to get improved error estimates for each term in the mixed formulation by decoupling them. In this subsection, we show how this more refined analysis can be carried out for the mixed finite element approximation of the Hodge Laplacian. In particular, we show that for any Lipschitz polyhedral domain Ω and any $f \in L^2\Lambda^k(\Omega)$, we have

$$\|\sigma - \sigma_h\| + \|u - u_h\|_{H\Lambda} + \|p - p_h\| = \mathcal{O}(h^{1/2}).$$

Without the assumption of additional regularity on f, such an estimate can not be obtained from the quasi-optimal result stated above, since the error in that estimate also depends on the approximation of $d\sigma = f_d$ and this requires more regularity than just $f_d \in L^2\Lambda^k(\Omega)$ to achieve a positive rate of convergence. Higher-order improved rates of convergence with less regularity can also be obtained by using this more refined analysis.

We will assume throughout the discussion below that the domain Ω is s-regular. As above, we will use the notation $P_{\mathfrak{H}_h^k}$, $P_{\mathfrak{B}_h^k}$, and $P_{\mathfrak{Z}_h^k}$ to denote the L^2-projection onto the spaces $\mathfrak{H}_h^k$, $\mathfrak{B}_h^k$, and $\mathfrak{Z}_h^k$, respectively. We further introduce P_h as the L^2-projection onto Λ_h^k. To obtain improved error estimates, we will break up the solutions of problems (7.1) and (7.13) into the three subproblems corresponding to the Hodge decomposition of the right-hand side f. Our error analysis will be based on a separate analysis for each of these pieces.

We start with the almost trivial case when $f \in \mathfrak{H}^k$.

Lemma 7.6. Assume that $f \in \mathfrak{H}^k$ and that $(\sigma, u, p) \in H\Lambda^{k-1}(\Omega) \times H\Lambda^k(\Omega) \times \mathfrak{H}^k$ is the solution of problem (7.1). Then $\sigma = 0$, $u = 0$, and $p = f$. The corresponding solution $(\sigma_h, u_h, p_h) \in \Lambda_h^{k-1} \times \Lambda_h^k \times \mathfrak{H}_h^k$ of (7.13) satisfies $\sigma_h = 0$, $u_h = u_{\delta,h} \in \mathfrak{Z}_h^{k\perp}$, $p_h = P_{\mathfrak{H}_h^k} f$ and there is a constant C, independent of h, such that
$$\|p - p_h\| + \|u_h\|_{H\Lambda} \le C\|(I - \tilde{\Pi}_h)f\|.$$

Proof. It is straightforward to check that the given solutions satisfy the corresponding systems. In particular, $u_h \in \mathfrak{Z}_h^{k\perp}$ satisfies
$$\langle du_h, dv \rangle = \langle (I - P_{\mathfrak{H}_h^k})f, v \rangle, \quad v \in \Lambda_h^k.$$

The discrete Poincaré inequality therefore implies $\|u_h\|_{H\Lambda} \le C\|(I - P_{\mathfrak{H}_h^k})f\|$. In addition, Theorem 5.8 implies that
$$\|p - p_h\| = \|(I - P_{\mathfrak{H}_h^k})f\| \le \|(I - \tilde{\Pi}_h)f\|. \qquad \square$$

Remark. Note that for $f \in \mathfrak{H}^k$, it follows from (7.19) that $f \in H^s\Lambda^k(\Omega)$ and $\|f\|_{H^s\Lambda} \le C\|f\|$. Hence, we can conclude from Theorem 5.6 that
$$\|(I - \tilde{\Pi}_h)f\| \le Ch^s\|f\|, \quad f \in \mathfrak{H}^k.$$

Lemma 7.7. Assume that $f \in \mathfrak{B}^k$ and that $(\sigma, u, p) \in H\Lambda^{k-1}(\Omega) \times H\Lambda^k(\Omega) \times \mathfrak{H}^k$ is the solution of problem (7.1). Then $d\sigma = f$, $u = u_d \in \mathfrak{B}^k$, and $p = 0$. The corresponding solution $(\sigma_h, u_h, p_h) \in \Lambda_h^{k-1} \times \Lambda_h^k \times \mathfrak{H}_h^k$ of (7.13) satisfies $d\sigma_h = P_{\mathfrak{B}_h^k} f$, and there is a constant C, independent of h, such that
$$\|\sigma - \sigma_h\| + \|P_{\mathfrak{B}_h^k} u - u_h\| + \|du_h\| + \|p_h\|$$
$$\le C(h^s\|(I - \tilde{\Pi}_h)f\| + \|(I - \tilde{\Pi}_h)\sigma\|).$$

Proof. Note that since $\tilde{\Pi}_h(\mathfrak{B}^k) \subset \mathfrak{B}_h^k$ it follows that

$$\|(I - P_{\mathfrak{B}_h^k})f\| \le \|(I - \tilde{\Pi}_h)f\|.$$

To establish the error estimate, we will decompose the right-hand side $f \in \mathfrak{B}^k$ into $(I - P_{\mathfrak{B}_h^k})f$ and $P_{\mathfrak{B}_h^k}f$. Let (σ^1, u^1, p^1) be the solution of (7.1) with right-hand side $(I - P_{\mathfrak{B}_h^k})f$ and let $(\sigma_h^1, u_h^1, p_h^1)$ be the corresponding discrete solution. Since $(I - P_{\mathfrak{B}_h^k})f \in \mathcal{B}^k$, it follows that $d\sigma^1 = (I - P_{\mathfrak{B}_h^k})f$, $u^1 = u_d^1$, and $p^1 = 0$. Furthermore, from the system (7.1) and Theorem 5.6 we obtain

$$\|\sigma^1\|^2 = \langle d\sigma^1, u^1 \rangle = \langle (I - P_{\mathfrak{B}_h^k})f, (I - \tilde{\Pi}_h)u^1 \rangle \le Ch^s \|u^1\|_{H^s\Lambda}\|(I - \tilde{\Pi}_h)f\|.$$

Furthermore, since $\sigma^1 = \delta u^1$ and $\mathrm{Tr}\star u^1 = 0$, we conclude from (7.19) that

$$\|u^1\|_{H^s\Lambda} \le C\|\sigma^1\|.$$

Collecting the estimates above, and stating only the weaker result we will need, we have

$$\|\sigma^1\|, \|u^1\| \le Ch^s\|(I - \tilde{\Pi}_h)f\|. \tag{7.20}$$

On the other hand, since $(I - P_{\mathfrak{B}_h^k})f$ is orthogonal to $\mathfrak{B}_h^k$, we must have $\sigma_h^1 = 0$ and $u_h^1 = u_{\delta,h}^1$. Furthermore,

$$\|du_h^1\|^2 = \langle (I - P_{\mathfrak{B}_h^k})f, u_h^1 - w \rangle \le \|(I - \tilde{\Pi}_h)f\|\,\|u_h^1 - w\|,$$

where $w \in \mathfrak{Z}^{k\perp}$ is arbitrary. In particular, by Lemma 5.10 and (7.19), we can choose w such that $dw = du_h$ and

$$\|u_h^1 - w\| \le \|(I - \tilde{\Pi}_h)w\| \le Ch^s\|w\|_{H^s\Lambda} \le Ch^s\|du_h^1\|.$$

As a consequence, we have from the discrete Poincaré inequality that

$$\|u_h^1\|_{H\Lambda} \le Ch^s\|(I - \tilde{\Pi}_h)f\|.$$

In a similar manner, we have

$$\|p_h^1\|^2 = \langle (I - P_{\mathfrak{B}_h^k})f, p_h^1 - r \rangle \le \|(I - \tilde{\Pi}_h)f\|\,\|p_h^1 - r\|,$$

where $r \in \mathfrak{H}^k$ can be chosen such that $\|p_h^1 - r\| \le \|(I - \tilde{\Pi}_h)r\|$, and $\|r\| \le \|p_h^1\|$ (*cf.* Lemma 5.9). In particular, since $r \in \mathfrak{H}^k$, we have by (7.19) that

$$\|(I - \tilde{\Pi}_h)r\| \le Ch^s\|r\|_{H^s\Lambda} \le Ch^s\|r\| \le Ch^s\|p_h^1\|,$$

which implies

$$\|p_h^1\| \le Ch^s\|(I - \tilde{\Pi}_h)f\|.$$

By collecting (7.20) and the estimates just obtained for u_h^1 and p_h^1, we obtain

$$\|\sigma^1\| + \|u^1\| + \|u_h^1\|_{H\Lambda} + \|p_h^1\| \leq Ch^s \|(I - \tilde{\Pi}_h)f\|. \tag{7.21}$$

We will use a superscript 2 to indicate the solutions of (7.1) and (7.13) with right-hand side $P_{\mathfrak{B}_h^k} f \in \mathfrak{B}_h^k \subset \mathfrak{B}^k$. Then, $d\sigma^2 = d\tilde{\Pi}_h \sigma^2 = d\sigma_h^2 = P_{\mathfrak{B}_h^k} f$, $u_\delta^2 = u_{\delta,h}^2 = 0$, and $p^2 = p_h^2 = 0$. By combining these properties with the error equation

$$\langle \sigma^2 - \sigma_h^2, \tau \rangle = \langle d\tau, u^2 - u_h^2 \rangle, \quad \tau \in \Lambda_h^{k-1}, \tag{7.22}$$

we obtain

$$\|\sigma^2 - \sigma_h^2\|^2 = \langle d(\tilde{\Pi}_h \sigma^2 - \sigma_h^2), u^2 - u_h^2 \rangle + \langle \sigma^2 - \tilde{\Pi}_h \sigma^2, \sigma^2 - \sigma_h^2 \rangle$$
$$\leq \|(I - \tilde{\Pi}_h)\sigma^2\| \, \|\sigma^2 - \sigma_h^2\|,$$

or

$$\|\sigma^2 - \sigma_h^2\| \leq \|(I - \tilde{\Pi}_h)\sigma^2\|. \tag{7.23}$$

Note that the error equation (7.22) implies that

$$\langle \sigma^2 - \sigma_h^2, \tau \rangle = \langle d\tau, P_{\mathfrak{B}_h^k} u^2 - u_h^2 \rangle, \quad \tau \in \Lambda_h^{k-1}.$$

Choosing $\tau \in \mathfrak{Z}_h^{(k-1)\perp}$ such that $d\tau = P_{\mathfrak{B}_h^k} u^2 - u_h^2$, we obtain from the discrete Poincaré inequality that

$$\|P_{\mathfrak{B}_h^k} u^2 - u_h^2\| \leq C\|\sigma^2 - \sigma_h^2\| \leq C\|(I - \tilde{\Pi}_h)\sigma^2\|. \tag{7.24}$$

Finally, the uniform boundedness of the operators $\tilde{\Pi}_h$ implies that

$$\|(I - \tilde{\Pi}_h)\sigma^2\| \leq \|(I - \tilde{\Pi}_h)\sigma\| + C\|\sigma^1\|.$$

Together with (7.21), (7.23), and (7.24), this implies the desired estimate. $\qquad \square$

Lemma 7.8. Assume that $f \in \mathfrak{Z}^{k\perp}$ and that $(\sigma, u, p) \in H\Lambda^{k-1}(\Omega) \times H\Lambda^k(\Omega) \times \mathfrak{H}^k$ is the solution of problem (7.1), and $(\sigma_h, u_h, p_h) \in \Lambda_h^{k-1} \times \Lambda_h^k \times \mathfrak{H}_h^k$ is the corresponding solution of (7.13). Then $\sigma = \sigma_h = 0$, $p = p_h = 0$, and $u \in \mathfrak{Z}^{k\perp}$ and $u_h \in \mathfrak{Z}_h^{k\perp}$ satisfy

$$\|d(u - u_h)\| \leq \|(I - \tilde{\Pi}_h)\,du\|.$$

Furthermore, there is a constant C, independent of h, such that

$$\|(I - P_{\mathfrak{Z}_h^k})(\tilde{\Pi}_h u - u_h)\| \leq C\|(I - \tilde{\Pi}_h)\,du\|.$$

Proof. Since $\sigma = \sigma_h = 0$ and $p = p_h = 0$, we have

$$\langle d(u - u_h),\, dv \rangle = 0, \quad v \in \Lambda_h^k,$$

which implies

$$\|d(u - u_h)\|^2 + \|d(\tilde{\Pi}_h u - u_h)\|^2 = \|(I - \tilde{\Pi}_h)\,du\|^2.$$

Hence, the first bound is established. Since $d(I - P_{\mathfrak{Z}_h^k})(\tilde{\Pi}_h u - u_h) = d(\tilde{\Pi}_h u - u_h)$ the second bound follows from the discrete Poincaré inequality.   □

Combining these lemmas, we obtain the following error bounds.

Theorem 7.9.   For $f \in L^2\Lambda^k(\Omega)$, let $(\sigma, u, p) \in H\Lambda^{k-1}(\Omega) \times H\Lambda^k(\Omega) \times \mathfrak{H}^k$ be the solution of problem (7.1), and let $(\sigma_h, u_h, p_h) \in \Lambda_h^{k-1} \times \Lambda_h^k \times \mathfrak{H}_h^k$ be the corresponding solution of (7.13). Then

$$\|\sigma - \sigma_h\| + \|u - u_h\|_{H\Lambda} + \|p - p_h\| \le C(\|(I - \tilde{\Pi}_h)f_{\mathfrak{H}}\|$$
$$+ h^s\|(I - \tilde{\Pi}_h)f_{\mathrm{d}}\| + \|(I - \tilde{\Pi}_h)\sigma\| + \|(I - \tilde{\Pi}_h)u\|_{H\Lambda} + \|P_{\mathfrak{H}_h^k}u\|) \quad (7.25)$$

and

$$\|d(\sigma - \sigma_h)\| \le \|(I - \tilde{\Pi}_h)f_{\mathrm{d}}\|. \quad (7.26)$$

Proof.   Note that if we apply the Hodge decomposition $f = f_{\mathrm{d}} + f_{\mathfrak{H}} + f_\delta$ to a general right-hand side $f \in L^2\Lambda^k(\Omega)$, we obtain from Lemmas 7.6–7.8 that

$$\|\sigma - \sigma_h\| + \|d(u - u_h)\| + \|p - p_h\|$$
$$\le C(\|(I - \tilde{\Pi}_h)f_{\mathfrak{H}}\| + h^s\|(I - \tilde{\Pi}_h)f_{\mathrm{d}}\| + \|(I - \tilde{\Pi}_h)\sigma\| + \|(I - \tilde{\Pi}_h)\,du\|)$$

and

$$\|d(\sigma - \sigma_h)\| = \|(I - P_{\mathfrak{B}_h^k})f_{\mathrm{d}}\| \le \|(I - \tilde{\Pi}_h)f_{\mathrm{d}}\|.$$

Furthermore, we can conclude from Lemma 7.7 that

$$\|P_{\mathfrak{B}_h^k}(u - u_h)\| \le C(h^s\|(I - \tilde{\Pi}_h)f_{\mathrm{d}}\| + \|(I - \tilde{\Pi}_h)\sigma\|), \quad (7.27)$$

and from Lemmas 7.6 and 7.8 and the proof of Lemma 7.7 that

$$\|(I - P_{\mathfrak{Z}_h^k})(\tilde{\Pi}_h u - u_h)\| \le C(\|(I - \tilde{\Pi}_h)f_{\mathfrak{H}}\| + \|(I - \tilde{\Pi}_h)\,du\| + h^s\|(I - \tilde{\Pi}_h)f_{\mathrm{d}}\|).$$

However, from the last bound, we obtain

$$\|(I - P_{\mathfrak{Z}_h^k})(u - u_h)\|$$
$$\le \|(I - P_{\mathfrak{Z}_h^k})(u - \tilde{\Pi}_h u)\| + \|(I - P_{\mathfrak{Z}_h^k})(\tilde{\Pi}_h u - u_h)\| \quad (7.28)$$
$$\le C(\|(I - \tilde{\Pi}_h)f_{\mathfrak{H}}\| + \|(I - \tilde{\Pi}_h)u\|_{H\Lambda} + h^s\|(I - \tilde{\Pi}_h)f_{\mathrm{d}}\|).$$

Since $P_h = P_{\mathfrak{B}_h^k} + P_{\mathfrak{H}_h^k} + (I - P_{\mathfrak{Z}_h^k})$, we can conclude from (7.27) and (7.28) that

$$\|P_h u - u_h\| \le C(\|(I - \tilde{\Pi}_h)f_{\mathfrak{H}}\| + h^s\|(I - \tilde{\Pi}_h)f_{\mathrm{d}}\|$$
$$+ \|(I - \tilde{\Pi}_h)\sigma\| + \|(I - \tilde{\Pi}_h)u\|_{H\Lambda} + \|P_{\mathfrak{H}_h^k}u\|).$$

Hence, since

$$\|u - u_h\| \le \|(I - P_h)u\| + \|P_h u - u_h\|$$
$$\le \|(I - \tilde{\Pi}_h)u\| + \|P_h u - u_h\|,$$

we obtain

$$\|u - u_h\| \le C(\|(I - \tilde{\Pi}_h)f_{\mathfrak{H}}\| + h^s\|(I - \tilde{\Pi}_h)f_{\mathrm{d}}\|$$
$$+ \|(I - \tilde{\Pi}_h)\sigma\| + \|(I - \tilde{\Pi}_h)u\|_{H\Lambda} + \|P_{\mathfrak{H}_h^k} u\|).$$

Combining these results establishes the theorem. □

7.9. *Order of convergence estimates*

Using the regularity estimate (7.19), it is straightforward to check that if $(\sigma, u, p) \in H\Lambda^{k-1}(\Omega) \times H\Lambda^k(\Omega) \times \mathfrak{H}^k$ solves problem (7.1) for a given $f \in L^2\Lambda^k(\Omega)$, then $f_{\mathfrak{H}}$, σ, u, u_{d}, and $\mathrm{d}u$, are all in $H^s\Lambda^k(\Omega)$, with

$$\|f_{\mathfrak{H}}\|_{H^s\Lambda}, \|\sigma\|_{H^s\Lambda}, \|u\|_{H^s\Lambda}, \|u_{\mathrm{d}}\|_{H^s\Lambda}, \|\mathrm{d}u\|_{H^s\Lambda} \le C\|f\|. \tag{7.29}$$

Furthermore, from (7.18) it follows that

$$\|P_{\mathfrak{H}_h^k} u\| \le Ch^s\|(I - \tilde{\Pi}_h)u_{\mathrm{d}}\| \le Ch^{2s}\|f\|. \tag{7.30}$$

Using these results, together with the approximation properties of the operators $\tilde{\Pi}_h$, we can immediately conclude that the right-hand side of (7.25) is of order h^s. Furthermore, additional smoothness of the right-hand side and the solution will lead to higher-order convergence. More precisely, we have the following result.

Theorem 7.10. Suppose that the domain Ω is *s-regular* for some $0 < s \le 1$. Let $(\sigma, u, p) \in H\Lambda^{k-1}(\Omega) \times H\Lambda^k(\Omega) \times \mathfrak{H}^k$ be the solution of problem (7.1) and let $(\sigma_h, u_h, p_h) \in \Lambda_h^{k-1} \times \Lambda_h^k \times \mathfrak{H}_h^k$ be the solution of problem (7.13), where $\Lambda_h^{k-1} \times \Lambda_h^k$ is one of the choices given in (7.14). Then, for $f \in L^2\Lambda^k(\Omega)$,

$$\|\sigma - \sigma_h\| + \|u - u_h\|_{H\Lambda} + \|p - p_h\| \le Ch^s\|f\|.$$

If $f_{\mathrm{d}}, u_{\mathrm{d}} \in H^{t-s}\Lambda^k(\Omega)$ and $f_{\mathfrak{H}}, \sigma, u, \mathrm{d}u \in H^t\Lambda^k(\Omega)$, for $s \le t \le r$, then

$$\|\sigma - \sigma_h\| + \|u - u_h\|_{H\Lambda} + \|p - p_h\| \le Ch^t(\|f_{\mathrm{d}}\|_{H^{t-s}\Lambda} + \|f_{\mathfrak{H}}\|_{H^t\Lambda}$$
$$+ \|\sigma\|_{H^t\Lambda} + \|u\|_{H^t\Lambda} + \|u_{\mathrm{d}}\|_{H^{t-s}\Lambda} + \|\mathrm{d}u\|_{H^t\Lambda}).$$

If $\mathfrak{H}_h^k \subset \mathfrak{H}^k$, then the term $\|u_{\mathrm{d}}\|_{H^{t-s}\Lambda}$ and the corresponding regularity hypothesis can be dropped. Finally, if $f_{\mathrm{d}} \in H^t\Lambda^k(\Omega)$, then

$$\|\mathrm{d}(\sigma - \sigma_h)\| \le Ch^t\|f_{\mathrm{d}}\|_{H^t\Lambda}.$$

Proof. The desired estimates follow directly from the basic error bounds (7.25) and (7.26), combined with (7.29), (7.30) and the approximation properties of $\tilde{\Pi}_h$ (*cf.* Theorem 5.6). □

8. Eigenvalue problems

The purpose of this section is to discuss approximations of the eigenvalue
problem for the Hodge Laplacian using the same finite element spaces
$\Lambda_h^k \subset H\Lambda^k(\Omega)$ used for the boundary value problems discussed in the previous section. In fact, there is a vast literature on the approximation of eigenvalue problems for mixed systems, and it is well known that the standard
Brezzi stability conditions for linear saddle point systems are not sufficient
to guarantee convergence of the corresponding eigenvalue approximations.
Typically, spurious eigenvalues/eigenfunctions can occur, even if the Brezzi
conditions are fulfilled. However, by now the proper conditions for guaranteeing convergence and no spurious eigenmodes are well understood, and
discrete Hodge (or Helmholtz) decompositions seem to be a useful tool for
verifying these conditions. For the main results in this direction, we refer to
Osborn (1979), Mercier, Osborn, Rappaz and Raviart (1981), Babuška and
Osborn (1991), Boffi, Brezzi and Gastaldi (1997, 2000), and Boffi (2000,
2001, 2006).

Here, we shall use this theory to show that the finite element spaces
defined in this paper lead to convergence of the corresponding finite element
approximations of the eigenvalue problem for the Hodge Laplacian. In the
final subsection, we present some results on convergence rates for the special
case of a simple eigenvalue.

8.1. The eigenvalue problem for the Hodge Laplacian

We shall only consider the eigenvalue problem with boundary conditions
which are natural with respect to the mixed formulation, $i.e.$, the boundary
conditions are $\mathrm{Tr}(\star u) = 0$ and $\mathrm{Tr}(\star du) = 0$. The eigenvalue problem for the
Hodge Laplacian then takes the following form.

Find $\lambda \in \mathbb{R}$ and $(\sigma, u) \in H\Lambda^{k-1}(\Omega) \times H\Lambda^k(\Omega)$ such that

$$
\begin{aligned}
\langle \sigma, \tau \rangle - \langle d\tau, u \rangle &= 0, & \tau \in H\Lambda^{k-1}(\Omega), \\
\langle d\sigma, v \rangle + \langle du, dv \rangle &= \lambda \langle u, v \rangle, & v \in H\Lambda^k(\Omega),
\end{aligned}
\tag{8.1}
$$

where (σ, u) is not identically zero. We remark that standard symmetry
arguments show that only real eigenvalues are possible for this problem:
see (8.5) below. Also, the identity

$$\|\sigma\|^2 + \|du\|^2 = \lambda\|u\|^2$$

implies that all eigenvalues are nonnegative.

The first equation above expresses that $\delta u = \sigma \in L^2\Lambda^k(\Omega)$ and that
$\mathrm{Tr}(\star u) = 0$. Hence, $u \in H\Lambda^k(\Omega) \cap \mathring{H}^*\Lambda^k(\Omega)$. The eigenvalue problem (8.1)
can therefore alternatively be written as follows.

Find $u \in H\Lambda^k(\Omega) \cap \mathring{H}^*\Lambda^k(\Omega)$ such that

$$\langle \delta u, \delta v \rangle + \langle du, dv \rangle = \lambda \langle u, v \rangle, \quad v \in H\Lambda^k(\Omega) \cap \mathring{H}^*\Lambda^k(\Omega).$$

In general, $\lambda = 0$ will be an eigenvalue for this problem with the space of harmonic k-forms, $\mathfrak{H}^k$, as the corresponding eigenspace. In fact, just as we split the source problem into independent problems, we may also split the eigenvalue problem (8.1) into three independent eigenvalue problems. To see this, assume that (λ, σ, u) satisfies (8.1) with (σ, u) not identically equal to zero. Decompose

$$(\sigma, u) = (\sigma, u_{\mathrm{d}}) + (0, u_{\mathfrak{H}}) + (0, u_\delta),$$

where u_{d} and $u_{\mathfrak{H}}$ are the projections of u onto $\mathfrak{B}^k$ and $\mathfrak{H}^k$, respectively, with respect to the inner product of $L^2\Lambda^k(\Omega)$. If $\lambda = 0$ then $\sigma = 0$ and $u = u_{\mathfrak{H}}$. On the other hand, if $\lambda > 0$ then $\lambda\|u_{\mathfrak{H}}\|^2 = 0$, and therefore $u_{\mathfrak{H}} = 0$. Furthermore, $u_\delta \in \mathfrak{Z}^{k\perp}$ satisfies

$$\langle du_\delta, dv \rangle = \lambda \langle u_\delta, v \rangle, \quad v \in \mathfrak{Z}^{k\perp}. \tag{8.2}$$

Finally, (σ, u_{d}) satisfies the system

$$\begin{aligned}
\langle \sigma, \tau \rangle - \langle d\tau, u_{\mathrm{d}} \rangle &= 0, & \tau &\in H\Lambda^{k-1}(\Omega), \\
\langle d\sigma, v \rangle &= \lambda \langle u_{\mathrm{d}}, v \rangle, & v &\in \mathfrak{B}^k.
\end{aligned} \tag{8.3}$$

Observe that (σ, u_{d}), $u_{\mathfrak{H}}$, or u_δ may very well be identically equal to zero, even if (σ, u) is nonzero. On the other hand, any eigenvalue/eigenvector of (8.2) or (8.3) corresponds to an eigenvalue/eigenvector of (8.1).

We will introduce K as the solution operator for the Hodge Laplacian. More precisely, for a given $f \in L^2\Lambda^k(\Omega)$, consider the following problem. Find $(\sigma, u, p) \in H\Lambda^{k-1}(\Omega) \times H\Lambda^k(\Omega) \times \mathfrak{H}^k$ such that

$$\begin{aligned}
\langle \sigma, \tau \rangle - \langle d\tau, u \rangle &= 0, & \tau &\in H\Lambda^{k-1}(\Omega), \\
\langle d\sigma, v \rangle + \langle du, dv \rangle + \langle v, p \rangle &= \langle f, v \rangle, & v &\in H\Lambda^k(\Omega), \\
\langle u, q \rangle &= 0, & q &\in \mathfrak{H}^k.
\end{aligned} \tag{8.4}$$

The solution operator $K : L^2\Lambda^k(\Omega) \to H\Lambda^k(\Omega) \cap \mathring{H}^*\Lambda^k(\Omega)$ is given by

$$K : f \mapsto Kf = u + p.$$

Note that in (8.4), p is the L^2-projection of $u + p$ onto $\mathfrak{H}^k$. Therefore, an alternative characterization of the operator K is $Kf = u'$, where $u' \in H\Lambda^k(\Omega) \cap \mathring{H}^*\Lambda^k(\Omega)$ solves the system

$$\langle \delta u', \delta v \rangle + \langle du', dv \rangle + \langle u'_{\mathfrak{H}}, v_{\mathfrak{H}} \rangle = \langle f, v \rangle, \quad v \in H\Lambda^k(\Omega) \cap \mathring{H}^*\Lambda^k(\Omega),$$

where, as above, $v_{\mathfrak{H}}$ is the L^2-projection of v onto $\mathfrak{H}^k$. The operator K is the identity on $\mathfrak{H}^k$, and if $f_{\mathfrak{H}} = 0$ then $(Kf)_{\mathfrak{H}} = 0$. Observe also that the operator K is symmetric and positive definite on $L^2\Lambda^k(\Omega)$ since

$$\langle f, Kg \rangle = \langle \delta Kf, \delta Kg \rangle + \langle dKf, dKg \rangle + \langle (Kf)_{\mathfrak{H}}, (Kg)_{\mathfrak{H}} \rangle \qquad (8.5)$$

for all $f, g \in L^2\Lambda^k(\Omega)$. Furthermore, as a consequence of Theorem 2.1, the operator K is a compact operator in $\mathcal{L}(L^2\Lambda^k(\Omega), L^2\Lambda^k(\Omega))$.

If $\lambda > 0$ and $u \in H\Lambda^k(\Omega) \cap \mathring{H}^*\Lambda^k(\Omega)$ corresponds to an eigenvalue/eigenvector for (8.1) then

$$Ku = \lambda^{-1}u, \qquad (8.6)$$

and $u_{\mathfrak{H}} = 0$. On the other hand, if $u \in \mathfrak{H}^k$ then u and $\lambda = 1$ satisfy (8.6). In fact, the two eigenvalue problems (8.1) and (8.6) are equivalent, if we just recall that the eigenvalue $\lambda = 0$ in (8.1), corresponding to the eigenspace $\mathfrak{H}^k$, is shifted to $\lambda = 1$ in (8.6). Since K is compact in $\mathcal{L}(L^2\Lambda^k(\Omega), L^2\Lambda^k(\Omega))$, and not of finite rank, we conclude that the Hodge–Laplace problem (8.1) has a countable set of nonnegative eigenvalues

$$0 \leq \lambda_1 \leq \lambda_2 \leq \cdots \leq \lambda_j \leq \cdots$$

such that $\lim_{j\to\infty} \lambda_j = \infty$.

8.2. The discrete eigenvalue problem

In order to approximate the eigenvalue problem for the Hodge Laplacian (8.1), we need to introduce finite element spaces Λ_h^{k-1} and Λ_h^k, which are subspaces of the corresponding Sobolev spaces $H\Lambda^{k-1}(\Omega)$ and $H\Lambda^k(\Omega)$ occurring in the formulation of (8.1). We will continue to assume that these discrete spaces satisfy (7.10), (7.11) and (7.12). Given the discrete spaces, the corresponding discrete eigenvalue problem takes the following form.

Find $\lambda_h \in \mathbb{R}$ and $(\sigma_h, u_h) \in \Lambda_h^{k-1} \times \Lambda_h^k$, (σ_h, u_h) not identically zero, such that

$$\langle \sigma_h, \tau \rangle - \langle d\tau, u_h \rangle = 0, \qquad\qquad \tau \in \Lambda_h^{k-1},$$
$$\langle d\sigma_h, v \rangle + \langle du_h, dv \rangle = \lambda_h \langle u_h, v \rangle, \quad v \in \Lambda_h^k. \qquad (8.7)$$

If we define a discrete coderivative operator $d_h^* : \Lambda_h^k \to \Lambda_h^{k-1}$ by

$$\langle d_h^*\omega, \tau \rangle = \langle d\tau, \omega \rangle, \qquad \tau \in \Lambda_h^{k-1},$$

then the first equation states that $\sigma_h = d_h^* u_h$. Note that this identity also contains the information that $\mathrm{Tr}(\star u_h)$ is 'weakly zero', i.e., the boundary condition $\mathrm{Tr}(\star u)$ is approximated as a natural boundary condition.

As in the continuous case above, it follows by a straightforward energy argument that if $(\lambda_h, \sigma_h, u_h)$ solves (8.7), then

$$\|\sigma_h\|^2 + \|du_h\|^2 = \lambda_h\|u_h\|^2.$$

Therefore, all discrete eigenvalues λ_h are nonnegative. Furthermore, the eigenspace for the eigenvalue $\lambda_h = 0$ corresponds exactly to the space $\mathfrak{H}_h^k$ of discrete harmonic k-forms. We can assume that the discrete eigenvalues are ordered such that

$$0 \leq \lambda_{1,h} \leq \lambda_{2,h} \leq \cdots \leq \lambda_{N(h),h},$$

where $N(h)$ is the dimension of the space Λ_h^k.

An alternative characterization of the eigenvalue problem is obtained by introducing a discrete solution operator $K_h : L^2\Lambda^k(\Omega) \to \Lambda_h^k$. In parallel to the discussion in the continuous case, we define $K_h f = u_h + p_h$ if $(\sigma_h, u_h, p_h) \in \Lambda_h^{k-1} \times \Lambda_h^k \times \mathfrak{H}_h^k$ is the solution of the problem

$$\langle \sigma_h, \tau \rangle - \langle \mathrm{d}\tau, u_h \rangle = 0, \qquad \tau \in \Lambda_h^{k-1},$$

$$\langle \mathrm{d}\sigma_h, v \rangle + \langle \mathrm{d}u_h, \mathrm{d}v \rangle + \langle v, p_h \rangle = \langle f, v \rangle, \quad v \in \Lambda_h^k,$$

$$\langle u_h, q \rangle = 0, \qquad q \in \mathfrak{H}_h^k.$$

Here $\sigma_h = \mathrm{d}_h^* u_h = \mathrm{d}_h^* K_h f$. The operator K_h is equivalently characterized by $K_h f = u_h'$, where $u_h' \in \Lambda_h^k$ solves

$$\langle \delta u_h', \delta v \rangle + \langle \mathrm{d}u_h', \mathrm{d}v \rangle + \langle u_{\mathfrak{H},h}', v_{\mathfrak{H},h} \rangle = \langle f, v \rangle, \quad v \in \Lambda_h^k,$$

where $v_{\mathfrak{H},h}$ denotes the L^2-projection of v onto $\mathfrak{H}_h^k$.

The eigenvalue problem (8.7) is equivalent to the corresponding eigenvalue problem for the operator K_h given by

$$K_h u_h = \lambda_h^{-1} u_h, \tag{8.8}$$

with the interpretation that the eigenvalue $\lambda_h = 0$ in (8.7) is shifted to $\lambda_h = 1$ in (8.8). The discrete operator K_h is again symmetric on $L^2\Lambda^k(\Omega)$ since it is straightforward to verify that

$$\langle f, K_h g \rangle = \langle \mathrm{d}_h^* K_h f, \mathrm{d}_h^* K_h g \rangle + \langle \mathrm{d}K_h f, \mathrm{d}K_h g \rangle + \langle (K_h f)_{\mathfrak{H},h}, (K_h g)_{\mathfrak{H},h} \rangle$$

for all $f, g \in L^2\Lambda^k(\Omega)$.

8.3. Convergence of the discrete approximations

For every positive integer j, we let $m(j)$ be the dimension of the eigenspace spanned by the first j distinct eigenvalues of the Hodge–Laplace problem (8.1). We let $\mathfrak{E}_i$ denote the eigenspace associated to the eigenvalue λ_i, while $\mathfrak{E}_{i,h}$ is the corresponding discrete eigenspace associated to $\lambda_{i,h}$.

The discrete eigenvalue problem (8.7) is said to *converge* to the exact eigenvalue problem (8.1) if, for any $\epsilon > 0$ and integer $j > 0$, there exists a mesh parameter $h_0 > 0$ such that, for all $h \leq h_0$, we have

$$\max_{1 \leq i \leq m(j)} |\lambda_i - \lambda_{i,h}| \leq \epsilon,$$

$$\mathrm{gap}\left(\bigoplus_{i=1}^{m(j)} \mathfrak{E}_i, \bigoplus_{i=1}^{m(j)} \mathfrak{E}_{i,h} \right) \leq \epsilon.$$

Here gap $= \mathrm{gap}(E, F)$ is the *gap* between two subspaces E and F of a Hilbert space H given by

$$\mathrm{gap}(E, F) = \max\left(\sup_{\substack{u \in E \\ \|u\|_H=1}} \inf_{v \in F} \|u - v\|_H, \sup_{\substack{v \in F \\ \|v\|_H=1}} \inf_{u \in E} \|u - v\|_H \right).$$

This is a reasonable concept of convergence since, besides convergence of the eigenmodes, it also contains the information that no spurious eigenmodes pollute the spectrum. Furthermore, for eigenvalue problems of the form (8.6) and (8.8), which are equivalent to (8.1) and (8.7), convergence will follow if the operators K_h converge to K in the operator norm. In other words, if

$$\|K_h - K\|_{\mathcal{L}(L^2\Lambda^k(\Omega), L^2\Lambda^k(\Omega))} \longrightarrow 0 \quad \text{as } h \to 0, \tag{8.9}$$

then the discrete eigenvalue problem (8.7) converges to the eigenvalue problem (8.1) in the sense specified above: see Kato (1995, Chapter IV). Here, the gap between the subspaces is defined with respect to the Hilbert space $L^2\Lambda^k(\Omega)$. In fact, it was established in Boffi, Brezzi and Gastaldi (2000) that this operator convergence is both sufficient and necessary for obtaining convergence of the eigenvalue approximations.

As a consequence, in the present case it only remains to estimate $\|K_h - K\|_{\mathcal{L}(L^2\Lambda^k(\Omega), L^2\Lambda^k(\Omega))}$. However, it is a direct consequence of Theorem 7.10 and the definitions of the operator K and K_h above that

$$\|K_h - K\|_{\mathcal{L}(L^2\Lambda^k(\Omega), L^2\Lambda^k(\Omega))} \leq ch^s, \tag{8.10}$$

where $s \geq 1/2$ and the constant c is independent of h. Therefore, the convergence property (8.9) holds and convergence of the eigenvalues and eigenvectors are guaranteed.

Convergence
As a consequence of (8.10) and the discussion above, the following theorem is obtained.

Theorem 8.1. The discrete eigenvalue problem (8.7) converges to the eigenvalue problem (8.1) in the sense defined above.

In particular, this theorem implies the following result on the approximation of the k-harmonic forms.

Corollary 8.2.

$$\lim_{h \to 0} \operatorname{gap}(\mathfrak{H}^k, \mathfrak{H}_h^k) = 0.$$

Recall that for the continuous problem (8.1), all eigenfunctions corresponding to eigenvalues $\lambda > 0$ are L^2-orthogonal to $\mathfrak{H}^k$, and for such eigenfunctions, the problem (8.1) can be split into two independent problems (8.2) and (8.3). The corresponding property is valid for the discrete problem (8.7) as well. If $\lambda_h > 0$, then all eigenfunctions are L^2-orthogonal to $\mathfrak{H}_h^k$, and if $u_h = u_{\mathrm{d},h} + u_{\delta,h}$ is the discrete Hodge decomposition of an eigenfunction u_h in Λ_h^k then

$$\langle \mathrm{d} u_{\delta,h}, \mathrm{d} v \rangle = \lambda_h \langle u_{\delta,h}, v \rangle, \quad v \in \mathfrak{Z}_h^{k\perp}, \tag{8.11}$$

and

$$\langle \sigma_h, \tau \rangle - \langle \mathrm{d}\tau, u_{\mathrm{d},h} \rangle = 0, \qquad \tau \in \Lambda_h^{k-1},$$
$$\langle \mathrm{d}\sigma_h, v \rangle = \lambda_h \langle u_{\mathrm{d},h}, v \rangle, \quad v \in \mathfrak{B}_h^k. \tag{8.12}$$

Here $\sigma_h = \mathrm{d}_h^* u_h$. The converse also holds, *i.e.*, any eigenvalue/eigenfunction of (8.11) or (8.12) is an eigenvalue/eigenfunction of (8.7). In fact the discrete problems (8.11) and (8.12) converge separately to the eigenvalue problems (8.2) and (8.3) in the sense specified above. This is basically a consequence of Theorem 8.1 and the orthogonality property of the Hodge decompositions. To see this, the following result is useful.

Lemma 8.3. If $u \in \mathfrak{B}^k$ and $v \in \operatorname{span}(\mathfrak{Z}^{k\perp}, \mathfrak{Z}_h^{k\perp})$ then we obtain the bounds

$$\|u - v\| \geq \|v\| - \|(I - \tilde{\Pi}_h)u\| \quad \text{and} \quad \|u - v\| \geq \|u\| - 2\|(I - \tilde{\Pi}_h)u\|.$$

Proof. Since $\tilde{\Pi}_h u \in \mathfrak{B}_h^k \subset \mathfrak{B}^k$ and v are orthogonal we have

$$\|u - v\| \geq \|\tilde{\Pi}_h u - v\| - \|(I - \tilde{\Pi}_h)u\|$$
$$\geq \max(\|\tilde{\Pi}_h u\|, \|v\|) - \|(I - \tilde{\Pi}_h)u\|.$$

This gives the first inequality, and the second bound follows since $\|\tilde{\Pi}_h u\| \geq \|u\| - \|(I - \tilde{\Pi}_h)u\|$. □

We now have the following result, which is a strengthening of Theorem 8.1.

Theorem 8.4. The discrete eigenvalue problems (8.11) and (8.12) converge separately to the corresponding problems (8.2) and (8.3) in the sense defined above.

Proof. Let $\lambda > 0$ be an eigenvalue for the problem (8.1) with corresponding eigenspace $\mathfrak{E}$, and let λ_h be the corresponding eigenvalue for (8.7), with eigenspace $\mathfrak{E}_h$, such that $\lambda_h \to \lambda$ and $\operatorname{gap}(\mathfrak{E}, \mathfrak{E}_h) \to 0$ as h tends to zero.

Let

$$\mathfrak{E} = \mathfrak{E}_d \oplus \mathfrak{E}_\delta \quad \text{and} \quad \mathfrak{E}_h = \mathfrak{E}_{d,h} \oplus \mathfrak{E}_{\delta,h}$$

be the corresponding Hodge decompositions (continuous and discrete) of the spaces, $i.e.$, $\mathfrak{E}_d = \mathfrak{E} \cap \mathfrak{B}^k$ and $\mathfrak{E}_{d,h} = \mathfrak{E}_h \cap \mathfrak{B}_h^k$. The desired result will follow if we can show that $\mathrm{gap}(\mathfrak{E}_d, \mathfrak{E}_{d,h}) \to 0$ and $\mathrm{gap}(\mathfrak{E}_\delta, \mathfrak{E}_{\delta,h}) \to 0$.

However, if $u \in \mathfrak{E}_d$, with $\|u\| = 1$, then by Lemma 8.3,

$$\inf_{v \in \mathfrak{E}_{d,h}} \|u - v\| \le \inf_{v \in \mathfrak{E}_h} \|u - v\| + 2\|(I - \tilde{\Pi}_h)u\|$$

$$\le \mathrm{gap}(\mathfrak{E}, \mathfrak{E}_h) + 2\|(I - \tilde{\Pi}_h)u\|,$$

and since the space $\mathfrak{E}_d$ is finite-dimensional, the right-hand side converges to zero uniformly in u. On the other hand, for a given $v \in \mathfrak{E}_{d,h}$, with $\|v\| = 1$, we have

$$\inf_{u \in \mathfrak{E}_d} \|u - v\| = \inf_{u \in \mathfrak{E}} \|u - v\| \le \mathrm{gap}(\mathfrak{E}, \mathfrak{E}_h).$$

This shows that $\mathrm{gap}(\mathfrak{E}_d, \mathfrak{E}_{d,h}) \to 0$, and a corresponding argument will show that $\mathrm{gap}(\mathfrak{E}_\delta, \mathfrak{E}_{\delta,h}) \to 0$.   □

Convergence rates

The results of Section 7 can also be combined with the standard theory for eigenvalue approximation to obtain rates of convergence. To simplify the presentation, we do this only for the simplest case, where we assume we have only simple eigenvalues. Then, from Theorem 7.3 of Babuška and Osborn (1991), if $Ku = \lambda^{-1}u$, $\|u\| = 1$, then

$$|\lambda^{-1} - \lambda_h^{-1}| \le C(|\langle (K - K_h)u, u \rangle| + \|(K - K_h)u\|^2).$$

For the second term, we can use the error estimates of Section 7. To get a similar squaring of the error from the first term, we need some additional analysis, which is a slight modification of Theorem 11.1 of Babuška and Osborn (1991). To use the results of Section 7, and avoid confusing the terminology, we estimate $|\langle (K - K_h)f, f \rangle|$, where now $\|f\| = 1$ and $(K - K_h)f = u - u_h + p - p_h$, $i.e.$, we estimate $|\langle u - u_h + p - p_h, f \rangle|$.

We first note that from the definitions of σ, u, and p, and their finite element approximations, we get the error equations

$$\langle \sigma - \sigma_h, \tau \rangle = \langle d\tau, u - u_h \rangle, \qquad \tau \in \Lambda_h^{k-1}, \qquad (8.13)$$

$$\langle d(\sigma - \sigma_h), v \rangle + \langle d(u - u_h), dv \rangle + \langle p - p_h, v \rangle = 0, \qquad v \in \Lambda_h^k. \qquad (8.14)$$

Also, from the definitions of σ, u, and p, we have

$$\langle f, u - u_h + p - p_h \rangle = \langle d\sigma, u - u_h + p - p_h \rangle + \langle du, d(u - u_h) \rangle$$

$$+ \langle p, u - u_h + p - p_h \rangle,$$

$$\langle \sigma, \sigma - \sigma_h \rangle = \langle d(\sigma - \sigma_h), u \rangle.$$

Since $\langle d\sigma, p\rangle = \langle d\sigma_h, p\rangle = 0$, we obtain

$$\langle f, u - u_h + p - p_h\rangle = \langle d\sigma, u - u_h + p - p_h\rangle - \langle \sigma, \sigma - \sigma_h\rangle$$
$$+ \langle d(\sigma - \sigma_h), u + p\rangle + \langle du, d(u - u_h)\rangle + \langle p - p_h, p + u\rangle$$
$$+ \langle p, u - u_h\rangle - \langle p - p_h, u\rangle.$$

Since $\langle u, q\rangle = 0, q \in \mathfrak{H}^k$ and $\langle u_h, q_h\rangle = 0, q_h \in \mathfrak{H}^k_h$,

$$\langle p, u - u_h\rangle - \langle p - p_h, u\rangle = \langle p - p_h, u - u_h\rangle + 2\langle p_h, u\rangle.$$

Using the Hodge decomposition $u = u_\mathrm{d} + u_{\mathfrak{H}} + u_\delta$, and observing that $u_{\mathfrak{H}} = 0$, $\langle p_h, u_\delta\rangle = 0$, and $\tilde{\Pi}_h u_\mathrm{d} \in \mathfrak{B}^k_h \subset \mathfrak{B}^k$, we get

$$\langle p_h, u\rangle = \langle p_h, u_\mathrm{d}\rangle = \langle p_h, u_\mathrm{d} - \Pi_h u_\mathrm{d}\rangle = \langle p_h - p, u_\mathrm{d} - \Pi_h u_\mathrm{d}\rangle.$$

Choosing $\tau = \sigma_h$ and $v = u_h + p_h$ in (8.13) and (8.14), and combining these results, we get

$$\langle f, u - u_h + p - p_h\rangle = \langle d(\sigma - \sigma_h), u - u_h + p - p_h\rangle - \langle \sigma - \sigma_h, \sigma - \sigma_h\rangle$$
$$+ \langle d(\sigma - \sigma_h), u - u_h + p - p_h\rangle + \langle d(u - u_h), d(u - u_h)\rangle$$
$$+ \langle p - p_h, u - u_h + p - p_h\rangle + \langle p - p_h, u - u_h\rangle + 2\langle p_h - p, u_\mathrm{d} - \tilde{\Pi}_h u_\mathrm{d}\rangle.$$

Hence,

$$|\langle f, u - u_h + p - p_h\rangle| \leq C(\|\sigma - \sigma_h\|_{H\Lambda} + \|u - u_h\|_{H\Lambda}$$
$$+ \|p - p_h\|)^2 + 2\|p_h - p\|\|u_\mathrm{d} - \tilde{\Pi}_h u_\mathrm{d}\|.$$

Applying our approximation theory results and Theorem 7.10 in the case when $s = 1$ and the eigenfunction is sufficiently smooth, we obtain

$$|\langle f, u - u_h + p - p_h\rangle| \leq Ch^{2r}$$

and hence it follows that

$$|\lambda - \lambda_h| \leq Ch^{2r}.$$

9. The $H\Lambda$ projection and Maxwell's equations

In this section, we consider the approximation of variations of the following problem.

Given $f \in L^2\Lambda^k(\Omega)$, find $u \in H\Lambda^k(\Omega)$ satisfying

$$\langle u, v\rangle_{H\Lambda} := \langle du, dv\rangle + \langle u, v\rangle = \langle f, v\rangle, \quad v \in H\Lambda^k(\Omega). \qquad (9.1)$$

Note that when $k = 0$, this corresponds to a perturbation of the Hodge–Laplace problem we have considered previously, by adding a lower-order term. Problems of this form will also arise later in this paper in Section 10, on preconditioning. The time-harmonic Maxwell equations can be written as a variation of this form, where $k = 1$ and the term $\langle u, v \rangle$ is replaced by $-m^2 \langle u, v \rangle$. We will study this case below, when m^2 is not an eigenvalue, i.e., when there is no nonzero $u \in H\Lambda^k(\Omega)$ for which $\langle du, dv \rangle = m^2 \langle u, v \rangle$ for all $v \in H\Lambda^k(\Omega)$.

A simple approximation scheme for (9.1) is to seek $u_h \in \Lambda_h^k$ satisfying

$$\langle u_h, v \rangle_{H\Lambda} = \langle f, v \rangle, \quad v \in \Lambda_h^k. \tag{9.2}$$

The error analysis in the $H\Lambda$-norm of such a problem is straightforward. By subtracting (9.2) from (9.1), we get the error equation

$$\langle u - u_h, v \rangle_{H\Lambda} = 0, \quad v \in \Lambda_h^k. \tag{9.3}$$

Hence, u_h is the $H\Lambda$-projection of u into Λ_h^k and therefore satisfies the optimal error estimate

$$\|u - u_h\|_{H\Lambda} \leq \inf_{v \in \Lambda_h^k} \|u - v\|_{H\Lambda}.$$

If we modify the problem by replacing the term $\langle u, v \rangle$ by $m^2 \langle u, v \rangle$, then the analysis is essentially the same, since $\langle du, dv \rangle + m^2 \langle u, v \rangle$ defines an equivalent inner product on $H\Lambda^k(\Omega)$.

9.1. Maxwell-type problems

We shall also consider the modification of the problem (9.1) obtained by replacing $\langle u, v \rangle$ by $-m^2 \langle u, v \rangle$. That is, we consider the following problem.
Find $u \in H\Lambda^k(\Omega)$ such that

$$\langle du, dv \rangle - m^2 \langle u, v \rangle = \langle f, v \rangle, \quad v \in H\Lambda^k(\Omega). \tag{9.4}$$

This problem is *indefinite*, and its analysis more complicated. We will assume that m^2 is not an eigenvalue, since otherwise the problem is not well-posed.

Problems of the form (9.4) arise, for example, in the study of Maxwell's equations. In the cavity problem for the time-harmonic Maxwell equations subject to a perfect conducting boundary condition, one seeks the electric field E satisfying

$$\mathrm{curl}(\mathrm{curl}\, E) - m^2 E = F \text{ in } \Omega, \qquad \nu \times E = 0 \text{ on } \Gamma = \partial\Omega.$$

Here F is a given function related to the imposed current sources and the parameter m is the wave number, assumed to be real and positive, and the permittivity and permeability coefficients are set equal to one. This problem corresponds to the problem (9.4) with $k = 1$ and $f \in \mathfrak{B}^{*k}$, but with boundary conditions given as $\mathrm{Tr}(u) = 0$ and $\mathrm{Tr}(\delta u) = 0$. For an extensive treatment of Maxwell's equation using finite element exterior calculus, see the paper by Hiptmair (2002). A comprehensive treatment in standard finite element notation can be found in the book by Monk (2003a).

As in the case of the time-harmonic Maxwell equation, problems of the form (9.4) arise by considering the steady state problem obtained from the ansatz of a time-harmonic solution to the following second-order time-dependent differential equation related to the Hodge Laplacian:

$$\frac{\partial^2 \omega}{\partial t^2} + (\delta d + d\delta)\omega = g.$$

If we assume that g is time-harmonic, $i.e.$, can be expressed as the product of a function which is independent of time with e^{imt} for some positive real number m, and then we seek ω of the same form, we are led to a time-independent equation of the form

$$(\delta d + d\delta)u - m^2 u = f.$$

With the boundary conditions $\text{Tr}(\star u) = 0$ and $\text{Tr}(\star du) = 0$, a weak formulation of this problem is as follows.

Find $u \in H\Lambda^k(\Omega)$ such that

$$\langle du, dv \rangle + \langle \delta u, \delta v \rangle - m^2 \langle u, v \rangle = \langle f, v \rangle, \quad v \in H\Lambda^k(\Omega).$$

If we consider the special case $f \in \mathfrak{B}^{*k}$, and split u by its Hodge decomposition, $u = u_d + u_{\mathfrak{H}} + u_\delta$, then the subproblem for u_δ may be written as follows.

Find $u \in \mathfrak{B}^{*k}$ such that

$$\langle du, dv \rangle - m^2 \langle u, v \rangle = \langle f, v \rangle, \quad v \in \mathfrak{B}^{*k}.$$

Note that for $f \in \mathfrak{B}^{*k}$, this problem is equivalent to (9.4) (in the sense that the unique solution to this problem is the unique solution of (9.4)).

To approximate (9.4), we use the obvious extension of the method defined in (9.2). That is, we seek $u_h \in \Lambda_h^k$ such that

$$\langle du_h, dv \rangle - m^2 \langle u_h, v \rangle = \langle f, v \rangle, \quad v \in \Lambda_h^k. \tag{9.5}$$

Note that since we have assumed that m^2 is not an eigenvalue for the continuous problem (9.4), it follows from Theorem 8.4 that m^2 is not an eigenvalue for the corresponding discrete problem (9.5) if the mesh parameter h is sufficiently small. In fact the following error bound can be established.

Theorem 9.1. Assume that $f \in L^2\Lambda^k(\Omega)$ and a real number $m > 0$ are fixed. Furthermore, let $u \in H\Lambda^k(\Omega)$ and $u_h \in \Lambda_h^k$ be the corresponding solutions of (9.4) and (9.5), respectively. If the domain Ω is s-regular, with

$0 < s \le 1$, then for h sufficiently small,

$$\|u - u_h\|_{H\Lambda} \le \frac{1}{1 - Ch^s} \inf_{v \in \Lambda_h^k} \|u - v\|_{H\Lambda},$$

where the constant C is independent of h.

Proof. To obtain this result, we follow the proof of Monk (2003b), but for the case of natural, rather than essential boundary conditions; see also Monk (1992), Monk and Demkowicz (2001) and Boffi and Gastaldi (2002). First, subtracting (9.5) from (9.4), and setting $e_h = u - u_h$, we obtain the error equation

$$\langle de_h, dv \rangle - m^2 \langle e_h, v \rangle = 0, \quad v \in \Lambda_h^k. \tag{9.6}$$

If we choose $v = dq$ for $q \in \Lambda_h^{k-1}$, we get

$$\langle e_h, dq \rangle = 0, \quad q \in \Lambda_h^{k-1}. \tag{9.7}$$

Letting Q_h denote the $H\Lambda$-projection into Λ_h^k, we then obtain from (9.6)

$$\begin{aligned}
\|e_h\|_{H\Lambda}^2 &= \langle de_h, de_h \rangle - m^2 \langle e_h, e_h \rangle + (1 + m^2)\langle e_h, e_h \rangle \tag{9.8}\\
&= \langle de_h, d(u - Q_h u) \rangle - m^2 \langle e_h, u - Q_h u \rangle + (1 + m^2)\langle e_h, e_h \rangle\\
&= \langle de_h, d(u - Q_h u) \rangle + \langle e_h, u - Q_h u \rangle + (1 + m^2)\langle e_h, Q_h u - u_h \rangle\\
&= \langle de_h, d(u - Q_h u) \rangle + \langle e_h, u - Q_h u \rangle + (1 + m^2)\langle e_h, Q_h e_h \rangle\\
&\le \|e_h\|_{H\Lambda} \|u - Q_h u\|_{H\Lambda} + (1 + m^2)\langle e_h, Q_h e_h \rangle.
\end{aligned}$$

The main work of the proof is then to estimate the term $\langle e_h, Q_h e_h \rangle$. To do so, we use the Hodge decomposition to write

$$e_h = d\rho + \psi, \quad \rho \in H\Lambda^{k-1}, \ \psi \in \mathfrak{B}^{k\perp}, \quad \|e_h\|^2 = \|d\rho\|^2 + \|\psi\|^2,$$

and the discrete Hodge decomposition to write

$$Q_h e_h = d\rho_h + \psi_h, \quad \rho_h \in \Lambda_h^{k-1}, \ \psi_h \in \mathfrak{B}_h^{k\perp}, \quad \|Q_h e_h\|^2 = \|d\rho_h\|^2 + \|\psi_h\|^2.$$

Then by (9.7),

$$\langle e_h, Q_h e_h \rangle = \langle e_h, d\rho_h + \psi_h \rangle = \langle e_h, \psi_h \rangle = \langle d\rho, \psi_h \rangle + \langle \psi, \psi_h \rangle. \tag{9.9}$$

We next obtain a bound on $\|\psi\|_0$. To do so, we define the adjoint variable $z \in H\Lambda^k(\Omega)$ satisfying

$$\langle d\phi, dz \rangle - m^2 \langle z, \phi \rangle = \langle \psi, \phi \rangle, \quad \phi \in H\Lambda^k(\Omega). \tag{9.10}$$

Since m is not an eigenvalue, z is well-defined, and there is a constant C such that $\|z\|_{H\Lambda} \le C\|\psi\|$. Furthermore, since $\langle dq, \psi \rangle = 0$ for $q \in \Lambda^{k-1}$, it follows that $\langle dq, z \rangle = 0$ for $q \in \Lambda^{k-1}$. Hence, $\delta z = 0$. Since $\mathrm{Tr}\,{\star}z = 0$ and Ω is s-regular, $z \in H^s\Lambda^k(\Omega)$ for some $s \in (0,1]$ and from (7.19), we have the estimate $\|z\|_{H^s\Lambda} \le C\|dz\| \le C\|\psi\|$. In addition, we see that dz satisfies

$$\delta dz = m^2 z + \psi, \quad \mathrm{Tr}\,{\star}dz = 0.$$

Again using (7.19), we obtain $dz \in H^s \Lambda^{k+1}(\Omega)$ and

$$\|dz\|_{H^s \Lambda} \leq C\|\delta dz\|_0 \leq C\|m^2 z + \psi\| \leq C\|\psi\|.$$

Using the approximation properties of $\tilde{\Pi}_h$, we obtain the estimate

$$\|z - \tilde{\Pi}_h z\|_{H\Lambda} \leq Ch^s(\|z\|_s + \|dz\|_s) \leq Ch^s \|\psi\|.$$

Choosing $\phi = \psi$ in (9.10), and using the fact that $\delta z = 0$, and (9.6), we get

$$\|\psi\|^2 = \langle d\psi, dz \rangle - m^2 \langle z, \psi \rangle = \langle de_h, dz \rangle - m^2 \langle z, e_h \rangle$$
$$= \langle de_h, d(z - \tilde{\Pi}_h z) \rangle - m^2 \langle z - \tilde{\Pi}_h z, e_h \rangle$$
$$\leq C\|e_h\|_{H\Lambda} \|z - \tilde{\Pi}_h z\|_{H\Lambda} \leq Ch^s \|e_h\|_{H\Lambda} \|\psi\|.$$

Hence, $\|\psi\| \leq Ch^s \|e_h\|_{H\Lambda}$, and so

$$|\langle \psi, \psi_h \rangle| \leq \|\psi\| \|\psi_h\| \leq \|\psi\| \|Q_h e_h\| \leq Ch^s \|e_h\|_{H\Lambda}^2. \qquad (9.11)$$

It thus remains to bound the term $\langle d\rho, \psi_h \rangle$. To do so, we first write $\psi_h = \psi_{h,\delta} + \psi_{h,\mathfrak{H}}$, with $\psi_{h,\delta} \in \mathfrak{Z}_h^{k\perp}$, and $\psi_{h,\mathfrak{H}} \in \mathfrak{H}_h^k$. Then, by Lemmas 5.10 and 5.9, we can find $\chi_\delta \in \mathfrak{Z}^{k\perp}$ and $\chi_\mathfrak{H} \in \mathfrak{H}^k$ satisfying $d\chi_\delta = d\psi_h$, $\|\chi_\mathfrak{H}\| \leq \|\psi_{h,\mathfrak{H}}\|$ and (also using (7.19))

$$\|\psi_{h,\delta} - \chi_\delta\| \leq \|(I - \tilde{\Pi}_h)\chi_\delta\| \leq Ch^s \|\chi_\delta\|_s \leq Ch^s \|d\chi_\delta\| \leq Ch^s \|d\psi_h\|,$$
$$\|\psi_{h,\mathfrak{H}} - \chi_\mathfrak{H}\| \leq \|(I - \tilde{\Pi}_h)\chi_\mathfrak{H}\| \leq Ch^s \|\chi_\mathfrak{H}\|_s \leq Ch^s \|\chi_\mathfrak{H}\| \leq Ch^s \|\psi_{h,\mathfrak{H}}\|.$$

Hence,

$$|\langle d\rho, \psi_h \rangle| = |\langle d\rho, \psi_{h,\delta} - \chi_\delta + \psi_{h,\mathfrak{H}} - \chi_\mathfrak{H} \rangle|$$
$$\leq Ch^s \|d\rho\| (\|d\psi_h\| + \|\psi_{h,\mathfrak{H}}\|) \leq Ch^s \|d\rho\| (\|d\psi_h\| + \|\psi_h\|) \quad (9.12)$$
$$\leq Ch^s \|e_h\| \|Q_h e_h\|_{H\Lambda} \leq Ch^s \|e_h\|_{H\Lambda}^2.$$

Combining (9.8), (9.9), (9.11) and (9.12), we obtain

$$\|e_h\|_{H\Lambda}^2 \leq \|e_h\|_{H\Lambda} \|u - Q_h u\|_{H\Lambda} + (1 + m^2)Ch^s \|e_h\|_{H\Lambda}^2.$$

The theorem follows immediately from this result, provided h is sufficiently small. $\qquad \square$

9.2. Refined estimates

In the error estimates derived above, the estimates for $\|u - u_h\|$ and $\|d(u - u_h)\|$ are tied together in the $H\Lambda$-norm. It is well known, however, that for the standard Galerkin method (corresponding to the case $k = 0$), the error bounds for $\|u - u_h\|$ can be up to one power higher in h than the error for $\|d(u - u_h)\|$.

To understand the type of improvement we might hope to get in this more general case, we use the Hodge decomposition to write the error $u - u_h = e_d + e_\mathfrak{H} + e_\delta = dg + e_\mathfrak{H} + e_\delta$, where $e_d \in \mathfrak{B}^k$, $e_\mathfrak{H} \in \mathfrak{H}^k$, $e_\delta \in \mathring{\mathfrak{B}}^{*k}$, and

$g \in \mathring{\mathfrak{B}}^{*(k-1)}$. Since $\|e_d + e_{\mathfrak{H}}\|_{H\Lambda} = \|e_d + e_{\mathfrak{H}}\|$, we cannot expect to improve on this part of the error. However, we will establish the following result.

Theorem 9.2. Suppose $u \in H\Lambda^k(\Omega)$ is the solution of problem (9.1) and $u_h \in \Lambda_h^k$ is the solution of problem (9.2). Further, suppose that the domain Ω is 1-regular and that $u - u_h = dg + e_{\mathfrak{H}} + e_\delta$, where $g \in \mathring{\mathfrak{B}}^{*(k-1)}$, $e_{\mathfrak{H}} \in \mathfrak{H}^k$, and $e_\delta \in \mathfrak{B}^{*k}$. Then there is a constant C independent of h such that

$$\|g\| \le Ch\|u - u_h\|, \qquad \|e_\delta\| \le Ch\|u - u_h\|_{H\Lambda}.$$

Proof. To obtain the first bound, we observe that by (9.3)

$$\langle dg, dz_h \rangle = \langle u - u_h, dz_h \rangle = 0, \quad z_h \in \Lambda_h^{k-1}.$$

Since $g \in \mathring{\mathfrak{B}}^{*(k-1)}$, we can define (see (7.3)) $z^g \in \mathring{\mathfrak{B}}^{*(k-1)}$ such that

$$\langle dz^g, d\mu \rangle = \langle g, \mu \rangle, \quad \mu \in H\Lambda^{k-1}(\Omega).$$

Hence, choosing $\mu = g$, we obtain

$$\|g\|^2 = \langle dz^g, dg \rangle = \langle d(z^g - z_h), dg \rangle \le \|d(z^g - z_h)\|\|u - u_h\|.$$

Choosing $z_h = \tilde{\Pi}_h z^g$, using (7.19), and noting that $\delta dz^g = g$ and $\text{Tr} \star dz^g = 0$, we get that

$$\|d(z^g - z_h)\| = \|(I - \tilde{\Pi}_h)dz^g\| \le Ch\|dz^g\|_1 \le Ch\|\delta dz^g\| \le Ch\|g\|.$$

This establishes the first estimate. To get the second estimate, we now define $z^\delta \in \mathfrak{B}^{*k}$ such that

$$\langle z^\delta, \mu \rangle_{H\Lambda} = \langle e_\delta, \mu \rangle, \quad \mu \in H\Lambda^k(\Omega).$$

Then, using (9.3), we get

$$\begin{aligned}
\|e_\delta\|^2 &= \langle e_\delta, u - u_h \rangle = \langle z^\delta, u - u_h \rangle_{H\Lambda} \\
&= \langle z^\delta - \tilde{\Pi}_h z^\delta, u - u_h \rangle_{H\Lambda} \le \|z^\delta - \tilde{\Pi}_h z^\delta\|_{H\Lambda}\|u - u_h\|_{H\Lambda} \\
&\le Ch(\|z^\delta\|_{H^1\Lambda} + \|dz^\delta\|_{H^1\Lambda})\|u - u_h\|_{H\Lambda}.
\end{aligned}$$

Again applying (7.19), we find that

$$\|z^\delta\|_{H^1\Lambda} \le C\|dz^\delta\| \le C\|e_\delta\|,$$
$$\|dz^\delta\|_{H^1\Lambda} \le C\|\delta dz^\delta\| \le C(\|e_\delta\| + \|z^\delta\|) \le C\|e_\delta\|.$$

The result follows by combining these estimates. $\square$

We close by remarking that essentially the same result holds, with essentially the same proof, in the case where u is the solution of (9.4) and u_h is defined by (9.5).

10. Preconditioning

The purpose of this section is to discuss preconditioned iterative methods for the discrete Hodge–Laplace problem. Such solvers have been considered previously for some of the PDE problems that can be derived from the Hodge–Laplace problem, but to our knowledge not in the full generality presented here. We will first give a quick review of Krylov space iterations and block diagonal preconditioners. Thereafter, we will illustrate how the discrete Hodge decomposition enters as a fundamental tool in the construction of the diagonal blocks of the preconditioner.

The discrete Hodge–Laplace problem can be written as a linear system

$$\mathcal{A}_h x_h = f_h, \tag{10.1}$$

where the operator $\mathcal{A}_h$ is a self-adjoint operator from $X_h = \Lambda_h^{k-1} \times \Lambda_h^k \times \mathfrak{H}_h^k$ onto its dual X_h^* and $f_h \in X_h^*$ is given. In the discussion below, we will consider $X_h^* = X_h$ as a set, but with different norms and inner products.

If systems of the form (10.1) are solved by an iterative method, then the convergence of the iteration depends on the conditioning of the coefficient operator $\mathcal{A}_h$. In particular, the convergence rate can be bounded by the spectral condition number of the operator $\mathcal{A}_h$, cond$(\mathcal{A}_h)$, given by

$$\mathrm{cond}(\mathcal{A}_h) = \frac{\sup |\lambda|}{\inf |\lambda|}, \tag{10.2}$$

where the supremum and infimum are taken over the spectrum of $\mathcal{A}_h$. For linear operators arising as discretization of operators such as the Hodge Laplacian, this condition number will typically tend to infinity as the mesh parameter h tends to zero, leading to slow convergence of the iterative methods for fine triangulations $\mathcal{T}_h$. The standard way to overcome this problem is to introduce preconditioners.

10.1. Krylov space iterations

Consider a linear equation of the form

$$\mathcal{A}x = f, \tag{10.3}$$

where $f \in X$ is given, $x \in X$ is the unknown, and $\mathcal{A} \in \mathcal{L}(X, X)$ is a symmetric operator mapping a Hilbert space X into itself. Furthermore, assume that $\mathcal{A}$ is invertible with $\mathcal{A}^{-1} \in \mathcal{L}(X, X)$. Solutions of equations of the form (10.3) can be approximated by a Krylov space iteration. For each integer $m \geq 1$ we define $K_m = K_m(f)$ as the finite-dimensional subspace of X spanned by the elements $f, \mathcal{A}f, \ldots, \mathcal{A}^{m-1}f$. If $\mathcal{A}$ is positive definite then the approximations $x_m \in K_m$ can be generated by *the conjugate gradient method*, i.e., x_m is uniquely determined by

$$\langle \mathcal{A}x_m, y \rangle_X = \langle f, y \rangle_X, \quad y \in X,$$

or, equivalently, x_m is the minimizer of the quadratic functional

$$F(y) = \frac{1}{2}\langle Ay, y\rangle_X - \langle f, y\rangle_X$$

over K_m. Here, $\langle \cdot, \cdot \rangle_X$ is the inner product in X, and $\|\cdot\|_X$ is the corresponding norm. If the coefficient operator A is indefinite, we can instead use *the minimum residual method*, where $x_m \in K_m$ is uniquely characterized by

$$\|Ax_m - f\|_X = \inf_{y \in X} \|Ay - f\|_X.$$

These optimal characterizations of the approximations x_m ensure that the iterative method converges and that the reduction factor

$$\frac{\|x_m - x\|_X}{\|x_0 - x\|_X}$$

can be bounded, *a priori*, by a function only depending on the number of iterations, m, and the condition number of the coefficient operator A given by

$$\text{cond}(A) = \|A\|_{\mathcal{L}(X,X)} \cdot \|A^{-1}\|_{\mathcal{L}(X,X)}.$$

However, from a computational point of view, it is also important that the approximations x_m can be cheaply computed by a recurrence relation. Typically, only one or two evaluations of the operator A, in addition to a few calculations of the inner products $\langle \cdot, \cdot \rangle_X$, are required to compute x_m from x_{m-1}. For more details on Krylov space methods, we refer, for example, to Hackbusch (1994).

10.2. Block diagonal preconditioners

Our goal is to design effective iterations for the discrete Hodge–Laplace problem. However, we will motivate the approach by first studying the corresponding continuous problem. For simplicity we will assume throughout this subsection that there are no nontrivial harmonic k-forms, *i.e.*, $\mathfrak{H}^k = \{0\}$, since the inclusion of such a finite-dimensional space of multipliers is relatively insignificant for the design of preconditioners. Furthermore, the boundary conditions are taken to be $\text{Tr}\star u = 0$ and $\text{Tr}\star du = 0$.

The continuous problem

Recall that the Hodge–Laplace problem in mixed form is as follows. Find $(\sigma, u) \in H\Lambda^{k-1}(\Omega) \times H\Lambda^k(\Omega)$ such that

$$\begin{aligned}
\langle \sigma, \tau\rangle - \langle d\tau, u\rangle &= 0, & \tau \in H\Lambda^{k-1}(\Omega), \\
\langle d\sigma, v\rangle + \langle du, dv\rangle &= \langle f, v\rangle, & v \in H\Lambda^k(\Omega),
\end{aligned} \tag{10.4}$$

where $f \in L^2\Lambda^k(\Omega)$ is given. In order to align the description of this system with the framework above, we let $X = H\Lambda^{k-1}(\Omega) \times H\Lambda^k(\Omega)$, and $H = L^2\Lambda^{k-1}(\Omega) \times L^2\Lambda^k(\Omega)$. Note that any element $g = (g^{k-1}, g^k) \in H$ gives rise to an element of X^* by the identification

$$x = (\tau, v) \mapsto \langle g, x \rangle = \langle g^{k-1}, \tau \rangle + \langle g^k, v \rangle, \quad x = (\tau, v) \in X.$$

In fact, X^* can be identified with the completion of H in the dual norm

$$\|g\|_{X^*} = \sup_{x \in X, \|x\|_X = 1} \langle g, x \rangle.$$

In this way we obtain *a Gelfand triple*

$$X \subset H \subset X^*,$$

where X is dense in H and H is dense in X^*. Furthermore, the duality pairing between X^* and X can be defined as an extension of the H-inner product, still denoted $\langle \cdot, \cdot \rangle$.

The system (10.4) can now be formally written in the form

$$\mathcal{A}x = f, \tag{10.5}$$

where $x = (\sigma, -u) \in X$, $f \in X^*$ is the functional associated to $(0, f) \in H$ and $\mathcal{A} : X \to X^*$ is the operator

$$\mathcal{A} = \begin{pmatrix} I & \mathrm{d}^* \\ \mathrm{d} & -\mathrm{d}^*\mathrm{d} \end{pmatrix}.$$

Here $\mathrm{d}^* = \mathrm{d}_k^* : L^2\Lambda^{k+1}(\Omega) \to (H\Lambda^k(\Omega))^*$ is the adjoint of $\mathrm{d}_k : H\Lambda^k(\Omega) \to L^2\Lambda^{k+1}(\Omega)$, *i.e.*,

$$\langle \mathrm{d}^*\tau, \omega \rangle = \langle \tau, \mathrm{d}\omega \rangle, \quad \tau \in L^2\Lambda^{k+1}(\Omega), \ \omega \in H\Lambda^k(\Omega)$$

where, as above, $\langle \cdot, \cdot \rangle$ is the extension of the inner product on $L^2\Lambda^k(\Omega)$ to the duality pairing between $(H\Lambda^k(\Omega))^*$ and $H\Lambda^k(\Omega)$.

Note that the operator $\mathcal{A}$ is H-symmetric in the sense that if $x = (\sigma, u) \in X$ and $y = (\tau, v) \in X$, then

$$\langle \mathcal{A}x, y \rangle = \langle \sigma, \tau \rangle + \langle \mathrm{d}\sigma, v \rangle + \langle \mathrm{d}\tau, u \rangle - \langle \mathrm{d}u, \mathrm{d}v \rangle = \langle x, \mathcal{A}y \rangle.$$

On the other hand, the operator $\mathcal{A}$ is not positive definite in any sense.

It is a consequence of the discussion in Section 7 (see Theorem 7.2), that $\mathcal{A}$ defines an isomorphism between X and X^*, *i.e.*,

$$\mathcal{A} \in \mathcal{L}(X, X^*) \quad \text{and} \quad \mathcal{A}^{-1} \in \mathcal{L}(X^*, X).$$

However, since the operator $\mathcal{A}$ maps the Hilbert space X into a larger space X^*, a Krylov space iteration for the linear system (10.5) is not well-defined. Instead we will consider an equivalent preconditioned system of the form

$$\mathcal{B}\mathcal{A}x = \mathcal{B}f, \tag{10.6}$$

where $\mathcal{B} \in \mathcal{L}(X^*, X)$ is an isomorphism, *i.e.*, we also have $\mathcal{B}^{-1} \in \mathcal{L}(X, X^*)$.

Hence the new coefficient operator $\mathcal{B}\mathcal{A}$ is an element of $\mathcal{L}(X,X)$ with a bounded inverse. In order to preserve the symmetry of the system, we will also require that $\mathcal{B}$ is symmetric in the sense that

$$\langle \mathcal{B}f, g \rangle = \langle f, \mathcal{B}g \rangle,$$

and positive definite in the sense that there is a constant $C > 0$ such that

$$\langle \mathcal{B}f, f \rangle \geq C\|\mathcal{B}f\|_X, \quad f \in X^*.$$

If these conditions are satisfied, then the bilinear form $\langle \mathcal{B}^{-1} \cdot, \cdot \rangle$ defines a new inner product on X, and the coefficient operator $\mathcal{B}\mathcal{A}$ of (10.6) is symmetric in this inner product since

$$\langle \mathcal{B}^{-1}\mathcal{B}\mathcal{A}x, y \rangle = \langle \mathcal{A}x, y \rangle = \langle \mathcal{B}^{-1}x, \mathcal{B}\mathcal{A}y \rangle, \quad x, y \in X.$$

Therefore the preconditioned system (10.6) fulfils all the properties required such that the minimum residual method can be applied. But observe that in order to do so, the preconditioner $\mathcal{B}$ has to be evaluated at least once for each iteration.

One possible operator $\mathcal{B}$ with all the required properties for a preconditioner is the block diagonal operator operator given by

$$\begin{pmatrix} (I + \mathrm{d}^*\mathrm{d})^{-1} & 0 \\ 0 & (I + \mathrm{d}^*\mathrm{d})^{-1} \end{pmatrix} = \begin{pmatrix} (I + \mathrm{d}_{k-1}^*\mathrm{d}_{k-1})^{-1} & 0 \\ 0 & (I + \mathrm{d}_k^*\mathrm{d}_k)^{-1} \end{pmatrix}.$$

Here the operator $(I + \mathrm{d}^*\mathrm{d})^{-1} \in \mathcal{L}((H\Lambda^k(\Omega))^*, H\Lambda^k(\Omega))$ is the solution operator defined from the inner product on $H\Lambda^k(\Omega)$, i.e., $(I + \mathrm{d}^*\mathrm{d})^{-1}f = \omega$ if

$$\langle \omega, \tau \rangle_{H\Lambda} = \langle f, \tau \rangle, \quad \tau \in H\Lambda^k(\Omega).$$

Furthermore, any block diagonal operator of the form

$$\mathcal{B} = \begin{pmatrix} B_{k-1} & 0 \\ 0 & B_k \end{pmatrix},$$

where $B = B_k \in \mathcal{L}(H\Lambda^k(\Omega)^*, H\Lambda^k(\Omega))$ is a preconditioner for the operator $I + \mathrm{d}^*\mathrm{d}$, will also work. Here an operator $B \in \mathcal{L}(H\Lambda^k(\Omega)^*, H\Lambda^k(\Omega))$ is referred to as a preconditioner for $I + \mathrm{d}^*\mathrm{d}$ if it is an isomorphism which is symmetric and positive definite with respect to $\langle \cdot, \cdot \rangle$.

The discrete problem

The main reason for including the discussion of preconditioners for the continuous problem above was to use it to motivate the corresponding discussion for the discrete case. Of course, in real computations we need to apply preconditioners for the discrete problem. In fact, there exist several approaches to the design of preconditioners for mixed problems such as the discrete Hodge–Laplace problem, and not all of them are of the form discussed here. As an example, we mention the positive definite reformulation,

performed in Bramble and Pasciak (1988), and the use of triangular preconditioners, as discussed in Klawonn (1998). However, here, in complete
analogy with the discussion above, we restrict the attention to block diagonal preconditioners, as discussed, for example, in Rusten and Winther
(1992), Silvester and Wathen (1994), Arnold, Falk and Winther (1997a),
and Bramble and Pasciak (1997).

Recall that the discrete Hodge–Laplace problem takes the following form.
For a given $f \in L^2 \Lambda^k(\Omega)$, find $(\sigma_h, u_h) \in \Lambda_h^{k-1} \times \Lambda_h^k$ such that

$$\begin{aligned}
\langle \sigma_h, \tau \rangle - \langle d\tau, u_h \rangle &= 0, & \tau &\in \Lambda_h^{k-1}, \\
\langle d\sigma_h, v \rangle + \langle du_h, dv \rangle &= \langle f, v \rangle, & v &\in \Lambda_h^k.
\end{aligned}$$
(10.7)

We assume throughout this section that the discrete spaces Λ_h^{k-1} and Λ_h^k
are chosen such that they are part of one of the discrete de Rham complexes
introduced in Section 5 above.

Let $X_h = \Lambda_h^{k-1} \times \Lambda_h^k$ and define an operator $\mathcal{A}_h : X_h \to X_h^*$ by

$$\langle \mathcal{A}_h x, y \rangle = \langle \sigma, \tau \rangle + \langle d\sigma, v \rangle + \langle d\tau, u \rangle - \langle du, dv \rangle,$$

where $x = (\sigma, u)$ and $y = (\tau, v)$, and, as above, $\langle \cdot, \cdot \rangle$ is the inner product
on $H = L^2 \Lambda^{k-1}(\Omega) \times L^2 \Lambda^k(\Omega)$. The space X_h^* will be equal to X_h as a set,
but with norm given by

$$\|g\|_{X_h^*} = \sup_{x \in X_h, \|x\|_X = 1} \langle g, x \rangle.$$
(10.8)

With these definitions, the system (10.7) can be alternatively written as

$$\mathcal{A}_h x_h = f_h,$$
(10.9)

where $x_h = (\sigma_h, -u_h)$ and where $f_h \in X_h^*$ represents the data.

Under the present conditions, the operator $\mathcal{A}_h$ is invertible and symmetric
with respect to the inner product $\langle \cdot, \cdot \rangle$. Furthermore, since $\mathcal{A}_h$ maps X_h
into itself, a Krylov space iteration, like the minimum residual method, is
well-defined for the system (10.9). On the other hand, it can be shown that
the spectral condition number of $\mathcal{A}_h$, defined by (10.2), will grow asymptotically like h^{-2} as the mesh parameter h tends to zero. Therefore, such an
iteration will not be very efficient for fine triangulations, and it is also necessary in the discrete case to introduce a preconditioner in order to obtain
an efficient iteration.

If we let X_h inherit the norm of $X = H\Lambda^{k-1}(\Omega) \times H\Lambda^k(\Omega)$, and the
norm of X_h^* is given by (10.8), then it follows from results of Section 7
(see Theorem 7.3), that the norms

$$\|\mathcal{A}_h\|_{\mathcal{L}(X_h, X_h^*)} \quad \text{and} \quad \|\mathcal{A}_h^{-1}\|_{\mathcal{L}(X_h^*, X_h)}$$

are both bounded independently of the mesh parameter h. Therefore, if $\mathcal{B}_h$

is another operator on X_h such that the norms

$$\|\mathcal{B}_h\|_{\mathcal{L}(X_h^*,X_h)} \quad \text{and} \quad \|\mathcal{B}_h^{-1}\|_{\mathcal{L}(X_h,X_h^*)} \tag{10.10}$$

are bounded independently of h, then the spectral condition number of the composition $\mathcal{B}_h\mathcal{A}_h$ is bounded independently of h. As above, we can also argue that if the preconditioner $\mathcal{B}_h$ is symmetric and positive definite with respect to the inner product $\langle \cdot, \cdot \rangle$, then the operator $\mathcal{B}_h\mathcal{A}_h$ is symmetric with respect to the inner product $\langle \mathcal{B}_h^{-1} \cdot, \cdot \rangle$ on X_h. Hence, the preconditioned system

$$\mathcal{B}_h\mathcal{A}_h x_h = \mathcal{B}_h f_h,$$

fulfils all the necessary conditions required for applying the minimum residual method. Furthermore, since the condition numbers of the operators $\mathcal{B}_h\mathcal{A}_h$ are bounded independently of h, the convergence rate will not deteriorate as the mesh becomes finer.

As in the continuous case, it seems canonical to construct block diagonal preconditioners of the form

$$\mathcal{B}_h = \begin{pmatrix} B_{k-1,h} & 0 \\ 0 & B_{k,h} \end{pmatrix}, \tag{10.11}$$

where the operator $B_h = B_{k,h} : \Lambda_h^k \to \Lambda_h^k$ is constructed such that it is symmetric with respect to the L^2-inner product $\langle \cdot, \cdot \rangle$, and such that there exists constants $c_1, c_2 > 0$, independent of h, satisfying

$$c_1 \langle B_h^{-1}w, w \rangle \le \langle w, w \rangle_{H\Lambda} \le c_2 \langle B_h^{-1}w, w \rangle \quad w \in \Lambda_h^k. \tag{10.12}$$

If both operators $B_{k-1,h}$ and $B_{k,h}$ occurring in (10.11) satisfy such a spectral equivalence relation, the corresponding operator $\mathcal{B}_h$ will satisfy all requirements given above. In particular, the norms given in (10.10) will be bounded independently of h.

Finally, we observe that the preconditioner $\mathcal{B}_h$, and hence the operators $B_{k-1,h}$ and $B_{k,h}$, have to be evaluated at least once for each iteration. Therefore, the operators $B_{k-1,h}$ and $B_{k,h}$ have to be defined such that they can be evaluated effectively. The real challenge is to design preconditioners $\mathcal{B}_h$ which satisfy (10.12) and which, at the same time, can be evaluated cheaply. The construction of the operators $\mathcal{B}_h$ will be further discussed below.

10.3. Constructing effective preconditioners

Motivated by the development above, we will now discuss the design of preconditioners for the discrete operators $I + d_h^*d_h$, i.e., we will construct $B_h =: \Lambda_h^k \to \Lambda_h^k$ such that the spectral equivalence relation (10.12) holds and such that B_h can be evaluated cheaply. Before doing so, we note the correspondence between the operators $I + d^*d = I + d_k^*d_k$ and well-known

differential operators in the three-dimensional case, if the k-forms are identified with the associated proxy fields.

The operators $I + \mathrm{d}^*\mathrm{d}$

Consider the case $n = 3$, and identify k-forms by their associated proxy
fields, as described in Section 2.3 above. If we consider the case $k = 0$, then
0 forms correspond to real-valued functions and the inner product $\langle \omega, \mu \rangle_{H\Lambda}$
to the standard inner product of $H^1(\Omega)$ given by

$$\langle \omega, \mu \rangle + \langle \mathrm{grad}\, \omega, \mathrm{grad}\, \mu \rangle.$$

Furthermore, $I + \mathrm{d}_0^*\mathrm{d}_0$ corresponds to $I - \mathrm{div}\,\mathrm{grad} = I - \Delta$. Here Δ is the
standard Laplace operator on scalar functions, and $\langle \cdot, \cdot \rangle$ the inner product
in $L^2(\Omega)$. If $k = 1$ and $\omega, \mu \in H\Lambda^1(\Omega)$ are identified with vector fields, then
$\langle \omega, \mu \rangle_{H\Lambda}$ becomes

$$\langle \omega, \mu \rangle + \langle \mathrm{curl}\, \omega, \mathrm{curl}\, \mu \rangle,$$

and $I + \mathrm{d}_1^*\mathrm{d}_1$ is represented by $I + \mathrm{curl}\,\mathrm{curl}$. For $k = 2$, the inner product
corresponds to

$$\langle \omega, \mu \rangle + \langle \mathrm{div}\, \omega, \mathrm{div}\, \mu \rangle,$$

and $I + \mathrm{d}_2^*\mathrm{d}_2$ is represented by $I - \mathrm{grad}\,\mathrm{div}$. Finally, for $k = 3$ the inner
product is simply the L^2-inner product and $I + \mathrm{d}_3^*\mathrm{d}_3$ reduces to the identity
operator.

Note that $I + \mathrm{d}_0^*\mathrm{d}_0$ corresponds to a standard second-order elliptic differential operator. In fact, the modern preconditioning techniques built
on space decompositions, such as domain decomposition and multigrid algorithms, are tailored to operators of this form. On the other hand, the
differential operators corresponding to $I + \mathrm{d}_k^*\mathrm{d}_k$, for $k = 1$ and $k = 2$, are
not standard second-order elliptic operators. For example, the operator
$I + \mathrm{curl}\,\mathrm{curl}$ behaves like a second-order differential operator on curl fields,
but degenerates to the identity on gradient fields. The operator $I - \mathrm{grad}\,\mathrm{div}$
has similar properties. These properties just mirror the fact that the oper-
ator $I + \mathrm{d}_k^*\mathrm{d}_k$ is the identity operator on $\mathfrak{B}^k$, but acts like a second-order
differential operator on $\mathfrak{Z}^{k\perp}$.

It is well known that the simplest preconditioners constructed for discretizations of standard second-order elliptic operators will not work for the
operators $I + \mathrm{curl}\,\mathrm{curl}$ and $I - \mathrm{grad}\,\mathrm{div}$. In fact, it seems that a necessary
condition for the construction of optimal preconditioners is the existence
of proper Hodge decompositions for the discrete spaces. Early work on
this topic in two and three space dimensions can be found in Vassilevski
and Wang (1992), Hiptmair (1997, 1999b), Hiptmair and Toselli (2000) and
Arnold, Falk and Winther (1997b, 2000). Here we shall indicate how such
a construction can be carried out for general n and k using the spaces Λ_h^k
introduced above.

The discretization of the operator $I + \mathrm{d}^*\mathrm{d}$

As approximations of the differential operator $I + \mathrm{d}^*\mathrm{d}$, we define the operator $A_h : \Lambda_h^k \to (\Lambda_h^k)^* = \Lambda_h^k$ by

$$\langle A_h \omega, \tau \rangle = \langle \omega, \tau \rangle_{H\Lambda}, \quad \tau \in \Lambda_h^k. \tag{10.13}$$

The operator A_h is symmetric and positive definite with respect to the L^2-inner product $\langle \cdot, \cdot \rangle$. Note that the inequalities (10.12) are equivalent to the statement that the spectrum of the operator $A_h^{-1} B_h^{-1}$ is contained in the interval $[c_2^{-1}, c_1^{-1}]$. This is again equivalent to the property that the spectrum of $B_h A_h$ is contained in the interval $[c_1, c_2]$, or

$$c_1 \|\omega\|_{H\Lambda}^2 \leq \langle B_h A_h \omega, \omega \rangle_{H\Lambda} \leq c_2 \|\omega\|_{H\Lambda}^2, \quad \omega \in \Lambda_h^k.$$

A two-level preconditioner

Let $\mathcal{T}_h$ be the triangulation of Ω used to construct the finite element space Λ_h^k. Assume further that $\mathcal{T}_h$ is obtained from a refinement of a coarser triangulation $\mathcal{T}_{h_0}$, where the mesh parameter $h_0 > h$. If $\Lambda_{h_0}^k$ is the corresponding subspace of $H\Lambda^k(\Omega)$, constructed from the same polynomial functions as Λ_h^k, then $\Lambda_{h_0}^k \subset \Lambda_h^k$. In fact, we obtain a commuting diagram of the form

$$
\begin{array}{ccccccccc}
0 \to & \Lambda_h^0 & \xrightarrow{\mathrm{d}} & \Lambda_h^1 & \xrightarrow{\mathrm{d}} & \cdots & \xrightarrow{\mathrm{d}} & \Lambda_h^n & \to 0 \\
& \downarrow{\scriptstyle \Pi_{h_0}^0} & & \downarrow{\scriptstyle \Pi_{h_0}} & & & & \downarrow{\scriptstyle \Pi_{h_0}^n} & \\
0 \to & \Lambda_{h_0}^0 & \xrightarrow{\mathrm{d}} & \Lambda_{h_0}^1 & \xrightarrow{\mathrm{d}} & \cdots & \xrightarrow{\mathrm{d}} & \Lambda_{h_0}^n & \to 0,
\end{array}
$$

where the projection operators $\Pi_{h_0}^k$ are the canonical interpolation projections onto the spaces $\Lambda_{h_0}^k$.

The inverse of the discrete operator $A_{h_0} : \Lambda_{h_0}^k \to \Lambda_{h_0}^k$, defined by (10.13) with Λ_h^k replaced by $\Lambda_{h_0}^k$, will be used to define the two-level preconditioner B_h for the operator A_h. Furthermore, we will assume that the coarse mesh $\mathcal{T}_{h_0}$ is not too coarse, in the sense that there is a constant $C > 0$, independent of h, such that

$$h_0 \leq Ch. \tag{10.14}$$

The following two-level version of the dual estimate of Theorem 9.2, which is a generalization to n dimensions of Propositions 4.3 and 4.4 of Arnold, Falk and Winther (2000), will be useful below.

Theorem 10.1. Suppose that the domain is convex or, more generally, 1-regular (see (7.19)), and that (10.14) holds. Suppose further that $\omega \in \Lambda_h^k$ satisfies

$$\langle \omega, \tau \rangle_{H\Lambda} = 0, \quad \tau \in \Lambda_{h_0}^k$$

and has the discrete Hodge decomposition

$$\omega = \omega_{d,h} + \omega_{\delta,h} = d\sigma_h + \omega_{\delta,h},$$

where $\sigma_h \in \mathfrak{Z}_h^{(k-1)\perp}$, and $\omega_{\delta,h} \in \mathfrak{Z}_h^{k\perp}$. Then there is a constant c, independent of h, such that

$$\|\sigma_h\| \le ch\|\omega\| \quad \text{and} \quad \|\omega_{\delta,h}\| \le ch\|\omega\|_{H\Lambda}.$$

Proof. The orthogonality condition implies that

$$\langle d\sigma_h, d\mu \rangle = 0, \quad \mu \in \Lambda_{h_0}^{k-1}.$$

Introduce $u \in \mathfrak{Z}^{(k-1)\perp}$ as the solution of

$$\langle du, dv \rangle = \langle \sigma_h, v \rangle, \quad v \in \mathfrak{Z}^{(k-1)\perp}.$$

Then $du \in H\Lambda^k(\Omega) \cap \mathring{H}^*\Lambda^k(\Omega) = H^1\Lambda^k(\Omega)$, with $\delta du = \sigma_h$. Therefore,

$$\|\sigma_h\|^2 = \langle d\sigma_h, du \rangle = \langle d\sigma_h, (I - \tilde{\Pi}_{h_0}) du \rangle$$
$$\le ch\|d\sigma_h\|\|du\|_{H^1\Lambda} \le ch\|\omega\|\|\sigma_h\|,$$

which shows the first bound. For the second bound, we compare the discrete Hodge decomposition of ω, given above, with the corresponding continuous decomposition

$$\omega = d\sigma + \omega_\delta, \quad \sigma \in \mathfrak{Z}^{(k-1)\perp}, \ \omega_\delta \in \mathfrak{Z}^{k\perp}.$$

Then $\omega_\delta \in H^1\Lambda^k(\Omega)$ with $\|\omega_\delta\|_{H^1\Lambda} \le c\|\omega\|_{H\Lambda}$. Furthermore, from Theorem 9.2, we obtain

$$\|\omega_\delta\| \le Ch\|\omega\|_{H\Lambda}.$$

Since $d\omega_\delta = d\omega_{\delta,h}$, we get by Lemma 5.10 that

$$\|\omega_{\delta,h} - \omega_\delta\| \le Ch\|\omega_\delta\|_{H^1\Lambda} \le Ch\|\omega\|_{H\Lambda}. \qquad \Box$$

In order to simplify the notation, we will simply write A_0 instead of A_{h_0}, and Λ_0^k instead of $\Lambda_{h_0}^k$. Furthermore, we let $P_0 = P_{h_0} : H\Lambda^k(\Omega) \to \Lambda_0^k$ and $Q_0 = Q_{h_0} : L^2\Lambda^k(\Omega) \to \Lambda_0^k$ be the orthogonal projections with respect to the inner products $\langle \cdot, \cdot \rangle_{H\Lambda}$ and $\langle \cdot, \cdot \rangle$, respectively. With this notation it is straightforward to verify that

$$A_0 P_0 = Q_0 A_h.$$

In order to construct the preconditioner B_h, we will need a smoothing operator $R_h : \Lambda_h^k \to \Lambda_h^k$. (Note that in the context of preconditioners, a smoothing operator refers to an approximate inverse, typically computed by a classical iteration, and is not related to the mollification procedure discussed in Section 5.4.) Typically, the operator R_h will approximate A_h^{-1}, but with high accuracy only on a part of the space. More precisely, the operator

R_h is assumed to be symmetric and positive definite with respect to $\langle \cdot , \cdot \rangle$. In addition, we assume that there is a positive constant c_0, independent of h, such that

$$\langle R_h^{-1} w, w \rangle \le c_0 \| w \|_{H\Lambda}^2, \quad w \in (I - P_0)\Lambda_h^k, \tag{10.15}$$

and that R_h satisfies

$$\langle R_h A_h w, w \rangle_{H\Lambda} \le \| w \|_{H\Lambda}^2, \quad w \in \Lambda_h^k. \tag{10.16}$$

The first condition states that the smoothing operator R_h approximates A_h well for highly oscillating functions (orthogonal to the coarse space Λ_0^k), while the second condition states that R_h is properly scaled.

The two-level preconditioner $B_h : \Lambda_h^k \to \Lambda_h^k$ is defined as $w \mapsto B_h w = \sigma_3$ where $\sigma_1, \sigma_2, \sigma_3$ is defined by the following algorithm:

(i) $\sigma_1 = R_h w$,
(ii) $\sigma_2 = \sigma_1 - A_0^{-1} Q_0 (A_h \sigma_1 - w)$,
(iii) $\sigma_3 = \sigma_2 - R_h (A_h \sigma_2 - w)$.

In steps (i) and (iii), the smoothing operator R_h is evaluated, while in step (ii), we must find the solution $\mu_0 \in \Lambda_0^k$ of the discrete problem

$$\langle \mu_0, \tau \rangle_{H\Lambda} = \langle \sigma_1, \tau \rangle_{H\Lambda} - \langle w, \tau \rangle, \quad \tau \in \Lambda_0^k.$$

Therefore, in order to evaluate the preconditioner B_h, approximating A_h^{-1}, R_h is evaluated twice, and A_h and A_0^{-1} once.

Such a two-level algorithm as presented here may not be very efficient, since it is usually necessary to assume that condition (10.14) holds. As a consequence, the linear system associated to A_0 is not much smaller than the system associated to A_h. Therefore, in most practical computations a multi-level algorithm should be used. Such an algorithm is defined by using the algorithm above recursively, where the operator A_0^{-1} is replaced by the use of the two-level algorithm at the coarser level, and repeated until a sufficiently small system is obtained. However, for the theoretical discussion here, where the main purpose is to discuss how the discrete Hodge decomposition for the spaces Λ_h^k enters into the construction of the smoothing operators R_h, the two-level algorithm is sufficient. We will therefore restrict the discussion here to the two-level algorithm above.

Properties of the preconditioner B_h

In the rest of this section, only the discrete spaces will occur. For simplicity of notation, we will therefore drop the subscript h, and just write Λ^k, A, B, and R instead of Λ_h^k, A_h, B_h, and R_h. It is a simple computation to show that the two-level preconditioner B satisfies the identity

$$I - BA = (I - RA)(I - P_0)(I - RA). \tag{10.17}$$

Furthermore, all the eigenvalues of BA are contained in the interval $[1/c_0, 1]$, where c_0 is the positive constant appearing in (10.15). This is a consequence of the theorem below, which is a two-level version of a pioneering result first proved in Braess and Hackbusch (1983).

Theorem 10.2. If the assumptions (10.15) and (10.16) hold, then the two-level preconditioner $B : \Lambda^k \mapsto \Lambda^k$ defined above satisfies

$$0 \leq \langle (I - BA)\omega, \omega \rangle_{H\Lambda} \leq \left(1 - \frac{1}{c_0}\right) \|\omega\|_{H\Lambda}, \quad \omega \in \Lambda^k.$$

Proof. The left inequality follows since (10.17) implies that

$$\langle (I - BA)\omega, \omega \rangle_{H\Lambda} = \|(I - P_0)(I - RA)\omega\|_{H\Lambda}^2 \geq 0.$$

Let $\tau \in (I - P_0)\Lambda^k$. To obtain the right inequality, we note that it follows from (10.15) and the Cauchy–Schwarz inequality that

$$\langle RA\tau, \tau \rangle_{H\Lambda} \geq c_0^{-1}\|\tau\|_{H\Lambda}^2. \tag{10.18}$$

This follows since

$$\|\tau\|_{H\Lambda}^2 = \langle R^{-1/2}\tau, R^{1/2}A\tau \rangle$$

$$\leq \langle R^{-1}\tau, \tau \rangle^{1/2} \langle RA\tau, \tau \rangle_{H\Lambda}^{1/2}$$

$$\leq c_0 \|\tau\|_{H\Lambda} \langle RA\tau, \tau \rangle_{H\Lambda}^{1/2}.$$

Note that the operator $I - RA$ is symmetric with respect to the inner product $\langle \cdot, \cdot \rangle_{H\Lambda}$, and as a consequence of (10.16), positive semidefinite. Therefore, we obtain from (10.18) that

$$\|(I - RA)\tau\|_{H\Lambda}^2 = \|\tau\|_{H\Lambda}^2 - \langle (2I - RA)RA\tau, \tau \rangle_{H\Lambda}$$

$$\leq \|\tau\|_{H\Lambda}^2 - \langle RA\tau, \tau \rangle_{H\Lambda}$$

$$\leq \left(1 - \frac{1}{c_0}\right)\|\tau\|_{H\Lambda}^2.$$

This final inequality further implies that

$$\|(I - RA)(I - P_0)\omega\|_{H\Lambda} \leq \left(1 - \frac{1}{c_0}\right)\|\omega\|_{H\Lambda}, \quad \omega \in \Lambda^k.$$

In other words, the operator norm $\|(I - RA)(I - P_0)\|_{\mathcal{L}(H\Lambda^k(\Omega), H\Lambda^k(\Omega))}$ is bounded by $1 - 1/c_0$. Hence, the corresponding norm of the adjoint operator, $(I - P_0)(I - RA)$ admits the same bound, and this is equivalent to the lower bound. $\quad\square$

Schwarz smoothers

The main challenge in obtaining effective preconditioners for the discrete operator $A : \Lambda^k \mapsto \Lambda^k$ is the construction of the smoothing operator R. In fact, it is only in this part of the construction that the special properties of the operator A and the discrete spaces Λ^k will enter. We will consider so called Schwarz smoothers. These smoothers will be defined from a collection of subspaces $\{\Lambda_j^k\}_{j=1}^m$ of Λ^k. We will assume that

$$\Lambda^k = \sum_{j=1}^m \Lambda_j^k$$

in the sense that all elements $\omega \in \Lambda^k$ can be written as $\omega = \sum_j \omega_j$, where $\omega_j \in \Lambda_j^k$. The decomposition is not required to be unique. The spaces Λ_j^k should be of low dimension, independent of the triangulation $\mathcal{T}_h$. In fact, frequently they are taken to be generated by a single basis function.

From the family of subspaces $\{\Lambda_j^k\}$, we can construct two different smoothing operators, usually referred to as the multiplicative and the additive Schwarz smoother. The multiplicative Schwarz smoother is given by the following algorithm. Define $R\omega = \sigma_{2m}$, where $\sigma_0 = 0$,

$$\sigma_j = \sigma_{j-1} - P_j(\sigma_{j-1} - A^{-1}\omega), \qquad j = 1, 2, \ldots, m,$$
$$\sigma_j = \sigma_{j-1} - P_{2m+1-j}(\sigma_{j-1} - A^{-1}\omega), \quad j = m+1, m+2, \ldots, 2m.$$

Here P_j is the $H\Lambda$-orthogonal projection onto Λ_j^k. Note that if A_j is the representation of A on Λ_j^k, defined from (10.13) with Λ^k replaced by Λ_j^k, and Q_j is the L^2-orthogonal projection onto Λ_j^k, then $P_j A^{-1} = A_j^{-1} Q_j$. This shows that that in each step of the algorithm above, only the 'local operator' A_j has to be inverted.

The corresponding additive smoother is given as a suitable scaling of the operator $R_a = \sum_{j=1}^m P_j A^{-1}$. For simplicity, we will restrict the discussion here to the multiplicative Schwarz smoother. This smoothing operator will always satisfy the condition (10.16). In fact, it is straightforward to verify that

$$\langle (I - BA)\omega, \omega \rangle_{H\Lambda} = \langle (I - P_m) \cdots (I - P_1)\omega, (I - P_m) \cdots (I - P_1)\omega \rangle_{H\Lambda} \geq 0.$$

Furthermore, the first condition can be verified by applying the following result.

Theorem 10.3. Suppose that the subspaces $\{\Lambda_j^k\}$ of Λ^k satisfy the two conditions

$$\sum_{i=1}^m \sum_{j=1}^m |\langle \omega_i, \tau_j \rangle_{H\Lambda}| \leq a_1 \left(\sum_{i=1}^m \|\omega_i\|_{H\Lambda}^2 \right)^{1/2} \left(\sum_{j=1}^m \|\tau_j\|_{H\Lambda}^2 \right)^{1/2}, \qquad (10.19)$$

where $\omega_i \in \Lambda_i^k$ and $\tau_j \in \Lambda_j^k$, and

$$\inf_{\substack{\omega_i \in \Lambda_i^k \\ \omega = \sum \omega_i}} \sum_{i=1}^m \|\omega_i\|_{H\Lambda}^2 \leq a_2 \|\omega\|_{H\Lambda}^2, \quad \omega \in (I - P_0)\Lambda^k, \tag{10.20}$$

for some positive constants a_1 and a_2. Then the corresponding multiplicative Schwarz smoother satisfies condition (10.15) with $c_0 = a_1^2 a_2$.

The second condition here is frequently referred to as a *stable decomposition property*. This condition is in fact closely related to the additive Schwarz operator R_a by the relation

$$\langle R_a^{-1}\omega, \omega \rangle = \inf_{\substack{\omega_i \in \Lambda_i^k \\ \omega = \sum \omega_i}} \sum_{i=1}^m \|\omega_i\|_{H\Lambda}^2.$$

Furthermore, the first condition can be used to obtain the bound

$$\langle R_a\omega, \omega \rangle \leq a_1^2 \langle R\omega, \omega \rangle.$$

In fact, results of this type can be found in many places, *i.e.*, in Bramble (1993), Dryja and Widlund (1995), Smith, Bjørstad and Gropp (1996), Xu (1992): see also Section 3 of Arnold *et al.* (2000). We will therefore not give a proof here.

The choice of space decomposition
Let $\{\Omega_j\}$ be the subsets of Ω defined by

$$\Omega_j = \{x \in \Omega \mid \operatorname{supp}\omega \subset \bar{\Omega}_j, \, \omega \in \Lambda_j^k\}.$$

Note that if Λ_j^k is generated by a single basis function, then any point of Ω will only be contained in a finite number of the subdomains Ω_j, and the number of overlapping subdomains will not grow with h. The same property will hold if the spaces Λ_j^k are generated by a fixed number of neighbouring basis functions. Furthermore, the constant a_1 appearing in (10.19) can be bounded by the sum of the characteristic functions of the domains Ω_j. So for a locally defined Schwarz smoother, the constant a_1 will be *a priori* bounded independently of h.

The stable decomposition property, (10.20), is harder to fulfil. For the rest of this section, let Γ_i^k be the one-dimensional subspaces of Λ^k generated by each basis function. Then, for any $\omega \in \Lambda^k$ the decomposition

$$\omega = \sum_i \omega_i, \quad \omega_i \in \Gamma_i^k,$$

is unique. Furthermore, by using a scaling argument, we have

$$\sum_i \|\omega_i\|^2 \leq C_1 \|\omega\|^2 \quad \text{and} \quad \|d\omega_i\| \leq C_1 h^{-1} \|\omega_i\|, \tag{10.21}$$

for each $\omega \in \Lambda^k$, where the constant C_1 is independent of h.

Consider first the case $k = 0$ and assume that the hypotheses of Theorem 10.1 hold. It follows from the standard duality argument of finite element theory that for all $\omega \in (I - P_0)\Lambda^0$, which are orthogonal to constants, we have the estimate

$$\|\omega\| \le ch\|d\omega\|. \tag{10.22}$$

In fact, this result is also contained in Theorem 10.1. The property (10.22) reflects the fact that the elements of $(I - P_0)\Lambda^0$ are highly oscillating functions. However, estimate (10.21) gives

$$\sum_{i=1}^{m} \|\omega_i\|_{H\Lambda}^2 \le C_1 h^{-2}\|\omega\|^2,$$

and therefore (10.22) implies that (10.20) holds.

Consider next the case when $1 \le k < n$. Of course, in this case we can not expect the bound (10.22) to hold in general, since the L^2-norm and the $H\Lambda$-norm coincide on 3^k. Instead we have to rely on the estimates given in Theorem 10.1. Hence, it seems necessary that the discrete Hodge decomposition will enter the construction.

The smoothers introduced by Vassilevski and Wang (1992) in two space dimensions, and Hiptmair (1997) in three dimensions use the decomposition

$$\sum_i d\Gamma_i^{k-1} + \sum_j \Gamma_j^k \tag{10.23}$$

of Λ^k. This decomposition is stable. In order to see this let $\omega \in (I - P_0)\Lambda^k$ be decomposed as

$$\omega = d\sigma + \omega_\delta = \sum_i d\sigma_i + \sum_j \omega_{\delta,j},$$

where σ and ω_δ are as in Theorem 10.1, $\sigma_i \in \Gamma_i^{k-1}$ and $\omega_{\delta,j} \in \Gamma_j^k$. Using the estimates from (10.21) and Theorem 10.1, we now have

$$\sum_i \|d\sigma_i\|^2 + \sum_j \|\omega_{\delta,j}\|_{H\Lambda}^2 \le ch^{-2}(\|\sigma\|^2 + \|\omega_\delta\|^2)$$

$$\le c\|\omega\|_{H\Lambda}^2,$$

which shows (10.20).

As an alternative to the decomposition (10.23), which employs the basis functions of both the spaces Λ^{k-1} and Λ^{k-1}, the construction carried out in Arnold *et al.* (2000) only utilizes the basis functions of the space Λ^k. However, the subspaces will then not be one-dimensional, since one has to make sure that all the spaces $d\Gamma_i^{k-1}$ are contained in at least one subspace.

This will typically lead to the property that a low (< 10)-dimensional linear system has to be solved for each step of the Schwarz algorithm. We refer to Arnold *et al.* (2000) for further details.

11. The elasticity equations

11.1. Introduction

The equations of linear elasticity can be written as a system of equations of the form

$$A\sigma = \epsilon\, u, \qquad \operatorname{div} \sigma = f \quad \text{in } \Omega. \tag{11.1}$$

Here the unknowns σ and u denote the stress and displacement fields engendered by a body force f acting on a linearly elastic body which occupies a region $\Omega \subset \mathbb{R}^n$, with boundary $\partial\Omega$. Then σ takes values in the space $\mathbb{S} = \mathbb{R}^{n\times n}_{\mathrm{sym}}$ of symmetric second-order tensors and u takes values in $V = \mathbb{R}^n$. The differential operator ϵ is the symmetric part of the gradient, div denotes the divergence operator taking tensor fields to vector fields, and the fourth order compliance tensor $A = A(x) : \mathbb{S} \to \mathbb{S}$ is a bounded and symmetric, uniformly positive definite operator reflecting the properties of the material at each point. If the body is clamped on the boundary $\partial\Omega$ of Ω, then the proper boundary condition for the system (11.1) is $u = 0$ on $\partial\Omega$. For simplicity, this boundary condition will be assumed throughout most of the discussion here. However, there are issues that arise when other boundary conditions are assumed (*e.g.*, traction boundary conditions $\sigma n = 0$). The modifications needed to deal with such boundary conditions are discussed in Section 11.7.

The system (11.1) can be formulated weakly in a number of ways. One is to seek $\sigma \in H(\operatorname{div}, \Omega; \mathbb{S})$, the space of square-integrable symmetric tensor fields with square-integrable divergence, and $u \in L^2(\Omega; V)$, satisfying

$$\int_\Omega (A\sigma : \tau + \operatorname{div} \tau \cdot u)\, dx = 0, \qquad \tau \in H(\operatorname{div}, \Omega; \mathbb{S}),$$

$$\int_\Omega \operatorname{div} \sigma \cdot v\, dx = \int_\Omega fv\, dx, \quad v \in L^2(\Omega; V). \tag{11.2}$$

A second weak formulation, that enforces the symmetry weakly, seeks $\sigma \in H(\operatorname{div}, \Omega; \mathbb{M})$, $u \in L^2(\Omega; V)$, and $p \in L^2(\Omega; \mathbb{K})$ satisfying

$$\int_\Omega (A\sigma : \tau + \operatorname{div} \tau \cdot u + \tau : p)\, dx = 0, \qquad \tau \in H(\operatorname{div}, \Omega; \mathbb{M}),$$

$$\int_\Omega \operatorname{div} \sigma \cdot v\, dx = \int_\Omega fv\, dx, \quad v \in L^2(\Omega; V), \tag{11.3}$$

$$\int_\Omega \sigma : q\, dx = 0, \qquad q \in L^2(\Omega; \mathbb{K}),$$

where $\mathbb{M}$ is the space of second-order tensors, $\mathbb{K}$ is the subspace of skew-symmetric tensors, and the compliance tensor $A(x)$ is now considered as a symmetric and positive definite operator mapping $\mathbb{M}$ into $\mathbb{M}$. In the isotropic case, the mapping $\sigma \mapsto A\sigma$ has the form

$$A\sigma = \frac{1}{2\mu} \left(\sigma - \frac{\lambda}{2\mu + n\lambda} \operatorname{tr}(\sigma) I \right),$$

where $\lambda(x), \mu(x)$ are positive scalar coefficients, the Lamé coefficients.

Stable finite element discretizations with reasonable computational complexity based on the variational formulation (11.2) have proved very difficult to construct. In two space dimensions, the first stable finite elements with polynomial shape functions were presented in Arnold and Winther (2002). For the lowest-order element, the approximate stress space is composed of certain piecewise cubic functions, while the displacement space consists of piecewise linear functions. In three dimensions, a piecewise quartic stress space is constructed with 162 degrees of freedom on each tetrahedron (Adams and Cockburn 2005). Another approach which has been discussed in the two-dimensional case is the use of composite elements, in which the approximate displacement space consists of piecewise polynomials with respect to one triangulation of the domain, while the approximate stress space consists of piecewise polynomials with respect to a different, more refined, triangulation (Fraeijs de Veubeke 1965, Watwood and Hartz 1968, Johnson and Mercier 1978, Arnold, Douglas and Gupta 1984b).

Because of the lack of suitable mixed elasticity elements that strongly impose the symmetry of the stresses, a number of authors have developed approximation schemes based on the weak symmetry formulation (11.3): see Fraeijs de Veubeke (1965), Amara and Thomas (1979), Arnold, Brezzi and Douglas (1984a), Stenberg (1986, 1988a, 1988b), Arnold and Falk (1988), Morley (1989), Stein and Rolfes (1990) and Farhloul and Fortin (1997). Although (11.2) and (11.3) are equivalent on the continuous level, an approximation scheme based on (11.3) may not produce a symmetric approximation to the stress tensor, depending on the choices of finite element spaces.

In this section of the paper, we build on the techniques derived in the previous sections to develop and analyse finite element approximations of the equations of linear elasticity based on the mixed formulation (11.3) with weak symmetry. The basic ideas first appeared in Arnold, Falk and Winther (2006c) in two dimensions and Arnold, Falk and Winther (2005) in three dimensions.

In order to write (11.3) in the language of exterior calculus, we will use the spaces of vector-valued differential forms presented in Section 6. The domain Ω is a bounded open set in $\mathbb{R}^n$, $V = \mathbb{R}^n$ denotes its tangent space at any point, and $\mathbb{K} = V \wedge V$ the space of bivectors defined in Section 2.1, which is identified with the space of skew-symmetric linear operators $V \to V$. As explained in Section 6, it is natural to seek the stress σ in the space $\Lambda^{n-1}(\Omega; V)$, so that if Γ is an $(n-1)$-dimensional surface embedded in $\bar{\Omega}$, e.g., a portion of the boundary of a subdomain of Ω, then $\int_\Gamma \sigma$ is a vector representing force. The Hodge star operator then represents the stress by $\star\sigma \in \Lambda^1(\Omega; V)$, which means it defines a linear operator $V \to V$ (i.e., a second-order tensor or matrix) at every point: this is the classical representation of stress. In Proposition 6.1, we showed that the operator $S = S_{n-1} : \Lambda^{n-1}(\Omega; V) \to \Lambda^n(\Omega; \mathbb{K})$ defined in (6.2) corresponds (up to a factor of ± 2) to taking the skew-symmetric part of its argument. Thus the elasticity problem (11.3) becomes: find $(\sigma, u, p) \in H\Lambda^{n-1}(\Omega; V) \times L^2\Lambda^n(\Omega; V) \times L^2\Lambda^n(\Omega; \mathbb{K})$ such that

$$\langle A\sigma, \tau\rangle + \langle d\tau, u\rangle - \langle S\tau, p\rangle = 0, \qquad \tau \in H\Lambda^{n-1}(\Omega; V),$$
$$\langle d\sigma, v\rangle = \langle f, v\rangle, \quad v \in L^2\Lambda^n(\Omega; V), \qquad (11.4)$$
$$\langle S\sigma, q\rangle = 0, \qquad q \in L^2\Lambda^n(\Omega; \mathbb{K}).$$

This problem is well-posed in the sense that, for each $f \in L^2\Lambda^n(\Omega; V)$, there exists a unique solution $(\sigma, u, p) \in H\Lambda^{n-1}(\Omega; V) \times L^2\Lambda^n(\Omega; V) \times L^2\Lambda^n(\Omega; \mathbb{K})$, and the solution operator is a bounded operator

$$L^2\Lambda^n(\Omega; V) \to H\Lambda^{n-1}(\Omega; V) \times L^2\Lambda^n(\Omega; V) \times L^2\Lambda^n(\Omega; \mathbb{K}).$$

This will follow from the general theory of such saddle point problems (Brezzi 1974) once we establish two conditions:

(W1) $\|\tau\|_{H\Lambda}^2 \leq c_1\langle A\tau, \tau\rangle$ whenever $\tau \in H\Lambda^{n-1}(\Omega; V)$ satisfies $\langle d\tau, v\rangle = 0$
$\forall v \in L^2\Lambda^n(\Omega; V)$ and $\langle S\tau, q\rangle = 0 \ \forall q \in L^2\Lambda^n(\Omega; \mathbb{K})$,

(W2) for all nonzero $(v, q) \in L^2\Lambda^n(\Omega; V) \times L^2\Lambda^n(\Omega; \mathbb{K})$, there exists nonzero
$\tau \in H\Lambda^{n-1}(\Omega; V)$ with $\langle d\tau, v\rangle - \langle S\tau, q\rangle \geq c_2\|\tau\|_{H\Lambda}(\|v\| + \|q\|)$,

for some positive constants c_1 and c_2. The first condition is obvious (and does not even utilize the orthogonality of $S\tau$). However, the second condition is more subtle. We will verify it in Theorem 11.1 below.

We next consider a finite element discretizations of (11.4). For this, we choose families of finite-dimensional subspaces $\Lambda_h^{n-1}(V) \subset H\Lambda^{n-1}(\Omega; V)$, $\Lambda_h^n(V) \subset L^2\Lambda^n(\Omega; V)$, and $\Lambda_h^n(\mathbb{K}) \subset L^2\Lambda^n(\Omega; \mathbb{K})$, indexed by h, and seek the discrete solution $(\sigma_h, u_h, p_h) \in \Lambda_h^{n-1}(V) \times \Lambda_h^n(V) \times \Lambda_h^n(\mathbb{K})$ such that

$$\langle A\sigma_h, \tau\rangle + \langle d\tau, u_h\rangle - \langle S\tau, p_h\rangle = 0, \qquad \tau \in \Lambda_h^{n-1}(V),$$
$$\langle d\sigma_h, v\rangle = \langle f, v\rangle, \quad v \in \Lambda_h^n(V), \qquad (11.5)$$
$$\langle S\sigma_h, q\rangle = 0, \qquad q \in \Lambda_h^n(\mathbb{K}).$$

In analogy with the well-posedness of the problem (11.4), the stability of the saddle point system (11.5) will be ensured by the Brezzi stability conditions:

(S1) $\|\tau\|_{H\Lambda}^2 \le c_1(A\tau, \tau)$ whenever $\tau \in \Lambda_h^{n-1}(\mathbb{V})$ satisfies $\langle d\tau, v \rangle = 0$
$\forall v \in \Lambda_h^n(\mathbb{V})$ and $\langle S\tau, q \rangle = 0 \ \forall q \in \Lambda_h^n(\mathbb{K})$,

(S2) for all nonzero $(v, q) \in \Lambda_h^n(\mathbb{V}) \times \Lambda_h^n(\mathbb{K})$, there exists nonzero
$\tau \in \Lambda_h^{n-1}(\mathbb{V})$ with $\langle d\tau, v \rangle - \langle S\tau, q \rangle \ge c_2\|\tau\|_{H\Lambda}(\|v\| + \|q\|)$,

where now the constants c_1 and c_2 must be independent of h. The difficulty is, of course, to design finite element spaces satisfying these conditions.

Just as there is a close relation between the construction of stable mixed finite element methods for the approximation of the Hodge Laplacian and discrete versions of the de Rham complex, there is also a close relation between mixed finite elements for linear elasticity and discretization of an associated complex, the elasticity complex, which will be derived in the next subsection. The importance of this complex for the stability of discretizations of elasticity was first recognized in Arnold and Winther (2002), where mixed methods for elasticity in two space dimensions were discussed. It turns out that there is also a close, but nonobvious, connection between the elasticity complex and the de Rham complex. This connection is described in Eastwood (2000) and is related to a general construction given in Bernšteĭn, Gel'fand and Gel'fand (1975), called the BGG resolution (see also Čap, Slovák and Souček (2001)). In Arnold et al. (2006c) (two dimensions) and Arnold et al. (2005) (three dimensions), we developed a discrete version of the BGG construction, which allowed us to derive stable mixed finite elements for elasticity in a systematic manner based on the finite element subcomplexes of the de Rham complex described earlier. The resulting elements in both two and three space dimensions are simpler than any derived previously. For example, as we shall see, the simple choice of $\mathcal{P}_1\Lambda^{n-1}(\mathcal{T}_h; \mathbb{V})$ for stress, $\mathcal{P}_0\Lambda^n(\mathcal{T}_h; \mathbb{V})$ for displacement, and $\mathcal{P}_0\Lambda^n(\mathcal{T}_h; \mathbb{K})$ for the multiplier results in a stable discretization of the problem (11.5). In Figure 11.1, this element is depicted in two dimensions with a conventional finite element diagram that portrays the degrees of freedom on a triangle: for stress the first two moments of its trace on the edges, and for the displacement and multiplier, their integrals on the triangle (two components for displacement, one for the multiplier). Moreover, this element is the lowest order of a family of stable elements in n dimensions utilizing $\mathcal{P}_r\Lambda^{n-1}(\mathcal{T}_h; \mathbb{V})$ for stress, $\mathcal{P}_{r-1}\Lambda^n(\mathcal{T}_h; \mathbb{V})$ for displacement, and $\mathcal{P}_{r-1}\Lambda^n(\mathcal{T}_h; \mathbb{K})$ for the multiplier.

In this section we shall basically follow the approach of Arnold et al. (2006c) and Arnold et al. (2005), but with some simplifications, and in a manner which works in n dimensions. In Section 11.2, we show how an elasticity complex with weakly imposed symmetry can be derived from the de Rham complex. For the convenience of readers more familiar with the classical notation for elasticity, in Section 11.3 we specialize to the cases

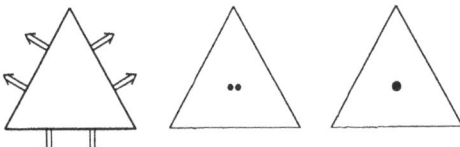

Figure 11.1. Approximation of stress, displacement, and multiplier for the simplest element in two dimensions.

$n = 2$ and $n = 3$ and translate our results from the language of exterior calculus to the classical notation. In Section 11.4, we give a proof of the well-posedness of the mixed formulation of elasticity with weak symmetry for the continuous problem, as a guide for establishing a similar result for the discrete problem. In Section 11.5, we derive discrete analogues of the elasticity complex obtained in Section 11.2, beginning from discrete analogues of the de Rham complex, and identify the required properties of the discrete spaces necessary for this construction. These results are then used to establish the main stability result for weakly symmetric mixed finite element approximations of the equations of elasticity in Section 11.6. In Sections 11.7 and 11.8, we show how our results can be extended to the equations of elasticity with traction boundary conditions and also to obtain some simplified elements.

11.2. From the de Rham complex to the elasticity complex

In this subsection we derive the elasticity complex in n dimensions starting with the de Rham complex. Our derivation is strongly influenced by the derivation given in Eastwood (2000) in three dimensions, although we have rearranged it substantially and, moreover, obtain a different complex which is the appropriate one when the symmetry of the stress is imposed weakly. The derivation is such that in Section 11.5 we are able to mimic it on the discrete level and so obtain finite element subcomplexes of the elasticity complex from corresponding finite element subcomplexes of the de Rham complex.

We start with the Cartesian product of two vector-valued de Rham complexes, as in (6.1), one with values in $\mathbb{V}$ and one with values in $\mathbb{K}$. Letting $\mathbb{W} := \mathbb{K} \times \mathbb{V}$ and writing $\Lambda^k(\mathbb{W})$ for the more cumbersome $\Lambda^k(\Omega; \mathbb{K}) \times \Lambda^k(\Omega; \mathbb{V})$, we have the starting complex:

$$0 \to \Lambda^0(\mathbb{W}) \xrightarrow{\left(\begin{smallmatrix} d & 0 \\ 0 & d \end{smallmatrix}\right)} \Lambda^1(\mathbb{W}) \xrightarrow{\left(\begin{smallmatrix} d & 0 \\ 0 & d \end{smallmatrix}\right)} \cdots \xrightarrow{\left(\begin{smallmatrix} d & 0 \\ 0 & d \end{smallmatrix}\right)} \Lambda^n(\mathbb{W}) \to 0. \tag{11.6}$$

Here the two d operators in the diagonal matrices represent $d \otimes \mathrm{id}_{\mathbb{K}}$ and $d \otimes \mathrm{id}_{\mathbb{V}}$, respectively. The elasticity complex will be realized as a subcomplex of an isomorphic image of this complex.

Define $K = K_k : \Lambda^k(\Omega; \mathbb{V}) \to \Lambda^k(\Omega; \mathbb{K})$ by

$$(K\omega)_x(v_1, \ldots, v_k) = X(x) \wedge \omega_x(v_1, \ldots, v_k).$$

where $X(x) \in \mathbb{V}$ is the vector corresponding to x as in Section 3.2. Set $S = \mathrm{d}K - K\mathrm{d}$, or, more precisely, $S_k = \mathrm{d}_{k+1}K_k - K_{k+1}\mathrm{d}_k : \Lambda^k(\Omega; \mathbb{V}) \to \Lambda^{k+1}(\Omega; \mathbb{K})$. Since $\mathrm{d}^2 = 0$, it follows that

$$\mathrm{d}S + S\mathrm{d} = 0. \tag{11.7}$$

Next, we define an isomorphism $\Phi = \Phi_k : \Lambda^k(\mathbb{W}) \to \Lambda^k(\mathbb{W})$ by

$$\Phi(\omega, \mu) = (\omega + K\mu, \mu),$$

with inverse given by

$$\Phi^{-1}(\omega, \mu) = (\omega - K\mu, \mu),$$

and an operator $\mathcal{A} = \mathcal{A}_k : \Lambda^k(\mathbb{W}) \to \Lambda^{k+1}(\mathbb{W})$ by $\mathcal{A} = \Phi\mathrm{d}\Phi^{-1}$. The operator $\mathcal{A}$ has a simple form. Using the definition of Φ, we obtain for $(\omega, \mu) \in \Lambda^k(\mathbb{W})$,

$$\mathcal{A}(\omega, \mu) = \Phi \circ \mathrm{d}(\omega - K\mu, \mu) = \Phi(\mathrm{d}\omega - \mathrm{d}K\mu, \mathrm{d}\mu) = (\mathrm{d}\omega - S\mu, \mathrm{d}\mu).$$

By construction, $\mathcal{A} \circ \mathcal{A} = 0$, and Φ is a cochain isomorphism from (11.6) to the complex

$$0 \to \Lambda^0(\mathbb{W}) \xrightarrow{\mathcal{A}} \Lambda^1(\mathbb{W}) \xrightarrow{\mathcal{A}} \cdots \xrightarrow{\mathcal{A}} \Lambda^n(\mathbb{W}) \to 0. \tag{11.8}$$

Using the definition of the exterior derivative, the definition of K, and the Leibniz rule, we obtain

$$(S_k\omega)_x(v_1, \ldots, v_{k+1}) = \sum_{j=1}^{k+1}(-1)^{j+1}v_j \wedge \omega_x(v_1, \ldots, \hat{v}_j, \ldots, v_{k+1}),$$

so S_k is precisely the operator defined in (6.2) of Section 6, applied pointwise. As shown in that section, the operator $S_{n-2} : \Lambda^{n-2}(\Omega; \mathbb{V}) \to \Lambda^{n-1}(\Omega; \mathbb{K})$ is an isomorphism. This will be crucial to the construction.

To proceed, we define

$$\Gamma^{n-2} = \{ (\omega, \mu) \in \Lambda^{n-2}(\mathbb{W}) \mid \mathrm{d}\omega = S\mu \},$$
$$\Gamma^{n-1} = \{ (\omega, \mu) \in \Lambda^{n-1}(\mathbb{W}) \mid \omega = 0 \},$$

with projections $\pi^{n-2} : \Lambda^{n-2}(\mathbb{W}) \to \Gamma^{n-2}$ and $\pi^{n-1} : \Lambda^{n-1}(\mathbb{W}) \to \Gamma^{n-1}$ given by

$$\pi^{n-2}(\omega, \mu) = (\omega, S_{n-2}^{-1}\mathrm{d}\omega), \quad \pi^{n-1}(\omega, \mu) = (0, \mu + \mathrm{d}S_{n-2}^{-1}\omega).$$

Using (11.7), it is straightforward to check that $\mathcal{A}$ maps $\Lambda^{n-2}(\mathbb{W})$ into Γ^{n-2} and Γ^{n-2} into Γ^{n-1}, so that

$$0 \to \Lambda^0(\mathbb{W}) \xrightarrow{\mathcal{A}} \cdots \xrightarrow{\mathcal{A}} \Lambda^{n-3}(\mathbb{W}) \xrightarrow{\mathcal{A}} \Gamma^{n-2} \xrightarrow{\mathcal{A}} \Gamma^{n-1} \xrightarrow{\mathcal{A}} \Lambda^n(\mathbb{W}) \to 0$$

is a subcomplex of (11.8). Moreover, the diagram

$$\cdots \xrightarrow{\mathcal{A}} \Lambda^{n-3}(\mathbb{W}) \xrightarrow{\mathcal{A}} \Lambda^{n-2}(\mathbb{W}) \xrightarrow{\mathcal{A}} \Lambda^{n-1}(\mathbb{W}) \xrightarrow{\mathcal{A}} \Lambda^{n}(\mathbb{W}) \to 0$$

$$\downarrow \mathrm{id} \qquad \downarrow \pi^{n-2} \qquad \downarrow \pi^{n-1} \qquad \downarrow \mathrm{id} \qquad (11.9)$$

$$\cdots \xrightarrow{\mathcal{A}} \Lambda^{n-3}(\mathbb{W}) \xrightarrow{\mathcal{A}} \Gamma^{n-2} \xrightarrow{\mathcal{A}} \Gamma^{n-1} \xrightarrow{\mathcal{A}} \Lambda^{n}(\mathbb{W}) \to 0$$

commutes, and as each of the vertical maps is a projection, they induce a surjective map on cohomology. Now we introduce the obvious isomorphisms

$$\Lambda^{n-2}(\Omega; \mathbb{K}) \cong \Gamma^{n-2}, \quad \omega \mapsto (\omega, S_{n-2}^{-1} d\omega),$$

and

$$\Lambda^{n-1}(\Omega; \mathbb{V}) \cong \Gamma^{n-1}, \quad \mu \mapsto (0, \mu).$$

Then the bottom row of (11.9) becomes

$$0 \to \Lambda^0(\mathbb{W}) \xrightarrow{\mathcal{A}} \cdots \xrightarrow{\mathcal{A}} \Lambda^{n-3}(\mathbb{W}) \xrightarrow{(d, -S_{n-3})} \Lambda^{n-2}(\Omega; \mathbb{K})$$

$$\xrightarrow{d \circ S_{n-2}^{-1} \circ d} \Lambda^{n-1}(\Omega; \mathbb{V}) \xrightarrow{(-S_{n-1}, d)^T} \Lambda^{n}(\mathbb{W}) \to 0. \quad (11.10)$$

We refer to the complex (11.10), or at least terms of degree $n - 3$ through n, as the *elasticity complex*. Since the highest-order de Rham cohomology group of Ω vanishes, it follows from the construction that the same holds true for the highest-order cohomology group of the elasticity complex. The same is true for the L^2 version of the elasticity complex, and this is what is needed to show the well-posedness of the problem (11.4). We show this in Section 11.4.

11.3. Connections to 2D and 3D elasticity complexes

In this subsection, we consider the special cases $n = 2$ and $n = 3$ and identify the elasticity complex with a complex consisting of spaces of scalar-, vector-, and tensor-valued proxy fields, *i.e.*, fields with values in $\mathbb{R}$, $\mathbb{V} = \mathbb{R}^n$, and $\mathbb{M} := \mathbb{V} \otimes \mathbb{V}$, and mappings which are differential operators.

Our tools for making the identifications are simple.

(1) Algebraic 1-forms may be identified with vectors via the Riesz map j induced by the inner product in $\mathbb{V}$. In this way $\mathrm{Alt}^1 \mathbb{V} \cong \mathbb{V}$ and $\mathrm{Alt}^1(\mathbb{V}; \mathbb{V}) \cong \mathbb{M}$.

(2) In 2D the Hodge star operation provides isomorphisms $\mathrm{Alt}^2 \mathbb{V} \xrightarrow{\star} \mathbb{R}$ and $\mathrm{Alt}^1 \mathbb{V} \xrightarrow{\star} \mathrm{Alt}^1 \mathbb{V}$. In 3D it provides isomorphisms $\mathrm{Alt}^3 \mathbb{V} \xrightarrow{\star} \mathbb{R}$ and $\mathrm{Alt}^2 \mathbb{V} \xrightarrow{\star} \mathrm{Alt}^1 \mathbb{V}$.

(3) In 2D the Hodge star operation for multivectors (which in this subsection we denote $\circledast$) provides an isomorphism from $\mathbb{K} \xrightarrow{\circledast} \mathbb{R}$. In 3D the isomorphism is from $\mathbb{K} \xrightarrow{\circledast} \mathbb{V}$.

If we use a positively oriented orthonormal basis e_i and dual basis x_i, these isomorphisms are as follows.

(1) $dx_i \overset{j}{\mapsto} e_i$ and $e_j dx_i \overset{j}{\mapsto} e_i \otimes e_j$.

(2) 2D: $dx_1 \wedge dx_2 \overset{\star}{\mapsto} 1$ and $dx_1 \overset{\star}{\mapsto} dx_2$, $dx_2 \overset{\star}{\mapsto} -dx_1$;
   3D: $dx_1 \wedge dx_2 \wedge dx_3 \overset{\star}{\mapsto} 1$ and $dx_1 \wedge dx_2 \overset{\star}{\mapsto} dx_3$, $dx_1 \wedge dx_3 \overset{\star}{\mapsto} -dx_2$, $dx_2 \wedge dx_3 \overset{\star}{\mapsto} dx_1$.

(3) 2D: $e_1 \wedge e_2 \overset{\circledast}{\mapsto} 1$ 3D: $e_1 \wedge e_2 \overset{\circledast}{\mapsto} e_3$, etc.

As an example, we compute the operator $\mathbb{V} \to \mathbb{V}$ in two dimensions which corresponds to the map $S_0 : \mathbb{V} \to \mathrm{Alt}^1(\mathbb{V}; \mathbb{K})$ after identifying the last space with $\mathbb{V}$ via $\circledast$, $\star$, and j. Now

$$(S_0 e_1)(v) = v \wedge e_1 = [dx_1(v)e_1 + dx_2(v)e_2] \wedge e_1 = -e_1 \wedge e_2 \, dx_2(v),$$

so $S_0 e_1 = -e_1 \wedge e_2 \, dx_2$. Thus, after the identifications, we compute the image of e_1 as

$$e_1 \overset{S_0}{\longrightarrow} -e_1 \wedge e_2 \, dx_2 \overset{\circledast}{\longrightarrow} -dx_2 \overset{\star}{\mapsto} dx_1 \overset{j}{\mapsto} e_1.$$

Similarly $e_2 \mapsto e_2$. Thus, modulo these isomorphisms, the map S_0 is simply the identify $\mathbb{V} \to \mathbb{V}$.

The elasticity complex in two dimensions
In the case $n = 2$, the elasticity complex is

$$0 \to \Lambda^0(\Omega; \mathbb{K}) \xrightarrow{d_0 \circ S_0^{-1} \circ d_0} \Lambda^1(\Omega; \mathbb{V}) \xrightarrow{(-S_1, d_1)^T} \Lambda^2(\Omega; \mathbb{K}) \times \Lambda^2(\Omega; \mathbb{V}) \to 0.$$

Using the listed isomorphisms, we can convert the spaces of differential forms to function spaces:

$$\Lambda^0(\Omega; \mathbb{K}) \overset{\circledast}{\to} \Lambda^0(\Omega) = C^\infty(\Omega),$$

$$\Lambda^1(\Omega; \mathbb{V}) \overset{\star}{\to} \Lambda^1(\Omega; \mathbb{V}) \overset{j}{\to} C^\infty(\Omega; \mathbb{M}),$$

$$\Lambda^2(\Omega; \mathbb{K}) \overset{\star}{\to} \Lambda^0(\Omega; \mathbb{K}) = C^\infty(\Omega; \mathbb{K}),$$

$$\Lambda^2(\Omega; \mathbb{V}) \overset{\star}{\to} \Lambda^0(\Omega; \mathbb{V}) = C^\infty(\Omega; \mathbb{V}).$$

In this way, the elasticity complex becomes

$$0 \to C^\infty(\Omega) \overset{J}{\to} C^\infty(\Omega; \mathbb{M}) \xrightarrow{(\mathrm{skw}, \mathrm{div})^T} C^\infty(\Omega; \mathbb{K}) \times C^\infty(\Omega; \mathbb{V}) \to 0.$$

Indeed, we have already seen that the operator S_1 corresponds to a constant multiple of skw, and that the operator d_1 corresponds to the tensor divergence (*i.e.*, the map $v \otimes w \mapsto (\mathrm{div}\, v) \otimes w$). So we need only clarify the operator J which corresponds to $d_0 \circ S_0^{-1} \circ d_0$. Identifying $\Lambda^1(\Omega; \mathbb{K})$ with $C^\infty(\Omega; \mathbb{V})$ via $\circledast$, $\star$, and j, we find that $d_0 : \Lambda^0(\Omega; \mathbb{K}) \to \Lambda^1(\Omega; \mathbb{K})$

corresponds, as usual, to the operator curl : $C^\infty(\Omega) \to C^\infty(\Omega; \mathbb{V})$ and the operator $d_0 : \Lambda^0(\Omega; \mathbb{V}) \to \Lambda^1(\Omega; \mathbb{V})$ to the corresponding operator on vectors curl : $C^\infty(\Omega; \mathbb{V}) \to C^\infty(\Omega; \mathbb{M})$ given by curl$(\phi e_i) = (\text{curl } \phi) \otimes e_i$. Also, we have seen, modulo these identifications, that S_0 corresponds to the identity operator on $C^\infty(\Omega; \mathbb{V})$. Thus we conclude that the operator J corresponding to $d_0 \circ S_0^{-1} \circ d_0$ is curl curl : $C^\infty(\Omega) \to C^\infty(\Omega; \mathbb{M})$. Written in terms of the usual basis on $\mathbb{V}$, it is given by

$$J\phi = \begin{pmatrix} \dfrac{\partial^2 \phi}{\partial x_2{}^2} & -\dfrac{\partial^2 \phi}{\partial x_1 \partial x_2} \\ -\dfrac{\partial^2 \phi}{\partial x_1 \partial x_2} & \dfrac{\partial^2 \phi}{\partial x_1{}^2} \end{pmatrix}.$$

The elasticity complex in three dimensions
When $n = 3$, the elasticity complex (11.10) is

$$0 \to \Lambda^0(\Omega; \mathbb{K}) \times \Lambda^0(\Omega; \mathbb{V}) \xrightarrow{(d_0, -S_0)} \Lambda^1(\Omega; \mathbb{K})$$

$$\xrightarrow{d_1 \circ S_1^{-1} \circ d_1} \Lambda^2(\Omega; \mathbb{V}) \xrightarrow{(-S_2, d)^T} \Lambda^3(\Omega; \mathbb{K}) \times \Lambda^3(\Omega; \mathbb{V}) \to 0. \quad (11.11)$$

We shall show that this corresponds to

$$0 \to C^\infty(\Omega; \mathbb{V}) \times C^\infty(\Omega; \mathbb{K}) \xrightarrow{(\text{grad}, i)} C^\infty(\Omega; \mathbb{M})$$

$$\xrightarrow{J} C^\infty(\Omega; \mathbb{M}) \xrightarrow{(\text{skw}, \text{div})^T} C^\infty(\Omega; \mathbb{K}) \times C^\infty(\Omega; \mathbb{V}) \to 0, \quad (11.12)$$

where the operator i represents the inclusion of $\mathbb{K}$ into $\mathbb{M}$, and the operator J is a second-order differential operator which will be defined below. More precisely, if the spaces in (11.11) are mapped isomorphically onto the corresponding spaces in (11.12) using the three classes of isomorphisms listed at the start of this subsection, then the maps in (11.11) correspond via composition to the maps shown in (11.12), up to nonzero constant factors.

The correspondence between the last parts of the sequences

$$\Lambda^2(\Omega; \mathbb{V}) \xrightarrow{(-S_2, d_2)^T} \Lambda^3(\Omega; \mathbb{K}) \times \Lambda^3(\Omega; \mathbb{V})$$

and

$$C^\infty(\Omega; \mathbb{M}) \xrightarrow{(\text{skw}, \text{div})^T} C^\infty(\Omega; \mathbb{K}) \times C^\infty(\Omega; \mathbb{V})$$

is straightforward. We have already seen that S_2 corresponds to a multiple of skw and that d_2 corresponds to the tensor divergence.

The correspondence between the first parts of the sequences

$$\Lambda^0(\Omega; \mathbb{K}) \times \Lambda^0(\Omega; \mathbb{V}) \xrightarrow{(d_0, -S_0)} \Lambda^1(\Omega; \mathbb{K})$$

and

$$C^\infty(\Omega; \mathbb{V}) \times C^\infty(\Omega; \mathbb{K}) \xrightarrow{(\text{grad}, i)} C^\infty(\Omega; \mathbb{M})$$

is similar. It is easy to see that d_0 corresponds to the vector gradient, and we can follow the basis elements through the mappings to see that S_0 corresponds to the inclusion i. For example, on the basis element $e_1 \wedge e_2 = e_1 \otimes e_2 - e_2 \otimes e_1$, we get

$$e_1 \wedge e_2 \overset{\circledast}{\mapsto} e_3 \overset{S_0}{\mapsto} e_1 \wedge e_3 \, dx_1 + e_2 \wedge e_3 \, dx_2 \overset{\circledast}{\mapsto} -e_2 \, dx_1 + e_1 \, dx_2 \overset{j}{\mapsto} e_1 \otimes e_2 - e_2 \otimes e_1.$$

In order to identify the operator J corresponding to $d_1 \circ S_1^{-1} \circ d_1$, consider the diagram

$$
\begin{array}{ccccccc}
\Lambda^1(\Omega; \mathbb{K}) & \overset{d_1}{\longrightarrow} & \Lambda^2(\Omega; \mathbb{K}) & \overset{S_1^{-1}}{\longrightarrow} & \Lambda^1(\Omega; V) & \overset{d_1}{\longrightarrow} & \Lambda^2(\Omega; V) \\
{\scriptstyle j\circledast}\big\downarrow & & {\scriptstyle j\star\circledast}\big\downarrow & & {\scriptstyle j}\big\downarrow & & {\scriptstyle j\star}\big\downarrow \\
C^\infty(\Omega; \mathbb{M}) & \overset{\mathrm{curl}}{\longrightarrow} & C^\infty(\Omega; \mathbb{M}) & \overset{\Xi^{-1}}{\longrightarrow} & C^\infty(\Omega; \mathbb{M}) & \overset{\mathrm{curl}}{\longrightarrow} & C^\infty(\Omega; \mathbb{M}).
\end{array}
$$

We have indicated the operators corresponding to the two occurrences of d_1. As is easy to check, they are both occurrences of the tensor curl, $f \otimes e_i \mapsto \mathrm{curl}\, f \otimes e_i$ for any smooth vector field f. We have also denoted the operator corresponding to S_1, namely $j \star \circledast S_1 j^{-1}$, by Ξ. We now compute Ξ using a basis. Since

$$S_1(e_1 \, dx_1)(v, w) = v \wedge e_1 \, dx_1(w) - w \wedge e_1 \, dx_1(v)$$

$$= e_2 \wedge e_1 \, dx_2(v) dx_1(w) + e_3 \wedge e_1 \, dx_3(v) dx_1(w)$$

$$- e_2 \wedge e_1 \, dx_2(w) dx_1(v) - e_3 \wedge e_1 \, dx_3(w) dx_1(v)$$

$$= (e_1 \wedge e_2 \, dx_1 \wedge dx_2 + e_1 \wedge e_3 \, dx_1 \wedge dx_3)(v, w),$$

we have

$$S_1(e_1 \, dx_1) = e_1 \wedge e_2 \, dx_1 \wedge dx_2 + e_1 \wedge e_3 \, dx_1 \wedge dx_3.$$

A similar computation gives

$$S_1(e_1 \, dx_2) = e_1 \wedge e_3 \, dx_2 \wedge dx_3.$$

Thus $\Xi(e_1 \otimes e_1)$ is the composition

$$e_1 \otimes e_1 \overset{j^{-1}}{\longmapsto} e_1 \, dx_1 \overset{S_1}{\longmapsto} e_1 \wedge e_2 \, dx_1 \wedge dx_2 + e_1 \wedge e_3 \, dx_1 \wedge dx_3$$

$$\overset{\circledast\star}{\longmapsto} e_3 \, dx_3 + e_2 \, dx_2 \overset{j}{\mapsto} e_2 \otimes e_2 + e_3 \otimes e_3,$$

and $\Xi(e_2 \otimes e_1) = -e_1 \otimes e_2$. Since similar expressions apply to the other basis functions, we find that

$$\Xi(e_i \otimes e_j) = \delta_{ij} \sum_k e_k \otimes e_k - e_j \otimes e_i$$

for all i, j, or, equivalently, $\Xi F = \mathrm{tr}(F)I - F^T$ for $F \in \mathbb{M}$. We then have $\Xi^{-1} F = (1/2) \mathrm{tr}(F)I - F^T$, and

$$JF = \mathrm{curl}(\Xi \, \mathrm{curl}\, F), \quad F \in C^\infty(\Omega; \mathbb{M}).$$

It is worth remarking that if $F = F^T$, then $JF = \mathrm{curl}(\mathrm{curl}\, F)^T$, and if $F = -F^T$ then $JF = 0$.

There are also elasticity complexes corresponding to the case of strongly imposed symmetry. In two dimensions, this complex takes the form

$$0 \to C^\infty \xrightarrow{J} C^\infty(\Omega; \mathbb{S}) \xrightarrow{\mathrm{div}} C^\infty(\Omega; \mathbb{V}) \to 0, \qquad (11.13)$$

where $\mathbb{S} \subset \mathbb{M}$ is the space of symmetric tensors. The complex (11.13) can be obtained from the complex (11.13) by performing a projection step. To see this, consider the diagram

$$
\begin{array}{ccccccc}
0 \to C^\infty(\Omega) & \xrightarrow{J} & C^\infty(\Omega; \mathbb{M}) & \xrightarrow{(\mathrm{skw,div})^T} & C^\infty(\Omega; \mathbb{K}) \times C^\infty(\Omega; \mathbb{V}) & \to & 0 \\
\downarrow{\scriptstyle \mathrm{id}} & & \downarrow{\scriptstyle \mathrm{sym}} & & \downarrow{\scriptstyle \pi} & & \\
0 \to C^\infty(\Omega) & \xrightarrow{J} & C^\infty(\Omega; \mathbb{S}) & \xrightarrow{\mathrm{div}} & C^\infty(\Omega; \mathbb{V}) & & \to 0,
\end{array}
$$

$\pi(q, u) = u - \mathrm{div}\, q$. The vertical maps are projections onto subspaces and the diagram commutes, so define a cochain projection, and therefore induce a surjection on cohomology. The connection between the two versions of the elasticity complex is explored in more detail in Arnold et $al.$ (2006c), but will not be pursued here.

The corresponding elasticity complex in three dimensions with strongly imposed symmetry of the stress tensor is given by

$$0 \to C^\infty(\Omega; \mathbb{V}) \xrightarrow{\epsilon} C^\infty(\Omega; \mathbb{S}) \xrightarrow{J} C^\infty(\Omega; \mathbb{S}) \xrightarrow{\mathrm{div}} C^\infty(\Omega; \mathbb{V}) \to 0, \quad (11.14)$$

where $\epsilon\, u$ is the symmetric part of $\mathrm{grad}\, u$ for a vector field u. If we were to follow the program outlined previously for mixed methods for the Poisson equation, the construction of stable mixed finite elements for elasticity for strong symmetry would be based on extending the sequence (11.13) to a complete commuting diagram of the form

$$
\begin{array}{ccccccc}
0 \to C^\infty(\Omega) & \xrightarrow{J} & C^\infty(\Omega; \mathbb{S}) & \xrightarrow{\mathrm{div}} & C^\infty(\Omega; \mathbb{V}) & \to & 0 \\
\downarrow{\scriptstyle \pi_h} & & \downarrow{\scriptstyle \pi_h} & & \downarrow{\scriptstyle \pi_h} & & \qquad (11.15) \\
0 \to W_h & \xrightarrow{J} & \Sigma_h & \xrightarrow{\mathrm{div}} & V_h & \to & 0,
\end{array}
$$

where $W_h \subset H^2(\Omega)$, $\Sigma_h \subset H(\mathrm{div}, \Omega; \mathbb{S})$ and $V_h \subset L^2(\Omega; \mathbb{V})$ are suitable finite element spaces and the π_h^0 are corresponding projection operators defining a cochain projection. This is exactly the construction performed in Arnold and Winther (2002) in two dimensions. In particular, since the finite element space W_h is required to be a subspace of $H^2(\Omega)$ (the natural domain of J), we can conclude that the finite element space W_h must contain quintic polynomials, and therefore the lowest-order space Σ_h will at least involve piecewise cubics. In fact, for the lowest-order elements discussed in Arnold and Winther (2002), W_h is the classical Argyris space, while Σ_h consists of piecewise cubic symmetric tensor fields with a linear divergence.

The analogous approach in three dimensions would be based on developing finite element spaces approximating the spaces in the complex (11.14) and embedding (11.14) as the top row of a commuting diagram analogous to (11.15), with a corresponding finite element complex as the bottom row. However, as mentioned previously, when the symmetry constraint is enforced pointwise on the discrete stress space, this construction leads to intricate finite elements of high-order. In this paper, we instead pursue an approach based on the weak symmetry formulation.

11.4. Well-posedness of the continuous problem

As discussed in Section 11.1, to establish well-posedness of the elasticity problem with weakly imposed symmetry (11.4), it suffices to verify condition (W2) of that subsection. This follows from the following theorem, which says that the map

$$H\Lambda^{n-1}(\Omega; V) \xrightarrow{(-S, \mathrm{d})^T} H\Lambda^n(\Omega; \mathbb{K}) \times H\Lambda^n(\Omega; V)$$

is surjective, *i.e.*, that the highest-order cohomology of the L^2 elasticity sequence vanishes. We spell out the proof in detail, since it will give us guidance as we construct stable discretizations.

Theorem 11.1. Given $(\omega, \mu) \in L^2\Lambda^n(\Omega; \mathbb{K}) \times L^2\Lambda^n(\Omega; V)$, there exists $\sigma \in H\Lambda^{n-1}(\Omega; V)$ such that $\mathrm{d}\sigma = \mu$, $-S_{n-1}\sigma = \omega$. Moreover, we may choose σ so that

$$\|\sigma\|_{H\Lambda} \leq c(\|\omega\| + \|\mu\|),$$

for a fixed constant c.

Proof. The second sentence follows from the first by Banach's theorem, so we need only prove the first.

(1) By Theorem 2.4, we can find $\eta \in H^1\Lambda^{n-1}(\Omega; V)$ with $\mathrm{d}\eta = \mu$.

(2) Since $\omega + S_{n-1}\eta \in H\Lambda^n(\Omega; K)$, we can apply Theorem 2.4 a second time to find $\tau \in H^1\Lambda^{n-1}(\Omega; \mathbb{K})$ with $\mathrm{d}\tau = \omega + S_{n-1}\eta$.

(3) Since S_{n-2} is an isomorphism from $\mathrm{Alt}^{n-2}(V; V)$ to $\mathrm{Alt}^{n-1}(V; \mathbb{K})$, when applied pointwise, it gives an isomorphism of the space $H^1\Lambda^{n-2}(\Omega; V)$ onto $H^1\Lambda^{n-1}(\Omega; \mathbb{K})$, and so we have $\rho \in H^1\Lambda^{n-2}(\Omega; V)$ with $S_{n-2}\rho = \tau$.

(4) Define $\sigma = \mathrm{d}\rho + \eta \in H\Lambda^{n-1}(\Omega; V)$.

(5) From steps (1) and (4), it is immediate that $d\sigma = \mu$.

(6) From (4), $-S_{n-1}\sigma = -S_{n-1}d\rho - S_{n-1}\eta$. But, since $dS = -Sd$,

$$-S_{n-1}d\rho = dS_{n-2}\rho = d\tau = \omega + S_{n-1}\eta,$$

so $-S_{n-1}\sigma = \omega$.   □

We note a few points from the proof. First, although the elasticity problem (11.4) only involves the three spaces $H\Lambda^{n-1}(\Omega; \mathbb{V})$, $L^2\Lambda^n(\Omega; \mathbb{V})$, and $L^2\Lambda^n(\Omega; \mathbb{K})$, the proof brings in two additional spaces from the BGG construction: $H\Lambda^{n-2}(\Omega; \mathbb{V})$ and $H\Lambda^{n-1}(\Omega; \mathbb{K})$. Also, although S_{n-1} is the only S operator arising in the formulation, S_{n-2} plays a role in the proof. Note, however, that we do not fully use the fact that S_{n-2} is an isomorphism from $\mathrm{Alt}^{n-2}(\mathbb{V}; \mathbb{V})$ to $\mathrm{Alt}^{n-1}(\mathbb{V}; \mathbb{K})$, only the fact that it is a surjection. This will prove important in the next subsection, when we construct a discrete elasticity complex.

11.5. A discrete elasticity complex

In this subsection, we derive a discrete version of the elasticity sequence by adapting the construction of Section 11.2. To carry out the construction, we will use two discretizations of the de Rham sequence, in general different, one to discretize the $\mathbb{K}$-valued de Rham sequence and one the $\mathbb{V}$-valued de Rham sequence. For $k = 0, 1, \ldots, n$, let Λ_h^k define a finite-dimensional subcomplex of the L^2 de Rham complex with an associated cochain projection $\Pi_h^k : \Lambda^k \to \Lambda_h^k$. Thus the following diagram commutes:

$$
\begin{array}{ccccccccc}
0 \to \Lambda^0 & \xrightarrow{\mathrm{d}} & \cdots & \xrightarrow{\mathrm{d}} & \Lambda^{n-1} & \xrightarrow{\mathrm{d}} & \Lambda^n & \to 0 \\
& & & & & & & & \\
\downarrow{\Pi_h} & & & & \downarrow{\Pi_h} & & \downarrow{\Pi_h} & & \\
& & & & & & & & \\
0 \to \Lambda_h^0 & \xrightarrow{\mathrm{d}} & \cdots & \xrightarrow{\mathrm{d}} & \Lambda_h^{n-1} & \xrightarrow{\mathrm{d}} & \Lambda_h^n & \to 0.
\end{array}
\tag{11.16}
$$

We do not make a specific choice of the discrete spaces yet, but, as shown in Section 5, there exist many such complexes based on the spaces $\mathcal{P}_r\Lambda^k(\mathcal{T}_h)$ and $\mathcal{P}_r^-\Lambda^k(\mathcal{T}_h)$ for a simplicial decomposition $\mathcal{T}_h$ of Ω. In fact, for each polynomial degree $r \geq 0$ there exists 2^{n-1} such complexes with $\Lambda_h^n = \mathcal{P}_r\Lambda^n(\mathcal{T}_h)$.

Let $\bar{\Lambda}_h^k$ denote a second finite-dimensional subcomplex of the L^2 de Rham complex with a corresponding cochain projection $\bar{\Pi}_h$ enjoying the same properties. Supposing a compatibility condition between these two discretizations, which we describe below, we shall construct a discrete elasticity complex based on them, in close analogy with the BGG construction in Section 11.2.

Let $\Lambda_h^k(\mathbb{K}) = \Lambda_h^k \otimes \mathbb{K}$ and $\bar{\Lambda}_h^k(\mathbb{V}) = \bar{\Lambda}_h^k \otimes \mathbb{V}$. For brevity, we write $\Lambda_h^k(\mathbb{W})$

for $\Lambda_h^k(\mathbb{K}) \times \bar{\Lambda}_h^k(\mathbb{V})$. In analogy with (11.6), we start with the complex

$$0 \to \Lambda_h^0(\mathbb{W}) \xrightarrow{\begin{pmatrix} d & 0 \\ 0 & d \end{pmatrix}} \Lambda_h^1(\mathbb{W}) \xrightarrow{\begin{pmatrix} d & 0 \\ 0 & d \end{pmatrix}} \cdots \xrightarrow{\begin{pmatrix} d & 0 \\ 0 & d \end{pmatrix}} \Lambda_h^n(\mathbb{W}) \to 0. \qquad (11.17)$$

Since $\bar{\Lambda}_h^k$ may not equal Λ_h^k, the operator K may not map $\Lambda_h^k(\mathbb{K})$ into $\bar{\Lambda}_h^k(\mathbb{V})$. So we define $K_h : \bar{\Lambda}_h^k(\mathbb{V}) \to \Lambda_h^k(\mathbb{K})$ by $K_h = \Pi_h K$ where Π_h is the given projection operator onto $\Lambda_h^k(\mathbb{K})$.

Next we define $S_h = S_{k,h} : \bar{\Lambda}_h^k(\mathbb{V}) \to \Lambda_h^{k+1}(\mathbb{K})$ by $S_h = dK_h - K_h d$, for $k = 0, 1, \ldots, n - 1$. Observe that the discrete version of (11.7),

$$dS_h = -S_h d \qquad (11.18)$$

follows (exactly as in the continuous case) from $d^2 = 0$. From the commutative diagram (11.16), we see that

$$S_h = d\Pi_h K - \Pi_h K d = \Pi_h(dK - Kd) = \Pi_h S.$$

Continuing to mimic the continuous case, we define the automorphism Φ_h on $\Lambda_h^k(\mathbb{W})$ by

$$\Phi_h(\omega, \mu) = (\omega + K_h \mu, \mu),$$

and the operator $\mathcal{A}_h : \Lambda_h^k(\mathbb{W}) \to \Lambda_h^{k+1}(\mathbb{W})$ by $\mathcal{A}_h = \Phi_h d\Phi_h^{-1}$, which leads to

$$\mathcal{A}_h(\omega, \mu) = (d\omega - S_h \mu, d\mu).$$

Inserting the isomorphisms Φ_h into (11.17), we obtain the isomorphic complex

$$0 \to \Lambda_h^0(\mathbb{W}) \xrightarrow{\mathcal{A}_h} \Lambda_h^1(\mathbb{W}) \xrightarrow{\mathcal{A}_h} \cdots \xrightarrow{\mathcal{A}_h} \Lambda_h^n(\mathbb{W}) \to 0. \qquad (11.19)$$

As in the continuous case, the discrete elasticity complex will be realized as a subcomplex of this complex. We define

$$\Gamma_h^{n-2} = \{ (\omega, \mu) \in \Lambda_h^{n-2}(\mathbb{W}) \mid d\omega = S_{n-2,h}\mu \},$$
$$\Gamma_h^{n-1} = \{ (\omega, \mu) \in \Lambda_h^{n-1}(\mathbb{W}) \mid \omega = 0 \}.$$

Again, $\mathcal{A}_h$ maps $\Lambda_h^{n-2}(\mathbb{W})$ into Γ_h^{n-2} and Γ_h^{n-2} into Γ_h^{n-1}, so that

$$0 \to \Lambda_h^0(\mathbb{W}) \xrightarrow{\mathcal{A}_h} \cdots \xrightarrow{\mathcal{A}_h} \Lambda_h^{n-3}(\mathbb{W}) \xrightarrow{\mathcal{A}_h} \Gamma_h^{n-2} \xrightarrow{\mathcal{A}_h} \Gamma_h^{n-1} \xrightarrow{\mathcal{A}_h} \Lambda_h^n(\mathbb{W}) \to 0$$

is indeed a subcomplex of (11.19).

As in the continuous case, we could identify Γ_h^{n-1} with $\bar{\Lambda}_h^{n-1}(\mathbb{V})$, but, unlike in the continuous case, we cannot identify Γ_h^{n-2} with $\Lambda_h^{n-2}(\mathbb{K})$, since we do not require that $S_{n-2,h}$ be invertible (and it is in fact not invertible in the applications). However, we saw in the proof of Theorem 11.1 that the decisive property of S_{n-2} is that it be surjective, and surjectivity of $S_{n-2,h}$ is what we shall require in order to derive a cochain projection and obtain the analogue of the diagram (11.9). Thus we make the following surjectivity assumption.

Surjectivity assumption.

The operator $S_{n-2,h}$ maps $\bar{\Lambda}_h^{n-2}(\mathbb{V})$ onto $\Lambda_h^{n-1}(\mathbb{K})$.   (11.20)

Under this assumption, the operator $S_h = S_{n-2,h}$ has a right inverse $S_h^\dagger$ mapping $\Lambda_h^{n-1}(\mathbb{K})$ into $\Lambda_h^{n-2}(\mathbb{V})$. This allows us to define discrete counterparts of the projection operators π^{n-2} and π^{n-1} by

$$\pi_h^{n-2}(\omega,\mu) = (\omega, \mu - S_h^\dagger S_h \mu + S_h^\dagger d\omega), \quad \pi_h^{n-1}(\omega,\mu) = (0, \mu + dS_h^\dagger \omega),$$

and obtain the discrete analogue of (11.9):

$$
\begin{array}{ccccccccc}
\cdots \xrightarrow{\mathcal{A}_h} & \Lambda_h^{n-3}(\mathbb{W}) & \xrightarrow{\mathcal{A}_h} & \Lambda_h^{n-2}(\mathbb{W}) & \xrightarrow{\mathcal{A}_h} & \Lambda_h^{n-1}(\mathbb{W}) & \xrightarrow{\mathcal{A}_h} & \Lambda_h^n(\mathbb{W}) \to 0 \\
& \downarrow \mathrm{id} & & \downarrow \pi^{n-2} & & \downarrow \pi^{n-1} & & \downarrow \mathrm{id} \\
\cdots \xrightarrow{\mathcal{A}_h} & \Lambda_h^{n-3}(\mathbb{W}) & \xrightarrow{\mathcal{A}_h} & \Gamma_h^{n-2} & \xrightarrow{\mathcal{A}_h} & \Gamma_h^{n-1} & \xrightarrow{\mathcal{A}_h} & \Lambda_h^n(\mathbb{W}) \to 0.
\end{array}
$$

(11.21)

It is straightforward to check that this diagram commutes. For example, if $(\omega,\mu) \in \Lambda_h^{n-3}(\mathbb{W})$, then

$$
\begin{aligned}
\pi_h^{n-2}\mathcal{A}_h(\omega,\mu) &= \pi_h^{n-2}(d\omega - S_h\mu, d\mu) \\
&= (d\omega - S_h\mu, d\mu - S_h^\dagger S_h d\mu + S_h^\dagger d(d\omega - S_h\mu)) \\
&= (d\omega - S_h\mu, d\mu - S_h^\dagger(S_h d\mu + dS_h\mu)) = \mathcal{A}_h(\omega,\mu),
\end{aligned}
$$

where the last equality follows from (11.18). Thus the vertical maps in (11.21) indeed define a cochain projection.

Since $\mathcal{A}_h$ maps $\Lambda_h^{n-1}(\mathbb{W})$ onto $\Lambda_h^n(\mathbb{W})$, the diagram implies that $\mathcal{A}_h$ maps Γ_h^{n-1} onto $\Lambda_h^n(\mathbb{W})$, i.e., that $(-S_{n-1,h}, d)^T$ maps $\bar{\Lambda}_h^{n-1}(\mathbb{V})$ onto $\Lambda_h^n(\mathbb{K}) \times \bar{\Lambda}_h^n(\mathbb{V})$. This suggests that the choice of finite element spaces $\bar{\Lambda}_h^{n-1}(\mathbb{V})$ for stress, $\bar{\Lambda}_h^n(\mathbb{V})$ for displacements, and $\Lambda_h^n(\mathbb{K})$ for the multiplier will lead to a stable discretization of (11.5). We now make specific choices for the two sets of spaces Λ_h^k and $\bar{\Lambda}_h^k$ for $k = 0, 1, \ldots, n$ and verify the surjectivity assumption. Then in the next subsection we prove that they do, in fact, lead to a stable discretization.

Let $\mathcal{T}_h$ denote a family of shape-regular simplicial meshes of Ω indexed by h, the maximal diameter of the simplices in $\mathcal{T}_h$, and fix the degree $r \geq 0$. Our choices are then:

- $\Lambda_h^{n-1} = \mathcal{P}_{r+1}^-\Lambda^{n-1}(\mathcal{T}_h)$, $\Lambda_h^n = \mathcal{P}_r\Lambda^n(\mathcal{T}_h)$, and
- $\bar{\Lambda}_h^{n-2} = \mathcal{P}_{r+2}^-\Lambda^{n-2}(\mathcal{T}_h)$, $\bar{\Lambda}_h^{n-1} = \mathcal{P}_{r+1}\Lambda^{n-1}(\mathcal{T}_h)$, $\bar{\Lambda}_h^n = \mathcal{P}_r\Lambda_h^n(\mathcal{T}_h)$.

For the remaining spaces, we choose Λ_h^k and $\bar{\Lambda}_h^k$ as either $\mathcal{P}_{s+1}^-\Lambda^k(\mathcal{T}_h)$ or $\mathcal{P}_s\Lambda^k(\mathcal{T}_h)$, for appropriate degrees s, so as to obtain the commuting diagram (11.16). In all cases we use the canonical projection operator related to the degrees of freedom in the space, as defined at the end of Section 5.1. Note that in the lowest-order case $r = 0$, we are approximating the stresses by piecewise linear functions and the displacements and the multiplier by piecewise constants.

We now verify the surjectivity assumption for this choice.

Theorem 11.2. Let $\Pi_h^{n-1} : \Lambda_h^{n-1}(\Omega; \mathbb{K}) \to \mathcal{P}_{r+1}^-\Lambda^{n-1}(\mathcal{T}_h; \mathbb{K})$ and $\bar{\Pi}_h^{n-2} : \Lambda^{n-2}(\Omega; \mathbb{V}) \to \mathcal{P}_{r+2}^-\Lambda^{n-2}(\mathcal{T}_h; \mathbb{V})$ be the canonical projection operators defined in terms of the degrees of freedom (5.2). Then

$$\Pi_h^{n-1} S_{n-2} \bar{\Pi}_h^{n-2} = \Pi_h^{n-1} S_{n-2} \text{ on } \Lambda^{n-2}(\Omega; \mathbb{V}).$$

Consequently $S_{n-2,h} := \Pi_h^{n-1} S_{n-2}$ maps the space $\mathcal{P}_{r+2}^-\Lambda^{n-2}(\mathcal{T}_h; \mathbb{V})$ onto $\mathcal{P}_{r+1}^-\Lambda^{n-1}(\mathcal{T}_h; \mathbb{K})$.

Proof. Note that the second statement easily follows from the first, since Π_h^{n-1} and S_{n-2} are both surjective.

For the proof, we define the operator $K' : \Lambda^k(\Omega; \mathbb{K}) \to \Lambda^k(\Omega; \mathbb{V})$ by

$$(K'\omega)_x(v_1, \ldots, v_k) = \omega_x(v_1, \ldots, v_k) X(x)$$

where $X(x)$ is the element of $\mathbb{V}$ corresponding to x and the last product is the action of the skew-symmetric operator $\omega_x(v_1, \ldots, v_k)$ on the vector $X(x)$. We then have

$$K\omega \wedge \mu = \omega \wedge K'\mu, \quad \omega \in \Lambda^k(\Omega; \mathbb{V}), \ \mu \in \Lambda^j(\Omega; \mathbb{K}).$$

We next show that

$$S\omega \wedge \mu = (-1)^{k+1} \omega \wedge (K'd - dK')\mu, \quad \omega \in \Lambda^k(\Omega; \mathbb{V}), \ \mu \in \Lambda^j(\Omega; \mathbb{K}). \quad (11.22)$$

This follows from the Leibniz rule. We have

$$dK\omega \wedge \mu = d(K\omega \wedge \mu) - (-1)^k K\omega \wedge d\mu = d(\omega \wedge K'\mu) - (-1)^k \omega \wedge K'd\mu,$$

and

$$Kd\omega \wedge \mu = d\omega \wedge K'\mu = d(\omega \wedge K'\mu) - (-1)^k \omega \wedge dK'\mu.$$

Subtracting we get (11.22). Thus, if $\mu \in \mathcal{P}_r\Lambda^j(\Omega; \mathbb{K})$, there exists $\zeta \in \mathcal{P}_{r-1}\Lambda^{j+1}(\Omega; \mathbb{V})$ such that

$$S\omega \wedge \mu = \pm \omega \wedge \zeta, \quad \omega \in \Lambda^k(\Omega; \mathbb{K})$$

(namely, just take $\zeta = (K'd - dK')\mu$).

Now, to prove the theorem we must show that

$$(\Pi_h^{n-1} S_{n-2} - \Pi_h^{n-1} S_{n-2} \bar{\Pi}_h^{n-2})\sigma = 0$$

for all $\sigma \in \Lambda^{n-2}(\Omega; V)$. Defining $\omega = (I - \Pi_h^{n-2})\sigma$, the required condition becomes $\Pi_h^{n-1} S_{n-2}\omega = 0$. Since $\bar{\Pi}_h^{n-2}\omega = 0$, we have

$$\int_f \mathrm{Tr}_f\, \omega \wedge \zeta = 0, \quad \zeta \in \mathcal{P}_{r-d+n-1}\Lambda^{d-n+2}(f; V), \ f \in \Delta_d(T_h), \ n-1 \le d \le n,$$
(11.23)

(in fact (11.23) holds for $d = n - 2$ as well, but this is not used here). We must show that (11.23) implies

$$\int_f \mathrm{Tr}_f(S_{n-2}\omega) \wedge \mu = 0$$

for $\mu \in \mathcal{P}_{r-d+n-1}\Lambda^{d-n+1}(f; \mathbb{K})$, $f \in \Delta_d(T_h)$, $n-1 \le d \le n$. This follows in view of the result proved in the last paragraph (applied on the face f; note that d, K, K' and S commute with traces). $\square$

In the next subsection, we use this result to verify that this choice of spaces results in a stable finite element discretization of the variational formulation of elasticity with weak symmetry.

11.6. The main stability result for mixed finite elements for elasticity

We show in this subsection that the choices

$$\Lambda_h^{n-1}(V) = \mathcal{P}_{r+1}\Lambda^{n-1}(T_h; V),$$
$$\Lambda_h^n(V) = \mathcal{P}_r\Lambda^n(T_h; V),$$
$$\Lambda_h^n(\mathbb{K}) = \mathcal{P}_r\Lambda^n(T_h; \mathbb{K}),$$
(11.24)

give a stable finite element discretization of the system (11.5).

The first stability condition (S1) is obvious since, by construction,

$$\mathrm{d}\mathcal{P}_{r+1}\Lambda^{n-1}(T_h; V) \subset \mathcal{P}_r\Lambda^n(T_h; V).$$

The condition (S2) is more subtle. Our proof is inspired by the proof of the well-posedness result, Theorem 11.1, but involves a variety of projections from the continuous to the finite element spaces, and keeping track of norms. A technical difficulty arises because the canonical projection $\bar{\Pi}_h^{n-2} : \Lambda^{n-2}(\Omega; V) \to \mathcal{P}_{r+2}^-\Lambda^{n-2}(T_h; V)$ is not bounded on H^1, since its definition involves traces on subsimplices of codimension 2. On the other hand, we cannot use the smoothed projection operators introduced in Section 5.4, because these do not preserve the moments of traces on faces of codimension 0 and 1, which were required in the previous theorem to prove that $\Pi_h^{n-1} S_{n-2}\bar{\Pi}_h^{n-2} = \Pi_h^{n-1} S_{n-2}$. Hence we introduce a new operator, $\bar{P}_h : \Lambda^{n-2}(\Omega; V) \to \mathcal{P}_{r+2}^-\Lambda^{n-2}(T_h; V)$. Namely, as for the canonical projection, $\bar{P}_h\omega$ is defined in terms of the degrees of freedom in (5.2), but it is taken to be the element of $\mathcal{P}_{r+2}^-\Lambda^{n-2}(T_h; V)$ with the same moments as ω

on the faces of codimension 0 and 1, but with the moments of a smoothed approximation of ω on the faces of codimension 2. For more details see Arnold *et al.* (2006c). The properties we will need of this operator as well as the relevant canonical projections are summarized in the following lemma.

Lemma 11.3. Let

$$\Pi_h^{n-1} : \Lambda^{n-1}(\Omega; \mathbb{K}) \to \mathcal{P}_{r+1}^-\Lambda^{n-1}(\mathcal{T}_h; \mathbb{K}),$$

$$\Pi_h^n : \Lambda^n(\Omega; \mathbb{K}) \to \mathcal{P}_r\Lambda^n(\mathcal{T}_h; \mathbb{K}),$$

$$\bar{\Pi}_h^{n-1} : \Lambda^{n-1}(\Omega; \mathbb{V}) \to \mathcal{P}_{r+1}\Lambda^{n-1}(\mathcal{T}_h; \mathbb{V}),$$

$$\bar{\Pi}_h^n : \Lambda^n(\Omega; \mathbb{V}) \to \mathcal{P}_r\Lambda^n(\mathcal{T}_h; \mathbb{V})$$

be the canonical projections, and let $\bar{P}_h : \Lambda^{n-2}(\Omega; \mathbb{V}) \to \mathcal{P}_{r+2}^-\Lambda^{n-2}(\mathcal{T}_h; \mathbb{V})$ be the operator described above. Then

$$d\Pi_h^{n-1} = \Pi_h^n d, \quad d\bar{\Pi}_h^{n-1} = \bar{\Pi}_h^n d, \tag{11.25}$$

$$\Pi_h^{n-1} S_{n-2}\bar{P}_h = \Pi_h^{n-1} S_{n-2}, \tag{11.26}$$

$$\|\Pi_h^n\omega\| \le c\|\omega\|, \quad \omega \in \Lambda^n(\Omega; \mathbb{K}), \tag{11.27}$$

$$\|\bar{\Pi}_h^{n-1}\omega\| \le c\|\omega\|_1, \quad \omega \in \Lambda^{n-1}(\Omega; \mathbb{V}), \tag{11.28}$$

$$\|d\bar{P}_h\eta\| \le c\|\eta\|_1, \quad \eta \in \Lambda^{n-2}(\Omega; \mathbb{V}). \tag{11.29}$$

The constant c is uniform in the mesh size h (although it may depend on the shape regularity of the mesh).

Proof. The commutativity conditions in (11.25) are the standard ones. We proved (11.26) with $\bar{\Pi}_h^{n-2}$ in place of $\bar{P}_h$ in Theorem 11.2. Since the proof only depended on the fact that the projection preserved the appropriate moments on faces of codimension 0 or 1, the same proof works for $\bar{P}_h$. The L^2 bound (11.27) is obvious since Π_h^n is just the L^2-projection. The bound (11.28) is standard. Finally the bound in (11.29) can be proved using standard techniques; see Arnold *et al.* (2006c). □

We can now state the main stability result, following the outline of Theorem 11.1.

Theorem 11.4. Given $(\omega, \mu) \in \mathcal{P}_r\Lambda^n(\mathcal{T}_h; \mathbb{K}) \times \mathcal{P}_r\Lambda^n(\mathcal{T}_h; \mathbb{V})$, there exists $\sigma \in \mathcal{P}_{r+1}\Lambda^{n-1}(\mathcal{T}_h; \mathbb{V})$ such that $d\sigma = \mu$, $-\Pi_h^n S_{n-1}\sigma = \omega$, and

$$\|\sigma\|_{H\Lambda} \le c(\|\omega\| + \|\mu\|), \tag{11.30}$$

where the constant c is independent of ω, μ and h.

Proof.

(1) By Theorem 2.4, we can find $\eta \in H^1\Lambda^{n-1}(\Omega; \mathbb{V})$ with $d\eta = \mu$ and $\|\eta\|_1 \leq c\|\mu\|$.

(2) Since $\omega + \Pi_h^n S_{n-1}\bar{\Pi}_h^{n-1}\eta \in H\Lambda^n(\Omega; K)$, we can apply Theorem 2.4 a second time to find $\tau \in H^1\Lambda^{n-1}(\Omega; \mathbb{K})$ with $d\tau = \omega + \Pi_h^n S_{n-1}\bar{\Pi}_h^{n-1}\eta$ and $\|\tau\|_1 \leq c(\|\omega\| + \|\Pi_h^n S_{n-1}\bar{\Pi}_h^{n-1}\eta\|)$.

(3) Since S_{n-2} is an isomorphism from $H^1\Lambda^{n-2}(\Omega; \mathbb{V})$ to $H^1\Lambda^{n-1}(\Omega; \mathbb{K})$, we have $\rho \in H^1\Lambda^{n-2}(\Omega; \mathbb{V})$ with $S_{n-2}\rho = \tau$, and $\|\rho\|_1 \leq c\|\tau\|_1$.

(4) Define $\sigma = d\bar{P}_h\rho + \bar{\Pi}_h^{n-1}\eta \in \mathcal{P}_{r+1}\Lambda^{n-1}(\mathcal{T}_h; \mathbb{V})$.

(5) From step (4), (11.25), step (1), and the fact that $\bar{\Pi}_h^n$ is a projection, we have
$$d\sigma = d\bar{\Pi}_h^{n-1}\eta = \bar{\Pi}_h^n d\eta = \bar{\Pi}_h^n \mu = \mu.$$

(6) From step (4),
$$-\Pi_h^n S_{n-1}\sigma = -\Pi_h^n S_{n-1}d\bar{P}_h\rho - \Pi_h^n S_{n-1}\bar{\Pi}_h^{n-1}\eta.$$

Applying, in order, (11.7), (11.25), (11.26), step (3), (11.25), step (2), and the fact that Π_h^n is a projection, we obtain
$$\Pi_h^n S_{n-1}d\bar{P}_h\rho = -\Pi_h^n dS_{n-2}\bar{P}_h\rho = -d\Pi_h^{n-1}S_{n-2}\bar{P}_h\rho$$
$$= -d\Pi_h^{n-1}S_{n-2}\rho = -d\Pi_h^{n-1}\tau = -\Pi_h^n d\tau$$
$$= -\Pi_h^n(\omega + \Pi_h^n S_{n-1}\bar{\Pi}_h^{n-1}\eta) = -\omega - \Pi_h^n S_{n-1}\bar{\Pi}_h^{n-1}\eta.$$

Combining, we have $-\Pi_h^n S_{n-1}\sigma = \omega$.

(7) Finally, we prove the norm bound. From (11.27), the boundedness of S_{n-1} in L^2, (11.28), and step (1),
$$\|\Pi_h^n S_{n-1}\bar{\Pi}_h^{n-1}\eta\| \leq c\|S_{n-1}\bar{\Pi}_h^{n-1}\eta\| \leq c\|\bar{\Pi}_h^{n-1}\eta\| \leq c\|\eta\|_1 \leq c\|\mu\|.$$

Combining with the bounds in steps (3) and (2), this gives $\|\rho\|_1 \leq c(\|\omega\| + \|\mu\|)$. From (11.29), we then have $\|d\bar{P}_h\rho\| \leq c(\|\omega\| + \|\mu\|)$. From (11.28) and the bound in step (1), $\|\bar{\Pi}_h^{n-1}\eta\| \leq c\|\eta\|_1 \leq c\|\mu\|$. In view of the definition of σ, these two last bounds imply that $\|\sigma\| \leq c(\|\omega\| + \|\mu\|)$, while $\|d\sigma\| = \|\mu\|$, and thus we have the desired bound (11.30). $\square$

We have thus verified the stability conditions (S1) and (S2), and so may apply the standard theory of mixed methods (see Brezzi (1974), Brezzi and Fortin (1991), Douglas and Roberts (1985), Falk and Osborn (1980)) and standard results about approximation by finite element spaces to obtain convergence and error estimates.

Theorem 11.5. Suppose (σ, u, p) is the solution of the elasticity system (11.4) and (σ_h, u_h, p_h) is the solution of discrete system (11.5), where the finite element spaces are given by (11.24) for some integer $r \geq 0$. Then there is a constant C, independent of h, such that

$$\|\sigma - \sigma_h\|_{H\Lambda} + \|u - u_h\| + \|p - p_h\| \leq C \inf(\|\sigma - \tau\|_{H\Lambda} + \|u - v\| + \|p - q\|).$$

where the infimum is over all $\tau \in \Lambda_h^{n-1}(\mathbb{V})$, $v \in \Lambda_h^n(\mathbb{V})$, and $q \in \Lambda_h^n(\mathbb{K})$. If u and σ are sufficiently smooth, then

$$\|\sigma - \sigma_h\| + \|u - u_h\| + \|p - p_h\| \leq Ch^{r+1}\|u\|_{r+2},$$

$$\|d(\sigma - \sigma_h)\| \leq Ch^{r+1}\|d\sigma\|_{r+1}.$$

11.7. Traction boundary conditions

So far we have considered only the case of the Dirichlet boundary condition $u = 0$ on $\partial\Omega$. In this subsection, we consider the modifications that need to be made to deal with the case of the traction boundary condition $\sigma n = 0$ on $\partial\Omega$. For this boundary value problem, in order for a solution to exist, f must be orthogonal in $L^2(\Omega; \mathbb{V})$ to the space $\mathbb{T}$ of rigid motions, defined to be the restrictions to Ω of affine maps of the form $x \mapsto a + bx$ where $a \in \mathbb{V}$ and $b \in \mathbb{K}$. If f does satisfy this compatibility condition, then u is only unique up to addition of a rigid motion. One method of defining a well-posed weak formulations for this problem is to introduce a Lagrange multiplier to enforce the constraint on f. We are then led to the following weak formulation, analogous to (11.3).

Find $(\sigma, u, p, s) \in \mathring{H}(\mathrm{div}, \Omega; \mathbb{M}) \times L^2(\Omega; \mathbb{V}) \times L^2(\Omega; \mathbb{K}) \times \mathbb{T}$ satisfying

$$\int_\Omega (A\sigma : \tau + \mathrm{div}\,\tau \cdot u + \tau : p)\,dx = 0, \qquad\qquad \tau \in \mathring{H}(\mathrm{div}, \Omega; \mathbb{M}),$$

$$\int_\Omega (\mathrm{div}\,\sigma \cdot v + s \cdot v)\,dx = \int_\Omega f \cdot v\,dx, \quad v \in L^2(\Omega; \mathbb{V}),$$

$$\int_\Omega \sigma : q\,dx = 0, \qquad\qquad q \in L^2(\Omega; \mathbb{K}),$$

$$\int_\Omega u \cdot t\,dx = 0, \qquad\qquad t \in \mathbb{T},$$

where

$$\mathring{H}(\mathrm{div}, \Omega; \mathbb{M}) = \{\sigma \in H(\mathrm{div}, \Omega; \mathbb{M}) : \sigma n = 0 \text{ on } \partial\Omega\}.$$

We shall show below that this problem is well-posed.

To restate this in the language of differential forms we introduce $\mathbb{T}^\star = \star\mathbb{T} \subset \mathcal{P}_1\Lambda^n(\Omega; \mathbb{V})$. The problem then takes the form: given $f \in L^2\Lambda^n(\Omega; \mathbb{V})$, find $(\sigma, u, p, s) \in \mathring{H}\Lambda^{n-1}(\Omega; \mathbb{V}) \times L^2\Lambda^n(\Omega; \mathbb{V}) \times L^2\Lambda^n(\Omega; \mathbb{K}) \times \mathbb{T}^\star$ such that

$$
\begin{aligned}
\langle A\sigma, \tau \rangle + \langle d\tau, u \rangle - \langle S_{n-1}\tau, p \rangle &= 0, & \tau &\in \mathring{H}\Lambda^{n-1}(\Omega; \mathbb{V}), \\
\langle d\sigma, v \rangle + \langle s, v \rangle &= \langle f, v \rangle, & v &\in L^2\Lambda^n(\Omega; \mathbb{V}), \\
\langle S_{n-1}\sigma, q \rangle &= 0, & q &\in L^2\Lambda^n(\Omega; \mathbb{K}), \\
\langle u, t \rangle &= 0, & t &\in \mathbb{T}^\star.
\end{aligned}
\tag{11.31}
$$

We remark that taking $v \in \mathbb{T}^\star$ in the second equation and using the identity given in Lemma 11.8 below together with the third equation, implies that s is the L^2-projection of f into $\mathbb{T}^\star$.

We consider here the development of stable mixed finite elements for the linear elasticity problem with traction boundary conditions based on the variational formulation (11.31). To do so, we will follow the development for the Dirichlet problem. In particular, we will again use the link between stable mixed finite elements for elasticity and the existence of discrete versions of a corresponding elasticity complex and also the connection between the elasticity complex and the ordinary de Rham complex. Thus, the choice of stable finite element spaces for elasticity with traction boundary conditions will again have as its starting point discrete versions of an appropriate de Rham complex. Since the derivation is quite analogous to the case of Dirichlet boundary conditions, we will not provide all the details, but concentrate on the modifications that are needed. We will make use of finite element spaces of the form $\mathring{\Lambda}_h^k := \Lambda_h^k \cap \mathring{H}\Lambda^k$ (where $\Lambda_h^k = \mathcal{P}_r\Lambda^k(\mathcal{T}_h)$ or $\mathcal{P}_r^-\Lambda^k(\mathcal{T}_h)$). The canonical projection operator Π_h^k then maps $\mathring{H}\Lambda^k(\Omega)$ into $\mathring{\Lambda}_h^k$.

We begin with the BGG construction, parallel to Section 11.2. For the case of traction boundary conditions, the appropriate de Rham sequence is that with compact support, (2.13), and the corresponding L^2 complex (2.14). So our starting complex for the BGG construction is

$$
0 \to \mathring{\Lambda}^0(\mathbb{W}) \xrightarrow{\left(\begin{smallmatrix} d & 0 \\ 0 & d \end{smallmatrix}\right)} \mathring{\Lambda}^1(\mathbb{W}) \xrightarrow{\left(\begin{smallmatrix} d & 0 \\ 0 & d \end{smallmatrix}\right)} \cdots \xrightarrow{\left(\begin{smallmatrix} d & 0 \\ 0 & d \end{smallmatrix}\right)} \mathring{\Lambda}^n(\mathbb{W}) \to 0,
\tag{11.32}
$$

where $\mathbb{W} = \mathbb{K} \times \mathbb{V}$ and $\mathring{\Lambda}^k(\mathbb{W}) := \mathring{\Lambda}^k(\Omega; \mathbb{K}) \times \mathring{\Lambda}^k(\Omega; \mathbb{V})$. With Φ and $\mathcal{A}$ as before, Φ is a cochain isomorphism from (11.32) to

$$
0 \to \mathring{\Lambda}^0(\mathbb{W}) \xrightarrow{\mathcal{A}} \mathring{\Lambda}^1(\mathbb{W}) \xrightarrow{\mathcal{A}} \cdots \xrightarrow{\mathcal{A}} \mathring{\Lambda}^n(\mathbb{W}) \to 0.
\tag{11.33}
$$

Introducing the spaces $\mathring{\Gamma}^i$ in analogy to the spaces Γ^i of Section 11.2, we obtain the subcomplex

$$
0 \to \mathring{\Lambda}^0(\mathbb{W}) \xrightarrow{\mathcal{A}} \cdots \xrightarrow{\mathcal{A}} \mathring{\Lambda}^{n-3}(\mathbb{W}) \xrightarrow{\mathcal{A}} \mathring{\Gamma}^{n-2} \xrightarrow{\mathcal{A}} \mathring{\Gamma}^{n-1} \xrightarrow{\mathcal{A}} \mathring{\Lambda}^n(\mathbb{W}) \to 0,
$$

and a corresponding cochain projection. Identifying elements $(\omega, \mu) \in \mathring{\Gamma}^{n-2}$ with $\omega \in \mathring{\Lambda}^{n-2}(\Omega; \mathbb{K})$ and elements $(0, \mu) \in \mathring{\Gamma}^{n-1}$ with $\mu \in \mathring{\Lambda}^{n-1}(\Omega; \mathbb{V})$, we obtain the relevant elasticity complex

$$0 \to \mathring{\Lambda}^0(\mathbb{W}) \xrightarrow{A} \cdots \xrightarrow{A} \mathring{\Lambda}^{n-3}(\mathbb{W}) \xrightarrow{(d,-S_{n-3})} \mathring{\Lambda}^{n-2}(\Omega; \mathbb{K})$$

$$\xrightarrow{d \circ S_{n-2}^{-1} \circ d} \mathring{\Lambda}^{n-1}(\Omega; \mathbb{V}) \xrightarrow{(-S_{n-1}, d)^T} \mathring{\Lambda}^n(\mathbb{W}) \to 0. \quad (11.34)$$

The complex (11.32), and so the isomorphic complex (11.33) have a co-homology space of dimension $\dim \mathbb{V} + \dim \mathbb{K} = n(n+1)/2$ at the highest order. Thus the highest-order cohomology space for the elasticity complex (11.34) has dimension at most $n(n+1)/2$. In other words, solvability of the problem, given $(\omega, \mu) \in \mathring{\Lambda}^n(\mathbb{W})$, to find $\sigma \in \mathring{\Lambda}^{n-1}(\Omega; \mathbb{V})$ such that

$$(-S_{n-1}\sigma, d\sigma) = (\omega, \mu),$$

implies at most $n(n+1)/2$ constraints on the data (ω, μ). In fact, it implies exactly this many constraints, namely,

$$\int_\Omega \mu = 0, \quad \int_\Omega \omega = \int_\Omega K\mu.$$

Indeed, the first equation (n constraints) follows immediately from the equation $d\sigma = \mu$ and Stokes' theorem, while

$$\int_\Omega \omega = \int_\Omega (Kd - dK)\sigma = \int_\Omega K d\sigma = \int_\Omega K\mu.$$

Our next goal is to prove the well-posedness of (11.31). But first we prove a useful lemma.

Lemma 11.6. Given $a \in \mathbb{V}$ and $b \in \mathbb{K}$, there exists a unique $s \in \mathbb{T}^*$ such that $\int_\Omega s = a$, $\int_\Omega Ks = b$.

Proof. Since $\dim \mathbb{T}^* = \dim \mathbb{V} + \dim \mathbb{K}$, it is enough to show that if $\int_\Omega s$ and $\int_\Omega Ks = 0$, then $s = 0$. Now $s_x = (g + cx)\text{vol}_x$ for some $g \in \mathbb{V}$ and $c \in \mathbb{K}$ (in order to lighten the notation, we will henceforth not distinguish between the point x and the associated vector $X(x)$). From the vanishing of $\int s$ we can write g in terms of c and find that $s_x = c(x - \bar{x})\text{vol}_x$ where $\bar{x}$ is the barycentre of Ω. We have a simple linear algebra identity

$$\langle v \wedge bv, b \rangle = -2|bv|^2, \quad v \in \mathbb{V}, \ b \in \mathbb{K}, \quad (11.35)$$

where the inner product on the left is taken in $\mathbb{M}$ and the norm on the right is the norm in $\mathbb{V}$. Thus

$$-|c(x - \bar{x})|^2 = \frac{1}{2}\langle (x - \bar{x}) \wedge c(x - \bar{x}), c \rangle.$$

Integrating over Ω and using the fact that $\int s = 0$, we get

$$-\int_\Omega |c(x - \bar{x})|^2 \text{vol} = \frac{1}{2}\left\langle \int_\Omega x \wedge s_x, c \right\rangle = \frac{1}{2}\left\langle \int_\Omega Ks, c \right\rangle = 0.$$

Thus $c(x - \bar{x}) \equiv 0$, and so s vanishes.   □

We now turn to the proof of well-posedness, *i.e.*, the analogue of Theorem 11.1 for traction boundary conditions. The problem (11.31) is of the saddle point type to which Brezzi's theorem applies, with

$$a(\sigma, s; \tau, t) = \langle A\sigma, \tau \rangle,$$
$$b(\sigma, s; v, q) = \langle d\sigma, v \rangle - \langle S_{n-1}\sigma, q \rangle + \langle s, v \rangle.$$

The analogues of conditions (W1) and (W2) are:

(W1′) $\|\tau\|_{H\Lambda}^2 + \|t\|^2 \le c_1 \langle A\tau, \tau \rangle$ whenever $\tau \in \mathring{H}\Lambda^{n-1}(\Omega; V)$ and $t \in \mathbb{T}^\star$
   satisfy $\langle d\tau, v \rangle + \langle t, v \rangle = 0 \ \forall v \in L^2\Lambda^n(\Omega; V)$ and $\langle S_{n-1}\tau, q \rangle = 0$
   $\forall q \in L^2\Lambda^n(\Omega; \mathbb{K})$,

(W2′) for all nonzero $(v, q) \in \Lambda^n(\Omega; V) \times \Lambda^n(\Omega; \mathbb{K})$, there exists nonzero
   $\tau \in \mathring{\Lambda}^{n-1}(\Omega; V)$ and $t \in \mathbb{T}^\star$ with

$$\langle d\tau, v \rangle - \langle S_{n-1}\tau, q \rangle + \langle t, v \rangle \ge c_2(\|\tau\|_{H\Lambda} + \|t\|)(\|v\| + \|q\|),$$

for some positive constants c_1 and c_2. Again, the first condition is easy (since for such τ and t, the rigid motion t is an L^2-projection of $d\tau$). We now prove (W2′).

Theorem 11.7. Given $(\omega, \mu) \in L^2\Lambda^n(\Omega; \mathbb{K}) \times L^2\Lambda^n(\Omega; V)$, there exists $\sigma \in H\Lambda^{n-1}(\Omega; V)$, $s \in \mathbb{T}^\star$ such that $d\sigma + s = \mu$, $-S_{n-1}\sigma = \omega$. Moreover we may choose σ, s so that

$$\|\sigma\|_{H\Lambda} + \|s\| \le c(\|\omega\| + \|\mu\|),$$

for a fixed constant c.

Proof. Again, the norm bound is automatic once the existence is established.

(0) Define $s \in \mathbb{T}^*$ by

$$\int_\Omega s = \int_\Omega \mu, \quad \int_\Omega Ks = \int_\Omega (K\mu + \omega).$$

   By the lemma, this determines s.

(1) By Theorem 2.4, we can find $\eta \in \mathring{H}^1\Lambda^{n-1}(\Omega; V)$ with $d\eta = \mu - s$.

(2) Now $\omega + S_{n-1}\eta \in H\Lambda^n(\Omega; K)$ and has vanishing integral, since

$$\int_\Omega S_{n-1}\eta = \int_\Omega Kd\eta = \int_\Omega K(\mu - s) = -\int_\Omega \omega.$$

   Thus we can apply Theorem 2.4 another time to find $\tau \in \mathring{H}^1\Lambda^{n-1}(\Omega; K)$ with $d\tau = \omega + S_{n-1}\eta$.

(3) Take $\rho \in \mathring{H}^1\Lambda^{n-2}(\Omega; V)$ with $S_{n-2}\rho = \tau$.

(4) Define $\sigma = d\rho + \eta \in \mathring{H}\Lambda^{n-1}(\Omega; V)$.

(5) From steps (1) and (4), it is immediate that $d\sigma + s = \mu$.

(6) From (4),

$$-S_{n-1}\sigma = -S_{n-1}d\rho - S_{n-1}\eta = dS_{n-2}\rho - S_{n-1}\eta = d\tau - S_{n-1}\eta = \omega.$$

$\square$

We now turn to the discrete problem: find $(\sigma_h, u_h, p_h, s_h) \in \mathring{\Lambda}_h^{n-1}(V) \times \Lambda_h^n(V) \times \Lambda_h^n(\mathbb{K}) \times \mathbb{T}^*$ such that

$$\begin{aligned}
\langle A\sigma_h, \tau \rangle + \langle d\tau, u_h \rangle - \langle S_{n-1}\tau, p_h \rangle &= 0, & \tau &\in \mathring{\Lambda}_h^{n-1}(V), \\
\langle d\sigma_h, v \rangle + \langle s_h, v \rangle &= \langle f, v \rangle, & v &\in \Lambda_h^n(V), \\
\langle S_{n-1}\sigma_h, q \rangle &= 0, & q &\in \Lambda_h^n(\mathbb{K}), \\
\langle u_h, t \rangle &= 0, & t &\in \mathbb{T}^*.
\end{aligned} \tag{11.36}$$

The stability conditions for this system are then:

(S1') $\|\tau\|_{H\Lambda}^2 + \|s\|^2 \le c_1 \langle A\tau, \tau \rangle$ whenever $\tau \in \mathring{\Lambda}_h^{n-1}(V)$ and $s \in \mathbb{T}^*$ satisfy

$\langle d\tau, v \rangle + \langle s, v \rangle = 0 \ \forall v \in \Lambda_h^n(V)$ and $\langle S_{n-1}\tau, q \rangle = 0 \ \forall q \in \Lambda_h^n(\mathbb{K})$,

(S2') for all nonzero $(v, q) \in \Lambda_h^n(V) \times \Lambda_h^n(\mathbb{K})$, there exists nonzero $\tau \in \mathring{\Lambda}_h^{n-1}(V)$ and $s \in \mathbb{T}^*$ with

$$\langle d\tau, v \rangle - \langle S_{n-1}\tau, q \rangle + \langle s, v \rangle \ge c_2(\|\tau\|_{H\Lambda} + \|s\|)(\|v\| + \|q\|),$$

where c_1 and c_2 are positive constants independent of h.

We choose the same finite element spaces as before, except that the stress space now incorporates the boundary conditions:

$$\mathring{\Lambda}_h^{n-1}(V) = \mathring{\mathcal{P}}_{r+1}\Lambda^{n-1}(\mathcal{T}_h; V), \qquad \Lambda_h^n(V) = \mathcal{P}_r\Lambda^n(\mathcal{T}_h; V),$$
$$\Lambda_h^n(\mathbb{K}) = \mathcal{P}_r\Lambda^n(\mathcal{T}_h; \mathbb{K}).$$

We show that, for any $r \ge 0$, this choice gives a stable finite element discretization of the system (11.36). The case when $r = 0$ requires a bit of extra effort, because $\mathbb{T}^* \not\subseteq \Lambda_h^n(V)$ in this case. We begin by assuming that $r \ge 1$ and remark on the case $r = 0$ at the end.

The following simple identity will be useful in establishing stability.

Lemma 11.8. Let $s = (a + bx)\mathrm{vol} \in \mathbb{T}^*$, with $a \in V$ and $b \in \mathbb{K}$. Then

$$\langle d\tau, s \rangle = \frac{1}{2}\langle S_{n-1}\tau, b\,\mathrm{vol} \rangle, \qquad \tau \in \mathring{H}\Lambda^{n-1}(\Omega; V).$$

First we verify the stability (S1'). We have $\tau \in \mathring{\mathcal{P}}_{r+1}\Lambda^{n-1}(\mathcal{T}_h; V)$ and $s \in \mathbb{T}^*$ with

$$\langle d\tau, v \rangle + \langle s, v \rangle = 0, \ v \in \mathcal{P}_r\Lambda^n(\mathcal{T}_h; V), \quad \langle S_{n-1}\tau, q \rangle = 0, \ q \in \mathcal{P}_r\Lambda^n(\mathcal{T}_h; \mathbb{K}).$$

Taking $v = s$ in the first equation, applying the lemma, and then the second equation, we conclude that $s = 0$. Then we take $v = d\tau$ and conclude that $d\tau = 0$, so the bound in (S1') holds.

The proof of the second stability condition is very much as in the case of Dirichlet boundary conditions, with the minor extra complications which we have already seen in the continuous case in Theorem 11.7, so we just sketch the proof.

Theorem 11.9. Given $(\omega, \mu) \in \mathcal{P}_r \Lambda^n(\mathcal{T}_h; \mathbb{K}) \times \mathcal{P}_r \Lambda^n(\mathcal{T}_h; \mathbb{V})$, there exists $\sigma \in \mathring{\mathcal{P}}_{r+1}\Lambda^{n-1}(\mathcal{T}_h; \mathbb{V})$ and $s \in \mathbb{T}^\star$ such that

$$d\sigma + s = \mu, \qquad -\Pi_h^n S_{n-1}\sigma = \omega,$$

and

$$\|\sigma\|_{H\Lambda} + \|s\| \le c(\|\omega\| + \|\mu\|),$$

where the constant c is independent of ω, μ and h.

Proof. First define s and η as in steps (0) and (1) of Theorem 11.7. Note that

$$\int_\Omega (\omega + \Pi_h^n S_{n-1}\bar{\Pi}_h^{n-1}\eta) = \int_\Omega (\omega - Kd\bar{\Pi}_h^{n-1}\eta) = \int_\Omega (\omega - K\bar{\Pi}_h^n(\mu - s))$$

$$= \int_\Omega (\omega - K(\mu - s)) = 0.$$

Thus we can take $\tau \in \mathring{H}\Lambda^{n-1}(\Omega; \mathbb{K})$ with $d\tau = \omega + \Pi_h^n S_{n-1}\bar{\Pi}_h^{n-1}\eta$ and then $\rho = S_{n-2}^{-1}\tau$, and $\sigma = d\bar{P}_h + \bar{\Pi}_h^{n-1}\eta$. The remainder of the proof is just as for Theorem 11.4. $\qquad\square$

Finally we remark on the modifications that have to be made in the case $r = 0$. In the proof of (S1'), we cannot take $v = s$, since $\mathbb{T}^\star \not\subset \mathcal{P}_0\Lambda^n(\mathcal{T}_h; \mathbb{V})$. So we take $v = \bar{\Pi}_h^n s$, with $\bar{\Pi}_h^n$ simply the L^2-projection into the piecewise constant n-forms. Then

$$\|\bar{\Pi}_h^n s\|^2 = \langle s, v\rangle = -\langle d\tau, v\rangle = -\langle d\tau, s\rangle = 0,$$

with the last step following from the lemma, as before. The $\bar{\Pi}_h^n s = 0$. But for $s \in \mathbb{T}^\star$ it is easy to see that this implies that $s = 0$, at least for h sufficiently small.

A similar issue arises in the verification of (S2'). Now we want to find $\sigma \in \mathring{\mathcal{P}}_1\Lambda^{n-1}(\mathcal{T}_h; \mathbb{V})$, $s \in \mathbb{T}^\star$ with $d\sigma + \bar{\Pi}_h^n s = \mu$. This requires us to define s by

$$\int_\Omega s = \int_\Omega \mu, \qquad \int_\Omega K\bar{\Pi}_h^n s = \int_\Omega (K\mu + \omega).$$

The existence of such an s follows from a variant of Lemma 11.6 which replaces Ks by $K\bar{\Pi}_h^n s$. The variant lemma can be proved using the identity

(11.35), but taking $v = \bar{\Pi}_h(x - \bar{x})$, the L^2-projection of $x - \bar{x}$ into the piecewise constants, rather than $v = x - \bar{x}$ as before. It follows that $\bar{\Pi}_h^n s = 0$, which again implies that s vanishes for small h.

11.8. Simplified elements

The purpose of this subsection is to present a stable element for which the finite element spaces are slightly smaller than the simplest element derived so far, namely

$$\Lambda_h^{n-1}(\mathbb{V}) = \mathcal{P}_1\Lambda^{n-1}(\mathcal{T}_h; \mathbb{V}), \ \Lambda_h^n(\mathbb{V}) = \mathcal{P}_0\Lambda^n(\mathcal{T}_h; \mathbb{V}), \ \Lambda_h^n(\mathbb{K}) = \mathcal{P}_0\Lambda^n(\mathcal{T}_h; \mathbb{K}).$$

(We return to the Dirichlet problem for this.) In the new element, the spaces $\Lambda_h^n(\mathbb{V})$ and $\Lambda_h^n(\mathbb{K})$ are unchanged, but $\Lambda_h^{n-1}(\mathbb{V})$ will be reduced from a full space of piecewise linear elements to one where some of the components are only a reduced space of linears. Since the full details for the cases $n = 2$ and $n = 3$ are provided in Arnold et al. (2005) and (2006c), we only present the main ideas here.

By examining the proof of Theorem 11.4, we realize that we do not use the complete sequence (11.17) for the given spaces. We only use the sequences

$$\mathcal{P}_1^-\Lambda^{n-1}(\mathcal{T}_h; \mathbb{K}) \xrightarrow{\ d\ } \mathcal{P}_0\Lambda^n(\mathcal{T}_h; \mathbb{K}) \longrightarrow 0,$$

$$\mathcal{P}_2^-\Lambda^{n-2}(\mathcal{T}_h; \mathbb{V}) \xrightarrow{\ d\ } \mathcal{P}_1\Lambda^{n-1}(\mathcal{T}_h; \mathbb{V}) \xrightarrow{\ d\ } \mathcal{P}_0\Lambda^n(\mathcal{T}_h; \mathbb{V}) \longrightarrow 0. \tag{11.37}$$

The purpose here is to show that it is possible to choose subspaces of some of the spaces in (11.37) such that the desired properties still hold. More precisely, compared to (11.37), the spaces $\mathcal{P}_2^-\Lambda^{n-2}(\mathcal{T}_h; \mathbb{V})$ and $\mathcal{P}_1\Lambda^{n-1}(\mathcal{T}_h; \mathbb{V})$ are simplified, while the three others remain unchanged. If we denote by $\mathcal{P}_{2,-}^-\Lambda^{n-2}(\mathcal{T}_h; \mathbb{V})$ and $\mathcal{P}_{1,-}\Lambda^{n-1}(\mathcal{T}_h; \mathbb{V})$ the simplifications of the spaces $\mathcal{P}_2^-\Lambda^{n-2}(\mathcal{T}_h; \mathbb{V})$ and $\mathcal{P}_1\Lambda^{n-1}(\mathcal{T}_h; \mathbb{V})$, respectively, then the properties we need are that:

$$\mathcal{P}_{2,-}^-\Lambda^{n-2}(\mathcal{T}_h; \mathbb{V}) \xrightarrow{\ d\ } \mathcal{P}_{1,-}\Lambda^{n-1}(\mathcal{T}_h; \mathbb{V}) \xrightarrow{\ d\ } \mathcal{P}_0\Lambda^n(\mathcal{T}_h; \mathbb{V}) \longrightarrow 0 \tag{11.38}$$

is a complex and that the surjectivity assumption (11.20) holds, i.e., $S_h = S_{n-2,h}$ maps $\mathcal{P}_{2,-}^-\Lambda^{n-2}(\mathcal{T}_h; \mathbb{V})$ onto $\mathcal{P}_1^-\Lambda^{n-1}(\mathcal{T}_h; \mathbb{K})$. We note that if the space $\mathcal{P}_1^-\Lambda^{n-1}(\mathcal{T}_h; \mathbb{V}) \subset \mathcal{P}_{1,-}\Lambda^{n-1}(\mathcal{T}_h; \mathbb{V})$, then d maps $\mathcal{P}_{1,-}\Lambda^{n-1}(\mathcal{T}_h; \mathbb{V})$ onto $\mathcal{P}_0\Lambda^n(\mathcal{T}_h; \mathbb{V})$.

The key to this construction is to first show that a space $\mathcal{P}_{2,-}^-\Lambda^{n-2}(\mathcal{T}_h; \mathbb{V})$ can be constructed as a subspace of $\mathcal{P}_2^-\Lambda^{n-2}(\mathcal{T}_h; \mathbb{V})$, while still retaining the surjectivity assumption (11.20). This can be done locally on each simplex. We begin by recalling that the degrees of freedom on a face $f \in \Delta_{n-1}(T)$ of $\mathcal{P}_2^-\Lambda^{n-2}(T; \mathbb{V})$ have the form

$$\int_f \omega \wedge \mu, \quad \mu \in \mathcal{P}_0\Lambda^1(f, \mathbb{V}). \tag{11.39}$$

However, if we examine the proof of Theorem 11.2, we see that the only degrees of freedom that are used for an element $\omega \in \mathcal{P}_2^- \Lambda^{n-2}(T; \mathbb{V})$ are the subset of the $\Delta_{n-1}(T)$ face degrees of freedom given by

$$\int_f \omega \wedge \nu, \quad \nu \in \mathbb{K},$$

where in the integral we view $\nu \in \mathbb{K} \subset \mathbb{M} \cong \mathrm{Alt}^1(\mathbb{V}; \mathbb{V})$ as a 1-form with values in $\mathbb{V}$.

To classify the degrees of freedom that we need to retain to establish Theorem 11.2, we observe that the $n(n-1)$-dimensional space of test functions used in (11.39) can be decomposed into

$$\mathcal{P}_0 \Lambda^1(f; \mathbb{V}) = \mathcal{P}_0 \Lambda^1(f; T_f) + \mathcal{P}_0 \Lambda^1(f; N_f),$$

i.e., into forms with values in the tangent space to f, T_f, or its orthogonal complement, N_f. This is an $(n-1)^2 + (n-1)$-dimensional decomposition. Furthermore,

$$\mathcal{P}_0 \Lambda^1(f; T_f) = \mathcal{P}_0 \Lambda^1_{\mathrm{sym}}(f; T_f) + \mathcal{P}_0 \Lambda^1_{\mathrm{skw}}(f; T_f),$$

where $\mu \in \mathcal{P}_0 \Lambda^1(f; T_f)$ is in $\mathcal{P}_0 \Lambda^1_{\mathrm{sym}}(f; T_f)$ if and only if $\mu(s) \cdot t = \mu(t) \cdot s$ for orthonormal tangent vectors s and t. Note that when $n = 2$, this space is 1-dimensional, so there is only $\mathcal{P}_0 \Lambda^1_{\mathrm{sym}}(f; T_f)$. Finally, we obtain an $n(n-1)/2 + n(n-1)/2$-dimensional decomposition

$$\mathcal{P}_0 \Lambda^1(f; \mathbb{V}) = \mathcal{P}_0 \Lambda^1_{\mathrm{sym}}(f; T_f) + \mathcal{P}_0 \Lambda^1_{\mathrm{skw}}(f; \mathbb{V}),$$

where

$$\mathcal{P}_0 \Lambda^1_{\mathrm{skw}}(f; \mathbb{V}) = \mathcal{P}_0 \Lambda^1_{\mathrm{skw}}(f; T_f) + \mathcal{P}_0 \Lambda^1(f; N_f).$$

It can be shown that the degrees of freedom corresponding to $\mathcal{P}_0 \Lambda^1_{\mathrm{skw}}(f; \mathbb{V})$ are the ones that need to be retained, while those in $\mathcal{P}_0 \Lambda^1_{\mathrm{sym}}(f; T_f)$ are not needed.

The reduced space $\mathcal{P}_{2,-}^- \Lambda^{n-2}(T; \mathbb{V})$ that we now construct has two properties. The first is that it still contains the space $\mathcal{P}_1 \Lambda^{n-2}(T; \mathbb{V})$ and the second is that the unused face degrees of freedom are eliminated (by setting them equal to zero). We can achieve these conditions by first writing an element $\omega \in \mathcal{P}_2^- \Lambda^{n-2}(T; \mathbb{V})$ as $\omega = \bar{\Pi}_h \omega + (I - \bar{\Pi}_h)\omega$, where $\bar{\Pi}_h$ denotes the usual projection operator into $\mathcal{P}_1 \Lambda^{n-2}(T; \mathbb{V})$ defined by the moments on the faces $f \in \Delta_{n-2}(T)$. Then the elements in $(I - \bar{\Pi}_h)\mathcal{P}_2^- \Lambda^{n-2}(T; \mathbb{V})$ will have zero traces on these faces, so they are completely defined by their degrees of freedom on the faces $f \in \Delta_{n-1}(T)$:

$$\int_f \omega \wedge \mu, \quad \mu \in \mathcal{P}_0 \Lambda^1(f; \mathbb{V}), \quad f \in \Delta_{n-1}(T).$$

Thus, we henceforth denote $(I - \bar{\Pi}_h)\mathcal{P}_2^- \Lambda^{n-2}(T; \mathbb{V})$ by $\mathcal{P}_{2,f}^- \Lambda^{n-2}(T; \mathbb{V})$.

We then define our reduced space

$$\mathcal{P}_{2,-}^{-}\Lambda^{n-2}(T;\mathbb{V}) = \mathcal{P}_1\Lambda^{n-2}(T;\mathbb{V}) + \mathcal{P}_{2,f,-}^{-}\Lambda^{n-2}(T;\mathbb{V}),$$

where $\mathcal{P}_{2,f,-}^{-}\Lambda^{n-2}(T;\mathbb{V})$ denotes the set of forms $\omega \in \mathcal{P}_{2,f}^{-}\Lambda^{n-2}(T;\mathbb{V})$ satisfying

$$\int_f \omega \wedge \mu = 0, \quad \mu \in \mathcal{P}_0\Lambda_{\mathrm{sym}}^1(f;\mathbb{V}),$$

i.e., we have set the unused degrees of freedom to be zero. The space $\mathcal{P}_{2,-}^{-}\Lambda_h^2(\mathbb{V})$ can then be defined from the local spaces in the usual way. The degrees of freedom for this space are then given by

$$\int_f \omega \wedge \mu, \quad \mu \in \mathcal{P}_1\Lambda^0(f;\mathbb{V}), \qquad f \in \Delta_{n-2}(T),$$

$$\int_f \omega \wedge \mu, \quad \mu \in \mathcal{P}_0\Lambda_{\mathrm{skw}}^1(f;\mathbb{V}), \quad f \in \Delta_{n-1}(T).$$

When $n = 3$, the space $\mathcal{P}_{2,-}^{-}\Lambda^1(T;\mathbb{V})$ will have 48 degrees of freedom (36 edge degrees of freedom and 12 face degrees of freedom).

The motivation for this choice of the space $\mathcal{P}_{2,-}^{-}\Lambda_h^{n-2}(\mathbb{V})$ is that it easily leads to a definition of the space $\mathcal{P}_{1,-}\Lambda_h^{n-1}(\mathbb{V})$ that satisfies the property that (11.38) is a complex. We begin by defining

$$\mathcal{P}_{1,-}\Lambda^{n-1}(T;\mathbb{V}) = \mathcal{P}_1^{-}\Lambda^{n-1}(T;\mathbb{V}) + d\mathcal{P}_{2,f,-}^{-}\Lambda^{n-2}(T;\mathbb{V}).$$

When $n = 3$, this space will have 24 face degrees of freedom. The space $\mathcal{P}_{1,-}\Lambda_h^{n-1}(\mathbb{V})$ is then defined from the local spaces in the usual way. It is clear that $\mathcal{P}_1^{-}\Lambda_h^{n-1}(\mathbb{V}) \subset \mathcal{P}_{1,-}\Lambda_h^{n-1}(\mathbb{V})$ and easy to check that the complex (11.38) is exact.

We define appropriate degrees of freedom for the space $\mathcal{P}_{1,-}\Lambda^{n-1}(T;\mathbb{V})$ by using a subset of the degrees of freedom for $\mathcal{P}_1\Lambda^{n-1}(T;\mathbb{V})$, i.e., of $\int_f \omega \wedge \mu$, $\mu \in \mathcal{P}_1\Lambda^0(f;\mathbb{V})$, $f \in \Delta_{n-1}(T)$. In particular, we take as degrees of freedom for $\mathcal{P}_{1,-}\Lambda^{n-1}(T;\mathbb{V})$,

$$\int_f \omega \wedge \mu, \quad \mu \in \mathcal{P}_{1,\mathrm{skw}}\Lambda^0(f;\mathbb{V}), \quad f \in \Delta_{n-1}(T),$$

where $\mathcal{P}_{1,\mathrm{skw}}\Lambda^0(f;\mathbb{V})$ denotes the set of $\mu \in \mathcal{P}_1\Lambda^0(f;\mathbb{V})$ that satisfy $d\mu \in \mathcal{P}_0\Lambda_{\mathrm{skw}}^1(f;\mathbb{V})$.

Using an argument parallel to that used previously, it is straightforward to show that the simplified spaces also satisfy the surjectivity assumption (11.20). We can then complete the proof of stability and show that the convergence asserted in Theorem 11.5 for $r = 0$ holds also for the reduced spaces.

When $n = 3$, $\mathcal{P}_{1,\text{skw}}\Lambda^0(f; \mathbb{V})$ is a 6-dimensional space on each face, so the above quantities specify 24 degrees of freedom for the space $\mathcal{P}_{1,-}\Lambda^2(T; \mathbb{V})$. It is not difficult to check that these are a unisolvent set of degrees of freedom, and we can use the identification of an element $\omega \in \Lambda^2(\Omega; \mathbb{V})$ with a matrix F given by $\omega(v_1, v_2) = F(v_1 \times v_2)$ to describe the six degrees of freedom on a face:

$$\int_f Fn\,\mathrm{d}f, \qquad \int_f (x \cdot t)n^T Fn\,\mathrm{d}f,$$

$$\int_f (x \cdot s)n^T Fn\,\mathrm{d}f, \qquad \int_f [(x \cdot t)s^T - (x \cdot s)t^T]Fn\,\mathrm{d}f.$$

Acknowledgements

This work benefited substantially from the facilities and programming provided by the Institute for Mathematics and its Applications (IMA) at the University of Minnesota and by the Centre of Mathematics for Applications (CMA) at the University of Oslo. The workshop entitled *Compatible Spatial Discretizations for Partial Differential Equations* held in May 2004 at the IMA and the subsequent workshop entitled *Compatible Discretizations for Partial Differential Equations* held at the CMA in September 2005, were particularly important to the development of this field and this paper. We are grateful to the participants of these workshops for valuable discussions and presentations. We particularly acknowledge the help of Daniele Boffi, Snorre Christiansen and Joachim Schöberl, whose presentations at the workshops and subsequent discussions have influenced the presentations of parts of the paper. The first author is also grateful to Victor Reiner for his explanations of the representation theory used to characterize the affine-invariant spaces of polynomial differential form. We also gratefully acknowledge the support of this work by the National Science Foundation of the United States and the Norwegian Research Council.

REFERENCES

S. Adams and B. Cockburn (2005), 'A mixed finite element method for elasticity in three dimensions', *J. Sci. Comput.* **25**, 515–521.

M. Ainsworth and J. Coyle (2003), 'Hierarchic finite element bases on unstructured tetrahedral meshes', *Internat. J. Numer. Methods Engrg.* **58**, 2103–2130.

M. Amara and J. M. Thomas (1979), 'Equilibrium finite elements for the linear elastic problem', *Numer. Math.* **33**, 367–383.

C. Amrouche, C. Bernardi, M. Dauge and V. Girault (1998), 'Vector potentials in three-dimensional non-smooth domains', *Math. Methods Appl. Sci.* **21**, 823–864.

D. N. Arnold (2002), Differential complexes and numerical stability, in *Proceedings of the International Congress of Mathematicians*, Vol. I (Beijing, 2002), Higher Education Press, Beijing, pp. 137–157.

D. N. Arnold and R. S. Falk (1988), 'A new mixed formulation for elasticity', *Numer. Math.* **53**, 13–30.

D. N. Arnold and R. Winther (2002), 'Mixed finite elements for elasticity', *Numer. Math.* **92**, 401–419.

D. N. Arnold, P. B. Bochev, R. B. Lehoucq, R. A. Nicolaides and M. Shashkov, eds (2006*a*), *Compatible Spatial Discretizations*, Vol. 142 of *The IMA Volumes in Mathematics and its Applications*, Springer, Berlin.

D. N. Arnold, F. Brezzi and J. Douglas, Jr. (1984*a*), 'PEERS: a new mixed finite element for plane elasticity', *Japan J. Appl. Math.* **1**, 347–367.

D. N. Arnold, J. Douglas, Jr. and C. P. Gupta (1984*b*), 'A family of higher order mixed finite element methods for plane elasticity', *Numer. Math.* **45**, 1–22.

D. N. Arnold, R. S. Falk and R. Winther (1997*a*), 'Preconditioning discrete approximations of the Reissner-Mindlin plate model', *RAIRO Modél. Math. Anal. Numér.* **31**, 517–557.

D. N. Arnold, R. S. Falk and R. Winther (1997*b*), 'Preconditioning in $H(\mathrm{div})$ and applications', *Math. Comp.* **66**, 957–984.

D. N. Arnold, R. S. Falk and R. Winther (2000), 'Multigrid in $H(\mathrm{div})$ and $H(\mathrm{curl})$', *Numer. Math.* **85**, 197–217.

D. N. Arnold, R. S. Falk and R. Winther (2005), 'Mixed finite element methods for linear elasticity with weakly imposed symmetry', submitted to *Math. Comput.*

D. N. Arnold, R. S. Falk and R. Winther (2006*b*), Differential complexes and stability of finite element methods I: The de Rham complex, in *Compatible Spatial Discretizations*, Vol. 142 of *The IMA Volumes in Mathematics and its Applications*, Springer, Berlin, pp. 23–46.

D. N. Arnold, R. S. Falk and R. Winther (2006*c*), Differential complexes and stability of finite element methods II: The elasticity complex, in *Compatible Spatial Discretizations*, Vol. 142 of *The IMA Volumes in Mathematics and its Applications*, Springer, Berlin, pp. 47–68.

V. I. Arnold (1978), *Mathematical Methods of Classical Mechanics*, Springer, New York.

I. Babuška and A. K. Aziz (1972), Survey lectures on the mathematical foundations of the finite element method, in *The Mathematical Foundations of the Finite Element Method with Applications to Partial Differential Equations* (Proc. Sympos., Univ. Maryland, Baltimore, MD, 1972), Academic Press, New York, pp. 1–359.

I. Babuška and J. Osborn (1991), Eigenvalue problems, in *Handbook of Numerical Analysis*, Vol. II, North-Holland, Amsterdam, pp. 641–787.

I. N. Bernšteĭn, I. M. Gel'fand and S. I. Gel'fand (1975), Differential operators on the base affine space and a study of $\mathfrak{g}$-modules, in *Lie Groups and Their Representations* (Proc. Summer School, Bolyai János Math. Soc., Budapest, 1971), Halsted, New York, pp. 21–64.

P. Bochev and M. D. Gunzburger (2005), 'On least-squares finite element methods for the Poisson equation and their connection to the Dirichlet and Kelvin principles', *SIAM J. Numer. Anal.* **43**, 340–362.

P. B. Bochev and J. M. Hyman (2006), Principles of mimetic discretizations of differential operators, in *Compatible Spatial Discretizations*, Vol. 142 of *The IMA Volumes in Mathematics and its Applications*, Springer, Berlin, pp. 89–119.

D. Boffi (2000), 'Fortin operator and discrete compactness for edge elements', *Numer. Math.* **87**, 229–246.

D. Boffi (2001), 'A note on the de Rham complex and a discrete compactness property', *Appl. Math. Lett.* **14**, 33–38.

D. Boffi (2006), Compatible discretizations for eigenvalue problems, in *Compatible Spatial Discretizations*, Vol. 142 of *The IMA Volumes in Mathematics and its Applications*, Springer, Berlin, pp. 121–142.

D. Boffi and L. Gastaldi (2002), 'Edge finite elements for the approximation of Maxwell resolvent operator', *M2AN Math. Model. Numer. Anal.* **36**, 293–305.

D. Boffi, F. Brezzi and L. Gastaldi (1997), 'On the convergence of eigenvalues for mixed formulations', *Ann. Scuola Norm. Sup. Pisa Cl. Sci.* (4) **25**, 131–154 (1998).

D. Boffi, F. Brezzi and L. Gastaldi (2000), 'On the problem of spurious eigenvalues in the approximation of linear elliptic problems in mixed form', *Math. Comp.* **69**, 121–140.

A. Bossavit (1988), 'Whitney forms: A class of finite elements for three-dimensional computations in electromagnetism', *IEE Trans. Mag.* **135, Part A**, 493–500.

R. Bott and L. W. Tu (1982), *Differential Forms in Algebraic Topology*, Vol. 82 of *Graduate Texts in Mathematics*, Springer, New York.

D. Braess and W. Hackbusch (1983), 'A new convergence proof for the multigrid method including the V-cycle', *SIAM J. Numer. Anal.* **20**, 967–975.

J. H. Bramble (1993), *Multigrid Methods*, Vol. 294 of *Pitman Research Notes in Mathematics Series*, Longman Scientific & Technical, Harlow.

J. H. Bramble and J. E. Pasciak (1988), 'A preconditioning technique for indefinite systems resulting from mixed approximations of elliptic problems', *Math. Comp.* **50**, 1–17.

J. H. Bramble and J. E. Pasciak (1997), 'Iterative techniques for time dependent Stokes problems', *Comput. Math. Appl.* **33**, 13–30.

S. C. Brenner and L. R. Scott (2002), *The Mathematical Theory of Finite Element Methods*, Vol. 15 of *Texts in Applied Mathematics*, 2nd edn, Springer, New York.

F. Brezzi (1974), 'On the existence, uniqueness and approximation of saddle-point problems arising from Lagrangian multipliers', *Rev. Française Automat. Informat. Recherche Opérationnelle Sér. Rouge* **8**, 129–151.

F. Brezzi and M. Fortin (1991), *Mixed and Hybrid Finite Element Methods*, Vol. 15 of *Springer Series in Computational Mathematics*, Springer, New York.

F. Brezzi, J. Douglas, Jr. and L. D. Marini (1985), 'Two families of mixed finite elements for second order elliptic problems', *Numer. Math.* **47**, 217–235.

F. Brezzi, J. Douglas, Jr., R. Durán and M. Fortin (1987), 'Mixed finite elements for second order elliptic problems in three variables', *Numer. Math.* **51**, 237–250.

A. Čap, J. Slovák and V. Souček (2001), 'Bernstein–Gelfand–Gelfand sequences', *Ann. of Math.* (2) **154**, 97–113.

S. H. Christiansen (2005), Stability of Hodge decompositions of finite element spaces of differential forms in arbitrary dimensions. Preprint, Department of Mathematics, University of Oslo.

P. G. Ciarlet (1978), *The Finite Element Method for Elliptic Problems*, North-Holland, Amsterdam.

P. Clément (1975), 'Approximation by finite element functions using local regularization', *Rev. Française Automat. Informat. Recherche Opérationnelle Sér. Rouge, RAIRO Analyse Numérique* **9**, 77–84.

M. Costabel (1991), 'A coercive bilinear form for Maxwell's equations', *J. Math. Anal. Appl.* **157**, 527–541.

M. Desbrun, A. N. Hirani, M. Leok and J. E. Marsden (2005), Discrete exterior calculus. Available from `arXiv.org/math.DG/0508341`.

J. Douglas, Jr. and J. E. Roberts (1985), 'Global estimates for mixed methods for second order elliptic equations', *Math. Comp.* **44**, 39–52.

M. Dryja and O. B. Widlund (1995), 'Schwarz methods of Neumann–Neumann type for three-dimensional elliptic finite element problems', *Comm. Pure Appl. Math.* **48**, 121–155.

M. Eastwood (2000), 'A complex from linear elasticity', *Rend. Circ. Mat. Palermo* (2) *Suppl.* (63), 23–29.

R. S. Falk and J. E. Osborn (1980), 'Error estimates for mixed methods', *RAIRO Anal. Numér.* **14**, 249–277.

M. Farhloul and M. Fortin (1997), 'Dual hybrid methods for the elasticity and the Stokes problems: a unified approach', *Numer. Math.* **76**, 419–440.

H. Federer (1969), *Geometric Measure Theory*, Vol. 153 of *Die Grundlehren der Mathematischen Wissenschaften*, Springer, New York.

G. J. Fix, M. D. Gunzburger and R. A. Nicolaides (1981), 'On mixed finite element methods for first-order elliptic systems', *Numer. Math.* **37**, 29–48.

B. M. Fraeijs de Veubeke (1965), Displacement and equilibrium models in the finite element method, in *Stress Analysis* (O. C. Zienkiewicz and G. S. Holister, eds), Wiley, New York, pp. 145–197.

W. Fulton and J. Harris (1991), *Representation Theory*, Vol. 129 of *Graduate Texts in Mathematics*, Springer, New York.

M. P. Gaffney (1951), 'The harmonic operator for exterior differential forms', *Proc. Nat. Acad. Sci. USA* **37**, 48–50.

V. Girault and P.-A. Raviart (1986), *Finite element methods for Navier–Stokes equations*, Vol. 5 of *Springer Series in Computational Mathematics*, Springer, Berlin.

J. Gopalakrishnan, L. E. García-Castillo and L. F. Demkowicz (2005), 'Nédélec spaces in affine coordinates', *Comput. Math. Appl.* **49**, 1285–1294.

V. W. Guillemin and S. Sternberg (1999), *Supersymmetry and Equivariant de Rham Theory*, Mathematics Past and Present, Springer, Berlin.

W. Hackbusch (1994), *Iterative Solution of Large Sparse Systems of Equations*, Vol. 95 of *Applied Mathematical Sciences*, Springer, New York. Translated and revised from the 1991 German original.

P. J. Hilton and U. Stammbach (1997), *A Course in Homological Algebra*, Vol. 4 of *Graduate Texts in Mathematics*, 2nd edn, Springer, New York.

R. Hiptmair (1997), 'Multigrid method for **H**(div) in three dimensions', *Electron. Trans. Numer. Anal.* **6**, 133–152 (electronic). Special issue on Multilevel Methods (Copper Mountain, CO, 1997).

R. Hiptmair (1999*a*), 'Canonical construction of finite elements', *Math. Comp.* **68**, 1325–1346.

R. Hiptmair (1999*b*), 'Multigrid method for Maxwell's equations', *SIAM J. Numer. Anal.* **36**, 204–225 (electronic).

R. Hiptmair (2001), Higher order Whitney forms, in *Geometrical Methods in Computational Electromagnetics* (F. Teixeira, ed.), Vol. 32 of *PIER*, EMW Publishing, Cambridge, MA, pp. 271–299.

R. Hiptmair (2002), Finite elements in computational electromagnetism, in *Acta Numerica*, Vol. 11, Cambridge University Press, Cambridge, pp. 237–339.

R. Hiptmair and A. Toselli (2000), Overlapping and multilevel Schwarz methods for vector valued elliptic problems in three dimensions, in *Parallel Solution of Partial Differential Equations* (Minneapolis, MN, 1997), Vol. 120 of *IMA Vol. Math. Appl.*, Springer, New York, pp. 181–208.

K. Jänich (2001), *Vector Analysis*, Undergraduate Texts in Mathematics, Springer, New York. Translated from the second German (1993) edition by Leslie Kay.

C. Johnson and B. Mercier (1978), 'Some equilibrium finite element methods for two-dimensional elasticity problems', *Numer. Math.* **30**, 103–116.

T. Kato (1995), *Perturbation Theory for Linear Operators*, Classics in Mathematics, Springer, Berlin. Reprint of the 1980 edition.

A. Klawonn (1998), 'An optimal preconditioner for a class of saddle point problems with a penalty term', *SIAM J. Sci. Comput.* **19**, 540–552 (electronic).

S. Lang (1995), *Differential and Riemannian manifolds*, Vol. 160 of *Graduate Texts in Mathematics*, Springer, New York.

J.-L. Loday (1992), *Cyclic Homology*, Vol. 301 of *Grundlehren der Mathematischen Wissenschaften*, Springer, Berlin.

B. Mercier, J. Osborn, J. Rappaz and P.-A. Raviart (1981), 'Eigenvalue approximation by mixed and hybrid methods', *Math. Comp.* **36**, 427–453.

D. Mitrea, M. Mitrea and M. Taylor (2001), 'Layer potentials, the Hodge Laplacian, and global boundary problems in nonsmooth Riemannian manifolds', *Mem. Amer. Math. Soc.* **150**, # 713.

M. Mitrea (2001), 'Dirichlet integrals and Gaffney–Friedrichs inequalities in convex domains', *Forum Math.* **13**, 531–567.

P. Monk (1992), 'A finite element method for approximating the time-harmonic Maxwell equations', *Numer. Math.* **63**, 243–261.

P. Monk (2003*a*), *Finite Element Methods for Maxwell's Equations*, Numerical Mathematics and Scientific Computation, Oxford University Press, New York.

P. Monk (2003*b*), A simple proof of convergence for an edge element discretization of Maxwell's equations, in *Computational electromagnetics* (Kiel, 2001), Vol. 28 of *Lecture Notes in Computer Science and Engineering*, Springer, Berlin, pp. 127–141.

P. Monk and L. Demkowicz (2001), 'Discrete compactness and the approximation of Maxwell's equations in $\mathbb{R}^3$', *Math. Comp.* **70**, 507–523.

M. E. Morley (1989), 'A family of mixed finite elements for linear elasticity', *Numer. Math.* **55**, 633–666.

J.-C. Nédélec (1980), 'Mixed finite elements in $\mathbf{R}^3$', *Numer. Math.* **35**, 315–341.

J.-C. Nédélec (1986), 'A new family of mixed finite elements in $\mathbf{R}^3$', *Numer. Math.* **50**, 57–81.

R. A. Nicolaides and K. A. Trapp (2006), Covolume discretization of differential forms, in *Compatible Spatial Discretizations*, Vol. 142 of *The IMA Volumes in Mathematics and its Applications*, Springer, Berlin, pp. 89–119.

J. E. Osborn (1979), Eigenvalue approximation by mixed methods, in *Advances in Computer Methods for Partial Differential Equations III* (Proc. Third IMACS Internat. Sympos., Lehigh Univ., Bethlehem, PA, 1979), IMACS, New Brunswick, NJ, pp. 158–161.

R. Picard (1984), 'An elementary proof for a compact imbedding result in generalized electromagnetic theory', *Math. Z.* **187**, 151–164.

P.-A. Raviart and J. M. Thomas (1977), A mixed finite element method for 2nd order elliptic problems, in *Mathematical Aspects of Finite Element Methods* (Proc. Conf. CNR, Rome, 1975), Vol. 606 of *Lecture Notes in Mathematics*, Springer, Berlin, pp. 292–315.

T. Rusten and R. Winther (1992), 'A preconditioned iterative method for saddlepoint problems', *SIAM J. Matrix Anal. Appl.* **13**, 887–904. Special issue on Iterative Methods in Numerical Linear Algebra (Copper Mountain, CO, 1990).

J. Schöberl (2005), A posteriori error estimates for Maxwell equations. RICAM Report No. 2005-10.

D. Silvester and A. Wathen (1994), 'Fast iterative solution of stabilised Stokes systems II: Using general block preconditioners', *SIAM J. Numer. Anal.* **31**, 1352–1367.

B. F. Smith, P. E. Bjørstad and W. D. Gropp (1996), *Domain Decomposition: Parallel Multilevel Methods for Elliptic Partial Differential Equations*, Cambridge University Press, Cambridge.

E. Stein and R. Rolfes (1990), 'Mechanical conditions for stability and optimal convergence of mixed finite elements for linear plane elasticity', *Comput. Methods Appl. Mech. Engrg.* **84**, 77–95.

R. Stenberg (1986), 'On the construction of optimal mixed finite element methods for the linear elasticity problem', *Numer. Math.* **48**, 447–462.

R. Stenberg (1988a), 'A family of mixed finite elements for the elasticity problem', *Numer. Math.* **53**, 513–538.

R. Stenberg (1988b), Two low-order mixed methods for the elasticity problem, in *The Mathematics of Finite Elements and Applications VI* (Uxbridge, 1987), Academic Press, London, pp. 271–280.

M. E. Taylor (1996), *Partial Differential Equations I: Basic Theory*, Vol. 115 of *Applied Mathematical Sciences*, Springer, New York.

P. S. Vassilevski and J. P. Wang (1992), 'Multilevel iterative methods for mixed finite element discretizations of elliptic problems', *Numer. Math.* **63**, 503–520.

V. B. Watwood, Jr. and B. J. Hartz (1968), 'An equilibrium stress field model for finite element solution of two-dimensional elastostatic problems', *Internat. J. Solids Structures* **4**, 857–873.

J. P. Webb (1999), 'Hierarchal vector basis functions of arbitrary order for triangular and tetrahedral finite elements', *IEEE Trans. Antennas and Propagation* **47**, 1244–1253.

H. Whitney (1957), *Geometric Integration Theory*, Princeton University Press, Princeton, NJ.

J. Xu (1992), 'Iterative methods by space decomposition and subspace correction', *SIAM Rev.* **34**, 581–613.

Acta Numerica (2006), pp. 157–256 © Cambridge University Press, 2006
doi: 10.1017/S0962492906220014

General linear methods

J. C. Butcher
Department of Mathematics,
The University of Auckland, Auckland,
New Zealand
E-mail: butcher@math.auckland.ac.nz

General linear methods, as multistage multivalue methods, are the natural generalizations of linear multistep and Runge–Kutta methods. This survey contains a discussion of the traditional methods and a motivation for the general linear type of generalization. The new methods are introduced in terms of their formulation and the basic properties of consistency, stability and convergence. The order of general linear methods has to be looked at from a new point of view and it is shown how to use an algebraic structure (equivalent to B-series) to express conditions for a given order. Linear and non-linear stability for the new methods brings the theories for the classical methods into a comprehensive formulation and known results are outlined. Recently a number of subfamilies have been introduced and some of these are considered in detail. This applies in particular to methods with the property known as 'inherent Runge–Kutta stability'. These seem to have prospects of yielding useful and efficient methods, and some progress towards their practical implementation is outlined. Finally, the relationship between stability functions and order of methods is discussed in a setting wide enough to include general linear methods as well as multiderivative methods, such as Obreshkov methods. The classical barriers due to Ehle, Daniel–Moore and Dahlquist (second barrier) all fit into a common pattern and these are explored in a general setting.

CONTENTS

1. Introduction

The history of numerical methods for initial value problems up to 1965 was the history of Runge–Kutta methods and linear multistep methods. These seem to have been completely separate developments with the only meeting point being the existence of several low-order methods which simultaneously lie in each of the special classes.

Against this background, it must be asked why a more general type of method should be considered at all. Two reasons are proposed. First, the general linear method formulation is often the most natural framework for analysing the properties, even of traditional methods. Secondly, it is possible that new and potentially superior methods will arise, which could not possibly have been found as developments based on classical methods.

In this introductory section we will review the traditional methods, Euler, linear multistep and Runge–Kutta. Following this section, we will discuss some of the motivations for looking towards a more general type of method. In Section 3, we will consider the formulation of general linear methods and this is followed by a consideration of the meaning and significance of the order of a method. In the short Section 5 we will review the theories of linear and non-linear stability; a theme for this section is that the general linear method formulation is, for many questions, the most natural formulation, even in the case of classical linear multistep methods. The following two sections deal with some known new classes of methods, with a special emphasis on methods possessing the *inherent Runge–Kutta stability* (IRKS) structure. Finally, in Section 8, we study the interrelation between order and stability in the context of multivalue-multistage stability functions.

The bibliography is intended to be wider than references to the publications actually cited in this paper. There is no claim that it includes all work relevant to the development of general linear methods, but it is a start in this direction.

1.1. Initial value problems

The standard initial value problem is written in the form

$$y'(x) = f(x, y(x)), \qquad y(x_0) = y_0,$$

where $f : \mathbb{R} \times \mathbb{R}^N \to \mathbb{R}^N$, although it will sometimes be more convenient to use an autonomous form of this problem,

$$y'(x) = f(y(x)), \qquad y(x_0) = y_0, \tag{1.1}$$

where $f : \mathbb{R}^N \to \mathbb{R}^N$. The individual components will sometimes need to be written out in full:

$$y_i'(x) = f_i(y_1(x), y_2(x), \ldots, y_N(x)), \qquad y_i(x_0) = y_{0i}, \quad i = 1, 2, \ldots, N,$$

where $y_{01}, y_{02}, \ldots, y_{0N}$ are the components of y_0

Even though many practical problems are conveniently presented in non-autonomous form, it is a simple matter to rewrite these problems as an autonomous system, possibly with N increased to $N + 1$. For example, if

$$y_i'(x) = f_i(x, y_1(x), y_2(x), \ldots, y_N(x)), \qquad y_i(x_0) = y_{0i}, \quad i = 1, 2, \ldots, N,$$

then an equivalent autonomous system would be

$$\overline{y}_i'(x) = \overline{f}_i(\overline{y}_0(x), \overline{y}_1(x), \ldots, \overline{y}_N(x)), \qquad \overline{y}_i(x_0) = \overline{y}_{0i}, \quad i = 0, 1, 2, \ldots, N,$$

where

$$\overline{f}_0(\overline{y}_0(x), \overline{y}_1(x), \ldots, \overline{y}_N(x)) = 1,$$
$$\overline{f}_i(\overline{y}_0(x), \overline{y}_1(x), \ldots, \overline{y}_N(x)) = f_i(\overline{y}_0(x), \overline{y}_1(x), \ldots, \overline{y}_N(x)),$$
$$\overline{y}_{0i} = \begin{cases} x_0, & i = 0, \\ y_{0i}, & i = 1, 2, \ldots, N. \end{cases}$$

The autonomous form has significant advantages in that the theory of Runge–Kutta methods is much simpler with this formulation.

It is often convenient to consider an integrated form of the basic initial value problem, that is,

$$y(x) = y_0 + \int_{x_0}^{x} f(x, y(x)) \, dx,$$

so that the process of numerical solution consists in approximating the integral appearing in this formulation.

1.2. The Euler method

If an approximation to the solution to an initial value problem is known at $x = x_{n-1}$, then the solution at $x_n = x_{n-1} + h$ can be written as

$$y(x_n) = y(x_{n-1}) + \int_{x_{n-1}}^{x_n} f(x, y(x)) \, dx, \tag{1.2}$$

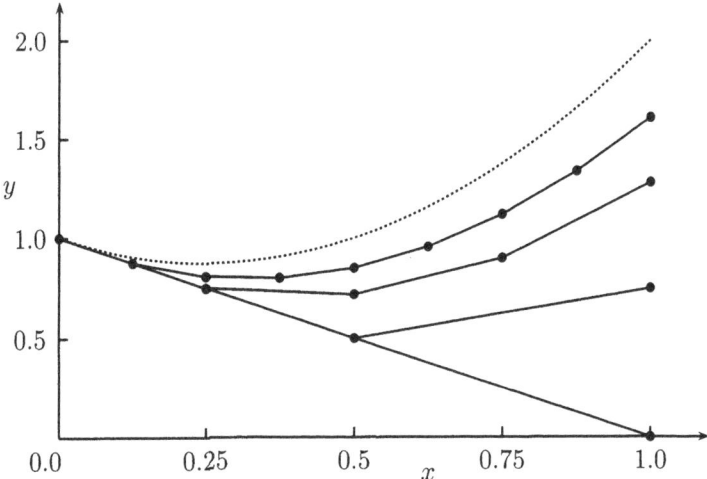

Figure 1.1. Euler method with various step-sizes and exact solution (······).

where $y(x)$ is the trajectory defined by the initial data $y(x_{n-1}) = y_{n-1}$. Approximate the integral by the left Riemann sum

$$\int_a^b \phi(x)\,dx \approx (b - a)\phi(a), \tag{1.3}$$

and we have the approximation

$$y(x_n) \approx y_{n-1} + hf(x_{n-1}, y_{n-1}).$$

Hence we obtain the basic form of the Euler method,

$$y_n = y_{n-1} + hf(x_{n-1}, y_{n-1}).$$

Because of its simplicity, the Euler method is a suitable prototype for discussing a range of questions which can also be asked about more complicated methods. Central to these considerations is the question as to when we can rely on a numerical scheme to provide arbitrarily accurate approximations, provided that sufficient computational effort is extended. This is the question of convergence. There are aspects of stability also associated with the Euler method which provide insights into corresponding questions for more general methods.

Discussion of convergence

In the computation shown in Figure 1.1, the problem

$$y'(x) = y - 2 + 5x - 2x^2, \qquad y(0) = 1,$$

is solved by the Euler method on the interval $[0, 1]$ using n steps with step-size $h = 1/n$, for $n = 1, 2, 4, 8$. It is seen that the approximations for $y(1)$ become steadily more accurate as n increases. This phenomenon is known as 'convergence' and is a necessary property for a numerical method to possess if it is to be used in practical computation. The precise definition and criteria for convergence is best subsumed under the corresponding theory for linear multistep methods which will be informally discussed in Section 1.4. In keeping with the intentions of this paper, we will in turn regard the linear multistep convergence theory as included in the general linear method formulation in Section 3.1.

The implicit Euler method

If instead of the left Riemann sum approximation (1.3) to (1.2), we use the *right* Riemann sum,

$$\int_a^b \phi(x)\,dx \approx (b-a)\phi(b),$$

we arrive at the numerical method

$$y_n = y_{n-1} + hf(x_n, y_n).$$

This is *implicit* because y_n is not given by an explicit formula but is defined as the solution to this algebraic equation.

1.3. Stability of the Euler and implicit Euler methods

The justification for linear stability analysis is argued along the following lines. Consider an autonomous differential equation system

$$y'(x) = f(y(x)), \tag{1.4}$$

and ask how the introduction of a perturbation into the solution carries through to later times. This perturbation may be thought of as the result of computational errors caused by the inaccuracy of a numerical method, or simply as an imprecision in the initial data for the problem. Suppose the perturbation is expressed as a function $\eta(x)$ and that we can assume that η takes on sufficiently small values for the approximation

$$f(y(x) + \eta(x)) \approx f(y(x)) + f'(y(x))\eta(x)$$

to be realistic. If $y(x) + \eta(x)$ is supposed to satisfy the original differential equation system (1.4), then the development of $\eta(x)$ as time passes is approximately as the solution to the problem

$$\eta'(x) = f'(y(x))\eta(x).$$

If the Jacobian matrix $f'(y(x))$ has an eigenvalue q, assumed to be approximately constant over a (possibly small) range of x values, then we are faced with the need to consider the linear differential equation

$$Y'(x) = qY(x), \tag{1.5}$$

as representing some aspect of the behaviour of the perturbation.

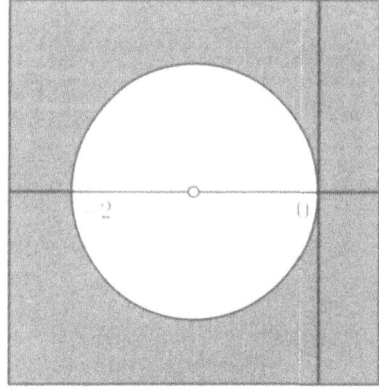
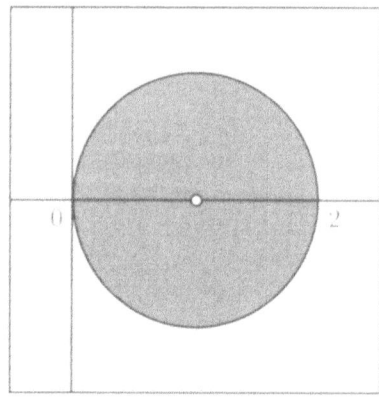

Euler method   Implicit Euler method

Figure 1.2. Stability regions of Euler and implicit Euler methods.

A differential equation in which q can be large, with negative real part, based on a time-scale that might seem appropriate to the numerical modelling of (1.4) may cause serious difficulties, and is described as 'stiff'. The exact solution of (1.5) is a multiple of $\exp(qx)$ and dies away as x increases. However, the result computed by the Euler method, increases in magnitude by a factor $|1 + hq|$ in each time-step. Write $z = hq$ and refer to $1 + z$ as the stability function for the Euler method. The value of h that will permit stable computations to be carried out by the Euler method must be such that $|1 + z| \leq 1$. On the other hand, if it is somehow possible to compute a sequence of numerical approximations using the implicit Euler method, a similar analysis leads to the formula $(1 - z)^{-1}$ for the stability function and the set of points satisfying $|1 - z| \geq 1$ for the stability region.

The stability regions for the Euler and implicit Euler methods are shown in Figure 1.2. From this figure, we see that solving stiff problems is likely to be more successful with the implicit Euler than with the Euler method itself. Generalizations of these methods will need to take this phenomenon into account.

A method, such as the implicit Euler method, for which the stability region includes the left half-plane is said to be 'A-stable'. We will return to this concept in the context of Runge–Kutta, linear multistep and, of course, general linear methods.

Because of its good stability properties in handling linear stiff problems, it might be possible to ask how the implicit Euler methods might be expected to cope with non-linear stiff problems. We consider a test problem introduced in Dahlquist (1976) in his study of linear multistep methods. In this model problem the function $f(x, y)$ is assumed to satisfy the constraint

$$\langle u - v, f(x, u) - f(x, v) \rangle \leq 0, \qquad x \in \mathbb{R}, \quad u, v \in \mathbb{R}^N.$$

The significance of this assumption is that two particular solutions, y and $\widehat{y}$, cannot drift apart, because

$$\frac{d}{dx}\|y(x) - \widehat{y}(x)\|^2 = 2\langle y(x) - \widehat{y}(x), f(x, y(x)) - f(x, \widehat{y}(x))\rangle \leq 0.$$

A corresponding property for two numerical approximations y_n and $\widehat{y}_n$ would be that

$$\|y_n - \widehat{y}_n\|^2 \leq \|y_{n-1} - \widehat{y}_{n-1}\|^2,$$

and this is actually the case for the implicit Euler method because

$$\begin{aligned}
\|y_n - \widehat{y}_n\|^2 - \|y_{n-1} - \widehat{y}_{n-1}\|^2 &+ \|(y_n - \widehat{y}_n) - (y_{n-1} - \widehat{y}_{n-1})\|^2 \\
&= 2\langle y_n - \widehat{y}_n, (y_n - \widehat{y}_n) - (y_{n-1} - \widehat{y}_{n-1})\rangle \\
&= 2\langle y_n - \widehat{y}_n, hf(x_n, y_n) - hf(x_n, \widehat{y}_n)\rangle \\
&\leq 0.
\end{aligned}$$

This model problem is the basis for separate studies of non-linear stability for Runge–Kutta methods, as well as for linear multistep methods. We will return to this question in Section 5, but in the more comprehensive context of general linear methods.

1.4. Linear multistep methods

Given existing approximations $y_i \approx y(x_{n-i})$, $f_i \approx y'(x_{n-i})$, $i = 1, 2, \ldots, k$, a linear k-step method is an algorithm for computing y_n and f_n so that $f_n = f(x_n, y_n)$ and

$$\sum_{i=0}^{k} \alpha_i y_{n-i} = h \sum_{i=0}^{k} \beta_i f_{n-i}. \tag{1.6}$$

In this formulation, $\alpha_0 \neq 0$, because we will want to compute the new approximation value y_n from (1.6). It is possible to rescale by multiplying (1.6) by an arbitrary nonzero factor, so that we could always assume for convenience that $\alpha_0 = 1$. However, different normalizations often lead to simplifications and we will keep the scaling of α_0 open.

Introduction of characteristic polynomials

Following the fundamental ideas of Dahlquist (1956), we introduce polynomials

$$\rho(w) = \sum_{i=0}^{k} \alpha_i w^{k-i}, \tag{1.7}$$

$$\sigma(w) = \sum_{i=0}^{k} \beta_i w^{k-i}. \tag{1.8}$$

It is customary to identify the polynomial pair (ρ, σ) with the linear multi-step method it represents. The polynomials ρ and σ will be assumed to have no common polynomial factor because if $\rho = \phi\hat{\rho}$ and $\sigma = \phi\hat{\sigma}$, where ϕ has nonzero degree $k - \hat{k}$, then numerical results computed using $(\hat{\rho}, \hat{\sigma})$ would also satisfy results computed using (ρ, σ). In particular we can assume that α_k and β_k are not both zero.

Consistency, stability and convergence

A linear multistep method (ρ, σ) is said to be consistent if

$$\rho(1) = 0, \tag{1.9}$$

$$\rho'(1) = \sigma(1). \tag{1.10}$$

The significance of these assumptions is that for a consistent method, the method is able to solve any problem of the form $y'(x) = 1$ exactly over a single step, assuming that exact values of previous step values are used.

The method (ρ, σ) is said to be stable if ρ has all its zeros in the closed unit disc and repeated zeros are in the open unit disc. The significance of this assumption is that the method can not only solve problems of the form $y'(x) = 0$ exactly over many steps but it can do so even with slightly perturbed initial data. This leads to a main theorem in Dahlquist (1956) which relates convergence of a method to the method being both consistent and stable. A precise definition of convergence and further details can be found in standard textbooks. We will return to these ideas again, in the context of general linear methods, in Section 3.1.

Order of methods

A method has order p if it is capable of solving any differential equation exactly if its solution is a polynomial of degree not exceeding p. Put another way, this means that if the expression

$$\sum_{i=0}^{k} \alpha_i y_{n-i} - h \sum_{i=0}^{k} \beta_i f_{n-i}$$

is evaluated, with all y and f values replaced by the quantities they are supposed to approximate, then its formal Taylor series vanishes up to and including terms in h^p. Evaluate this series, expanding about x_{n-k}, and we obtain an expression of the form

$$\sum_{i=0}^{k} \alpha_i y(x_{n-i}) - h \sum_{i=0}^{k} \beta_i y'(x_{n-i}) = \sum_{i=0}^{\infty} C_i h^i y^{(i)}(x_{n-k}). \tag{1.11}$$

From the known Taylor expansions of $y(x_{n-i})$ and $y'(x_{n-i})$ we find the following formulae for $C_0, C_1, \ldots$:

$$C_0 = \sum_{j=0}^{k} \alpha_j, \qquad (1.12)$$

$$C_1 = \sum_{j=0}^{k-1} (k-j)\alpha_j - \sum_{j=0}^{k} \beta_j \qquad (1.13)$$

$$C_i = \frac{1}{i!}\left(\sum_{j=0}^{k-1} (k-j)^i \alpha_j - i\sum_{j=0}^{k-1} (k-j)^{i-1}\beta_j\right), \quad i = 2, 3, \ldots. \qquad (1.14)$$

Evaluate (1.11) for the special case $y(x) = \exp(z(x - x_{n-k})/h)$ (so that $y'(x) = (z/h)\exp(z(x-x_{n-k})/h))$, where z is an arbitrary complex number, and we find

$$\sum_{i=0}^{k} \alpha_i \exp((k-i)z) - z\sum_{i=0}^{k} \beta_i \exp((k-i)z) = C_0 + C_1 z + C_2 z^2 + \cdots,$$

so that

$$\rho(\exp(z)) - z\sigma(\exp(z)) = C_0 + C_1 z + C_2 z^2 + \cdots.$$

This enables us to state a convenient criterion for order.

Theorem 1.1.   A linear multistep method (ρ, σ) has order p if and only if

$$\rho(\exp(z)) - z\sigma(\exp(z)) = \mathcal{O}(z^{p+1}). \qquad (1.15)$$

By substituting $\log(1 + z) = z - \frac{1}{2}z^2 + \frac{1}{3}z^3 - \cdots$ for z in (1.15) and rearranging we find

$$\frac{1}{\log(1+z)}\rho(1+z) - \sigma(1+z) = \mathcal{O}(z^p),$$

where, because of consistency, $\rho(1 + z)$ is a multiple of z. Hence, we have the following result.

Corollary 1.2.   A linear multistep method (ρ, σ) has order p if and only if

$$\frac{1}{\log(1+z)/z}\frac{\rho(1+z)}{z} - \sigma(1+z) = \mathcal{O}(z^p). \qquad (1.16)$$

Adams and BDF methods

By writing $\rho(z) = z^k - z^{k-1}$ and defining σ to be the degree $k-1$ polynomial satisfying (1.16), with $p = k$, Adams–Bashforth methods are derived. By increasing the degree of σ to k and p to $k + 1$, Adams–Moulton methods are found.

Table 1.1. Adams methods.

	Adams–Bashforth				Adams–Moulton				
k	β_1	β_2	β_3	β_4	β_0	β_1	β_2	β_3	β_4
1	1				$\frac{1}{2}$	$\frac{1}{2}$			
2	$\frac{3}{2}$	$-\frac{1}{2}$			$\frac{5}{12}$	$\frac{2}{3}$	$-\frac{1}{12}$		
3	$\frac{23}{12}$	$-\frac{4}{3}$	$\frac{5}{12}$		$\frac{3}{8}$	$\frac{19}{24}$	$-\frac{5}{24}$	$\frac{1}{24}$	
4	$\frac{55}{24}$	$-\frac{59}{24}$	$\frac{37}{24}$	$-\frac{3}{8}$	$\frac{251}{720}$	$\frac{323}{360}$	$-\frac{11}{30}$	$\frac{53}{360}$	$-\frac{19}{720}$

Table 1.2. BDF methods.

k	β_0	α_1	α_2	α_3	α_4
1	1	1			
2	$\frac{2}{3}$	$\frac{4}{3}$	$-\frac{1}{3}$		
3	$\frac{6}{11}$	$\frac{18}{11}$	$-\frac{9}{11}$	$\frac{2}{11}$	
4	$\frac{12}{25}$	$\frac{48}{25}$	$-\frac{36}{25}$	$\frac{16}{25}$	$-\frac{3}{25}$

The series for $z/\log(1+z)$, occurring in (1.16) is

$$\frac{z}{\log(1+z)} =$$
$$1 + \tfrac{1}{2}z - \tfrac{1}{12}z^2 + \tfrac{1}{24}z^3 - \tfrac{19}{720}z^4 + \tfrac{3}{160}z^5 - \tfrac{863}{60480}z^6 + \tfrac{275}{24192}z^7 - \tfrac{33953}{3628800}z^8 + \cdots,$$

and we readily find the first few Adams–Bashforth (order k) and Adams–Moulton (order $k+1$) methods as shown in Table 1.1. In this table the values of β_0 (for the AM method only), β_1, β_2, ..., β_k, are given, assuming the scaling $\alpha_0 = -\alpha_1 = 1$.

The 'backward difference formulae' (BDF) are approximations to the derivative of $y(x)$ at x_n in terms of the values of $y(x_{n-i})$, $i = 0, 1, \ldots, k$. The corresponding linear multistep methods are referred to as BDF methods. To derive such methods of order $p = k$, write $\sigma(z) = z^k$, and find ρ of degree k from

$$\rho(1+z) = \log(1+z)(1+z)^k + \mathcal{O}(z^{p+1}).$$

For the first few BDF methods, the coefficients are shown in Table 1.2, scaled so that $\alpha_0 = 1$.

Stability regions and A-stability
For a linear differential equation $y' = qy$, the difference equation (1.6) simplifies to

$$\sum_{i=0}^{k}(\alpha_i - z\beta_i)y_{n-i} = 0,$$

where here $z = hq$. The stability region is the set of points in the complex plane for which this difference equation has only bounded solution sequences. This means that, for z in the stability region,

$$\rho(w) - z\sigma(w),$$

regarded as a polynomial in w, has all its zeros in the closed unit disc and has all repeated zeros in the interior. Generalizing the discussion in Section 1.3, we will refer to a method as being A-stable if the stability region includes the left half-plane.

Dahlquist barriers
The barriers of Dahlquist (1956, 1963) state fundamental limitations on achievable order for linear multistep methods which satisfy various stability properties. The second barrier is concerned with A-stability and we will discuss this in Section 8. The first barrier is stated in the following result.

Theorem 1.3. The order p of a stable k-step method is bounded by

$$p \leq \begin{cases} k+1, & k \text{ odd,} \\ k+2, & k \text{ even.} \end{cases}$$

Proof. Substitute $\log(\frac{1+z}{1-z})$ for z in (1.15), to obtain

$$\rho\left(\frac{1+z}{1-z}\right) - \log\left(\frac{1+z}{1-z}\right)\sigma\left(\frac{1+z}{1-z}\right) = \mathcal{O}(z^{p+1}), \qquad (1.17)$$

Let

$$r(z) = (1-z)^k\rho\left(\frac{1+z}{1-z}\right) = \sum_{i=0}^{k}a_iz^i, \qquad (1.18)$$

$$s(z) = (1-z)^k\sigma\left(\frac{1+z}{1-z}\right) = \sum_{i=0}^{k}b_iz^i. \qquad (1.19)$$

If z_0 is a zero of r, then $\rho((1+z_0)/(1-z_0)) = 0$. Hence, $(1+z_0)/(1-z_0)$ is in the closed unit disc, implying that z_0 is in the closed left half-plane. Because 1 is a zero of ρ, 0 is a zero of r and hence, $a_0 = 0$. However, because 1 is not a *repeated* zero of ρ, 0 is not a repeated zero of r. Hence, $a_1 \neq 0$. Without loss of generality, assume that $a_1 > 0$. Because all zeros of r are in the left half-plane, and because these are all real or exist in conjugate pairs, r is a constant multiplied by products of the form $z+\alpha$, where $\alpha \geq 0$, or of the form $z^2 + \alpha z + \beta$, where α and β are each nonnegative. Hence, no two coefficients in r can have opposite signs and therefore $a_i \geq 0$, for $i = 1, 2, \ldots, k$.

Multiply (1.17) by $(1-z)^k$ and divide by $\log(\frac{1+z}{1-z})$ and we find

$$(c_0 + c_2 z^2 + c_4 z^4 + \cdots)(a_1 + a_2 z + \cdots + a_k z^{k-1})$$
$$= b_0 + b_1 z + \cdots + b_k z^k + \mathcal{O}(z^p), \quad (1.20)$$

where

$$\left(c_0 + c_2 z^2 + c_4 z^4 + \cdots\right)\left(\tfrac{1}{z} \log\left(\tfrac{1+z}{1-z}\right)\right) = 1. \quad (1.21)$$

From the known Taylor series for $\log(\frac{1+z}{1-z})$, we see that $c_0 = \frac{1}{2}$ and $c_2 = -\frac{1}{6}$. We now prove by induction that $c_{2n} < 0$ for $n \geq 1$. From the z^{2n} and z^{2n-2} terms in the expansion of (1.21), we find

$$c_{2n} + \frac{1}{3}c_{2n-2} + \cdots + \frac{1}{2n+1}c_0 = 0, \quad (1.22)$$

$$c_{2n} + \frac{1}{3}c_{2n-4} + \cdots + \frac{1}{2n-1}c_0 = 0. \quad (1.23)$$

Multiply (1.22) by $2n+1$, subtract the result of multiplying (1.23) by $2n-1$, and rearrange to find c_{2n} as a positive linear combination of $c_2, \ldots, c_{2n-2}$, completing the proof that $c_{2n} < 0$ for all positive n.

We now need to prove that an order $p > k+1$ is impossible for k odd. If it were possible, then the coefficient of z^{k+1} in (1.20) would be zero. However, this equals

$$a_k c_2 + a_{k-2} c_4 + \cdots + a_1 c_{k+1},$$

which cannot be zero unless all the terms are zero; but this would imply $a_1 = 0$, which is impossible. If k is even and $p > k+2$, then the coefficient of z^{k+2} in (1.20) would be zero. This implies

$$a_{k-1} c_4 + a_{k-3} c_6 + \cdots + a_1 c_{k+2} = 0,$$

which again would lead to the impossible conclusion that $a_1 = 0$. $\square$

1.5. Runge–Kutta methods

Formulation of methods

The well-known methods of Runge (1895), Heun (1900) and Kutta (1901), generalize the classical Euler method by allowing for additional functional evaluations in each time-step. Write $Y_1, Y_2, \ldots, Y_s$ for the arguments of these evaluations, and $F_1, F_2, \ldots, F_s$ for the corresponding derivative approximations. For explicit methods, as in the cited works, each Y_i is a linear combination of the hF_j values, for $j < i$ added on to the input approximation, which we will write as y_{n-1} for step number n. Once the s stages and corresponding derivative approximations have been computed, the output value for the step is computed as a further linear combination of the hF_i values, for $i = 1, 2, \ldots, s$.

Putting all this together we formulate the method as

$$Y_i = y_{n-1} + h \sum_{j<i} a_{ij} F_j, \quad F_i = f(Y_i), \quad i = 1, 2, \ldots, s, \qquad (1.24)$$

$$y_n = y_{n-1} + h \sum_{i=1}^{s} b_i F_i, \qquad (1.25)$$

where the numbers a_{ij}, b_i are characteristic of a specific method.

Because we will also want to consider 'implicit' methods in which the sums in (1.24) extend beyond $j < i$, we will conventionally introduce a full matrix of a_{ij} coefficients where, for the explicit case we have so far considered, $a_{ij} = 0$ if $j \geq i$.

This formulation is for an autonomous problem. In the case of a non-autonomous problem, we need to take account of the point within the step to which each stage corresponds. Suppose that stage number i evaluates an approximation at $x_{n-1} + hc_i$. By considering the simple differential equation $y' = 1$ we can evaluate c_i as the sum of the elements in row number i of the a_{ij} table of coefficients. That is,

$$c_i = \sum_{j=i}^{s} a_{ij}. \qquad (1.26)$$

The modification to (1.24) and (1.25) to handle the non-autonomous case is

$$Y_i = y_{n-1} + h \sum_{j<i} a_{ij} F_j, \quad F_i = f(x_{n-1} + hc_i, Y_i), \quad i = 1, 2, \ldots, s, \quad (1.27)$$

$$y_n = y_{n-1} + h \sum_{i=1}^{s} b_i F_i. \qquad (1.28)$$

It is customary to write the collection of coefficients a_{ij}, b_i and c_i in a tableau thus:

0				
c_2	a_{21}			
c_3	a_{31}	a_{32}		
$\vdots$	$\vdots$	$\vdots$	$\ddots$	
c_s	a_{s1}	a_{s2}	$\cdots$	$a_{s,s-1}$
	b_1	b_2	$\cdots$	b_{s-1} $\quad b_s$

where we note that $c_1 = 0$ and that we have omitted those elements of the a_{ij} array which are necessarily zero.

Sometimes we will need to introduce the full matrix of coefficients and we denote this by A. The two vectors b^T and c are also introduced. Later we will need to consider implicit methods and in this case A will be a full matrix. Using these notations, the tableau of coefficients will be written as

$$\begin{array}{c|c} c & A \\ \hline & b^T \end{array}.$$

Order conditions

Order p linear multistep methods are constructed using approximations to $y(x_n)$ in terms of y evaluated at x_{n-i}, $i = 1, 2, \ldots, k$ and hy' evaluated at x_{n-i}, $i = 0, 1, \ldots, k$. To achieve the required order, the approximation must be exactly satisfied whenever y is a polynomial of degree less than p. If the values on which the approximations are based are themselves approximations found in previous steps, we can still interpret the current approximation as having order p, because the total error in $y(x_n)$ is made up from the truncation error in this approximation, together with inherited errors, all of which can be estimated in terms of $O(h^{p+1})$.

For Runge–Kutta methods, the situation is more complicated because even though the approximation (1.28) is based on the integral

$$y(x_n) = y(x_{n-1}) + h \int_0^1 y'(x_{n-1} + h\xi) \, d\xi \approx y(x_{n-1}) + h \sum_{i=1}^s b_i y'(x_{n-1} + hc_i),$$

the values of F_i are not accurate approximations to $y'(x_{n-1} + hc_i)$, $i = 1, 2, \ldots, s$.

We deal with this complication by carrying out three steps. First we find the Taylor expansion of the exact solution; secondly we find the Taylor expansion for the approximation computed using a Runge–Kutta method. Finally, by comparing these two Taylor expansions term by term, we arrive at conditions for the difference between them to equal $O(h^{p+1})$.

To commence the first step of finding the formal Taylor expansion of y satisfying (1.1), we need formulae for the second, third, $\ldots$, derivatives for this function:

$$y'(x) = f(y(x)),$$
$$y''(x) = f'(y(x))y'(x)$$
$$\quad\ = f'(y(x))f(y(x)),$$
$$y'''(x) = f''(y(x))(f(y(x)), y'(x)) + f'(y(x))f'(y(x))y'(x)$$
$$\quad\ = f''(y(x))(f(y(x)), f(y(x))) + f'(y(x))f'(y(x))f(y(x)).$$

This sequence of expressions becomes increasingly complicated as we evaluate higher derivatives and we look for a systematic pattern.

Table 1.3. Tree-like structure of terms appearing in derivative formulae.

$$y'(x) = \mathbf{f} \qquad\qquad \bullet\, \mathbf{f}$$

$$y''(x) = \mathbf{f'f}$$

$$y'''(x) = \mathbf{f''(f,f)}$$

$$+\,\mathbf{f'f'f}$$

Write $\mathbf{f} = f(y(x))$, $\mathbf{f'} = f'(y(x))$, $\mathbf{f''} = f''(y(x))$, ... and consider Table 1.3 where the expressions for y' and y'' and the two terms occurring in y''' are shown together with their tree-like structures.

Motivated by this structure, we introduce the set of all rooted trees and the corresponding derivative terms.

Trees and elementary differentials

Let T denote the set of rooted trees:

$$T = \left\{ \begin{matrix}\end{matrix} \;,\; \; , \; \; , \; \; , \; \; , \; \; , \; \; , \; \; , \; \; \dots \right\}. \tag{1.29}$$

It is convenient to introduce some notation, including a tree-building structure. We will usually omit 'rooted' and refer only to trees.

The tree $\bullet$ will be denoted by τ. Given trees $t_1, t_2, \ldots, t_m$ we consider the tree formed by joining the roots of each of these trees to a new root. This will be written as $[t_1 t_2 \ldots t_m]$. Furthermore the notation will be made more compact by denoting repeated trees within $[\cdot]$ using exponents. Repeated use of the $[\cdot]$ operation will be denoted using subscripts.

For example the first 8 trees in the sequence, listed in (1.29) are written in terms of this new notation as follows:

$$T = \left\{ \tau, [\tau], [\tau^2], [_2\tau]_2, [\tau^3], [\tau[\tau]], [_2\tau^2]_2, [_3\tau]_3, \dots \right\}.$$

Trees of the form $[\tau^n]$ are sometimes referred to as 'bushy trees' and trees of the form $[_n\tau]_n$ as tall trees.

We can now define the elementary differentials.

Definition 1.4. The elementary differential associated with the tree t, the function f and the evaluation point y_0 is defined by

$$F(\tau)(y_0) = f(y_0),$$

$$F([t_1, t_2, \ldots, t_m])(y_0) = f^{(m)}(y_0)(F(t_1)(y_0), F(t_2)(y_0), \ldots, F(t_m)(y_0)).$$

Table 1.4. Some functions on trees and an example.

Function	Name (and meaning)	Example	Construction
$r(t)$	order of t (number of vertices)	7	
$\sigma(t)$	symmetry of t (order of automorphism group)	8	
$\gamma(t)$	density of t	63	
$\alpha(t)$	(number of ways of labelling t with an ordered set)	10	$\dfrac{r(t)!}{\sigma(t)\gamma(t)}$
$\beta(t)$	(number of ways of labelling t with an unordered set)	630	$\dfrac{r(t)!}{\sigma(t)}$
$F(t)(y_0)$	elementary differential	$\mathbf{f}'\left(\mathbf{f}'(\mathbf{f},\mathbf{f}),\mathbf{f}'(\mathbf{f},\mathbf{f})\right)$	
$\Phi(t)$	elementary weight	$\displaystyle\sum_{i,j,k=1}^{s} b_i a_{ij} c_j^2 a_{ik} c_k^2$	

Functions on trees

The various functions we will need are summarized in Table 1.4, with a more detailed explanation available in Butcher (2003). In Table 1.4, t denotes a typical tree. Also given are examples of these functions based on the tree $t = $, which, in terms of the notation we have introduced, can also be written as $[[\tau^2]^2]$.

The function $\Phi(t)$ will be explained below. The remaining functions are easy to compute up to order 4 trees and are shown in Table 1.5.

Taylor expansions and order conditions

The formal Taylor expansion of the solution at $x_0 + h$ is

$$y(x_0 + h) = y_0 + \sum_{t \in T} \frac{\alpha(t) h^{r(t)}}{r(t)!} F(t)(y_0).$$

Using the known formula for $\alpha(t)$, we can write this as

$$y(x_0 + h) = y_0 + \sum_{t \in T} \frac{h^{r(t)}}{\sigma(t)\gamma(t)} F(t)(y_0). \qquad (1.30)$$

Our aim will now be to find a corresponding formula for the result computed by one step of a Runge–Kutta method. By comparing these formulae term by term, we will obtain conditions for a specific order of accuracy.

Table 1.5. Various functions on trees.

t	$r(t)$	$\sigma(t)$	$\gamma(t)$	$\alpha(t)$	$\beta(t)$	$F(t)$	$\Phi(t)$
•	1	1	1	1	1	$\mathbf{f}$	$\sum b_i$
	2	1	2	1	2	$\mathbf{f'f}$	$\sum b_i c_i$
	3	2	3	1	3	$\mathbf{f''(f,f)}$	$\sum b_i c_i^2$
	3	1	6	1	6	$\mathbf{fff}$	$\sum b_i a_{ij} c_j$
	4	6	4	1	4	$\mathbf{f^{(3)}(f,f,f)}$	$\sum b_i c_i^3$
	4	1	8	3	24	$\mathbf{f''(f,f'f)}$	$\sum b_i c_i a_{ij} c_j$
	4	2	12	1	12	$\mathbf{f'f''(f,f)}$	$\sum b_i a_{ij} c_j^2$
	4	1	24	1	24	$\mathbf{f'f'f'f}$	$\sum b_i a_{ij} a_{jk} c_k$

We need to evaluate various expressions, known as 'elementary weights', which depend on the tableau for a particular method. First we use the example tree we have already considered to illustrate the construction of the elementary weight $\Phi(t)$ for this tree t:

$$t = \quad \text{(tree diagram with nodes labelled } l, m, n, o, j, k, i\text{)}$$

The elementary weight for this tree is

$$\Phi(t) = \sum_{i,j,k,l,m,n,o=1}^{s} b_i a_{ij} a_{ik} a_{jl} a_{jm} a_{kn} a_{ko},$$

which can be simplified by summing over l, m, n, o and using (1.26):

$$\Phi(t) = \sum_{i,j,k=1}^{s} b_i a_{ij} c_j^2 a_{ik} c_k^2.$$

It is now possible to write down the formal Taylor expansion of the solution at $x_0 + h$ in the form

$$y_1 = y_0 + \sum_{t \in T} \frac{\beta(t) h^{r(t)}}{r(t)!} \Phi(t) F(t)(y_0).$$

Using the known formula for $\beta(t)$, this can be re-written as

$$y_1 = y_0 + \sum_{t \in T} \frac{h^{r(t)}}{\sigma(t)} \Phi(t) F(t)(y_0). \tag{1.31}$$

If the Taylor series (1.31) is to match the Taylor series (1.30), up to h^p terms, we need to ensure that

$$\Phi(t) = \frac{1}{\gamma(t)},$$

for all trees such that

$$r(t) \le p.$$

These are the 'order conditions'.

Low-order explicit methods

We will attempt to construct methods of order $p = s$ with s stages for $s = 1, 2, \ldots$. We will find that this is possible up to order 4 but not for $p \ge 5$. The usual approach will be to first choose $c_2, c_3, \ldots, c_s$ and then solve for $b_1, b_2, \ldots, b_s$. After this solve for those of the a_{ij} which can be found as solutions to linear equations.

Order 2. The order conditions are

$$b_1 + b_2 = 1,$$

$$b_2 c_2 = \tfrac{1}{2},$$

with solution, for arbitrary nonzero c_2,

$$
\begin{array}{c|cc}
0 & & \\
c_2 & c_2 & \\
\hline
& 1 - \frac{1}{2c_2} & \frac{1}{2c_2}
\end{array}.
$$

Choose $c_2 = \tfrac{1}{2}$ and $c_2 = 1$, respectively, and we obtain the two well-known special cases

$$
\begin{array}{c|cc}
0 & & \\
\frac{1}{2} & \frac{1}{2} & \\
\hline
& 0 & 1
\end{array},
\qquad
\begin{array}{c|cc}
0 & & \\
1 & 1 & \\
\hline
& \frac{1}{2} & \frac{1}{2}
\end{array}.
$$

Order 3. The order conditions are

$$b_1 + b_2 + b_3 = 1,$$

$$b_2 c_2 + b_3 c_3 = \tfrac{1}{2},$$

$$b_2 c_2^2 + b_3 c_3^2 = \tfrac{1}{3},$$

$$b_3 a_{32} c_2 = \tfrac{1}{6}.$$

Three representative special cases are

$$
\begin{array}{c|ccc}
0 \\
\frac{1}{2} & \frac{1}{2} \\
1 & -1 & 2 \\
\hline
& \frac{1}{6} & \frac{2}{3} & \frac{1}{6}
\end{array}
\quad,\quad
\begin{array}{c|ccc}
0 \\
\frac{2}{3} & \frac{2}{3} \\
\frac{2}{3} & 0 & \frac{2}{3} \\
\hline
& \frac{1}{4} & \frac{3}{8} & \frac{3}{8}
\end{array}
\quad,\quad
\begin{array}{c|ccc}
0 \\
\frac{2}{3} & \frac{2}{3} \\
0 & -1 & 1 \\
\hline
& 0 & \frac{3}{4} & \frac{1}{4}
\end{array}\;.
$$

Order 4. The order conditions are

$$b_1 + b_2 + b_3 + b_4 = 1, \tag{1.32}$$

$$b_2 c_2 + b_3 c_3 + b_4 c_4 = \tfrac{1}{2}, \tag{1.33}$$

$$b_2 c_2^2 + b_3 c_3^2 + b_4 c_4^2 = \tfrac{1}{3}, \tag{1.34}$$

$$b_3 a_{32} c_2 + b_4 a_{42} c_2 + b_4 a_{43} c_3 = \tfrac{1}{6}, \tag{1.35}$$

$$b_2 c_2^3 + b_3 c_3^3 + b_4 c_4^3 = \tfrac{1}{4}, \tag{1.36}$$

$$b_3 c_3 a_{32} c_2 + b_4 c_4 a_{42} c_2 + b_4 c_4 a_{43} c_3 = \tfrac{1}{8}, \tag{1.37}$$

$$b_3 a_{32} c_2^2 + b_4 a_{42} c_2^2 + b_4 a_{43} c_3^2 = \tfrac{1}{12}, \tag{1.38}$$

$$b_4 a_{43} a_{32} c_2 = \tfrac{1}{24}. \tag{1.39}$$

To solve these equations, treat c_2, c_3, c_4 as parameters, and solve for b_1, b_2, b_3, b_4 from (1.32), (1.33), (1.34), (1.36). Now solve for a_{32}, a_{42}, a_{43} from (1.35), (1.37) and (1.38). Finally, use (1.39) to obtain a consistency condition on c_2, c_3, c_4. This consistency condition is found to be $c_4 = 1$.

We will prove a stronger result in another way.

Theorem 1.5. If an explicit Runge–Kutta method with $s = 4$ has order 4, then

$$\sum_{i=j+1}^{s} b_i a_{ij} = b_j (1 - c_j), \qquad j = 1, 2, 3, 4$$

and, in particular, $c_4 = 1$.

Proof. The result $c_4 = 1$ is proved in Lemma 1.6 below. Hence, $v_4 = 0$, where $v_j = \sum_{i=j+1}^{s} b_i a_{ij} - b_j (1 - c_j)$, $j = 1, 2, 3, 4$. Also $v_3 = 0$ because $\sum_{j,k} v_j a_{jk} c_k = 0$, which we find by expanding and using the order conditions. Finally, $\sum_j v_j c_j = 0$, implying $v_2 = 0$, and $\sum_j v_j = 0$, implying $v_1 = 0$. $\qquad \square$

Lemma 1.6. If an explicit Runge–Kutta method has order p where $s = p \geq 4$, then $c_4 = 1$.

Proof. Consider the matrix formed as the result of the product

$$
\begin{bmatrix} 1 & 0 & 0 \\ 0 & 1 & -c_4 \end{bmatrix}
\begin{bmatrix} b^T A^{p-3} \\ b^T A^{p-4} C \\ b^T A^{p-4} \end{bmatrix}
\begin{bmatrix} Ac & Cc & c \end{bmatrix}
\begin{bmatrix} 1 & 0 \\ 0 & 1 \\ 0 & -c_2 \end{bmatrix}.
$$

This matrix has rank one because $b^T A^{p-3}$ and $b^T A^{p-4} C - c_4 b^T A^{p-4}$ are each zero except for components number $1, 2, 3$ and because Ac and $Cc - c_2 c$ are each zero in components $1, 2$. Hence, multiplying the middle two factors, we see that the product can be written as

$$
\begin{bmatrix} 1 & 0 & 0 \\ 0 & 1 & -c_4 \end{bmatrix}
\begin{bmatrix} b^T A^{p-2} c & b^T A^{p-3} C c & b^T A^{p-3} c \\ b^T A^{p-4} C A c & b^T A^{p-4} C^2 c & b^T A^{p-4} C c \\ b^T A^{p-3} c & b^T A^{p-4} C c & b^T A^{p-4} c \end{bmatrix}
\begin{bmatrix} 1 & 0 \\ 0 & 1 \\ 0 & -c_2 \end{bmatrix}.
$$

Evaluate the second factor by the order conditions and we obtain the result

$$
\begin{bmatrix} 1 & 0 & 0 \\ 0 & 1 & -c_4 \end{bmatrix}
\begin{bmatrix} \frac{1}{p!} & \frac{2}{p!} & \frac{p}{p!} \\ \frac{3}{p!} & \frac{6}{p!} & \frac{2p}{p!} \\ \frac{p}{p!} & \frac{2p}{p!} & \frac{p(p-1)}{p!} \end{bmatrix}
\begin{bmatrix} 1 & 0 \\ 0 & 1 \\ 0 & -c_2 \end{bmatrix}
$$

$$
= \frac{1}{p!} \begin{bmatrix} 1 & 2 - pc_2 \\ 3 - pc_4 & 6 - 2pc_2 - 2pc_4 + p(p-1)c_2 c_4 \end{bmatrix}.
$$

Because this matrix has rank not exceeding 1, its determinant is zero. This gives the result $c_2(1 - c_4) = 0$. The possibility that $c_2 = 0$ has to be rejected because this would lead to the contradiction

$$
0 = b^T A^{p-2} c = \frac{1}{p!}.
$$

Hence, for any Runge–Kutta method with $s = p \geq 4$, c_4 necessarily equals 1. □

As a result of Theorem 1.5, the construction of fourth-order Runge–Kutta methods now becomes straightforward. Kutta (1901) classified all solutions to the fourth-order conditions.

In particular, we have the famous method:

$$
\begin{array}{c|cccc}
0 & & & & \\
\frac{1}{2} & \frac{1}{2} & & & \\
\frac{1}{2} & 0 & \frac{1}{2} & & \\
1 & 0 & 0 & 1 & \\
\hline
& \frac{1}{6} & \frac{1}{3} & \frac{1}{3} & \frac{1}{6}
\end{array}.
$$

An order barrier

We will review what is achievable up to order 8. In Table 1.6, N_p is the number of order conditions to achieve this order. $M_s = s(s+1)/2$ is the

Table 1.6. Minimum number of stages s to achieve order p.

p	N_p	s	M_s
1	1	1	1
2	2	2	3
3	4	3	6
4	8	4	10
5	17	6	21
6	37	7	28
7	115	9	45
8	200	11	66

number of free parameters to satisfy the order conditions for the required s stages.

According to Table 1.6, it is suggested that, for $p \geq 5$, it is necessary that $s > p$. We will now prove this result.

Theorem 1.7. There does not exist an explicit Runge–Kutta method with order $p = s \geq 5$.

Proof. Recall from Lemma 1.6 that $c_4 = 1$. If $s = p \geq 5$, repeat the argument but starting from the product

$$\begin{bmatrix} 1 & 0 & 0 \\ 0 & 1 & -c_5 \end{bmatrix} \begin{bmatrix} b^T A^{p-4} \\ b^T A^{p-5} C \\ b^T A^{p-5} \end{bmatrix} \begin{bmatrix} A^2 c & ACc & Ac \end{bmatrix} \begin{bmatrix} 1 & 0 \\ 0 & 1 \\ 0 & -c_2 \end{bmatrix}$$

and we now find that $c_5 = 1$. If $s = p \geq 5$ and $c_4 = c_5 = 1$, we obtain the contradiction

$$0 = b^T A^{p-5}(I - C)A^2 c = \frac{1}{p!}. \qquad \square$$

1.6. Algebraic theory and B-series

We will review work first presented in Butcher (1972a); it has since become known under the name B-series (Hairer and Wanner 1974). A more recent account is given in Butcher (2003).

As a first step, make a slight generalization to the formulation of Runge–Kutta methods, by inserting a factor b_0 in the term y_{n-1} in (1.25). In such a generalized Runge–Kutta method, the extra coefficient can be conveniently inserted into its tableau:

$$\frac{\begin{array}{c|c} c & A \end{array}}{\begin{array}{c|c} b_0 & b^T \end{array}}. \qquad (1.40)$$

We will conventionally add an additional 'empty tree', denoted by $\emptyset$, to the set of rooted trees to form the augmented set

$$T^{\#} = T \cup \{\emptyset\}.$$

For the generalized Runge–Kutta tableau (1.40), define the corresponding elementary weight as

$$\Phi(\emptyset) = b_0,$$

so that, for a standard Runge–Kutta method, $\Phi(\emptyset) = 1$.

We will denote by X the set of mappings $T^{\#} \to \mathbb{R}$ and by X_1 the subset for which $\emptyset \mapsto 1$. Thus, to each Runge–Kutta method, we can associate a member of X_1, such that $t \mapsto \Phi(t)$.

The order conditions can be written in terms of elementary weights, and it is natural to ask in what sense the elementary weights characterize a Runge–Kutta method. The answer is that if two Runge–Kutta methods have the same sequence of elementary weights, then they are equivalent methods in a very natural sense. For example, if two methods are equivalent then they give the same numerical result when applied to the same problem. Furthermore they are equivalent also in the sense that if unused stages are eliminated and sets of stages which give identical results are collapsed into a single stage, then the two methods have equivalent tableaux, except for the ordering of the stages. For a more detailed explanation of equivalences amongst Runge–Kutta methods, see, for example, Butcher (1996b).

Given that we can represent equivalence classes of Runge–Kutta methods using the sequence of elementary weights, we might ask: What is the significance of the right-hand sides of the order conditions? We will give an interpretation of these quantities in terms of a limiting type of Runge–Kutta method which can be thought of as having arbitrarily high order. For convenience we will consider a step of size h starting from $y(x_0) = y_0$.

The exact solution at the end of this step, and at points within the step, is given by the Picard integral equation

$$y(x_0 + h\xi) = y_0 + h \int_0^{\xi} f(y(x_0 + h\eta)) \, \mathrm{d}\eta.$$

We can regard this as an idealized Runge–Kutta method in which the finite index set for the stages, $\{1, 2, \ldots, s\}$, is replaced by an interval $[0, 1]$. This means that for any $\xi \in [0, 1]$, we can associate a 'stage value', $Y_{\xi} = y(x_0 + h\xi)$, with corresponding stage derivative $f(Y_{\xi})$. In this limiting interpretation, the matrix $A : \mathbb{R}^s \to \mathbb{R}^s$ is replaced by a linear operator on the set of continuous functions on $[0, 1]$. At the same time, the vector b^T, is replaced by a linear functional on the continuous functions on $[0, 1]$. More specifically, the idealized A and b^T are given by

$$(A\phi)_\xi = \int_0^\xi \phi_\eta \, d\eta, \quad (b^T\phi) = \int_0^1 \phi_\eta \, d\eta = (A\phi)_1.$$

The elementary weights for this idealized method are found from the formula for $\Phi(t)$ and replacing various sums by integrals. For example, for the example tree used in Table 1.4, the calculation of the limiting elementary weight is as follows:

$$\int_0^1 \left(\int_0^x x^2 \, dx \right)^2 dx = \int_0^1 \left(\tfrac{1}{3}x^3 \right)^2 dx = \tfrac{1}{63} = \tfrac{1}{\gamma(t)}.$$

The representation of this method as a member of X_1 will be denoted by E. Hence, $E(t) = 1/\gamma(t)$.

If we consider equivalence classes of Runge–Kutta method as basic objects of study, then we might ask: What is the significance of the composition of two methods, one from each class? If the two methods are denoted by M_1 and M_2, with s_1 and s_2 stages, respectively, then $M_1 M_2$ will denote the combined operation of calculating the stages of the first method and the output value so that it now becomes possible to write the stages of the second method as though they were additional stages appended to the first method. Thus $M_1 M_2$ is also a Runge–Kutta method but the equivalence class to which it belongs is independent of how the representative methods M_1 and M_2 were chosen from within their classes. Furthermore we can compute the elementary weights for the product class directly from those of the classes containing M_1 and M_2.

For convenience, we will write the function on trees to elementary weights corresponding to two specific methods as α and β. For tree number i we will write $\alpha_i = \alpha(t_i)$ and $\beta_i = \beta(t_i)$. The value of $(\alpha\beta)(t_i)$ will be a function of the α and β values and formulae for these are shown in Table 1.7, up to order 4 trees. The value of β_0 which appears in this table is equal to $\beta(\emptyset)$. Restricted to $X_1 \times X_1$, the operation defined by this table generates a group. However, X also has a vector space structure and left-multiplication by a member of X_1 is a linear operator on this vector space.

We now discuss an important example of the vector space structure. The output from the Euler method, starting from initial value $y(x_0) = y_0$, gives a result $y_0 + hy'(x_0)$. Subtract y_0 from this and we obtain exactly the scaled derivative $hy'(x_0) = hf(y_0)$. We will regard the elementary weights for this scaled derivative as being exactly the same as for the Euler method, but with β_0 set equal to zero. We will denote this special generalized Runge–Kutta method by D so that $D(\emptyset) = 0$, $D(\tau) = 1$, $D(t) = 0$ $(r(t) > 1)$.

It is quite convenient to build up elementary weights, and more complicated objects, using the D operation. If 1 is used to represent the identity element of the group. We can then write the group elements representing

Table 1.7. Runge–Kutta group operation, with $\beta_0 = 1$.

i	t_i	$\alpha(t_i)$	$\beta(t_i)$	$(\alpha\beta)(t_i)$
1	.	α_1	β_1	$\alpha_1\beta_0 + \beta_1$
2	!	α_2	β_2	$\alpha_2\beta_0 + \alpha_1\beta_1 + \beta_2$
3	V	α_3	β_3	$\alpha_3\beta_0 + \alpha_1^2\beta_1 + 2\alpha_1\beta_2 + \beta_3$
4	!	α_4	β_4	$\alpha_4\beta_0 + \alpha_2\beta_1 + \alpha_1\beta_2 + \beta_4$
5	V	α_5	β_5	$\alpha_5\beta_0 + \alpha_1^3\beta_1 + 3\alpha_1^2\beta_2 + 3\alpha_1\beta_3 + \beta_5$
6	V	α_6	β_6	$\alpha_6\beta_0 + \alpha_1\alpha_2\beta_1 + (\alpha_1^2 + \alpha_2)\beta_2 + \alpha_1(\beta_3 + \beta_4) + \beta_6$
7	Y	α_7	β_7	$\alpha_7\beta_0 + \alpha_3\beta_1 + \alpha_1^2\beta_2 + 2\alpha_1\beta_4 + \beta_7$
8	!	α_8	β_8	$\alpha_8\beta_0 + \alpha_4\beta_1 + \alpha_2\beta_2 + \alpha_1\beta_4 + \beta_8$

the stages by η_i, $i = 1, 2, \ldots, s$ which satisfy

$$\eta_i = 1 + \sum_{j=1}^{s} a_{ij}\eta_j D.$$

The output at the end of a Runge–Kutta method will then be

$$1 + \sum_{i=1}^{s} b_i\eta_i D.$$

Collocation methods and implicit Runge–Kutta methods

A possible approach to the solution of an initial value problem, on an interval $[x_0, x_0 + h]$, is to assume an approximation of the form

$$y(x_0 + \xi h) = P(\xi),$$

and to define the polynomial P, assumed to be of degree s, by the conditions

$$P(0) = y_0,$$
$$P'(c_i) = f(P(c_i)), \quad i = 1, 2, \ldots, s.$$

In these conditions, the 'collocation points' $c_1, c_2, \ldots, c_s$, are distinct and nonzero. To obtain a step-by-step sequence of approximations, define $y_1 = P(1)$, and compute y_2 in a similar way from y_1, as the next step in the process. An attraction of such methods is the fact that an interpolated approximation is automatically available between step values.

    It was pointed out in Wright (1970) that collocation methods are equivalent to implicit Runge–Kutta methods, with the abscissae identical to the collocation points. In the implicit Runge–Kutta representation, the elements of A and b^T are defined by

$$\sum_{j=1}^{s} a_{ij} c_j^{k-1} = \frac{1}{k} c_i^k, \qquad k = 1, 2, \ldots, s, \quad i = 1, 2, \ldots s, \tag{1.41}$$

$$\sum_{j=1}^{s} b_j c_j^{k-1} = \frac{1}{k}, \qquad k = 1, 2, \ldots, s. \tag{1.42}$$

For example, if $c = [0, \frac{1}{2}, 1]^T$, we obtain the method

$$\begin{array}{c|ccc}
0 & 0 & 0 & 0 \\
\frac{1}{2} & \frac{5}{24} & \frac{1}{3} & -\frac{1}{24} \\
1 & \frac{1}{6} & \frac{2}{3} & \frac{1}{6} \\
\hline
& \frac{1}{6} & \frac{2}{3} & \frac{1}{6}
\end{array} \tag{1.43}$$

The condition (1.41) expresses the fact that not only does the output from a step have a specific order but the stage values themselves are computed to a high precision, as measured in terms of an asymptotic error $\mathcal{O}(h^{q+1})$. Even for a method which is not based on collocation, the stage order, which we denote by q, can be close to s for implicit methods. This is regarded as an advantage in terms of the ability of the method to solve stiff problems reliably (Prothero and Robinson 1974). The next example method has stage order 2 and order 3:

$$\begin{array}{c|cc}
\frac{1}{3} & \frac{5}{12} & -\frac{1}{12} \\
1 & \frac{3}{4} & \frac{1}{4} \\
\hline
& \frac{3}{4} & \frac{1}{4}
\end{array} \tag{1.44}$$

A-stable Runge–Kutta methods

As for the Euler method and linear multistep methods, the first step in assessing the stability properties of a Runge–Kutta method is to investigate its behaviour with the linear problem $y' = qy$. For this problem, the stages and final output in a single step are given by

$$Y = zAY + y_{n-1}\mathbf{1},$$

$$y_n = zb^T Y + y_{n-1},$$

where $z = hq$ and $\mathbf{1} \in \mathbb{R}^s$ has every component equal to 1. Eliminate Y and the result is

$$y_n = R(z)y_{n-1},$$

where the 'stability function' is

$$R(z) = 1 + zb^T (I - zA)^{-1}\mathbf{1}.$$

The set of z values for which $|R(z)| \le 1$ is the 'stability region'. If this includes the left half-plane then the method is A-stable. Two examples of

A-stable methods can be found in (1.43) and (1.44) for which the stability functions are

$$R(z) = \frac{1 + \frac{1}{2}z + \frac{1}{12}z^2}{1 - \frac{1}{2}z + \frac{1}{12}z^2},$$

$$R(z) = \frac{1 + \frac{1}{3}z}{1 - \frac{2}{3}z + \frac{1}{12}z^2},$$

respectively. These stability functions are examples of Padé approximations and proof of the A-stability, in these particular cases, is included in Theorem 8.8.

Runge–Kutta methods for stiff problems

For stiff problems, it is not satisfactory to use explicit methods, and we need to consider methods in which A has a more complicated structure. We will consider five levels of implicitness, in terms of restrictions on the coefficients in the $s \times s$ matrix A and the vector b^T:

(i) $a_{ij} = 0$ if $j > i$,
(ii) $a_{ij} = 0$ if $j > i$; $a_{ii} = \lambda$, $i = 1, 2, \ldots, s$,
(iii) $a_{11} = 0$; $a_{ij} = 0$, $j > i$; $a_{ii} = \lambda$, $i = 2, 3, \ldots, s$; $a_{sj} = b_j$, $j = 1, 2, \ldots, s$,
(iv) $\sigma(A) = \{\lambda\}$,
(v) A an arbitrary full matrix.

The use of fully implicit methods (v) was proposed by Ceschino and Kuntzmann (1963) and Butcher (1964), with the abscissae based on Gauss–Legendre integration points. The Gauss methods and related methods based on other high-order quadrature formulae, have an important role in the solution of stiff problems. In Butcher (1964) so-called semi-implicit methods (i) were introduced but without a specific application in mind.

For efficiency reasons, there is also an interest in diagonally implicit (or DIRK) methods included within the (ii) and (iii) families. Finally, singly implicit (SIRK) methods (Burrage 1978a) were introduced to yield methods which not only have efficient implementation properties but have high stage order.

The following example of (i) has order 5; this would have required 6 stages if the method had been explicit. For example, the following method has order 5:

$$
\begin{array}{c|ccccc}
0 & & & & & \\
\frac{1}{4} & \frac{1}{8} & \frac{1}{8} & & & \\
\frac{7}{10} & -\frac{1}{100} & \frac{14}{25} & \frac{3}{20} & & \\
1 & \frac{2}{7} & 0 & \frac{5}{7} & & \\
\hline
 & \frac{1}{14} & \frac{32}{81} & \frac{250}{567} & \frac{5}{54} &
\end{array}
$$

An example of (ii) is the following order 3 method:

$$
\begin{array}{c|ccc}
\lambda & \lambda \\
\frac{1}{2}(1+\lambda) & \frac{1}{2}(1-\lambda) & \lambda \\
1 & \frac{1}{4}(-6\lambda^2+16\lambda-1) & \frac{1}{4}(6\lambda^2-20\lambda+5) & \lambda \\
\hline
& \frac{1}{4}(-6\lambda^2+16\lambda-1) & \frac{1}{4}(6\lambda^2-20\lambda+5) & \lambda
\end{array}
\, ,
$$

where $\lambda \approx 0.4358665215$ satisfies $\frac{1}{6}-\frac{3}{2}\lambda+3\lambda^2-\lambda^3 = 0$. This method is A-stable but its stage order is only 1, making it of limited value in the solution of stiff problems.

The next method is an example of (iv) and has order and stage order 2:

$$
\begin{array}{c|cc}
3-2\sqrt{2} & \frac{5}{4}-\frac{3}{4}\sqrt{2} & \frac{7}{4}-\frac{5}{4}\sqrt{2} \\
1 & \frac{1}{4}+\frac{1}{4}\sqrt{2} & \frac{3}{4}-\frac{1}{4}\sqrt{2} \\
\hline
& \frac{1}{4}+\frac{1}{4}\sqrt{2} & \frac{3}{4}-\frac{1}{4}\sqrt{2}
\end{array}
\, .
$$

The method is A-stable, as are similar methods up to order 8 (with the exception of 7). Their major disadvantage is the fact that, for $s > 2$, not all the abscissae lie in $[0, 1]$.

It is implemented using a transformation which makes it effectively like DIRK methods, in terms of cost, at least for large problems where the overheads due to the transformations are relatively insignificant.

Finally, a simple example of (v). This is one of the Gauss–Legendre family of methods and it has order 4 and stage order 2:

$$
\begin{array}{c|cc}
\frac{1}{2}-\frac{1}{6}\sqrt{3} & \frac{1}{4} & \frac{1}{4}-\frac{1}{6}\sqrt{3} \\
\frac{1}{2}+\frac{1}{6}\sqrt{3} & \frac{1}{4}+\frac{1}{6}\sqrt{3} & \frac{1}{4} \\
\hline
& \frac{1}{2} & \frac{1}{2}
\end{array}
\, .
$$

Similar methods exist for all positive values of s with order $2s$ and stage order s. Although they are A-stable, they are difficult to implement efficiently.

2. Motivations for general linear methods

We will describe some of the circumstances and events which have led to an interest in more general methods.

The traditional methods can be regarded as generalizations, in one way or another, of the Euler method. In the terminology of general linear methods, this is a one-value ($r = 1$), one-stage ($s = 1$) method. Increasing the value of the integer r leads to linear multistep methods and increasing s leads to Runge–Kutta methods. It seems natural to consider methods in which both $r > 1$ and $s > 1$ are possible.

Thus the first motivation for studying general linear methods is that this generalization is natural and there seems to be no reason for not adopting it. Indeed even some existing methods are more naturally formulated in a general linear method ansatz.

We will look at some of the limitations of existing methods to see that the general linear method generalization is not only natural but also potentially useful. We will pursue this point of view by looking at some simple modifications of existing methods and ultimately by attempting to find some new and potentially efficient methods which do not arise naturally in any other way.

2.1. Limitations of linear multistep methods

Even though there are $2k+1$ free parameters in the specification of a linear k-step method, so that order $2k$ would seem to be possible, in fact practical methods are limited in order to $k + 2$ (if k is even) and $k + 1$ (if k is odd), because of the stability condition. This result, Dahlquist's first barrier (Dahlquist 1956), which we reviewed in Theorem 1.3, is coupled with the second Dahlquist barrier (Dahlquist 1963), which limits the order of A-stable methods to exactly 2. A proof, using order arrows, of the second barrier is given in Theorem 8.10. In spite of this barrier, if A(α)-stability, with a reasonably large angle α, is regarded as acceptable, it is possible to go to at least order 4.

This applies in particular to BDF methods where we note that BDF4 is A(0.4π)-stable. For any p, it is possible to replace the factor 0.4 by a number arbitrarily close to $\frac{1}{2}$ (Widlund 1967, Grigorieff and Schroll 1978) but at the cost of impractically high error constants (Jeltsch and Nevanlinna 1982).

In addition to stability constraints, another type of limitation is the complication associated with change of step-size and change of order. Each of these requires considerable overheads.

2.2. Limitations of Runge–Kutta methods

While explicit order p Runge–Kutta methods exist with p stages, for $p = 1, 2, 3, 4$, no such methods exist for $p > 4$. Furthermore, if the minimal number of stages to achieve order p is $s(p)$, then $s(p) - p$ increases steadily as p increases, as we recall from Table 1.6.

Variable step-size and order are made difficult by the need to estimate local truncation errors in a reliable way. This is an increasingly expensive requirement as the order increases.

In the case of implicit methods, the achievable order is exactly $2s$, and methods which achieve this maximum are A-stable. This seems to be a satisfactory situation but the actual methods have two serious handicaps.

The first of these is that they suffer from error reduction and the second is the very high implementation cost.

Even though the global truncation error is asymptotically $O(h^p)$, for step-sizes which often arise in practice, the error behaves more like $O(h^q)$, where q is the stage order.

Solving the non-linear equations defining the stage values involves a process based on the Newton method. This is much more expensive than for implicit linear multistep methods, unless the coefficient matrix A has a special structure, such as the DIRK structure. Unfortunately DIRK methods necessarily have low stage order. In the opinion of this author the only way to overcome this difficulty is to use SIRK methods, in the modified form discussed in Butcher and Chen (1998). But there seem to be better algorithms within the larger family of general linear methods.

2.3. Modifications of linear multistep methods

Many examples are known of modifications to standard methods, which somehow acquire enhanced properties. For example, by adding one or more offstep points, it is possible to give a linear multistep method a little closer to that of Runge–Kutta methods. This can break the Dahlquist barrier by permitting methods to have order greater than $2k$ and still remain stable. A class of methods in this hybrid family takes the idea of predictor–corrector pairs based on Adams–Bashforth and Adams–Moulton methods further, by including a single offstep predictor as well as the usual predictor and corrector at the end of the step. Thus for $k = 2$ the k-step PECE[1] method,

$$y_n^* = y_{n-1} + \tfrac{3}{2} h f_{n-1} - \tfrac{1}{2} h f_{n-2},$$
$$y_n = y_{n-1} + \tfrac{1}{2} h f_n^* + \tfrac{1}{2} h f_{n-1},$$

generalizes to

$$y_{n-\frac{1}{2}}^* = y_{n-2} + \tfrac{9}{8} h f_{n-1} + \tfrac{3}{8} h f_{n-2},$$
$$y_n^* = \tfrac{28}{5} y_{n-1} - \tfrac{23}{5} y_{n-2} + \tfrac{32}{15} h f_{n-\frac{1}{2}}^* - 4 h f_{n-1} - \tfrac{26}{15} h f_{n-2},$$
$$y_n = \tfrac{32}{31} y_{n-1} - \tfrac{1}{31} y_{n-2} + \tfrac{5}{31} h f_n^* + \tfrac{64}{93} h f_{n-\frac{1}{2}}^* + \tfrac{4}{31} h f_{n-1} - \tfrac{1}{93} h f_{n-2}.$$

Note that in this discussion f_n^* denotes $f(x_n, y_n^*)$ and $f_{n-\frac{1}{2}}^*$ denotes

$$f\left(x_n - \frac{1}{2} h, y_{n-\frac{1}{2}}^* \right).$$

Even though the two predictors generate approximations only of order 3, the overall result has order 5.

---

[1] PECE denotes 'Predict–Evaluate–Correct–Evaluate'.

'Hybrid' methods, as Gear named them, were introduced in Butcher (1965), Gear (1965) and Gragg and Stetter (1964).

A completely different generalization of linear multistep methods is that of cyclic composite methods, first proposed by Donelson and Hansen (1971). If we are given m linear multistep methods

$$y_n = \sum_{i=1}^{k} \alpha_i^{[j]} y_{n-i} + \sum_{i=0}^{k} \beta_i^{[j]} h f_{n-i}, \quad j = 1, \ldots, m,$$

the idea is to apply them cyclically. That is, in a sequence of m steps, use method number 1 followed by method number 2 and so on until method number m has been applied. For steps after this, the cycle is repeated.

We present just two examples. In the first we consider two methods, each based on open Newton–Cotes quadrature formulae:

$$y_n = y_{n-2} + 2h f_{n-1},$$

$$y_n = y_{n-3} + \tfrac{3}{2} h f_{n-1} + \tfrac{3}{2} h f_{n-2}.$$

Taken alone, each of these methods is 'weakly stable'. That is, regarded for convenience as 3-step methods, their zero stability matrices are

$$M_1 = \begin{bmatrix} 0 & 1 & 0 \\ 1 & 0 & 0 \\ 0 & 1 & 0 \end{bmatrix} \quad \text{and} \quad M_2 = \begin{bmatrix} 0 & 0 & 1 \\ 1 & 0 & 0 \\ 0 & 1 & 0 \end{bmatrix},$$

respectively. For M_1 the eigenvalues are $\{1, -1, 0\}$ and for M_2 the eigenvalues are $\{1, \exp(2\pi i/3), \exp(-2\pi i/3)\}$. Weak stability is a consequence of the existence of eigenvalues on the unit disc, in addition to the principal eigenvalue at 1 in each case. However, when the two methods are used in alternation then the stability matrix over the pair of steps becomes

$$M_2 M_1 = \begin{bmatrix} 0 & 1 & 0 \\ 0 & 1 & 0 \\ 1 & 0 & 0 \end{bmatrix},$$

with eigenvalues $\{1, 0, 0\}$.

It is possible to go even further and to construct cycles of explicit methods which overcome the first Dahlquist barrier. For example, consider the two methods

$$y_n = -\tfrac{8}{11} y_{n-1} + \tfrac{19}{11} y_{n-2} + \tfrac{10}{11} h f_n + \tfrac{19}{11} h f_{n-1} + \tfrac{8}{11} h f_{n-2} - \tfrac{1}{33} h f_{n-3},$$

$$y_n = \tfrac{449}{240} y_{n-1} + \tfrac{19}{30} y_{n-2} - \tfrac{361}{240} y_{n-3} + \tfrac{251}{720} h f_n + \tfrac{19}{30} h f_{n-1} - \tfrac{449}{240} h f_{n-2}$$

$$- \tfrac{35}{72} h f_{n-3}.$$

Each of these methods has order 5 and each is unstable, but we will see that the corresponding cyclic method has perfect stability. To verify this remark, analyse stability using $y' = 0$:

$$y_n = -\tfrac{8}{11}y_{n-1} + \tfrac{19}{11}y_{n-2}, \tag{2.1}$$

$$y_n = \tfrac{449}{240}y_{n-1} + \tfrac{19}{30}y_{n-2} - \tfrac{361}{240}y_{n-3}. \tag{2.2}$$

The difference equation for $y_n - y_{n-1}$ is

$$\begin{bmatrix} y_n - y_{n-1} \\ y_{n-1} - y_{n-2} \end{bmatrix} = X \begin{bmatrix} y_{n-1} - y_{n-2} \\ y_{n-2} - y_{n-3} \end{bmatrix},$$

where X is

$$\begin{bmatrix} -\tfrac{19}{11} & 0 \\ 1 & 0 \end{bmatrix}$$

for (2.1), and

$$\begin{bmatrix} \tfrac{209}{240} & \tfrac{361}{240} \\ 1 & 0 \end{bmatrix}$$

for (2.2). Neither matrix is power-bounded but their product, corresponding to the cyclic use of the two methods, is nilpotent.

By applying the cyclic composite idea to implicit methods it is also possible to overcome the second Dahlquist barrier (Bickart and Picel 1973).

2.4. Modifications of Runge–Kutta methods

The following family of fourth-order methods is one of several such families found by Kutta:

0				
c_2	c_2			
$\tfrac{1}{2}$	$\tfrac{1}{2} - \tfrac{1}{8c_2}$	$\tfrac{1}{8c_2}$		
1	$\tfrac{1}{2c_2} - 1$	$-\tfrac{1}{2c_2}$	2	
	$\tfrac{1}{6}$	0	$\tfrac{2}{3}$	$\tfrac{1}{6}$

If we substitute $c_2 = -1$, it is found that

0				
-1	-1			
$\tfrac{1}{2}$	$\tfrac{5}{8}$	$-\tfrac{1}{8}$		
1	$-\tfrac{3}{2}$	$\tfrac{1}{2}$	2	
	$\tfrac{1}{6}$	0	$\tfrac{2}{3}$	$\tfrac{1}{6}$

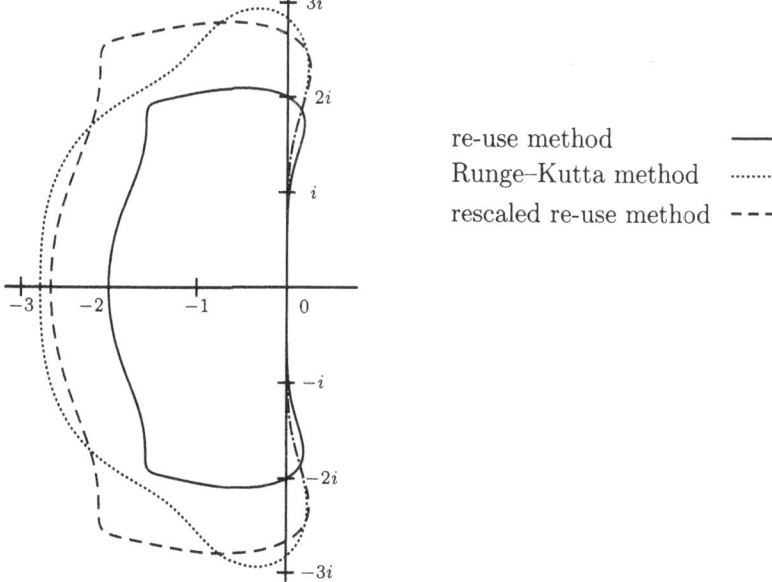

re-use method   ———
Runge–Kutta method   ·········
rescaled re-use method   – – –

Figure 2.1. Stability region for the re-use method
compared with the classical Runge–Kutta method.

We can interpret the abscissa at -1 as re-use of the derivative found at the
beginning of the previous step. We then have the method

$$Y_1 = y_{n-1} + \tfrac{5}{8}hf(y_{n-1}) - \tfrac{1}{8}hf(y_{n-2}), \qquad\qquad F_1 = f(Y_1),$$

$$Y_2 = y_{n-1} - \tfrac{3}{2}hf(y_{n-1}) + \tfrac{1}{2}hf(y_{n-2}) + 2hF_1, \qquad F_2 = f(Y_2),$$

$$y_n = y_{n-1} + \tfrac{1}{6}hf(y_{n-1}) + \tfrac{2}{3}hF_1 + \tfrac{1}{6}hF_2.$$

Like the Runge–Kutta method on which it is based, this method retains
order 4, even though it evaluates f only 3 times per time-step compared
with 4 for the original method.

We can understand something about the behaviour of the new method by
plotting its stability region. This is shown in Figure 2.1, with the classical
fourth-order method included for comparison. Because $s = 3$ for the re-use
method, rather than $s = 4$ for the Runge–Kutta method, a more appropriate
comparison is achieved by the rescaling $z \mapsto \tfrac{4}{3}z$ in the case of the re-use
method; this is also shown in the figure. Based on this comparison, there
seems to be little advantage in either the Runge–Kutta method or the re-use
method.

As a general linear method, using a notation we will introduce in Section 3, the re-use method has the following matrix representation:

$$
\begin{bmatrix} A & U \\ B & V \end{bmatrix} = \left[\begin{array}{ccc|cc} 0 & 0 & 0 & 1 & 0 \\ \frac{5}{8} & 0 & 0 & 1 & -\frac{1}{8} \\ -\frac{3}{2} & 2 & 0 & 1 & \frac{1}{2} \\ \hline \frac{1}{6} & \frac{2}{3} & \frac{1}{6} & 1 & 0 \\ 1 & 0 & 0 & 0 & 0 \end{array} \right].
\tag{2.3}
$$

3. Formulations

In the formulation of general linear methods given in Butcher (1966), a collection of approximations $y_i^{[n]}$, $i = 1, 2, \ldots, r$, together with derivative approximations $F_i^{[n]} = f(y_i^{[n]})$ is computed at the end of step number n in terms of the corresponding quantities available as input to the step. Making use of three matrices of coefficients, A, B, C, which characterize a specific method, the step n approximations are given by

$$
y_i^{[n]} = \sum_{j=1}^{r} a_{ij} y_j^{[n-1]} + \sum_{j=1}^{r} b_{ij} h F_j^{[n]} + \sum_{j=1}^{r} c_{ij} h F_j^{[n-1]}.
$$

This method was referred to as (A, B, C). It is easy to see, by raising the value of r if necessary, that C can be removed from the formulation.

An equivalent, but in many ways more convenient, formulation was introduced in Burrage and Butcher (1980) and this is now the standard way of representing general linear methods.

Denote the output approximations from step number n by $y_i^{[n]}$, $i = 1, 2, \ldots, r$, the stage values by Y_i, $i = 1, 2, \ldots, s$ and the stage derivatives by F_i, $i = 1, 2, \ldots, s$.

For convenience, write

$$
y^{[n-1]} = \begin{bmatrix} y_1^{[n-1]} \\ y_2^{[n-1]} \\ \vdots \\ y_r^{[n-1]} \end{bmatrix}, \quad y^{[n]} = \begin{bmatrix} y_1^{[n]} \\ y_2^{[n]} \\ \vdots \\ y_r^{[n]} \end{bmatrix}, \quad Y = \begin{bmatrix} Y_1 \\ Y_2 \\ \vdots \\ Y_s \end{bmatrix}, \quad F = \begin{bmatrix} F_1 \\ F_2 \\ \vdots \\ F_s \end{bmatrix}.
$$

It is assumed that Y and F are related by a differential equation.

The computation of the stages and the output from step number n is carried out according to the formulae

$$Y_i = \sum_{j=1}^{s} a_{ij} h F_j + \sum_{j=1}^{r} u_{ij} y_j^{[n-1]}, \qquad i = 1, 2, \ldots, s,$$

$$y_i^{[n]} = \sum_{j=1}^{s} b_{ij} h F_j + \sum_{j=1}^{r} v_{ij} y_j^{[n-1]}, \qquad i = 1, 2, \ldots, r,$$

where the matrices $A = [a_{ij}]$, $U = [u_{ij}]$, $B = [b_{ij}]$, $V = [v_{ij}]$ are characteristic of a specific method.

We can write these relations more compactly in the form

$$\begin{bmatrix} Y \\ y^{[n]} \end{bmatrix} = \begin{bmatrix} A \otimes I & U \otimes I \\ B \otimes I & V \otimes I \end{bmatrix} \begin{bmatrix} hF \\ y^{[n-1]} \end{bmatrix},$$

which we can simplify by making a harmless abuse of notation in the form

$$\begin{bmatrix} Y \\ y^{[n]} \end{bmatrix} = \begin{bmatrix} A & U \\ B & V \end{bmatrix} \begin{bmatrix} hF \\ y^{[n-1]} \end{bmatrix}. \tag{3.1}$$

An alternative formulation, of the closely related A-methods, is given in Albrecht (1985).

3.1. Consistency, stability and convergence

An idea that will be developed in Section 4 is that there is always a starting method associated with each method. For our present purpose it is enough to ask what quantity the numerical solution is supposed to approximate, at least to a first-order approximation. As a very basic requirement we ask if it is possible to approximate the solution to the problem $y'(x) = 0$, exactly. This condition has two parts. First, we want to ensure that quantities input to step number n are capable of remaining unchanged at the end of the step. Secondly we want to be able to guarantee long-term adherence to this solution, even if a slight perturbation is introduced. The first requirement will be written in terms of the existence of a 'pre-consistency vector' u which is unchanged when acted upon by V and the second that V is power-bounded. Note that for the differential equation $y' = 0$, input component number i is assumed to have the form $u_i y(x_{n-1})$ so that $Vu = u$ is the first of our conditions. In addition to this property of u we will require that $Uu = 1$ so that each stage gives an approximation close to $y(x_{n-1})$.

For a pre-consistent stable method, we also want to guarantee that the solution of the problem $y'(x) = 1$ has correctly advanced one step forward. We summarize these remarks with a series of definitions.

Definition 3.1. A general linear method (A, U, B, V) is 'stable' if there exists a constant C such that $\|V^n\| \le C$ for any positive integer n.

Definition 3.2. A general linear method (A, U, B, V) is 'pre-consistent' if there exists a 'pre-consistency vector' u such that

$$Vu = u,$$
$$Uu = \mathbf{1}.$$

Definition 3.3. A general linear method (A, U, B, V) is consistent if it is pre-consistent with pre-consistency vector u and furthermore, there exists a vector v such that

$$B\mathbf{1} + Vv = u + v.$$

Given the properties embodied in these definitions it is possible to guarantee that the approximation computed by a general linear method can be found arbitrarily close to the exact solution. We express this in terms of a definition, followed by a theorem, which will not be proved in this survey.

Definition 3.4. Consider an initial value problem

$$y'(x) = f(x, y(x)), \qquad y(x_0) = y_0,$$

where $f : [x_0, \bar{x}] \times \mathbb{R}^N \to \mathbb{R}^N$ is continuous in its first variable and satisfies a Lipschitz condition in its second variable. Let (A, U, B, V) be a consistent, stable general linear method with pre-consistency and consistency vectors u and v. Let $S(h)$ denote a starting approximation which depends on h in such a way that $\lim_{h \to 0} S_i(h) = u_i y_0$, $i = 1, 2, \ldots, r$. Let $\eta(n)$ denote the value of $y^{[n]}$, computed using the given method for the given problem, with starting value defined by $y^{[0]} = S((\bar{x} - x_0)/n)$. The method is 'convergent' if, for any choice of initial value problem and starting approximation S, $\lim_{n \to \infty} \eta_i(n) = u_i y(\bar{x})$, $i = 1, 2, \ldots, r$.

Theorem 3.5. Any stable and consistent general linear method is convergent.

This result includes the theories for convergence for special methods, such as Runge–Kutta and linear multistep methods. In practice, methods will be designed to have a stage order equal to at least 0 and an order equal to at least 1. Such order conditions imply consistency so the crucial question to ask in the search for acceptable methods, is whether the method is or is not stable.

The proof of Theorem 3.5 is technical and is given in Butcher (2003), for example.

3.2. Representation of standard methods

For a linear multistep method with input and output

$$
y^{[n-1]} = \begin{bmatrix} y_{n-1} \\ y_{n-2} \\ \vdots \\ y_{n-k} \\ \hline hf(x_{n-1}, y_{n-1}) \\ hf(x_{n-2}, y_{n-2}) \\ \vdots \\ hf(x_{n-k}, y_{n-k}) \end{bmatrix}, \qquad
y^{[n]} = \begin{bmatrix} y_{n} \\ y_{n-1} \\ \vdots \\ y_{n-k+1} \\ \hline hf(x_{n}, y_{n}) \\ hf(x_{n-1}, y_{n-1}) \\ \vdots \\ hf(x_{n-k+1}, y_{n-k+1}) \end{bmatrix},
$$

the single stage Y_1 will be identical with the first output component, and we have the method

$$
\left[
\begin{array}{ccccc|ccccc}
\beta_0 & \alpha_1 & \alpha_2 & \cdots & \alpha_{k-1} & \alpha_k & \beta_1 & \beta_2 & \cdots & \beta_{k-1} & \beta_k \\
\beta_0 & \alpha_1 & \alpha_2 & \cdots & \alpha_{k-1} & \alpha_k & \beta_1 & \beta_2 & \cdots & \beta_{k-1} & \beta_k \\
0 & 1 & 0 & \cdots & 0 & 0 & 0 & 0 & \cdots & 0 & 0 \\
\vdots & \vdots & \vdots & & \vdots & \vdots & \vdots & \vdots & & \vdots & \vdots \\
0 & 0 & 0 & \cdots & 1 & 0 & 0 & 0 & \cdots & 0 & 0 \\
\hline
1 & 0 & 0 & \cdots & 0 & 0 & 0 & 0 & \cdots & 0 & 0 \\
0 & 0 & 0 & \cdots & 0 & 1 & 0 & 0 & \cdots & 0 & 0 \\
0 & 0 & 0 & \cdots & 0 & 0 & 1 & 0 & \cdots & 0 & 0 \\
\vdots & \vdots & \vdots & & \vdots & \vdots & \vdots & \vdots & & \vdots & \vdots \\
0 & 0 & 0 & \cdots & 0 & 0 & 0 & 0 & \cdots & 1 & 0
\end{array}
\right].
\tag{3.2}
$$

3.3. Transformation of methods

Because the data imported at the start of a step and exported at the end of the step is capable of being repackaged in a different way, we consider two methods (A, U, B, V) and $(A, \widehat{U}, \widehat{B}, \widehat{V})$, so related that

$$
\begin{bmatrix} A & \widehat{U} \\ \widehat{B} & \widehat{V} \end{bmatrix} = \begin{bmatrix} I & 0 \\ 0 & T^{-1} \end{bmatrix} \begin{bmatrix} A & U \\ B & V \end{bmatrix} \begin{bmatrix} I & 0 \\ 0 & T \end{bmatrix},
\tag{3.3}
$$

where T is an arbitrary $r \times r$ non-singular matrix.

If $\widehat{y}^{[n]}$ is the output from the transformed method and $y^{[n]}$ is the output from the original method then

$$
\widehat{y}^{[n-1]} = (T \otimes I) y^{[n-1]}, \qquad \widehat{y}^{[n]} = (T^{-1} \otimes I) y^{[n]}.
$$

The relationship between the basic properties of the two methods is expressed in the following result which is proved in a routine way.

Theorem 3.6. Let (A, U, B, V) and $(A, \widehat{U}, \widehat{B}, \widehat{V})$ be two methods related by (3.3), then if either is consistent, then so is the other, with preconsistency and consistency vectors related by

$$\widehat{u} = T^{-1}u,$$
$$\widehat{v} = T^{-1}v.$$

Furthermore, if either method is stable, then so is the other.

The significance of transformations predates by many years the introduction of general linear methods. In traditional formulations of Adams methods, for example, the input data for step number n may consist of approximations to $y(x_{n-1})$, $hy'(x_{n-1})$, $hy'(x_{n-2})$, ..., $hy'(x_{n-k})$. On the other hand, an alternative representation, popular in the days of hand computation, is to use $y(x_{n-1})$, together with backward differences, from order 0 to $k-1$, of the derivative information. The use of approximations to scaled derivatives was proposed by Nordsieck (1962) and promoted in Gear (1967, 1971).

Given a method $(A, \widehat{U}, \widehat{B}, \widehat{V})$ with $\widehat{r}$ input and output values, it may happen that a method (A, U, B, V) with $r < \widehat{r}$ input and output values might be related to it by the existence of an $r \times \widehat{r}$ matrix T, with rank r, such that

$$\widehat{U} = UT, \quad T\widehat{B} = B, \quad T\widehat{V} = VT.$$

In this case, it might be asked to what extent the method (A, U, B, V) carries out essentially the same task as the original method $(A, \widehat{U}, \widehat{B}, \widehat{V})$. To understand this question, introduce an arbitrary $(\widehat{r} - r) \times \widehat{r}$ matrix $\dot{T}$ whose rows, together with the rows of T, constitute a basis for $\mathbb{R}^{\widehat{r}}$. Now write

$$\begin{bmatrix} y^{[n]} \\ \dot{y}^{[n]} \end{bmatrix} = \begin{bmatrix} T \otimes I \\ \dot{T} \otimes I \end{bmatrix} \widehat{y}^{[n]},$$

where it is assumed that the $\widehat{y}$ sequence satisfies the original method. Transform the original ⌢ method to give the method

$$\left[\begin{array}{c|cc} A & U & 0 \\ \hline B & V & 0 \\ \dot{B} & \dot{V} & \ddot{V} \end{array}\right] = \begin{bmatrix} T \\ \dot{T} \end{bmatrix} \begin{bmatrix} A & \widehat{U} \\ \widehat{B} & \widehat{V} \end{bmatrix} \begin{bmatrix} T \\ \dot{T} \end{bmatrix}^{-1}.$$

It is apparent that the $y^{[n]}$ is generated by the method (A, U, B, V), without any reference to the the the $\dot{y}^{[n]}$ sequence. Hence, the transformation of methods can also have the effect of reducing a method to a simpler 'reduced method'. Assuming that A has no unused stages, we refer to a method that is not capable of further reduction as irreducible.

The most readily available example of a reducible method is for a linear multistep method written in the form (3.2). Define

$$
T = \left[
\begin{array}{cccc|cccc}
\alpha_1 & \alpha_2 & \cdots & \alpha_{k-1} & \alpha_k & \beta_1 & \beta_2 & \cdots & \beta_{k-1} & \beta_k \\
\alpha_2 & \alpha_3 & \cdots & \alpha_k & 0 & \beta_2 & \beta_3 & \cdots & \beta_k & 0 \\
\vdots & \vdots & & \vdots & \vdots & \vdots & \vdots & & \vdots & \vdots \\
\alpha_k & 0 & \cdots & 0 & 0 & \beta_k & 0 & \cdots & 0 & 0
\end{array}
\right],
$$

which has rank k because $|\alpha_k| + |\beta_k| \neq 0$.

Hence, we arrive at the r input method

$$
\begin{bmatrix} A & U \\ B & V \end{bmatrix} =
\left[
\begin{array}{c|ccccc}
\beta_0 & 1 & 0 & 0 & \cdots & 0 & 0 \\
\hline
\beta_0\alpha_1 + \beta_1 & \alpha_1 & 1 & 0 & \cdots & 0 & 0 \\
\beta_0\alpha_2 + \beta_2 & \alpha_2 & 0 & 1 & \cdots & 0 & 0 \\
\beta_0\alpha_3 + \beta_3 & \alpha_3 & 0 & 0 & \cdots & 0 & 0 \\
\vdots & \vdots & \vdots & \vdots & & \vdots & \vdots \\
\beta_0\alpha_{k-1} + \beta_{k-1} & \alpha_{k-1} & 0 & 0 & \cdots & 0 & 1 \\
\beta_0\alpha_k + \beta_k & \alpha_k & 0 & 0 & \cdots & 0 & 0
\end{array}
\right]. \qquad (3.4)
$$

4. Order conditions

4.1. General definition of order

In the formulation of a general linear method, there is not always a natural meaning that can be given to the quantities $y^{[n]}$ output at the end of step number n. Hence we will introduce a 'starting method', to represent the quantity we are trying to approximate. Write this as a modified type of general linear method with only a single input but with r outputs. It can also be thought of as a Runge–Kutta method with multiple outputs.

If the starting method is applied to a given initial value the output can be used as input to the first step of the main method. If $\mathcal{S}$ denotes the starting method and $\mathcal{M}$ the main method then $\mathcal{S}\mathcal{M}$ will denote the combined operation. Similarly, $\mathcal{E}$ denotes the exact solution evaluated after a time-step h, the same as the step-size for $\mathcal{M}$ and $\mathcal{S}$, and $\mathcal{E}\mathcal{S}$ denotes the result of applying $\mathcal{S}$ to the exact solution evaluated after a time h.

If we compare the result computed by $\mathcal{S}\mathcal{M}$ and compare it with the result of the computation $\mathcal{E}\mathcal{S}$, that is, two members of $\mathbb{R}^{rN}$, we have a measure of how closely $\mathcal{M}$ is able to preserve approximations to $\mathcal{S}$ applied to the exact trajectory. If $\mathcal{S}$ can be chosen so that the norm of the difference between these two results can be estimated in terms of h^{p+1}, then we say that 'the method $\mathcal{M}$ has order p relative to $\mathcal{S}$'. This enables us to state:

Definition 4.1. The method $\mathcal{M}$ has order p if there exists a starting method $\mathcal{S}$ such that $\mathcal{M}$ has order p relative to $\mathcal{S}$.

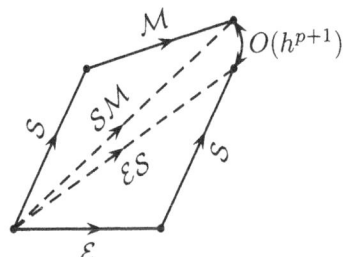

Figure 4.1. Order of general linear method.

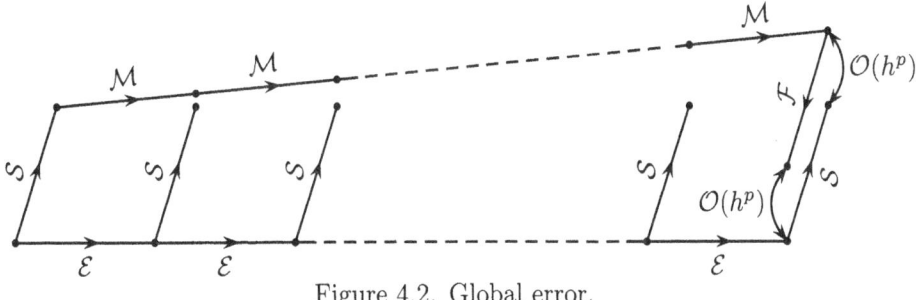

Figure 4.2. Global error.

The relationship between $\mathcal{M}$, $\mathcal{S}$ and $\mathcal{E}$ is illustrated in Figure 4.1.

According to this view of order, the method under consideration cannot be looked at in isolation but will always be related to the starting method. Looked at another way, the order is related to the *interpretation* of the quantities the method is intended to approximate. This is actually a classical point of view and reflects the fact that multivalue methods typically input approximations to specific quantities known in advance. The most well-known example is linear k-step methods in which the data input to step number n consists of approximations to $y(x_{n-i})$ and $hy'(x_{n-i})$ for $i = 1, 2, \ldots, k$. From the classical point of view, a Runge–Kutta method treats its single item of input as an approximation to $y(x_{n-1})$. However, there is no fundamental reason why we should restrict ourselves to this interpretation and it might be possible to regard $\mathcal{S}$ as an arbitrary Runge–Kutta method. Does this lead to any sort of enhancement of the concept of order, for Runge–Kutta methods? The answer is yes, as we will see in Section 4.2.

In addition to the mapping $\mathcal{S}$, we postulate the existence of a 'finishing method' $\mathcal{F}$, defined as a right-sided inverse of $\mathcal{S}$. That is, if $y^{[0]}$ is found by applying $\mathcal{S}$ to y_0, then y_0 is found by applying $\mathcal{F}$ to $y^{[0]}$.

Consider a long term integration consisting of n steps with step-size $h = (\bar{x} - x_0)/n$. Once n steps have been carried out to give the result $y^{[n]}$, $\mathcal{F}$ is applied to this to obtain an approximation to $y(\bar{x})$. Assuming that the method is stable, the errors in each of the steps combine to give an overall error, that is a global error, $n\mathcal{O}(h^{p+1}) = \mathcal{O}(h^p)$. The way this works is shown in Figure 4.2.

Table 4.1. Analysis for effective order 5.

i	t_i	$(\Psi\Phi)(t_i) - (E\Psi)(t_i)$
1	.	$\phi_1 - 1$
2	ı	$\phi_2 - \frac{1}{2}$
3	v	$\phi_3 - \frac{1}{3} - 2\psi_2$
4	ı	$\phi_4 + \psi_2\phi_1 - \frac{1}{6} - \psi_2$
5	v	$\phi_5 - \frac{1}{4} - 3\psi_2 - 3\psi_3$
6	√	$\phi_6 + \phi_2\psi_2 - \frac{1}{8} - \frac{3}{2}\psi_2 - \psi_3 - \psi_4$
7	Y	$\phi_7 + \phi_1\psi_3 - \frac{1}{12} - \psi_2 - 2\psi_4$
8	ı	$\phi_8 + \phi_1\psi_4 + \phi_2\psi_2 - \frac{1}{24} - \frac{1}{2}\psi_2 - \psi_4$
9	✶	$\phi_9 - \frac{1}{5} - 4\psi_2 - 6\psi_3 - 4\psi_5$
10	✶	$\phi_{10} + \phi_3\psi_2 - \frac{1}{10} - 2\psi_2 - \frac{5}{2}\psi_3 - \psi_4 - \psi_5 - 2\psi_6$
11	✶	$\phi_{11} + \phi_2\psi_3 - \frac{1}{15} - \frac{4}{3}\psi_2 - \psi_3 - 2\psi_4 - 2\psi_6 - \psi_7$
12	✶	$\phi_{12} + \phi_2\psi_4 + \phi_3\psi_2 - \frac{1}{30} - \frac{2}{3}\psi_2 - \frac{1}{2}\psi_3 - \psi_4 - \psi_6 - \psi_8$
13	✶	$\phi_{13} + \phi_1\psi_2^2 + 2\phi_4\psi_2 - \frac{1}{20} - \psi_2 - \psi_3 - \psi_4 - 2\psi_6$
14	Y	$\phi_{14} + \phi_1\psi_5 - \frac{1}{20} - \psi_2 - 3\psi_4 - 3\psi_7$
15	✶	$\phi_{15} + \phi_1\psi_6 + \phi_4\psi_2 - \frac{1}{40} - \frac{1}{2}\psi_2 - \frac{3}{2}\psi_4 - \psi_7 - \psi_8$
16	Y	$\phi_{16} + \phi_1\psi_7 + \phi_2\psi_3 - \frac{1}{60} - \frac{1}{3}\psi_2 - \psi_4 - 2\psi_8$
17	ı	$\phi_{17} + \phi_1\psi_8 + \phi_2\psi_4 + \phi_4\psi_2 - \frac{1}{120} - \frac{1}{6}\psi_2 - \frac{1}{2}\psi_4 - \psi_8$

4.2. Effective order of Runge–Kutta methods

In the case of a Runge–Kutta method, the starting method must itself be a Runge–Kutta method because it accepts a single input and yields a single output. For a given tree t, let $\Phi(t)$ denote the elementary differential associated with the main method and $\Psi(t)$ the elementary differential associated with the starting method. For convenience, we will write $\Phi(t_i) = \phi_i$ and $\Psi(t_i) = \psi_i$. Using this notation, the numbered trees are shown, together with expressions for $(\Psi\Phi)(t_i) - (E\Psi)(t_i)$ in Table 4.1. For simplification, ψ_1 has been assigned the value 0. This turns out not to limit the generality of the conditions on Ψ.

Because ψ_2, ψ_3, ψ_4, ψ_5, ψ_6 and ψ_7 can have arbitrary values, there is much more freedom on Φ for effective order 5 than for classical order. The following is a possible solution:

$$[\phi_1, \phi_2, \ldots, \phi_{17}] = [1, \tfrac{1}{2}, \tfrac{1}{3}, \tfrac{1}{6}, \tfrac{1}{4}, \tfrac{1}{8}, \tfrac{1}{12}, \tfrac{1}{24}, \tfrac{31}{150}, \tfrac{31}{300}, \tfrac{1}{15}, \tfrac{1}{30}, \tfrac{31}{600}, \tfrac{13}{300}, \tfrac{13}{600}, \tfrac{1}{60}, \tfrac{1}{120}],$$

$$[\psi_1, \psi_2, \ldots, \psi_8] = [0, 0, 0, 0, \tfrac{1}{600}, \tfrac{1}{1200}, -\tfrac{1}{600}, -\tfrac{1}{1200}],$$

and the tableau for a method yielding the given Φ values is

$$
\begin{array}{c|ccccc}
0 & & & & & \\
\frac{2}{5} & \frac{2}{5} & & & & \\
\frac{2}{5} & \frac{1}{5} & \frac{1}{5} & & & \\
\frac{3}{5} & \frac{3}{20} & -\frac{3}{10} & \frac{3}{4} & & \\
1 & \frac{9}{44} & \frac{5}{22} & -\frac{15}{44} & \frac{10}{11} & \\
\hline
 & \frac{11}{72} & 0 & \frac{25}{72} & \frac{25}{72} & \frac{11}{72}
\end{array}
$$

Effective order of singly implicit Runge–Kutta methods

Singly implicit Runge–Kutta methods represent an attempt to achieve the combined goals of L-stability, stage order and order equal to s and efficient implementation. It is possible to satisfy all these requirements up to order 8, with 7 the only exception, or slightly weakened requirements (A(α)-stability for α close to $\frac{1}{2}\pi$ and zero stability function at infinity) for s much higher. Unfortunately, these methods have a disadvantage that abscissae lie outside the interval $[0,1]$, if $s > 2$. This can be overcome by applying the principle of effective order. Even the difficulty associated with variable step-size can be overcome for this type of method because the high stage order makes it possible to correct the implied starting perturbation as the solution develops. Furthermore, no finishing method is required for individual steps because one of the stages can be used for output.

4.3. Algebraic criterion for order

Let $\xi \in X^r$ represent the starting method, where X is the algebraic structure introduced in Section 1.6. Then the vector of stages, as represented by a member of X_1^s, is found from the relation

$$\eta = A(\eta D) + U\xi,$$

and the result computed at the end of the step is represented by

$$B(\eta D) + V\xi.$$

If this is to agree with $E\xi$ up to trees with order p then we have a convenient criterion for this order. We formalize this as follows.

Theorem 4.2. The general linear method (A, U, B, V) has order p if there exists $\xi \in X^r$ and $\eta \in X_1^s$, such that, for every tree t satisfying $r(t) \le p$,

$$\eta(t) = A(\eta D)(t) + U\xi(t),$$
$$(E\xi)(t) = B(\eta D)(t) + V\xi(t).$$

Table 4.2. Calculation of order for method (4.1).

t	ξ_1	ξ_2	η_1	$\eta_1 D$	η_2	$\eta_2 D$	η_3	$\eta_3 D$	$\hat{\xi}_1$	$\hat{\xi}_2$	$E\xi_1$	$E\xi_2$
1	0	-1	0	1	$\frac{1}{2}$	1	1	1	1	0	1	0
2	0	$\frac{1}{2}$	0	0	$\frac{1}{8}$	$\frac{1}{2}$	$\frac{1}{2}$	1	$\frac{1}{2}$	0	$\frac{1}{2}$	0
3	0	$-\frac{1}{3}$	0	0	$-\frac{1}{12}$	$\frac{1}{4}$	$\frac{5}{6}$	1	$\frac{1}{3}$	0	$\frac{1}{3}$	0
4	0	$-\frac{1}{6}$	0	0	$-\frac{1}{24}$	$\frac{1}{8}$	$\frac{5}{12}$	$\frac{1}{2}$	$\frac{1}{6}$	0	$\frac{1}{6}$	0
5	0	$\frac{1}{4}$	0	0	$\frac{1}{16}$	$\frac{1}{8}$	0	1	$\frac{1}{4}$	0	$\frac{1}{4}$	0
6	0	$\frac{1}{8}$	0	0	$\frac{1}{32}$	$\frac{1}{16}$	0	$\frac{1}{2}$	$\frac{1}{8}$	0	$\frac{1}{8}$	0
7	0	$\frac{1}{12}$	0	0	$\frac{1}{48}$	$-\frac{1}{12}$	$-\frac{1}{4}$	$\frac{5}{6}$	$\frac{1}{12}$	0	$\frac{1}{12}$	0
8	0	$\frac{1}{24}$	0	0	$\frac{1}{96}$	$-\frac{1}{24}$	$-\frac{1}{8}$	$\frac{5}{12}$	$\frac{1}{24}$	0	$\frac{1}{24}$	0
9	0	$-\frac{1}{5}$	0	0	$-\frac{1}{20}$	$\frac{1}{16}$	$\frac{13}{40}$	1	$\frac{5}{24}$	0	$\frac{1}{5}$	0
10	0	$-\frac{1}{10}$	0	0	$-\frac{1}{40}$	$\frac{1}{32}$	$\frac{13}{80}$	$\frac{1}{2}$	$\frac{5}{48}$	0	$\frac{1}{10}$	0
11	0	$-\frac{1}{15}$	0	0	$-\frac{1}{60}$	$-\frac{1}{24}$	$-\frac{1}{60}$	$\frac{5}{6}$	$\frac{1}{9}$	0	$\frac{1}{15}$	0
12	0	$-\frac{1}{30}$	0	0	$-\frac{1}{120}$	$-\frac{1}{48}$	$-\frac{1}{120}$	$\frac{5}{12}$	$\frac{1}{18}$	0	$\frac{1}{30}$	0
13	0	$-\frac{1}{20}$	0	0	$-\frac{1}{80}$	$\frac{1}{64}$	$\frac{13}{160}$	$\frac{1}{4}$	$\frac{5}{96}$	0	$\frac{1}{20}$	0
14	0	$-\frac{1}{20}$	0	0	$-\frac{1}{80}$	$\frac{1}{16}$	$\frac{7}{40}$	0	$\frac{1}{24}$	0	$\frac{1}{20}$	0
15	0	$-\frac{1}{40}$	0	0	$-\frac{1}{160}$	$\frac{1}{32}$	$\frac{7}{80}$	0	$\frac{1}{48}$	0	$\frac{1}{40}$	0
16	0	$-\frac{1}{60}$	0	0	$-\frac{1}{240}$	$\frac{1}{48}$	$\frac{7}{120}$	$-\frac{1}{4}$	$-\frac{1}{36}$	0	$\frac{1}{60}$	0
17	0	$-\frac{1}{120}$	0	0	$-\frac{1}{480}$	$\frac{1}{96}$	$\frac{7}{240}$	$-\frac{1}{8}$	$-\frac{1}{72}$	0	$\frac{1}{120}$	0

An example of order calculation

Consider the general linear method

$$
\left[\begin{array}{c|c} A & U \\ \hline B & V \end{array}\right] =
\left[\begin{array}{ccc|cc}
0 & 0 & 0 & 1 & 0 \\
\frac{3}{4} & 0 & 0 & \frac{3}{4} & \frac{1}{4} \\
-2 & 2 & 0 & 2 & -1 \\
\hline
\frac{1}{6} & \frac{2}{3} & \frac{1}{6} & 1 & 0 \\
0 & 0 & 0 & 1 & 0
\end{array}\right]. \tag{4.1}
$$

Let η_1, η_2, η_3 denote the group elements representing the stages and assume the starting method is given by 1 for the first component and E^{-1} for the second component. The calculation of the various quantities needed to establish order are given in Table 4.2. The columns headed $\hat{\xi}_1$ and $\hat{\xi}_2$ represent the output computed by the method. For order 5, these would equal the $E\xi_1$ and $E\xi_2$ columns, respectively. Note that the trees are numbered in the same order as for Table 4.1.

Because of differences between $\hat{\xi}_1$ and $E\xi_1$, the method has order only 4.

4.4. Methods with high stage order

In the formal definition of order based on Figure 4.1, the existence of S so that this diagram commutes to within $\mathcal{O}(h^{p+1})$ was required for the method to have order p. If this starting methods exists so that, in addition, the stages approximate y at specific points related to the current step to within $\mathcal{O}(h^{q+1})$, where $q \leq p$, then the method is said to have stage order q. If $q \geq p - 1$, the combined criteria for order and stage order become much simpler.

Denote the tall tree $[_k\tau]_k$ by t_k. In the special case $k = 0$, t_k will be the empty tree. Suppose that, for a starting method which gives order p and stage order q, $\xi(t_k) = w_k$. The order conditions now give

$$\eta(t_k) = A\eta_{k-1} + Uw_k, \qquad\qquad k = 1, 2, \ldots, q,$$

$$\sum_{l=0}^{k} \frac{1}{l!} w_{k-l} = B\eta_{k-1} + Vw_k, \qquad\qquad k = 1, 2, \ldots, p.$$

By the stage order conditions, $\eta(t_k)$ is the vector $c^k/k!$, where c^k denotes the component-by-component power. Furthermore, $(\eta D)(t_k) = c^{k-1}/(k-1)!$.

Write

$$C = \begin{bmatrix} 1 & c_1 & \frac{1}{2!}c_1^2 & \cdots & \frac{1}{p!}c_1^p \\ 1 & c_2 & \frac{1}{2!}c_2^2 & \cdots & \frac{1}{p!}c_2^p \\ 1 & c_3 & \frac{1}{2!}c_3^2 & \cdots & \frac{1}{p!}c_3^p \\ \vdots & \vdots & \vdots & & \vdots \\ 1 & c_s & \frac{1}{2!}c_s^2 & \cdots & \frac{1}{p!}c_s^p \end{bmatrix}, \quad K - \begin{bmatrix} 0 & 1 & 0 & \cdots & 0 \\ 0 & 0 & 1 & \cdots & 0 \\ 0 & 0 & 0 & \cdots & 0 \\ \vdots & \vdots & \vdots & & \vdots \\ 0 & 0 & 0 & \cdots & 1 \\ 0 & 0 & 0 & \cdots & 0 \end{bmatrix},$$

and the *necessary* order and stage order p conditions become

$$C = ACK + UW, \tag{4.2}$$

$$WE = BCK + VW, \tag{4.3}$$

where W and E are the matrices

$$W = \begin{bmatrix} w_0 & w_1 & w_2 & \cdots & w_p \end{bmatrix}, \quad E = \begin{bmatrix} 1 & 1 & \frac{1}{2!} & \frac{1}{3!} & \cdots & \frac{1}{p!} \\ 0 & 1 & 1 & \frac{1}{2!} & \cdots & \frac{1}{(p-1)!} \\ 0 & 0 & 1 & 1 & \cdots & \frac{1}{(p-2)!} \\ 0 & 0 & 0 & 1 & \cdots & \frac{1}{(p-3)!} \\ \vdots & \vdots & \vdots & \vdots & & \vdots \\ 0 & 0 & 0 & 0 & \cdots & 1 \end{bmatrix}.$$

If the stage order is only required to be $q = p - 1$, then the last column is ignored in (4.2). We will show that these are actually *sufficient* order conditions. First, however, we remark that (4.2) and (4.3) can be expressed in a different form.

Theorem 4.3. Necessary conditions for order p and stage order $q \geq p-1$ conditions are the existence of a polynomial-valued vector $\phi(z)$ such that

$$\exp(cz) = zA\exp(cz) + U\phi(z) + \mathcal{O}(z^{q+1}), \qquad (4.4)$$
$$\exp(z)\phi(z) = zB\exp(cz) + V\phi(z) + \mathcal{O}(z^{p+1}). \qquad (4.5)$$

Proof. Let $\phi(z) = \sum_{i=0}^{p} z^i w_i$. Add the columns of (4.2) and (4.3), in each case multiplying by the sequence of factors, $1, z, z^2, \ldots, z^p$ and the result follows. $\qquad\square$

Theorem 4.4. The conditions given in Theorem 4.3 are sufficient for order p and stage order q.

Note that, because of the equivalence of (4.2) (with the last column ignored if $q = p - 1$) and (4.3) with (4.4) and (4.5), we will actually prove the sufficiency of the former conditions.

Proof of Theorem 4.4. Define the starting method so that

$$y^{[0]} = \sum_{i=0}^{p} w_i h^i y^{(i)}(x_0).$$

We will first show the stage order property. That is, the stage values are

$$Y_i = \sum_{k=0}^{q} h^k c_i^k y^{(k)}(x_0)/k! + \mathcal{O}(h^{q+1}), \qquad i = 1, 2, \ldots, s.$$

This is verified by noting that these stage values imply that

$$hF_j = \sum_{k=1}^{q} h^k c_j^{k-1} y^{(k-1)}(x_0)/(k-1)! + \mathcal{O}(h^{q+2}), \qquad j = 1, 2, \ldots, s,$$

and that, because of (4.2),

$$Y_i - \sum_{j=1}^{s} a_{ij} hF_j = \sum_{k=0}^{q}(C - ACK)_{ik} h^k y^{(k)}(x_0) + \mathcal{O}(h^{q+1})$$

$$= \sum_{k=0}^{q}(UW)_{ik} h^k y^{(k)}(x_0) + \mathcal{O}(h^{q+1})$$

$$= \sum_{j=1}^{r} u_{ij} y_j^{[0]} + \mathcal{O}(h^{q+1}).$$

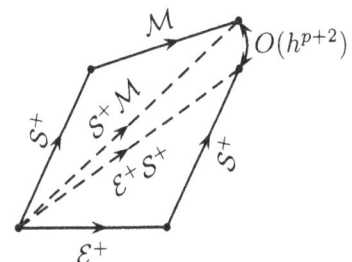

Figure 4.3. Iterative structure
of underlying one-step method.

Multiplying $y^{[0]}$ by E is equivalent to replacing x_0 by $x_0 + h$ in the formula for the starting method components, to within $\mathcal{O}(h^{p+1})$. The order result for the method now follows in a similar way to the stage order result. $\square$

4.5. The underlying one-step method

In Figure 4.1 denote the error term by ϕ. This is the Taylor expansion of $\mathcal{S}$ applied to $y(x_0 + h)$ minus the composition $\mathcal{S}\mathcal{M}$ applied to $y(x_0)$. Resolve ϕ into two terms,

$$\phi = \epsilon u + (I - V)\delta, \tag{4.6}$$

where u is the preconsistency vector which satisfies $(I - V)u = 0$. Now construct a new diagram in which $\mathcal{E}$ is replaced by $\mathcal{E}^+ = \mathcal{E} - \epsilon$ and $\mathcal{S}$ is replaced by $\mathcal{S}^+ = \mathcal{S} - \delta$.

The meanings of $\mathcal{E}^+$ and $\mathcal{S}^+$ require some explanation. In the case of $\mathcal{E}^+$, this represents a perturbation of the flow of the differential equation in which the value of ϵ, evaluated at $y(x_0)$, is subtracted from $y(x_0 + h)$. Similarly, $\mathcal{S}^+$ represents the unperturbed starting method $\mathcal{S}$ applied to $y(x_0)$, with the vector-valued error term δ, evaluated at $y(x_0)$, subtracted from the result.

The diagram in Figure 4.1 is now replaced by Figure 4.3 so that the order, in the sense of this diagram, has been increased to $p+1$. This process can be repeated to obtain a sequence of one-step methods which we can denote by $\mathcal{E}_p = \mathcal{E}, \mathcal{E}_{p+1} = \mathcal{E}^+, \mathcal{E}_{p+2}, \dots$, together with corresponding starting methods $\mathcal{S}_p = \mathcal{S}, \mathcal{S}_{p+1} = \mathcal{S}^+, \mathcal{S}_{p+2}, \dots$. According to the construction of these various operations, the two compositions

$$\mathcal{S}_i\mathcal{M} \quad \text{and} \quad \mathcal{E}_i\mathcal{S}_i$$

commute to within order i for $i = p, p + 1, p + 2, \dots$. The underlying one-step method is the notional limit of the $\mathcal{E}_i$ sequence. This construction was proposed, in the case of linear multistep methods by Kirchgraber (1986) and extended to the case of general linear methods by Stoffer (1993).

Denote the underlying one-step method by $\mathcal{E}^\star$ and the limit of the sequence of iterated starting methods by $\mathcal{S}^\star$ and we have a new diagram corresponding to Figure 4.1 given by Figure 4.4. Now the diagram exactly

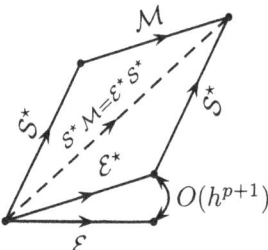

Figure 4.4. Underlying one-step method.

commutes but $\mathcal{E}$ is replaced by $\mathcal{E}^\star$. Thus, we can interpret this to mean that the general linear method behaves exactly like a one-step method. Consequently, error analysis is reduced to understanding how well $\mathcal{E}^\star$ approximates $\mathcal{E}$.

As an example of the computations involved in the analysis of the underlying one-step method, return to the method given in (4.1) and the information given in Table 4.2. From this table we see that the coefficients of $\mathcal{E}_i S_i - S_i M$ of the elementary differentials associated with the fifth-order trees $t_9, t_{10}, \ldots, t_{17}$ are

$$[\tfrac{1}{5}, \tfrac{1}{10}, \tfrac{1}{15}, \tfrac{1}{30}, \tfrac{1}{20}, \tfrac{1}{20}, \tfrac{1}{40}, \tfrac{1}{60}, \tfrac{1}{120}] - [\tfrac{5}{24}, \tfrac{5}{48}, \tfrac{1}{9}, \tfrac{1}{18}, \tfrac{5}{96}, \tfrac{1}{24}, \tfrac{1}{48}, -\tfrac{1}{36}, -\tfrac{1}{72}]$$
$$= [-\tfrac{1}{120}, -\tfrac{1}{240}, -\tfrac{2}{45}, -\tfrac{1}{45}, -\tfrac{1}{480}, \tfrac{1}{120}, \tfrac{1}{240}, \tfrac{2}{45}, \tfrac{1}{45}].$$

Let $\phi_1 = \sum_{i=9}^{17} \sigma(t_i)^{-1} C_i h^5 F(t_i)(y_0)$ denote the corresponding error terms, where C_i is found from this array. Because there are no error terms of this order in the second output component, we find ϵ and δ by solving the equation

$$\begin{bmatrix} \phi_1 \\ 0 \end{bmatrix} = \epsilon \begin{bmatrix} 1 \\ 1 \end{bmatrix} + \begin{bmatrix} 0 & 0 \\ -1 & 1 \end{bmatrix} \begin{bmatrix} \delta_1 \\ \delta_2 \end{bmatrix},$$

where we have inserted the actual values of u and $I - V$. A possible solution is $\epsilon = \delta_1 = \phi_1$, $\delta_2 = 0$.

5. Linear and non-linear stability

5.1. Linear stability

The linear stability analysis of numerical methods, exemplified in Figure 1.2 is based on the test problem

$$y'(x) = qy(x). \tag{5.1}$$

The aim of this type of analysis is to investigate the existence of bounds on the step-size to achieve stable numerical behaviour for a stable problem. We will want to identify methods for which there is never any such restriction because stable behaviour will occur whenever $\operatorname{Re} hq \leq 0$.

For a general linear method (A, U, B, V), the linear problem (5.1) will produce output which satisfies the equations

$$Y = hqAY + Uy^{[n-1]}, \tag{5.2}$$

$$y^{[n]} = hqBY + Vy^{[n-1]}. \tag{5.3}$$

For convenience write $z = hq$ and solve for Y from (5.2), to give $Y = (I - zA)^{-1}Uy^{[n-1]}$. Substitute into (5.3) to give $y^{[n]} = M(z)y^{[n-1]}$, where

$$M(z) = V + zB(I - zA)^{-1}U. \tag{5.4}$$

The matrix-valued function $M(z)$ is the 'stability function' and its characteristic polynomial

$$\Phi(w, z) = \det(wI - M(z)),$$

determines its linear stability properties.

Definition 5.1. A general linear method (A, U, B, V) is A-stable if $M(z)$ is power bounded whenever $\mathrm{Re}\, z < 0$.

The test problem on which this definition is based can be made more realistic, but of course more difficult to analyse if q is allowed to be time-dependant. That is, we might consider the problem

$$y'(x) = q(x)y(x). \tag{5.5}$$

In this case, we need to evaluate $z = hq(x)$ at each stage value and we write the collection of all values of this quantity that occur in the step in the form of a diagonal matrix:

$$Z = \mathrm{diag}(hq(x_{n-1} + hc_1),\, hq(x_{n-1} + hc_2),\, \ldots,\, hq(x_{n-1} + hc_s)).$$

For simplicity we will assume that the c components are distinct so that there is no reason to suppose that two diagonal elements of Z are necessarily equal. The stage derivative vector is now ZY and the stage values and output values are now

$$Y = (I - AZ)^{-1}Uy^{[n-1]}, \qquad y^{[n]} = (V + BZ(I - AZ)^{-1}U)y^{[n-1]}. \tag{5.6}$$

Thus for a problem of the form (5.5), it is natural in stability considerations to replace use of $M(z)$ given by (5.4) by the matrix-valued function of r complex variables, given by

$$\widetilde{M}(Z) = V + BZ(I - AZ)^{-1}U.$$

This leads to the following definition.

Definition 5.2. A general linear method (A, U, B, V) is AN-stable if $\widetilde{M}(Z)$ given by (5.6) is power-bounded for $Z = \mathrm{diag}(z_1, z_2, \ldots, z_s)$ if

$$\mathrm{Re}\, z_i < 0, \qquad i = 1, 2, \ldots, s.$$

5.2. Non-linear stability

To analyse stable behaviour for non-linear problems it is necessary to find a problem class for which stability of the exact solution is assured. We reintroduce (from Section 1.3) a problem made famous in Dahlquist (1976):

$$y'(x) = f(x, y(x)), \quad \langle u - v, f(x, u) - f(x, v) \rangle \leq 0. \tag{5.7}$$

Consideration of this problem led to the construction of 'one-leg methods' and to the definition of G-stability.

For a linear multistep method (ρ, σ), normalized so that $\sigma(1) = 1$, in which $\widehat{y}_n$ is computed as the solution to the equation

$$\sum_{i=0}^{k} \alpha_i \widehat{y}_{n+i} = h \sum_{i=0}^{k} \beta_i f(\widehat{x}_{n+i}, \widehat{y}_{n+i}), \tag{5.8}$$

the one-leg counterpart defines y_n as the solution to the equation

$$\sum_{i=0}^{k} \alpha_i y_{n-i} = h f\left(x_n, \sum_{i=0}^{k} \beta_i y_{n-i}\right). \tag{5.9}$$

From a linear stability point of view, these two methods have the same stability function:

$$\rho(w) - z\sigma(w) = 0.$$

Furthermore, stable behaviour, even for solutions to the non-linear problem (5.7), is closely related in the sense that if the y sequence satisfies (5.9) and

$$\widehat{x}_n = \sum_{i=0}^{k} \beta_i x_{n+i}, \qquad \widehat{y}_n = \sum_{i=0}^{k} \beta_i y_{n+i},$$

then $\widehat{y}$ satisfies (5.8). The crucial result in Dahlquist (1976) is a condition on (ρ, σ) such that, if two sequences y and $\bar{y}$ satisfy (5.9), then, for a norm $\|\cdot\|_G$, defined in the paper,

$$\|y^{[n]} - \bar{y}^{[n]}\|_G \leq \|y^{[n-1]} - \bar{y}^{[n-1]}\|_G, \tag{5.10}$$

if f satisfies the condition (5.7).

Why, it might be asked, was it necessary to introduce one-leg methods, rather than carry out an analysis of non-linear stability directly in terms of linear multistep methods?

From the general linear methods point of view, the answer is simple. A linear multistep method, in its standard formulation, is reducible, as we discussed in Section 3.3. With the irreducible formulation given by (3.4), non-linear stability for linear multistep methods can be analysed in their own right (Butcher and Hill 2006).

Non-linear stability of Runge–Kutta methods was introduced in Butcher (1975) where it was shown that certain implicit methods, such as Gauss methods and Radau IIA methods have the property known as B-stability. A method is B-stable if, for a method satisfying (5.7),

$$\|y_n - \bar{y}_n\| \leq \|y_{n-1} - \bar{y}_{n-1}\|,$$

for two approximation sequences y and $\bar{y}$. This definition was originally introduced for an autonomous version of (5.7), but BN-stability, introduced in Burrage and Butcher (1979) made use of the general non-autonomous version of this model problem. At the same time AN-stability, where a non-autonomous linear problem is used, was introduced.

The necessary and sufficient conditions for BN-stability (and incidentally for AN-stability), at least for non-confluent methods, was shown to hinge on a matrix M given by

$$M = \mathrm{diag}(b)A + A^T \, \mathrm{diag}(b) - bb^T.$$

It was shown in Burrage and Butcher (1979) and Crouzeix (1979) that a method is B-stable if M and $\mathrm{diag}(b)$ are each positive semi-definite. In an unpublished report, Dahlquist and Jeltsch showed that the elements of b must actually be positive for B-stability, or the method can be reduced to a simpler method, with fewer stages.

Linear and non-linear stability were analysed and inter-related in a series of papers (Burrage and Butcher 1980, Burrage 1980, Butcher 1981b, 1987b). The criterion, referred to as algebraic stability, which generalizes Dahlquist's G-stability criterion and the Runge–Kutta criterion based on M, makes use of the matrix

$$\hat{M} = \begin{bmatrix} DA + A^T D - B^T GB & DU - B^T GV \\ U^T D - V^T GB & G - V^T GV \end{bmatrix}. \tag{5.11}$$

In (5.11), G is a positive-definite matrix and D is a diagonal matrix of positive numbers. Under various conditions, it was shown in Butcher (1987c) that $\hat{M}$ positive semi-definite is equivalent to AN-stability and to stable behaviour in the sense of (5.10), for problems satisfying (5.7).

An interesting example of an algebraically stable method is the following, discovered by Dekker (1981), and shown by him to have this property:

$$\begin{bmatrix} A & U \\ B & V \end{bmatrix} = \left[\begin{array}{cc|cc} \frac{2}{3} & 0 & 1 & -\frac{7}{6} \\ \frac{2}{3} & \frac{2}{3} & 1 & \frac{1}{6} \\ \hline \frac{1}{2} & \frac{1}{2} & 1 & 0 \\ -\frac{1}{11} & \frac{7}{11} & 0 & \frac{5}{11} \end{array} \right]. \tag{5.12}$$

To verify algebraic stability, substitute $G = \mathrm{diag}(1, \frac{11}{12})$, $D = \mathrm{diag}(\frac{1}{2}, \frac{1}{2})$ in (5.11). Like Runge–Kutta methods, but unlike linear multistep methods,

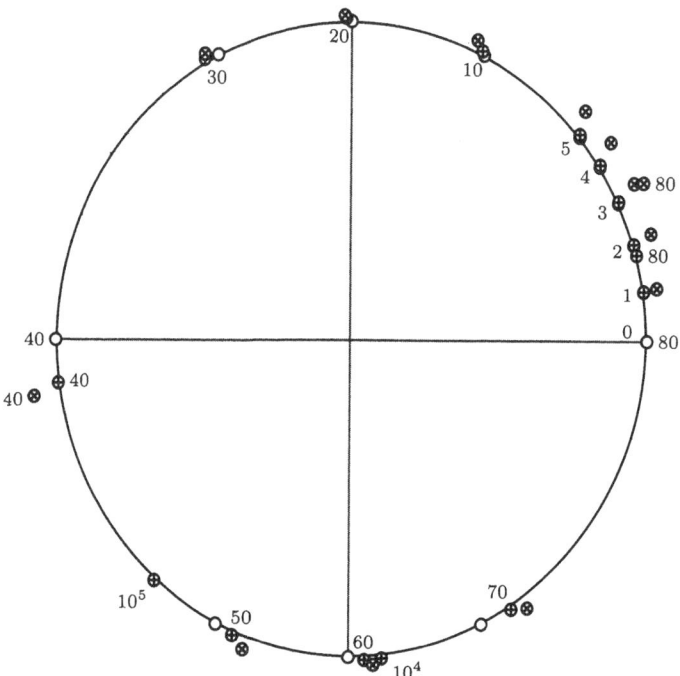

Figure 5.1. Solution of (5.15) using (5.12) (⊕) and
(5.13) (⊗), compared with the exact solution (○).

algebraic stability is *not* equivalent to A-stability. For example, compare
with (5.12) the method

$$
\begin{bmatrix} A & U \\ B & V \end{bmatrix} =
\left[
\begin{array}{cc|cc}
\frac{2}{3} & 0 & 1 & -\frac{7}{6} \\
\frac{2}{3} & \frac{2}{3} & 1 & \frac{1}{6} \\
\hline
\frac{179}{88} & -\frac{19}{88} & 1 & -\frac{9}{11} \\
\frac{23}{44} & \frac{1}{44} & 0 & \frac{5}{11}
\end{array}
\right].
\tag{5.13}
$$

The two methods have the same stability functions, and are thus each A-
stable. Furthermore, the stage abscissa vector is $[-\frac{1}{2}, \frac{3}{2}]$ in each case. However, (5.13) is not algebraically stable.

As an example, it should be expected that a problem of the form

$$
y'(x) = L(x, y(x))y(x),
\tag{5.14}
$$

where L takes values on the set of $N \times N$ matrices such that the symmetric part of $-L$ is positive semi-definite, will exhibit stable behaviour when
solved using (5.12) but not when solved using (5.13).

In particular consider the initial value problem

$$
\begin{bmatrix} y_1'(x) \\ y_2'(x) \end{bmatrix} = \left(y_1(x)^2 + \tfrac{1}{4} y_2(x)^2 \right) \begin{bmatrix} -y_2(x) \\ y_1(x) \end{bmatrix}, \quad \begin{bmatrix} y_1(0) \\ y_2(0) \end{bmatrix} = \begin{bmatrix} 1 \\ 0 \end{bmatrix}.
\tag{5.15}
$$

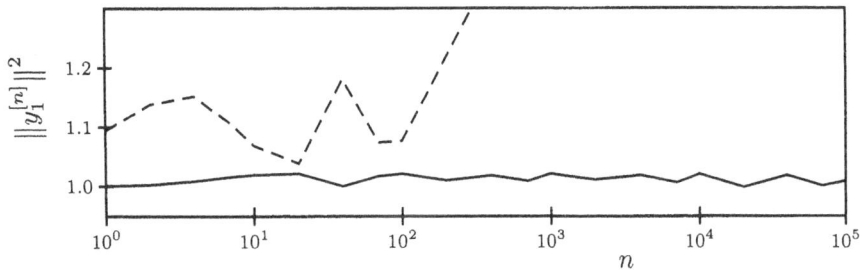

Figure 5.2. $\|y_1^{[n]}\|^2$ after n steps,
using (5.12) (——) and (5.13) (---).

For this problem, L takes on skew-symmetric values, so that $\|y(x)\|^2$ is not simply bounded, but is invariant.

To solve (5.15), using either (5.12) or (5.13), an appropriate starting value is $y_1^{[0]} = [1,0]^T$, $y_2^{[0]} = [0,h]^T$. Results found from these methods using constant step-size $h = \pi/20$ are presented in Figure 5.1 for n steps up to $n = 80$, which corresponds to a single period in the exact solution. In addition, $n = 10^4$ and 10^5 are also given in the case of (5.12). Even though there is a considerable phase shift, in accordance with the low accuracy of the numerical approximation, it is seen that the solutions using (5.12) adhere closely to the $\|y(x)\|^2 = 1$ manifold. This is explored further in Figure 5.2, where $\|y_1^{[n]}\|^2$ is evaluated for (5.12) up to $n = 10^5$ and for (5.13) until the value of this quantity drifts too far away. Also computed, but not explicitly shown, is the value of $\|y_1^{[n]}\|^2 + \frac{11}{12}\|y_2^{[n]}\|^2$, which is virtually constant.

We will now give a partial explanation of this phenomenon. First we note that, for (5.12), with the values of G and D that have been proposed, the partitioned matrix in (5.11) has the value

$$\widehat{M} = \frac{8}{11} \begin{bmatrix} \frac{3}{4} \\ \frac{1}{4} \\ 0 \\ -1 \end{bmatrix} \begin{bmatrix} \frac{3}{4} & \frac{1}{4} & 0 & -1 \end{bmatrix},$$

so that we can evaluate the following inner product:

$$\begin{bmatrix} hF^T & (y^{[n-1]})^T \end{bmatrix} \widehat{M} \begin{bmatrix} hF \\ y^{[n-1]} \end{bmatrix} = \frac{8}{11}\|\frac{3}{4}hF_1 + \frac{1}{4}hF_2 - y_2^{[n-1]}\|^2. \tag{5.16}$$

Doing the calculation another way we find that (5.16) can be written as

$$hF^T D(hAF + Uy^{[n-1]}) + (hAF + Uy^{[n-1]})^T DhF$$
$$- (hBF + Vy^{[n-1]})^T G(hBF + Vy^{[n-1]})$$
$$+ (y^{[n-1]})^T G(y^{[n-1]}). \tag{5.17}$$

Write $v^T G v$ as $\|v\|_G^2$ with the corresponding inner product equal to $\langle u, v \rangle = u^T G v$ and (5.17) simplifies to

$$2 \sum_{i=1}^{s} d_i \langle hF_i, Y_i \rangle + \|y^{[n]}\|_G^2 - \|y^{[n-1]}\|_G^2.$$

With the assumed form of the differential equation, we deduce

$$\|y^{[n]}\|_G = \|y^{[n-1]}\|_G - \tfrac{8}{11}\|\tfrac{3}{4}hF_1 + \tfrac{1}{4}hF_2 - y_2^{[n-1]}\|^2,$$

so that we cannot expect $\|y^{[n]}\|_G$ to be invariant. However, for the problem we are considering, $\tfrac{3}{4}hF_1 + \tfrac{1}{4}hF_2 - y_2^{[n-1]}$ has a small norm which decreases rapidly if $\|y^{[n]}\|_G$ becomes small. If we replace the method (5.12) by one in which, for an appropriate D and G, $\widehat{\mathsf{M}}$ is the zero matrix, then we can expect precise invariance of $\|y^{[n]}\|_G$. For example,

$$\begin{bmatrix} A & U \\ B & V \end{bmatrix} = \left[\begin{array}{cc|cc} \tfrac{1}{2} & 0 & 1 & -\tfrac{1}{2} \\ 1 & \tfrac{1}{2} & 1 & -\tfrac{1}{2} \\ \hline \tfrac{1}{2} & \tfrac{1}{2} & 1 & 0 \\ 1 & 1 & 0 & -1 \end{array} \right], \tag{5.18}$$

for which we need to use $G = \operatorname{diag}(1, \tfrac{1}{4})$, $D = \operatorname{diag}(\tfrac{1}{2}, \tfrac{1}{2})$. The invariant behaviour of $\|y^{[n]}\|_G^2 = (y_1^{[n]})^2 + \tfrac{1}{4}(y_2^{[n]})^2$ is verified by numerical experiment.

This method has order only 2 and does not seem to have any real advantages over the implicit mid-point rule method given by

$$\begin{bmatrix} A & U \\ B & V \end{bmatrix} = \left[\begin{array}{c|c} \tfrac{1}{2} & 1 \\ \hline 1 & 1 \end{array} \right],$$

However, it is possible to construct more accurate methods such as

$$\begin{bmatrix} A & U \\ B & V \end{bmatrix} = \left[\begin{array}{cc|cc} \tfrac{3+\sqrt{3}}{6} & 0 & 1 & -\tfrac{3+2\sqrt{3}}{3} \\ -\tfrac{\sqrt{3}}{3} & \tfrac{3+\sqrt{3}}{6} & 1 & \tfrac{3+2\sqrt{3}}{3} \\ \hline \tfrac{1}{2} & \tfrac{1}{2} & 1 & 0 \\ \tfrac{1}{2} & -\tfrac{1}{2} & 0 & -1 \end{array} \right], \tag{5.19}$$

which has order 4, as we see below. As for (5.18), $\widehat{\mathsf{M}} = 0$; but in this case we use $G = \operatorname{diag}(1, 1 + \tfrac{2\sqrt{3}}{3})$, $D = \operatorname{diag}(\tfrac{1}{2}, \tfrac{1}{2})$.

Because (5.19) is symplectic, in a slightly more general sense than applies to Runge–Kutta methods, it has a potential role in structure-preserving algorithms. Before we discuss this question, we verify its order.

Theorem 5.3. The order of (5.19) is 4.

Proof. Given an input approximation

$$y^{[0]} = \begin{bmatrix} y(x_0) \\ \frac{\sqrt{3}}{12}h^2 y''(x_0) - \frac{\sqrt{3}}{108}h^4 y^{(4)}(x_0) + \frac{9+5\sqrt{3}}{216}h^4 \frac{\partial f}{\partial y} y^{(4)}(x_0) \end{bmatrix}, \qquad (5.20)$$

we need to verify that the output is

$$y^{[1]} = \begin{bmatrix} y(x_0)+hy'(x_0)+\frac{1}{2}h^2 y''(x_0)+\frac{1}{6}h^3 y^{(3)}+\frac{1}{24}h^4 y^{(4)}+\mathcal{O}(h^5) \\ \frac{\sqrt{3}}{12}h^2 y''(x_0) + \frac{\sqrt{3}}{12}h^3 y^{(3)}(x_0)+ \\ \frac{7\sqrt{3}}{216}h^4 y^{(4)}(x_0) + \frac{9+5\sqrt{3}}{216}h^4 \frac{\partial f}{\partial y} y^{(4)}(x_0) + \mathcal{O}(h^5) \end{bmatrix}, \qquad (5.21)$$

found by replacing x_0 by $x_1 = x_0 + h$ and expanding about x_0. By Taylor expansions we find

$$Y_1 = y\left(x_0 + h\tfrac{3+\sqrt{3}}{6}\right) + \tfrac{9+5\sqrt{3}}{108}h^3 y^{(3)}(x_0) + \mathcal{O}(h^4),$$

$$hF_1 = hy'\left(x_0 + h\tfrac{3+\sqrt{3}}{6}\right) + \tfrac{9+5\sqrt{3}}{108}h^4 \frac{\partial f}{\partial y} y^{(3)}(x_0) + \mathcal{O}(h^5), \qquad (5.22)$$

$$Y_2 = y\left(x_0 + h\tfrac{3-\sqrt{3}}{6}\right) - \tfrac{9+5\sqrt{3}}{108}h^3 y^{(3)}(x_0) + \mathcal{O}(h^4),$$

$$hF_2 = hy'\left(x_0 + h\tfrac{3-\sqrt{3}}{6}\right) - \tfrac{9+5\sqrt{3}}{108}h^4 \frac{\partial f}{\partial y} y^{(3)}(x_0) + \mathcal{O}(h^5). \qquad (5.23)$$

Using (5.22) and (5.23), evaluate $y^{[1]} = hAF + Vy^{[0]}$ by Taylor expansions, to obtain agreement with (5.21). □

5.3. Experiments with a Hamiltonian problem

Consider the simple-pendulum problem

$$\begin{aligned} \dot{p} &= -\sin(q), & p(0) &= 1, \\ \dot{q} &= p, & q(0) &= 0. \end{aligned} \qquad (5.24)$$

This is based on the Hamiltonian $H(p,q) = \frac{1}{2}p^2 - \cos(q)$, where we note that

$$\dot{p} = -\frac{\partial H}{\partial q}, \qquad \dot{q} = \frac{\partial H}{\partial p}.$$

Because of the initial values assigned to (5.24), the dependent variables lie in the intervals

$$p \in [-1,1], \quad q \in [-\tfrac{1}{3}\pi, \tfrac{1}{3}\pi]$$

and the period is calculated to be $T = 6.743001419251$.

Attempts to solve this problem using the Euler and implicit Euler methods, are shown in Figure 5.3, with the exact solution also given for

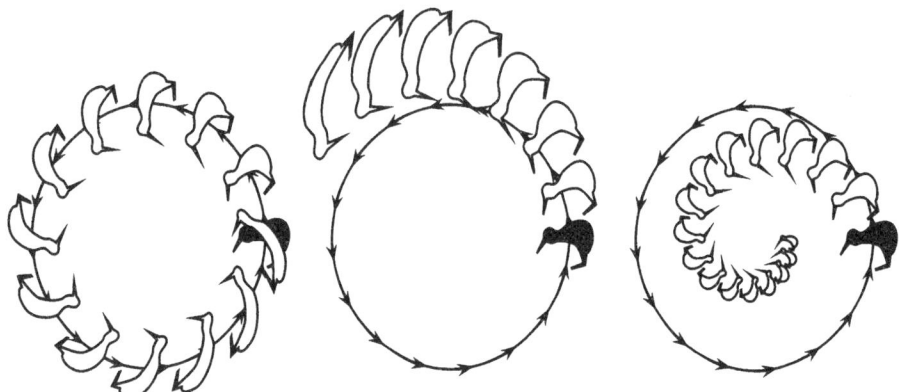

Figure 5.3. Left to right: exact solution,
Euler method, implicit Euler method.

comparison. A step-size $\frac{1}{16}T$ is used and the computations are confined to
the interval $[0,T]$, except for the Euler case which leaves the field of view
after only 7 time-steps.

To illustrate symplectic behaviour for the exact solution, and to indicate
that it does not occur for the two numerical approximations, a set of initial
points, shown in black, is used at time zero. For the exact symplectic result,
even though the set of points has its shape distorted, the area remains
unchanged. For the Euler and implicit Euler methods, however, not only
do the computed results drift rapidly away from the correct trajectory, but
the areas change in size.

We will consider the use of three alternative methods. These are:

(i) the order 4 Gauss Runge–Kutta method with defining matrices

$$
\begin{bmatrix} A & U \\ B & V \end{bmatrix} =
\left[\begin{array}{cc|c}
\frac{1}{4} & \frac{1}{4}-\frac{1}{6}\sqrt{3} & 1 \\
\frac{1}{4}+\frac{1}{6}\sqrt{3} & \frac{1}{4} & 1 \\
\hline
\frac{1}{2} & \frac{1}{2} & 1
\end{array} \right] ;
$$

(ii) the order 2 Gauss method, usually referred to as 'the mid-point rule
method', defined by

$$
\begin{bmatrix} A & U \\ B & V \end{bmatrix} =
\left[\begin{array}{c|c}
\frac{1}{2} & 1 \\
\hline
1 & 1
\end{array} \right] ;
$$

(iii) the general linear method given by (5.19).

The general linear method requires a starting procedure to produce input

Figure 5.4. Left to right: exact solution,
mid-point rule, symplectic general linear method.

for the first step of the method and it is proposed to use

$$
\begin{bmatrix} A & U \\ B & V \end{bmatrix} =
\left[
\begin{array}{cc|c}
\frac{3+\sqrt{3}}{6} & 0 & 1 \\
-\frac{3+\sqrt{3}}{3} & \frac{3+\sqrt{3}}{6} & 1 \\
\hline
0 & 0 & 1 \\
\frac{\sqrt{3}-1}{8} & \frac{1-\sqrt{3}}{8} & 0
\end{array}
\right].
$$

This simply passes through the initial value $y(x_0)$ as the first component $y_1^{[0]}$ and produces an approximation to $\frac{1}{12}\sqrt{3}h^2 y''(x_0)$ as the value of $y_2^{[0]}$.

To evaluate the behaviour of these methods, a greater step-size than was used in Figure 5.3 is needed, otherwise the errors will be imperceptible; we will use a step-size $\frac{1}{8}T$. Even with steps this large, it is impossible to distinguish method (i) from the exact solution and we therefore omit this case from the results we present. The exact result and methods (ii) and (iii) are shown in Figure 5.4. We see that the mid-point rule exhibits inexact behaviour but, because it is symplectic, the areas of the solution sets do not change. Similarly, the general linear method performs very well and also appears to preserve areas.

6. Special families of methods

6.1. Re-use methods and two-step Runge–Kutta methods

The idea of using derivative approximations, computed in a previous step, as contributing to the computation of the current step, was proposed in Butcher (1966). It has recently been developed under the name 'two-step Runge–Kutta methods'. Although we will not attempt to survey this large body of work in detail, we will derive order conditions for a family of these methods.

As a general linear method, the inputs to step n are the computed approximations to $y(x_{n-2})$ and $y(x_{n-1})$ together with the s scaled stage derivatives computed within step number $n - 1$. Hence we write

$$
y^{[n-1]} = \begin{bmatrix} y_{n-1} \\ y_{n-2} \\ hF_1^{[n-1]} \\ hF_2^{[n-1]} \\ \vdots \\ hF_s^{[n-1]} \end{bmatrix}, \quad
y^{[n]} = \begin{bmatrix} y_n \\ y_{n-1} \\ hF_1^{[n]} \\ hF_2^{[n]} \\ \vdots \\ hF_s^{[n]} \end{bmatrix}, \quad
Y^{[n]} = \begin{bmatrix} Y_1^{[n]} \\ Y_2^{[n]} \\ \vdots \\ Y_s^{[n]} \end{bmatrix}, \quad
F^{[n]} = \begin{bmatrix} F_1^{[n]} \\ F_2^{[n]} \\ \vdots \\ F_s^{[n]} \end{bmatrix}
$$

and we write the coefficient matrix in the partitioned form

$$
\left[\begin{array}{c|c} A & U \\ \hline B & V \end{array}\right] =
\left[\begin{array}{c|ccc}
A & u & e-u & \bar{A} \\ \hline
b^T & \theta & 1-\theta & \bar{b}^T \\
0 & 1 & 0 & 0 \\ \hline
I & 0 & 0 & 0
\end{array}\right].
$$

This method is also conveniently written using an extension of the standard tableau for Runge–Kutta methods,

$$
\begin{array}{c|c|c|c}
c & u & \bar{A} & A \\ \hline
 & \theta & \bar{b}^T & b^T
\end{array},
$$

indicating that $Y_i^{[n]}$, $i = 1, 2, \ldots, s$ and y_n are computed using the formulae

$$
Y_i^{[n]} = u_i y_{n-2} + (1 - u_i)y_{n-1} + h \sum_{j=1}^{s}(\bar{a}_{ij}F_j^{[n-1]} + a_{ij}F_j^{[n]}), \tag{6.1}
$$

$$
F_i^{[n]} = f(Y_i^{[n]}), \tag{6.2}
$$

$$
y_n = \theta y_{n-2} + (1 - \theta)y_{n-1} + h \sum_{i=1}^{s}(\bar{b}_i F_i^{[n-1]} + b_i F_i^{[n]}). \tag{6.3}
$$

To find the order conditions, write $\eta \in X_1^s$, to represent the stage values. We then have

$$
\eta = uE^{-1} + (1 - u) + \bar{A}E^{-1}\eta D + A\eta D. \tag{6.4}
$$

The values of $\eta_i(t)$, $i = 1, 2, \ldots, s$ are found recursively and these are then substituted into the order equation

$$
E(t) = \theta E^{-1}(t) + \bar{b}^T(E^{-1}\eta D)(t) + b^T(\eta D)(t), \qquad r(t) \leq p.
$$

If we are going to require that $\eta(\tau) = c$, with c prescribed in advance, then

Table 6.1. Analysis of re-use method.

$\eta(\cdot)$	c
$(\eta D)(\cdot)$	1
$(E^{-1}\eta)(\cdot)$	$c - 1$
$(E^{-1}\eta D)(\cdot)$	1
$\eta(\mathfrak{t})$	$\frac{1}{2}u + \overline{A}(c-1) + Ac$
$(\eta D)(\mathfrak{t})$	c
$(E^{-1}\eta)(\mathfrak{t})$	$\frac{1}{2}u + \overline{A}(c-1) + Ac - c + \frac{1}{2}1$
$(E^{-1}\eta D)(\mathfrak{t})$	$c - 1$
$\eta(\mathsf{V})$	$-\frac{1}{3}u + \overline{A}(c-1)^2 + Ac^2$
$(\eta D)(\mathsf{V})$	c^2
$(E^{-1}\eta)(\mathsf{V})$	$2c - \frac{1}{3}1 - \frac{4}{3}u + \overline{A}(c^2 - 4c + 31) + A(c^2 - 2c)$
$(E^{-1}\eta D)(\mathsf{V})$	$(c-1)^2$
$\eta(\mathfrak{t})$	$-\frac{1}{6}u + \overline{A}(\frac{1}{2}u + \overline{A}(c-1) + Ac - c + \frac{1}{2}1)$ $\quad + A(\frac{1}{2}u + \overline{A}(c-1) + Ac)$
$(\eta D)(\mathfrak{t})$	$\frac{1}{2}u + \overline{A}(c-1) + Ac$
$(E^{-1}\eta)(\mathfrak{t})$	$\frac{1}{2}c - \frac{1}{6}1 - \frac{2}{3}u + \overline{A}(\frac{1}{2}u + \overline{A}(c-1) + Ac - 2c + \frac{3}{2}1)$ $\quad + A(\frac{1}{2}u + \overline{A}(c-1) + Ac - c)$
$(E^{-1}\eta D)(\mathfrak{t})$	$\frac{1}{2}u + \overline{A}(c-1) + Ac - c + \frac{1}{2}1$

the conditions become simpler. The additional condition is equivalent to

$$c = -u + (\overline{A} + A)1 \qquad (6.5)$$

and this can always be satisfied by the choice of u.

We will present in Table 6.1 formulae for $\eta(t)$ and associated quantities up to order 3. It will be assumed throughout that c is defined by (6.5). It is now a routine task to compute the order conditions up to order 4. However, we will avoid the full generality of this task and settle for the case defined by $s = 2$, $u = 0$, $\theta = 0$ and $c = [\frac{1}{2}, 1]^T$. The order conditions associated with the trees $\cdot$, $\mathfrak{t}$, V, V enable b^T and $\overline{b}^T$ to be found. These are

$$\overline{b}^T = [0, \frac{1}{6}], \qquad b^T = [\frac{2}{3}, \frac{1}{6}].$$

The order conditions associated with the trees $\mathfrak{t}$, V and Y become

$$-\frac{1}{3}\overline{a}_{11} - \frac{1}{6}\overline{a}_{21} + \frac{1}{6}a_{21} = \frac{1}{4},$$
$$-\frac{1}{6}\overline{a}_{11} - \frac{1}{12}\overline{a}_{21} + \frac{1}{12}a_{21} = \frac{1}{8},$$
$$-\frac{1}{6}\overline{a}_{11} - \frac{1}{4}\overline{a}_{21} + \frac{1}{12}a_{21} = \frac{7}{36}.$$

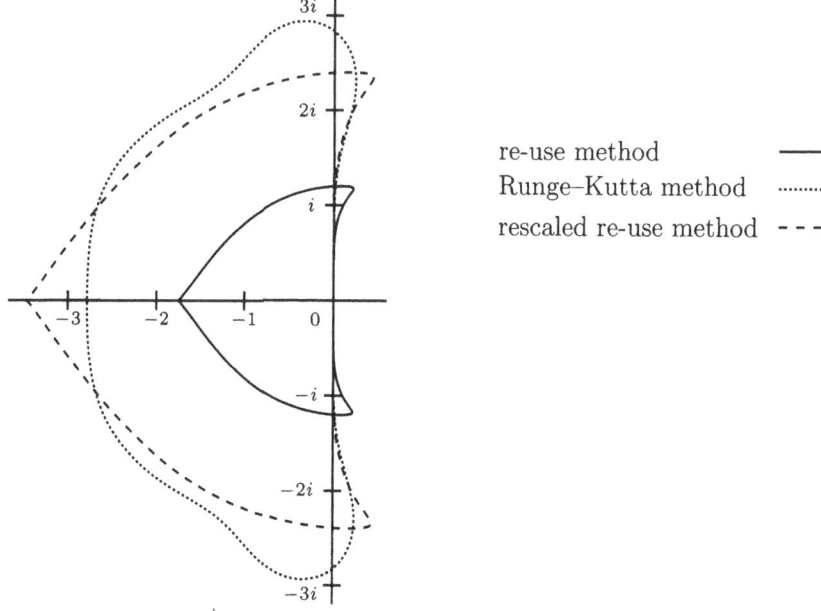

Figure 6.1. Stability region for re-use method.

These have solution

$$\bar{a}_{11} = \tfrac{1}{2}a_{21} - \tfrac{13}{24}, \qquad \bar{a}_{21} = -\tfrac{5}{12},$$

and the order condition associated with the remaining tree $\mathbf{I}$ gives $a_{21} = \tfrac{13}{12}$ or $a_{21} = \tfrac{11}{12}$. Choose the second of these and we find the tableau for the method to be

$$
\begin{array}{c|c|cc|cc}
\tfrac{1}{2} & 0 & -\tfrac{1}{12} & \tfrac{7}{12} & 0 & 0 \\
1 & 0 & -\tfrac{5}{12} & \tfrac{1}{2} & \tfrac{11}{12} & 0 \\
\hline
 & 0 & 0 & \tfrac{1}{6} & \tfrac{2}{3} & \tfrac{1}{6}
\end{array}.
$$

The stability region for this method is shown in Figure 6.1. As in Figure 2.1, we also give a rescaled stability region to allow for the fact that this method has only two stages, compared with four for the Runge–Kutta method.

Recently methods based on re-use of quantities computed in the previous step have been investigated under the name 'two-step Runge–Kutta methods'. Basic references on these methods are Jackiewicz and Tracogna (1995, 1996) and Bartoszewski and Jackiewicz (1998).

6.2. ARK methods

The aim of 'almost Runge–Kutta' or ARK methods is impose on re-use methods the additional requirement expressed in the following.

Definition 6.1. A general linear method is RK-stable (possesses Runge–Kutta stability) if its stability matrix has only a single nonzero eigenvalue.

This means that, for a method with r values passed from step to step and stability matrix $M(z)$,

$$\det(wI - M(z)) = w^{r-1}(w - R(z)).$$

The function $R(z)$, because it equals the trace of $M(z)$, is a rational function, and in the case of explicit methods a polynomial, and will be referred to as the stability function for the method.

Before formulating ARK methods, we will make a brief remark about traditional Runge–Kutta methods. The classical Kutta method, as a general linear method, is

$$\left[\begin{array}{cccc|c}
0 & 0 & 0 & 0 & 1 \\
\frac{1}{2} & 0 & 0 & 0 & 1 \\
0 & \frac{1}{2} & 0 & 0 & 1 \\
0 & 0 & 1 & 0 & 1 \\
\hline
\frac{1}{6} & \frac{1}{3} & \frac{1}{3} & \frac{1}{6} & 1
\end{array}\right]. \tag{6.6}$$

It is also possible to formulate this as a multivalue method with $r = 2$:

$$\left[\begin{array}{cccc|cc}
0 & 0 & 0 & 0 & 1 & \frac{1}{2} \\
\frac{1}{2} & 0 & 0 & 0 & 1 & 0 \\
0 & 1 & 0 & 0 & 1 & 0 \\
\frac{1}{3} & \frac{1}{3} & \frac{1}{6} & 0 & 1 & \frac{1}{6} \\
\hline
\frac{1}{3} & \frac{1}{3} & \frac{1}{6} & 0 & 1 & \frac{1}{6} \\
0 & 0 & 0 & 1 & 1 & 0
\end{array}\right], \tag{6.7}$$

where the derivative from the first stage in (6.6) is now computed as the *last* stage of (6.7). The two output quantities are approximations to the first terms in the Taylor series at the start of the following step:

$$y_1^{[n]} \approx y(x_n), \qquad y_2^{[n]} \approx hy'(x_n).$$

Make a similar change to the re-use method (2.3) and renumber the existing $y_2^{[n]}$ as $y_3^{[n]}$:

$$\left[\begin{array}{ccc|ccc}
0 & 0 & 0 & 1 & \frac{5}{8} & -\frac{1}{8} \\
2 & 0 & 0 & 1 & -\frac{3}{2} & \frac{1}{2} \\
\frac{2}{3} & \frac{1}{6} & 0 & 1 & \frac{1}{6} & 0 \\
\hline
\frac{2}{3} & \frac{1}{6} & 0 & 1 & \frac{1}{6} & 0 \\
0 & 0 & 1 & 0 & 0 & 0 \\
0 & 0 & 0 & 0 & 1 & 0
\end{array}\right]. \tag{6.8}$$

Instead of passing on $y_3^{[n]} \approx hy'(x_{n-1})$, in addition to $y(x_n)$, and $hy'(x_n)$, it is possible to replace $y_3^{[n]}$ by an approximation to $h^2 y''(x_n)$ which can be found as the difference between approximations to $hy'(x_n)$ and $hy'(x_{n-1})$. This gives the equivalent but more convenient formulation

$$
\left[
\begin{array}{ccc|ccc}
0 & 0 & 0 & 1 & \frac{1}{2} & \frac{1}{8} \\
2 & 0 & 0 & 1 & -1 & -\frac{1}{2} \\
\frac{2}{3} & \frac{1}{6} & 0 & 1 & \frac{1}{6} & 0 \\
\hline
\frac{2}{3} & \frac{1}{6} & 0 & 1 & \frac{1}{6} & 0 \\
0 & 0 & 1 & 0 & 0 & 0 \\
0 & 0 & 1 & 0 & -1 & 0
\end{array}
\right].
\tag{6.9}
$$

Even though we now pass on approximations to $y(x), hy'(x), h^2 y''(x)$ from step to step, the third of these is accurate only to within $\mathcal{O}(h^3)$. This is not a serious handicap because the consequence of this inaccuracy cancels out to within $\mathcal{O}(h^5)$ because $y_3^{[n]}$ appears only within the arguments of the first and second scaled stage derivatives and because the first row of B is orthogonal to the last column of U. Properties like this are referred to as 'annihilation conditions' and are crucial to the design of ARK methods.

This method cannot possibly possess RK stability but if we restore the value of s to 4, this does become possible. The derivation of methods up to order 4 is given in Butcher (1997b). The following example is based on the abscissae $c = [1, \frac{1}{2}, 1, 1]$:

$$
\left[
\begin{array}{cccc|ccc}
0 & 0 & 0 & 0 & 1 & 1 & \frac{1}{2} \\
\frac{1}{16} & 0 & 0 & 0 & 1 & \frac{7}{16} & \frac{1}{16} \\
-\frac{1}{4} & 2 & 0 & 0 & 1 & -\frac{3}{4} & -\frac{1}{4} \\
0 & \frac{2}{3} & \frac{1}{6} & 0 & 1 & \frac{1}{6} & 0 \\
\hline
0 & \frac{2}{3} & \frac{1}{6} & 0 & 1 & \frac{1}{6} & 0 \\
0 & 0 & 0 & 1 & 0 & 0 & 0 \\
-\frac{1}{3} & 0 & -\frac{2}{3} & 2 & 0 & -1 & 0
\end{array}
\right].
$$

In a variable step-size implementation, when h changes to rh between steps $n-1$ and n, a simple rescaling of $y_2^{[n-1]} \approx hy'(x_{n-1})$ by a factor r and $y_3^{[n-1]} \approx h^2 y''(x_{n-1})$ by a factor r^2 is adequate to preserve fourth-order behaviour.

It is possible to find a five-stage fourth-order ARK method with the special property that its error constants exactly vanish. Unfortunately, the annihilation conditions satisfied by this method are not sufficient for this method to act like a fifth-order method when implemented in a manner in which variable step-size is dealt with by simple rescaling. However, it is possible to adjust things with negligible additional work so that variable h fifth-order behaviour is achieved. Methods with this effectively fifth-order behaviour are presented in Butcher and Moir (2003) and Rattenbury (2005). Along with methods of effective order five, they present a means of breaking the order barrier on explicit Runge–Kutta methods.

Although ARK methods were originally designed for non-stiff problems, it has recently been found how to adapt them for the solution of stiff problems (Butcher and Rattenbury 2005, Rattenbury 2005). Extensive numerical testing shows this type of method to be very competitive for many stiff problems.

6.3. DIMSIM methods

In the search for practical general linear methods a systematic family was sought such that $p = q = r = s$ and such that, if possible, they possessed RK stability. Because the structure of the matrix A plays a crucial role in the implementation cost in both sequential and parallel environments, it seemed to be a good design choice to consider only lower triangular matrices in this role. Furthermore there are often advantages in forcing the diagonal elements to be equal and we will assume this to be the case. Methods designed with these considerations in mind are referred to as a *diagonally implicit multi-stage integration method* or DIMSIM (Butcher 1995). Because applications are needed for both stiff and non-stiff problems and because we will want to consider parallel as well as sequential architectures, four types of methods, determined by the structure of A are considered and these are summarized in Table 6.2 (overleaf).

Type 1 and 2 methods
The following is an example of a type 1 DIMSIM with $p = q = r = s = 2$:

$$\left[\begin{array}{cc|cc} 0 & 0 & 1 & 0 \\ 2 & 0 & 0 & 1 \\ \hline \frac{5}{4} & \frac{1}{4} & \frac{1}{2} & \frac{1}{2} \\ \frac{3}{4} & -\frac{1}{4} & \frac{1}{2} & \frac{1}{2} \end{array} \right].$$

Even though this method has the same stability region as the classical Runge–Kutta methods of Runge, it has advantages associated with stage order $q = 2$. In particular, at no additional cost it yields interpolated results, suitable for dense output or application to certain delay differential equations. Furthermore, asymptotically correct local error estimates are available. Variable step-size, of course, presents complications which are not present for the corresponding Runge–Kutta methods. However, there are satisfactory ways round these complications.

Table 6.2. The four DIMSIM types.

	Structure of A	Stiffness type	Architecture
type 1	$\begin{bmatrix} 0 & 0 & 0 & \cdots & 0 \\ a_{21} & 0 & 0 & \cdots & 0 \\ a_{31} & a_{32} & 0 & \cdots & 0 \\ \vdots & \vdots & \vdots & & \vdots \\ a_{s1} & a_{s2} & a_{s3} & \cdots & 0 \end{bmatrix}$	nonstiff	sequential
type 2	$\begin{bmatrix} \lambda & 0 & 0 & \cdots & 0 \\ a_{21} & \lambda & 0 & \cdots & 0 \\ a_{31} & a_{32} & \lambda & \cdots & 0 \\ \vdots & \vdots & \vdots & & \vdots \\ a_{s1} & a_{s2} & a_{s3} & \cdots & \lambda \end{bmatrix}$	stiff	sequential
type 3	$\begin{bmatrix} 0 & 0 & 0 & \cdots & 0 \\ 0 & 0 & 0 & \cdots & 0 \\ 0 & 0 & 0 & \cdots & 0 \\ \vdots & \vdots & \vdots & & \vdots \\ 0 & 0 & 0 & \cdots & 0 \end{bmatrix}$	nonstiff	parallel
type 4	$\begin{bmatrix} \lambda & 0 & 0 & \cdots & 0 \\ 0 & \lambda & 0 & \cdots & 0 \\ 0 & 0 & \lambda & \cdots & 0 \\ \vdots & \vdots & \vdots & & \vdots \\ 0 & 0 & 0 & \cdots & \lambda \end{bmatrix}$	stiff	parallel

A similar method, but of order and stage order 3, has the coefficient matrix (Butcher and Jackiewicz 1996)

$$\left[\begin{array}{ccc|ccc} 0 & 0 & 0 & 1 & 0 & 0 \\ 1 & 0 & 0 & 0 & 1 & 0 \\ \frac{1}{4} & 1 & 0 & 0 & 0 & 1 \\ \hline \frac{5}{4} & \frac{1}{3} & \frac{1}{6} & -\frac{2}{3} & \frac{4}{3} & \frac{1}{3} \\ \frac{35}{24} & -\frac{1}{3} & \frac{1}{8} & -\frac{2}{3} & \frac{4}{3} & \frac{1}{3} \\ -\frac{17}{12} & 0 & \frac{1}{12} & -\frac{2}{3} & \frac{4}{3} & \frac{1}{3} \end{array}\right].$$

The construction of higher-order type 1 DIMSIMs becomes increasingly complicated and numerical searches have to be made (Butcher and Jacki-ewicz 2004, Butcher, Jackiewicz and Mittelmann 1997). However, order 4 methods have been found by Wright (2001).

Type 2 methods with $p = q = r = s = 2$ are easy to find, for example, an L-stable method:

$$\begin{bmatrix} A & U \\ B & V \end{bmatrix} = \left[\begin{array}{cc|cc} \frac{2-\sqrt{2}}{2} & 0 & 1 & 0 \\ \frac{6+2\sqrt{2}}{7} & \frac{2-\sqrt{2}}{2} & 0 & 1 \\ \hline \frac{73-34\sqrt{2}}{28} & \frac{4\sqrt{2}-5}{4} & \frac{3-\sqrt{2}}{2} & \frac{\sqrt{2}-1}{2} \\ \frac{87-48\sqrt{2}}{28} & \frac{34\sqrt{2}-45}{28} & \frac{3-\sqrt{2}}{2} & \frac{\sqrt{2}-1}{2} \end{array} \right].$$

For higher orders, L-stable type 2 methods are also increasingly difficult to construct. However, the following method with $p = q = r = s = 3$ is A-stable:

$$\begin{bmatrix} A & U \\ B & V \end{bmatrix} = \left[\begin{array}{ccc|ccc} \frac{1}{2} & 0 & 0 & 1 & 0 & 0 \\ \frac{5}{4} & \frac{1}{2} & 0 & 0 & 1 & 0 \\ \frac{7}{5} & \frac{4}{5} & \frac{1}{2} & 0 & 0 & 1 \\ \hline \frac{14}{15} & \frac{1}{5} & -\frac{1}{12} & \frac{5}{6} & \frac{1}{3} & -\frac{1}{6} \\ \frac{17}{20} & \frac{7}{60} & -\frac{1}{6} & \frac{5}{6} & \frac{1}{3} & -\frac{1}{6} \\ \frac{23}{30} & \frac{2}{15} & -\frac{1}{20} & \frac{5}{6} & \frac{1}{3} & -\frac{1}{6} \end{array} \right].$$

Type 3 and 4 methods

It is impossible to obtain RK stability for high-order methods in these families. However, reasonable stability regions are possible for type 3 methods, as in the example with $p = q = r = s = 2$:

$$\begin{bmatrix} A & U \\ B & V \end{bmatrix} = \left[\begin{array}{cc|cc} 0 & 0 & 1 & 0 \\ 0 & 0 & 0 & 1 \\ \hline -\frac{3}{8} & -\frac{3}{8} & -\frac{3}{4} & \frac{7}{4} \\ -\frac{7}{8} & \frac{9}{8} & -\frac{3}{4} & \frac{7}{4} \end{array} \right].$$

The error constant for this method has magnitude $\frac{19}{24}$, which is abnormally large, even allowing for any possible gain due to parallelism.

An example of a type 4 method also with $p = q = r = s = 2$ is

$$\begin{bmatrix} A & U \\ B & V \end{bmatrix} = \left[\begin{array}{cc|cc} \frac{3-\sqrt{3}}{2} & 0 & 1 & 0 \\ 0 & \frac{3-\sqrt{3}}{2} & 0 & 1 \\ \hline \frac{18-11\sqrt{3}}{4} & -\frac{12+7\sqrt{3}}{4} & \frac{3}{2}-\sqrt{3} & \sqrt{3}-\frac{1}{2} \\ \frac{22-13\sqrt{3}}{4} & -\frac{12+9\sqrt{3}}{4} & \frac{3}{2}-\sqrt{3} & \sqrt{3}-\frac{1}{2} \end{array} \right].$$

In this case the stability polynomial is

$$\left(1 - z\frac{3-\sqrt{3}}{2}\right)^2 w^2 - \left(1 - z\frac{3-\sqrt{3}}{2}\right) w + \frac{1-\sqrt{3}}{2} z,$$

and it is possible to verify that the method is A-stable and that it has zero spectral radius at infinity.

An experimental implementation of methods of type 4 is reported in Singh (1999).

7. Methods with inherent RK-stability

Even though DIMSIM methods of types 1 and 2 cannot be constructed in a systematic manner, it is possible, by increasing both r and s, to $p+1 = q+1$ to derive methods which possess Runge–Kutta stability purely as a result of their structure. These methods, which are said to possess inherent RK stability, can be constructed in various ways but it seems most convenient to use 'doubly companion matrices', and this is the approach we will use.

7.1. Doubly companion matrices

Consider a matrix of the form

$$X(\alpha, \beta) = \begin{bmatrix} -\alpha_1 & -\alpha_2 & -\alpha_3 & \cdots & -\alpha_{n-1} & -\alpha_n - \beta_n \\ 1 & 0 & 0 & \cdots & 0 & -\beta_{n-1} \\ 0 & 1 & 0 & \cdots & 0 & -\beta_{n-2} \\ \vdots & \vdots & \vdots & & \vdots & \vdots \\ 0 & 0 & 0 & \cdots & 0 & -\beta_2 \\ 0 & 0 & 0 & \cdots & 1 & -\beta_1 \end{bmatrix},$$

where

$$\alpha(z) = 1 + \alpha_1 z + \cdots + \alpha_n z^n,$$
$$\beta(z) = 1 + \beta_1 z + \cdots + \beta_n z^n.$$

If $\beta(z) = 1$, so that $\beta_1 = \beta_2 = \cdots = \beta_n$, or similarly if $\alpha(z) = 1$, then the characteristic polynomials can be found from

$$\det(I - zX(\alpha, 1)) = \alpha(z),$$
$$\det(I - zX(1, \beta)) = \beta(z).$$

We now consider the general case.

Theorem 7.1. The characteristic polynomial of $X(\alpha, \beta)$ is given by

$$\det(I - zX(\alpha, \beta)) = \alpha(z)\beta(z) + \mathcal{O}(z^{n+1}). \tag{7.1}$$

In (7.1), the effect of the term $\mathcal{O}(z^{n+1})$ is to simply remove from the expanded product $\alpha(z)\beta(z)$ all terms with degree greater than n. If we denote the usual characteristic polynomial $\det(zI - X(\alpha, \beta))$ by $\phi(z)$, then (7.1) can be interpreted to mean

$$\phi(z) = [z^{-n} \det(zI - X(\alpha, 1)) \det(zI - X(1, \beta))],$$

where, in this formula, $[\cdot]$ means that negative powers of z are omitted.

Proof. Define the vector-valued function $P(z)$ by

$$P(z) = \begin{bmatrix} \vdots \\ z^2 + \beta_1 z + \beta_2 \\ z + \beta_1 \\ 1 \end{bmatrix}.$$ (7.2)

A simple calculation shows that

$$X(\alpha, \beta)P(z) = zP(z) + \phi(z)e_1,$$ (7.3)

showing that $(\lambda, P(\lambda))$ are an eigenvalue–eigenvector pair if $\phi(\lambda) = 0$. □

In applications of this result, we will be given $\beta(z)$ and the characteristic polynomial of $X(\alpha, \beta)$, and we will need to find α. In particular we will need to consider the case $\det(I - zX(\alpha, \beta)) = (1 - \lambda z)^n$ and we find, in this case,

$$\alpha(z) = (1 - \lambda z)^n / \beta(z) + \mathcal{O}(z^{n+1}).$$

It is possible to find explicit formulae for transformation matrices Ψ^{-1} and Ψ so that $\Psi^{-1}X\Psi$ is in Jordan canonical form. We will specialize this to the case that $X(\alpha, \beta)$ has a one-point spectrum $\sigma(X(\alpha, \beta)) = \{\lambda\}$. In addition to $P(z)$ given by (7.2), we will need the vector-valued function $Q(z)$ given by

$$Q(z) = \begin{bmatrix} 1 & z + \alpha_1 & z^2 + \alpha_1 z + \alpha_2 & \cdots \end{bmatrix}.$$

For the remainder of this paper, we will write X in place of $X(\alpha, \beta)$, unless there is the possibility of ambiguity. For the trivial case in which all α_i and β_i are zero, we will write J for this value of X.

Theorem 7.2. Define

$$\Psi = \begin{bmatrix} \frac{1}{(n-1)!}P^{(n-1)}(\lambda) & \cdots & \frac{1}{2!}P''(\lambda) & P'(\lambda) & P(\lambda) \end{bmatrix},$$

then, if the characteristic polynomial of X is $(z - \lambda)^n$,

$$\Psi^{-1}X\Psi = \lambda I + J,$$ (7.4)

and Ψ^{-1} is given by

$$\Psi^{-1} = \begin{bmatrix} Q(\lambda) \\ Q'(\lambda) \\ \frac{1}{2!}Q''(\lambda) \\ \vdots \\ \frac{1}{(n-1)!}Q^{(n-1)}(\lambda) \end{bmatrix}.$$

Proof. From the Taylor expansion of (7.3) about $z = \lambda$, it is found that

$$XP(\lambda) = \lambda P(\lambda),$$

$$X \frac{1}{i!} P^{(i)} = \lambda \frac{1}{i!} P^{(i)} + \frac{1}{(i-1)!} P^{(i-1)}, \quad i = 1, 2, \ldots, n-1,$$

and (7.4) follows. Similar formulae are found for $Q(\lambda)X$ and $\frac{1}{i!} Q^{(i)}(\lambda)X$ and the fact that both Ψ and the formula given for Ψ^{-1} are unit upper triangular, completes the proof. $\qquad\square$

7.2. Formulation of IRKS methods

We will consider the construction of methods with $p = q$, and $r = s = p+1$. Without loss of generality, we can assume that the starting method corresponds to the evaluation of the scaled derivatives $h^i y^{(i)}(x_0)$, $i = 0, 1, 2, \ldots, p$. This means that the vector $\phi(z)$ in Theorem 4.3 is equal to Z given by

$$Z = \begin{bmatrix} 1 \\ z \\ z^2 \\ \vdots \\ z^p \end{bmatrix}.$$

A consequence of this assumption is that V necessarily has the form

$$V = \begin{bmatrix} 1 & v^T \\ 0 & \dot{V} \end{bmatrix},$$

and stability requires that $\rho(\dot{V}) \le 1$. We will want to go further than this and actually assume that $\rho(\dot{V}) = 0$.

Because we want to minimize computational cost, we will consider only methods in which A has a lower triangular structure with constant diagonals:

$$A = \begin{bmatrix} \lambda & 0 & 0 & \cdots & 0 \\ a_{21} & \lambda & 0 & \cdots & 0 \\ a_{31} & a_{32} & \lambda & \cdots & 0 \\ \vdots & \vdots & \vdots & & \vdots \\ a_{s1} & a_{s2} & a_{s3} & \cdots & \lambda \end{bmatrix}.$$

For RK-stable methods, using A with this structure will result in a stability function of the form

$$R(z) = \frac{N(z)}{(1 - \lambda z)^{p+1}},$$

where, because the order is p, $N(z)$ is given by

$$N(z) = \exp(z)(1 - \lambda z)^{p+1} - \mathrm{const}\, z^{p+1},$$

where const is the 'error constant'. If n_{p+1} is the coefficient of z^{p+1} in $N(z)$ then

$$n_{p+1} = \frac{1}{(p+1)!} - \frac{1}{p!}\binom{p+1}{1}\lambda + \frac{1}{(p-1)!}\binom{p+1}{2}\lambda^2 - \cdots + (-\lambda)^{p+1} - \mathrm{const}.$$

In the construction of a specific method, the values of λ and either n_{p+1} or const are available as design choices. For example we may want to choose λ to achieve A-stability and we might require that $n_{p+1} = 0$ to obtain the additional property of L-stability. For a non-stiff option, λ would be chosen as zero, to obtain explicit methods, and const $= n_{p+1} - \frac{1}{(p+1)!}$ would be chosen to balance the requirements of accuracy and stability.

In the formulation of these new methods we will want to find the remaining (strictly lower triangular) elements of A and the elements of B as starting points, and evaluate U and V from

$$U = C - ACK, \tag{7.5}$$
$$V = E - BCK, \tag{7.6}$$

where C, K and E have the meanings introduced in Section 4.4. It will be a constraint on the choice of the elements in B to make sure that V has the correct form.

Definition 7.3. A general linear method (A, U, B, V) is said to possess inherent Runge–Kutta stability (IRKS) if it satisfies the assumptions introduced in Section 7.2 and there exists a doubly companion matrix X such that $\alpha(z)\beta(z) = (1 - \lambda z)^{p+1} + \mathcal{O}(z^{s+1})$, and a vector ξ^T, such that

$$BA = XB, \tag{7.7}$$
$$BU = XV - VX + e_1\xi^T. \tag{7.8}$$

The value of the vector ξ^T will be explored below.

The significance of Definition 7.3 is summed up in the following result.

Theorem 7.4. The characteristic polynomial of a general linear method possessing the IRKS property has only a single nonzero eigenvalue.

Proof. The stability matrix is

$$M(z) = V + zB(I - zA)^{-1}U,$$

and the characteristic polynomial of $M(z)$ is the same as for the matrix formed by similarity, using the transformation matrix $(I - zX)$. Evaluate this as follows:

$$
\begin{aligned}
(I - zX)M(z)(I - zX)^{-1} &= (I - zX)(V + zB(I - zA)^{-1}U)(I - zX)^{-1} \\
&= (I - zX)(V + z(I - zX)^{-1}BU)(I - zX)^{-1} \\
&= (V - zXV + z(XV - VX + e_1\xi^T))(I - zX)^{-1} \\
&= V + e_1\xi^T(I - zX)^{-1}.
\end{aligned}
\tag{7.9}
$$

This matrix has the same form as V except for the first row. Hence, p of the zeros of the characteristic polynomial are equal to zero.   □

This result makes it possible to determine ξ^T. Write $\xi(z) = \xi_1 z + \xi_2 z^2 + \cdots + \xi_{p+1} z^{p+1}$ and use (7.9) to give

$$
R(z) = 1 + z\xi^T(I - zX)^{-1}e_1
\tag{7.10}
$$

$$
= \frac{\det(I + z(e_1\xi^T - X))}{\det(I - zX)}
\tag{7.11}
$$

$$
= \frac{(\alpha(z) + \xi(z))\beta(z)}{\alpha(z)\beta(z)} + O(z^{p+2}).
\tag{7.12}
$$

Because $R(z) = N(z)(1 - \lambda z)^{-p-1}$, it follows that

$$
\xi(z) = (N(z) - (1 - \lambda z)^{p+1})\beta(z)^{-1} + O(z^{p+2}),
$$

and the coefficients in $\xi(z)$ are found as the components of ξ^T.

7.3. Construction of specific methods

We first explore consequences of (7.7) and (7.8). Substitute U and V from (7.5) and (7.6) into (7.8) and use (7.7) to find:

$$
BC(I - KX) = XE - EX + e_1\xi^T.
\tag{7.13}
$$

It is found that both $I - KX$ and $XE - EX + e_1\xi^T$ are zero except for their final columns. Deleting the irrelevant columns of (7.13) we find, after some manipulation,

$$
BC \begin{bmatrix} \beta_p \\ \beta_{p-1} \\ \vdots \\ \beta_1 \\ 1 \end{bmatrix} = \begin{bmatrix} \frac{1}{(p+1)!} + \sum_{i=1}^{p} \frac{1}{(p+1-i)!}\beta_i - \text{const} \\ \frac{1}{p!} + \sum_{i=1}^{p-1} \frac{1}{(p-i)!}\beta_i \\ \vdots \\ \frac{1}{2!} + \frac{1}{1!}\beta_1 \\ \frac{1}{1!} \end{bmatrix}.
\tag{7.14}
$$

Define $\tilde{B} = \Psi^{-1}B$ and rewrite (7.7) in the form

$$
\tilde{B}(A - \lambda I) = J\tilde{B}.
$$

It follows from this equation that $\widetilde{B}$ is lower triangular and we can write (7.14) in the form

$$
\widetilde{B}C \begin{bmatrix} \beta_p \\ \beta_{p-1} \\ \vdots \\ \beta_1 \\ 1 \end{bmatrix} = \Psi^{-1} \begin{bmatrix} \frac{1}{(p+1)!} + \sum_{i=1}^{p} \frac{1}{(p+1-i)!}\beta_i - \text{const} \\ \frac{1}{p!} + \sum_{i=1}^{p-1} \frac{1}{(p-i)!}\beta_i \\ \vdots \\ \frac{1}{2!} + \frac{1}{1!}\beta_1 \\ \frac{1}{1!} \end{bmatrix}. \tag{7.15}
$$

The condition that $\rho(\dot{V}) = 0$ can be written as a linear constraint on B and therefore of $\widetilde{B}$.

The construction of methods now consists of the following steps.

(i) Select suitable values of λ, c_1, $\ldots$, c_{p+1}, β_1, $\ldots$, β_p and const.

(ii) Find α_1, $\ldots$, α_{p+1} from

$$
\alpha(z) = (1 - \lambda z)^{p+1}/\beta(z) + \mathcal{O}(z^{p+2}).
$$

(iii) Construct X, Ψ and other related matrices.

(iv) Choose $\widetilde{B}$ so that (7.15) and so that $\rho(\dot{V}) = 0$.

(v) Find the coefficient matrices using the formulae

$$
B = \Psi\widetilde{B},
$$
$$
A = B^{-1}XB,
$$
$$
U = C - ACK,
$$
$$
V = E - BCK.
$$

To illustrate this procedure we construct an A-stable method of order 3. In step (i) we make the following choices.

- $\lambda = \frac{1}{2}$. This was chosen for simplicity, taking into account the need for A-stable behaviour and for a reasonably small absolute error constant. Assuming that L-stability is sought, then the error constant is given by

$$
\text{const} = \lambda^4 - 4\lambda^3 + 3\lambda^2 - \frac{2}{3}\lambda + \frac{1}{24},
$$

and is shown in Figure 7.1 for the interval $[0.223647801, 0.572816062]$, which is approximately the set of λ values yielding A-stable methods.

- The stage abscissae are chosen as $[\frac{1}{3}, \frac{2}{3}, 1, 1]$. This choice is obviously chosen for simplicity and convenience. By imposing an additional constraint, it is possible to force the first row of B to be identical to the last row of A and at the same time forcing the second row of B to be $[0, 0, 0, 1]$. This enables the method to have properties similar to the

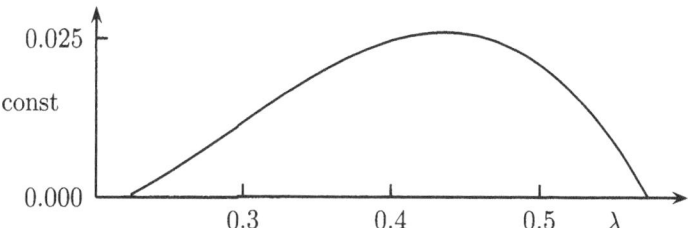

Figure 7.1. Error constant for third-order method.

so-called FSAL Runge–Kutta methods. In particular, $y_2^{[n]} = hf(y_1^{[n]})$ so that in step number $n+1$, we have available what is effectively a further stage derivative for use in interpolation and similar purposes.

- The values of $[\beta_1, \beta_2, \beta_3]$ are chosen as $[-1, \frac{1}{3}, 0]$. The zero value of β_3 is a consequence of the FSAL condition whereas β_1 and β_2 are chosen to ensure that the coefficients given as elements of $[A, U, B, V]$ have reasonably small magnitudes and are reasonably simple numbers.

- Because we want L-stability, we choose const as we have described above.

There are choices to be made in how step (iv) is to be carried out. In the present construction, we have forced $\dot V$ to be strictly lower triangular.

The coefficients for the method described under these choices are given by

$$
\begin{bmatrix} A & U \\ B & V \end{bmatrix} =
\left[
\begin{array}{cccc|ccccc}
\frac{1}{2} & 0 & 0 & 0 & 1 & -\frac{1}{6} & -\frac{1}{9} & -\frac{7}{324} \\[4pt]
\frac{36}{295} & \frac{1}{2} & 0 & 0 & 1 & \frac{79}{1770} & -\frac{403}{2655} & -\frac{1637}{23895} \\[4pt]
-\frac{705}{472} & \frac{177}{160} & \frac{1}{2} & 0 & 1 & \frac{8377}{9440} & -\frac{1131}{4720} & -\frac{581}{2360} \\[4pt]
\frac{39}{160} & \frac{15}{128} & \frac{5}{24} & \frac{1}{2} & 1 & -\frac{133}{1920} & \frac{353}{960} & \frac{109}{480} \\[4pt]
\hline
\frac{39}{160} & \frac{15}{128} & \frac{5}{24} & \frac{1}{2} & 1 & -\frac{133}{1920} & \frac{353}{960} & -\frac{109}{480} \\[4pt]
0 & 0 & 0 & 1 & 0 & 0 & 0 & 0 \\[4pt]
-\frac{45}{2} & 18 & \frac{5}{2} & -6 & 0 & 8 & 0 & 0 \\[4pt]
\frac{171}{4} & -\frac{531}{16} & 0 & 12 & 0 & -\frac{345}{16} & -\frac{33}{8} & 0
\end{array}
\right].
$$

$$(7.16)$$

As a start towards approximating the underlying one-step method and finding an improved starting method, we evaluate the asymptotic error, for step number n, in the sense of Figure 4.1. This is found to be

$$
\begin{bmatrix}
-\frac{241}{2880} h^4 y^{(4)}(x_n) + \mathcal{O}(h^5) \\[4pt]
\mathcal{O}(h^5) \\[4pt]
\frac{1}{3} h^4 y^{(4)}(x_n) + \mathcal{O}(h^5) \\[4pt]
\frac{3}{8} h^4 y^{(4)}(x_n) + \mathcal{O}(h^5)
\end{bmatrix}.
$$

Denote this by ϕ and carry out the decomposition in (4.6) to obtain a solution

$$\epsilon = \frac{1}{48}h^4 y^{(4)}(x_n) + \mathcal{O}(h^5), \qquad \delta = \begin{bmatrix} 0 \\ \mathcal{O}(h^5) \\ \frac{1}{3}h^4 y^{(4)}(x_n) + \mathcal{O}(h^5) \\ -h^4 y^{(4)}(x_n) + \mathcal{O}(h^5) \end{bmatrix}.$$

Note that the coefficients of $h^4 y^{(4)}(x_n)$, in the last three components of δ, are equal to β_3, β_2, β_1, an example of a result by Wright (2002b). We can now construct the underlying one-step method and the corresponding modified starting method, to within $\mathcal{O}(h^5)$. For the underlying one-step method, with input $y(x_{n-1})$, simply evaluate the Taylor expansion to within this accuracy, with ϵ subtracted from it. The starting method is now a modified version of the Nordsieck vector with δ subtracted from it. Thus,

$$y^{[n-1]} = \begin{bmatrix} y(x_{n-1}) \\ hy'(x_{n-1}) + \mathcal{O}(h^5) \\ h^2 y''(x_{n-1}) - \frac{1}{3}h^4 y^{(4)}(x_{n-1}) + \mathcal{O}(h^5) \\ h^3 y'''(x_{n-1}) + h^4 y^{(4)}(x_{n-1}) + \mathcal{O}(h^5) \end{bmatrix}.$$

In Section 7.4, we will discuss the use of the modified starting method in the estimation of error information. However, the very existence of an underlying one-step method hinges on the use of constant step-size. Two approaches for dealing with variable step-size have been considered (Butcher and Jackiewicz 2002, 2003). In this paper we will emphasize the second of these, which is to 'correct' the drift away from the correct starting approximation caused by unmodified Nordsieck scaling.

7.4. Implementation issues

In the practical implementation of any method, or family of methods, it is desirable to have available an asymptotically correct error estimator together with a mechanism for adjusting the step-size. Most well-known methods have these but usually at a computational cost. Our aim in the design of general linear methods is to keep any overhead costs as low as possible. For local error estimation, the secret seems to be to insist on methods with high stage order, and for variation of step-size, the essential idea is to use the Nordsieck representation of the data passed between steps. However, a simple rescaling of the Nordsieck vector by powers of the step-size ratio is not always a satisfactory way of adjusting for a new step-size.

There are two reasons for this. The first is that a method which might exhibit stable behaviour for constant step-size might act unstably when the step-size is varied, especially if large variations are permitted. The second is that we will not only want to estimate errors for a method currently in use, but we will also want to estimate errors for an alternative method of *higher* order, which is in contention as a possibly more efficient method for succeeding steps.

This leads to the idea of a 'scale and modify' scheme for step-size control. Suppose that, at the end of step n, a step-size change $h \mapsto rh$ is to be made. Also assume that in the underlying one-step method, the quantities being approximated at step number n are given by

$$
\begin{bmatrix}
y(x_n) + \mathcal{O}(h^{p+2}) \\
hy'(x_n) - \delta_1 h^{p+1} y^{(p+1)}(x_n) + \mathcal{O}(h^{p+2}) \\
h^2 y''(x_n) - \delta_2 h^{p+1} y^{(p+1)}(x_n) + \mathcal{O}(h^{p+2}) \\
\vdots \\
h^p y^{(p)}(x_n) - \delta_p h^{p+1} y^{(p+1)}(x_n) + \mathcal{O}(h^{p+2})
\end{bmatrix}. \tag{7.17}
$$

When this quantity is computed as the output to step n and an unmodified Nordsieck scaling is performed, we have as intended input for step number $n + 1$, the quantities

$$
\begin{bmatrix}
y(x_n) + \mathcal{O}(h^{p+2}) \\
(rh)y'(x_n) - r\delta_1 h^{p+1} y^{(p+1)}(x_n) + \mathcal{O}(h^{p+2}) \\
(rh)^2 y''(x_n) - r^2 \delta_2 h^{p+1} y^{(p+1)}(x_n) + \mathcal{O}(h^{p+2}) \\
\vdots \\
(rh)^p y^{(p)}(x_n) - r^p \delta_p h^{p+1} y^{(p+1)}(x_n) + \mathcal{O}(h^{p+2})
\end{bmatrix},
$$

which differs from what is required by

$$
\begin{bmatrix}
\mathcal{O}(h^{p+2}) \\
(r - r^{p+1})\delta_1 h^{p+1} y^{(p+1)}(x_n) + \mathcal{O}(h^{p+2}) \\
(r^2 - r^{p+1})\delta_2 h^{p+1} y^{(p+1)}(x_n) + \mathcal{O}(h^{p+2}) \\
\vdots \\
(r^p - r^{p+1})\delta_p h^{p+1} y^{(p+1)}(x_n) + \mathcal{O}(h^{p+2})
\end{bmatrix}.
$$

The scale and modify scheme requires us to add to the scaled Nordsieck vector, an approximation to this quantity. However, we have a choice of possible approximations to $h^{p+1} y^{(p+1)}(x_n) + \mathcal{O}(h^{p+2})$ and the choice we make, which might differ from component to component, needs to take account of stability requirements.

We will illustrate how this is done using the example method (7.16). By matching Taylor expansions, we find a family of linear combinations of various quantities which give asymptotically correct approximations to $h^4 y^{(4)}$. These quantities are $hF_i = hy'(x_{n-1} + hc_i) + \mathcal{O}(h^5)$, $i = 1, 2, 3, 4$, together with $y_2^{[n-1]} = hy'(x_{n-1}) + \mathcal{O}(h^5)$, $y_3^{[n-1]} = h^2 y''(x_{n-1}) - \frac{1}{3}h^4 y^{(4)}(x_{n-1}) + \mathcal{O}(h^5)$. Note that we do not involve $y_4^{[n-1]}$ in the error estimate because we want the modified and scaled B and V matrices to have an unchanged sparsity pattern. This will guarantee a variable step analogue of zero stability.

We have two free parameters, which we denote by C_2, C_3 (C_1 will be introduced below) and the suggested approximation is

$$(81 + 18C_3)hF_1 + (-81 - \tfrac{45}{2}C_3)hF_2 + (27 + 8C_3 - C_2)hF_3 + C_2 hF_4$$
$$+ (-27 - \tfrac{7}{2}C_3)y_2^{[n-1]} + C_3 y_3^{[n-1]}$$
$$= h^4 y^{(4)}(x_n) + \mathcal{O}(h^5).$$

We use this approximation in two places, in the modification of the scaled $y_3^{[n]}$, with C_3 replaced by zero and C_2 replaced by C_1, and in the modification to the scaled $y_4^{[n]}$. If we write $B(r)$ and $V(r)$ for the scaled and modified versions of B and V respectively, then $y^{[n]}$, as input to step number $n+1$ with step-size rh, is given in a modified version of (3.1),

$$y^{[n]} = B(r)hF + V(r)y^{[n-1]},$$

where $B(r)$ and $V(r)$ are each given as the sum of the simply scaled version plus the modifier terms:

$$B(r) = \begin{bmatrix} \frac{39}{160} & \frac{15}{128} & \frac{5}{24} & \frac{1}{2} \\ 0 & 0 & 0 & r \\ -\frac{45}{2}r^2 & 8r^2 & \frac{5}{2}r^2 & -6r^2 \\ \frac{171}{4}r^3 & -\frac{531}{16}r^3 & 0 & 12r^3 \end{bmatrix}$$

$$+ \operatorname{diag}\left(0, 0, \tfrac{r^2-r^4}{3}, -(r^3-r^4)\right) \begin{bmatrix} 0 & 0 & 0 & 0 \\ 0 & 0 & 0 & 0 \\ 81 & -81 & 27-C_1 & C_1 \\ 81+18C_3 & -81-\tfrac{45}{2}C_3 & 27+8C_3-C_2 & C_2 \end{bmatrix},$$

$$V(r) = \begin{bmatrix} 1 & -\frac{133}{1920} & -\frac{353}{960} & -\frac{109}{480} \\ 0 & 0 & 0 & 0 \\ 0 & 8r^2 & 0 & 0 \\ 0 & -\frac{345}{16}r^3 & -\frac{33}{8}r^3 & 0 \end{bmatrix}$$

$$+ \operatorname{diag}\left(0, 0, \tfrac{r^2-r^4}{3}, -(r^3-r^4)\right) \begin{bmatrix} 0 & 0 & 0 & 0 \\ 0 & 0 & 0 & 0 \\ 0 & -27 & 0 & 0 \\ 0 & -27-\tfrac{7}{2}C_3 & C_3 & 0 \end{bmatrix}.$$

For arbitrary (but bounded) step-size ratios, the products of matrices $V(r)$ over many steps acts in a stable manner, simply because these are all strictly lower triangular. However, we will also seek stable behaviour for infinitely stiff problems. That is, we will want to form products of matrices like

$$V(r) - B(r)A^{-1}U. \qquad (7.18)$$

Although we have described C_1, C_2 and C_3 as constants, there is no reason why they should not depend on r. If we choose $C_3 = -2.6$ and define $C_1(r)$ and $C_2(r)$ so that the characteristic equation of (7.18) has only a single non-zero root, then this root is bounded in magnitude by 1 at least for $r \in [\frac{1}{2}, 2]$. This is, of course, not sufficient for acceptable variable step stability for stiff problems, but is at least an encouragement to explore this aspect of L-stable general linear methods further.

The availability of asymptotically correct error estimators, and the variable step-size adjustments we have described, provides all the equipment that is needed for reliable step-size control. However, we also want variable order. Although we will not consider adjustments to Nordsieck vector approximations when order is increased, we will discuss as the final detail on implementation, the estimation in step number n of $h^{p+2}y^{(p+2)}(x_n)$, because the asymptotic error of a method of order $p+1$ will be proportional to this quantity.

The key to estimating $h^{p+2}y^{(p+2)}(x_n)$ is the fact that

$$hF_i = hy'(x_{n-1} + hc_i) + \mathcal{O}(h^{p+2}), \quad i = 1, 2, \dots, p+1. \qquad (7.19)$$

We will consider only methods which, like the example method (7.16), have 'Property F', otherwise known as the FSAL property.

Definition 7.5. A general linear method with the IRKS property has Property F if

(i) $c_s = 1$,

(ii) $b_{1j} = a_{sj}$, $j = 1, 2, \dots, s$,

(iii) $v_{1j} = u_{sj}$, $j = 1, 2, \dots, r$,

(iv) $b_{2j} = \delta_{sj}$, $j = 1, 2, \dots, s$,

(v) $v_{2j} = 0$, $j = 1, 2, \dots, r$.

For a method with Property F, we effectively have an additional accurate derivative approximation, $hF_0 = hy'(x_0 + hc_0) + \mathcal{O}(h^{p+2})$, where $c_0 = 0$. Hence, we will regard (7.19) as holding for $i = 0, 1, \dots, p+1$.

The FSAL property, on which Definition 7.5 is modelled, was made popular in the design of Runge–Kutta methods, by the work of Dormand and Prince (1980) because it gives an additional apparently free derivative approximation to widen options for error estimators. This is exactly how we will use Property F. For the remainder of this section we will assume this property, just as we will always assume that the scale and multiply technique is used when the step-size varies.

Suppose that the error coefficients δ_1, δ_2, are as in (7.17). Then the errors introduced into the stage values are

$$y(x_{n-1} + hc_i) - h \sum_{j=1}^{p+1} a_{ij} y'(x_{n-1} + hc_j)$$

$$- \sum_{j=1}^{p+1} u_{ij} \left(h^{j-1} y^{(j-1)} - \delta_{j-1} h^{p+1} y^{(p+1)}(x_{n-1}) \right),$$

$$i = 1, 2, \ldots, p+1.$$

By Taylor's theorem this equals

$$\sigma_i h^{p+1} y^{(p+1)} y(x_{n-1}) + \mathcal{O}(h^{p+2}),$$

where

$$\sigma = \frac{1}{(p+1)!} c^{p+1} - \frac{1}{p!} A c^p + U \delta,$$

and we find, for the corresponding error in the stage derivatives,

$$\sigma_i h^{p+2} \frac{\partial f}{\partial y} y^{(p+1)}(x_{n-1}) + \mathcal{O}(h^{p+3}).$$

If we contemplate using linear combinations of hF_i, $i = 0, 1, \ldots, p+1$ to estimate quantities related to errors, to within $\mathcal{O}(h^{p+3})$, we need to make use of the matrix

$$L = \begin{bmatrix} \sigma_0 & 1 & c_0 & \frac{1}{2} c_0^2 & \cdots & \frac{1}{p!} c_0^p & \frac{1}{(p+1)!} c_0^{p+1} \\ \sigma_1 & 1 & c_1 & \frac{1}{2} c_1^2 & \cdots & \frac{1}{p!} c_1^p & \frac{1}{(p+1)!} c_1^{p+1} \\ \vdots & \vdots & \vdots & \vdots & & \vdots & \vdots \\ \sigma_{p+1} & 1 & c_{p+1} & \frac{1}{2} c_{p+1}^2 & \cdots & \frac{1}{p!} c_{p+1}^p & \frac{1}{(p+1)!} c_{p+1}^{p+1} \end{bmatrix}.$$

We distinguish three cases:

(i) the c components are distinct and the first $p+2$ columns of L are linearly independent,

(ii) the c components are distinct and the first $p+2$ columns of L are linearly dependent,

(iii) $c_p = c_{p+1} = 1$ and $\sigma_p \neq \sigma_{p+1}$.

A final case, in which $c_p = 1$ but $\sigma_p = \sigma_{p+1}$, will not be explored because there does not seem to be a simple error estimator of the type we want in this case. Note that the example method (7.16) is in case (iii).

In case (i), construct a coefficient vector $\xi^T = [\xi_0, \xi_1, \ldots, \xi_{p+1}]$ such that

$$\xi^T L = e_{p+2} + \theta e_{p+3},$$

so that $\sum_{i=0} \xi_i h F_i$ equals

$$\Psi_n = h^{p+1} y^{(p+1)}(x_{n-1}) + \theta h^{p+2} y^{(p+2)}(x_{n-1}) + \mathcal{O}(h^{p+3})$$
$$= h^{p+1} y^{(p+1)}(x_{n-1} + h\theta) + \mathcal{O}(h^{p+3}).$$

If the step-size used in step number $n-1$ was $r^{-1}h$, then an asymptotically correct estimate of $h^{p+2} y^{(p+2)}(x_{n-1})$ can be found from

$$\frac{r}{1+\theta(r-1)} \left(\Psi_n - r^{p+1} \Psi_{n-1} \right).$$

In cases (ii), it is possible to give in a single step an approximation

$$\sum_{i=0} \xi_i h F_i = h^{p+2} y^{(p+2)}(x_{n-1}) + \mathcal{O}(h^{p+3})$$

by choosing ξ^T to satisfy

$$\xi^T L = e_{p+3}.$$

Case (iii) is similar to case (i) and we will illustrate this using the example method (7.16). For this method L is found to be

$$\begin{bmatrix} 0 & 1 & 0 & 0 & 0 & 0 \\ -\frac{35}{1944} & 1 & \frac{1}{3} & \frac{1}{18} & \frac{1}{162} & \frac{1}{1944} \\ \frac{10}{14337} & 1 & \frac{2}{3} & \frac{2}{9} & \frac{4}{81} & \frac{2}{243} \\ \frac{187}{2360} & 1 & 1 & \frac{1}{2} & \frac{1}{6} & \frac{1}{24} \\ \frac{1}{48} & 1 & 1 & \frac{1}{2} & \frac{1}{6} & \frac{1}{24} \end{bmatrix},$$

leading to the approximation

$$-27 y_2^{[n-1]} + 81 h F_1 - 81 h F_2 + \frac{13485}{827} h F_3 + \frac{8844}{827} h F_4$$
$$\approx h^4 y^{(4)}(x_{n-1}) + \frac{1}{2} h^5 y^{(5)}(x_{n-1}).$$

The estimation of local truncation errors is discussed in Butcher and Podhaisky (2006); this includes the estimation of $h^{p+2} y^{(p+2)}$ using the method described in this section.

8. Order and stability barriers

This discussion is relevant to multi-derivative methods, otherwise known as Obreshkov methods, as well as to general linear methods. The essential question concerns polynomial functions in two complex variables and the extent to which they can represent high-order approximations to exp and at the same time represent A-stable behaviour.

Given positive integers r, s, we consider a polynomial function of two complex variables $\Phi(w, z)$, which has degree r in w and degree s in z. Given a general linear method (A, U, B, V), the stability matrix is

$$M(z) = V + zB(I - zA)^{-1}U$$

and its linear stability properties are defined in terms of the characteristic polynomial

$$\det(wI - M(z)). \tag{8.1}$$

Only when A is nilpotent, such as in an explicit method, will this expression be a polynomial. It is, however, always a polynomial in w and z divided by a polynomial in z. We can define Φ for this method by using the numerator of (8.1).

Two special cases correspond to classical methods. If $s = 1$ we will write

$$\Phi(w, z) = \rho(w) - z\sigma(w),$$

using the standard notation for linear multistep methods. On the other hand, if $r = 1$ then we will write

$$\Phi(w, z) = wD(z) - N(z),$$

corresponding to the stability function $R(z) = N(z)/D(z)$ of a Runge–Kutta method.

As much as possible, we will distance ourselves from an actual method but will consider properties of Φ in its own right.

Definition 8.1.   A stability function has order p if

$$\Phi(\exp(z), z) = \mathcal{O}(z^{p+1}).$$

Note that this definition does not necessarily coincide with the order of the underlying general linear method. However, the actual order of the method cannot exceed p in Definition 8.1.

Definition 8.2.   A stability function is A-stable if for every complex numbers z such that $\operatorname{Re} z \le 0$,

(i) if w satisfies $\Phi(w, z) = 0$, then $|w| \le 1$,

(ii) if w satisfies $\Phi(w, z) = \frac{\partial}{\partial w}\Phi(w, z) = 0$, then $|w| < 1$.

This definition is not the usual one. However, our aim will be to understand the conflict between order and stable behaviour for stiff problems and we want consistent conclusions which reasonably well make it possible to decide between suitable and unsuitable methods. Consider the approximation

$$\Phi(w, z) = \left(1 - \tfrac{5}{8}z + \tfrac{1}{8}z^2\right)w^2 - 2w + 1 + \tfrac{5}{8}z + \tfrac{1}{8}z^2.$$

According to Definition 8.1, this method has order 5 but it is not possible in a numerical computation to realize the corresponding accuracy. On the other hand, according to Definition 8.2 it is not A-stable even though, for every z satisfying $\operatorname{Re} z < 0$, the corresponding w values are in the open unit disc. Hence, its properties are consistent with Theorem 8.11, which it should be.

For a particular approximation we might wish to refer to the polynomials in z arising as coefficients of various powers of w. Write

$$\Phi(w, z) = P_0(z)w^r + P_1(z)w^{r-1} + \cdots + P_{r-1}(z)w + P_r(z). \tag{8.2}$$

We will refer to the zeros of $P_0(z)$ as the 'poles of Φ' and the zeros of P_r as the 'zeros of Φ'.

Theorem 8.3. The approximation Φ is A-stable if and only if

(i) Φ has no poles in the left half-plane,

(ii) there do not exist complex numbers w and z such that $\operatorname{Re} z = 0$, $|w| > 1$ and $\Phi(w, z) = 0$,

(iii) there do not exist complex numbers w and z such that $\operatorname{Re} z = 0$, $|w| = 1$ and $\Phi(w, z) = \frac{\partial}{\partial w}\Phi(w, z) = 0$.

Proof. (i) is necessary because if z is near a pole there are arbitrarily high values of w satisfying $\Phi(w, z) = 0$. (ii) and (iii) are necessary because the imaginary axis is a subset of the closed left half-plane. Sufficiency follows from the maximum-modulus theorem. $\square$

Associated with a given approximation Φ is the Riemann surface defined by $\Phi(\widehat{w}\exp(z), z) = 0$. The use of this 'relative stability function' was made famous by its use in the theory of order stars. We will use the closely related 'order arrows' in this paper to achieve many of the same goals.

In the search for A-stable methods of high order, we will consider (8.2) with the degrees

$$n_i = \deg(P_i), \qquad i = 0, 1, 2, \ldots, r, \tag{8.3}$$

specified and the coefficients chosen to maximize the order of the approximation.

Definition 8.4. Given a sequence of degrees, $n_i \geq -1$, $i = 0, 1, \ldots, r$ a generalized Padé approximation (to exp) is a sequence of polynomials $P_0, P_1, \ldots, P_r$, satisfying (8.3) with order $p = \sum_{i=0}^{r} n_i + r - 1$.

We will always assume that $n_0 \geq 0$ and we will usually assume that $n_r \geq 0$ (otherwise, r could be reduced to $r-1$). In the interpretation Definition 8.4, we will regard a polynomial of degree -1 as being the zero polynomial.

8.1. Padé approximations

The Padé approximations, that is the generalized Padé approximations in the case $r = 1$, arise as the stability functions of certain implicit Runge–Kutta methods. If n_0 the degree of D, is written as d and n_1, the degree of N is written as n, then the (d, n) Padé approximation is given by

$$D(z)\exp(z) - N(z) = (-1)^d \frac{C}{(n+d+1)!} z^{n+d+1} + \mathcal{O}(z^{n+d+2}), \qquad (8.4)$$

where the constant C is an arbitrary nonzero scale factor.

Theorem 8.5. The polynomials N and D in (8.4) are given by

$$N(z) = C \sum_{i=0}^{n} \frac{z^{n-i}}{(n-i)!} \binom{d+i}{i}, \qquad (8.5)$$

$$D(z) = C \sum_{i=0}^{d} \frac{(-z)^{d-i}}{(d-i)!} \binom{n+i}{i}. \qquad (8.6)$$

Proof. Operate on (8.4) by $(\frac{d}{dz})^{n+1}$ to obtain

$$\exp(z)(1 + \tfrac{d}{dz})^{n+1} D(z) = (-1)^d C \frac{z^d}{d!} + \mathcal{O}(z^{d+1}).$$

Multiply by $\exp(-z)$ and the right-hand side is unchanged. However, the left-hand side is a polynomial of degree exactly d and the $\mathcal{O}(z^{d+1})$ can be omitted. It now follows that

$$D(z) = (-1)^d C (1 + \tfrac{d}{dz})^{-(n+1)} \frac{z^d}{d!},$$

and (8.6) follows. Similarly, multiply (8.4) by $\exp(-z)$ and operate on the resulting equation by $(\frac{d}{dz})^{d+1}$, leading to (8.5). $\qquad \square$

A convenient choice of C is $n!d!/(n+d)!$ leading to formulae in which $N(0) = D(0) = 1$:

$$N(z) = \sum_{i=0}^{n} \frac{n!(n+d-i)!}{(n-i)!(n+d)!i!} z^i, \qquad (8.7)$$

$$D(z) = \sum_{i=0}^{d} \frac{d!(n+d-i)!}{(d-i)!(n+d)!i!} (-z)^i. \qquad (8.8)$$

A partial table of Padé approximations to the exponential function is given in Table 8.1 (overleaf).

Recurrence relations

We will write $V_{dn}(z)$ to denote the two-dimensional vector whose first and second components are $N_{dn}(z)$ and $D_{dn}(z)$, respectively. Many relationships

Table 8.1. Padé approximations to exp of degrees $[n,d]$.

d \ n	0	1	2	3
0	$\frac{1}{1}$	$\frac{1+z}{1}$	$\frac{1+z+\frac{1}{2}z^2}{1}$	$\frac{1+z+\frac{1}{2}z^2+\frac{1}{6}z^3}{1}$
1	$\frac{1}{1-z}$	$\frac{1+\frac{1}{2}z}{1-\frac{1}{2}z}$	$\frac{1+\frac{2}{3}z+\frac{1}{6}z^2}{1-\frac{1}{3}z}$	$\frac{1+\frac{3}{4}z+\frac{1}{4}z^2+\frac{1}{24}z^3}{1-\frac{1}{4}z}$
2	$\frac{1}{1-z+\frac{1}{2}z^2}$	$\frac{1+\frac{1}{3}z}{1-\frac{2}{3}z+\frac{1}{6}z^2}$	$\frac{1+\frac{1}{2}z+\frac{1}{12}z^2}{1-\frac{1}{2}z+\frac{1}{12}z^2}$	$\frac{1+\frac{3}{5}z+\frac{3}{20}z^2+\frac{1}{60}z^3}{1-\frac{2}{5}z+\frac{1}{20}z^2}$
3	$\frac{1}{1-z+\frac{1}{2}z^2-\frac{1}{6}z^3}$	$\frac{1+\frac{1}{4}z}{1-\frac{3}{4}z+\frac{1}{4}z^2-\frac{1}{24}z^3}$	$\frac{1+\frac{2}{5}z+\frac{1}{20}z^2}{1-\frac{3}{5}z+\frac{3}{20}z^2-\frac{1}{60}z^3}$	$\frac{1+\frac{1}{2}z+\frac{1}{10}z^2+\frac{1}{120}z^3}{1-\frac{1}{2}z+\frac{1}{10}z^2-\frac{1}{120}z^3}$

exist between adjacent members of the Padé table. We will here consider just one of these because of its application in this work.

Theorem 8.6. If $n \geq 2$ then

$$V_{n,n}(z) = V_{n-1,n-1}(z) + \frac{z^2}{4(2n-1)(2n-3)} V_{n-2,n-2}(z).$$

Proof. Because $D_{nn}(z) = N_{nn}(-z)$, it follows that $N_{nn}(z)\exp(-z/2) - \exp(z/2)D_{nn}(z)$ is an odd function; combining this with the fact that

$$\frac{N_{nn}(z)}{D_{nn}(z)} = \exp(z) + \mathcal{O}(z^{2n+1}),$$

we find

$$\left[\exp(-z/2) - \exp(z/2)\right] V_{nn}(z) = \frac{(-1)^{n+1}n!^2}{(2n)!(2n+1)!}z^{2n+1} + \mathcal{O}(z^{2n+3}).$$

Hence,

$$\left[\exp(-z/2) - \exp(z/2)\right]\left(V_{n-1,n-1}(z) + \frac{z^2}{4(2n-1)(2n-3)} V_{n-2,n-2}(z)\right)$$
$$= (-1)^n \theta z^{2n-1} + \mathcal{O}(z^{2n+1}),$$

where

$$\theta = -\frac{1}{4(2n-1)(2n-3)}\frac{(n-2)!^2}{(2n-4)!(2n-3)!} + \frac{(n-1)!^2}{(2n-2)!(2n-1)!} = 0.$$

It follows that $V_{n-1,n-1}(z)+\frac{z^2}{4(2n-1)(2n-3)}V_{n-2,n-2}(z)$ is the unique, correctly scaled vector V_{nn}.   □

A-stable Padé approximations

The A-stable members of the Padé table for the exponential functions are those for which $d-n \in \{0,1,2\}$. The fact that approximations with $d > n+2$ cannot be A-stable will be proved in Theorem 8.9 and the corresponding result for $d < n$ is covered by the following result.

Theorem 8.7. A Padé approximation to exp with $d > n$ is not A-stable.

Proof. For $|z|$ large, we find from (8.7) and (8.8)

$$\left|\frac{N(z)}{D(z)}\right| = \frac{d!}{n!}|z|^{n-d} + \mathcal{O}(|z|^{n-d-1}),$$

and is greater than 1 for $z \in \mathbb{C}^-$ with $|z|$ sufficiently large. $\square$

In remains to prove the result.

Theorem 8.8. If $d - n \in \{0, 1, 2\}$, the $[d, n]$ Padé approximation to exp is A-stable.

Proof. In the case $n = d$, let $\zeta_n = D_{nn}(z)/zD_{n-1,n-1}(z)$, where $\mathrm{Re}\,(z) < 0$ and deduce from the second component of (8.6) that

$$\zeta_n = \frac{1}{z} + \frac{1}{4(2n-1)(2n-3)\zeta_{n-1}}.$$

Starting from $\zeta_1 = z^{-1} - \frac{1}{2}$, we deduce that each of $\zeta_1, \zeta_2, \dots$ is in the left half-plane, where we use the fact that the inverse of a number in the left half-plane, and the sum of two such numbers, are each also in the left half-plane. It follows that ζ_n is not zero and neither is

$$D_{nn}(z) = z^n \zeta_n \zeta_{n-1} \cdots \zeta_1.$$

Since D_{nn} has no zeros in the left half-plane, we use the maximum modulus principle to deduce that $|N_{nn}z/D_{nn}(z)|$ is bounded by 1, because the value is achieved exactly on the imaginary axis.

In the case $n = d - 1$, define the approximation $\tilde{N}(z)/\tilde{D}(z)$ as the components of the vector

$$\tilde{V}(z) = (1 - t)V_{nn}(z) + tV_{n,n-1}(z),$$

where the homotopy variable t moves from 0 (diagonal approximation) to 1 (sub-diagonal approximation). Because $\tilde{N}(z) = \exp(z)\tilde{D}(z) + \mathcal{O}(z^{2n})$ it follows that

$$|\tilde{D}(iy)|^2 - |\tilde{N}(iy)|^2 = C(t)y^{2n},$$

where $C(t) > 0$ for $t > 0$. Hence, $D(iy) > 0$. As t increases in $[0, 1]$, the zeros of $\tilde{D}(z)$ move continuously and can never cross the imaginary axis, because $D(iy)$ never vanishes.

In the case $n = d - 2$ carry out a similar homotopy from $V_{n,n-1}(z)$ to $V_{n,n-2}(z)$ and obtain a similar result. $\square$

8.2. Quadratic Padé approximations

The derivation of Padé approximations we have given can be easily generalized to the quadratic case $r = 2$.

Table 8.2. Some quadratic Padé approximations to exp.

p	$[n_0, n_1, n_2]$	$P_0(z)$	$P_1(z)$	$P_2(z)$
2	$[1,0,0]$	$1 - \frac{2}{3}z$	$-\frac{4}{3}$	$\frac{1}{3}$
3	$[2,0,0]$	$1 - \frac{6}{7}z + \frac{2}{7}z^2$	$-\frac{8}{7}$	$\frac{1}{7}$
4	$[2,1,0]$	$1 - \frac{10}{17}z + \frac{2}{17}z^2$	$-\frac{16}{17} - \frac{8}{17}z$	$-\frac{1}{17}$
4	$[2,0,1]$	$1 - \frac{8}{11}z + \frac{2}{11}z^2$	$-\frac{16}{11}$	$\frac{5}{11} + \frac{2}{11}z$
4	$[3,0,0]$	$1 - \frac{14}{15}z + \frac{2}{5}z^2 - \frac{4}{45}z^3$	$-\frac{16}{15}$	$\frac{1}{15}$
5	$[3,1,0]$	$1 - \frac{34}{49}z + \frac{10}{49}z^2 - \frac{4}{147}z^3$	$-\frac{48}{49} - \frac{16}{49}z$	$-\frac{1}{49}$
5	$[3,0,1]$	$1 - \frac{11}{13}z + \frac{4}{13}z^2 - \frac{2}{39}z^3$	$-\frac{16}{13}$	$\frac{3}{13} + \frac{1}{13}z$
5	$[4,0,0]$	$1 - \frac{30}{31}z + \frac{14}{31}z^2 - \frac{4}{31}z^3 + \frac{2}{93}z^4$	$-\frac{32}{31}$	$\frac{1}{31}$

Suppose $n_0 \geq 0$, $n_1, n_2 \geq -1$, $\min(n_1, n_2) \geq 0$ and

$$\Phi(w, z) = w^2 P_0(z) + w P_1(z) + P_2(z), \qquad \deg(P_i) = n_i, \quad i = 0, 1, 2,$$

and that $\Phi(\exp(z), z) = \mathcal{O}(z^{p+1})$, with $p = n_0 + n_1 + n_2 + 1$.
Assume for some C

$$\exp(2z) P_0(z) + \exp(z) P_1(z) + P_2(z) = C \frac{z^{p+1}}{(p+1)!} + \mathcal{O}(z^{p+2}).$$

To find P_0, multiply by $\exp(-2z)$ and apply $(1 + \frac{d}{dz})^{n_1+1}(2 + \frac{d}{dz})^{n_2+1}$ to both sides. The result is

$$\left(1 + \tfrac{d}{dz}\right)^{n_1+1}\left(2 + \tfrac{d}{dz}\right)^{n_2+1} P_0(z) = C \frac{z^{n_0}}{n_0!},$$

where $\mathcal{O}(z^{n_0+1})$ is omitted on the right-hand side because the left-hand side is a polynomial of degree n_0. Find similar expressions involving P_1 and P_2 and rearrange to obtain

$$P_0(z) = C\left(1 + \tfrac{d}{dz}\right)^{-(n_1+1)}\left(2 + \tfrac{d}{dz}\right)^{-(n_2+1)} \frac{z^{n_0}}{n_0!},$$
$$P_1(z) = C\left(-1 + \tfrac{d}{dz}\right)^{-(n_0+1)}\left(1 + \tfrac{d}{dz}\right)^{-(n_2+1)} \frac{z^{n_1}}{n_1!},$$
$$P_2(z) = C\left(-2 + \tfrac{d}{dz}\right)^{-(n_0+1)}\left(-1 + \tfrac{d}{dz}\right)^{-(n_1+1)} \frac{z^{n_2}}{n_2!}.$$

A number of quadratic approximations are given in Table 8.2.

8.3. Order stars and order arrows

The famous theory of order stars (Wanner, Hairer and Nørsett 1978) was introduced as a means of settling some outstanding open questions. The idea is based on the observation that a rational approximation $R(z) = N(z)/D(z)$

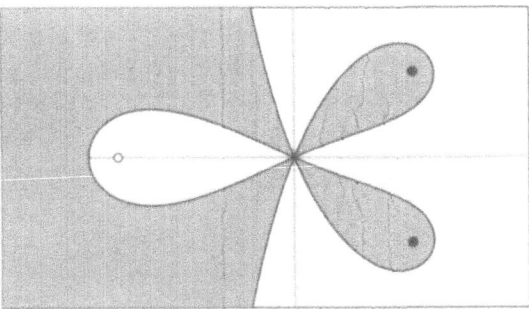

Figure 8.1. Order star for $[2,1]$ Padé approximation.

(N and D having no common factor) of order p is A-stable if and only if

(i) D has no zeros in the open left half-plane,

and

(ii) $|R(z)| \leq 1$, when $\mathrm{Re}\,(z) = 0$,

and the further observation that the criteria still hold if, in (ii), $R(z)$ is replaced by $R(z)\exp(-z)$.

The advantage of the modified form of this criterion is that the behaviour of $R(z)\exp(-z)$ is known in considerable detail when $|z|$ is small. In fact,

$$R(z)\exp(-z) = 1 - C z^{p+1} + \mathcal{O}(z^{p+2}),$$

where the error constant C is defined by $R(z) = \exp(z) - C z^{p+1} + \mathcal{O}(z^{p+2})$. Write $z = r\exp t\theta$ and we find

$$|R(z)\exp(-z)| = 1 - C r^{p+1}\cos((p+1)\theta) + \mathcal{O}(r^{p+2}).$$

For arguments θ such that $C\cos((p+1)\theta) < 0$, $R(z)\exp(-z) > 1$ for sufficiently small $|z|$, and conversely, $R(z)\exp(-z) < 1$ if $C\cos((p+1)\theta) > 0$ and $|z|$ is sufficiently small.

The order star corresponding to this approximation is defined as the set of points in the complex plane such that $|R(z)\exp(-z)| > 1$ and the dual star is defined as the set of points for which $|R(z)\exp(-z)| < 1$. We have seen which points near zero lie in each of these sets. The components of the order star (respectively, dual star) close to zero are referred to as fingers (respectively, dual fingers). Further details are available in Wanner, Hairer and Nørsett (1978) and in other expositions of the theory.

The criterion for A-stability, that $|R(z)\exp(-z)| \leq 1$ on the imaginary axis, translates, in order star language, to the requirement that a finger cannot intersect the imaginary axis. Two examples are presented; first Figure 8.1 for the $[2,1]$ Padé approximation. Here two 'bounded fingers' enclose the poles and a single 'bounded dual finger' encloses the zero. The

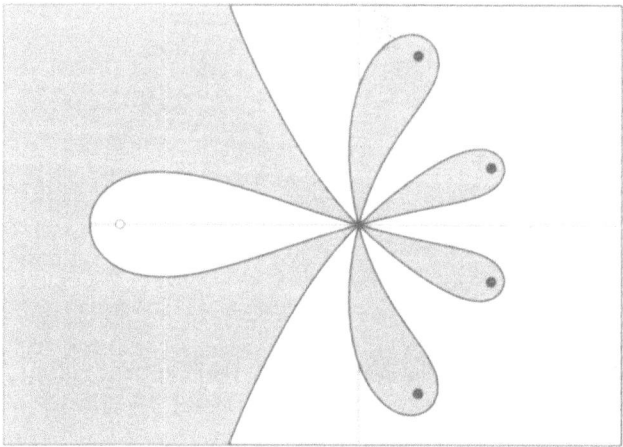

Figure 8.2. Order star for [4, 1] Padé approximation.

unbounded finger and the unbounded dual finger divide the parts of the complex plane distant from zero into two parts. The underlying approximation is A-stable because the two poles are in the right half-plane and there is no intersection between the order star (the shaded region) and the imaginary axis.

In contrast, we present Figure 8.2 for the [4, 1] Padé approximation. This is *not* A-stable, because in this case the order star intersects the imaginary axis. This is known to be the case because there are too many bounded fingers containing poles for all of them to lie entirely in the right half-plane.

As an alternative to the order star technique, 'order arrows' have been proposed. For an approximation $R(z) = N(z)/D(z)$, this also uses the modified function formed by dividing by $\exp(z)$, but considers the set of points in the complex plane for which $R(z)\exp(-z)$ is real and positive. These emanate from zero as 'up-arrows' which terminate at poles or at $-\infty$, or down-arrows which terminate at zeros or at $+\infty$. A-stability does not hold if an up-arrow leaves zero (with magnitude 1) and either crosses the imaginary axis or is tangential to it. We present the order star diagrams for the two approximations already considered. First, Figure 8.3 corresponds to the A-stable approximation [2, 1]. In contrast, the arrow diagram for the [4, 1] approximation is shown in Figure 8.4. This approximation cannot be A-stable for the following reasons.

(i) Exactly four up-arrows terminate at poles, otherwise some up-arrows would cross down-arrows.

(ii) The angle subtended by the tangents at zero to two of these up-arrows is at least $4 \times \pi/4 = \pi$.

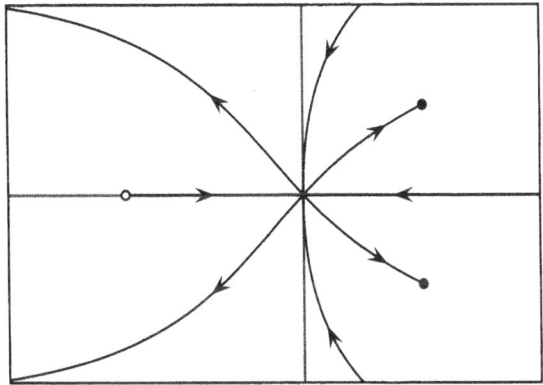

Figure 8.3. Order arrows for $[2, 1]$ Padé approximation.

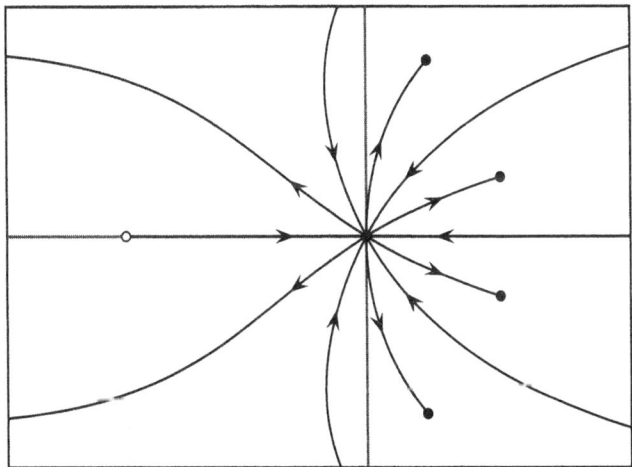

Figure 8.4. Order arrows for $[4, 1]$ Padé approximation.

(iii) Hence, at least one up-arrow is either tangential to the imaginary axis or else it emanates into the left half-plane and terminates at a pole.

(iv) Hence, there is either a pole in the left half-plane or this up-arrow crosses the imaginary axis before terminating at a pole in the right half-plane.

If we define order arrows from their basic property that $\Phi(\widehat{w}\exp(z), z) = 0$ with $\widehat{w}$ real and positive, then, in addition to those arrows emanating from zero, there is an infinite family of arrows spaced approximately $2\pi i$ apart, as illustrated in Figure 8.5, for the $[1, 0]$ case,

$$\Phi(w, z) = w(1 - z) - 1.$$

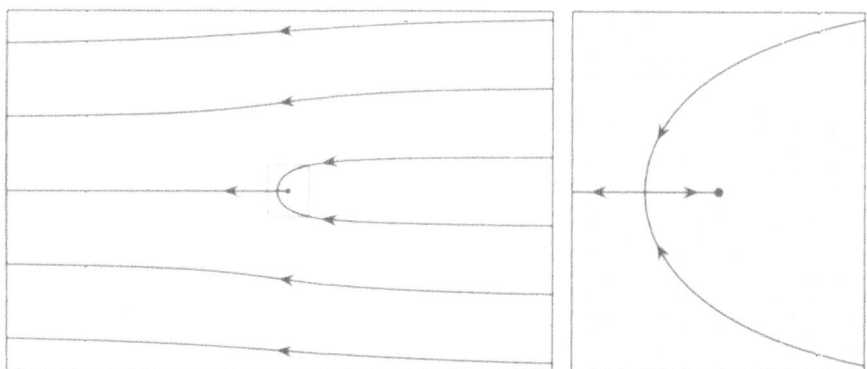

Figure 8.5. Order arrows for $[1,0]$ Padé approximation,
together with magnified detail near $z = 0$.

8.4. The Ehle barrier

The result formerly known as the Ehle conjecture was one of the first successes of the order star theory (Wanner, Hairer and Nørsett 1978). Here we will present an alternative proof using order arrows.

Theorem 8.9. Let $R(z) = N(z)/D(z)$ denote the $[d, n]$ Padé approximation to exp. Then, if $d > n + 2$, this approximation is not A-stable.

Proof. There are $n+d+1$ up-arrows emanating from zero, alternating with $n+d+1$ down-arrows. Suppose that $\tilde{d}$ up-arrows terminate at poles so that $(n + d + 1) - \tilde{d}$ up-arrows terminate at $-\infty$. Suppose that $\tilde{n}$ down-arrows terminate at zeros. These must fit into the $(n+d+1) - \tilde{d} - 1$ gaps between the up-arrows which terminate at $-\infty$. Hence

$$\tilde{n} + \tilde{d} \le n + d.$$

Because $\tilde{d} \le d$ and $\tilde{n} \le n$ it follows that $\tilde{d} = d$ and $\tilde{n} = n$. Since d up-arrows terminate at poles, there must be at least one up-arrow emanating from zero with tangent making an angle to the the positive real axis at least equal to $\frac{d-1}{2} \times 2\pi/(n + d + 1)$. For A-stability either (i) this angle must me less than $\pi/2$ or (ii) at least one of the up-arrows terminating at a pole emanates from zero with an argument greater than $\pi/2$. Hence, in case (i),

$$\frac{d - 1}{2} \cdot \frac{2\pi}{n + d + 1} < \frac{\pi}{2},$$

implying $2d - 2 < n + d + 1$ so that $d < n + 3$. In case (ii), the up-arrow referred to either terminates at a pole in the left half-plane, or crosses the imaginary axis, each of which is impossible. □

8.5. Order arrows on Riemann surfaces

To generalize the use of relative stability regions, by inserting the factor $\exp(-z)$ in $R(z)\exp(-z)$, we consider the modification of a generalized approximation $\Phi(w, z)$ by considering the function $\widehat{\Phi}$ defined by

$$\widehat{\Phi}(\widehat{w}, z) = \Phi(\widehat{w}\exp(z), z). \tag{8.9}$$

The order star theory for this type of generalization is developed in Wanner, Hairer and Nørsett (1978) and we will discuss here only the order arrow approach.

The Riemann surface for (8.9) is the subset of $\mathbb{C} \times \mathbb{C}$ for which $\widehat{\Phi}(\widehat{w}, z) = 0$. It is usual to think of the Riemann surface as a multivalued function for which values of z are the arguments, and the corresponding values of $\widehat{w}$ which satisfy the equation $\widehat{\Phi}(\widehat{w}, z) = 0$ are the values of this function. Except at isolated points at which $(\partial\widehat{\Phi}/\partial\widehat{w}) = 0$, an open set in the z plane exists so that, in this open set, each value of $\widehat{w}$ acts like a function of a complex variable and satisfies the Cauchy–Riemann conditions. Except in trivial cases in which the sheets of the Riemann surface do not interact with each other, analytic extension leads to a migration onto other sheets.

We superimpose order arrows onto the Riemann surface by considering the subset for which the value $\widehat{w}$ is real and positive. Starting at a specific point on the Riemann surface for which $\widehat{w}$ has this property, up-arrows and down-arrows can be traced out. In particular, if the approximation has order p, then at $z = 0$, there are $p + 1$ up-arrows and $p + 1$ down-arrows emanating from this point.

8.6. The Dahlquist second barrier

The famous second barrier result of Dahlquist (1963), states that an A-stable linear multistep method cannot have order greater than 2. In our context this means the following theorem.

Theorem 8.10.   Let $\Phi(w, z)$ denote an $(r, 1)$ A-stable approximation with order p. Then $p \le 2$.

Although many proofs exist, we will here use an order arrow approach, if only as an example of the use of this technique. Note that the approximation is not assumed to be Padé.

Proof of Theorem 8.10.   Because $p > 2$, and because up-arrows cannot be tangential to the imaginary axis, there are at least 2 up-arrows leaving the origin with directions in $[-\frac{1}{2}\pi, \frac{1}{2}\pi]$. These arrows cannot terminate at $-\infty$ without crossing the imaginary axis. Hence there are at least 2 poles, contrary to the assumption that $s = 1$.   □

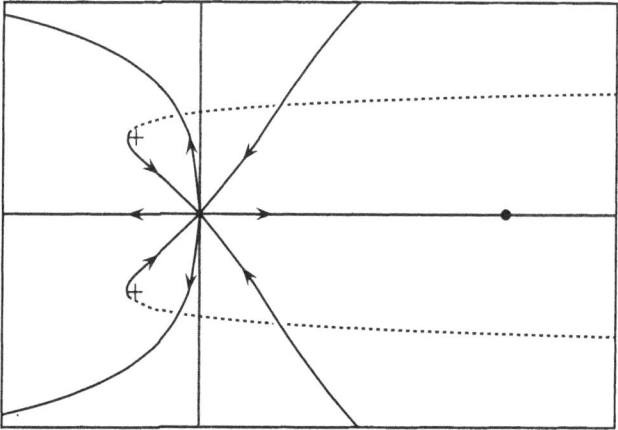

Figure 8.6. Order arrows for BDF3 (pole: • , branch points: +).

This result is illustrated in Figure 8.6 where it is seen that only a single up-arrow emanates from zero in a positive direction but two up-arrows which terminate at $-\infty$ are tangential to the imaginary axis. Note that two down-arrows which emanate in the negative direction terminate at $+\infty$ on lower sheets.

8.7. The Daniel–Moore barrier

It was conjectured in Daniel and Moore (1970) that an order $2s$, which is achieved for Gauss–Legendre Runge–Kutta methods, cannot be exceeded, except at the expense of A-stability. This was eventually proved in Wanner, Hairer and Nørsett (1978) using order stars. The proof given here uses order arrows.

Theorem 8.11. Let $\Phi(w, z)$ denote an (r, s) A-stable approximation with order p. Then $p \leq 2s$.

Proof. If the approximation is A-stable, at most s up-arrows emanate from zero in the positive direction and terminate at poles. The next up-arrow in the anticlockwise direction and the next up-arrow in the clockwise direction do not terminate at poles and must emanate in the negative direction. Because the angle between up-arrows is $2\pi/(p+1)$, it follows that

$$(s+1)\frac{2\pi}{p+1} > \pi,$$

implying that $2s \geq p$.   □

We present two order arrow diagrams to illustrate this result. First the A-stable, $[2, 0, 1]$ approximation with order 4. This is given in Figure 8.7. Note that, because of the complicated behaviour on the real axis in which

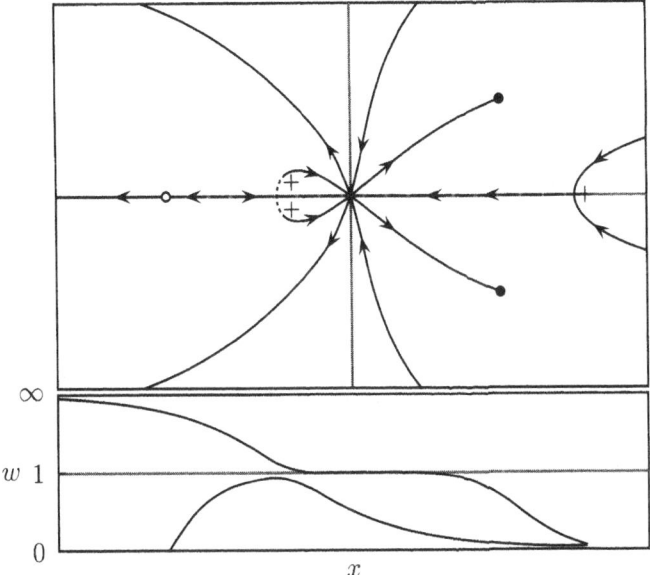

Figure 8.7. Order arrows for $[2, 0, 1]$ approximation
(poles: •, zero: ∘, branch points: +).

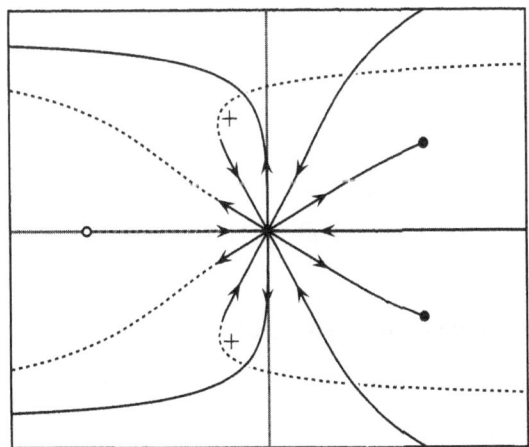

Figure 8.8. Order arrows for $[2, 1, 1]$ approximation
(poles: •, zero: ∘, branch points: +).

arrows on both sheets of the Riemann surface overlap, an additional view is given, showing w as a function of x (the real part of z). Secondly, in Figure 8.8, the $[2, 1, 1]$ approximation with order 5 is given. That this cannot be A-stable is seen from the up-arrows tangential to the imaginary axis at zero.

The $[2, 0, 1]$ and $[2, 1, 1]$ approximations are given respectively by

$$\Phi(w, z) = w^2\left(1 - \tfrac{8}{11}z + \tfrac{2}{11}z^2\right) - \tfrac{16}{11}w + \tfrac{5}{11} + \tfrac{2}{11}z,$$

$$\Phi(w, z) = w^2\left(1 - \tfrac{12}{23}z + \tfrac{2}{23}z^2\right) - w\left(\tfrac{16}{23} + \tfrac{16}{23}z\right) - \tfrac{7}{23} - \tfrac{2}{23}z.$$

8.8. The Butcher–Chipman conjecture

For an $[n_0, n_1, n_2]$ approximation, the value of $2n_0 - p$ seems to be related to possible A-stability. If $2n_0 - p \leq 0$ then, from Theorem 8.11, A-stability is impossible. On the other hand, if $2n_0 - p > 2$, there is no known case in which A-stability occurs. For $2n_0 - p \in \{0, 1, 2\}$, most methods easily analysed are A-stable but not always; for example $[7, 0, 4]$ is not A-stable; for this approximation,

$$\Phi(w, z) = w^2\left(1 - \tfrac{1486}{1651}z + \tfrac{638}{1651}z^2 - \tfrac{512}{4953}z^3 + \tfrac{31}{1651}z^4 - \tfrac{58}{24765}z^5 + \tfrac{14}{74295}z^6 - \tfrac{4}{520065}z^7\right)$$
$$- \tfrac{2048}{1651}w + \tfrac{397}{1651} + \tfrac{232}{1651}z + \tfrac{56}{1651}z^2 + \tfrac{20}{4953}z^3 + \tfrac{1}{4953}z^4. \quad (8.10)$$

Write this as $A(z)w^2 + B(z)w + C(z)$ then, because $|A(0)| > |C(0)|$, we can use the Schur criterion to find a simple necessary and sufficient condition for A-stability. This is that

$$\left(A(iy)A(-iy) - C(iy)C(-iy)\right)^2$$
$$- \left(A(iy)B(-iy) - C(-iy)B(iy)\right)\left(A(-iy)B(iy) - C(iy)B(-iy)\right)$$
$$\geq 0, \quad (8.11)$$

for all real y. In the case of (8.10), (8.11) evaluates to

$$-\frac{3424256}{245746955354703075}y^{14} + \mathcal{O}(y^{16}).$$

After considerable numerical searching, the following statement was produced (Butcher and Chipman 1992).

Conjecture 8.12. A generalized Padé approximation with $2n_0 - p > 2$ is not A-stable.

Note that in the statement of this conjecture, the w-degree is not specified although most attempts at a proof have focused on the quadratic case. If this degree is 1, then the result is covered by Theorem 8.9. Furthermore, the method of proof for the linear case would easily generalize if it can be shown that, for each pole, there exists an up-arrow emanating from zero which terminates at this pole.

9. Conclusions and inconclusions

If there is a single theme to the ideas presented here, it is that the general linear method formulation has reached a reasonable level of maturity. Theoretically they are quite well understood and practical methods, as well as techniques for their implementation, are starting to be developed. However, they are not simply a new class of methods, because they represent a new way of looking at earlier and more established methods.

From many points of view, the general linear formulation is more natural than the traditional way of understanding even traditional methods. A first example concerns the order conditions for general linear methods which actually give fresh insight into the significance and meaning of order for traditional methods. This is especially true in the case of Runge–Kutta methods, for which effective order is a clear-cut and useful generalization which arises naturally from a general linear point of view.

A second example of new insight coming out of general linear methods concerns non-linear stability. The irreducible formulation of linear multistep methods can stand alongside one-leg methods as a valid way of understanding G-stability and algebraic stability.

Although order arrows are not specific to general linear methods, they are featured in this paper as a tool for studying the relationship between order and stability, especially for multivalue multistage methods. They provide alternative proofs to those made available by the use of order stars and give a slightly different insight into some problems. The author would like to see the two approaches used to examine new questions, with the expectation that each of them will sometimes turn out to be the more convenient.

There is still much to be done in some of the areas identified in this paper. It would be worthwhile to know more about the consequences of algebraic stability. In particular, it would be valuable to know the extent to which general linear methods can make worthwhile contributions towards the development of structure-preserving algorithms.

However, there are already sufficiently challenging questions arising in the construction of efficient new methods. The inherent Runge–Kutta stability ansatz is promising as a source of methods but it is not yet known where to search for the best methods in this already-large family.

Finally, more detailed information on the interplay between stability and order is needed. It is a simple matter to determine in particular cases what the order of an approximation is and whether or not it is A-stable. However, there are quite likely some general patterns that can be identified and verified. The simplest outstanding question is the so-called Butcher–Chipman conjecture and, in the view of this author, is an issue capable of resolution using known techniques.

Acknowledgements

The writing of this paper was assisted by a grant from the New Zealand Marsden Fund. I have worked, over the years, with many people on aspects of general linear methods and I wish to thank these colleagues for their collaborations. Especially I wish to acknowledge the opportunity to have worked in recent years with Zdzisław Jackiewicz, Will Wright and Helmut Podhaisky on the construction and implementation of practical methods. My interest in stability and related issues has been revived by a visit of Adrian Hill and I am grateful for the discussions he and I have had and for comments on early drafts of this paper. Steffen Voigtmann has also made constructive and helpful suggestions. Finally, I wish to thank Robert Chan, Allison Heard, and other members of the Auckland numerical analysis workshop, who have been a constant resource of support and encouragement.

REFERENCES

P. Albrecht (1978a) 'On the order of composite multistep methods for ordinary differential equations', *Numer. Math.* **29**, 381–396.

P. Albrecht (1978b) 'Explicit, optimal stability functionals and their application to cyclic discretization methods', *Computing* **19**, 233–249.

P. Albrecht (1979), *Die Numerische Behandlung Gewöhnlicher Differentialgleichungen: Eine Einführung unter Besonderer Berücksichtigung Zyklischer Verfahren*, Carl Hanser Verlag, Munich.

P. Albrecht (1985), 'Numerical treatment of ODEs: The theory of A-methods', *Numer. Math.* **47**, 59–87.

P. Albrecht (1988), 'The extension of the theory of A-methods to RK-methods', in *Numerical Treatment of Differential Equations: Halle, 1987*, Vol. 104 of *Teubner-Texte Math.*, Teubner, Leipzig, pp. 8–18.

P. Albrecht (1989), 'Elements of a general theory of composite integration methods', in *Numerical Ordinary Differential Equations: Albuquerque, NM, 1986*, *Appl. Math. Comput.* **31**, 1–17.

P. Albrecht (1996), 'The common basis of the theories of linear cyclic methods and Runge–Kutta methods', *Appl. Numer. Math.* **22**, 3–21.

R. Alexander (1977), 'Diagonally implicit Runge–Kutta methods for stiff ODEs', *SIAM J. Numer. Anal.* **14**, 1006–1021.

Z. Bartoszewski and Z. Jackiewicz (1998), 'Construction of two-step Runge–Kutta methods of high order for ordinary differential equations', *Numer. Algorithms* **18**, 51–70.

T. A. Bickart and Z. Picel (1973), 'High order stiffly stable composite multistep methods for numerical integration of stiff differential equations', *BIT* **13**, 272–286.

D. G. Brush, J. J. Kohfeld and G. T. Thompson (1967), 'Solution of ordinary differential equations using two off-step points', *J. Assoc. Comput. Mach.* **14**, 769–784.

K. Burrage (1978a), 'A special family of Runge–Kutta methods for solving stiff differential equations', *BIT* **18**, 22–41.

K. Burrage (1978*b*) 'High order algebraically stable Runge–Kutta methods', *BIT* **18**, 373–383.

K. Burrage (1980), 'Non-linear stability of multivalue multiderivative methods', *BIT* **20**, 316–325.

K. Burrage (1988), 'Order properties of implicit multivalue methods for ordinary differential equations', *IMA J. Numer. Anal.* **8**, 43–69.

K. Burrage and J. C. Butcher (1979), 'Stability criteria for implicit Runge–Kutta methods', *SIAM J. Numer. Anal.* **16**, 46–57.

K. Burrage and J. C. Butcher (1980), 'Non-linear stability of a general class of differential equation methods', *BIT* **20**, 185–203.

K. Burrage and F. H. Chipman (1985), 'The stability properties of singly-implicit general linear methods', *IMA J. Numer Anal.* **5**, 287–295.

K. Burrage and F. H. Chipman (1989), 'Construction of A-stable diagonally implicit multivalue methods', *SIAM J. Numer. Anal.* **26**, 397–413.

K. Burrage and P. Moss (1980), 'Simplifying assumptions for the order of partitioned multivalue methods', *BIT* **20**, 452–465.

K. Burrage and P. W. Sharp (1994), 'A class of variable-step explicit Nordsieck multivalue methods', *SIAM J. Numer. Anal.* **31**, 1434–1451.

J. C. Butcher (1964), 'Implicit Runge–Kutta processes', *Math. Comp.* **18**, 50–64.

J. C. Butcher (1965), 'A modified multistep method for the numerical integration of ordinary differential equations', *J. Assoc. Comput. Mach.* **12**, 124–135.

J. C. Butcher (1966), 'On the convergence of numerical solutions of ordinary differential equations', *Math. Comp.* **20**, 1–10.

J. C. Butcher (1967), 'A multistep generalization of Runge–Kutta methods with four or five stages', *J. Assoc. Comput. Mach.* **14**, 84–89.

J. C. Butcher (1969), 'The effective order of Runge–Kutta methods', in *Proc. Conference on the Numerical Solution of Differential Equations: Dundee 1969* (J. L. Morris, ed.), Vol. 109 of *Lecture Notes in Mathematics*, Springer, pp. 133–139.

J. C. Butcher (1972*a*) 'An algebraic theory of integration methods', *Math. Comp.* **26**, 79–106.

J. C. Butcher (1972*b*) 'A convergence criterion for a class of integration methods', *Math. Comp.* **26**, 107–117.

J. C. Butcher (1973*a*) 'The order of numerical methods for ordinary differential equations', *Math. Comp.* **27**, 793–806.

J. C. Butcher (1973*b*) 'Order conditions for a general class of numerical methods for ordinary differential equations', in *Topics in Numerical Analysis* (J. J. H. Miller, ed.), Academic Press, London, pp. 35–40.

J. C. Butcher (1974*a*) 'The order of differential equation methods', in *Proc. Conference on the Numerical Solution of Ordinary Differential Equations: Austin 1972* (D. G. Bettis, ed.), Vol. 362 of *Lecture Notes in Mathematics*, Springer, pp. 72–75.

J. C. Butcher (1974*b*) 'Order conditions for general linear methods for ordinary differential equations', in *Numerische Methoden bei Differentialgleichungen und mit funktionalanalytischen Hilfsmitteln: Oberwolfach 1972*, Vol. 19 of *International Series of Numerical Mathematics*, Birkhäuser, pp. 77–81.

J. C. Butcher (1975), 'A stability property of implicit Runge–Kutta methods', *BIT*
15, 358–361.

J. C. Butcher (1981*a*) 'A generalization of singly-implicit methods', *BIT* **21**, 175–
189.

J. C. Butcher (1981*b*) 'Stability properties for a general class of methods for ordinary differential equations', *SIAM J. Numer. Anal.* **18**, 37–44.

J. C. Butcher (1984), 'An application of the Runge–Kutta space', *BIT* **24**, 425–440.

J. C. Butcher (1985), 'General linear methods: A survey', *Appl. Numer. Math.* **1**,
273–284.

J. C. Butcher (1987*a*) *The Numerical Analysis of Ordinary Differential Equations:
Runge–Kutta and General Linear Methods*, Wiley, Chichester.

J. C. Butcher (1987*b*) 'Linear and non-linear stability for general linear methods',
BIT **27**, 182–189.

J. C. Butcher (1987*c*) 'The equivalence of algebraic stability and AN-stability',
BIT **27**, 510–533.

J. C. Butcher (1988), 'On a class of matrices with real eigenvalues', *Linear Algebra
Appl.* **103**, 1–12.

J. C. Butcher (1992), 'Some new hybrid methods for initial value problems', in
Computational Ordinary Differential Equations (J. R. Cash and I. Gladwell,
eds), Clarendon Press, Oxford, pp. 29–46.

J. C. Butcher (1993*a*) 'Diagonally-implicit multi-stage integration methods', *Appl.
Numer. Math.* **11**, 347–363.

J. C. Butcher (1993*b*) 'General linear methods for the parallel solution of ordinary
differential equations', *World Sci. Ser. Appl. Anal.* **2**, 99–111.

J. C. Butcher (1994*a*) 'The parallel solution of ordinary differential equations and
some special functions', in *Approximation and Computation: West Lafayette
1993* (R. V. M. Zahar, ed.), Vol. 119 of *International Series of Numerical
Mathematics*, Birkhäuser, pp. 67–76.

J. C. Butcher (1994*b*) 'A transformation for the analysis of DIMSIMs', *BIT* **34**,
25–32.

J. C. Butcher (1994*c*), 'Laguerre polynomials: Applications in numerical ordinary
differential equations', in *Proc. Cornelius Lanczos International Centenary
Conference* (D. Brown *et al.*, eds), SIAM, Philadelphia, PA, pp. 371–373.

J. C. Butcher (1995), 'An introduction to DIMSIMs', *Mat. Apl. Comput.* **14**, 59–72.

J. C. Butcher (1996*a*) 'General linear methods', *Comput. Math. Appl.* **31**, 105–112.

J. C. Butcher (1996*b*) 'Runge–Kutta methods as mathematical objects', in *Numerical Analysis: A. R. Mitchell 75th Birthday* (D. F. Griffiths and G. A. Watson,
eds), World Scientific, Singapore, pp. 39-56.

J. C. Butcher (1997*a*) 'Order and stability of parallel methods for stiff problems',
Adv. Comput. Math. **7**, 79–96.

J. C. Butcher (1997*b*) 'An introduction to 'Almost Runge–Kutta' methods', *Appl.
Numer. Math.* **24**, 331–342.

J. C. Butcher (1998), 'ARK methods up to order five', *Numer. Algorithms* **17**,
193–221.

J. C. Butcher (2000), 'Numerical methods for ordinary differential equations in
the 20th century', in *Numerical Analysis 2000*, Vol. VI: *Ordinary Differential
Equations and Integral Equations, J. Comput. Appl. Math.* **125**, 1–29.

J. C. Butcher (2001), 'General linear methods for stiff differential equations', *BIT*, **41**, 240–264.

J. C. Butcher (2002*a*) 'The A-stability of methods with Padé and generalized Padé stability functions', *Numer. Algorithms* **31**, 47–58.

J. C. Butcher (2002*b*) 'Software issues for ordinary differential equations', *Numer. Algorithms* **31**, 401–418.

J. C. Butcher (2003), *Numerical Methods for Ordinary Differential Equations*, Wiley, Chichester.

J. C. Butcher and J. Cash (1989), 'Some recent developments on numerical initial value problems: A survey', in *Recent Theoretical Results in Numerical Ordinary Differential Equations, Appl. Numer. Math.* **5**, 3–18.

J. C. Butcher and P. Chartier (1994), The construction of DIMSIMs for stiff ODEs and DAEs. Report Series, University of Auckland, New Zealand.

J. C. Butcher and P. Chartier (1995), 'Parallel general linear methods for stiff ordinary differential and differential algebraic equations', *Appl. Numer. Math.* **17**, 213–222.

J. C. Butcher and P. Chartier (1997), 'A generalization of singly-implicit Runge–Kutta methods', *Appl. Numer. Math.* **24**, 343–350.

J. C. Butcher and D. J. L. Chen (1998), 'ESIRK methods and variable stepsize', *Appl. Numer. Math.* **28**, 193–207.

J. C. Butcher and D. J. L. Chen (2001), 'On the implementation of ESIRK methods for stiff IVPs', *Numer. Algorithms* **26**, 201–218.

J. C. Butcher and J. H. Chipman (1992), 'Generalized Padé approximations to the exponential function', *BIT* **32**, 118–130.

J. C. Butcher and A. D. Heard (2002), 'Stability of numerical methods for ordinary differential equations', *Numer. Algorithms* **31**, 59–73.

J. C. Butcher and A. T. Hill (2006), 'Linear multistep methods as irreducible general linear methods', to appear in *BIT*.

J. C. Butcher and Z. Jackiewicz (1993), 'Diagonally implicit general linear methods for ordinary differential equations', *BIT* **33**, 452–472.

J. C. Butcher and Z. Jackiewicz (1996), 'Construction of diagonally implicit general linear methods of type 1 and 2 for ordinary differential equations', *Appl. Numer. Math.* **21**, 385–415.

J. C. Butcher and Z. Jackiewicz (1997*a*) 'Implementation of diagonally implicit multistage integration methods for ordinary differential equations', *SIAM J. Numer. Anal.* **34**, 2119–2141.

J. C. Butcher and Z. Jackiewicz (1997*b*) 'Construction of high order diagonally implicit multistage integration methods for ordinary differential equations', *Appl. Numer. Math.* **27**, 1–12.

J. C. Butcher and Z. Jackiewicz (2001), 'A reliable error estimation for diagonally implicit multistage integration methods', *BIT* **41**, 656–665.

J. C. Butcher and Z. Jackiewicz (2002), 'Error estimation for Nordsieck methods', *Numer. Algorithms* **31**, 75–85.

J. C. Butcher and Z. Jackiewicz (2003), 'A new approach to error estimation for general linear methods', *Numer. Math.* **95**, 487–502.

J. C. Butcher and Z. Jackiewicz (2004), 'Construction of general linear methods with Runge–Kutta stability properties', *Numer. Algorithms* **36**, 53–72.

J. C. Butcher and N. Moir (2003), 'Experiments with a new fifth order method', *Numer. Algorithms* **33**, 137–151.

J. C. Butcher and A. E. O'Sullivan (2002), 'Nordsieck methods with an off-step point', *Numer. Algorithms* **31**, 87–101.

J. C. Butcher and H. Podhaisky (2006), 'On error estimation in general linear methods for stiff ODEs', *Appl. Numer. Math.* **56**, 345–357.

J. C. Butcher and N. Rattenbury (2005), 'ARK methods for stiff problems', *Appl. Numer. Math.* **53**, 165–181.

J. C. Butcher and A. D. Singh (2000), 'The choice of parameters in parallel general linear methods for stiff problems', *Appl. Numer. Math.* **34**, 59–84.

J. C. Butcher and S. Tracogna (1997), 'Order conditions for two-step Runge–Kutta methods', *Appl. Numer. Math.* **24**, 351–364.

J. C. Butcher and W. M. Wright (2003*a*) 'A transformation relating explicit and diagonally-implicit general linear methods', *Appl. Numer. Math.* **44**, 313–327.

J. C. Butcher and W. M. Wright (2003*b*) 'The construction of practical general linear methods', *BIT* **43**, 695–721.

J. C. Butcher, J. R. Cash and M. T. Diamantakis (1996), 'DESI methods for stiff initial value problems', *ACM Trans. Math. Software* **22**, 401–422.

J. C. Butcher, P. Chartier and Z. Jackiewicz (1997), 'Nordsieck representation of DIMSIMs', *Numer. Algorithms* **16**, 209–230.

J. C. Butcher, P. Chartier and Z. Jackiewicz (1999), 'Experiments with a variable-order type 1 DIMSIM code', *Numer. Algorithms* **22**, 237–261.

J. C. Butcher, Z. Jackiewicz and H. D. Mittelmann (1997), 'A nonlinear optimization approach to the construction of general linear methods of high order', *J. Comput. Appl. Math.* **81**, 181–196.

G. D. Byrne and R. J. Lambert (1966), 'Pseudo-Runge–Kutta methods involving two points', *J. Assoc. Comput. Mach.* **13**, 114–123.

R. Caira, C. Costabile and F. Costabile (1990), 'A class of pseudo Runge–Kutta methods', *BIT* **30**, 642–649.

J. Cash (1980), 'On the integration of stiff systems of ODEs using extended backward differentiation formulae', *Numer. Math.* **34**, 235–246.

J. Cash (1981), 'Second derivative extended backward differentiation formulas for the numerical integration of stiff systems', *SIAM J. Numer. Anal.* **18**, 21–36.

F. Ceschino and J. Kuntzmann (1963), *Problèmes Différentiels de Conditions Initiales*, Dunod, Paris.

T. M. H. Chan (1998), Algebraic structures for the analysis of numerical methods. PhD thesis, University of Auckland, New Zealand.

P. E. Chartier (1994), 'L-stable parallel one-block methods for ordinary differential equations', *SIAM J. Numer. Anal.* **31**, 552–571.

P. Chartier (1998), 'The potential of parallel multi-value methods for the simulation of large real-life problems, solving differential equations on parallel computers', *CWI Quarterly* **11**, 7–32.

G. J. Cooper (1978), 'The order of convergence of general linear methods for ordinary differential equations', *SIAM J. Numer. Anal.* **15**, 643–661.

G. J. Cooper (1981), 'Error estimates for general linear methods for ordinary differential equations', *SIAM J. Numer. Anal.* **18**, 65–82.

M. Crouzeix (1979), 'Sur la B-stabilité des méthodes de Runge–Kutta', *Numer. Math.* **32**, 75–82.

G. Dahlquist (1956), 'Convergence and stability in the numerical integration of ordinary differential equations', *Math. Scand.* **4**, 33–53.

G. Dahlquist (1963), 'A special stability problem for linear multistep methods', *BIT* **3**, 27–43.

G. Dahlquist (1975), On stability and error analysis for stiff non-linear problems 1. Report NA 75.08, Department of Information Processing, Royal Institute of Technology, Stockholm.

G. Dahlquist (1976), 'Error analysis of a class of methods for stiff nonlinear initial value problems', in *Numerical Analysis, Dundee*, Vol. 506 of *Lecture Notes in Mathematics*, pp. 60–74.

G. Dahlquist (1978), 'G-stability is equivalent to A-stability', *BIT* **18**, 384–401.

J. W. Daniel and R. E. Moore (1970), *Computation and Theory in Ordinary Differential Equations*, Freeman.

K. Dekker (1981), Algebraic stability of general linear methods. Technical Report No. 25, Computer Science Department, University of Auckland, New Zealand.

K. Dekker (1982), Reducibility of algebraically stable general linear methods. Preprint No. NW 131/82, Mathematics Centre Amsterdam, Numerical Mathematics.

J. Donelson and E. Hansen (1971), 'Cyclic composite multistep predictor-corrector methods', *SIAM J. Numer. Anal.* **8**, 137–157.

J. R. Dormand and P. J. Prince (1980), 'A family of embedded Runge–Kutta formulae', *J. Comput. Appl. Math.* **6**, 19–26.

B. L. Ehle (1969), On Padé approximation to the exponential function and A-stable methods for the numerical solution of initial value problems. Research Report CSRR 2010, Department AACS, University of Waterloo, Canada.

B. L. Ehle (1973), 'A-stable methods and Padé approximations to the exponential', *SIAM J. Math. Anal.* **4**, 671–680.

R. Frank, J. Schneid and C. W. Ueberhuber (1981), 'The concept of B-convergence', *SIAM J. Numer. Anal.* **18**, 753–780.

C. W. Gear (1965), ' Hybrid methods for initial value problems in ordinary differential equations', *SIAM J. Numer. Anal.* **2**, 69–86.

C. W. Gear (1967), 'The numerical integration of ordinary differential equations', *Math. Comp.* **21**, 146–156.

C. W. Gear (1971), *Numerical Initial Valve Problems in Ordinary Differential Equations*, Prentice-Hall, Englewood Cliffs, NJ.

W. B. Gragg and H. J. Stetter (1964), 'Generalized multistep predictor-corrector methods', *J. Assoc. Comput. Mach.* **11**, 188–209.

R. D. Grigorieff and J. Schroll (1978), 'Über $A(\alpha)$-stabile Verfahren hoher Konsistenzordnung', *Computing* **20**, 343–350.

N. Guglielmi and N. Zennaro (2001), 'On the zero-stability of variable stepsize multistep methods: The spectral radius approach', *Numer. Math.* **88**, 445–4548.

E. Hairer and G. Wanner (1973), 'Multistep-multistage-multiderivative methods of ordinary differential equations', *Computing* **11**, 287–303.

E. Hairer and G. Wanner (1974), 'On the Butcher group and general multi-value methods', *Computing* **13**, 1–15.

E. Hairer and G. Wanner (1996), *Solving Ordinary Differential Equations Numerically II: Stiff Problems and Differential-Algebraic Equations*, Springer, Berlin.

E. Hairer and G. Wanner (1997), 'Order conditions for general two-step Runge–Kutta methods', *SIAM J. Numer. Anal.* **34**, 2087–2089.

E. Hairer, C. Lubich and G. Wanner (2002), *Geometric Numerical Integration: Structure-Preserving Algorithms for Ordinary Differential Equations*, Springer, Berlin.

E. Hairer, S. P. Nørsett and G. Wanner (1993), *Solving Ordinary Differential Equations Numerically I: Nonstiff Problems*, Springer, Berlin.

A. D. Heard (1978), The solution of the order conditions for general linear methods. Thesis, University of Auckland, New Zealand.

K. Heun (1900), 'Neue Methoden zur approximativen Integration der Differentialgleichungen einer unabhängigen Veränderlichen', *Z. Math. Phys.* **45**, 23–38.

A. T. Hill (2005), 'Nonlinear stability of general linear methods'. Submitted for publication.

A. Iserles and S. P. Nørsett (1990), 'On the theory of parallel Runge–Kutta methods', *IMA J. Numer. Anal.* **10**, 463–488.

A. Iserles and S. P. Nørsett (1991), *Order Stars*, Vol. 2 of *Applied Mathematics and Mathematical Computation*, Chapman and Hall, London.

Z. Jackiewicz and H. D. Mittelmann (1999), 'Exploiting structure in the construction of DIMSIMs', *J. Comput. Appl. Math.* **107**, 233–239.

Z. Jackiewicz and S. Tracogna (1994), 'A representation formula for two-step Runge–Kutta methods', in *Hellenic European Research on Mathematics and Informatics: Athens 1994*, Vol. 1, 2, Hellenic Mathematical Society, Athens, pp. 111–120.

Z. Jackiewicz and S. Tracogna (1995), 'A general class of two-step Runge–Kutta methods for ordinary differential equations', *SIAM J. Numer. Anal.* **32**, 1390–1427.

Z. Jackiewicz and S. Tracogna (1996), 'Variable stepsize continuous two step Runge–Kutta methods for ordinary differential equations', *Numer. Algorithms* **12**, 347–368.

Z. Jackiewicz and R. Vermiglio (1996), 'General linear methods with external stages of different orders', *BIT* **36**, 688–712.

Z. Jackiewicz and M. Zennaro (1992), 'Variable stepsize explicit two-step Runge–Kutta methods', *Math. Comp.* **59**, 421–438.

Z. Jackiewicz, R. Renaut and A. Feldstein (1991), 'Two-step Runge–Kutta methods', *SIAM J. Numer. Anal.* **28**, 1165–1182.

Z. Jackiewicz, R. Renaut and M. Zennaro (1995), 'Explicit two-step Runge–Kutta methods', *Appl. Math.* **40**, 433–456.

Z. Jackiewicz, R. Vermiglio and M. Zennaro (1995), 'Variable stepsize diagonally implicit multistage integration methods for ordinary differential equations', *Appl. Numer. Math.* **16**, 343–367.

Z. Jackiewicz, R. Vermiglio and M. Zennaro (1997), 'Regularity properties of multistage integration methods', *J. Comput. Appl. Math.* **87**, 285–302.

R. Jeltsch (1976), 'A necessary condition for A-stability of multistep multiderivative methods', *Math. Comp.* **30**, 739–746.

R. Jeltsch and O. Nevanlinna, (1982), 'Stability and accuracy of time discretizations for initial value problems', *Numer. Math.* **40**, 245–296.

U. Kirchgraber (1986), 'Multi-step methods are essentially one-step methods', *Numer. Math.* **48**, 85–90.

J. J. Kohfeld and G. T. Thompson (1967), 'Multistep methods with modified predictors and correctors', *J. Assoc. Comput. Mach.* **14**, 155–166.

J. J. Kohfeld and G. T. Thompson (1968), 'A modification of Nordsieck's method using an off-step point', *J. Assoc. Comput. Mach.* **15**, 390–401.

W. Kutta (1901), 'Beitrag zur näherungsweisen Integration totaler Differentialgleichungen', *Z. Math. Phys.* **46**, 435–453.

M. A. Lopez-Marcos, J. M. Sanz-Serna and R. D. Skeel (1996), 'Cheap enhancement of symplectic integrators', in *Numerical Analysis*, Vol. 344 of *Pitman Res. Notes Math. Ser.*, Longman, Harlow, pp. 107–122.

M. Mihelcic (1977), 'Fast A-stable Donelson–Hansensche zyklische Verfahren zur numerischen Integration von stiff Differentialgleichungssystemen', *Ange. Inform.* **19**, 299–305.

A. Nordsieck (1962), 'On numerical integration of ordinary differential equations', *Math. Comp.* **16**, 22–49.

S. P. Nørsett (1969), 'An A-stable modification of the Adams–Bashforth methods', in *Proc. Conf. on the Numerical Solution of Differential Equations: Dundee, Scotland, June 1969*, Springer, Berlin, pp. 214–219.

N. Obreshkov (1940), 'Neue Quadraturformeln', *Abh. der Preuß. Akad. der Wiss., Math.-naturwiss. Klasse* **4**.

A. Prothero and A. Robinson (1974), 'On the stability and accuracy of one-step methods for solving stiff systems of ordinary differential equations', *Math. Comp.* **28**, 145–162.

N. Rattenbury (2005), Almost Runge–Kutta methods for stiff and non-stiff problems. PhD thesis, Department of Mathematics, University of Auckland, New Zealand.

R. Renaut (1990), 'Two step Runge–Kutta methods and hyperbolic partial differential equations', *Math. Comp.* **55**, 563–579.

C. Runge (1895), 'Über die numerische Auflösung von Differentialgleichungen', *Math. Ann.* **46**, 167–178.

A. D. Singh (1999), Parallel diagonally implicit multistage integration methods for stiff ordinary differential equations. PhD thesis, University of Auckland, New Zealand.

H. M. Sloate and T. A. Bickart (1973), 'A-stable composite multistep methods', *J. Assoc. Comput. Mach.* **20**, 7–26.

D. Stoffer (1993), ' General linear methods: Connection to one step methods and invariant curves', *Numer. Math.* **64**, 395–408.

S. Tracogna (1996), 'Implementation of two-step Runge–Kutta methods for ordinary differential equations', *J. Comput. Appl. Math.* **76**, 113–136.

S. Tracogna and B. Welfert (2000), 'Two-step Runge–Kutta methods: Theory and practice', *BIT* **40**, 775–799.

P. J. van der Houwen and B. P. Sommeijer (1982), 'A special class of multistep Runge–Kutta methods with extended real stability interval', *IMA J. Numer. Anal.* **2**, 183–209.

G. Wanner, E. Hairer and S. P. Nørsett (1978), 'Order stars and stability theorems', *BIT* **18**, 475–489.

O. B. Widlund (1967), 'A note on unconditionally stable linear multistep methods', *BIT* **7**, 65–70.

K. Wright (1970), 'Some relationships between implicit Runge–Kutta, collocation and Lanczos τ methods and their stability properties', *BIT* **10**, 217–227.

W. M. Wright (1999), General linear methods for ordinary differential equations. MSc thesis, University of Auckland, New Zealand.

W. M. Wright (2001), 'The construction of order 4 DIMSIMs for ordinary differential equations', *Numer. Algorithms* **26**, 123–130.

W. M. Wright (2002*a*) 'Explicit general linear methods with inherent Runge–Kutta stability', *Numer. Algorithms* **31**, 381–399.

W. M. Wright (2002*b*) General linear methods with inherent Runge–Kutta stability. PhD thesis, Department of Mathematics, University of Auckland, New Zealand.

M. Zennaro (1986), 'Natural continuous extensions of Runge–Kutta methods', *Math. Comp.* **46**, 119–133.

Acta Numerica (2006), pp. 257–325
doi: 10.1017/S0962492906230010

© Cambridge University Press, 2006

Modern statistical estimation
via oracle inequalities

Emmanuel J. Candès

Applied and Computational Mathematics,
California Institute of Technology,
Pasadena, CA 91125, USA
E-mail: `emmanuel@acm.caltech.edu`

A number of fundamental results in modern statistical theory involve thresholding estimators. This survey paper aims at reconstructing the history of how thresholding rules came to be popular in statistics and describing, in a not overly technical way, the domain of their application. Two notions play a fundamental role in our narrative: sparsity and oracle inequalities. Sparsity is a property of the object to estimate, which seems to be characteristic of many modern problems, in statistics as well as applied mathematics and theoretical computer science, to name a few. 'Oracle inequalities' are a powerful decision-theoretic tool which has served to understand the optimality of thresholding rules, but which has many other potential applications, some of which we will discuss.

Our story is also the story of the dialogue between statistics and applied harmonic analysis. Starting with the work of Wiener, we will see that certain representations emerge as being optimal for estimation. A leitmotif throughout our exposition is that efficient representations lead to efficient estimation.

CONTENTS

1. Introduction

1.1. Foreword

This paper is a survey article based on a series of lectures I gave at the
Institute of Mathematical Sciences at the National University of Singapore
in August 2004. The theme of these lectures was the interactions between
applied harmonic analysis and statistical estimation. I feel that it is important to state upfront that these lectures were by no means conceived
as an extended review of recent developments in the theory and practice
of nonparametric estimation but merely as an account of some important
ideas I had learned as a PhD student in the Department of Statistics at
Stanford University during the years 1995–1998. More to the point, these
lectures owe much to the scientific vision proposed by David Donoho and his
colleagues in a series of papers published in the early and mid-1990s, which
have influenced my thinking enormously, and continue to do so. I would
also like to acknowledge inspiration from a course I took called 'Function
Estimation in White Noise' taught by Iain Johnstone, and from a set of
notes written for this course, which have been updated since then, namely,
Johnstone (2002) in the reference section. This paper makes repeated references to Johnstone's unpublished manuscript, as the latter deals with many
of the topics we discuss here. I might have achieved something, should this
paper merely serve the purpose of encouraging the curious reader to take a
look at Donoho's papers and Johnstone's manuscript.

1.2. Interactions between statistical estimation and harmonic analysis

The interactions between harmonic analysis and statistical estimation have,
of course, a long history. Although it is amusing to note that Joseph Fourier,
the founding father of harmonic analysis, spent a significant fraction of
his research career studying statistical problems (see Stigler (1990) for an
excellent account of Fourier's contribution to early statistics), this history
cannot be traced quite that far back. Instead, the credit for bringing both
these topics together should probably go to Norbert Wiener where our story
begins. In the late 1930s and early 1940s, Wiener studied the problem of
filtering out noise (by statistical means) that has corrupted a time series.
He developed a solution by requiring information regarding the spectral
content of the original signal and the noise, and by creating a filter, which,
for stationary signals, filters selected frequencies. This filter was proposed
in the 1940s and first published in Wiener (1949). Since this fundamental
contribution, Fourier analysis has always played an important role in the
filtering literature and, more generally, in the analysis of time series.

Harmonic analysis and statistical estimation also remained connected via
the theory of splines (Wahba 1990), via the theory of estimation in statistical inverse problems and via key theoretical developments in function

estimation in the white noise model, to name a few examples. Having said that, it is nevertheless fair to say that the subject has been completely revitalized by Donoho and his colleagues. In the early 1990s, Donoho and his team realized that recent advances in applied harmonic analysis such as the theory of wavelets had very significant implications for statistical estimation. They developed *wavelet shrinkage* and established many of its spectacular properties, showing that, perhaps surprisingly, this algorithm has universal properties in the sense that it solves many statistical estimation problems simultaneously. I am sure that everyone reading this paper has heard about wavelet shrinkage as this has almost become a household word, and is perhaps the greatest application of wavelets to this date. But beyond wavelet shrinkage, Donoho also showed that efficient representations lead to efficient estimations, and that certain representations emerge as optimal. In doing so, he has linked statistical estimation and harmonic analysis in a durable and profound way. There is something remarkable about the timeliness of this discovery, since it occurred during a period marked and followed by intense research in computational harmonic analysis. On the one hand, applied mathematicians were energized by the prospect of new applications for the tools they were constructing, and on the other hand, statisticians had access to a brand new and powerful toolbox to refine and extend Donoho's ideas.

1.3. Our preoccupations

Such a broad subject imposes a selection of topics that will be covered and others that will not. As emphasized earlier, we will focus on ideas that have shaped my thinking; our focus is on the key structures and tools that bind statistical estimation and harmonic analysis. For example, we will explore the consequences of sparsity and emphasize the key role played by oracle inequalities – a new, fruitful and enlightening concept with an almost unlimited range of applications.

Our focus on sparsity and oracle inequalities serves a simple purpose: we wish to provide the reader with the necessary ideas for understanding an important fraction of the literature on modern statistical estimation, and with tools for future research in this area. Our point of view is that both these notions are fundamental, and that many decision-theoretic results are, in fact, easy consequences from rather simple oracle inequalities. To make this point, the reader perhaps already knows that wavelet shrinkage, discussed above, is asymptotically optimal for recovering objects taken from certain functional classes, such as the so-called Besov spaces – a result which has attracted a lot of attention. In truth, this is an automatic consequence of the fact that (1) wavelets provide optimally sparse representations of such functional classes, and that (2) a fundamental oracle inequality relates the performance of thresholding rules to the sparsity of such wavelet representations. Although there exist other ways to comprehend these types of results – note that we are not saying that these alternatives are uninteresting – we have decided to shift focus away from these and, instead, discuss what we believe are more fundamental concepts.

Indeed, the concepts of sparsity and oracle inequalities have already had a significant impact and everything suggests that this impact will last for a very long time. For example, 'sparsity' has become a true paradigm in many fields (not only statistics) including applied mathematics, theoretical computer science, signal and image processing, inverse problems, scientific computing and so on. While the potential for sparsity has been understood for a while now, there were relatively very few papers on this subject twenty years ago. In contrast, it is startling to see that the number of research papers and talks with 'sparsity' as a central theme has been exploding over the last few years. An oracle inequality, on the other hand, is a decision-theoretic tool and its use has thus far been confined to the field of statistical estimation. There are many forms of oracle inequalities and, as we will see, they have proved extremely successful in addressing the performance of many new estimation strategies 'post-wavelet shrinkage'. Without a doubt, oracle inequalities will continue to play a vital role in years to come.

1.4. Organization of the paper

We begin our survey with early important ideas in linear estimation, which are presented in Section 2. What is interesting here is that these ideas make explicit the connection between the estimation problem and the representation problem (the subject of applied harmonic analysis). Section 3 motivates the need for nonlinear estimation procedures. Section 4 introduces nonlinear estimation (nonlinear shrinkage to be more exact) and the powerful concept of oracle inequality. Section 5 introduces the notion of sparsity and shows that thresholding rules are very accurate for estimating sparse objects, e.g., parameter vectors with only a few significant entries. Section 6 argues that the problems of efficiently estimating, approximating, or compressing a signal (or a function) are all related and all linked to the fundamental problem of finding efficient signal representations. In Section 7, we consider extensions of thresholding ideas when there is no orthobasis (i.e., orthonormal basis) in which the object is sparse. Section 8 revisits some topics in model selection and introduces the Dantzig selector, a new effective and computationally tractable estimation strategy for estimating signals from undersampled data. Section 9 explores the possibility of adaptive basis estimation. Finally, we close the paper by discussing further topics, essentially inverse problems and false discovery rate thresholding rules in Section 10.

Because the intended audience is wide-ranging, we also include a Glossary, on page 324, where the reader will find definitions or explanations of the main statistical terms or concepts. The words or expressions to be found in the Glossary are marked by a superscript star, $\star$.

2. Linear estimation

2.1. The Wiener filter

We start with a classical estimation problem known as 'Wiener filtering' in the electrical engineering literature. This example is primarily of historical significance and the author would otherwise have been guilty of omission. But more importantly, this example will play a pedagogical purpose as it wonderfully introduces some of the key ideas surveyed in this paper.

We wish to recover a Gaussian signal$\star$ $X = (X_1, X_2, \ldots, X_n)$ from noisy data Y of the form

$$Y_t = X_t + Z_t, \quad t = 1, \ldots, n; \tag{2.1}$$

here, Y is the observed process, X is the signal, which is assumed to be a Gaussian process with mean zero and covariance matrix Σ, $i.e.$, $X \sim N(0, \Sigma)$, and Z is Gaussian white noise, $i.e.$, $Z \sim N(0, \sigma^2 I)$, and independent of the signal X. One may want to view this as a Bayesian estimation$\star$ problem where the prior on the unknown signal is Gaussian. The goal is to reconstruct the signal by producing an estimator $\hat{X} = g(Y)$ which can be computed from the data, and which has small mean-squared error

$$\mathrm{MSE}(X, \hat{X}) = \mathbb{E}\|X - \hat{X}\|_2^2 = \mathbb{E}\sum_{t=1}^{n}(X_t - \hat{X}_t)^2. \tag{2.2}$$

As is well known in Bayesian statistics ($e.g.$, see Lehmann (1997)), the estimator which achieves the minimum MSE is the conditional expectation of X given the observed process Y:

$$\hat{X} = \mathbb{E}(X \mid Y). \tag{2.3}$$

In detail, the tth component is given by

$$\hat{X}_t = \int_{\mathbb{R}^n} z_t\, p_{X|Y}(z)\, \mathrm{d}z,$$

where $p_{X|Y}$ is the conditional density of the random vector X. At first glance, the analytical evaluation of the conditional expectation might seem a little delicate. Having said that, a detour by way of principal components greatly simplifies things.

Recall that the principal components of a process $(X_t)_{1 \leq t \leq n}$ are the orthonormal eigenvectors φ_k, $1 \leq k \leq n$, which diagonalize the covariance matrix

Σ of X. In matrix notation, the matrix of principal components Φ is the n by n orthonormal matrix obeying

$$\Sigma = \Phi D \Phi^T, \quad D = \mathrm{diag}(d_k^2). \tag{2.4}$$

We will assume that the eigenvalues are arranged in decreasing order of magnitude $d_1^2 \geq d_2^2 \geq \cdots \geq d_n^2$. (We use the notation d_k^2 to emphasize that the eigenvalues of Σ are nonnegative since Σ is positive semidefinite.) The interpretation is that, if X is Gaussian, then the level sets of the joint density of the vector X are concentric ellipsoids, and the principal components are simply the (normalized) principal axes of these ellipsoids. A more general interpretation, which holds for general stochastic processes (not necessarily Gaussian), is that the first principal component is a projection with maximal variance; φ_1 is a unit vector obeying

$$\mathrm{Var}(u^T X) \leq \mathrm{Var}(\varphi_1^T X), \quad \text{for all } u \in \mathbb{R}^n : \|u\| = 1.$$

The second principal component φ_2 is then a projection with maximal variance among all projections orthogonal to φ_1

$$\mathrm{Var}(u^T X) \leq \mathrm{Var}(\varphi_2^T X), \quad \text{for all } u \in \mathbb{R}^n : \|u\| = 1, u \perp \varphi_1,$$

and so on for $\varphi_3, \varphi_4, \ldots, \varphi_n$.

With this in mind, principal component analysis is the action of decomposing a process X as a superposition of its principal components. It consists of two steps.

(1) The analysis step finds the orthonormal eigenvectors φ_k and projects X onto this basis, i.e.,

$$X' = \Phi^T X.$$

(2) The synthesis step reconstructs the process from the principal components using the orthonormal eigenvectors by $X = \Phi X'$, i.e.,

$$X_t = \sum_{k=1}^n X_k' \varphi_k(t). \tag{2.5}$$

This formula is also known as the Karhunen–Loeve decomposition: see Leon-Garcia (1994).

By definition, the coefficients X_k' in the expansion (2.5) are uncorrelated – the covariance matrix of X' is the diagonal matrix D – and are therefore also independent in the case where X is Gaussian since $X' \sim N(0, D)$. Hence, the Karhunen–Loeve decomposition provides a representation of Gaussian stochastic processes as a superposition of *independent* components.

We now return to the estimation problem and 'rotate' the observation vector Y in the orthonormal basis of principal components by applying Φ^T

on both sides of (2.1)

$$\langle Y, \phi_k \rangle = \langle X, \phi_k \rangle + \langle Z, \phi_k \rangle,$$
$$Y_k' = X_k' + Z_k'.$$

The coordinates $X_k' \sim N(0, d_k^2)$ are independent; the Z_k' are i.i.d.$\star$ $N(0, \sigma^2)$ and independent of X'. Obviously, the problem has not changed and we are merely looking at it from a different perspective . In particular, to estimate X, we may just as well estimate its coefficient sequence X' with $\widehat{X'}$: that is, with any estimator with minimum mean-squared error. The synthesis step would then provide the reconstruction $\hat{X} = \Phi\widehat{X'}$,

$$\hat{X}_t = \sum_{k=1}^{n} \widehat{X_k'}\, \varphi_k(t),$$

and owing to the isometry

$$\|X - \hat{X}\|^2 = \|X' - \widehat{X'}\|^2,$$

this would be exactly the estimator with minimum MSE: $\hat{X} = \mathbb{E}(X \mid Y)$. The point of all this is that $\widehat{X'}$ is now easy to compute since

$$\widehat{X_k'} = \mathbb{E}(X_k' \mid Y') = \mathbb{E}(X_k' \mid Y_k'),$$

where the second equality uses the fact that X_k' is independent of all the components Y_j' with $j \neq k$. Now the pair (X_k', Y_k') follows a bivariate normal distribution with mean zero and covariance matrix $\mathrm{Var}(X_k') = d_k^2 = \mathrm{Cov}(X_K', Y_k')$, and $\mathrm{Var}(Y_k') = d_k^2 + \sigma^2$. It is a classical exercise in regression analysis to show that the conditional distribution of X_k' is Gaussian with conditional mean

$$\mathbb{E}(X_k' \mid Y_k') = \frac{\lambda_k^2}{\lambda_k^2 + \sigma^2} Y_k', \tag{2.6}$$

so that the Wiener estimator is given by

$$\hat{X}_t = \sum_{k=1}^{n} w_k \langle Y, \phi_k \rangle \varphi_k(t), \quad w_k = \frac{\lambda_k^2}{\lambda_k^2 + \sigma^2}. \tag{2.7}$$

In short, the Wiener filter transforms the data with respect to the orthobasis of principal components, and downweights each coefficient as a function of the signal-to-noise ratio since one can think of the coordinates of w as the ratio between the expected signal power and the expected signal + noise power. Note that downweighting and the whole estimation procedure are linear, and that one can write $\hat{X}$ as

$$\hat{X} = \Phi W \Phi^T Y,$$

where $W = \mathrm{diag}(w_k)$.

It is interesting to consider special instances of Wiener filtering. Suppose for example that the process process X is stationary (and periodic) in the sense that the covariance between X_s and X_t only depends on the time lag

$$\Sigma_{s,t} = \text{Cov}(X_s, X_t) = \gamma(s - t), \quad 1 \le s, t \le n,$$

where it is understood that subtraction operates modulo n. This property says that the statistical properties of the signal are invariant with respect to time shifts, which conveys the idea that the process is spatially homogeneous. Because Σ is a circulant matrix, the basis of principal components is the Fourier basis which, for even sample sizes n, takes the form

$$\varphi_1(t) = 1/\sqrt{n},$$
$$\varphi_{2k}(t) = \sqrt{2/n}\cos(2\pi kt/n), \quad k = 1, 2, \ldots, n/2 - 1,$$
$$\phi_{2k+1}(t) = \sqrt{2/n}\sin(2\pi kt/n), \quad k = 1, 2, \ldots, n/2 - 1,$$
$$\varphi_n(t) = (-1)^t/\sqrt{n},$$

and the eigenvalues are the Fourier coefficients of the vector $(\gamma(0), \ldots, \gamma(n - 1))$. Hence, Bayes' rule or Wiener's solution exhibit the following key structure:

(1) Bayes' rule transforms the data in the frequency domain,

(2) Bayes' rule shrinks the noisy Fourier coefficients towards zero using a specially selected frequency-dependent factor,

(3) finally, Bayes' rule reconstructs the signal by inverting the Fourier transform.

As we shall see, this transformation–shrinkage–inverse transformation structure is a recurrent theme in modern statistical estimation. What is interesting here is that the estimation problem makes no reference to any particular basis, nor to any particular shrinkage rule, and yet this structure naturally emerges as the optimal strategy.

In conclusion, the Wiener filter is optimal for Gaussian signal priors. In the case where X is non-Gaussian, however, the estimator (2.7) is only guaranteed to have minimum mean-squared error among all *linear* estimators; see Leon-Garcia (1994).

2.2. Kernel methods

In contemporary nonparametric statistics, there are other models which do not assume a prior distribution on the signals or functions of interest. The so-called frequentist viewpoint assumes a model of the form

$$y_i = f(t_i) + z_i, \quad 1 \le i \le n, \tag{2.8}$$

where again y is a vector of observations, the function $f(t)$ is the object we wish to recover, and z is a vector of stochastic and independent errors. In nonparametrics, the object f is completely unknown and does not depend upon a few parameters. The goal is to estimate f from the data y. Note that to develop a fruitful methodology, one would need to restrict the classes of objects f of interest, since to extract the object, one would need to be able to distinguish it from noise. Examples of common assumptions include imposing a bounded total variation, a bounded curvature, or bounded higher-order derivatives.

One of the first developed and most frequently discussed approaches for estimating the regression function f is the kernel method: see Silverman (1986) and Scott (1992) for an introduction. The idea is to estimate the response $f(t)$ by a local averaging of the data y_i with 'time indices near' the point t under consideration. To do this, one selects a kernel K, usually a symmetric density function, which is nonnegative and integrates up to one. Typical examples include the boxcar kernel $K(t) = 1$ if $-1/2 \leq t \leq 1/2$ and zero otherwise, the Gaussian kernel $K(t) = (2\pi)^{-1/2}e^{-t^2/2}$, and the 'spline' kernel or Epanechnikov kernel equal to $\frac{3}{4}(1 - t^2)_+$, where here and below x_+ is the positive part of the scalar x. With such a kernel, the kernel regression sets

$$\hat{f}(t) = \frac{\sum_{i=1}^n w_i y_i}{\sum_{i=1}^n w_i},$$
(2.9)

where the weights are given by the formula

$$w_i = K(h^{-1}(t - t_i)).$$
(2.10)

Hence, the estimator is a weighted average and closer points naturally receive larger weights since typical kernels $K(t)$ decay as $|t|$ increases. The parameter h is the window width, or the bandwidth, and essentially determines which observations are averaged together. A small bandwidth averages over very few points, while a very large bandwidth may average over a significant fraction of the data set.

To connect kernel regression with our earlier discussion, suppose that the t_is are equispaced in $[0, 1]$, e.g., $t_i = i/n$ with $1 \leq i \leq n$ and that the estimand $f(t)$ is periodic. These assumptions are only useful for getting simple results. In the equispaced design, the Priestley–Chao kernel smoother is of the form

$$\hat{f}(t) = \frac{1}{nh} \sum_{i=1}^n K(h^{-1}(t - t_i))y_i,$$
(2.11)

where the subtraction is understood modulo $[0, 1]$. The estimator is then a convolution in the time domain or, equivalently, a multiplication in the Fourier domain. Let $(w_k(h))_{k \in \mathbb{Z}}$ be the sequence of Fourier coefficients of the density $h^{-1}K(\cdot/h)$

$$w_k(h) = \int_0^1 h^{-1}K(h^{-1}t)e^{-i2\pi kt}\,dt$$

and let $(\tilde{y}_k)_{k\in\mathbb{Z}}$ be those of the vector y

$$\tilde{y}_k = \int_0^1 n^{-1}\sum_{i=1}^n y_i\delta(t - t_i)\,e^{-i2\pi kt}\,dt = \frac{1}{n}\sum_{j=1}^n e^{-i2\pi kt_j}y_j$$

(note that $\tilde{y}_{k+n} = \tilde{y}_k$). We also denote the coefficient sequence of f by $(\theta_k)_{k\in\mathbb{Z}}$. In the frequency domain, the estimator (2.11) obeys

$$\hat{\theta}_k = w_k(h)\cdot\tilde{y}_k, \tag{2.12}$$

where we observe that $0 \le |w_k(h)| \le 1$. In short, the kernel method estimates the Fourier coefficients of f by shrinking those of the observations y, and hence the structure of this procedure is similar to that of the Wiener filter: the estimation combines the transformation of the data in the Fourier domain with frequency-by-frequency dumping. If W is the Fourier transform of K,

$$W(\omega) = \int K(t)e^{-i2\pi\omega t}\,dt,$$

then $w_k(h) \approx W(kh)$ and $|w_k(h)|$ typically decreases as the frequency index $|k|$ increases. For example, if K is the Gaussian kernel, $W(kh) = e^{-(kh)^2/2}$. The bandwidth h controls the decay of the weights $w_k(h)$; the larger h, the faster the decay and hence the greater the amount of smoothing.

Whereas the Wiener filter gives an explicit formula for the weights, here the sequence $w_k(h)$ depends upon the kernel and above all upon the bandwidth. Automatic selection of the bandwidth h – i.e., how much to smooth – is the topic of an immense literature. There are theoretical rules based on asymptotics which guarantee good MSEs for estimating smooth functions together with more practical rules for finite samples, e.g., based on cross-validation: see Green and Silverman (1994).

2.3. Smoothing splines

Another popular approach for estimating the regression function is based on smoothing splines. The idea is to find an estimator $\hat{f}$ which minimizes the trade-off between the goodness of fit and the complexity of the estimator, as measured by the size of the second derivative of the fitted function. Quantitatively, we wish to find the function $\hat{f}(t)$ which minimizes the variational problem

$$\hat{f} = \operatorname{argmin}_g \sum_{i=1}^n (y_i - g(t_i))^2 + \lambda\int_0^1 |g''(u)|^2\,du. \tag{2.13}$$

Like the bandwidth, the parameter $\lambda > 0$ controls the smoothness of the fit. The larger λ, the smoother the fit (in the limit where λ goes to infinity, the regression function is the regression line). It is not difficult to show that the solution $\hat{f}(t)$ to (2.13) is a cubic spline with knots at the sampled points t_i – hence the name of the method. The problem of fitting the data is then a finite-dimensional problem, which can be solved efficiently on a computer.

As before, we wish to develop an understanding of the structure of the solution by making some useful simplifying assumptions. Suppose that the points $t_i = i/n$, $1 \le i \le n$ are equispaced and that the estimand f is periodic. We approximate the second term of (2.13) by finite differences so that one is interested in finding the vector $g \in \mathbb{R}^n$ minimizing

$$\min \sum_{1 \le i \le n} (y_i - g_i)^2 + \lambda \sum_{1 \le i \le n} |(D^2 g)_i|^2, \tag{2.14}$$

with

$$(D^2 g)_i = \frac{g_{i+1} - 2g_i + g_{i-1}}{n^2}.$$

(Because f is assumed periodic, we set $g_0 = g_n$ in the above formula so that the matrix D^2 is circulant.) Let $\tilde{y}_k$ (resp. $\tilde{g}_k$) be the discrete Fourier coefficients of y (resp. g)

$$\tilde{y}_k = \sum_{1 \le i \le n} y_i \phi_k(i/n),$$

where $(\phi_k(t))_{1 \le k \le n}$ is the sequence of sines and cosines introduced in Section 2.1. Since D^2 is diagonal with eigenvalues $d_1 = 0$, $d_{2k} = d_{2k+1} = 4n^{-2} \sin^2(\pi k/n)$ for $1 \le k \le n/2 - 1$ and $d_n = 4n^{-2}$, then owing to the Fourier isometry, the minimization problem is equivalent to

$$\min \sum_{1 \le k \le n} [(\tilde{y}_k - \tilde{g}_k)^2 + \lambda \cdot d_k^2 \tilde{g}_k^2]. \tag{2.15}$$

The solution is now readily available; namely, the discrete Fourier coefficients $(\hat{\theta}_k)$ of the fitted vector $\hat{f}(i/n)$ are given by

$$\hat{\theta}_k = \frac{\tilde{y}_k}{1 + \lambda d_k^2}. \tag{2.16}$$

Once again, a familiar structure emerges. Spline smoothing rotates the data in the frequency domain and linearly shrinks the high-frequency components towards zero, i.e.,

$$\hat{f}_\lambda := \Phi W_\lambda \Phi^T y, \qquad W_\lambda = \text{diag}((1 + \lambda d_k^2)^{-1}).$$

The larger λ, the greater the shrinkage. A small value of λ does not imply a lot of smoothing and yields a low bias* but a large variance. Conversely, a large value of λ gives a fit with large bias and small variance. An important topic in spline smoothing is then how to select the parameter λ. In other words, how best to trade off between bias and variance.

To understand the trade-off, we examine the mean-squared error of the fit

$$\mathbb{E}\|f - \hat{f}_\lambda\|^2 = \sum_{i=1}^{n} \mathbb{E}(f(t_i) - \hat{f}_\lambda(t_i))^2,$$

where $\|f - \hat{f}_\lambda\|^2$ is the Euclidean norm $\sum_{i=1}^{n}(f(t_i) - f(t))^2$ and $\hat{f}_\lambda$ is the solution to (2.14); that is, $\hat{f}_\lambda = S_\lambda y$ where we put $S_\lambda = \Phi W_\lambda \Phi^T$ for short. The classical bias variance decomposition gives

$$\mathbb{E}\|f - \hat{f}_\lambda\|^2 = \|f - \mathbb{E}\hat{f}_\lambda\|^2 + \mathbb{E}\|\hat{f}_\lambda - \mathbb{E}\hat{f}_\lambda\|^2;$$

the bias term obeys $f_\lambda - \mathbb{E}\hat{f}_\lambda = (I - S_\lambda)f$ while the 'variance term' is given by

$$\mathbb{E}\|\hat{f}_\lambda - \mathbb{E}\hat{f}_\lambda\|^2 = \mathbb{E}\|S_\lambda z\|^2 = \sigma^2 \cdot \mathrm{Tr}(S_\lambda^T S_\lambda) = \sigma^2 \cdot \sum_k w_k^2(\lambda),$$

where $w_k(\lambda) = (1 + \lambda d_k^2)^{-1}$. The squared bias increases as λ increases whereas the variance decreases so that the optimal value of λ trades off between the source of errors. Suppose that the sequence $(\theta_k)_{1 \leq k \leq n}$ is the discrete Fourier coefficient sequence of $(f(t_i))_{1 \leq i \leq n}$; then the MSE obeys

$$\mathbb{E}\|f - \hat{f}_\lambda\|^2 = \sum_{1 \leq k \leq n} [(1 - w_k(\lambda))^2 \theta_k^2 + \sigma^2 w_k^2(\lambda)]. \qquad (2.17)$$

The best value of the smoothing parameter is that value λ^* which minimizes the above mean-squared error. Expressed in a different way, an 'omniscient' procedure knowing in advance λ^* would automatically answer the fundamental question: how much to smooth? This information is, of course, not available in practice, and this is why we used the word 'omniscient' to qualify the procedure. In practice, the best one can hope for is to select a smoothing parameter $\hat{\lambda}$ – based on the data – close to the optimal one. An interesting question is then whether it is possible to find $\hat{\lambda}$ such that the performance of the resulting estimator is close to that of the ideal one. As we will see, such issues will form a recurring theme of this paper.

We conclude this short overview of smoothing splines by pointing out that the solution to (2.13) has the exact same structure as that discussed above even in the case where the design points t_i are unequispaced. In short, there is an orthonormal basis $(\varphi_k)_{1 \leq k \leq n}$ known as the Demmler–Reinsch system (Wahba 1990) which – like the discrete Fourier basis – diagonalizes the minimization problem (2.13) so that the solution in that basis is given by

$$\hat{f}(t) = \sum_{1 \leq k \leq n} \hat{\theta}_k \phi_k(t),$$

where the coefficients $\hat{\theta}_k$ are given by the same relation as (2.16) with, of course, slightly different eigenvalues. The Demmler–Reinsch functions are boundary-adapted sinusoidal waveforms.

2.4. Statistical theory

On the theoretical side, there is a large literature showing that if the shrinkage parameters are chosen appropriately, the corresponding linear estimators are, in an asymptotic sense, optimal for recovering objects assumed to belong to certain types of functional classes. These results are perhaps best presented in the so-called 'white noise model', that is,

$$Y(dt) = f(t)\,dt + \varepsilon W(dt), \quad t \in [0,1]. \tag{2.18}$$

Here $W(t)$ denotes a Wiener process (*i.e.*, the primitive of white noise); ε is a noise level; and f is the object to be recovered. Formally, this model says that if we take a finite numbers of projections of the data Y and define

$$y_k := \langle Y, \varphi_k \rangle = \langle f, \varphi_k \rangle + \varepsilon\, z_k, \quad 1 \le k \le n$$

where the $\varphi_k(t)$s are any functions bounded in L_2, then $z = (z_1, \ldots, z_n)$ is a Gaussian vector with mean 0 and covariance matrix $\mathrm{Cov}(z_k, z_\ell) = \langle \varphi_k, \varphi_\ell \rangle$, the Gram matrix of the waveforms φ_k. In particular, if the φ_ks are orthogonal, the coordinates of z are independent. This explains why the white noise model should be understood as the large sample limit of the discrete model (2.8) where the errors z_i are i.i.d. $N(0, \sigma^2)$ under the calibration $\varepsilon = \sigma/\sqrt{n}$. To see why this is so, consider averaging (2.18) over intervals of the form $I_i := [(i-1)/n, i/n]$. This gives

$$y_i := n\langle Y, 1_{I_i} \rangle = \bar{f}_i + \varepsilon\sqrt{n}\, z_i,$$

where $\bar{f}_i = \mathrm{Ave}_{I_i} f$ and the z_is are i.i.d. $N(0,1)$. For sufficiently nice functions, $\bar{f}_i$ is close to $f(i/n)$ when n is large, which justifies the claim. In summary, the asymptotics in the continuous white noise model as $\varepsilon \to 0$ have similar characteristics to the asymptotics in the discrete model as $n \to \infty$. In fact, although the model is continuous and real data are typically discretely sampled, the asymptotic theory deriving from the white noise model has typically been found to lead directly to comparable asymptotic theory in a sampled data model. We do not wish to elaborate on this point, and refer the reader to Brown and Low (1996), Nussbaum (1996) for general theory, and to Efroĭmovich and Pinsker (1981, 1982), Nussbaum (1996), Donoho and Nussbaum (1990), Donoho and Liu (1991) and Donoho and Johnstone (1999) for examples of translations of optimal solutions in the white noise model to corresponding solutions in the sampled data model. The advantage is that the white noise model is more homogeneous than sampled data models, and since estimation in the white noise model is in general neither easier nor harder than in sampled models, it has proved to be a fruitful theoretical tool.

Decision theory develops a mathematical theory for making decisions in the face of uncertainty. In the theory of estimation, for example, suppose we wish to estimate a function θ on the basis of a sample $Y = (Y_1, \ldots, Y_n)$, where the distribution of the Y_is depend on θ. Then, by choosing an estimator $\hat{\theta} = g(Y)$, the decision maker incurs a loss $\ell(\theta, \hat{\theta})$ whose expected value is called the risk* function

$$R(\theta, \hat{\theta}) = \mathbb{E}\ell(\theta, \hat{\theta}).$$

In the set-up of interest here, the parameter f is the unknown regression function and the observations follow the white noise model (2.18). If we take as a loss the L_2-squared error $\ell(f, \hat{f}) = \|f - \hat{f}\|_{L_2}^2$, the risk is the integrated mean-squared error

$$\mathrm{MSE}(f, \hat{f}) = \mathbb{E}\|f - \hat{f}\|_{L_2}^2.$$

Decision theory is concerned with finding good decisions, $i.e.$, decision functions with small risk. Note that the risk depends on f which is not known. Some decisions may be good for certain values of the parameters and poor for others. Consider for instance, two estimators $\hat{f}_i$, $i = 1, 2$, which are constant and equal to f_i. Suppose f_1 and f_2 are wildly different. When the true state of nature is f_1 the first estimator has vanishing risk, but a very large risk when the true state is f_2, and $vice~versa$ for the second estimator. The two dominating viewpoints for getting around this difficulty are the minimax and Bayesian paradigms.

(1) The minimax* point of view defines a functional class $\mathcal{F}$ and searches for an estimator $\hat{f}$ which exactly or approximately attains the minimax risk (here the minimax mean-squared error):

$$M^*(\varepsilon, \mathcal{F}) = \inf_{\hat{f}} \sup_{f \in \mathcal{F}} \mathrm{MSE}(f, \hat{f}).$$

In other words, one is interested in the estimator with minimum worst-case error. The minimax approach puts no restriction on the estimator; all measurable procedures – $i.e.$, all measurable functions of Y – are allowed.

(2) The Bayesian point of view assumes a prior process π about f (so that $\pi(A)$ is the probability that the object f belongs to the set A) and searches for the estimator achieving the minimum average mean-squared error, the so-called Bayes risk

$$B(\pi) = \mathbb{E}_\pi \mathrm{MSE}(f, \hat{f}).$$

Here one averages the MSE over the prior distribution π. This is the viewpoint of the Wiener filter which assumes a Gaussian prior process.

If one is given a functional class, as in the minimax framework, then a possible approach is to select a prior on $\mathcal{F}$, a probability distribution on the elements $f \in \mathcal{F}$ obeying $\pi(\mathcal{F}) = 1$.

A key result of statistical decision theory is that the minimax risk is lower-bounded by the Bayes risk for any choice of prior π obeying $\pi(\mathcal{F}) = 1$.

$$\inf_{\hat{f}} \sup_{f \in \mathcal{F}} \mathrm{MSE}(f, \hat{f}) \geq B(\pi). \tag{2.19}$$

Under mild conditions, a famous result due to Wald proves the existence of prior distributions satisfying inequality (2.19); such distributions are called *least favourable priors*.

A splendid result in the minimax theory of linear estimation is due to Pinsker. We wish to recover an object f which is assumed to lie in a Sobolev ball

$$\mathcal{F} = \{f : \|f\|_{W_2^m} \leq R\},$$

where $\| \cdot \|_{W_2^m}$ is the Sobolev norm

$$\|f\|_{W_2^m}^2 := \int_{[0,1]} |f(t)|^2 + |f^{(m)}(t)|^2 \, dt, \tag{2.20}$$

in which $f^{(m)}$ is the mth derivative of the function f. In short, the mth derivative of f is assumed to be bounded in an L_2-sense. Pinsker's solution applies linear shrinkage in the Fourier domain, and is given by

$$\hat{f}(t) = \sum_{k \geq 0} w_{k,\varepsilon} \langle Y, \varphi_k \rangle \varphi_k(t). \tag{2.21}$$

Because we are now studying continuous-time models, $(\varphi_k(t))_{k \geq 0}$ is the continuous-time orthonormal Fourier basis of $L_2(0, 1)$

$$\phi_0(t) = 1,$$
$$\phi_{2k}(t) = \sqrt{2} \cos(2\pi k t), \quad k \geq 1,$$
$$\phi_{2k-1}(t) = \sqrt{2} \sin(2\pi k t), \quad k \geq 1,$$

and the weights are given by

$$W_{k,\epsilon} = (1 - \lambda k^m)_+;$$

in the above expression, the scalar λ actually depends on ε and R: see (2.24). It is important to take note that the weights depend on the parameters that define the functional class: the degree of smoothness m and the modulus of smoothness R. The result is that $\hat{f}(t)$ is asymptotically minimax.

Theorem 2.1. (Pinsker's theorem) The estimator (2.21) is asymptotically minimax

$$\sup_{\mathcal{F}} \mathrm{MSE}(f, \hat{f}) = M^*(\varepsilon, \mathcal{F})(1 + o(1)),$$

where $o(1)$ is a term tending to zero as ε tends to zero.

To give a geometric interpretation of Pinsker's theorem, introduce the empirical Fourier coefficients

$$\langle Y, \varphi_k \rangle = \langle f, \varphi_k \rangle + \varepsilon \langle W, \varphi_k \rangle,$$
$$y_k = \theta_k + \varepsilon z_k.$$

By the Parseval theorem, the condition imposing a size constraint on the size of the mth derivative is equivalent to a weighted-ℓ_2 size estimate on the Fourier coefficient sequence of f:

$$f \in \mathcal{F} \quad \Leftrightarrow \quad \theta \in \Theta,$$

where Θ is the infinite-dimensional ellipsoid

$$\Theta := \left\{ \theta : \sum_{k \geq 0} (1 + k^{2m})(|\theta_{2k-1}|^2 + |\theta_{2k}|^2) \leq R^2 \right\}. \tag{2.22}$$

The problem is then to recover $\theta \in \Theta$ from the infinite Gaussian sequence model $y \sim N(\theta, \varepsilon^2 I)$. The idea is that for ellipsoids, least favourable priors are essentially Gaussian. Consider a general ellipsoid

$$\Theta(R) := \left\{ \theta : \sum_k a_k^2 \theta_k^2 \leq R^2 \right\}$$

in which $a_k > 0$ tends to infinity as k tends to infinity. Note that in the case of the Sobolev ball, $a_{2k-1} = a_{2k} = k^m$, or $(1 + |k|^{2m})^{1/2}$ to be more exact. The least favourable prior over the ellipsoid nearly has Gaussian independent components given by

$$\theta_k \sim N(0, \tau_k^2), \quad \tau_k^2 = \varepsilon^2 \lambda^{-1}(a_k^{-1} - \lambda)_+, \tag{2.23}$$

where the scalar λ is that appearing in Pinsker's weights. This scalar is chosen as the smallest real number with $\sum_k a_k^2 \tau_k^2 \leq R^2$, i.e., λ is the solution to

$$\varepsilon^2 \lambda^{-1} \sum_k a_k (1 - \lambda a_k)_+ = R^2. \tag{2.24}$$

The careful reader will notice that $\pi(\Theta(R)) < 1$ but it is possible to consider small perturbations of this prior which asymptotically concentrate on $\Theta(R)$. We leave out the details and refer to Johnstone (2002). For Gaussian priors, one can calculate Bayes' rule, which takes the form

$$\hat{\theta}_k := (1 - \lambda a_k)_+ \, y_k.$$

This is none other than Pinsker's estimate with weights $w_k = (1 - \lambda a_k)_+$, and the MSE of this estimator obeys

$$\text{MSE}(\theta, \hat{\theta}) = \sum_k (1 - w_k)^2 \theta_k^2 + w_k^2 \varepsilon^2,$$

which actually simplifies to $\varepsilon^2 \sum_k w_k$.

3. Why nonlinear estimation?

Linear estimation is well suited for estimating Gaussian processes, or objects taken from functional classes which are ellipsoids when viewed in the right basis. The problem is that many stochastic processes of scientific interest are not Gaussian and that many functional classes are not ellipsoids. Unfortunately, linear estimation is very often of poor quality in such circumstances. We give a few examples.

3.1. Non-Gaussian processes

We follow Yves Meyer and introduce the *Ramp process* $X(t)$, $t \in [0, 1)$, with periodic sample paths defined by

$$X(t) = t - 1(t \geq \tau), \tag{3.1}$$

where τ is drawn uniformly at random in $[0, 1)$. The sample path increases linearly from 0 to τ in the interval $[0, \tau)$, is decreased by 1 at $t = \tau$, and increases linearly from $\tau - 1$ to 0 in the interval $[\tau, 1)$. This process is very simple, and estimating X from noisy data is an exercise in parametric statistics. Without calculating Bayes' rule, one could recover X by simply estimating the location of the discontinuity.

The best linear estimator is given by the Wiener filter. To calculate the Karhunen–Loève decomposition of X, Meyer observes that the covariance matrix is given by

$$\text{Cov}(X(s)X(t)) = \min(s, t) - st,$$

and is the same as that of the Brownian bridge $B(t) = W(t) - tW(1)$ where W is a Brownian motion. Since $(\sqrt{2}\sin(\pi kt))_{k \geq 1}$ are the eigenfunctions of the covariance matrix of the Brownian bridge with eigenvalues $d_k^2 = (\pi k)^{-2}$, the best linear estimator would operate by linearly shrinking the Fourier coefficients of $Y(dt) = X(t)\,dt + \varepsilon W(dt)$. Obviously, this is a poor estimation strategy since, to achieve a small MSE, partial Fourier series would need to give very good approximations of the sample paths of the process X with just a few terms (we will elaborate on this later). But the slow decay of the eigenvalues of the covariance matrix says that this is not the case. This is an instance of the well-known Gibbs phenomenon, which asserts that partial Fourier series provide poor reconstructions of otherwise smooth signals with isolated singularities. Quantitatively, the MSE of the Wiener filter is given by

$$\text{MSE}(X, \hat{X}) = \sum_{k \geq 1} \frac{d_k^2 \varepsilon^2}{d_k^2 + \varepsilon^2} \geq \frac{1}{2} \sum_{k \geq 1} \min(d_k^2, \varepsilon^2), \tag{3.2}$$

since $a^2 b^2/(a^2 + b^2) \geq \frac{1}{2}\min(a^2, b^2)$ for all $a, b \in \mathbb{R}$, with equality when $a = b$. With $d_k^2 = (\pi k)^{-2}$, this gives

$$\text{MSE}(X, \hat{X}) = \sum_{k \geq 1} \frac{d_k^2 \varepsilon^2}{d_k^2 + \varepsilon^2} \geq \varepsilon/\pi.$$

To drive the point home, recall the asymptotic calibration $\varepsilon = 1/\sqrt{n}$, which says that if we were to think about this estimation in the sampled data model, the MSE would scale like $1/\sqrt{n}$, where n is the sample size. This is substandard since we are dealing with a parametric problem for which there are estimators converging at the parametric rate of about $1/n$ (or about ε^2).

3.2. Other functional classes

Suppose now that we are interested in estimating objects with bounded variations. A function with finite bounded variations is a function whose first derivative is a signed measure with finite mass. Then it turns out that, for this functional class, any estimator which asymptotically achieves or nearly achieves the minimax risk *must be nonlinear*. There are many such examples. Suppose the functional class is defined via

$$\mathcal{F} = \{f : \|f\|_{W_p^m} \leq R\},$$

where $\| \cdot \|_{W_p^m}$ is the L_p-Sobolev norm

$$\|f\|_{W_p^m}^2 := \int_{[0,1]} |f(t)|^p + |f^{(m)}(t)|^p \, dt. \tag{3.3}$$

When $m = 1$ and $p = 1$, this definition is close to the bounded variation norm (with the proviso that the first derivative may not be an integrable function). In Section 2.4, we have seen that if $p = 2$, there is a clean solution which achieves the minimax risk and that this solution is linear. When $p < 2$, however, any estimator whose risk scales like the minimax risk as $\varepsilon \to 0$ *must be nonlinear*. In other words, linear estimators achieve markedly suboptimal rates of convergence.

Geometrically suppose that one is interested in estimating the mean vector θ from the data $y_k = \theta_k + \sigma z_k$, where the z_ks are i.i.d. $N(0, 1)$. Then, if Θ

is an ellipsoid, linear estimation is all-powerful! But suppose Θ is the body

$$\Theta := \left\{ \theta : \sum_{k \geq 1} |\theta_k| \leq R \right\}.$$

This is a convex body – an octahedron to be precise – but not an ellipsoid, and this causes a substantial modification in what constitutes an optimal or near-optimal estimation strategy.

3.3. Spatial adaptivity

Suppose that the function we wish to recover has a few isolated singularities but is otherwise smooth, and that we employ a linear kernel smoother. Suppose, further, that we have available an *oracle* which supplies the best bandwidth, in the sense that it tells us which h yields the smallest MSE. This optimal choice of the bandwidth comes from the classical bias/variance trade-off: the smaller the bandwidth, the smaller the bias but the greater the variance. On the one hand, to keep the bias low we would need to use a small bandwidth, as otherwise the estimation error would be large, since one would smooth away the discontinuities. But on the other hand, to keep the variance low we would need to use a large bandwidth, as otherwise the error would be large, since one would undersmooth the flat part of the object f.

To get out of this dead end, one would like to use, instead, a spatially varying bandwidth. That is, one would like to be able to use a small bandwidth when the estimand is rough or discontinuous and a larger bandwidth when it is smooth or flat. That is, one could imagine using a spatially adaptive bandwidth which we would estimate from the data. This would turn the overall estimation strategy into a nonlinear procedure. And if we could somehow find the right bandwidth at every point, we could in principle obtain much better MSEs.

3.4. Adaptive estimation

The asymptotically optimal estimator (2.21) is sensitive to the parameters m and R which define the class $\mathcal{F} := \{f : \|f\|_{W_2^m} \leq R\}$. Should these parameters be mis-specified, statistical optimality would no longer hold. In practice, however, one must confess that we would rarely know in advance the exact degree of smoothness or the object we wish to estimate. And even if we did, we would not know the exact size of the radius of the ball. Such practical considerations suggest abandoning the idea of an asymptotically exact estimator for a particular class in favour of estimators with nearly optimal asymptotic properties *simultaneously* over a wide range of classes of interest. Admittedly, this may seem like an overly ambitious goal. Perhaps surprisingly, this is, however, possible in many interesting cases. The upshot is that such estimators are nonlinear.

4. Shrinkage estimators and oracle inequalities

In this section and the next, we consider the problem of estimating a (possibly infinite) vector $\theta \in \mathbb{R}^d$ from observations $y \sim N(\theta, 1)$, and focus on the statistical underpinnings of this problem. Only much later shall we identify θ with the coefficient sequence of a function f in an appropriate basis, and translate some of the decision-theoretic results in the language of nonparametric function estimation. The importance of this section relies upon the fact that it introduces the idea of an oracle inequality.

4.1. The James–Stein estimator

We wish to estimate $\theta \in \mathbb{R}^d$ from $y \sim N(\theta, I)$, and use the mean-squared error to measure performance

$$\mathrm{MSE}(\hat{\theta}, \theta) = \mathbb{E}\|\hat{\theta} - \theta\|^2$$

(here and below $\|\cdot\|$ denotes the Euclidean norm). The maximum-likelihood estimate (MLE) is of course given by $\hat{\theta}^{\mathrm{MLE}} = y$ and obeys

$$\mathrm{MSE}(\hat{\theta}^{\mathrm{MLE}}, \theta) = d.$$

Everybody would agree that the MLE is a good estimator. After all, what other estimator could we use in the absence of any additional information about the parameter θ? The surprising discovery of James and Stein (1961) is that when $d > 2$, the MLE is not admissible. That is, there exist estimators which are more accurate than the MLE (or better than the sample mean in the case where one gets independent copies of y). Consider, for example, the estimator

$$\hat{\theta}^{\mathrm{JS}} = w(y) \cdot y, \quad w(y) = \left(1 - \frac{d-2}{\|y\|^2}\right)_+ \tag{4.1}$$

which shrinks the data y towards the origin. James and Stein proved that $\hat{\theta}^{\mathrm{JS}}$ obeys

$$\mathrm{MSE}(\hat{\theta}^{\mathrm{JS}}, \theta) < \mathrm{MSE}(\hat{\theta}^{\mathrm{MLE}}, \theta), \quad \text{for all } \theta \in \mathbb{R}^d.$$

In words, the performance of the shrinkage estimator is superior to that of the sample mean *for all values of the parameter θ*. This is surprising, because y may measure seemingly unrelated quantities such as the taste of clams and the age of the universe, to paraphrase Le Cam (2000). It is therefore surprising that by mixing information about completely disconnected problems, one can obtain an estimator with a total mean-squared error that is smaller than that one would obtain by considering each problem separately.

   This result has had an enormous influence on the field and is still difficult to comprehend, although, by now, there are many papers that provide some explanations for this strange phenomenon: see, for example, the empirical Bayes interpretation of Efron and Morris (1971). We will not attempt to summarize this literature and, instead, merely note that nonlinear shrinkage improves performance.

4.2. Ideal linear shrinkage estimator and oracle inequalities

It is time to revisit the main issue discussed thus far – although in an abstract setting: how much should we smooth or, rather, how much should we shrink? To estimate $\theta \in \mathbb{R}^d$ from $y \sim N(\theta, I)$, consider the family of diagonal estimators

$$\hat{\theta}^c = c \cdot y$$

where c is a scalar. For each coordinate, recall that the bias $\hat{\theta}^c_k$ is given by $\theta_k - \mathbb{E}\hat{\theta}^c_k = (1 - c)\theta_k$ and the variance obeys $\mathrm{Var}(\hat{\theta}^c_k) = c^2$ so that $\mathbb{E}(\theta_k - \hat{\theta}^c_k)^2 = (1 - c^2)\theta_k^2 + c^2$. Summing over coordinates gives

$$\mathrm{MSE}(\hat{\theta}^c, \theta) = (1 - c)^2\|\theta\|^2 + c^2 d.$$

We now search for an *ideal estimator* which selects that estimator $\hat{\theta}^{c^*}$ from the family $(\hat{\theta}^c)_{c \in \mathbb{R}}$ with minimal MSE: that is, c^* is the solution to

$$\min_{c \in \mathbb{R}}(1 - c)^2\|\theta\|^2 + c^2 d.$$

Analytically, c^* is given by

$$c^* = \frac{\|\theta\|^2}{\|\theta\|^2 + d},$$

and the ideal MSE obeys

$$\mathrm{MSE}(\hat{\theta}^{c^*}, \theta) = \frac{\|\theta\|^2 d}{\|\theta\|^2 + d}.$$

This estimator is ideal because we would of course not know which estimator $\hat{\theta}^c$ is best; that is, to achieve the ideal MSE, one would need an *oracle* that would tell us which shrinkage factor to choose. The difference from the James–Stein estimate is that $\hat{\theta}^{\mathrm{JS}}$ is estimating the shrinkage factor from the data y, while in the ideal scenario, the ideal shrinkage factor which depends on $\|\theta\|$ is simply given to us. Obviously,

$$\inf_c \mathrm{MSE}(\hat{\theta}^c, \theta) \leq \mathrm{MSE}(\hat{\theta}^{\mathrm{JS}}, \theta).$$

But the interesting fact is that there is an equality in the other direction.

Theorem 4.1.   The James–Stein estimate obeys

$$\mathrm{MSE}(\hat{\theta}^{\mathrm{JS}}, \theta) \leq 2 + \inf_c \mathrm{MSE}(\hat{\theta}^c, \theta). \tag{4.2}$$

In other words, the James–Stein estimator is almost as good as the ideal estimator in a mean-squared error sense. When the dimension d is large, the additive factor is small compared to the MSE of the MLE, which is equal to d. The inequality (4.2) is an *oracle inequality*. An oracle inequality relates the performance of a real estimator with that of an ideal estimator which relies on perfect information supplied by an oracle, and which is not available in practice. Oracle inequalities are a powerful concept that we shall use extensively in the remainder of this paper.

To prove (4.2), one needs to come up with a formula, or at least with an estimate for the MSE of the James–Stein estimate. Perhaps the most elegant derivation is based on the *Stein unbiased risk estimate*, due to Stein (1981), which goes as follows. Let $Y \sim N(\theta, I)$ and consider the estimator $\theta = Y + g(Y)$ where $g : \mathbb{R}^d \to \mathbb{R}^d$ is a weakly differentiable function. Then, under mild integrability assumptions,

$$\mathbb{E}\|Y + g(Y) - \theta\|^2 = \mathbb{E}[d + 2\nabla \cdot g(Y) + \|g(Y)\|^2], \qquad (4.3)$$

where $\nabla \cdot g(Y)$ is the divergence of g, $\nabla \cdot g(Y) := \sum_{k=1}^{d} \partial_k g_k(Y)$. To see why this is so, observe that

$$\mathbb{E}\|Y + g(Y) - \theta\|^2 = \mathbb{E}\|Y - \theta\|^2 + 2\mathbb{E}(Y - \theta)^T g(Y) + \mathbb{E}\|g(Y)\|^2.$$

Since $\mathbb{E}\|Y - \theta\|^2 = d$, we only need to argue that

$$\mathbb{E}(Y - \theta)^T g(Y) = \mathbb{E}\nabla \cdot g(Y).$$

This follows from an integration by parts. Let $\phi(y)$ be the density function of the standard multivariate normal distribution $\phi(y) = (2\pi)^{-n/2} e^{-\|y\|^2/2}$, and recall that $\partial_k \phi(y - \theta) = -(y_k - \theta) \phi(y - \theta)$. Then, assuming that g is sufficiently smooth,

$$\mathbb{E}(Y_k - \theta_k) g_k(Y) = \int_{\mathbb{R}^d} (y_k - \theta_k) g_k(y) \phi(y - \theta) \, dy$$

$$= \int_{\mathbb{R}^d} \partial_k g_k(y) \phi(y - \theta) \, dy.$$

The idea is now to use the relation (4.3) to compute the MSE of the James–Stein estimate. To avoid unnecessary technicalities due to the non-differentiability of $\hat{\theta}^{JS}$, we prove (4.2) with the slightly modified estimator $\hat{\theta} = \tilde{w}(y)y$, where $\tilde{w}(y) = (1 - (d-2)/\|y\|^2)$; that is, we remove the positive part. It seems intuitively clear that $\mathrm{MSE}(\hat{\theta}^{JS}, \theta) \le \mathrm{MSE}(\hat{\theta}, \theta)$, which is true. With this notation, $\hat{\theta} = Y + g(Y)$, where

$$g(Y) = -\frac{d-2}{\|Y\|^2} Y.$$

Since

$$\nabla \cdot g(Y) = -\frac{(d-2)^2}{\|Y\|^2},$$

the Stein unbiased risk formula reads

$$\mathbb{E}\|Y + g(Y) - \theta\|^2 = d - (d-2)^2 \cdot \mathbb{E}\frac{1}{\|Y\|^2}.$$

Set $X = \|Y\|^2$, then $\mathbb{E}X = \|\theta\|^2 + d$, and since the function $1/x$ is convex, Jensen's inequality yields

$$\mathbb{E}\frac{1}{X} \geq \frac{1}{\mathbb{E}X} = \frac{1}{\|\theta\|^2 + d}.$$

In other words, this would give

$$\mathbb{E}\|\hat{\theta}^{JS} - \theta\|^2 \leq d - \frac{(d-2)^2}{\|\theta\|^2 + d} \leq 4 + \inf_c \mathbb{E}\|\theta^c - \theta\|^2.$$

This is not exactly the content of (4.2) since we have an additive factor of 4 instead of 2. To improve on this, we need a sharper lower bound on $\mathbb{E}\|Y\|^{-2}$. More work would show that

$$\mathbb{E}\frac{1}{\|Y\|^2} \geq \frac{1}{d - 2 + \|\theta\|^2},$$

where the equality holds if $\theta = 0$. This sharper estimate would give (4.2). We refer the reader to Johnstone (2002) for details.

4.3. Ideal shrinkage and adaptive estimation

Returning to the theme of nonparametric estimation, there is a beautiful application of such oracle inequalities. We have seen that one can find asymptotically minimax estimators for L_2-Sobolev balls of the form $\mathcal{F}^m(R) = \{f : \|f\|_{W_2^m} \leq R\}$. Pinsker's solution requires knowledge of m and R, but in practice these are unknown. Is it possible to achieve asymptotic minimaxity over $\mathcal{F}^m(R)$, simultaneously for each value of m and $R > 0$?

Taking the sequence space viewpoint, the problem is equivalent to that of estimating the Fourier coefficients (θ_k) of f from the Gaussian sequence model

$$y_k = \theta_k + \varepsilon z_k, \tag{4.4}$$

where the infinite-dimensional vector θ belongs to the ellipsoid

$$\Theta =: \left\{\theta : \sum_{j \geq 0}\sum_{k \in B_j}(1 + k^{2m})|\theta_k|^2 \leq R^2\right\}. \tag{4.5}$$

In the above expansion, we have partitioned the sum into blocks which we assume are dyadic sub-bands

$$B_j := \{k \geq 0 : 2^j \leq k < 2^{j+1}\}.$$

That is, the block B_j is the family of all those Fourier coefficients with frequency indices in the dyadic interval $[2^j, 2^{j+1})$. This partitioning goes back a long way in harmonic analysis and was first introduced by Littlewood and Paley (see Frazier, Jawerth and Weiss (1991)) to study the property of functions and of their Fourier series.

Let $d_j = 2^j$ be the size of the jth block B_j. With this notation, we introduce the block James–Stein estimator defined by

$$\hat{\theta}_j^{\mathrm{BJS}}(y) = \begin{cases} y_j, & j < J_0, \\ \left(1 - \frac{(d_j-2)\varepsilon^2}{\|y_j\|^2}\right)_+ y_j, & J_0 \leq j < J_\varepsilon, \\ 0, & j \geq J_\varepsilon. \end{cases} \tag{4.6}$$

For example, one can set $J_0 = 2$, and J_ε to be the nearest integer to $\log_2(1/\varepsilon^2)$. The interpretation is that the very low-frequency components are untouched, the intermediate-frequency components are shrunk towards zero, and the high-frequency components are thrown away. In summary, the function $f(t)$ is estimated by (1) taking the data in the frequency domain, (2) applying the James–Stein estimator to each dyadic sub-band B_j, and (3) returning to the original time domain.

A remarkable result due to Efroïmovich and Pinsker (1984) shows that the block James–Stein estimator is asymptotically minimax over all Sobolev ellipsoids.

Theorem 4.2. For all ellipsoids of the form (4.5), the MSE of the block James–Stein estimator (4.6) obeys

$$\sup_{\theta \in \Theta} \mathrm{MSE}(\hat{\theta}^{\mathrm{BJS}}, \theta) \leq 2^{2m} M^*(\varepsilon, \Theta)(1 + o(1)), \tag{4.7}$$

where $o(1)$ is a term tending to zero as $\varepsilon \to 0$. In fact it is possible to get asymptotic minimaxity, namely,

$$\sup_{\theta \in \Theta} \mathrm{MSE}(\hat{\theta}^{\mathrm{BJS}}, \theta) = M^*(\varepsilon, \Theta)(1 + o(1)),$$

by choosing shorter (but not too short) blocks $B_j = \{k : \ell_j \leq k \leq \ell_{j+1}\}$ obeying $\ell_{j+1}/\ell_j \to 1$.

The intuition is as follows. Suppose that we have a block $B_j = \{k : \ell_j \leq k \leq \ell_{j+1}\}$ obeying $\ell_{j+1}/\ell_j \to 1$, and let θ^j be the vector $(\theta_k)_{k \in B_j}$. The key point is that to estimate the coordinates of θ_j, an estimator of the form

$$\hat{\theta}_k^j = c_j \cdot y_k,$$

with weights depending on the block index, but not on the individual coefficients within a block, is almost as efficient as any other estimator. To
understand this, one can check that Pinsker's (optimal) weights are nearly
constant on each block for sufficiently large j. With the notation of Sec-
tion 2.4, this is indeed a consequence of $\sup_{k,k' \in B_j} a_k/a_{k'} = (\ell_{j+1}/\ell_j)^m \to 1$.
Continuing at this informal level of discussion, it follows that if we could
find the best block-dependent shrinkage factor, then we would do very well.
But we have seen that this is precisely what the James–Stein estimate does
(Theorem 4.1). Thus $\hat{\theta}^{\mathrm{BJS}}$ is efficient and provably asymptotically minimax:
see Johnstone (2002) for a rigorous argument. When one uses dyadic blocks,
$\ell_{j+1}/\ell_j \to 2$ and the weights are not nearly constant but vary within a factor
2^m. Replacing these variable weights with a constant weight is responsible
for the slight loss in precision; compare (4.7).

5. Ideal shrinkage and thresholding rules

All of the estimators we have encountered so far are based on the belief that
large coefficients occur at low frequencies. As a consequence, high-frequency
components are systematically shrunk toward zero. We remarked earlier
that signals of interest may exhibit significant high-frequency components
because of singularities or otherwise. Why should we then enforce shrinkage
if the data provide evidence that some special high-frequency components
are statistically significant or unlikely to be noise?

To makes things concrete, consider an extreme example, where $\theta \in \mathbb{R}^n$ is
of the form

$$\theta = (0, \ldots, 0, \mu, 0, \ldots, 0),$$

where $\mu \neq 0$ and the location of the nonzero coordinate is not known in
advance. Then it is clear that linear estimators would be highly ineffective
in this setting. The James–Stein estimator, which is essentially a linear
estimator – albeit with a nonlinear data-dependent shrinkage factor – would
also be very ineffective. This section introduces thresholding rules which are
true nonlinear estimation procedures, and which perform very well in this
setting and, of course, in much more complicated settings as well.

5.1. Ideal shrinkage

We consider the same Gaussian sequence model (4.4), where we think of
$(\theta_k)_{1 \leq k \leq n}$ as the coefficient sequence of f in a fixed basis $(\psi_k(t))_{1 \leq k \leq n}$. To
recover $\theta \in \mathbb{R}^n$ from $y \sim N(0, \varepsilon^2 I)$, we now consider the family of diagonal
shrinkage estimators

$$\hat{\theta}^w = Wy \quad \Leftrightarrow \quad \hat{\theta}_k = w_k y_k$$

where $W = \text{diag}(w_k)$. Just as before, we consider the ideal estimator θ^* which minimizes the MSE among all diagonal shrinkage estimators

$$\theta^* = \text{argmin}_{w \in \mathbb{R}^n} \mathbb{E}\|\hat{\theta}^w - \theta\|^2.$$

Note that we have already computed θ^*, since for each coordinate k, the optimal weight w_k^* minimizes the trade-off between the squared bias and the variance

$$\mathbb{E}(w_k y_k - \theta_k)^2 = (1 - w_k)^2 \theta_k^2 + w_k^2 \varepsilon^2$$

whose solution is given by

$$w_k^* = \frac{\theta_k^2}{\theta_k^2 + \varepsilon^2}, \quad \text{and} \quad E(\hat{\theta}_k^* - \theta_k) = \frac{\theta_k^2 \varepsilon^2}{\theta_k^2 + \varepsilon^2}.$$

Closely related is the ideal projection estimator θ^I, where we additionally require that W be a projection matrix. This condition simply says that the weights w_k are either 0 or 1,

$$\theta^I = \text{argmin}_{w \in \{0,1\}^n} \mathbb{E}\|\hat{\theta}^w - \theta\|^2.$$

A simple calculation then shows that

$$\theta_k^I = w_k y_k, \quad w_k = \begin{cases} 0, & |\theta_k| < \varepsilon, \\ 1, & |\theta_k| \geq \varepsilon. \end{cases}$$

This is a keep-or-kill estimate. The interpretation is that, for $w_k = 1$, $w_k y_k$ has vanishing bias and a variance equal to ε^2, while for $w_k = 0$, $w_k y_k$ has bias θ_k and vanishing variance. The optimal choice then minimizes between the squared bias and the variance and, therefore, the risk of the ideal projection is given by

$$\mathbb{E}(\theta_k^I - \theta_k)^2 = \min(\theta_k^2, \varepsilon^2).$$

We have already seen that for $a, b \geq 0$, $ab/(a + b) \leq 2\min(a, b)$ and thus

$$\mathbb{E}(\theta_k^I - \theta_k)^2 \leq 2\min(\theta_k^2, \varepsilon^2),$$

which gives

$$\text{MSE}(\theta^*, \theta) \leq \text{MSE}(\theta^I, \theta) \leq 2\,\text{MSE}(\theta^*, \theta).$$

In short, the risk of the ideal projection comes within a factor of 2 of that of the ideal shrinkage estimator. From now on, it will be convenient to compare the risk of any real estimator with that of the ideal projection which obeys

$$\text{MSE}(\theta^I, \theta) = \sum_k \min(\theta_k^2, \varepsilon^2). \tag{5.1}$$

We then ask the question: is it possible to find estimators whose risk comes close to that of the ideal projection?

5.2. Thresholding rules

In the spirit of the ideal projection, we consider thresholding rules for estimating the mean of a Gaussian distribution. There are many such rules, and we focus on the most commonly studied rules, namely the so-called hard-thresholding and soft-thresholding rules. For other types of thresholding rules, consider the garrote method of Gao (1998), for example. A hard-thresholding rule is of the form

$$\hat{\theta}_k = \begin{cases} y_k, & |y_k| \geq \lambda, \\ 0, & |y_k| < \lambda, \end{cases} \tag{5.2}$$

where λ is a some positive scalar parameter. A hard-thresholding rule yields a keep-or-kill estimate. Observations which pass the threshold are considered significant and untouched, while all observations below the threshold are set to zero. A soft-thresholding rule is similar but performs additional shrinkage:

$$\hat{\theta}_k = \begin{cases} y_k - \lambda, & y_k \geq \lambda, \\ 0, & |y_k| < \lambda, \\ y_k + \lambda, & y_k < -\lambda. \end{cases} \tag{5.3}$$

That is, the significant observations are also pulled towards zero by an amount equal to λ. We note that a soft-thresholding $\hat{\theta}(y)$ rule is a continuous function of y while the hard-thresholding rule is not. In this sense, the soft-thresholding rule is a smoother rule, hence the name.

The hard- and soft-thresholding rules also have an interpretation as minimum complexity estimates for complexity penalties which are not quadratic. For example, the hard thresholding rule at level λ is the solution to

$$\min_{\tau \in \mathbb{R}} \ (y_k - \tau)^2 + \lambda^2 \cdot 1(\tau \neq 0),$$

while the soft-thresholding rule solves

$$\min_{\tau \in \mathbb{R}} \ (y_k - \tau)^2 + 2\lambda \cdot |\tau|.$$

For n-dimensional problems, hard-thresholding each coordinate at level λ solves the variational problem

$$\min_{\tau \in \mathbb{R}^n} \ \|y - \theta\|^2 + \lambda^2 \cdot \|\tau\|_{\ell_0},$$

where $\|\tau\|_{\ell_0} := \sum_{1 \leq k \leq n} 1(\tau_k \neq 0)$ is the number of nonzero components of τ. Similarly, soft-thresholding each coordinate at level λ solves the variational problem

$$\min_{\tau \in \mathbb{R}^n} \ \|y - \theta\|^2 + 2\lambda \cdot \|\tau\|_{\ell_1},$$

where $\|\tau\|_{\ell_1} := \sum_{1 \leq k \leq n} |\tau_k|$. Hence, thresholding rules may be thought of

as a complexity-penalized estimation procedure where the complexity of the
fit is nonquadratic and given either by the ℓ_0 or the ℓ_1-norm.

5.3. Oracle inequalities

A foundational result in modern estimation is that correctly tuned thresholding rules nearly achieve the risk of ideal projections.

Theorem 5.1. (Donoho and Johnstone) Suppose that $n \geq 2$ and set
$\lambda = \epsilon\sqrt{2 \log n}$. Assume that $y \sim N(\theta, \varepsilon^2 I_n)$ and let $\hat{\theta}$ be either a hard- or
soft-thresholding estimate with parameter λ. Then

$$\mathbb{E}\|\theta - \hat{\theta}\|^2 \leq (2 \log n + 1) \cdot \left(\varepsilon^2 + \sum_{k=1}^{n} \min(\theta_k^2, \varepsilon^2) \right). \tag{5.4}$$

To sum up, the risk of a thresholding estimator is at most $2 \log n$ times
larger than the ideal mean-squared error. Further, what is interesting here
is that the oracle inequality (5.4) is nonasymptotic and holds for any finite
sample size $n \geq 2$. Finally, we have seen somewhat sharper oracle inequalities where the multiplicative factor is actually equal to one (see (4.2)), and
it is therefore legitimate to ask whether the logarithmic factor is sharp. It
turns out that without any further assumptions on the parameter θ, the
logarithmic factor is optimal – in an asymptotic sense.

Theorem 5.2. (Donoho and Johnstone) Consider the class of diago-
nal estimators obeying $\hat{\theta}_k = \hat{\theta}_k(y_k)$. Under the same assumptions as before,

$$\inf_{\hat{\theta} \text{ diagonal}} \sup_{\theta \in \mathbb{R}^n} \frac{\mathbb{E}\|\theta - \hat{\theta}\|^2}{\varepsilon^2 + \sum_k \min(\theta_k^2, \varepsilon^2)} \to 2 \log n \quad \text{as} \quad n \to \infty. \tag{5.5}$$

The above result says that when the parameter space of interest is $\mathbb{R}^n$,
then from a minimax point of view, no diagonal estimator can essentially
do better, at least asymptotically.

5.4. Risk of thresholding rules

This section gives a proof of Theorem 5.1 for the soft-thresholding rule. The
proof for the hard-thresholding rule is similar and is only more technical.
We may also just assume that $\varepsilon = 1$ as the general case follows from a
simple rescaling argument.

We need to develop a formula for the risk of a scalar soft-thresholding
rule and introduce some notation. We let η_S be the scalar nonlinearity
$\eta_S(y) = \text{sgn}(y)(y - \lambda)_+$ and let $r_S(\lambda, \mu)$ be the risk of the soft-thresholding
rule η_S, i.e.,

$$r_S(\lambda, \mu) = \mathbb{E}(\eta_S(y) - \mu)^2, \quad y \sim N(\mu, 1).$$

Because soft-thresholding rules treat each coordinate separately, the idea of the proof is to develop an upper bound on the accuracy of scalar thresholding rules for $\mu = 0$ in a first step, and to use the bound to deduce a bound for all values of $\mu \in \mathbb{R}$ in a second step. This strategy uses the following lemma.

Lemma 5.3. The risk of the soft-thresholding rule obeys

$$r_S(\lambda, \mu) \leq r_S(\lambda, 0) + \min(\mu^2, 1 + \lambda^2). \tag{5.6}$$

Proof. The proof is an exercise in calculus. By symmetry, we may just as well assume that $\mu \geq 0$. Note that

$$r_S(\lambda, \mu) = \int (\eta_S(y) - \mu)^2 \, \phi(y - \mu) \, dy$$

$$= \mu^2 \int_{|y| \leq \lambda} \phi(y - \mu) \, dy + \int_{y > \lambda} (y - \lambda - \mu)^2 \, \phi(y - \mu) \, dy$$

$$+ \int_{y < -\lambda} (y + \lambda - \mu)^2 \, \phi(y - \mu) \, dy,$$

where $\phi(y) = (2\pi)^{-1/2} e^{-y^2/2}$. A change of variables then gives

$$r_S(\lambda, \mu) = \mu^2 \int_{-\lambda-\mu}^{\lambda-\mu} \phi(z) \, dz + \int_{\lambda-\mu}^{\infty} (z - \lambda)^2 \, \phi(z) \, dz + \int_{-\infty}^{-\lambda-\mu} (z + \lambda)^2 \, \phi(z) \, dz,$$

which shows that the derivative with respect to μ obeys

$$\partial_\mu r_S(\lambda, \mu) = 2\mu \int_{-\lambda-\mu}^{\lambda-\mu} \phi(z) \, dz \leq 2\mu.$$

Therefore, $r_S(\lambda, \mu)$ is increasing in μ, and on the one hand

$$r_S(\lambda, \mu) \leq \lim_{\mu \to \infty} r_S(\lambda, \mu) = 1 + \lambda^2.$$

On the other hand,

$$r_S(\lambda, \mu) - r_S(\lambda, 0) \leq \int_0^\mu 2u \, du = \mu^2,$$

and we conclude that

$$r_S(\lambda, \mu) \leq \min(r_S(\lambda, 0) + \mu^2, 1 + \lambda^2),$$

which proves the lemma.   □

It is interesting to note that we established an estimate which is slightly better than (5.6). The quantity $\min(r(\lambda, 0) + \mu^2, 1 + \lambda^2)$ is of interest because one can prove that this is a proxy for the risk of the soft-thresholding rule since there is an inequality in the other direction:

$$r_S(\lambda, \mu) \geq \frac{1}{2} \min(r_S(\lambda, 0) + \mu^2, 1 + \lambda^2). \tag{5.7}$$

In other words, the risk of soft-thresholding is just about $\min(r_S(\lambda, 0) + \mu^2, 1 + \lambda^2)$.

The second lemma develops a bound on $r_S(\lambda, 0)$.

Lemma 5.4. The risk of the soft-thresholding rule obeys

$$r_S(\lambda, 0) \leq \frac{2\phi(\lambda)}{\lambda}. \tag{5.8}$$

Proof. By symmetry of the Gaussian distribution, the risk $r_S(\lambda, 0)$ obeys

$$r_S(\lambda, 0) = 2 \int_{y>\lambda} (y - \lambda)^2 \, \phi(y) \, dy,$$

and an integration by parts shows that

$$\int_{y>\lambda} (y - \lambda)^2 \, \phi(y) \, dy = -\lambda\phi(\lambda) + (1 + \lambda^2)\Phi([\lambda, \infty)),$$

where $\Phi([\lambda, \infty)) = \int_{y \in [\lambda, \infty)} \phi(y) \, dy$. The claim then follows from

$$\Phi([\lambda, \infty)) \leq \int_{\lambda}^{\infty} \phi(y) \, dy \leq \int_{\lambda}^{\infty} \frac{y}{\lambda} \phi(y) \, dy = \frac{\phi(\lambda)}{\lambda}. \qquad \square$$

We now specialize (5.6) and (5.8) to $\lambda = \sqrt{2 \log n}$, which gives

$$r_S(\sqrt{2 \log n}, 0) \leq \frac{1}{n \sqrt{\pi \cdot \log n}} \leq \frac{2 \log n + 1}{n},$$

as soon as $n \geq 2$. This proves Theorem 5.1 since

$$\mathbb{E}\|\theta - \hat{\theta}\|^2 \leq n \cdot r_S(\sqrt{2 \log n}, 0) + \sum_k \min(\theta_k^2, 1 + 2 \log n)$$

$$\leq (1 + 2 \log n) + \sum_k \min(\theta_k^2, 1 + 2 \log n)$$

$$\leq (2 \log n + 1)\left(1 + \sum_k \min(\theta_k^2, 1)\right),$$

as claimed.

5.5. *Choice of threshold*

Besides the fact that $\lambda = \sqrt{2 \log n}$ allows proving sharp estimation results, there is a large literature arguing why this is intuitively the correct threshold for the Gaussian model. One explanation is as follows. Suppose that θ is identically equal to zero, *i.e.*, $\theta_i = 0$ for all *is*. In the language of signal estimation, this assumption states that there is no signal and that y is just white noise, $y \sim N(0, I_n)$. Then one would like to declare that there is no signal, *i.e.*, we would like to have an estimator obeying $\hat{\theta}_i = 0$ for all *is* with large probability. In the language of tests of hypotheses, we would like to accept the null hypothesis (which postulates that there is no signal) with large probability whenever the null is true. From this standpoint, one should select a threshold λ so that

$$P(\max_i |z_i| > \lambda) \leq \alpha, \qquad z_i \text{ i.i.d. } N(0,1),$$

where α is a tolerance set in advance. In other words, λ should be a quantile of the distribution of the maximum absolute value of n i.i.d. standard normal random variables. It is well known (Williams 1991), however, that

$$\lim_{n \to \infty} \frac{\max_{1 \leq i \leq n} |z_i|}{\sqrt{2 \log n}} = 1 \quad \text{almost surely,}$$

which justifies the choice of threshold in an asymptotic sense.

This can be made a little more quantitative. In fact, it is possible to show that

$$\lim_{n \to \infty} \mathbb{P}(\max_{1 \leq i \leq n} |z_i| > \sqrt{2 \log n}) = 0,$$

which shows that asymptotically $\mathbb{P}(\hat\theta = 0) \to 1$ as $n \to \infty$ whenever $\theta = 0$. Introduce the indicator variables

$$I_k(\lambda) = \begin{cases} 1, & |z_k| \geq \lambda, \\ 0, & |z_k| < \lambda. \end{cases}$$

Then

$$\mathbb{P}(\max_k |z_k| > \lambda) \leq \sum_k \mathbb{E}[I_k(\lambda)] = n \cdot \mathbb{P}(|z_1| > \lambda) \leq 2n \frac{\phi(\lambda)}{\lambda},$$

which gives

$$\mathbb{P}(\max_k |z_k| > \sqrt{2 \log n}) \leq \frac{1}{\sqrt{\pi \cdot \log n}},$$

and the right-hand side tends to zero as n tends to infinity. Conversely, for a fixed threshold λ, the expected number of observations above λ in absolute value obeys

$$\sum_k \mathbb{E}[I_k(\lambda)] = n \cdot \mathbb{E}[I_1(\lambda)] = n \cdot \Phi([\lambda, \infty)) \geq 2n \cdot \frac{\phi(\lambda)}{\lambda} \cdot \left(1 - \frac{1}{\lambda^2}\right).$$

This shows that for λ slightly smaller than $\sqrt{2 \log n}$, *i.e.*, $\lambda = (1-\delta) \cdot \sqrt{2 \log n}$ for some $\delta > 0$, the number of expected white noise coordinates above threshold tends to infinity as n increases.

Having said all this, one still needs to keep in mind that the $\sqrt{2 \log n}$ threshold is driven by asymptotic considerations. In practice, this choice tends to be a little too conservative, in the sense that its bias has a tendency to be a little too large. That is, many coordinates in which the value of θ_k is potentially large are set to zero. In statistical terms, the burden of proof to be deemed 'estimable' is perhaps not as reasonable as one would want. We shall later discuss more flexible and adaptive choices of threshold.

5.6. Example: estimating a very sparse vector

Thresholding is very effective for estimating sparse vectors $\theta \in \mathbb{R}^n$, *i.e.*, vectors which only have a few significant coordinates with unknown *a priori* locations. We illustrate this with a simple toy example. We observe

$$y_k = \theta_k + z_k, \quad z_k \text{ i.i.d. } N(0,1), \quad k = 1, \ldots, n,$$

and suppose that all the coefficients are zero except for two spikes, each of size $\mu = \sqrt{n/2}$. (We have adjusted the heights of the spikes so that $\|\theta\|^2 = n = \mathbb{E}\|z\|^2$, so that the signal to noise ratio is one.) The James–Stein estimate is highly ineffective in this setting since the risk of the ideal shrinkage estimator $\hat{\theta}^* = c^* y$ studied in Section 4 obeys

$$\mathbb{E}\|\theta - \theta^*\|^2 \geq n/2. \tag{5.9}$$

Note that the risk of the MLE is n.

In contrast, consider the risk of a hard-thresholding rule with $\lambda = \sqrt{2 \log n}$.

(1) The two observations corresponding to the spikes pass the threshold with overwhelming probability; for each coordinate, the risk is thus about equal to the variance which is one. Formally, for any such coordinate, the risk is equal to

$$\mu^2 \mathbb{E}1\{|Z + \mu| < \lambda\} + \mathbb{E}[Z^2 1\{|Z + \mu| > \lambda\}] \leq \mu^2 \mathbb{E}1\{|Z + \mu| < \lambda\} + 1,$$

where Z is a standard normal random variable. Now, because $\mu = \sqrt{n/2}$ and $\mathbb{E}1\{|Z + \mu| < \lambda\}$ is ridiculously small, *i.e.*, exponentially decaying in n, the risk is about 1.

(2) In all other coordinates, the estimator sets all the data to zero except for a possibly minuscule fraction of noise realizations exceeding the threshold. For each such coordinate, the risk obeys

$$\mathbb{E}[Z^2 1\{|Z| > \lambda\}] \leq 2(\lambda + \lambda^{-1})\phi(\lambda) = \frac{2}{\sqrt{\pi}} \cdot \frac{\sqrt{\log n}}{n}.$$

In conclusion, the risk of the hard-thresholding rule is about

$$\mathbb{E}\|\hat{\theta} - \theta\|^2 \lesssim 2 + (n-2)\frac{1.13\sqrt{\log n}}{n} \approx 2 + 1.13\sqrt{\log n},$$

which is far better than (5.9).

More generally, the oracle inequality guarantees that if the mean vector θ is sparse in the sense that it has S nonzero and 'significant coordinates', then the mean-squared error of the thresholding rule obeys

$$\mathbb{E}\|\hat{\theta} - \theta\|^2 \leq (2\log n + 1) \cdot (S + 1),$$

which, ignoring the log-factor, is the MSE one would obtain if one had an oracle supplying perfect information about the location of those significant coordinates. In conclusion, thresholding is very effective when the mean vector is sparse – when there is a comparably small number of large coefficients at unpredictable locations so that one cannot say *a priori* where the 'significant coefficients' will be.

6. Interactions with modern harmonic analysis

We have seen that thresholding comes close to the ideal risk (5.1) so that one can think of the ideal risk as a proxy for the performance of thresholding estimators in the white noise model.

6.1. Interpretation of the ideal risk

We now give an interpretation of the ideal risk which links statistical estimation to other contemporary topics. We rearrange the coefficient sequence $(\theta_1, \ldots, \theta_n)$ in decreasing order of magnitude $|\theta|_{(1)} \geq |\theta|_{(2)} \geq \cdots \geq |\theta|_{(n)}$ and let $N(\varepsilon)$ be the number of those coefficients whose absolute value exceeds the noise level ε:

$$N(\varepsilon) = \#\{k : |\theta_k| \geq \varepsilon\}.$$

With this notation, one can express the ideal risk as

$$\sum_k \min(\theta_k^2, \varepsilon^2) = N(\varepsilon) \cdot \varepsilon^2 + \sum_{k > N(\varepsilon)} |\theta|_{(k)}^2$$

$$= N(\varepsilon) \cdot \varepsilon^2 + e_{N(\varepsilon)}^2(\theta),$$

where for a fixed number B, $e_B^2(\theta)$ is the approximation obtained by keeping the B largest coefficients of θ:

$$e_B(\theta)^2 = \|\theta - \theta_B\|^2;$$

θ_B is the truncated vector equal to the B-largest value of θ and zero otherwise. In other words, the proxy for the risk is simply equal to the number of terms above the noise level times the squared noise level plus the approximation error.

The interpretation is now self-evident. Suppose we are interested in estimating an object f and that θ is the coefficient sequence of f in an orthobasis $\mathcal{B}$. Then the mean-squared error of the thresholding estimator in this basis is small if and if the signal f is *compressible* in this basis. That is, if and only if it is possible to obtain an accurate approximation of the signal f with a superposition of just a few selected elements from the basis $\mathcal{B}$. This links nonparametric estimation with nonlinear approximation theory, a subject concerned with methods for finding good approximations to various classes of functions.

It is also interesting to compare the ideal risk with the risk of a *linear* projection

$$\hat{\theta}_k^L = \begin{cases} y_k, & k \in \mathcal{M}, \\ 0, & \text{otherwise}, \end{cases}$$

where the set $\mathcal{M}$ would be set in advance (for example, a set corresponding to low-frequency waveforms). The MSE of this projection obeys

$$\mathbb{E}\|\theta^L - \theta\| \leq \#\mathcal{M}\varepsilon^2 + \sum_{k \notin \mathcal{M}} |\theta_k|^2,$$

where the second term of the right-hand side is of course the linear approximation error. The performance of linear projection procedure depends on the precision of linear approximation, while that of thresholding depends on that of nonlinear approximation. Because nonlinear approximation is in general much more precise than linear approximation, thresholding rules are usually far more accurate than the linear estimation strategies we discussed earlier.

There is also a connection to the problem of data compression in information theory. Consider encoding a function $f \in \mathbb{R}^n$ (a digital signal or a digital image) by the method of wavelet transform coding. First, one quantizes its wavelet coefficients $\theta_k = \langle f, \psi_k \rangle$ into integers n_k using a uniform quantum q: for example, one rounds up the coefficients to the nearest multiple of $2q$. One encodes the positions and values of the nonzero coefficients as bit strings by standard devices (run-length coding and so forth). Later, an approximate reconstruction of f can be obtained from $f^q = 2q \sum_k n_k \psi_k$. Here we retain the index q to remind us that the quantization stepsize q controls the behaviour of the algorithm. This coding method has distortion $\delta(q)$ obeying

$$\delta(q) \leq N(q)q^2 + \sum_{k>N(q)} |\theta|_{(k)}^2 = N(q) \cdot q^2 + e_{N(q)}^2(\theta), \qquad (6.1)$$

and is the ideal risk with the quantum playing the role of the noise level.

6.2. Sparsity

From a certain viewpoint, statistical estimation, nonlinear approximation, and data compression are closely related. For example, the quality of estimation by thresholding rules depends on the sparsity of the coefficient sequence $(\theta_k)_{k \geq 1}$. One measure of sparsity is the Marcinkiewicz weak-ℓ_p norm defined by

$$\|\theta\|_{w\ell_p} := \sup_{k \geq 1} k^{1/p} |\theta|_{(k)}. \tag{6.2}$$

(In all rigour, $\| \cdot \|_{w\ell_p}$ is only a quasi-norm in the sense that it does not obey the triangle inequality, but only $\|\theta^0 + \theta^1\|_{w\ell_p} \leq c_p \cdot (\|\theta^0\|_{w\ell_p} + \|\theta^1\|_{w\ell_p})$ where c_p is a constant which can be calculated explicitly.) Suppose that $\|\theta\|_{w\ell_p} < \infty$, then the reordered entries of the possibly infinite sequence $(\theta_k)_{k \geq 1}$ decay at least as fast as $k^{-1/p}$; the smaller p, the faster the decay. We will be interested in bounded sequences in the weak-ℓ_p norm

$$w\ell_p(R) = \{(\theta_k) : |\theta|_{(k)} \leq R \cdot k^{-1/p}, \quad \text{for all } k \geq 1\},$$

which are those sequences that exhibit a special power law decay. Note that weak-ℓ_p balls are slightly larger than corresponding ℓ_p balls

$$\ell_p(R) \subset w\ell_p(R), \qquad \ell_p(R) := \left\{ (\theta_k), \sum_k |\theta_k|^p \leq R^p \right\}.$$

Weak-ℓ_p norms are useful because the decay of the ideal risk, as $\varepsilon \to 0$, or of the approximation error $e_B(\theta)$, as $B \to \infty$, are simply deduced from membership of $w\ell_p(R)$. We follow Donoho (1993), and introduce norms which measure the precision of nonlinear approximation and the size of the ideal risk. To measure the asymptotics of approximation/compression, define the quasi-norm

$$\|\theta\|_{c,m} = \sup_{k \geq 1} k^m \cdot e_k(\theta),$$

which says that $\|\theta\|_{c,m}$ is finite if and only if the approximation error $e_k(\theta)$ obeys $e_k(\theta) = O(k^{-m})$. In a similar fashion, we introduce a quasi-norm to measure the scaling of the ideal risk

$$\|\theta\|_{e,r} = \sup_{\varepsilon > 0} \left(\varepsilon^{-2r} \cdot \sum_k \min(\theta_k^2, \varepsilon^2) \right)^{1/2},$$

which says that $\|\theta\|_{e,r}$ is finite if and only the ideal risk is $O(\varepsilon^{2r})$.

Lemma 6.1. (Donoho 1993) Let $p > 0$ and set $m = 1/p - 1/2$ and $r = \frac{2m}{2m+1}$. Then all these quasi-norms are equivalent: there exist positive finite constants $c_i(p)$ such that

$$c_0(p)\|\theta\|_{c,m} \leq \|\theta\|_{w\ell_p} \leq c_1(p)\|\theta\|_{c,m},$$
$$c_2(p)\|\theta\|_{e,r} \leq \|\theta\|_{w\ell_p} \leq c_3(p)\|\theta\|_{e,r}.$$

The assertions that $|\theta_{(k)}| = O(k^{-1/p})$, or $e_k(\theta) = O(k^{-m})$, or the ideal risk is $O(\varepsilon^{2r})$ are, therefore, all roughly equivalent. Sparsity implies good compressibility, which in turn implies good estimation.

6.3. Minimax estimation of weak-ℓ_p balls

Consider the infinite Gaussian model (4.4) and suppose $\theta \in \Theta \subset w\ell_p(R)$. Lemma 6.1 shows that the ideal risk obeys

$$\sum_k \min(\theta_k^2, \epsilon^2) = O((\epsilon^2)^{\frac{2m}{2m+1}}), \quad 1/p =: m + 1/2.$$

If one further makes an extra assumption on Θ, which roughly says that the large coefficients of $\theta \in \Theta$ do not occur at infinity, thresholding achieves the ideal risk up to a multiplicative logarithmic factor scaling like $O(\log \varepsilon)$. For example, assume that

$$\sum_{k>n_\varepsilon} |\theta_k|^2 = O(\varepsilon^{2r}), \tag{6.3}$$

where n_ε grows at most polynomially in ε. Then set

$$\hat{\theta}_k = \begin{cases} \eta(y_k), & k \le n_\varepsilon, \\ 0, & k \ge n_\varepsilon, \end{cases}$$

where η is a thresholding rule at $\lambda = \varepsilon \cdot \sqrt{2 \log n_\varepsilon}$; we threshold the coefficients in the zone $k \in [1, n_\varepsilon]$ and throw out the others. Then the oracle inequality (5.4) together with (6.3) give

$$\mathbb{E}\|\hat{\theta} - \theta\|^2 \le O(\log \varepsilon) \cdot (\epsilon^2)^{\frac{2m}{2m+1}}. \tag{6.4}$$

To develop lower bounds, we use a standard argument, which consists in embedding large hypercubes or hyper-rectangles in Θ. Suppose that

$$\ell_{p,+}(R) \subset \Theta,$$

where this means that Θ contains n-dimensional hyper-rectangles of the form $[0, Rn^{-1/p}]^n$ for arbitrary large n. Then the minimax risk obeys

$$\inf_{\hat{\theta}} \sup_{\Theta} \mathbb{E}\|\hat{\theta} - \theta\|^2 \ge \inf_{\hat{\theta}} \sup_{\ell_{p,+}(R)} \mathbb{E}\|\hat{\theta} - \theta\|^2,$$

and we will show that the minimax risk over the hyper-rectangle is bounded below by

$$\inf_{\hat{\theta}} \sup_{\ell_{p,+}(R)} \mathbb{E}\|\hat{\theta} - \theta\|^2 \ge c \cdot R^p \cdot (\epsilon^2)^{\frac{2m}{2m+1}}, \tag{6.5}$$

for some positive constant $c > 0$.

To establish (6.5), we choose a prior π which is supported on the vertices of the hyper-rectangle

$$\mathcal{H} := \prod_k [0, \tau_k] \subset \Theta,$$

and defined by

$$\theta_k = \begin{cases} 0, & \text{with probability } 1/2, \\ \tau_k, & \text{with probability } 1/2, \end{cases}$$

with independent coordinates so that informally $\pi(\theta) = \prod_k \pi(\theta_k)$. Since the coordinates are independent, any given coordinate does not give any information about any other and, therefore, good procedures treat each coordinate individually. In fact, we have already seen that Bayes' rule is indeed given by

$$\hat{\theta}_{\pi,k} = \mathbb{E}(\theta_k \mid y_k).$$

Suppose that the rectangle is tuned so that the sidelength is about equal to the noise level, *i.e.*, we pick n_ε as the largest integer obeying

$$R\,n_\varepsilon^{-1/p} \leq \varepsilon,$$

so that $n_\varepsilon \approx R^p \varepsilon^{-p}$. It follows from the choice of parameters that $\theta_k = 0$ with probability $1/2$ and $\theta_k \approx \varepsilon$ with probability $1/2$. Assume for simplicity that $\theta_k = \varepsilon$ with probability $1/2$. A simple rescaling argument shows that

$$\mathbb{E}(\hat{\theta}_{\pi,k} - \theta_k)^2 = B \cdot \varepsilon^2,$$

where B is the Bayes risk of estimating $\theta_k \in \{0, 1\}$ from $y_k \sim N(\theta, 1)$ with a prior which puts equal probability on both outcomes. Therefore, with this choice of prior on the hyper-rectangle, the Bayes risk obeys

$$B(\pi) \geq B \cdot n_\varepsilon \cdot \varepsilon^2 \approx B \cdot R^p \cdot \varepsilon^{2-p}$$

$$= B \cdot R^p \cdot (\varepsilon^2)^{\frac{2m}{2m+1}},$$

as claimed.

In closing, we have thus established that the minimax risk of weak-ℓ_p balls with the tail property (6.3) is at most within a logarithmic factor of the ideal risk, and that thresholding rules are nearly minimax since they are also within a logarithmic factor of the ideal risk.

6.4. Statistical estimation and harmonic analysis

The consequence of these results is that the problem of finding efficient representations becomes central now that the benefits of sparsity are well understood. The goal is then (1) to identify problems and object classes of scientific interest, and (2) to find efficient representations (orthobases) for those classes. Once such orthobases are constructed, one simply transforms the data into those bases, applies thresholding, and inverts the transformation to separate signal from noise. The best basis to use is of course that in which the objects considered have the sparsest representation. Additionally, one might be interested in representations with fast algorithms for computational efficiency. These are the areas of preoccupation of modern harmonic analysis and this is the reason why, over the last decade or so, there has been, and still is, significant interaction between these two communities.

One such important development is that the program outlined above has been perfectly executed when the functional classes under study belong either to the L_2-Sobolev scale, the L_p-Sobolev scale, or the Besov and Triebel–Lizorkin scales. All these spaces admit *unconditional bases* which are especially well adapted to the estimation problem.

6.5. Optimality of unconditional bases

Assume we are given a function space with a norm $\|f\|_{\mathcal{F}}$. Then an orthonormal basis $(\phi_k)_k$ is said to be *unconditional* for the normed space $\mathcal{F}$ if, for all choices of signs,

$$\left\|\sum \pm_k \theta_k(f)\,\varphi_k\right\|_{\mathcal{F}} \le C \cdot \|f\|_{\mathcal{F}},$$

where $(\theta_k(f))$ are the coefficients of f in the basis (ϕ_k). This says that arbitrary changes of signs in the expansion do not change the norm by much. Another way to put it is that there is an equivalent norm $\|\theta\|_{\mathbf{f}}$ in the sequence space

$$\|f\|_{\mathcal{F}} \sim \|\theta(f)\|_{\mathbf{f}}$$

obeying

$$\|(\pm_i \theta_i)\|_{\mathbf{f}} = \|\theta\|_{\mathbf{f}}$$

for all choices of signs.

Define Θ as the image of the unit ball in the sequence space

$$\Theta = \{\theta(f) : \|f\|_{\mathcal{F}} \le 1\},$$

and its critical exponent

$$p^*(\Theta) := \inf\{p : \ \Theta \subset w\ell_p\}.$$

Then, for any orthogonal transform U, Donoho (1993) shows that

$$p^*(U\Theta) \ge p^*(\Theta). \tag{6.6}$$

For a fixed U, one should think of $U\Theta$ as the body of coefficients of the unit ball in another basis. With this in mind, the interpretation is that, among all orthobases, the unconditional basis is that which provides the sparsest coefficient sequence. As a consequence, if there is an unconditional basis, this is the best orthonormal basis to use for nonlinear approximation and for diagonal estimation, in the sense that it provides optimal rates of approximation/estimation.

Fortunately, harmonic analysts have constructed unconditional bases for some important cases of function spaces. Some notable examples are as follows (Meyer 1992).

- Fourier bases are unconditional bases for L_2-Sobolev spaces in any dimension.

- Wavelet bases are unconditional bases for L_p-Sobolev spaces in any dimension.

- Wavelet bases are unconditional bases for Besov and Triebel spaces in any dimension. These spaces depend on 3 parameters (m, p, q) and are extensions of L_p-Sobolev spaces which depend on the pair (m, p): see Triebel (1992) for a definition.

6.6. The wavelet shrinkage

Suppose we wish to recover objects taken from a Besov or a Triebel body from the data

$$Y(dt) = f(t)\, dt + \varepsilon W(dt),$$

and seek an estimator $\hat{f}$ which nearly achieves the minimax risk. Then the answer is simply given by the celebrated wavelet shrinkage algorithm of Donoho. We take a nice wavelet basis $\psi_{j,k}(t)$, where $j \geq j_0$ indexes the scale of the wavelet and $k = 0, 1, \ldots, 2^j - 1$ indexes the location of the wavelet, go into the wavelet domain, and estimate the coefficients of f in the wavelet basis via

$$\hat{\theta}_{j,k}(y) = \begin{cases} y_{j,k}, & j = j_0, \\ \eta(y_{j,k}), & j_0 < j < j_\varepsilon, \\ 0, & j \geq j_\varepsilon; \end{cases} \tag{6.7}$$

in the above equation, the $y_{j,k}$s are the noisy coefficients, and η is a hard- or soft-thresholding rule at the level $\lambda = \varepsilon \cdot \sqrt{2 \log n_\varepsilon}$, where n_ε is the number of coefficients to which the scalar nonlinearity applies. For example, one can set j_ε to be the nearest integer to $\log_2(1/\varepsilon^2)$ so that $n_\varepsilon \approx 1/\varepsilon^2$. Inverting the wavelet transforms gives the estimate

$$\hat{f}(t) = \sum_{j,k} \hat{\theta}_{j,k} \psi_{j,k}(t). \tag{6.8}$$

This estimator has a simple structure since we just take the data in the wavelet domain and throw out the small coefficients.

As an example, suppose we are interested in the space of two-dimensional functions on $[0,1]^2$ of bounded variation,

$$\mathcal{F} := \{f : \|f\|_{\mathrm{BV}} \leq 1\}.$$

We recall that the bounded variation norm is given by $\|f\|_{\mathrm{BV}} = \int |df|$. Technically speaking, the space of functions of bounded variations does not admit an unconditional basis, although it is tightly bracketed between two Besov spaces with wavelet orthobases as unconditional bases. Letting $\Theta = \{\theta(f),\ f \in \mathcal{F}\}$ be the coefficient sequence in a sufficiently nice wavelet basis, it is possible to use embeddings of Besov spaces to show that

$$\ell_{1,+}(R) \subset \Theta,$$

for some positive $R > 0$. As we have seen earlier, this immediately gives

$$\inf_{\hat{f}} \sup_{\mathcal{F}} \mathrm{MSE}(f, \hat{f}) \geq c \cdot \varepsilon.$$

The minimax risk of two-dimensional functions with controlled bounded variations goes to zero as least as slowly as ε. In the other direction, a result of Cohen, DeVore, Petrushev and Xu (1999) shows that the wavelet sequence of a function with bounded variations belong to the weak-ℓ_1 ball, which gives that the ideal risk in our wavelet basis obeys

$$\mathbb{E}\|\theta^I - \theta\|^2 \leq C \cdot \varepsilon.$$

Since the wavelet shrinkage estimate $\hat{f}$ (6.7)–(6.8) in a 2-dimensional basis comes within a logarithmic factor of the ideal risk, we have

$$\sup_{\mathcal{F}} \mathbb{E}\|f - \hat{f}\|^2 = O(\log \epsilon^{-1}) \cdot \inf_{\hat{f}} \sup_{\mathcal{F}} \mathrm{MSE}(f, \hat{f})$$

and it is, therefore, asymptotically near-optimal.

6.7. Adaptive minimaxity

The wavelet shrinkage algorithm does not really depend upon the parameters of the functional class one wishes to estimate, which in practice are not known. To guarantee near-optimality, we simply need to work with a basis which is unconditional for the functional class and correctly set the thresholding zone. Seen a little bit differently, suppose first that we settle on a nice wavelet basis. Our basis may not be an unconditional basis for *all* L_p-Sobolev spaces or *all* Besov spaces, but it will be an unconditional basis for many of them, *e.g.*, for all L_p-Sobolev space with $m \leq m_1$ and $p \geq 1$. (For the specialist, the regularity of the wavelet limits the smoothness range over which the fixed wavelet basis is unconditional.) Second, suppose that we ignore small-scale coefficients, *e.g.*, exceeding a fixed scale $j_\varepsilon = \log_2(1/\varepsilon^2)$ which only depends upon the noise level. Then Donoho, Johnstone, Kerkyacharian and Picard (1995) show that the wavelet shrinkage nearly achieves the asymptotic minimax risk for each value of the parameter $m \in [m_0, m_1]$, p, and $R > 0$ (R is the radius of the ball). This is another example of adaption by an oracle inequality.

This universal aspect of wavelet shrinkage should not be understated. The *same* algorithm is near-optimal simultaneously over a wide range of functional classes and the performance automatically adapts to that one would expect if one knew the functional class in advance. The wavelet shrinkage may not be an exact solution to a tightly specified minimax problem but it is an approximate solution for many interesting problems.

6.8. Challenges and limitations

In summary, we have seen that efficient representations lead to efficient estimations, and that certain representations emerge as optimal. In addition, the same representation may very well solve many estimation problems (adaptivity). The challenge is, therefore, to find optimal representations for models of scientific interest. For those models, unconditional bases are, however, unlikely ...

7. Empirical model selection

We have just learned that thresholding in an unconditional basis is statistically near-optimal. Arguably, such results are very satisfying except for the fact that, more often than not, unconditional bases are simply not available. For example, a commonly discussed and interesting model of images without an unconditional is the class of functions $f(x_1, x_2) \in L_2([0, 1]^2)$, which are twice differentiable away from edges with bounded curvature. To say this slightly differently, our class is composed of objects that are discontinuous along smooth curves, *i.e.*, edges, but otherwise smooth so that one can think about such objects as cartoon-like images. This class and many others do not admit unconditional bases and, therefore, one needs to extend the tools for adaptive estimation to deal with these more common situations. This section has two goals: (1) to develop more flexible estimation strategies which go beyond coefficient estimation in a single basis, and (2) to show that it is possible to deal with classes other than the traditional smoothness classes.

7.1. Estimation with general dictionaries

Instead of being sparse in an orthobasis, a signal $f(t)$ might be sparse in a general dictionary $\mathcal{D}$ of waveforms denoted by $\mathcal{D} = (\varphi_i(t))_{i \in I}$, where I is a finite or countable set. The elements $\varphi_i(t)$ of $\mathcal{D}$ may not be orthogonal or even linearly independent. Given such a dictionary, we will assume that one can write $f(t)$ as the linear combination

$$f(t) = \sum_i \theta_i \varphi_i(t),$$

where this expansion is not unique in the case where the dictionary $\mathcal{D}$ is overcomplete (meaning that the φ_is are linearly dependent). As before, we wish to recover an object from the sampled data model (2.8) or from the continuous white noise model (2.18), and seek an estimator of the form

$$\hat{f}(t) = \sum_i \hat{\theta}_i \varphi_i(t). \tag{7.1}$$

This problem is central in statistics since this is none other than the classical multivariate regression problem, which we discuss next.

7.2. Model selection

To simplify matters, suppose that we have a finite problem and let $\Phi \in \mathbb{R}^{n \times p}$ denote the matrix whose columns are the individual waveforms $\varphi_i(t)$, $t = 1, \ldots, n$, so that the sampled model assumes the form

$$y = \Phi \theta + z,$$

where y is an n-dimensional vector of observations, and $z \sim N(0, \sigma^2 I_n)$ is white noise. Note that when the dictionary is overcomplete, one has $p > n$. We are interested in estimating the object $f = \Phi \theta$ and measure performance with the MSE

$$\mathbb{E}\|\Phi\theta - \Phi\hat{\theta}\|^2 = E\|f - \hat{f}\|^2,$$

where $\hat{f} = \Phi\hat{\theta}$ is our estimate.

We turn our attention to ideas which generalize ideal projection rules. Suppose we are given a subset $\mathcal{M} \subset \{1, \ldots, p\}$ of coordinates, and denote by $V(\mathcal{M})$ the span of $\mathcal{M}$, namely,

$$V(\mathcal{M}) := \{a \in \mathbb{R}^p : a_i = 0 \quad \text{for all} \ i \notin \mathcal{M}\}.$$

We then consider the least squares estimate which is the solution to

$$\hat{\theta}[\mathcal{M}] = \operatorname{argmin}_{a \in V(\mathcal{M})} \|y - \Phi a\|^2.$$

For example, in the case where Φ is the identity matrix as in Section 5, one would have $\hat{\theta}[\mathcal{M}]_i = y_i$ for $i \in \mathcal{M}$ and $\hat{\theta}[\mathcal{M}]_i = 0$ otherwise. What is the risk of $\hat{\theta}[\mathcal{M}]$? A classical computation which we shall not reproduce here (the reader should really make sure that this is okay!) shows that the MSE obeys

$$\mathbb{E}\|\Phi\theta - \Phi\hat{\theta}[\mathcal{M}]\|^2 = \inf_{a \in V(\mathcal{M})} \|\Phi\theta - \Phi a\|^2 + \sigma^2 |\mathcal{M}|. \tag{7.2}$$

Again, this has an interpretation in terms of the classical bias variance decomposition. The first term is the squared bias one gets by using only a subset of columns of Φ to approximate the true object $f = \Phi\theta$. The second term is the variance of the estimator and is simply proportional to the size of the model $\mathcal{M}$.

7.3. Ideal model selection

Just as we selected the ideal projection or keep-or-kill estimate in Section 5, we now introduce the ideal estimator $f^I = \Phi\theta^I$ which automatically selects the best model so that

$$\mathcal{R}^I(\theta, \Phi) := \inf_{\mathcal{M}} \mathbb{E}\|\Phi\theta - \Phi\hat{\theta}[\mathcal{M}]\|^2. \tag{7.3}$$

We will refer to this as the ideal risk. Note that in the case where Φ is the identity or, by extension, any orthonormal matrix, (7.3) is equal to $\sum_i \min(\theta_i^2, \sigma^2)$, which is the risk of the ideal projection we encountered earlier: compare (7.3). In the language of model selection, one would say that we have an oracle which would select for us the best model to use, $i.e.$, the best subset of explanatory variables.

Of course, if the 'true model' $f = \Phi\theta$ has coefficients θ which are very sparse, then the ideal estimator would do very well. For example, since

$$\mathcal{R}^I(\theta, \Phi) \le \mathbb{E}\|\Phi\theta - \Phi\hat{\theta}[\mathcal{M}^*]\|^2,$$

where $\mathcal{M}^*$ is the set of indices corresponding to the nonzero entries of θ, $\mathcal{M}^* := \{i : \theta_i \ne 0\}$, we have

$$\mathcal{R}^I(\theta, \Phi) \le \sigma^2 |\mathcal{M}^*|$$

(note that the estimator $\hat{\theta}[\mathcal{M}^*])$ is unbiased). In comparison, if one uses the MLE without model selection, the risk would be equal to $n\sigma^2$ and hence be much larger. The conclusion is that when there are only a few nonzero parameters and we know which ones they are, we can achieve substantial risk savings.

This extends to situations where most coefficients are nonzero but relatively small, so that there is a small subset $\mathcal{M}^*$ of cardinality much smaller than n with small bias, for instance such that

$$\inf_{a \in V(\mathcal{M}^*)} \|\Phi a - \Phi\theta\|^2 \approx \sigma^2 |\mathcal{M}^*|.$$

Then the ideal risk is bounded by

$$\inf_{a \in V(\mathcal{M}^*)} \|\Phi a - \Phi\theta\|^2 + \sigma^2 |\mathcal{M}^*| \ll n\sigma^2.$$

In other words, even though there are many parameters to estimate, we can, in principle, ignore the bulk of these to achieve substantial risk savings.

Finally, and just as before, the size of the ideal risk (7.3) quantifies the precision of nonlinear approximation. We let f_m be the best m-term approximation of f, i.e.,

$$\|f - f_m\|^2 = \inf_{a:\ \#\{i,\, a_i \neq 0\} \leq m} \|f - \Phi a\|^2;$$

that is, it is that linear combination of at most m columns of Φ which comes closest to the object f of interest. With this notation, one can rewrite the ideal risk as

$$\inf_m \|f - f_m\|^2 + m\sigma^2, \qquad (7.4)$$

which is exactly the same trade-off between the approximation error and the number of terms in the partial expansion.

7.4. Oracles and ideal risk

We have seen that one can achieve the ideal risk (7.4) with the help of an oracle and the real issue is how close one can get without. We follow Donoho and Johnstone (1995) and introduce

$$K(\Phi) = \inf_{\hat{\theta}} \sup_{\theta \in \mathbb{R}^p} \frac{\mathbb{E}\|\Phi\theta - \Phi\hat{\theta}\|^2}{\sigma^2 + \mathcal{R}^I(\theta, \Phi)}.$$

A value of $K(\Phi)$ close to one would indicate that one could mimic an oracle, while if $K(\Phi)$ were much greater than one, then one could not.

For orthonormal matrices Φ, we argued that $K(\Phi)$ obeys

$$K(\Phi) \approx 2\log n,$$

as shown by Donoho and Johnstone (1994a) and Foster and George (1994). For general $n \times p$ matrices ($p \geq n$), and not necessarily orthonormal, Foster and George (1994) and Donoho and Johnstone (1995) show that $K(\Phi)$ obeys

$$K(\Phi) = O(\log p). \qquad (7.5)$$

We also refer to Barron and Cover (1991), Barron (1994), Birgé and Massart (1997, 2001) and Baraud (2000) for similar results. Equation (7.5) is important because it asserts that it is possible to do nearly as well as someone using an oracle.

Which estimators then mimic the oracle up to at most a logarithmic multiplicative factor? To answer this question, we take a complexity-penalized fitting approach and consider an estimator $\hat{\theta}$ which minimizes the functional

$$\|y - \Phi a\|^2 + \lambda^2 \sigma^2 \cdot \|a\|_{\ell_0}, \qquad (7.6)$$

where we recall that $\|a\|_{\ell_0} = \#\{i : a_i \neq 0\}$. In other words, our estimator $\hat{\theta}$ is the solution of the complexity-penalized residual sum of squares

$$\min_{\mathcal{M}} \|y - \Phi\hat{\theta}[\mathcal{M}]\|^2 + \lambda^2 \sigma^2 \cdot |\mathcal{M}|.$$

Note that this a valid estimator since it can, at least in principle, be computed from the data y. This is the 'canonical selection procedure', to quote Foster and George (1994), and the estimator achieves the best trade-off between the goodness of fit and the complexity of the model. Popular selection procedures such as AIC, C_p, BIC and RIC are all of this form, with different values of the parameter: $\lambda^2 = 2$ in AIC (Akaike 1974, Mallows 1973), $\lambda^2 = \log n$ in BIC (Schwarz 1978), and $\lambda^2 = 2 \log p$ in RIC (Foster and George 1994).

In an unpublished manuscript, Donoho and Johnstone (1995) proved that the performance of this empirical model selection strategy obeys the oracle inequality below. A sharper version of this inequality is published in the authoritative reference on this subject, Birgé and Massart (2001, Theorem 2).

Theorem 7.1. (Donoho and Johnstone)   Select $\lambda^2 = A \cdot (1 + \sqrt{2 \log p})^2$ where $A > 8$, and let θ be the solution to (7.6). Then

$$E \| \Phi \theta - \Phi \hat{\theta} \|^2 \leq 6 \left(1 - 8/A\right)^{-1} \cdot \lambda^2 \cdot (\sigma^2 + \mathcal{R}^I(\theta, \Phi)). \qquad (7.7)$$

The oracle inequality (7.7) is valid for all $n \times p$ matrices Φ and all θ and, therefore, empirical model selection comes within a log factor of ideal model selection.

Proof.   We follow Donoho and Johnstone (1995) and sketch a proof based on complexity functionals. Without loss of generality, we may just assume the noise level $\sigma^2 = 1$ (the general follows by rescaling).

We introduce some notation and will call $K(\tilde{\theta}; y)$ the empirical complexity functional

$$K(\tilde{\theta}; y) = \| \Phi \tilde{\theta} - y \|^2 + \lambda^2 \| \tilde{\theta} \|_{\ell_0}.$$

We make the following observations.

(1) Consider a vector θ_0, which achieves the minimum *noiseless* complexity

$$\theta_0 = \operatorname{argmin} K(\tilde{\theta}; \Phi \theta).$$

Since $\hat{\theta}$ has minimum *noisy* complexity, $\hat{\theta}$ obeys

$$K(\hat{\theta}; y) \leq K(\theta_0; y). \qquad (7.8)$$

(2) It follows from the decomposition $y = \Phi \theta + z$ that

$$\begin{aligned} K(\hat{\theta}; y) &= \| \Phi \theta - \Phi \hat{\theta} \|^2 + 2 \langle z, \Phi \theta - \Phi \hat{\theta} \rangle + \| z \|^2 + \lambda^2 \| \hat{\theta} \|_{\ell_0} \\ &= K(\hat{\theta}; \Phi \theta) + 2 \langle z, \Phi \theta - \Phi \hat{\theta} \rangle + \| z \|^2. \end{aligned}$$

(3) We may develop a similar expression for $K(\theta_0; y)$, and plugging these equalities on both sides of (7.8) gives

$$K(\hat{\theta}; \Phi \theta) \leq K(\theta_0; \Phi \theta) + 2 \langle z, \Phi \hat{\theta} - \Phi \theta_0 \rangle. \qquad (7.9)$$

Put $\hat{K} = K(\hat{\theta}; \Phi\theta)$ and $K_0 = K(\theta_0; \Phi\theta)$ for convenience. We have

$$\|\Phi\theta - \Phi\hat{\theta}\|^2 \leq \hat{K}, \tag{7.10}$$

and it will therefore suffice to develop a bound on the expected value of $\hat{K}$. Now check (7.9). If we could somehow argue that the term $2\langle z, \Phi\hat{\theta} - \Phi\theta_0\rangle$ is small compared to $\hat{K}$, e.g., at least a fraction of $\hat{K}$, then we would be done. This is precisely the strategy we will employ.

To achieve this goal, we let $X(k)$ be the random variable defined by

$$X(k) = \sup_{\theta_1,\theta_2}\{\langle z, \Phi\theta_2 - \Phi\theta_1\rangle, \|\Phi\theta_j - \Phi\theta\|^2 \leq k, \lambda^2\|\theta_j\|_{\ell_0} \leq k\}. \tag{7.11}$$

The following lemma gives a bound on the size of $X(k)$.

Lemma 7.2. Define $k_j = 2^j (1-8/A)^{-1} \max(K_0, \lambda^2)$ for each $j \geq 0$. Then the event

$$B_j = \{X(k) \leq 4k/A\} \tag{7.12}$$

has probability at least $1 - 1/(2^j)!$.

Observe that on the event B_j, one cannot have $k \leq K_0 + 2X(k)$, which automatically implies that on this event

$$\hat{K} \leq k_j.$$

This property gives a bound on the expected value of $\hat{K}$ since

$$\mathbb{E}\hat{K} \leq k_0 \mathbb{P}(\hat{K} \leq k_0) + \sum_{j\geq 1} k_j \mathbb{P}(\hat{K} \geq k_{j-1})$$

$$\leq k_0 \cdot \left(1 + \sum_{j\geq 1} 2^j \mathbb{P}(B_{j-1}^c)\right).$$

It follows from $\mathbb{P}(B_j^c) \leq 1/(2^j)!$ that $\sum_{j\geq 1} 2^j \mathbb{P}(B_{j-1}^c) \leq 5$ and, therefore,

$$\mathbb{E}\hat{K} \leq 6k_0.$$

In conclusion,

$$\mathbb{E}\hat{K} \leq 6(1 - 8/A)^{-1} \max(\lambda^2, K_0),$$

which proves the claim since K_0 is no greater than λ^2 times the ideal risk. □

We only briefly discuss Lemma 7.2. We consider k in the range $[\ell\lambda^2, (\ell + 1)\lambda^2)$ where ℓ is a fixed positive integer. Note that each feasible element for the optimization problem is a linear combination of at most $\ell = \lfloor k/\lambda^2 \rfloor$ nonzero vectors, and therefore the difference $\theta_2 - \theta_1$ is a linear combination of at most 2ℓ distinct vectors from our dictionary; we let V be the linear space of dimension at most 2ℓ spanned by those vectors and denote by P_V the orthogonal projection onto V. The Cauchy–Schwarz inequality gives

$$|\langle z, \Phi\theta_2 - \Phi\theta_1 \rangle| \leq \|P_V z\| \cdot \|\Phi\theta_2 - \Phi\theta_1\| \leq 2\sqrt{k} \cdot \|P_V z\|,$$

since $\|\Phi\theta_2 - \Phi\theta_1\| \leq 2\sqrt{k}$ by assumption. The term $\|P_V z\|^2$ is a chi-squared$\star$ random variable with 2ℓ degrees of freedom. The claim essentially follows from large deviation bounds for such chi-squares. Because of space limitations, we do not dwell on this issue.

7.5. Serious limitations

Theorem 7.1 is of theoretical importance but highly impractical. Solving (7.6) is in general NP-hard (Natarajan 1995). To the best of our knowledge, solving this problem essentially requires exhaustive searches over all subsets of columns of Φ, a procedure which is clearly combinatorial in nature and has exponential complexity since, for p of size about n, there are about 2^p such subsets. (We are of course aware that in the special case where Φ is orthogonal, the solution is simply obtained by hard-thresholding the vector $\Phi^T y$ at the level $\sqrt{\lambda}\sigma$: see Section 5.)

In other words, and quoting from Candès and Tao (2005a), 'solving the model selection problem might be possible only when p ranges in the few dozens. This is especially problematic when one considers that we now live in a data-driven era marked by ever larger datasets.'

In some sense, Theorem 7.1 is merely a theoretical gadget. However, it is a very important one, since it shows what is achievable by a real estimator. A crucial issue is whether there are computationally more efficient estimators with similar properties. In Section 8, we will discuss a new breed of complexity-penalized estimators with surprising properties.

7.6. An example: recovering edges from noisy data

Despite its computational infeasibility, Theorem 7.1 gives a precise statement about the performance of a real estimator, and Donoho and Johnstone (1995) give an example of how this might be used. We consider an image model where one tries to recover the indicator function of a smooth set (a shape, if you will)

$$f(x) = 1_B(x), \tag{7.13}$$

where we assume that the second derivative or the edge curvature ∂B is bounded by some constant R, so that one can loosely express the class of objects of interest by

$$\mathcal{F}_2(R) := \{f = 1_B : \|\partial B\|_{C^2} \leq R\}.$$

Such models, also known as *boundary fragment* models, have been studied extensively by Korostelëv and Tsybakov (1993) and others. Note that this class of images is neither convex nor orthosymmetric, and does not admit an unconditional basis.

We will suppose that the observations come from the two-dimensional model

$$Y(\mathrm{d}x) = f(x)\,\mathrm{d}x + \varepsilon W(\mathrm{d}x),$$

where W is a two-dimensional Wiener sheet. The problem is to recover the edges of the unknown object from the noisy data and there are many known results about this: see Korostelëv and Tsybakov (1993) and Donoho (1999) and references therein.

It is well known that a good dictionary to represent elements in $\mathcal{F}_2(R)$ is the triangle dictionary

$$\mathcal{D} = \{1_T : (x, y, z) \in [0, 1]^6\},$$

where T denotes the triangle T with vertices x, y, z. The dictionary $\mathcal{D}$ is not countable and, in fact, we shall consider a finite version $\mathcal{D}_\varepsilon$ of $\mathcal{D}$ where one restricts the vertices to belong to a two-dimensional lattice with vertical and horizontal spacing equal to ε^2 so that the cardinality of $\mathcal{D}_\varepsilon$ is polynomial in ε.

It is not really difficult to show that, for objects $f = 1_B$ in the class of interest, there is a superposition of triangles, *i.e.*,

$$f_m = \sum_{i=1}^{m} 1_{T_i}, \qquad 1_{T_i} \in \mathcal{D}_\varepsilon,$$

whose approximation error obeys

$$\|f - f_m\|^2 \le C \cdot m^{-2},$$

at least in the range where the approximation error dominates the quantization error, *i.e.*, $m^{-2} \le \varepsilon^2$. This merely follows from a first-order Taylor approximation and we skip the details. Now it can be shown that there is no dictionary with size growing at most polynomially in m that would yield better rates of convergence: see Donoho (2001) and Candès and Donoho (2000), for example.

The approximation error allows us to derive a bound on the ideal risk in the triangle dictionary since

$$\inf_m \left(\|f - f_m\|^2 + m\epsilon^2 \right) \le \inf_m \left(C \cdot m^{-2} + \epsilon^2 m \right).$$

Optimizing over m gives that the ideal risk obeys

$$\text{ideal risk} \le C \cdot \epsilon^{4/3}.$$

We can then invoke the oracle inequality (7.7), together with the fact that the size of the dictionary is polynomial in ε, to show that the performance of empirical triangle selection obeys

$$\mathbb{E}\|\hat{f} - f\|^2 \le O(\log 1/\varepsilon) \cdot \varepsilon^{4/3}. \tag{7.14}$$

Now the risk of the empirical triangle selection is nearly optimal since one can show – by embedding appropriate hypercubes – that any estimator must obey

$$\inf_{\hat{f}} \sup_{f \in \mathcal{F}_2(R)} \mathbb{E}\|f - \hat{f}\|^2 \ge c \cdot \varepsilon^{4/3}$$

and, therefore, (7.14) comes within a logarithmic factor of the minimax risk.

In addition, one could also get similar results for other degrees of smoothness of the edge curve. For example, suppose that the boundary is C^s with $1 \le s \le 2$. A function g is bounded in C^s with $1 \le s \le 2$ if the first derivative obeys

$$\sup_{t, t'} \frac{|g'(t) - g'(t')|}{|t - t'|^{s-1}} < \infty.$$

(One can then define the modulus of smoothness as the supremum of this ratio.) Then the risk of empirical triangle selection obeys

$$\mathbb{E}\|\hat{f} - f\|^2 \le O(\log 1/\varepsilon) \cdot \varepsilon^{2s/(s+1)}$$

while the lower bound is at least of size $c \cdot \varepsilon^{2s/(s+1)}$. (To deal with smoother edges, one would need to employ dictionaries with higher-order polycurves.)

In conclusion, we have shown that statistical near-optimality and adaptivity can hold even though there are no unconditional bases.

8. The Dantzig selector

Model selection is an especially important topic in statistics in part because of the very large number of users who are routinely fitting large linear models or designing statistical experiments. Therefore, finding computationally feasible strategies whose predictive risk comes close to that of the ideal model selection would be likely to have a large impact. This section presents some new ideas by Candès and Tao which show that this is in fact possible, at least in some special settings.

This work is concerned with a more ambitious goal than that discussed earlier. Indeed, they seek to estimate the parameter vector $\theta \in \mathbb{R}^p$ from the data

$$y = \Phi\theta + z,$$

where Φ is an $n \times p$ matrix with $p \ge n$, and $z \sim N(0, \sigma^2 I_n)$. A typical problem of this nature might be the reconstruction of an image $\theta \in \mathbb{R}^p$ with p pixels from undersampled and noisy data, e.g., from its noisy and incomplete Fourier coefficients – a problem that frequently arises in medical imaging. Now, because $p \geq n$, one might wonder how this is possible. Indeed, suppose that we are in the noiseless case in which $\sigma = 0$; then, to recover θ, one would need to solve a system of linear equations *where there are more unknowns than equations*. Elementary linear algebra tells us that this is problematic. But suppose now that θ is sparse or has entries decaying like a power law, as explained in Section 6. Then this premise radically changes the problem, making the search for solutions feasible.

8.1. The noiseless case

In fact, Candès and Tao (2005*b*) showed that in the noiseless case, one could actually recover θ *exactly* by solving a linear program

$$(P_1) \qquad \min_{\tilde{\theta} \in \mathbb{R}^p} \|\tilde{\theta}\|_{\ell_1} \quad \text{subject to} \quad \Phi\tilde{\theta} = y, \qquad (8.1)$$

provided that the matrix $\Phi \in \mathbb{R}^{n \times p}$ obeys a so-called *uniform uncertainty principle* (recall $\|\tilde{\theta}\|_{\ell_1} := \sum_i |\theta_i|$). That is, ℓ_1-minimization finds without error both the location and amplitudes – which we emphasize are *a priori* completely unknown – of the nonzero components of the vector $\theta \in \mathbb{R}^p$.

In detail, Candès and Tao (2005*b*) show that exact reconstruction occurs provided that sparse subsets of columns of the data matrix Φ are approximately orthonormal. For each $\mathcal{M} \subset \{1, \ldots, p\}$, we let $\Phi[\mathcal{M}]$ be the $n \times |\mathcal{M}|$ submatrix obtained by extracting the columns of Φ corresponding to those indices in $\mathcal{M}$; then they define the number δ_S as the smallest quantity obeying

$$(1 - \delta_S) \|c\|^2 \leq \|\Phi[\mathcal{M}]c\|^2 \leq (1 + \delta_S) \|c\|^2 \qquad (8.2)$$

for all subsets $\mathcal{M}$ with $|\mathcal{M}| \leq S$ and coefficient sequences c. Small values of δ_S indicate that every set of columns with cardinality less than S approximately behaves like an orthonormal system. There is a related quantity $\gamma_{S,S'}$, which is the smallest quantity such that

$$|\langle \Phi[\mathcal{M}]c, \Phi[\mathcal{M}']c' \rangle| \leq \gamma_{S,S'} \|c\| \|c'\| \qquad (8.3)$$

holds for all *disjoint* sets $\mathcal{M}, \mathcal{M}' \subseteq \{1, \ldots, p\}$ of cardinality less or equal to S and S', respectively. Small values of γ indicate that disjoint subsets of covariates span nearly orthogonal subspaces.

Theorem 8.1. (Candès and Tao 2005*b*) Let S be the number of entries of $\theta \in \mathbb{R}^p$ that are nonzero, and suppose that $\delta_{2S} + \gamma_{S,2S} < 1$. Then the solution $\theta^\star$ to (8.1) is exact, *i.e.*, $\theta^\star = \theta$.

This theorem is remarkable since it says that one can solve underdetermined systems of linear equations by linear programming. For instance, together with Romberg (Candès and Tao 2004, Candès, Romberg and Tao

2006), they show that one can recover exactly all kinds of sparse signals in some fixed basis from undersampled Fourier data or other types of incomplete measurements, a phenomenon now known as *compressive sampling* and with far-reaching implications. But what is more surprising is that compressive sampling extends to noisy data.

8.2. Ideal model selection

To get a sense of what might be possible, let us consider as before the least squares estimate

$$\hat{\theta}[\mathcal{M}] = \text{argmin}_{a \in V(\mathcal{M})} \|y - \Phi a\|^2.$$

Since $\hat{\theta}[\mathcal{M}]$ vanishes outside $\mathcal{M}$, we have that

$$\mathbb{E}\|\theta - \hat{\theta}[\mathcal{M}]\|^2 = \|P\theta - P\hat{\theta}[\mathcal{M}]\|^2 + \sum_{i \notin \mathcal{M}} |\theta_i|^2,$$

where P is the projection on the coordinate subset $\mathcal{M}$. We then write

$$P\theta - P\hat{\theta}[\mathcal{M}] = H\,(g + z),$$

where $H = (\Phi[\mathcal{M}]^T \Phi[\mathcal{M}])^{-1} \Phi[\mathcal{M}]^T$ and $g = \Phi\theta - \Phi P\theta$. It follows that

$$\mathbb{E}\|P\theta - P\hat{\theta}[\mathcal{M}]\|^2 = \|Hg\|^2 + \sigma^2 \text{Tr}((\Phi[\mathcal{M}]^T \Phi[\mathcal{M}])^{-1}).$$

However, since all the eigenvalues of $\Phi[\mathcal{M}]^T \Phi[\mathcal{M}]$ belong to the interval $[1 - \delta_{|\mathcal{M}|}, 1 + \delta_{|\mathcal{M}|}]$, we have

$$\mathbb{E}\|P\theta - P\hat{\theta}[\mathcal{M}]\|^2 \geq \frac{1}{1 + \delta_{|\mathcal{M}|}} \cdot |\mathcal{M}| \cdot \sigma^2.$$

For each set $\mathcal{M}$ with $|\mathcal{M}| \leq S$ and $\delta_S < 1$, we have

$$\mathbb{E}\|\theta - \hat{\theta}[\mathcal{M}]\|^2 \geq \sum_{i \in \mathcal{M}^c} \theta_i^2 + \frac{1}{2} |\mathcal{M}| \cdot \sigma^2.$$

If we then define the ideal estimator θ^I as

$$\theta^I = \text{argmin}_{\mathcal{M}} \mathbb{E}\|\theta - \hat{\theta}[\mathcal{M}]\|^2,$$

we have shown that the ideal mean-squared error is bounded below by

$$\mathbb{E}\|\theta - \theta^I\|^2 \geq \frac{1}{2} \min_{\mathcal{M}} \|\theta - \hat{\theta}[\mathcal{M}]\|^2 + |\mathcal{M}| \cdot \sigma^2$$

$$= \frac{1}{2} \sum_i \min(\theta_i^2, \sigma^2).$$

We feel that we do not need to make further comment on the right-hand side! What we would like to know is whether there is a computationally efficient estimator which can mimic the ideal risk.

8.3. The noisy case

Assume for simplicity that the columns of Φ are normalized (there are variations to handle the general case). Then the Dantzig selector estimates θ by solving the convex program

$$\text{(DS)} \qquad \min_{\tilde{\theta}\in\mathbb{R}^p} \|\tilde{\theta}\|_{\ell_1} \quad \text{subject to} \quad \sup_{1\le i\le p} |(\Phi^T r)_i| \le \lambda\cdot\sigma \qquad (8.4)$$

for some $\lambda > 0$, and where r is the vector of residuals

$$r = y - \Phi\tilde{\theta}. \qquad (8.5)$$

The solution to this optimization problem is the minimum ℓ_1 vector which is consistent with the observations. The constraints impose that the residual vector is within the noise level and does not correlate too well with the columns of Φ. We would like to mention that there exist related, yet different proposals in the literature, and most notably the lasso introduced by Tibshirani (1996).

The program (DS) is convex and can be recast as a linear program (LP)

$$\min \sum_i u_i \qquad (8.6)$$

subject to

$$-u \le \tilde{\theta} \le u \quad -\lambda\sigma\, \mathbf{1} \le \Phi^T(y - \Phi\tilde{\theta}) \le \lambda\sigma\, \mathbf{1},$$

where the optimization variables are $u, \tilde{\theta} \in \mathbb{R}^p$, and $\mathbf{1}$ is a p-dimensional vector of ones. This is nice because linear programming is a very mature field with stable and efficient solvers. As a matter of fact, the paper by Candès and Tao (2005a) reports on experiments where p is in the hundreds of thousands.

The Dantzig selector is not only computationally tractable, it is also accurate.

Theorem 8.2. (Candès and Tao 2005a)   Set $\lambda := (1+t^{-1})\sqrt{2\log p}$ in (8.4) and suppose that θ has S nonzero terms with $\delta_{2S}+\gamma_{S,2S} < 1-t$. Then

$$\mathbb{E}\|\hat{\theta} - \theta\|^2 \le O(\log p)\cdot\left(\sigma^2 + \sum_i \min(\theta_i^2, \sigma^2)\right). \qquad (8.7)$$

The slogan is thus that linear programming can mimic the oracle. It is worth mentioning that the oracle inequality (8.7) is not exactly the statement contained in Candès and Tao (2005a) where it is only shown that $\|\hat{\theta} - \theta\|^2$ is bounded by the right-hand side of (8.7) with very large probability. A minor modification of their argument, however, gives the bound on the MSE.

The assumptions are here more restrictive than in Theorem 7.1, but this is to be expected since we are looking at a more difficult problem, namely, estimating θ rather than $\Phi\theta$. For example suppose that $\delta_{2S} = 0$, which may indicate that there is a matrix $\Phi[\mathcal{M}_1 \cup \mathcal{M}_2]$ with $2S$ columns ($|\mathcal{M}_1| = S$, $|\mathcal{M}_2| = S$), which is rank-deficient. If this is the case, there is a pair of vectors $\theta_1 \in V(\mathcal{M}_1)$, $\theta_2 \in V(\mathcal{M}_2)$ with the property

$$\Phi(\theta_2 - \theta_1) = 0, \quad \Leftrightarrow \quad \Phi\theta_2 = \Phi\theta_1.$$

This is why we need $\delta_{2S} < 1$. For, otherwise, the model may not be identifiable since both θ_1 and θ_2 have at most S nonzero entries. The condition $\delta_{2S} + \gamma_{S,2S} < 1$ (or less than $1 - t$) is only slightly stronger than the identifiability condition.

There are other versions of Theorem 8.2 which only require θ to be sparse in the sense that many of its entries are small but not necessarily zero, e.g., θ may belong to a weak-ℓ_p ball for some $p > 0$: see Candès and Tao (2005a) for details. In addition, the Dantzig selector is a kind of soft-thresholding estimator and therefore has the tendency to underestimate the true value of θ. The aforementioned reference details simple versions which correct for the bias and have better practical performance.

8.4. Comparison with the combinatorial search

For sufficiently sparse vectors the near-orthogonality property (8.2) of the matrix Φ shows that

$$\|\Phi(\theta - \hat{\theta})\| \asymp \|\theta - \hat{\theta}\|$$

where $\asymp$ means that the ratio is bounded above and below. Thus, one can recast (8.7) as

$$\mathbb{E}\|\Phi\theta - \Phi\hat{\theta}\|^2 \leq O(\log p) \cdot (\sigma^2 + \mathcal{R}^I(\theta, \Phi)). \tag{8.8}$$

Like the 'combinatorial search estimator' (7.6), the Dantzig selector comes within a logarithmic factor of the ideal risk (7.3).

The catch, however, is that although the hypotheses of Theorem 8.2 are in some sense necessary to estimate θ accurately, they are probably too restrictive when one is 'only' interested in estimating $\Phi\theta$. For instance, Theorem 7.7 does not assume anything about the matrix Φ and about the sparsity of the true vector $\theta \in \mathbb{R}^p$. It is likely that the Dantzig selector would also obey (8.8) under more general conditions. As a matter of fact, we regard as extremely significant the problem of deciding whether or not there is – under mild conditions – a computationally tractable estimator mimicking the oracle.

9. Frames and libraries

Getting back to the familiar framework of thresholding, it is important to realize that thresholding can be successful even outside the specific case where one is given a single orthobasis. In this section we discuss two cases in which thresholding is highly effective even though there is no (single) orthobasis.

9.1. Tight frames

In harmonic analysis, it is generally much easier to construct a tight frame than an orthobasis. In $\mathbb{R}^n$, a tight frame is a collection of vectors (φ_i) with the property

$$\|f\|^2 = \sum_i |\langle f, \varphi_i \rangle|^2. \tag{9.1}$$

If we arrange the vectors φ_i as the columns of a matrix Φ, then this property may be expressed as

$$\|\Phi^T f\|^2 = \|f\|^2,$$

which says that Φ^T is an isometry. The isometry property provides a simple reconstruction formula from the frame coefficients $(\langle f, \varphi_i \rangle)$ since $\Phi \Phi^T = I_n$, or equivalently

$$f = \sum_i \langle f, \varphi_i \rangle \varphi_i. \tag{9.2}$$

The only difference between (9.1)–(9.2) and an orthobasis is that the elements φ_i may not be linearly independent. In particular, we may have more elements than the dimension of the space. In general, a tight frame is a collection of vectors taken from a Hilbert space obeying (9.1). For example, we have tight frames in $L_2(\mathbb{R})$, $L_2(\mathbb{R}^2)$, and so on, where the inner product is of course the usual inner product over square integrable functions.

The exact orthogonality between elements is what can make the construction of orthobases extremely challenging. In contrast, one has more flexibility in constructing tight frames and this is why this is easier. For instance, while tight Gabor frames exist, Balian and Low have shown that it is impossible to find an orthonormal equivalent with nice time-frequency localization properties (there are orthobases of local cosines but this is somewhat different): see Mallat (1999). Also, Candès and Donoho (2004) have constructed nice tight frames of *curvelets* and it is not known whether one can construct an orthonormal equivalent with nice time-frequency localization properties.

Suppose that we observe $y \sim N(f, \sigma^2 I_n)$; then we can define the empirical frame coefficients $\tilde{y} = \Phi^T y$ which obey the Gaussian model

$$\tilde{y}_i = \theta_i + \tilde{z}_i, \tag{9.3}$$

where $\tilde{z}$ is a Gaussian process with zero mean and covariance matrix

$$\mathrm{Cov}(\tilde{z}_i, \tilde{z}_j) = \sigma^2 \langle \varphi_i, \varphi_j \rangle.$$

In particular, the variance of $\tilde{z}_i$ obeys $\mathrm{Var}(z_i) = \sigma^2 \|\varphi_i\|^2$ which we denote by σ_i^2. The situation is analogous in the continuous white-noise model where the empirical coefficients are defined by $\tilde{y}_i = \int \varphi_i(t)\, Y(dt)$ giving an infinite-dimensional version of the sequence model (9.3) (the covariance is $\varepsilon^2 \langle \varphi_i, \varphi_j \rangle$). Also note that, since $\|\varphi_i\| \leq 1$, we have $\sigma_i \leq \sigma$ and

$$\sum_i \sigma_i^2 = \mathbb{E}\|\tilde{z}\|^2 = \mathbb{E}\|\Phi^T z\|^2 = \mathbb{E}\|z\|^2 = n\sigma^2.$$

One can of course apply individual thresholding in a tight frame. Suppose we are in the sampled model with n observations. We have seen in Section 5 that the risk of a thresholding rule, with threshold $\sqrt{2\log n} \cdot \sigma_i$, obeys

$$\mathbb{E}\|\theta_i - \hat{\theta}_i\|^2 \leq (2\log n + 1) \cdot (\sigma_i^2/n + \min(\theta_i^2, \sigma_i^2))$$

and therefore

$$\mathbb{E}\|\theta - \hat{\theta}\|^2 \leq (2\log n + 1) \cdot \left(\sigma^2 + \sum_i \min(\theta_i^2, \sigma_i^2)\right).$$

Returning to the original domain gives an estimator $\hat{f} = \sum_i \hat{\theta}_i \varphi_i$ obeying

$$\mathbb{E}\|f - \hat{f}\|^2 = \mathbb{E}\|\Phi\theta - \Phi\hat{\theta}\|^2 \leq \mathbb{E}\|\theta - \hat{\theta}\|^2,$$

where we have used the fact that, for any vector h, $\|\Phi h\| \leq h$. It then follows that the performance of the shrinkage estimator is bounded by

$$\mathbb{E}\|f - \hat{f}\|^2 \leq (2\log n + 1) \cdot \left(\sigma^2 + \sum_i \min(\theta_i^2, \sigma_i^2)\right). \tag{9.4}$$

The message is of course that, *if the frame coefficient sequence is sparse*, then this strategy is highly effective.

We emphasized the 'frame coefficient sequence' for a reason. There are many ways to expand a signal or a vector in a frame, and depending upon the frame, the frame decomposition may be dense while there may exist other very sparse decompositions. We give an example. Suppose that the frame is composed of two orthobases $\Phi = [\Phi_1, \Phi_2]/\sqrt{2}$ where each Φ_j is an $n \times n$ orthonormal matrix. To make things concrete, suppose Φ is the time-frequency dictionary where Φ_1 is the identity matrix and Φ_2 is the unitary discrete Fourier matrix. Now consider a signal f made out of one spike

$$f = (\mu, 0, \dots, 0),$$

where μ is some large amplitude. Then f is a multiple of a single column of Φ and the ideal risk (Section 7) is simply equal to σ^2. Now, for each i, $|\Phi_2^T f|_i = \mu/\sqrt{n}$ and, if the amplitude of the spike is large enough, then all the Fourier coefficients will exceed the noise level. Applying the thresholding estimator and using the proxy (5.7), we would not expect anything substantially better than

$$\frac{2\log n + 1}{2} \cdot (n+1) \cdot \frac{\sigma^2}{2},$$

which is horrible since there is only one parameter to estimate!

9.2. The curvelet shrinkage

Candès and Donoho recently introduced tight frames of curvelets to overcome inherent limitations of traditional multiscale representations such as wavelets (Candès and Donoho 2000, Candès and Guo 2002, Candès and Donoho 2004). Conceptually, the curvelet transform is a multiscale pyramid with many directions and positions at each length scale, and needle-shaped elements at fine scales. This pyramid is nonstandard, however, as curvelets have useful geometric features that set them apart from wavelets and the like. For instance, curvelets obey a parabolic scaling relation which says that at scale 2^{-j}, each element has an envelope which is aligned along a 'ridge' of length $2^{-j/2}$ and width 2^{-j}. It is beyond the scope of this paper to discuss this new construction and we refer to Candès and Donoho (2004) for mathematical details and to Candès, Demanet, Donoho and Ying (2005) for the description of fast and accurate digital curvelet transform algorithms.

Curvelets are interesting because they efficiently address very important problems where wavelet ideas are far from ideal. Of interest here is that curvelets provide optimally sparse representations of objects which display *curve-punctuated smoothness* – smoothness except for discontinuity along a general curve with bounded curvature. Such representations are nearly as sparse as if the object were not singular and turn out to be far more sparse than the wavelet decomposition of the object.

Quantitatively speaking, let (θ_i) denote the curvelet coefficient sequence of a C^2 function with piecewise C^2 singularities (edges). Then Candès and Donoho (2004) showed that the nth largest entry $|\theta|_{(n)}$ in the sequence obeys

$$|\theta|_{(n)} \leq C \cdot n^{-3/2}(\log n)^{3/2}, \quad \text{for all } n > 0. \tag{9.5}$$

This decay is optimal: among all possible representations of objects with singularities, this is essentially the sparsest one. That is, there is no basis, tight frame, frames and so on in which the coefficients of a function f with piecewise C^2 edges would have a faster decay.

Of course, the enhanced sparsity shows that one can recover such objects from noisy data by simple curvelet shrinkage and obtain an MSE order of magnitude better than that achieved by more traditional methods, *e.g.*, wavelet shrinkage. Omitting details having to do with the definition of the thresholding zone (Candès and Donoho 2002), one can then plug the estimate (9.5) into the oracle inequality and obtain that the risk obeys

$$\mathbb{E}\|f - \hat{f}\|^2 \leq O(\log^2 \varepsilon^{-1}) \cdot \varepsilon^{4/3}.$$

(Recall that the minimax lower bound exceeds $c \cdot \varepsilon^{4/3}$.) It goes without saying that we do not need to solve an intractable problem (like empirical triangle selection) to recover a smooth image with edges from noisy data in an optimal fashion. Instead, one can just go into the curvelet domain (by means of the fast digital curvelet transform), throw out the small coefficients and invert the transform.

9.3. Statistical estimation in a library of bases

Suppose now that we are given a library $\mathcal{L}$ of orthonormal bases

$$\mathcal{L} = \{\mathcal{B}_1, \ldots, \mathcal{B}_L\},$$

where the $\mathcal{B}_i$s are L distinct orthonormal bases. For example, the library $\mathcal{L}$ might be a concatenation of several orthonormal bases, e.g., the canonical basis (or the spike basis, as it is called in signal processing), the Fourier basis, a wavelet basis, a spline basis, a ridgelet basis (Candès and Donoho 1999) and so on. Or the library $\mathcal{L}$ might be the cosine, the wavelet (Coifman and Meyer 1991) or the ridgelet packet library (Flesia, Hel-Or, Averbuch, Candès, Coifman and Donoho 2003). We would like to emphasize that we consider libraries of orthonormal bases for simplicity but the results extend to libraries of tight frames (see Candès (2002)), so that it is possible to include the aforementioned curvelets, contourlets, and many other recent interesting constructions in computational harmonic analysis.

We wish to explore the possibility of adaptive basis estimation. Suppose that we observe a signal in white noise. Adaptive basis estimation means that we would like to select, based on the data, the best basis in which to estimate the signal; that is, the basis in which the true unknown signal is in some sense the sparsest possible.

We let $y_i[\mathcal{B}]$ be the coordinates of the observations in the basis $\mathcal{B}$ and, likewise, we let $\theta_i[\mathcal{B}]$ and $z_i[\mathcal{B}]$ be the coordinates of the signal f and of the error vector in $\mathcal{B}$. In the basis $\mathcal{B}$, our statistical model is of the form

$$y_i[\mathcal{B}] = \theta_i[\mathcal{B}] + z_i[\mathcal{B}],$$

and the ideal risk in that basis $\mathcal{B}$ is

$$\mathcal{R}^I(\theta, \mathcal{B}) = \sum_i \min(|\theta_i[\mathcal{B}]|^2, \sigma^2).$$

We now introduce the ideal risk in the library as the minimum over all bases in the library

$$\mathcal{R}^I(\theta, \mathcal{L}) = \min_{\mathcal{B} \in \mathcal{L}} \mathcal{R}^I(\theta, \mathcal{B}). \tag{9.6}$$

This ideal risk is achievable with the aid of (1) a *basis* oracle which selects the best basis and (2) a *coordinate* oracle which tells us which coordinates in that basis are worth estimating.

The issue is then whether one can select a basis in a near-ideal fashion from the data alone. In order to do this, Donoho and Johnstone (1994b) introduce the entropy functional

$$\mathcal{E}_\lambda(y, \mathcal{B}) := \sum_i \min(|y_i[\mathcal{B}]|^2, \lambda^2 \sigma^2),$$

where λ is a parameter. This quantity is not surprising since this is none other than the empirical complexity functional (7.6) in the basis $\mathcal{B}$

$$\mathcal{E}_\lambda(y, \mathcal{B}) := \min_a \|y[\mathcal{B}] - a\|^2 + \lambda^2 \sigma^2 \|a\|_{\ell_0}.$$

It then seems sensible to choose the basis for estimation in which $\mathcal{E}_\lambda(y, \mathcal{B})$ is smallest. The estimation strategy consists of two simple stages.

(1) We select $\hat{\mathcal{B}}$ as the best orthobasis $\hat{\mathcal{B}}$ according to the entropy

$$\hat{\mathcal{B}} := \mathrm{argmin}\mathcal{E}_\lambda(y, \mathcal{B}).$$

(2) We then apply hard-thresholding (with level $\lambda\sigma$) in that basis so that

$$\hat{\theta}_i[\hat{\mathcal{B}}] = \begin{cases} y_i[\hat{\mathcal{B}}], & |y_i| > \lambda\sigma, \\ 0, & \text{otherwise.} \end{cases}$$

The result is that if λ is correctly tuned, empirical basis selection nearly achieves the performance of the ideal estimator.

Theorem 9.1. (Donoho and Johnstone 1994b) Let M_n be the number of distinct vectors in the library and set $\lambda^2 = A(1 + \sqrt{2 \log M_n})^2$ for some $A > 8$. Then

$$\mathbb{E}\|\hat{\theta}[\hat{\mathcal{B}}] - \theta[\mathcal{B}]\|^2 \leq 6(1 - 8/A)^{-1} \cdot \lambda^2 \cdot (\sigma^2 + \mathcal{R}^I(\theta, \mathcal{L})). \tag{9.7}$$

If there is an efficient basis for estimation, then empirical basis selection will find it and the error of estimation will be small.

The reader is right to suspect that the proof of Theorem 9.1 is based on minimum complexity functionals and is nearly identical to that of Theorem 7.1 and we will, therefore, not reproduce it.

An interesting example concerns denoising in a packet library such as cosine or wavelet packets. In a cosine packet library, for instance, there are about $n \log_2 n$ distinct elements where n is the number of samples, while the number of orthobases is equal to the number of dyadic trees of depth about $\log_2 n$, which is exponential in n. This looks daunting as one would naively think that one would need to evaluate exponentially many entropy functionals in order to find the best basis. Fortunately, because of the additivity property of the entropy functional and of the tree structure of the library of bases, there is a way to invoke dynamic programming to select the best basis. In particular, Coifman and Wickerhauser (1992) show that one can compute $\hat{\mathcal{B}}$ in $O(n)$. Since all the noisy coefficients in the library (there are about $n \log_2 n$ of them) can be computed in $O(n \log^2 n)$, the empirical best basis estimator can be rapidly computed.

10. Further topics

In this last section, we discuss a selection of other important problems and topics which we hope will give an idea of how broad the field really is.

10.1. From theory to practice

We have not talked much about the practical performance of shrinkage ideas in signal and image processing. Wavelet shrinkage ideas have indeed been deployed with great success in many applications, and are nowadays routinely used by researchers and engineers. We mention here a few topics which enhance the estimation.

Thresholding rules in a wavelet basis are known to produce some artifacts, some of which may be removed by applying a translation-invariant type of shrinkage. For example, a frequently discussed approach consists of applying cycle spinning. Cycle spinning is a kind of translation-invariant thresholding rule: this technique computes several individual reconstructions by applying shifts to the noisy data and averages them out, after applying the reverse shifts, of course. Another popular approach consists in applying thresholding in a redundant wavelet representation, such as the undecimated wavelet transform; see the 'à trous' algorithm in Starck, Murtagh and Bijaoui (1998). The basic idea underlying these methods is that an average of similar-looking estimators produces visually more pleasing results than any of the individual estimators taken individually.

Researchers have also developed the idea of 'block thresholding', which originates in Efroïmovich (1985). Instead of treating each coefficient individually, the idea is that the statistical properties of images may be used to group coefficients together to better inform the decision. For example, if a wavelet coefficient is large, it may indicate the presence of an edge and, therefore, some of the neighbouring coefficients are likely to be large as well. There are many variations on this theme and we will not attempt to define these strategies. We shall instead simply mention that block thresholding works well empirically and is also amenable to rigorous analysis. We refer the reader to Cai (1999) and Hall, Kerkyacharian and Picard (1999) for experimental and theoretical results in this direction.

In a different direction, several authors (Candès and Guo 2002, Malgouyres 2002, Durand and Froment 2003) have independently proposed an attractive alternative to single basis thresholding. The idea here is to combine basis function expansions with variational principles for the reconstruction of an image/signal whose coefficients (in some basis) are known only approximately: they might be noisy, quantized, and so on. In the denoising problem where one wishes to recover an object f from $y = f + z$, one could imagine solving the following problem:

$$\min \|g\|_{TV} \quad \text{subject to} \quad |\Phi^T(g - y)|_i \leq \lambda \sigma \quad \text{for all} \quad i, \tag{10.1}$$

where Φ^T is the transform of interest (e.g., the wavelet transform), $(\Phi^T f)_i = \langle f, \varphi_i \rangle$. Here, the total variation norm $\|g\|_{BV}$ measures the complexity of the fit and is roughly equal to the integral of the Euclidean norm of the gradient. The aforementioned references demonstrate that this procedure works extremely well. Thresholding rules tend to produce artificial oscillations near discontinuities even though the original signal/image may be flat on both sides of the discontinuity, a 'pseudo-Gibbs phenomenon'. Ideas like (10.1) are very effective at removing such artifacts while retaining other nice properties of shrinkage methods.

In closing, shrinkage methods have inspired a lot of activity and new methods have been tuned to achieve the best practical performance.

10.2. Inverse problems

Another interesting problem occurs when one cannot measure the object $f(t)$ directly, but can only make linearly distorted measurements. That is, we are only able to observe data about $g(u) = Kf(u)$, where K is a linear transform. Such problems arise in multiple scientific settings ranging from medical imaging to physical chemistry to extragalactic astronomy. For example, in the case where K is a convolution transform, the signal is blurred as one measures

$$g(t) = (k * f)(t),$$

where k is a convolution kernel. Recovering blurred images from noisy data is ubiquitous in science and engineering: see Bertero and Boccacci (1998) for a nice survey. Another problem which has received a lot of attention concerns the case where K is the Radon transform

$$g(t, \theta) = \int_{\mathcal{L}_{t,\theta}} f(x_1, x_2) \, dx_1 \, dx_2,$$

where for $\theta \in [0, 2\pi)$ and $t \in \mathbb{R}$, $\mathcal{L}_{t,\theta}$ is the line

$$\{x_1 \cos \theta + x_2 \sin \theta = t\}.$$

Recovering an image from its two-dimensional noisy projections (line integrals) is the subject of computed tomography, which has been and still is the focus of intense research. Most interesting problems are ill-posed in the sense that the singular values of K tend to zero (think about a deconvolution problem where the convolution kernel k 'blocks' the high-frequency content of the signal).

Suppose then that we observe y of the form

$$y = Kf + z, \qquad (10.2)$$

where z is white noise and f is the object we wish to recover. Suppose we are given an orthobasis or a tight frame (φ_i) for functions 'living' in the object space. Then, under certain conditions, one can define dual basis elements (ψ_i), which 'live' in the data space and obey the relation

$$[Kf, \psi_i] = \delta_i \langle f, \varphi_i \rangle, \qquad (10.3)$$

where, in the above display, we have used the notation $[\cdot, \cdot]$ to distinguish between the data and the object spaces. Here the δ_is are defined by properties of K and called quasi-singular values; if φ_i is an orthobasis, they are set in such way that $\|\psi_i\| = 1$ (if (φ_i) is a tight frame, we could impose $\|\varphi_i\| = \|\psi_i\|$). The quasi-singular value relation (10.3) expresses the idea that one can measure the coefficients of f from Kf. Suppose that the δ_is do not vanish, then a consequence of the identity $f = \sum \langle f, \varphi_i \rangle \varphi_i$ and (10.3) is the reconstruction formula

$$f = \sum_i \delta_i^{-1} [Kf, \psi_i] \varphi_i. \qquad (10.4)$$

This formula is what Donoho calls a biorthogonal decomposition of K; see Donoho (1995) or the wavelet–vaguelette decomposition (WVD) in the case when (ϕ_i) is a wavelet basis. It is an extension of the SVD decomposition which reads

$$f = \sum d_i^{-1} [Kf, h_i] e_i, \qquad (10.5)$$

where (d_i^2) and (e_i) are the eigenvalues and eigenfunctions of K^*K, $K^*Ke_i = d_i^2 e_i$, and where h_i is the image of e_i under K, $Ke_i = d_i h_i$. (The ill-posedness means that $d_i \to 0$.)

The point is that many of the tools and ideas we have seen before apply. To make this connection, consider the sequence space version of (10.2), namely,

$$[y, \psi_i] = [Kf, \psi_i] + [z, \psi_i],$$

which one can write as

$$y_i = \delta_i \theta_i + [z, \psi_i]$$

(recall that $\theta_i = \langle f, \varphi_i \rangle$ are the coordinates of f we wish to estimate).

Dividing the above display by δ_i shows that we wish to recover the mean of a Gaussian vector

$$\tilde{y}_i = \theta_i + \sigma_i z_i, \qquad (10.6)$$

where $\sigma_i = \sigma \|\psi_i\|$ and the z_is are $N(0,1)$ (the covariance matrix is given by $\mathrm{Cov}(z_i, z_j) = [\psi_i, \psi_j]/(\|\psi_i\| \, \|\psi_j\|)$). The only real difference is that the noise is now heteroscedastic* with σ_i increasing as the quasi-singular values are decreasing.

One can thus see that everything should generalize nicely. In particular, if we apply thresholding, the proxy for the mean-squared error will be

$$\sum_i \min(\theta_i^2, \sigma_i^2), \qquad (10.7)$$

and this approach will be very effective if the following two conditions hold: (1) the signal is sparse in the basis (φ_i) and (2) the z_is in (10.6) are not too correlated, so that treating each coefficient individually still makes sense. We note that the latter condition is equivalent to saying that the system (φ_i) nearly diagonalizes the Gram matrix K^*K; by near-diagonalization, we mean that the representation of K^*K in the system (φ_i) is sparse.

The challenge for applied harmonic analysts is then to construct representations which sparsely represent objects of scientific interest and, *at the same time*, sparsely represent the operators under study. This is precisely what multiscale systems such as wavelets and curvelets achieve. On the one hand, they provide sparse representations of convolutions, Radon transforms, and many other types of common operators, and on the other, they simultaneously provide sparse representations of objects allowing for point-like singularities (wavelets) and curve-like singularities (curvelets). This is the reason why they have proved to be useful for solving inverse problems (Donoho 1995, Candès and Donoho 2002). In two dimensions, for instance, there is a quantitative theory showing that, for certain kind of interesting models of images, simple algorithms based on the shrinkage of curvelet biorthogonal decompositions achieve near-optimal statistical rates of convergence (Candès and Donoho 2002).

On the other hand – and this is very important – if one employs instead the singular system (e_i) for estimation, as is common, then the MSE may be very large. The proxy (10.7) lets us understand why this is the case. For the MSE to be small, the signal must be concentrated in the coordinates where the eigenvalues are large. But this is not usually the case, and the MSE is large. For example, in deconvolution problems, tomography problems and many others, the eigenvectors e_i are sinusoids, at least roughly speaking. The problem is that sinusoids provide very poor partial reconstructions of the kinds of signals and images in which one is typically interested: *e.g.*, images of the brain or the interior of the earth all have edges and perhaps other types of singularities. As a consequence, SVD-based methods tend to underperform when the object we wish to image is not smooth.

10.3. FDR thresholding rules

The 'universal' threshold of $\sqrt{2 \log n}$ is often criticized because it is very conservative; it potentially sets to zero many coordinates where the signal is larger than the noise level. We close this paper by discussing innovative adaptive choices of thresholds which have their origin in the field of hypotheses testing – in multiple comparisons, to be more exact.

Consider the simpler problem of deciding, for each $i = 1, \ldots, n$, whether or not $\theta_i = 0$, given the data

$$y_i = \theta_i + z_i, \quad z_i \text{ i.i.d. } N(0, \sigma^2).$$

Formally, we wish to simultaneously test n hypotheses

$$\begin{aligned} H_{0,i} : \quad & \theta_i = 0, \\ H_{1,i} : \quad & \theta_i \neq 0. \end{aligned}$$

Then one could accept the ith null hypothesis if $|y_i| \leq \sigma \sqrt{2 \log n}$ and reject it otherwise. This would essentially correspond to the Bonferroni procedure which controls the so-called familywise error rate, defined as the probability of rejecting at least one hypothesis $H_{i,0}$ which is true. If we want a familywise error rate below α, the Bonferroni method would ask us to reject $H_{i,0}$ if and only if

$$|y_i| > \sigma \, z(\alpha/2n),$$

where $z(\alpha)$ is the upper quantile of the Gaussian distribution ($z(\alpha)$ is defined by $\mathbb{P}(N(0,1) > z(\alpha)) = \alpha$). For nearly all reasonable levels α and n large, $z(\alpha/2n)$ is nearly equal to $\sqrt{2 \log n}$.

In the problem of multiple comparisons, control of the familywise error rate yields very conservative decisions. Ten years ago, Benjamini and Hochberg (1995) introduced an alternative, and instead proposed to control the false discovery rate (FDR). The FDR is the expected ratio between the number of incorrectly rejected null hypotheses and the total number of rejections. The advantage is that FDR controlling procedures have greater power to detect alternatives. In our problem, we order the values by decreasing order of magnitude $|y|_{(1)} \geq |y|_{(2)} \geq \cdots \geq |y|_{(n)}$, and define i_{FDR} to be the largest index for which

$$|y|_{(i)} \geq \sigma \, z(q \, i/2n).$$

Then the procedure which rejects all the hypotheses corresponding to the i_{FDR} largest values of $|y_i|$ controls the FDR at level q (meaning that the expected proportion of false rejections is less than q).

A little later, Abramovich and Benjamini (1996) proposed applying FDR for estimation and introduced a new thresholding rule. The idea is simply to estimate the parameters corresponding to the rejected hypotheses (these are judged estimable) and set the others to zero. With $\lambda_{\mathrm{FDR}} = z(q\, i_{\mathrm{FDR}}/2n)$, the FDR thresholding rule is thus defined by

$$\hat{\theta}_i = \begin{cases} y_i, & |y_i| > \lambda_{\mathrm{FDR}}\, \sigma, \\ 0, & \text{else.} \end{cases} \tag{10.8}$$

This is interesting because (10.8) is a data-driven thresholding rule which adapts to the sparsity of the signal. The threshold is larger for sparser signals and smaller for denser ones.

To understand why FDR thresholding rules are a good thing, suppose that by looking at y we learn that many of the coordinates θ_i are nonzero. Then the FDR threshold will be lower than the universal threshold and the estimator will have a smaller bias. Of course, we will also occasionally estimate some θ_is which are close to zero, hence increasing the variance a little. But the proportion of 'erroneous' estimations is controlled, and in the bias + variance trade-off we will typically draw significantly ahead of universal thresholding rules. There are numerical experiments showing that FDR thresholding rules perform very well: see Abramovich and Benjamini (1996) and Abramovich, Benjamini, Donoho and Johnstone (2000). There is also a beautiful theory showing that, in some special set-ups where θ belongs to a weak-ℓ_p ball, for example, the estimator achieves adaptive asymptotic minimaxity (Abramovich et al. 2000).

FDR thresholding rules are a nice new chapter in the history of thresholding and we suspect that they will generate a lot of interest in the near future. There are also challenging questions that do not have satisfactory answers at the moment. For example, how would FDR thresholding rules adapt when the observations are correlated and how would one use them in more sophisticated estimation problems?

10.4. Last words

Near the beginning of this article, we emphasized that we would focus on a couple of key ideas that have had a very significant impact on my professional development and on the field in general. A large fraction of this paper is a write-up of a series of lectures I delivered in 2004, and the whole manuscript was conceived with the goal of teaching this material to nonspecialists. It is not an exhaustive survey of all the research that occurred in the field, and I hope that this personal selection of topics will not be found offensive.

Last but not least, I would like to thank Carl for encouraging me to write this article.

REFERENCES

F. Abramovich and Y. Benjamini (1996), 'Adaptive thresholding of wavelet coefficients', *Comput. Statist. Data Anal.* **22**, 351–361.

F. Abramovich, Y. Benjamini, D. L. Donoho and I. M. Johnstone (2000), Adapting to unknown sparsity by controlling the false discovery rate. Technical Report 2000-19, Department of Statistics, Stanford University. To appear in *Ann. Statist.*

H. Akaike (1974), 'A new look at the statistical model identification', *IEEE Trans. Automatic Control* **AC-19**, 716–723.

Y. Baraud (2000), Model selection for regression on a fixed design, *Probab. Theory Rel. Fields* **117**, 467–493.

A. R. Barron (1994), 'Approximation and estimation bounds for artificial neural networks', *Machine Learning* **14**, 113–143.

A. R. Barron and T. M. Cover (1991), 'Minimum complexity density estimation', *IEEE Trans. Inform. Theory* **37**, 1034–1054.

Y. Benjamini and Y. Hochberg (1995), 'Controlling the false discovery rate: A practical and powerful approach to multiple testing', *J. Roy. Statist. Soc. Ser. B* **57**, 289–300.

M. Bertero and P. Boccacci (1998), *Introduction to Inverse Problems in Imaging*, Institute of Physics Publishing, Bristol.

L. Birgé and P. Massart (1997), From model selection to adaptive estimation, in *Festschrift for Lucien Le Cam*, Springer, New York, pp. 55–87.

L. Birgé and P. Massart (2001), Gaussian model selection, *J. Eur. Math. Soc.* **3**, 203–268.

L. D. Brown and M. G. Low (1996), 'Asymptotic equivalence of nonparametric regression and white noise', *Ann. Statist.* **24**, 2384–2398.

T. T. Cai (1999), 'Adaptive wavelet estimation: A block thresholding and oracle inequality approach', *Ann. Statist.* **27**, 898–924.

E. J. Candès (2002), Multiscale chirplets and near-optimal recovery of chirps. Technical report, Stanford University.

E. J. Candès and D. L. Donoho (1999), 'Ridgelets: The key to higher-dimensional intermittency?', *Phil. Trans. R. Soc. Lond. A* **357**, 2495–2509.

E. J. Candès and D. L. Donoho (2000), Curvelets: A surprisingly effective nonadaptive representation for objects with edges, in *Curves and Surfaces* (C. R. A. Cohen and L. L. Schumaker, eds), Vanderbilt University Press, Nashville, TN, pp. 105–120.

E. J. Candès and D. L. Donoho (2002), 'Recovering edges in ill-posed inverse problems: Optimality of curvelet frames', *Ann. Statist.* **30**, 784 –842.

E. J. Candès and D. L. Donoho (2004), 'New tight frames of curvelets and optimal representations of objects with piecewise-C^2 singularities', *Comm. Pure Appl. Math.* **57**, 219–266.

E. J. Candès and F. Guo (2002), 'New multiscale transforms, minimum total variation synthesis: Applications to edge-preserving image reconstruction', *Signal Processing* **82**, 1519–1543.

E. J. Candès and T. Tao (2004) Near-optimal signal recovery from random projections and universal encoding strategies. Available on the ArXiv preprint server: math.CA/0410542. To appear in *IEEE Trans. Inform. Theory*.

E. J. Candès and T. Tao (2005a), The Dantzig selector: Statistical estimation when
    p is much larger than n. Technical report, California Institute of Technology,
    available on the ArXiv preprint server: math.ST/0506081. To appear in *Ann.
    Statist.*

E. J. Candès and T. Tao (2005b), 'Decoding by linear programming', *IEEE Trans.
    Inform. Theory* **51**, 4203–4215.

E. J. Candès, L. Demanet, D. L. Donoho and L. Ying (2005), Fast discrete curvelet
    transforms. Technical report, California Institute of Technology. To appear
    in *SIAM J. Multiscale Modeling and Simulations.*

E. J. Candès, J. Romberg and T. Tao (2006) 'Robust uncertainty principles: Exact
    signal reconstruction from highly incomplete frequency information', *IEEE
    Trans. Inform. Theory* **52**, 489–509.

A. Cohen, R. DeVore, P. Petrushev and H. Xu (1999), 'Nonlinear approximation
    and the space BV($\mathbf{R}^2$)', *Amer. J. Math.* **121**, 587–628.

R. R. Coifman and Y. Meyer (1991), 'Remarques sur l'analyse de Fourier à fenêtre',
    C. R. Acad. Sci. Paris Sér. I Math. **312**, 259–261.

R. R. Coifman and M. V. Wickerhauser (1992), 'Entropy-based algorithms for best
    basis selection', *IEEE Trans. Inform. Theory* **38**, 713–718.

D. L. Donoho (1993), 'Unconditional bases are optimal bases for data compression
    and for statistical estimation', *Appl. Comput. Harmon. Anal.* **1**, 100–115.

D. L. Donoho (1995), 'Nonlinear solution of linear inverse problems by wavelet-
    vaguelette decomposition', *Appl. Comput. Harmon. Anal.* **2**, 101–126.

D. L. Donoho (1999), 'Wedgelets: Nearly-minimax estimation of edges', *Ann.
    Statist.* **27**, 859–897.

D. L. Donoho (2001), 'Sparse components of images and optimal atomic decompo-
    sition', *Constr. Approx.* **17**, 353–382.

D. L. Donoho and I. Johnstone (1994a), 'Ideal spatial adaptation via wavelet
    shrinkage', *Biometrika* **81**, 425–455.

D. L. Donoho and I. M. Johnstone (1994b), 'Ideal denoising in an orthonormal
    basis chosen from a library of bases', *CR Acad. Sci. Paris Sér. I Math.* **319**,
    1317–1322.

D. L. Donoho and I. M. Johnstone (1995), Empirical atomic decomposition.
    Manuscript.

D. L. Donoho and I. M. Johnstone (1999), 'Asymptotic minimaxity of wavelet
    estimators with sampled data', *Statist. Sinica* **9**, 1–32.

D. L. Donoho and R. C. Liu (1991), 'Geometrizing rates of convergence II, III',
    Ann. Statist. **19**, 633–667, 668–701.

D. L. Donoho and M. Nussbaum (1990), 'Minimax quadratic estimation of a
    quadratic functional', *J. Complexity* **6**, 290–323.

D. L. Donoho, I. M. Johnstone, G. Kerkyacharian and D. Picard (1995), 'Wavelet
    shrinkage: Asymptopia?', *J. Roy. Statist. Soc. Ser. B* **57**, 301–369.

S. Durand and J. Froment (2003), 'Reconstruction of wavelet coefficients using total
    variation minimization', *SIAM J. Sci. Comput.* **24**, 1754–1767 (electronic).

S. Y. Efroïmovich (1985), 'Nonparametric estimation of a density of unknown
    smoothness', *Teor. Veroyatnost. i Primenen.* **30**, 524–534.

S. Y. Efroïmovich and M. S. Pinsker (1981), 'Estimation of square-integrable den-
    sity on the basis of a sequence of observations', *Problemy Peredachi Infor-
    matsii* **17**, 50–68.

S. Y. Efroĭmovich and M. S. Pinsker (1982), 'Estimation of square-integrable probability density of a random variable', *Problems Inform. Transmission* **18**, 175–189; translated from *Problemy Peredachi Informatsii* **18**, 19–38 (Russian).

S. Y. Efroĭmovich and M. S. Pinsker (1984), 'A self-training algorithm for non-parametric filtering', *Avtomat. i Telemekh.* (11), 58–65.

B. Efron and C. Morris (1971), 'Limiting the risk of Bayes and empirical Bayes estimators I: The Bayes case', *J. Amer. Statist. Assoc.* **66**, 807–815.

A. G. Flesia, H. Hel-Or, A. Averbuch, E. J. Candès, R. R. Coifman and D. L. Donoho (2003), Digital implementation of ridgelet packets, in *Beyond wavelets*, Vol. 10 of *Stud. Comput. Math.*, Academic Press/Elsevier, San Diego, CA, pp. 31–60.

D. P. Foster and E. I. George (1994), 'The risk inflation criterion for multiple regression', *Ann. Statist.* **22**, 1947–1975.

M. Frazier, B. Jawerth and G. Weiss (1991), *Littlewood–Paley Theory and the Study of Function Spaces*, Vol. 79 of *NSF-CBMS Regional Conf. Ser. in Mathematics*, AMS, Providence, RI.

H.-Y. Gao (1998), 'Wavelet shrinkage denoising using the non-negative garrote', *J. Comput. Graph. Statist.* **7**, 469–488.

P. J. Green and B. W. Silverman (1994), *Nonparametric Regression and Generalized Linear Models: A Roughness Penalty Approach*, Vol. 58 of *Monographs on Statistics and Applied Probability*, Chapman & Hall, London.

P. Hall, G. Kerkyacharian and D. Picard (1999), 'On the minimax optimality of block thresholded wavelet estimators', *Statist. Sinica* **9**, 33–49.

W. James and C. Stein (1961), Estimation with quadratic loss, in *Proc. 4th Berkeley Sympos. Math. Statist. and Prob.*, Vol. I, University of California Press, Berkeley, CA, pp. 361–379.

I. M. Johnstone (2002), Function estimation and Gaussian sequence models. Available at: http://www-stat.stanford.edu/~imj.

A. P. Korostelëv and Λ. B. Tsybakov (1993), *Minimax Theory of Image Reconstruction*, Vol. 82 of *Lecture Notes in Statistics*, Springer, New York.

L. Le Cam (2000), La statistique mathématique depuis 1950, in *Development of Mathematics 1950–2000*, Birkhäuser, Basel, pp. 735–761.

E. L. Lehmann (1997), *Theory of Point Estimation*, Springer, New York. Reprint of the 1983 original.

A. Leon-Garcia (1994), *Probability and Random Processes for Electrical Engineering*, 2nd edn, Addison-Wesley.

F. Malgouyres (2002), 'Minimizing the total variation under a general convex constraint for image restoration', *IEEE Trans. Image Process.* **11**, 1450–1456.

S. Mallat (1999), *A Wavelet Tour of Signal Processing*, 2nd edn, Academic Press, San Diego, CA.

C. L. Mallows (1973), 'Some comments on c_p', *Technometrics* **15**, 661–676.

Y. Meyer (1992), *Wavelets and Operators*, Cambridge University Press.

B. K. Natarajan (1995), 'Sparse approximate solutions to linear systems', *SIAM J. Comput.* **24**, 227–234.

M. Nussbaum (1996), 'Asymptotic equivalence of density estimation and Gaussian white noise', *Ann. Statist.* **24**, 2399–2430.

G. Schwarz (1978), 'Estimating the dimension of a model', *Ann. Statist.* **6**, 461–464.

D. W. Scott (1992), *Multivariate Density Estimation: Theory, Practice, and Visualization*, Wiley Series in Probability and Mathematical Statistics, Wiley, New York.

B. W. Silverman (1986), *Density Estimation for Statistics and Data Analysis*, Monographs on Statistics and Applied Probability, Chapman & Hall, London.

J.-L. Starck, F. Murtagh and A. Bijaoui (1998), *Image Processing and Data Analysis: The Multiscale Approach*, Cambridge University Press, Cambridge.

C. M. Stein (1981), 'Estimation of the mean of a multivariate normal distribution', *Ann. Statist.* **9**, 1135–1151.

S. M. Stigler (1990), *The History of Statistics: The Measurement of Uncertainty Before 1900*, The Belknap Press of Harvard University Press, Cambridge, MA. Reprint of the 1986 original.

R. Tibshirani (1996), 'Regression shrinkage and selection via the lasso', *J. Roy. Statist. Soc. Ser. B* **58**, 267–288.

H. Triebel (1992), *Theory of Function Spaces II*, Vol. 84 of *Monographs in Mathematics*, Birkhäuser, Basel.

G. Wahba (1990), *Spline Models for Observational Data*, Vol. 59 of *CBMS-NSF Regional Conference Series in Applied Mathematics*, SIAM, Philadelphia, PA.

N. Wiener (1949), *Extrapolation, Interpolation, and Smoothing of Stationary Time Series: With Engineering Applications*, The Technology Press of the Massachusetts Institute of Technology, Cambridge, MA.

D. Williams (1991), *Probability with Martingales*, Cambridge Mathematical Textbooks, Cambridge University Press, Cambridge.

Glossary

Bayesian estimation. In this paper, we often use the terms 'Bayesian estimator' or 'Bayes' rule' to denote any estimator which minimizes the so-called Bayes risk defined by

$$B(\pi) = E_\pi R(\theta, \hat{\theta}) = \int R(\theta, \hat{\theta}) \, \pi(\mathrm{d}\theta),$$

where π is the prior distribution on the parameter θ and $R(\theta, \hat{\theta})$ is the risk of $\hat{\theta}$; see below for a definition of the risk.

Bias. The bias of an estimator is defined as the difference between the true value of the parameter vector and the expected value of the estimator under the true distribution. Suppose Y is a vector with joint distribution f_θ, where $\theta \in \Theta$ is a parameter of interest, and let $\hat{\theta}$ be a function of Y used to estimate θ. Then the bias of $\hat{\theta}$ is given by

$$\mathrm{bias}(\hat{\theta}) = \theta - E_{f_\theta} \hat{\theta},$$

where E_{f_θ} is the expectation of $\hat{\theta}$ under the true distribution f_θ, $E_{f_\theta} \hat{\theta} = \int \hat{\theta}(y) \, f_\theta(\mathrm{d}y)$. We say that an estimator is unbiased if $\mathrm{bias}(\hat{\theta}) = 0$. For example, if $Y_1, Y_2, \ldots, Y_n$ are i.i.d. $N(\theta, 1)$, then $\hat{\theta} = (Y_1 + \cdots + Y_n)/n$ is unbiased for θ.

Chi-square distribution. The chi-square distribution is that of the sum of squares of independent standard normal random variables; if we let $Z_1, Z_2, \ldots, Z_d$ be i.i.d. $N(0,1)$, the random variable $Y := Z_1^2 + \cdots + Z_d^2$ follows the (central) chi-square distribution with d degrees of freedom.

Gaussian signal. A Gaussian signal is simply a Gaussian process. A Gaussian process $X = (X_1, X_2, \ldots, X_n)$ is a family of random variables whose joint distribution is multivariate normal. A random vector is said to be multivariate normal if every linear combination $a_1 X_1 + \cdots + a_n X_n$ (the a_is are nonrandom) is normally distributed. In the case where the covariance matrix is nonsingular, this is equivalent to saying that the joint density of the random vector is given by

$$f(x) = \frac{1}{(2\pi)^{n/2} (\det \Sigma)^{1/2}} e^{-(x-\mu)^T \Sigma^{-1} (x-\mu)/2},$$

where $\mu \in \mathbb{R}^n$ is the mean vector and $\Sigma \in \mathbb{R}^{n \times n}$ the covariance matrix.

i.i.d. 'i.i.d.' stands for independently and identically distributed. We say that the random variables $X_1, \ldots, X_n$ are i.i.d. when they are all independent and follow the same distribution.

Heteroscedasticity. A sequence or a vector of random variables is heteroscedastic when the variances of the random variables in the sequence are not all the same. The complement is homoscedasticity.

Minimax estimation. A minimax estimator is any estimator whose worst-case risk is minimal. In other words, a minimax estimator is the solution to

$$\inf_{\hat{\theta}} \sup_{\theta \in \Theta} R(\theta, \hat{\theta}),$$

where Θ is the parameter space and the infimum is taken over all measurable functions of the data.

Risk of an estimator. In decision theory, we measure the quality of an estimator by the nonnegative loss function $\ell(\theta, \hat{\theta})$. For example, the quadratic loss is given by $(\theta - \hat{\theta})^2$ for scalar-valued parameters or $\|\theta - \hat{\theta}\|_{\ell_2}^2$ for vector-valued parameters. The idea is that the loss is small when θ and $\hat{\theta}$ are close, and increases as they get far apart. The loss is a random variable since $\hat{\theta}$ is random, and the risk $R(\theta, \hat{\theta})$ is the expected value of the loss

$$R(\theta, \hat{\theta}) := E_{f_\theta} \ell(\theta, \hat{\theta}).$$

Again, E_{f_θ} is the expectation under the distribution f_θ (see the entry for 'bias').

Acta Numerica (2006), pp. 327–384
doi: 10.1017/S0962492906240017

Numerical linear algebra in data mining

Lars Eldén
Department of Mathematics,
Linköping University, SE-581 83 Linköping, Sweden
E-mail: `laeld@math.liu.se`

Ideas and algorithms from numerical linear algebra are important in several
areas of data mining. We give an overview of linear algebra methods in text
mining (information retrieval), pattern recognition (classification of handwritten digits), and PageRank computations for web search engines. The
emphasis is on rank reduction as a method of extracting information from a
data matrix, low-rank approximation of matrices using the singular value decomposition and clustering, and on eigenvalue methods for network analysis.

CONTENTS

1. Introduction

1.1. Data mining

In modern society huge amounts of data are stored in databases with the
purpose of extracting useful information. Often it is not known at the
occasion of collecting the data what information is going to be requested,
and therefore the database is often not designed for the distillation of

any particular information, but rather it is to a large extent unstructured. The science of extracting useful information from large data sets is usually referred to as 'data mining', sometimes along with 'knowledge discovery'.

There are numerous application areas of data mining, ranging from e-business (Berry and Linoff 2000, Mena 1999) to bioinformatics (Bergeron 2002), from scientific application such as astronomy (Burl, Asker, Smyth, Fayyad, Perona, Crumpler and Aubele 1998), to information retrieval (Baeza-Yates and Ribeiro-Neto 1999) and Internet search engines (Berry and Browne 2005).

Data mining is a truly interdisciplinary science, where techniques from computer science, statistics and data analysis, pattern recognition, linear algebra and optimization are used, often in a rather eclectic manner. Because of the practical importance of the applications, there are now numerous books and surveys in the area. We cite a few here: Christianini and Shawe-Taylor (2000), Cios, Pedrycz and Swiniarski (1998), Duda, Hart and Storck (2001), Fayyad, Piatetsky-Shapiro, Smyth and Uthurusamy (1996), Han and Kamber (2001), Hand, Mannila and Smyth (2001), Hastie, Tibshirani and Friedman (2001), Hegland (2001) and Witten and Frank (2000).

The purpose of this paper is not to give a comprehensive treatment of the areas of data mining, where linear algebra is being used, since that would be a far too ambitious undertaking. Instead we will present a few areas in which numerical linear algebra techniques play an important role. Naturally, the selection of topics is subjective, and reflects the research interests of the author.

This survey has three themes, as follows.

(1) *Information extraction from a data matrix by a rank reduction process.*

By determining the 'principal direction' of the data, the 'dominating' information is extracted first. Then the data matrix is deflated (explicitly or implicitly) and the same procedure is repeated. This can be formalized using the Wedderburn rank reduction procedure (Wedderburn 1934), which is the basis of many matrix factorizations.

The second theme is a variation of the rank reduction idea.

(2) *Data compression by low-rank approximation*: A data matrix $A \in \mathbb{R}^{m \times n}$, where m and n are large, will be approximated by a rank-k matrix,

$$A \approx WZ^T, \qquad W \in \mathbb{R}^{m \times k}, \quad Z \in \mathbb{R}^{n \times k},$$

where $k \ll \min(m, n)$.

In many applications the data matrix is huge, and difficult to use for storage and efficiency reasons. Thus, one evident purpose of compression is to obtain a representation of the data set that requires less memory than the original data set, and that can be manipulated more efficiently. Sometimes one wishes to obtain a representation that can be interpreted as the 'main directions of variation' of the data, the *principal components*. This is done by building the low-rank approximation from the left and right singular vectors of A that correspond to the largest singular values. In some applications, *e.g.*, information retrieval (see Section 4) it is possible to obtain better search results from the compressed representation than from the original data. There the low-rank approximation also serves as a 'denoising device'.

(**3**) *Self-referencing definitions that can be formulated mathematically as eigenvalue and singular value problems.*

The most well-known example is the Google PageRank algorithm, which is based on the notion that the importance of a web page depends on how many inlinks it has from other important pages.

2. Vectors and matrices in data mining

Often the data are numerical, and the data points can be thought of as belonging to a high-dimensional vector space. Ensembles of data points can then be organized as matrices. In such cases it is natural to use concepts and techniques from linear algebra.

Example 2.1. Handwritten digit classification is a sub-area of *pattern recognition*. Here vectors are used to represent digits. The image of one digit is a 16×16 matrix of numbers, representing grey-scale. It can also be represented as a vector in $\mathbb{R}^{256}$, by stacking the columns of the matrix.

A set of n digits (handwritten 3s, say) can then be represented by matrix $A \in \mathbb{R}^{256 \times n}$, and the columns of A can be thought of as a cluster. They also span a subspace of $\mathbb{R}^{256}$. We can compute an approximate basis of this subspace using the singular value decomposition (SVD) $A = U \Sigma V^T$. Three basis vectors of the '3-subspace' are illustrated in Figure 2.1. The digits are taken from the US Postal Service database (see, *e.g.*, Hastie *et al.* (2001)).

Let b be a vector representing an unknown digit, and assume that one wants to determine, automatically using a computer, which of the digits 0–9 the unknown digit represents. Given a set of basis vectors for 3s, $u_1, u_2, \ldots, u_k$, we may be able to determine whether b is a 3 or not, by checking if there is a linear combination of the k basis vectors, $\sum_{j=1}^{k} x_j u_j$, such that the residual $b - \sum_{j=1}^{k} x_j u_j$ is small. Thus, we determine the co-ordinates of b in the basis $\{u_j\}_{j=1}^{k}$, which is equivalent to solving a least squares problem with the data matrix $U_k = (u_1 \ldots u_k)$.

In Section 5 we discuss methods for classification of handwritten digits.

Figure 2.1. Handwritten digits from the US Postal Service
data base, and basis vectors for 3s (bottom).

It also happens that there is a natural or perhaps clever way of encoding
non-numerical data so that the data points become vectors. We will give
a couple of such examples from text mining (information retrieval) and
Internet search engines.

Example 2.2. *Term-document matrices* are used in information retrieval.
Consider the following set of five documents. Key words, referred to as
terms, are marked in boldface.[1]

Document 1:	The **Google matrix** P is a model of the **Internet.**
Document 2:	P_{ij} is nonzero if there is a **link** from **web page** j to i.
Document 3:	The **Google matrix** is used to **rank** all **web pages**
Document 4:	The **ranking** is done by solving a **matrix eigenvalue** problem.
Document 5:	**England** dropped out of the top 10 in the **FIFA ranking**.

Counting the frequency of terms in each document, we get the result shown
in Table 2.1. The total set of terms is called the *dictionary*. Each document

---

[1] To avoid making the example too large, we have ignored some words that would normally be considered as terms. Note also that only the stem of a word is significant:
'ranking' is considered the same as 'rank'.

Table 2.1.

Term	Doc. 1	Doc. 2	Doc. 3	Doc. 4	Doc. 5
eigenvalue	0	0	0	1	0
England	0	0	0	0	1
FIFA	0	0	0	0	1
Google	1	0	1	0	0
Internet	1	0	0	0	0
link	0	1	0	0	0
matrix	1	0	1	1	0
page	0	1	1	0	0
rank	0	0	1	1	1
web	0	1	1	0	0

is represented by a vector in $\mathbb{R}^{10}$, and we can organize the data as a *term-document matrix*,

$$
A = \begin{pmatrix}
0 & 0 & 0 & 1 & 0 \\
0 & 0 & 0 & 0 & 1 \\
0 & 0 & 0 & 0 & 1 \\
1 & 0 & 1 & 0 & 0 \\
1 & 0 & 0 & 0 & 0 \\
0 & 1 & 0 & 0 & 0 \\
1 & 0 & 1 & 1 & 0 \\
0 & 1 & 1 & 0 & 0 \\
0 & 0 & 1 & 1 & 1 \\
0 & 1 & 1 & 0 & 0
\end{pmatrix} \in \mathbb{R}^{10 \times 5}.
$$

Assume that we want to find all documents that are relevant with respect to the query 'ranking of web pages'. This is represented by a *query vector*, constructed in an analogous way to the term-document matrix, using the same dictionary,

$$
q = \begin{pmatrix}
0 \\
0 \\
0 \\
0 \\
0 \\
0 \\
0 \\
1 \\
1 \\
1
\end{pmatrix} \in \mathbb{R}^{10}.
$$

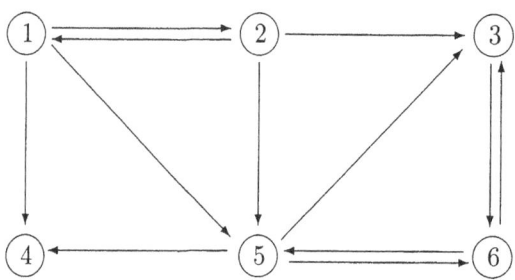

Figure 2.2.

Thus the query itself is considered as a document. The information retrieval task can now be formulated as a mathematical problem: *find the columns of A that are close to the vector q.* To solve this problem we use some distance measure in $\mathbb{R}^{10}$.

In information retrieval it is common that m is large, of the order 10^6, say. As most of the documents only contain a small fraction of the terms in the dictionary, the matrix is *sparse.*

In some methods for information retrieval, linear algebra techniques (*e.g.,* singular value decomposition (SVD)) are used for data compression and retrieval enhancement. We discuss vector space methods for information retrieval in Section 4.

The very idea of data mining is to extract useful information from large and often unstructured sets of data. Therefore it is necessary that the methods used are efficient and often specially designed for large problems. In some data mining applications huge matrices occur.

Example 2.3.   The task of extracting information from all the web pages available on the Internet is performed by *search engines.* The core of the Google search engine[2] is a matrix computation, probably the largest that is performed routinely (Moler 2002). The Google matrix P is assumed to be of dimension of the order billions (2005), and it is used as a model of (all) the web pages on the Internet.

In the Google PageRank algorithm the problem of assigning ranks to all the web pages is formulated as a matrix eigenvalue problem. Let all web pages be ordered from 1 to n, and let i be a particular web page. Then O_i will denote the set of pages that i is linked to, the *outlinks.* The number of outlinks is denoted $N_i = |O_i|$. The set of *inlinks*, denoted I_i, are the pages

[2] http://www.google.com.

that have an outlink to i. Now define Q to be a square matrix of dimension n, and let

$$
Q_{ij} = \begin{cases} 1/N_j, & \text{if there is a link from } j \text{ to } i, \\ 0, & \text{otherwise.} \end{cases}
$$

This definition means that row i has nonzero elements in those positions that correspond to inlinks of i. Similarly, column j has nonzero elements equal to $1/N_j$ in those positions that correspond to the outlinks of j.

The link graph in Figure 2.2 illustrates a set of web pages with outlinks and inlinks. The corresponding matrix becomes

$$
Q = \begin{pmatrix}
0 & \frac{1}{3} & 0 & 0 & 0 & 0 \\
\frac{1}{3} & 0 & 0 & 0 & 0 & 0 \\
0 & \frac{1}{3} & 0 & 0 & \frac{1}{3} & \frac{1}{2} \\
\frac{1}{3} & 0 & 0 & 0 & \frac{1}{3} & 0 \\
\frac{1}{3} & \frac{1}{3} & 0 & 0 & 0 & \frac{1}{2} \\
0 & 0 & 1 & 0 & \frac{1}{3} & 0
\end{pmatrix}.
$$

Define a vector r, which holds the ranks of all pages. The vector r is then defined[3] as the eigenvector corresponding to the eigenvalue $\lambda = 1$ of Q:

$$
\lambda r = Qr. \tag{2.1}
$$

We shall discuss some numerical aspects of the PageRank computation in Section 6.1.

3. Data compression: low-rank approximation

3.1. Wedderburn rank reduction

One way of measuring the information contents in a data matrix is to compute its rank. Obviously, linearly dependent column or row vectors are redundant, as they can be replaced by linear combinations of the other, linearly independent columns. Therefore, one natural procedure for extracting information from a data matrix is to systematically determine a sequence of linearly independent vectors, and deflate the matrix by subtracting rank-one matrices, one at a time. It turns out that this *rank reduction procedure* is closely related to *matrix factorization, data compression, dimension reduction*, and *feature selection/extraction*. The key link between the concepts is the *Wedderburn rank reduction theorem*.

---

[3] This definition is provisional since it does not take into account the mathematical properties of Q: as the problem is formulated so far, there is usually no unique solution of the eigenvalue problem.

Theorem 3.1. (Wedderburn 1934) Suppose $A \in \mathbb{R}^{m \times n}$, $f \in \mathbb{R}^{n \times 1}$, and $g \in \mathbb{R}^{m \times 1}$. Then

$$\operatorname{rank}(A - \omega^{-1} A f g^T A) = \operatorname{rank}(A) - 1,$$

if and only if $\omega = g^T A f \neq 0$.

Based on Theorem 3.1 a stepwise rank reduction procedure can be defined: Let $A^{(1)} = A$, and define a sequence of matrices $\{A^{(i)}\}$

$$A^{(i+1)} = A^{(i)} - \omega_i^{-1} A^{(i)} f^{(i)} g^{(i)^T} A^{(i)}, \tag{3.1}$$

for any vectors $f^{(i)} \in \mathbb{R}^{n \times 1}$ and $g^{(i)} \in \mathbb{R}^{m \times 1}$, such that

$$\omega_i = g^{(i)^T} A^{(i)} f^{(i)} \neq 0. \tag{3.2}$$

The sequence defined in (3.1) terminates in $r = \operatorname{rank}(A)$ steps, since each time the rank of the matrix decreases by one. This process is called a *rank-reducing process* and the matrices $A^{(i)}$ are called Wedderburn matrices. For details, see Chu, Funderlic and Golub (1995). The process gives a matrix *rank-reducing decomposition*,

$$A = \hat{F} \Omega^{-1} \hat{G}^T, \tag{3.3}$$

where

$$\hat{F} = \left(\hat{f}_1, \ldots, \hat{f}_r \right) \in \mathbb{R}^{m \times r}, \quad \hat{f}_i = A^{(i)} f^{(i)}, \tag{3.4}$$

$$\Omega = \operatorname{diag}(\omega_1, \ldots, \omega_r) \in \mathbb{R}^{r \times r}, \tag{3.5}$$

$$\hat{G} = \left(\hat{g}_1, \ldots, \hat{g}_r \right) \in \mathbb{R}^{n \times r}, \quad \hat{g}_i = A^{(i)^T} g^{(i)}. \tag{3.6}$$

Theorem 3.1 can be generalized to the case where the reduction of rank is larger than one, as shown in the next theorem.

Theorem 3.2. (Guttman 1957) Suppose $A \in \mathbb{R}^{m \times n}$, $F \in \mathbb{R}^{n \times k}$, and $G \in \mathbb{R}^{m \times k}$. Then

$$\operatorname{rank}(A - AFR^{-1} G^T A) = \operatorname{rank}(A) - \operatorname{rank}(AFR^{-1} G^T A), \tag{3.7}$$

if and only if $R = G^T A F \in \mathbb{R}^{k \times k}$ is nonsingular.

Chu *et al.* (1995) discuss Wedderburn rank reduction from the point of view of solving linear systems of equations. There are many choices of F and G that satisfy the condition (3.7). Therefore, various rank-reducing decompositions (3.3) are possible. It is shown that several standard matrix factorizations in numerical linear algebra are instances of the Wedderburn formula: Gram–Schmidt orthogonalization, singular value decomposition, QR and Cholesky decomposition, as well as the Lanczos procedure.

A complementary view is taken in data analysis[4] (see Hubert, Meulman and Heiser (2000)), where the Wedderburn formula and matrix factorizations are considered as tools for data analysis: 'The major purpose of a matrix factorization in this context is to obtain some form of lower-rank approximation to A for understanding the structure of the data matrix ...'.

One important difference in the way the rank reduction is treated in data analysis and in numerical linear algebra, is that in algorithm descriptions in data analysis the subtraction of the rank-one matrix $\omega^{-1}Afg^{T}A$ is often done explicitly, whereas in numerical linear algebra it is mostly implicit. One notable example is the Partial Least Squares method (PLS) that is widely used in chemometrics. PLS is equivalent to Lanczos bidiagonalization: see Section 3.4. This difference in description is probably the main reason why the equivalence between PLS and Lanczos bidiagonalization has not been widely appreciated in either community, even though it was pointed out quite early (Wold, Ruhe, Wold and Dunn 1984).

The application of the Wedderburn formula to data mining is further discussed in Park and Eldén (2005).

3.2. SVD, Eckart–Young optimality, and principal component analysis

We will here give a brief account of the SVD, its optimality properties for low-rank matrix approximation, and its relation to *principal component analysis* (PCA). For a more detailed exposition, see, *e.g.*, Golub and Van Loan (1996).

Theorem 3.3. Any matrix $A \in \mathbb{R}^{m \times n}$, with $m \geq n$, can be factorized

$$A = U\Sigma V^{T}, \qquad \Sigma = \begin{pmatrix} \Sigma_0 \\ 0 \end{pmatrix} \in \mathbb{R}^{m \times n}, \qquad \Sigma_0 = \operatorname{diag}(\sigma_1, \ldots, \sigma_n),$$

where $U \in \mathbb{R}^{m \times m}$ and $V \in \mathbb{R}^{n \times n}$ are orthogonal, and $\sigma_1 \geq \sigma_2 \geq \cdots \sigma_n \geq 0$.

The assumption $m \geq n$ is no restriction. The σ_i are the *singular values*, and the columns of U and V are *left and right singular vectors*, respectively. Suppose that A has rank r. Then $\sigma_r > 0$, $\sigma_{r+1} = 0$, and

$$A = U\Sigma V^{T} = \begin{pmatrix} U_r & \hat{U}_r \end{pmatrix} \begin{pmatrix} \Sigma_r & 0 \\ 0 & 0 \end{pmatrix} \begin{pmatrix} V_r^{T} \\ \hat{V}_r^{T} \end{pmatrix} = U_r \Sigma_r V_r^{T}, \qquad (3.8)$$

where $U_r \in \mathbb{R}^{m \times r}$, $\Sigma_r = \operatorname{diag}(\sigma_1, \ldots, \sigma_r) \in \mathbb{R}^{r \times r}$, and $V_r \in \mathbb{R}^{n \times r}$. From (3.8)

---

[4] It is interesting to note that several of the linear algebra ideas used in data mining were originally conceived in applied statistics and data analysis, especially in psychometrics.

we see that the columns of U and V provide bases for all four *fundamental* subspaces of A:

U_r gives an orthogonal basis for Range(A),
$\hat{V}_r$ gives an orthogonal basis for Null(A),
V_r gives an orthogonal basis for Range(A^T),
$\hat{U}_r$ gives an orthogonal basis for Null(A^T),

where Range and Null denote the range space and the null space of the matrix, respectively.

Often it is convenient to write the SVD in *outer product* form, *i.e.*, express the matrix as a sum of rank-one matrices,

$$A = \sum_{i=1}^{r} \sigma_i u_i v_i^T. \tag{3.9}$$

The SVD can be used to compute the rank of a matrix. However, in floating point arithmetic, the zero singular values usually appear as small numbers. Similarly, if A is made up from a rank-k matrix and additive noise of small magnitude, then it will have k singular values that will be significantly larger than the rest. In general, a large relative gap between two consecutive singular values is considered to reflect *numerical rank deficiency* of a matrix. Therefore, 'noise reduction' can be achieved via a *truncated SVD*. If trailing *small* diagonal elements of Σ are replaced by zeros, then a rank-k approximation A_k of A is obtained as

$$A = \begin{pmatrix} U_k & \hat{U}_k \end{pmatrix} \begin{pmatrix} \Sigma_k & 0 \\ 0 & \hat{\Sigma}_k \end{pmatrix} \begin{pmatrix} V_k^T \\ \hat{V}_k^T \end{pmatrix}$$

$$\approx \begin{pmatrix} U_k & \hat{U}_k \end{pmatrix} \begin{pmatrix} \Sigma_k & 0 \\ 0 & 0 \end{pmatrix} \begin{pmatrix} V_k^T \\ \hat{V}_k^T \end{pmatrix} = U_k \Sigma_k V_k^T =: A_k, \tag{3.10}$$

where $\Sigma_k \in \mathbb{R}^{k \times k}$ and $\|\hat{\Sigma}_k\| < \epsilon$ for a *small* tolerance ϵ.

The low-rank approximation of a matrix obtained in this way from the SVD has an optimality property specified in the following theorem (Eckart and Young 1936, Mirsky 1960), which is the foundation of numerous important procedures in science and engineering. An *orthogonally invariant matrix norm* is one, for which $\|QAP\| = \|A\|$, where Q and P are arbitrary orthogonal matrices (of conforming dimensions). The matrix 2-norm and the Frobenius norm are orthogonally invariant.

Theorem 3.4. Let $\|\cdot\|$ denote any orthogonally invariant norm, and let the SVD of $A \in \mathbb{R}^{m \times n}$ be given as in Theorem 3.3. Assume that an integer k is given with $0 < k \leq r = \mathrm{rank}(A)$. Then

$$\min_{\mathrm{rank}(B)=k} \|A - B\| = \|A - A_k\|,$$

where

$$A_k = U_k \Sigma_k V_k^T = \sum_{i=1}^{k} \sigma_i u_i v_i^T. \tag{3.11}$$

From the theorem we see that the singular values indicate how close a given matrix is to a matrix of lower rank.

The relation between the truncated SVD (3.11) and the Wedderburn matrix rank reduction process can be demonstrated as follows. In the rank reduction formula (3.7), define the error matrix E as

$$E = A - AF(G^T AF)^{-1} G^T A, \qquad F \in \mathbb{R}^{n \times k}, G \in \mathbb{R}^{m \times k}.$$

Assume that $k \leq \mathrm{rank}(A) = r$, and consider the problem

$$\min \|E\| = \min_{F \in \mathbb{R}^{n \times k}, G \in \mathbb{R}^{m \times k}} \|A - AF(G^T AF)^{-1} G^T A\|,$$

where the norm is orthogonally invariant. According to Theorem 3.4, the minimum error is obtained when

$$(AF)(G^T AF)^{-1}(G^T A) = U_k \Sigma_k V_k^T,$$

which is equivalent to choosing $F = V_k$ and $G = U_k$.

This same result can be obtained by a stepwise procedure, when k pairs of vectors $f^{(i)}$ and $g^{(i)}$ are to be found, where each pair reduces the matrix rank by 1.

The Wedderburn procedure helps to elucidate the equivalence between the SVD and *principal component analysis* (PCA) (Joliffe 1986). Let $X \in \mathbb{R}^{m \times n}$ be a data matrix, where each column is an observation of a real-valued random vector. The matrix is assumed to be centred, *i.e.*, the mean of each column is equal to zero. Let the SVD of X be $X = U\Sigma V^T$. The right singular vectors v_i are called *principal component directions* of X (Hastie *et al.* 2001, p. 62). The vector

$$z_1 = Xv_1 = \sigma_1 u_1$$

has the largest sample variance amongst all normalized linear combinations of the columns of X:

$$\mathrm{Var}(z_1) = \mathrm{Var}(Xv_1) = \frac{\sigma_1^2}{m}.$$

Finding the vector of maximal variance is equivalent, using linear algebra terminology, to maximizing the Rayleigh quotient:

$$\sigma_1^2 = \max_{v \neq 0} \frac{v^T X^T Xv}{v^T v}, \qquad v_1 = \arg\max_{v \neq 0} \frac{v^T X^T Xv}{v^T v}.$$

The normalized variable u_1 is called the *normalized first principal component* of X. The second principal component is the vector of largest sample variance of the deflated data matrix $X - \sigma_1 u_1 v_1^T$, and so on. Any subsequent principal component is defined as the vector of maximal variance subject to the constraint that it is orthogonal to the previous ones.

Example 3.5. PCA is illustrated in Figure 3.1. 500 data points from a correlated normal distribution were generated, and collected in a data matrix $X \in \mathbb{R}^{3 \times 500}$. The data points and the principal components are illustrated in the top plot. We then deflated the data matrix: $X_1 := X - \sigma_1 u_1 v_1^T$. The data points corresponding to X_1 are given in the bottom plot; they lie on a plane in $\mathbb{R}^3$, *i.e.*, X_1 has rank 2.

The concept of principal components has been generalized to *principal curves and surfaces*: see Hastie (1984) and Hastie *et al.* (2001, Section 14.5.2). A recent paper along these lines is Einbeck, Tutz and Evers (2005).

3.3. Generalized SVD

The SVD can be used for low-rank approximation involving one matrix. It often happens that two matrices are involved in the criterion that determines the dimension reduction: see Section 4.4. In such cases a generalization of the SVD to two matrices can be used to analyse and compute the dimension reduction transformation.

Theorem 3.6. (GSVD) Let $A \in \mathbb{R}^{m \times n}$, $m \geq n$, and $B \in \mathbb{R}^{p \times n}$. Then there exist orthogonal matrices $U \in \mathbb{R}^{m \times m}$ and $V \in \mathbb{R}^{p \times p}$, and a nonsingular $X \in \mathbb{R}^{n \times n}$, such that

$$U^T A X = C = \mathrm{diag}(c_1, \ldots, c_n), \quad 1 \geq c_1 \geq \cdots \geq c_n \geq 0, \qquad (3.12)$$

$$V^T B X = S = \mathrm{diag}(s_1, \ldots, s_q), \quad 0 \leq s_1 \leq \cdots \leq s_q \leq 1, \qquad (3.13)$$

where $q = \min(p, n)$ and

$$C^T C + S^T S = I.$$

A proof can be found in Golub and Van Loan (1996, Section 8.7.3); see also Van Loan (1976) and Paige and Saunders (1981).

The generalized SVD is sometimes called the *Quotient SVD*.[5] There is also a different generalization, called the *Product SVD*: see, *e.g.*, De Moor and Van Dooren (1992), Golub, Sølna and Van Dooren (2000).

3.4. Partial least squares: Lanczos bidiagonalization

Linear least squares (regression) problems occur frequently in data mining. Consider the minimization problem

$$\min_{\beta} \|y - X\beta\|, \qquad (3.14)$$

---

[5] Assume that B is square and nonsingular. Then the GSVD gives the SVD of AB^{-1}.

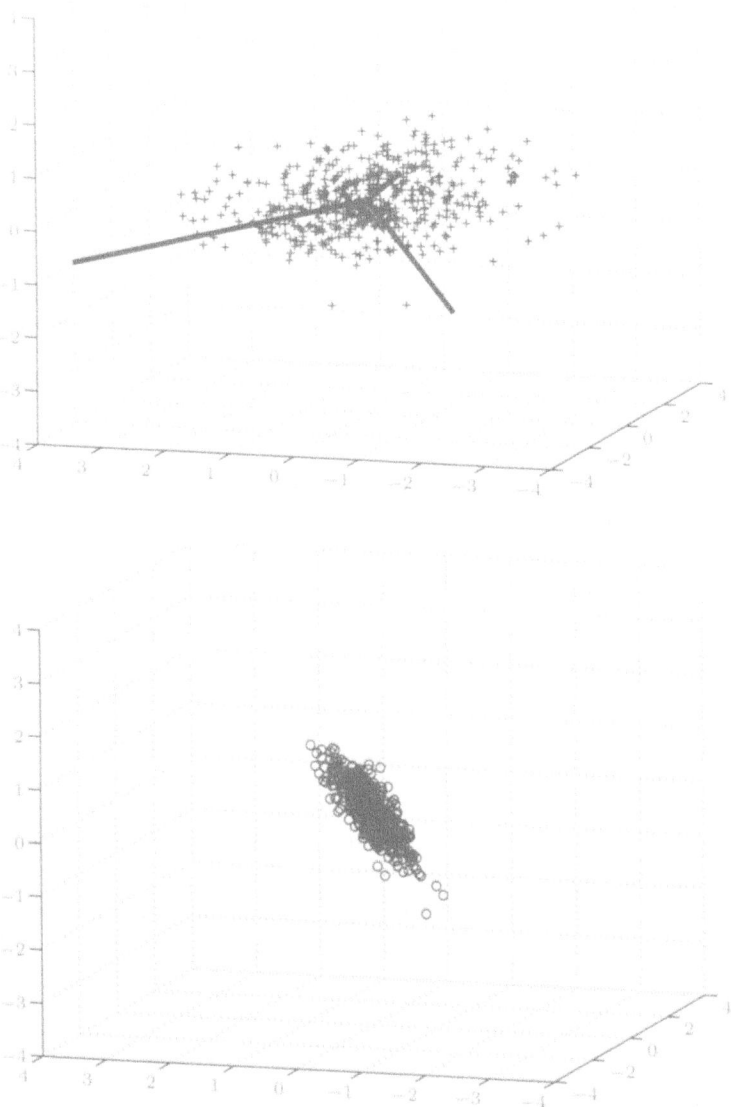

Figure 3.1. Cluster of points in $\mathbb{R}^3$ with (scaled) principal components (top). The same data with the contributions along the first principal component deflated (bottom).

where X is an $m \times n$ real matrix, and the norm is the Euclidean vector norm. This is the *linear least squares problem* in numerical linear algebra, and the *multiple linear regression problem* in statistics. Using regression terminology, the vector y consists of observations of a response variable, and the columns of X contain the values of the explanatory variables. Often the matrix is large and ill-conditioned: the column vectors are (almost) linearly dependent. Sometimes, in addition, the problem is under-determined, *i.e.*, $m < n$. In such cases the straightforward solution of (3.14) may be physically meaningless (from the point of view of the application at hand) and difficult to interpret. Then one may want to express the solution by projecting it onto a lower-dimensional subspace: let W be an $n \times k$ matrix with orthonormal columns. Using this as a basis for the subspace, one considers the approximate minimization

$$\min_{\beta} \|y - X\beta\| \approx \min_{z} \|y - XWz\|. \tag{3.15}$$

One obvious method for projecting the solution onto a low-dimensional subspace is *principal components regression* (*PCR*) (Massy 1965), where the columns of W are chosen as right singular vectors from the SVD of X. In numerical linear algebra this is called *truncated singular value decomposition* (TSVD). Another such projection method, the *partial least squares* (PLS) method (Wold 1975), is standard in chemometrics (Wold, Sjöström and Eriksson 2001). It has been known for quite some time (Wold *et al.* 1984) (see also Helland (1988), Di Ruscio (2000), Phatak and de Hoog (2002)) that PLS is equivalent to Lanczos (Golub–Kahan) bidiagonalization (Golub and Kahan 1965, Paige and Saunders 1982) (we will refer to this as LBD). The equivalence is further discussed in Eldén (2004*b*), and the properties of PLS are analysed using the SVD.

There are several variants of PLS: see, *e.g.*, Frank and Friedman (1993). The following is the so-called NIPALS version.

The NIPALS PLS algorithm

---

1 $X_0 = X$

2 for $i = 1, 2, \ldots, k$

    (a) $w_i = \frac{1}{\|X_{i-1}^T y\|} X_{i-1}^T y$

    (b) $t_i = \frac{1}{\|X_{i-1} w_i\|} X_{i-1} w_i$

    (c) $p_i = X_{i-1}^T t_i$

    (d) $X_i = X_{i-1} - t_i p_i^T$

In the statistics/chemometrics literature the vectors w_i, t_i, and p_i are called *weight*, *score*, and *loading vectors*, respectively.

It is obvious that PLS is a Wedderburn procedure. One advantage of PLS for regression is that the basis vectors in the solution space (the columns of W (3.15)) are influenced by the right-hand side.[6] This is not the case in PCR, where the basis vectors are singular vectors of X. Often PLS gives a higher reduction of the norm of the residual $y - X\beta$ for small values of k than does PCR.

The Lanczos bidiagonalization procedure can be started in different ways: see, e.g., Björck (1996, Section 7.6). It turns out that PLS corresponds to the following formulation.

Lanczos Bidiagonalization (LBD)
---

1 $v_1 = \frac{1}{\|X^T y\|} X^T y; \quad \alpha_1 u_1 = X v_1$

2 for $i = 2, \ldots, k$

    (a) $\gamma_{i-1} v_i = X^T u_{i-1} - \alpha_{i-1} v_{i-1}$

    (b) $\alpha_i u_i = X v_i - \gamma_{i-1} u_{i-1}$

The coefficients γ_{i-1} and α_i are determined so that $\|v_i\| = \|u_i\| = 1$.

Both algorithms generate two sets of orthogonal basis vectors: $(w_i)_{i=1}^k$ and $(t_i)_{i=1}^k$ for PLS, $(v_i)_{i=1}^k$ and $(u_i)_{i=1}^k$ for LBD. It is straightforward to show (directly using the equations defining the algorithm – see Eldén (2004b)) that the two methods are equivalent.

Proposition 3.7. The PLS and LBD methods generate the same orthogonal bases, and the same approximate solution, $\beta_{\text{pls}}^{(k)} = \beta_{\text{lbd}}^{(k)}$.

4. Text mining

By text mining we understand methods for extracting useful information from large and often unstructured collections of texts. A related term is *information retrieval*. A typical application is search in databases of abstract of scientific papers. For instance, in medical applications one may want to find all the abstracts in the database that deal with a particular syndrome. So one puts together a search phrase, a *query*, with key words that are relevant to the syndrome. Then the retrieval system is used to match the query to the documents in the database, and present to the user all the documents that are relevant, preferably ranked according to relevance.

---

[6] However, it is not always appreciated that the dependence of the basis vectors on the right-hand side is non-linear and quite complicated. For a discussion of these aspects of PLS, see Eldén (2004b).

Example 4.1. The following is a typical query:

9. *the use of induced hypothermia in heart surgery, neurosurgery, head injuries and infectious diseases.*

The query is taken from a test collection of medical abstracts, called Medline.[7] We will refer to this query as Q9 from here on.

Another well-known area of text mining is web search engines. There the search phrase is usually very short, and often there are so many relevant documents that it is out of the question to present them all to the user. In that application the ranking of the search result is critical for the efficiency of the search engine. We will come back to this problem in Section 6.1.

For overviews of information retrieval, see, *e.g.*, Korfhage (1997) and Grossman and Frieder (1998). In this section we will describe briefly one of the most common methods for text mining, namely the *vector space model* (Salton, Yang and Wong 1975). In Example 2.2 we demonstrated the basic ideas of the construction of a term-document matrix in the vector space model. Below we first give a very brief overview of the preprocessing that is usually done before the actual term-document matrix is set up. Then we describe a variant of the vector space model: *latent semantic indexing* (LSI) (Deerwester, Dumais, Furnas, Landauer and Harsman 1990), which is based on the SVD of the term-document matrix. For a more detailed account of the different techniques used in connection with the vector space model, see Berry and Browne (2005).

4.1. Vector space model: preprocessing and query matching

In information retrieval, key words that carry information about the contents of a document are called *terms*. A basic task is to create a list of all the terms in alphabetic order, a so-called *index*. But before the index is made, two preprocessing steps should be done: (1) eliminate all stop words, (2) perform stemming.

Stop words are extremely common words. The occurrence of such a word in a document does not distinguish it from other documents. The following is the beginning of one stop list:[8]

a, a's, able, about, above, according, accordingly, across, actually, after, afterwards, again, against, ain't, all, allow, allows, almost, alone, along, already, also, although, always, am, among, amongst, an, and, ...

---

[7] See, *e.g.*, `http://www.dcs.gla.ac.uk/idom/ir_resources/test_collections/`
[8] `ftp://ftp.cs.cornell.edu/pub/smart/english.stop`

Stemming is the process of reducing each word that is conjugated or has a suffix to its stem. Clearly, from the point of view of information retrieval, no information is lost in the following reduction:

$$
\left.
\begin{array}{l}
\text{computable} \\
\text{computation} \\
\text{computing} \\
\text{computed} \\
\text{computational}
\end{array}
\right\}
\quad \longrightarrow \quad \text{comput}
$$

Public domain stemming algorithms are available on the Internet.[9]

A number of pre-processed documents are parsed,[10] giving a term-document matrix $A \in \mathbb{R}^{m \times n}$, where m is the number of terms in the dictionary and n is the number of documents. It is common not only to count the occurrence of terms in documents but also to apply a *term-weighting scheme*, where the elements of A are weighted depending on the characteristics of the document collection. Similarly, document weighting is usually done. A number of schemes are described in Berry and Browne (2005, Section 3.2.1). For example, one can define the elements in A by

$$a_{ij} = f_{ij}\log(n/n_i), \tag{4.1}$$

where f_{ij} is the term frequency, the number of times term i appears in document j, and n_i is the number of documents that contain term i (inverse document frequency). If a term occurs frequently in only a few documents, then both factors are large. In this case the term discriminates well between different groups of documents, and it gets a large weight in the documents where it appears.

Normally, the term-document matrix is *sparse*: most of the matrix elements are equal to zero. Then, of course, one avoids storing all the zeros, and instead uses a sparse matrix storage scheme (see, *e.g.*, Saad (2003, Chapter 3) and Goharian, Jain and Sun (2003)).

Example 4.2. For the stemmed Medline collection (*cf.* Example 4.1) the matrix (including 30 query columns) is 4163×1063 with 48263 nonzero elements, *i.e.*, approximately 1%. The first 500 rows and columns of the matrix are illustrated in Figure 4.1.

The query (*cf.* Example 4.1) is parsed using the same dictionary as the documents, giving a vector $q \in \mathbb{R}^m$. *Query matching* is the process of finding

---

[9] http://www.tartarus.org/~martin/PorterStemmer/.

[10] Public domain text parsers are described in Giles, Wo and Berry (2003) and Zeimpekis and Gallopoulos (2005).

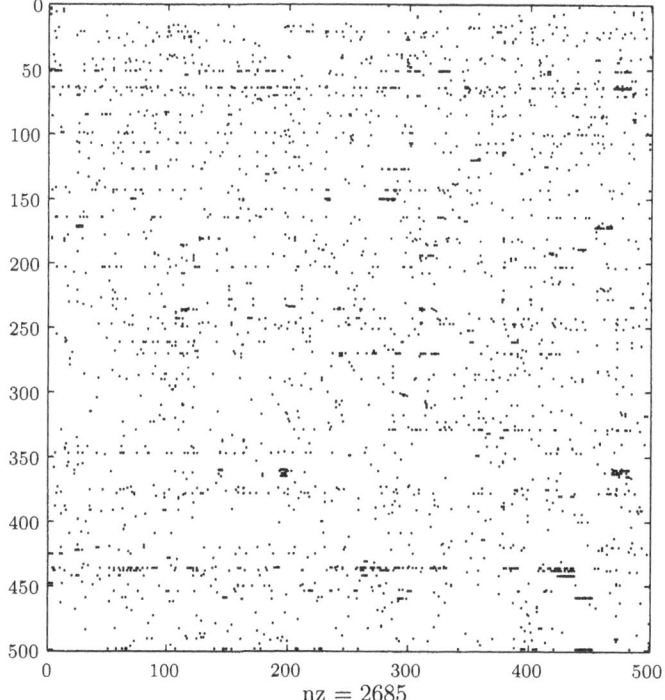

Figure 4.1. The first 500 rows and columns of the Medline matrix. Each dot represents a nonzero element.

all documents that are considered relevant to a particular query q. This is often done using the cosine distance measure: *all documents are returned for which*

$$\frac{q^T a_j}{\|q\|_2 \|a_j\|_2} > \text{tol}, \tag{4.2}$$

where tol is a predefined tolerance. If the tolerance is lowered, then more documents are returned, and then it is likely that more of the documents that are relevant to the query are returned. But at the same time there is a risk that more documents that are not relevant are also returned.

Example 4.3. We did query matching for query Q9 in the stemmed Medline collection. With tol = 0.19 only document 409 was considered relevant. When the tolerance was lowered to 0.17, then documents 409, 415, and 467 were retrieved.

We illustrate the different categories of documents in a query matching for two values of the tolerance in Figure 4.2. The query matching produces a good result when the intersection between the two sets of returned and

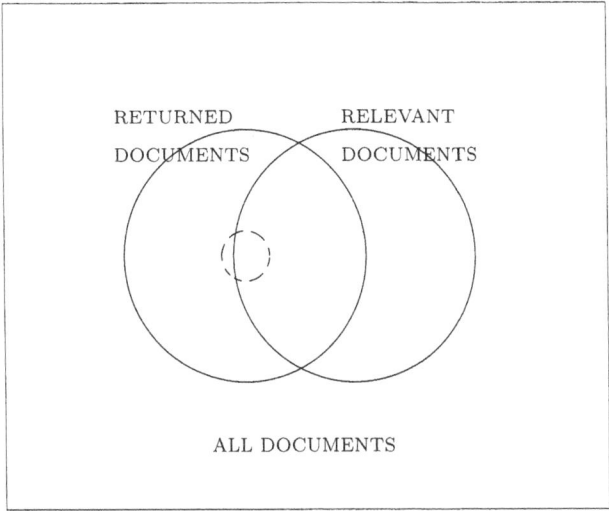

Figure 4.2. Returned and relevant documents for two values of the tolerance. The dashed circle represents the returned documents for a high value of the cosine tolerance.

relevant documents is as large as possible, and the number of returned irrelevant documents is small. For a high value of the tolerance, the retrieved documents are likely to be relevant (the small circle in Figure 4.2). When the cosine tolerance is lowered, then the intersection is increased, but at the same time, more irrelevant documents are returned.

In performance modelling for information retrieval we define the following measures:

$$precision \qquad P = \frac{D_r}{D_t},$$

where D_r is the number of relevant documents retrieved, and D_t the total number of documents retrieved; and

$$recall \qquad R = \frac{D_r}{N_r},$$

where N_r is the total number of relevant documents in the data base. With the cosine measure, we see that with a large value of tol we have high precision, but low recall. For a small value of tol we have high recall, but low precision.

In the evaluation of different methods and models for information retrieval usually a number of queries are used. For testing purposes all documents have been read by a human and those that are relevant to a certain query are marked.

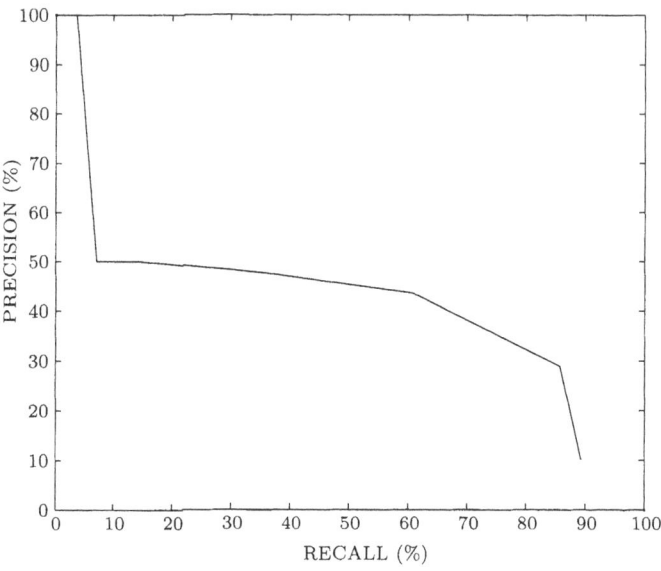

Figure 4.3. Query matching for Q9 using the
vector space method. Recall versus precision.

Example 4.4. We did query matching for query Q9 in the Medline collection (stemmed) using the cosine measure, and obtained recall and precision
as illustrated in Figure 4.3. In the comparison of different methods it is more
illustrative to draw the recall versus precision diagram. Ideally a method
has high recall at the same time as the precision is high. Thus, the closer
the curve is to the upper right corner, the better the method.

In this example and the following examples the matrix elements were
computed using term frequency and inverse document frequency weighting (4.1).

4.2. LSI: latent semantic indexing

Latent semantic indexing[11] (LSI) 'is based on the assumption that there is
some underlying latent semantic structure in the data . . . that is corrupted
by the wide variety of words used . . . ' (quoted from Park, Jeon and Rosen
(2001)) and that this semantic structure can be enhanced by projecting the
data (the term-document matrix and the queries) onto a lower-dimensional
space using the singular value decomposition. LSI is discussed in Deerwester
et al. (1990), Berry, Dumais and O'Brien (1995), Berry, Drmac and Jessup
(1999), Berry (2001), Jessup and Martin (2001) and Berry and Browne
(2005).

---

[11] Sometimes also called *latent semantic analysis* (LSA) (Jessup and Martin 2001).

Let $A = U \Sigma V^T$ be the SVD of the term-document matrix and approximate it by a matrix of rank k:

$$= U_k(\Sigma_k V_k) =: U_k D_k.$$

The columns of U_k live in the document space and are an orthogonal basis that we use to approximate all the documents: column j of D_k holds the coordinates of document j in terms of the orthogonal basis. With this k-dimensional approximation the term-document matrix is represented by $A_k = U_k D_k$, and in query matching we compute $q^T A_k = q^T U_k D_k = (U_k^T q)^T D_k$. Thus, we compute the coordinates of the query in terms of the new document basis and compute the cosines from

$$\cos \theta_j = \frac{q_k^T (D_k e_j)}{\|q_k\|_2 \, \|D_k e_j\|_2}, \qquad q_k = U_k^T q. \tag{4.3}$$

This means that the query matching is performed in a k-dimensional space.

Example 4.5. We did query matching for Q9 in the Medline collection, approximating the matrix using the truncated SVD of rank 100. The recall–precision curve is given in Figure 4.4. It is seen that for this query LSI improves the retrieval performance. In Figure 4.5 we also demonstrate a fact that is common to many term-document matrices: it is rather well conditioned, and there is no gap in the sequence of singular values. Therefore, we cannot find a suitable rank of the LSI approximation by inspecting the singular values: it must be determined by retrieval experiments.

Another remarkable fact is that with $k = 100$ the approximation error in the matrix approximation,

$$\frac{\|A - A_k\|_F}{\|A\|_F} \approx 0.8,$$

is large, and we still get *improved retrieval performance*. In view of the large approximation error in the truncated SVD approximation of the term-document matrix, one may question whether the 'optimal' singular vectors constitute the best basis for representing the term-document matrix. On the other hand, since we get such good results, perhaps a more natural conclusion may be that the Frobenius norm is not a good measure of the information contents in the term-document matrix.

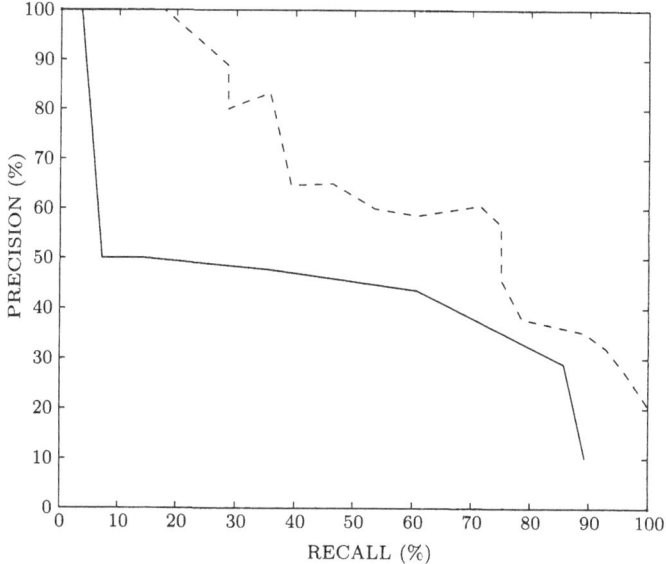

Figure 4.4. Query matching for Q9. Recall versus
precision for the full vector space model (solid
line) and the rank-100 approximation (dashed).

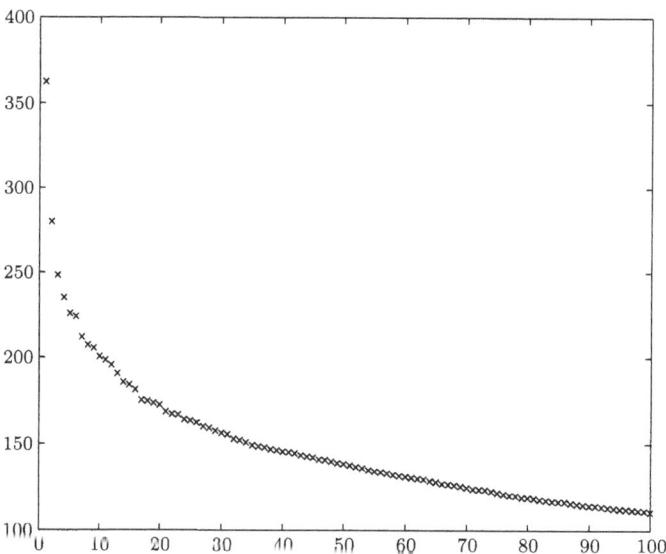

Figure 4.5. First 100 singular values
of the Medline (stemmed) matrix.

It is also interesting to see what are the most important 'directions' in the data. From Theorem 3.4 we know that the first few left singular vectors are the dominant directions in the document space, and their largest components should indicate what these directions are. The Matlab statements `find(abs(U(:,k))>0.13)`, combined with look-up in the dictionary of terms, gave the results shown in Table 4.1, for `k=1,2`.

Table 4.1.

U(:,1)	U(:,2)
cell	case
growth	cell
hormone	children
patient	defect
	dna
	growth
	patient
	ventricular

It should be said that LSI does not give significantly better results for all queries in the Medline collection: there are some where it gives results comparable to the full vector model, and some where it gives worse performance. However, it is often the average performance that matters.

Jessup and Martin (2001) made a systematic study of different aspects of LSI. They showed that LSI improves retrieval performance for surprisingly small values of the reduced rank k. At the same time the relative matrix approximation errors are large. It is probably not possible to prove any general results for LSI that explain how and for which data it can improve retrieval performance. Instead we give an artificial example (constructed using ideas similar to those of a corresponding example in Berry and Browne (2005)) that gives a partial explanation.

Example 4.6. Consider the term-document matrix from Example 2.2, and the query '**ranking** of **web pages**'. Obviously, Documents 1–4 are relevant with respect to the query, while Document 5 is totally irrelevant. However, we obtain the following cosines for query and the original data:

$$(0 \quad 0.6667 \quad 0.7746 \quad 0.3333 \quad 0.3333).$$

We then compute the SVD of the term-document matrix, and use a rank-two approximation. After projection to the two-dimensional subspace the

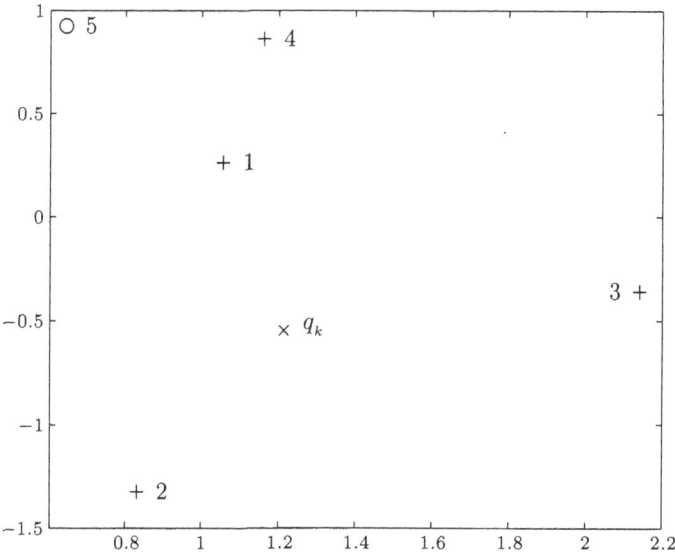

Figure 4.6. The documents and the query projected to
the coordinate system of the first two left singular vectors.

cosines, computed according to (4.3), are

$$(0.7857 \quad 0.8332 \quad 0.9670 \quad 0.4873 \quad 0.1819).$$

It turns out that Document 1, which was deemed totally irrelevant to the query in the original representation, is now highly relevant. In addition, the scores for the relevant Documents 2–4 have been reinforced. At the same time, the score for Document 5 has been significantly reduced. Thus, in this artificial example, the dimension reduction enhanced the retrieval performance. The improvement may be explained as follows.

   In Figure 4.6 we plot the five documents and the query in the coordinate system of the first two left singular vectors. Obviously, in this representation, the first document is is closer to the query than Document 5. The first two left singular vectors are

$$
u_1 = \begin{pmatrix} 0.1425 \\ 0.0787 \\ 0.0787 \\ 0.3924 \\ 0.1297 \\ 0.1020 \\ 0.5348 \\ 0.3647 \\ 0.4838 \\ 0.3647 \end{pmatrix}, \quad \begin{pmatrix} 0.2430 \\ 0.2607 \\ 0.2607 \\ -0.0274 \\ 0.0740 \\ -0.3735 \\ 0.2156 \\ -0.4749 \\ 0.4023 \\ -0.4749 \end{pmatrix},
$$

and the singular values are $\Sigma = \mathrm{diag}(2.8546, 1.8823, 1.7321, 1.2603, 0.8483)$. The first four columns in A are strongly coupled via the words *Google*, *matrix*, etc., and those words are the dominating contents of the document collection (*cf.* the singular values). This shows in the composition of u_1. So even if none of the words in the query is matched by Document 1, that document is so strongly correlated to the the dominating direction that it becomes relevant in the reduced representation.

4.3. Clustering and least squares

Clustering is widely used in pattern recognition and data mining. We give here a brief account of the application of clustering to text mining.

Clustering is the grouping together of similar objects. In the vector space model for text mining, similarity is defined as the distance between points in $\mathbb{R}^m$, where m is the number of terms in the dictionary. There are many clustering methods, *e.g.*, the k-means method, agglomerative clustering, self-organizing maps, and multi-dimensional scaling: see the references in Dhillon (2001), Dhillon, Fan and Guan (2001).

The relation between the SVD and clustering is explored in Dhillon (2001); see also Zha, Ding, Gu, He and Simon (2002) and Dhillon, Guan and Kulis (2005). Here the approach is graph-theoretic. The sparse term-document matrix represents a bi-partite graph, where the two sets of vertices are the documents $\{d_j\}$ and the terms $\{t_i\}$. An edge (t_i, d_j) exists if term t_i occurs in document d_j, *i.e.*, if the element in position (i, j) is nonzero. Clustering the documents is then equivalent to partitioning the graph. A *spectral partitioning* method is described, where the eigenvectors of a Laplacian of the graph are optimal partitioning vectors. Equivalently, the singular vectors of a related matrix can be used. It is of some interest that spectral clustering methods are related to algorithms for the partitioning of meshes in parallel finite element computations: see, *e.g.*, Simon, Sohn and Biswas (1998).

Clustering for text mining is discussed in Dhillon and Modha (2001) and Park, Jeon and Rosen (2003), and the similarities between LSI and clustering are pointed out in Dhillon and Modha (2001).

Given a partitioning of a term-document matrix into k clusters,

$$A = \begin{pmatrix} A_1 & A_2 & \cdots & A_k \end{pmatrix}, \tag{4.4}$$

where $A_j \in \mathbb{R}^{n_j}$, one can take the *centroid* of each cluster,[12]

$$c^{(j)} = \frac{1}{n_j} A_j e^{(j)}, \qquad e^{(j)} = \begin{pmatrix} 1 & 1 & \cdots & 1 \end{pmatrix}^T, \tag{4.5}$$

with $e^{(j)} \in \mathbb{R}^{n_j}$, as a representative of the class. Together the centroid

---

[12] In Dhillon and Modha (2001) normalized centroids are called *concept vectors*.

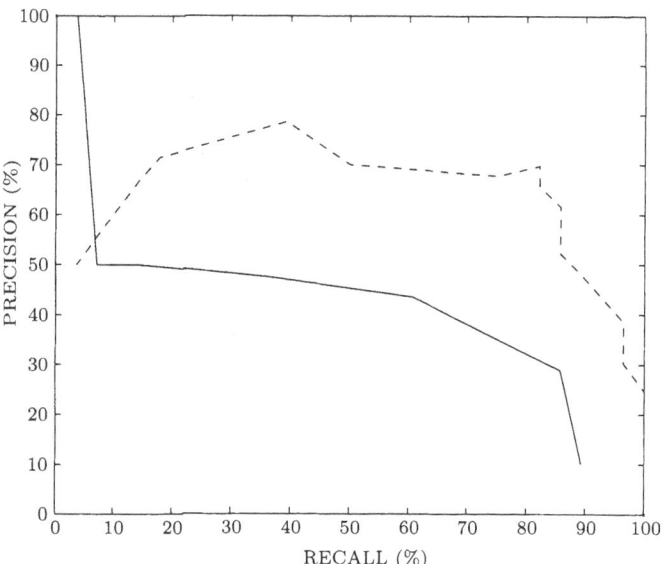

Figure 4.7. Query matching for Q9. Recall versus precision for the full vector space model (solid line) and the rank-50 centroid approximation (dashed).

vectors can be used as an approximate basis for the document collection, and the coordinates of each document with respect to this basis can be computed by solving the least squares problem

$$\min_{D} \|A - CD\|_F, \qquad C = \begin{pmatrix} c^{(1)} & c^{(2)} & \cdots & c^{(k)} \end{pmatrix}. \tag{4.6}$$

Example 4.7. We did query matching for Q9 in the Medline collection. Before computing the clustering we normalized the columns to equal Euclidean length. We approximated the matrix using the orthonormalized centroids from a clustering into 50 clusters. The recall–precision diagram is given in Figure 4.7. We see that for high values of recall, the centroid method is as good as the LSI method with double the rank: see Figure 4.4.

For rank 50 the approximation error in the centroid method,

$$\|A - CD\|_F / \|A\|_F \approx 0.9,$$

is even higher than for LSI of rank 100.

The improved performance can be explained in a similar way as for LSI. Being the 'average document' of a cluster, the centroid captures the main links between the dominant documents in the cluster. By expressing all documents in terms of the centroids, the dominant links are emphasized.

4.4. Clustering and linear discriminant analysis

When centroids are used as basis vectors, the coordinates of the documents are computed from (4.6) as

$$D := G^T A, \qquad G^T = R^{-1} Q^T,$$

where $C = QR$ is the thin QR decomposition[13] of the centroid matrix. The criterion for choosing G is based on approximating the term-document matrix A as well as possible in the Frobenius norm. As we have seen earlier (Examples 4.5 and 4.7), a good approximation of A is not always necessary for good retrieval performance, and it may be natural to look for other criteria for determining the matrix G in a dimension reduction.

   Linear discriminant analysis (LDA) is frequently used for classification (Duda *et al.* 2001). In the context of cluster-based text mining, LDA is used to derive a transformation G, such that the cluster structure is as well preserved as possible in the dimension reduction.

   In Howland, Jeon and Park (2003) and Howland and Park (2004) the application of LDA to text mining is explored, and it is shown how the GSVD (Theorem 3.6) can be used to extend the dimension reduction procedure to cases where the standard LDA criterion is not valid.

   Assume that a clustering of the documents has been made as in (4.4) with centroids (4.5). Define the overall centroid

$$c = Ae, \qquad e = \frac{1}{\sqrt{n}} \begin{pmatrix} 1 & 1 & \cdots & 1 \end{pmatrix}^T,$$

the three matrices[14]

$$\mathbb{R}^{m \times n} \ni H_w = \begin{pmatrix} A_1 - c^{(1)} e^{(1)T} & A_2 - c^{(2)} e^{(2)T} & \cdots & A_k - c^{(k)} e^{(k)T} \end{pmatrix},$$

$$\mathbb{R}^{m \times k} \ni H_b = \begin{pmatrix} \sqrt{n_1}(c^{(1)} - c) & \sqrt{n_2}(c^{(2)} - c) & \cdots & \sqrt{n_k}(c^{(k)} - c) \end{pmatrix},$$

$$\mathbb{R}^{m \times n} \ni H_m = A - c e^T,$$

and the corresponding *scatter matrices*

$$S_w = H_w H_w^T,$$
$$S_b = H_b H_b^T,$$
$$S_m = H_m H_m^T.$$

Assume that we want to use the (dimension-reduced) clustering for classifying new documents, *i.e.*, determine to which cluster they belong. The 'quality of the clustering' with respect to this task depends on how 'tight'

---

[13] The *thin QR decomposition* of a matrix $A \in \mathbb{R}^{m \times n}$, with $m \geq n$, is $A = QR$, where $Q \in \mathbb{R}^{m \times n}$ has orthonormal columns and R is upper triangular.

[14] Note: subscript w for 'within classes', b for 'between classes'.

or coherent each cluster is, and how well separated the clusters are. The overall tightness ('within-class scatter') of the clustering can be measured as

$$J_w = \mathrm{tr}(S_w) = \|H_w\|_F^2,$$

and the separateness ('between-class scatter') of the clusters by

$$J_b = \mathrm{tr}(S_b) = \|H_b\|_F^2.$$

Ideally, the clusters should be separated at the same time as each cluster is tight. Different quality measures can be defined. Often in LDA one uses

$$J = \frac{\mathrm{tr}(S_b)}{\mathrm{tr}(S_w)}, \tag{4.7}$$

with the motivation that if all the clusters are tight then S_w is small, and if the clusters are well separated then S_b is large. Thus the quality of the clustering with respect to classification is high if J is large. Similar measures are considered in Howland and Park (2004).

Now assume that we want to determine a dimension reduction transformation, represented by the matrix $G \in \mathbb{R}^{m \times d}$, such that the quality of the reduced representation is as high as possible. After the dimension reduction, the tightness and separateness are

$$J_b(G) = \|G^T H_b\|_F^2 = \mathrm{tr}(G^T S_b G),$$
$$J_w(G) = \|G^T H_w\|_F^2 = \mathrm{tr}(G^T S_w G).$$

Since $\mathrm{rank}(H_b) \le k - 1$, it is only meaningful to choose $d = k - 1$: see Howland et al. (2003).

The question arises whether it is possible to determine G so that, in a consistent way, the quotient $J_b(G)/J_w(G)$ is maximized. The answer is derived using the GSVD of H_w^T and H_b^T. We assume that $m > n$; see Howland et al. (2003) for a treatment of the general (but with respect to the text mining application more restrictive) case. We further assume

$$\mathrm{rank}\begin{pmatrix} H_b^T \\ H_w^T \end{pmatrix} = t.$$

Under these assumptions the GSVD has the form (Paige and Saunders 1981)

$$H_b^T = U^T \Sigma_b (Z\ 0) Q^T, \tag{4.8}$$
$$H_w^T = V^T \Sigma_w (Z\ 0) Q^T, \tag{4.9}$$

where U and V are orthogonal, $Z \in \mathbb{R}^{t \times t}$ is nonsingular, and $Q \in \mathbb{R}^{m \times m}$ is orthogonal. The diagonal matrices Σ_b and Σ_w will be specified shortly. We first see that, with

$$\tilde{G} = Q^T G = \begin{pmatrix} \tilde{G}_1 \\ \tilde{G}_2 \end{pmatrix}, \qquad \tilde{G}_1 \in \mathbb{R}^{t \times d},$$

we have

$$J_b(G) = \|\Sigma_b Z \tilde{G}_1\|_F^2, \qquad J_w(G) = \|\Sigma_w Z \tilde{G}_1\|_F^2. \qquad (4.10)$$

Obviously, we should not waste the degrees of freedom in G by choosing a nonzero $\tilde{G}_2$, since that would not affect the quality of the clustering after dimension reduction. Next we specify

$$\mathbb{R}^{(k-1) \times t} \ni \Sigma_b = \begin{pmatrix} I_b & 0 & 0 \\ 0 & D_b & 0 \\ 0 & 0 & 0_b \end{pmatrix},$$

$$\mathbb{R}^{n \times t} \ni \Sigma_w = \begin{pmatrix} 0_w & 0 & 0 \\ 0 & D_w & 0 \\ 0 & 0 & I_w \end{pmatrix},$$

where $I_b \in \mathbb{R}^{(t-s) \times (t-s)}$ and $I_w \in \mathbb{R}^{r \times r}$ are identity matrices with data-dependent values of r and s, and $0_b \in \mathbb{R}^{1 \times r}$ and $0_w \in \mathbb{R}^{(n-s) \times (t-s)}$ are zero matrices. The diagonal matrices satisfy

$$D_b = \text{diag}(\alpha_{r+1}, \ldots, \alpha_{r+s}), \quad \alpha_{r+1} \geq \cdots \geq \alpha_{r+s} > 0, \qquad (4.11)$$
$$D_w = \text{diag}(\beta_{r+1}, \ldots, \beta_{r+s}), \quad 0 < \beta_{r+1} \leq \cdots \leq \beta_{r+s}, \qquad (4.12)$$

and $\alpha_i^2 + \beta_i^2 = 1$, $i = r+1, \ldots, r+s$. Note that the column-wise partitionings of Σ_b and Σ_w are identical. Now we define

$$\hat{G} = Z\tilde{G}_1 = \begin{pmatrix} \hat{G}_1 \\ \hat{G}_2 \\ \hat{G}_3 \end{pmatrix},$$

where the partitioning conforms with that of Σ_b and Σ_w. Then we have

$$J_b(G) = \|\Sigma_b \hat{G}\|_F^2 = \|\hat{G}_1\|_F^2 + \|D_b \hat{G}_2\|_F^2,$$
$$J_w(G) = \|\Sigma_w \hat{G}\|_F^2 = \|D_w \hat{G}_2\|_F^2 + \|\hat{G}_3\|_F^2.$$

At this point we see that the maximization of

$$\frac{J_b(G)}{J_w(G)} = \frac{\text{tr}(\hat{G}^T \Sigma_b^T \Sigma_b \hat{G})}{\text{tr}(\hat{G}^T \Sigma_w^T \Sigma_w \hat{G})} \qquad (4.13)$$

is not a well-defined problem: We can make $J_b(G)$ large simply by choosing $\hat{G}_1$ large, without changing $J_w(G)$. On the other hand, (4.13) can be considered as the Rayleigh quotient of a generalized eigenvalue problem (see, e.g., Golub and Van Loan (1996, Section 8.7.2)), where the largest set of eigenvalues are infinite (since the first eigenvalues of $\Sigma_b^T \Sigma_b$ and $\Sigma_w^T \Sigma_w$ are 1 and 0, respectively), and the following are $\alpha_{r+i}^2 / \beta_{r+i}^2$, $i = 1, 2, \ldots, s$. With this in mind it is natural to constrain the data of the problem so that

$$\hat{G}^T \hat{G} = I. \qquad (4.14)$$

We see that, under this constraint,

$$\widehat{G} = \begin{pmatrix} I \\ 0 \end{pmatrix}, \tag{4.15}$$

is a (non-unique) solution of the maximization of (4.13). Consequently, the transformation matrix G is chosen as

$$G = Q \begin{pmatrix} Z^{-1}\widehat{G} \\ 0 \end{pmatrix} = Q \begin{pmatrix} Y_1 \\ 0 \end{pmatrix},$$

where Y_1 denotes the first $k-1$ columns of Z^{-1}.

LDA-based dimension reduction was tested in Howland *et al.* (2003) on data (abstracts) from the Medline database. Classification results were obtained for the compressed data, with much better precision than using the full vector space model.

4.5. Text mining using Lanczos bidiagonalization (PLS)

In LSI and cluster-based methods, the dimension reduction is determined completely from the term-document matrix, and therefore it is the same for all query vectors. In chemometrics it has been known for a long time that PLS (Lanczos bidiagonalization) often gives considerably more efficient compression (in terms of the dimensions of the subspaces used) than PCA (LSI/SVD), the reason being that the right-hand side (of the least squares problem) determines the choice of basis vectors.

In a series of papers (see Blom and Ruhe (2005)), the use of Lanczos bidiagonalization for text mining has been investigated. The recursion starts with the normalized query vector and computes two orthonormal bases[15] P and Q.

Lanczos Bidiagonalization

---

1 $q_1 = q/\|q\|_2, \quad \beta_1 = 0, \quad p_0 = 0.$

2 **for** $i = 2, \dots, k$

   (a) $\alpha_i p_i = A^T q_i - \beta_i p_{i-1}.$

   (b) $\beta_{i+1} q_{i+1} = Apk - \alpha_i q_i.$

The coefficients α_i and β_{i+1} are determined so that $\|p_i\|_2 = \|q_{i+1}\|_2 = 1.$

---

[15] We use a slightly different notation here to emphasize that the starting vector is different from that in Section 3.4.

Define the matrices

$$Q_i = \begin{pmatrix} q_1 & q_2 & \cdots & q_i \end{pmatrix},$$
$$P_i = \begin{pmatrix} p_1 & p_2 & \cdots & p_i \end{pmatrix},$$

$$B_{i+1,i} = \begin{pmatrix} \alpha_1 & & & \\ \beta_2 & \alpha_2 & & \\ & \ddots & \alpha_i & \\ & & \beta_{i+1} \end{pmatrix}.$$

The recursion can be formulated as matrix equations,

$$A^T Q_i = P_i B_{i,i}^T,$$
$$A P_i = Q_{i+1} B_{i+1,i}. \tag{4.16}$$

If we compute the thin QR decomposition of $B_{i+1,i}$,

$$B_{i+1,i} = H_{i+1,i+1} R,$$

then we can write (4.16)

$$A P_i = W_i R, \qquad W_i = Q_{i+1} H_{i+1,i+1},$$

which means that the columns of W_i are an approximate orthogonal basis of the document space (*cf.* the corresponding equation $A V_i = U_i \Sigma_i$ for the LSI approximation, where we use the columns of U_i as basis vectors). Thus we have

$$A \approx W_i D_i, \qquad D_i = W_i^T A, \tag{4.17}$$

and we can use this low-rank approximation in the same way as in the LSI method.

The convergence of the recursion can be monitored by computing the residual $\|A P_i z - q\|_2$. It is easy to show (see, *e.g.*, Blom and Ruhe (2005)) that this quantity is equal in magnitude to a certain element in the matrix $H_{i+1,i+1}$. When the residual is smaller than a prescribed tolerance, the approximation (4.17) is deemed good enough for this particular query.

In this approach the matrix approximation is recomputed for every query. This has the following advantages.

(1) Since the right-hand side influences the choice of basis vectors, only a very few steps of the bidiagonalization algorithm need be taken. Blom and Ruhe (2005) report tests for which this algorithm performed better, with $k = 3$, than LSI with subspace dimension 259.

(2) The computation is relatively cheap, the dominating cost being a small number of matrix-vector multiplications.

(3) Most information retrieval systems change with time, when new documents are added. In LSI this necessitates the updating of the SVD of the term-document matrix. Unfortunately, it is quite expensive to update an SVD. The Lanczos-based method, on the other hand, adapts immediately and at no extra cost to changes of A.

5. Classification and pattern recognition

5.1. Classification of handwritten digits using SVD bases

Computer classification of handwritten digits is a standard problem in pattern recognition. The typical application is automatic reading of zip codes on envelopes. A comprehensive review of different algorithms is given in LeCun, Bottou, Bengio and Haffner (1998).

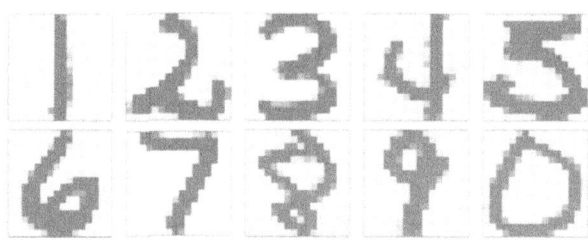

Figure 5.1. Handwritten digits from
the US Postal Service database.

In Figure 5.1 we illustrate handwritten digits that we will use in the examples in this section.

We will treat the digits in three different, but equivalent ways:

(1) 16×16 grey-scale images,

(2) functions of two variables,

(3) vectors in $\mathbb{R}^{256}$.

In the classification of an unknown digit it is necessary to compute the distance to known digits. Different distance measures can be used, perhaps the most natural is Euclidean distance: stack the columns of the image in a vector and identify each digit as a vector in $\mathbb{R}^{256}$. Then define the distance function

$$\text{dist}(x, y) = \|x - y\|_2.$$

An alternative distance function can be based on the cosine between two vectors.

In a real application of recognition of handwritten digits, *e.g.*, zip code reading, there are hardware and real time factors that must be taken into account. In this section we will describe an idealized setting. The problem is:

Given a set of of manually classified digits (the training set), classify a set of unknown digits (the test set).

In the US Postal Service database, the training set contains 7291 handwritten digits, and the test set has 2007 digits.

When we consider the training set digits as vectors or points, then it is reasonable to assume that all digits of one kind form a cluster of points in a Euclidean 256-dimensional vector space. Ideally the clusters are well separated and the separation depends on how well written the training digits are.

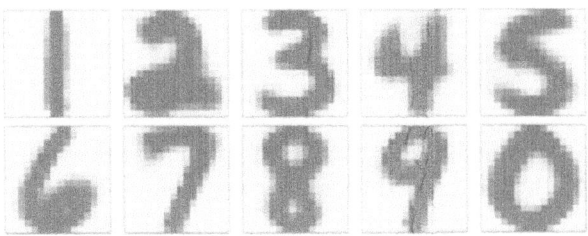

Figure 5.2. The means (centroids)
of all digits in the training set.

In Figure 5.2 we illustrate the means (centroids) of the digits in the training set. From this figure we get the impression that a majority of the digits are well written (if there were many badly written digits this would demonstrate itself as diffuse means). This means that the clusters are rather well separated. Therefore it is likely that a simple algorithm that computes the distance from each unknown digit to the means should work rather well.

A simple classification algorithm

Training. Given the training set, compute the mean (centroid) of all digits of one kind.

Classification. For each digit in the test set, compute the distance to all ten means, and classify as the closest.

Figure 5.3. Singular values (top), and the first
three singular images (vectors) computed
using the 131 3s of the training set (bottom).

It turns out that for our test set the success rate of this algorithm is
around 75%, which is not good enough. The reason is that the algorithm
does not use any information about the variation of the digits of one kind.
This variation can be modelled using the SVD.

Let $A \in \mathbb{R}^{m \times n}$, with $m = 256$, be the matrix consisting of all the training
digits of one kind, the 3s, say. The columns of A span a linear subspace of
$\mathbb{R}^m$. However, this subspace cannot be expected to have a large dimension,
because if it had, then the subspaces of the different kinds of digits would
intersect.

The idea now is to 'model' the variation within the set of training digits
of one kind using an orthogonal basis of the subspace. An orthogonal basis
can be computed using the SVD, and A can be approximated by a sum of
rank-one matrices (3.9),

$$A = \sum_{i=1}^{k} \sigma_i u_i v_i^T,$$

for some value of k. Each column in A is an image of a digit 3, and therefore the left singular vectors u_i are an orthogonal basis in the 'image space of 3s'. We will refer to the left singular vectors as 'singular images'. From the matrix approximation properties of the SVD (Theorem 3.4) we know that the first singular vector represents the 'dominating' direction of the data matrix. Therefore, if we fold the vectors u_i back to images, we expect the first singular vector to look like a 3, and the following singular images should represent the dominating variations of the training set around the first singular image. In Figure 5.3 we illustrate the singular values and the first three singular images for the training set 3s.

The SVD basis classification algorithm will be based on the following assumptions.

(1) Each digit (in the training and test sets) is well characterized by a few of the first singular images of its own kind. The more precise meaning of 'a few' should be investigated by experiment.

(2) An expansion in terms of the first few singular images discriminates well between the different classes of digits.

(3) If an unknown digit can be better approximated in one particular basis of singular images, the basis of 3s say, than in the bases of the other classes, then it is likely that the unknown digit is a 3.

Thus we should compute how well an unknown digit can be represented in the ten different bases. This can be done by computing the residual vector in *least squares problems* of the type

$$\min_{\alpha_i} \left\| z - \sum_{i=1}^{k} \alpha_i u_i \right\|,$$

where z represents an unknown digit, and u_i the singular images. We can write this problem in the form

$$\min_{\alpha} \| z - U_k \alpha \|_2,$$

where $U_k = \begin{pmatrix} u_1 & u_2 & \cdots & u_k \end{pmatrix}$. Since the columns of U_k are orthogonal, the solution of this problem is given by $\alpha = U_k^T z$, and the norm of the residual vector of the least squares problems is

$$\| (I - U_k U_k^T) z \|_2. \tag{5.1}$$

It is interesting to see how the residual depends on the number of terms in the basis. In Figure 5.4 we illustrate the approximation of a nicely written 3 in terms of the 3-basis with different numbers of basis images. In Figure 5.5 we show the approximation of a nice 3 in the 5-basis.

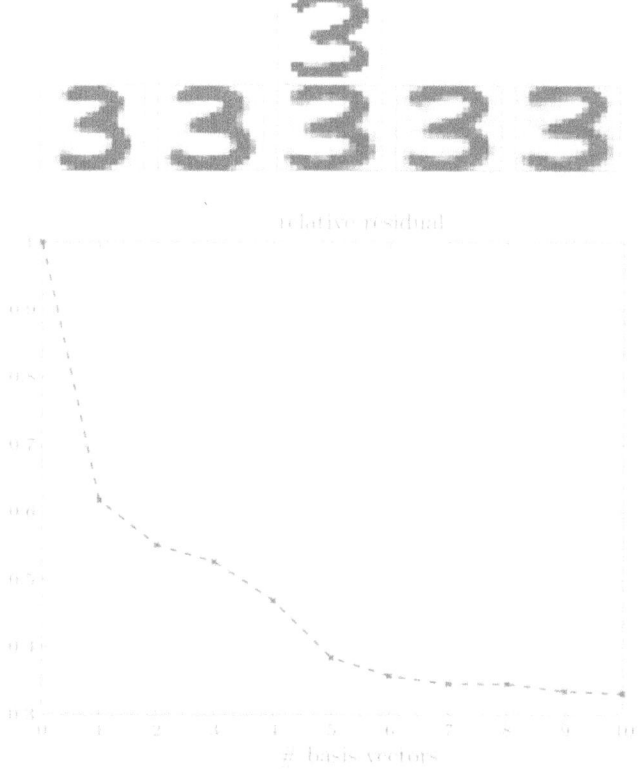

Figure 5.4. Unknown digit (nice 3) and approximations
using 1, 3, 5, 7, and 9 terms in the 3-basis (top).
Relative residual $\|(I - U_k U_k^T)z\|_2/\|z\|_2$ in least squares
problem (bottom).

From Figures 5.4 and 5.5 we see that the relative residual is considerably
smaller for the nice 3 in the 3-basis than in the 5-basis.

It is possible to devise several classification algorithm based on the model
of expanding in terms of SVD bases. Below we give a simple variant.

An SVD basis classification algorithm

Training. For the training set of known digits, compute the SVD of each
class of digits, and use k basis vectors for each class.

Classification. For a given test digit, compute its relative residual in all
ten bases. If one residual is significantly smaller than all the others,
classify as that. Otherwise give up.

The algorithm is closely related to the SIMCA method (Wold 1976,
Sjöström and Wold 1980).

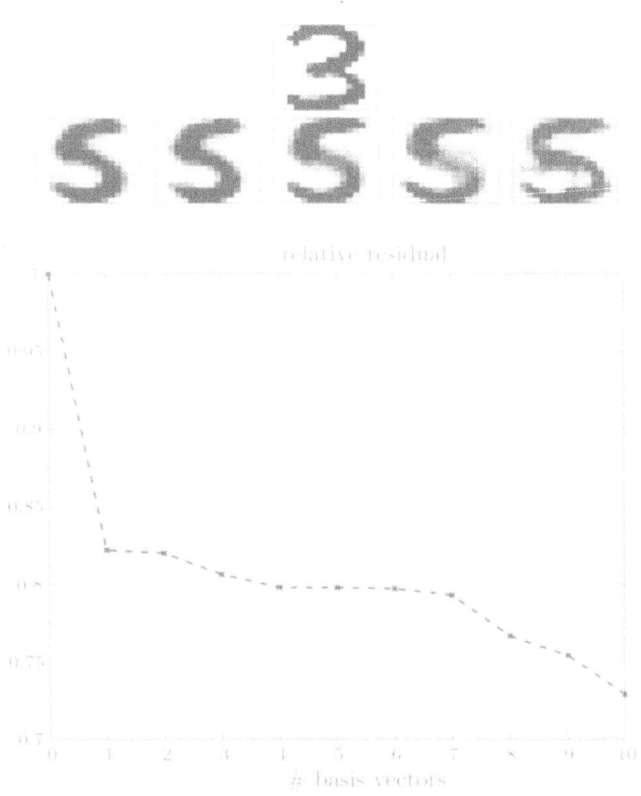

Figure 5.5. Unknown digit (nice 3) and approximations
using 1, 3, 5, 7, and 9 terms in the 5-basis (top).
Relative residual in least squares problem (bottom).

We next give some test results[16] for the US Postal Service data set, with
7291 training digits and 2007 test digits (Hastie *et al.* 2001). In Table 5.1
we give classification results as a function of the number of basis images for
each class.

Table 5.1. Correct classifications as a function of
the number of basis images (for each class).

# basis images	1	2	4	6	8	10
correct (%)	80	86	90	90.5	92	93

[16] From Savas (2002).

Even if there is a very significant improvement in performance compared to the method where one only used the centroid images, the results are not good enough, as the best algorithms reach about 97% correct classifications.

5.2. Tangent distance

Apparently it is the large variation in the way digits are written that makes it difficult to classify correctly. In the preceding subsection we used SVD bases to model the variation. An alternative model can be based on describing mathematically what are common and acceptable variations. We illustrate a few such variations in Figure 5.6. In the first column of modified

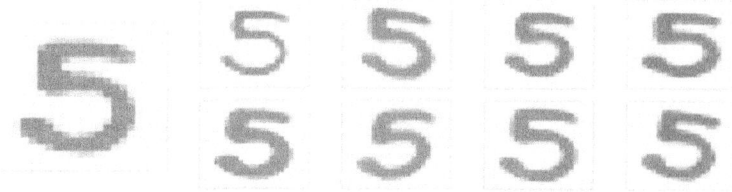

Figure 5.6. A digit and acceptable transformations.

digits, the digits appear to be written[17] with a thinner and thicker pen, in the second the digits have been stretched diagonal-wise, in the third they have been compressed and elongated vertically, and in the fourth they have been rotated. Such transformations constitute no difficulties for a human reader and ideally they should be easy to deal with in automatic digit recognition. A distance measure, *tangent distance*, that is invariant under small transformations of this type is described in Simard *et al.* (1993, 2001).

For now we interpret 16×16 images as points in $\mathbb{R}^{256}$. Let p be a fixed pattern in an image. We shall first consider the case of only one permitted transformation, diagonal stretching, say. The transformation can be thought of as moving the pattern along a curve in $\mathbb{R}^{256}$. Let the curve be parametrized by a real parameter α so that the curve is given by $s(p, \alpha)$, and in such a way that $s(p, 0) = p$. In general, such curves are nonlinear, and can be approximated by the first two terms in the Taylor expansion,

$$s(p, \alpha) = s(p, 0) + \frac{ds}{d\alpha}(p, 0)\, \alpha + O(\alpha^2) \approx p + t_p \alpha,$$

where $t_p = \frac{ds}{d\alpha}(p, 0)$ is a vector in $\mathbb{R}^{256}$. By varying α slightly around 0, we

---

[17] Note that the modified digits have not been written manually but using the techniques described later in this section. The presentation in this section is based on the papers by Simard, LeCun and Denker (1993) and Simard, LeCun, Denker and Victorri (2001), and the master's thesis of Savas (2002).

make a small movement of the pattern along the tangent at the point p on the curve. Assume that we have another pattern e that is approximated similarly,

$$s(e, \alpha) \approx e + t_e \alpha.$$

Since we consider small movements along the curves as allowed, such small movements should not influence the distance function. Therefore, ideally we would like to define our measure of closeness between p and e as the closest distance between the two curves.

In general we cannot compute the distance between the curves, but we can use the first-order approximations. Thus we move the patterns independently along their respective tangents, until we find the smallest distance. If we measure this distance in the usual Euclidean norm, we shall solve the least squares problem

$$\min_{\alpha_p, \alpha_e} \|p + t_p \alpha_p - e - t_e \alpha_e\|_2 = \min_{\alpha_p, \alpha_e} \left\| (p - e) - (-t_p \quad t_e) \begin{pmatrix} \alpha_p \\ \alpha_e \end{pmatrix} \right\|_2.$$

Consider now the case when we are allowed to move the pattern p along l different curves in $\mathbb{R}^{256}$, parametrized by $\alpha = (\alpha_1 \cdots \alpha_l)^T$. This is equivalent to moving the pattern on an l-dimensional surface (manifold) in $\mathbb{R}^{256}$. Assume that we have two patterns, p and e, each of which can move on its surface of allowed transformations. Ideally we would like to find the closest distance between the surfaces, but instead, since this is not possible to compute, we now define a distance measure where we compute the distance between the two *tangent planes* of the surface in the points p and e.

As before, the tangent plane is given by the first two terms in the Taylor expansion of the function $s(p, \alpha)$:

$$s(p, \alpha) = s(p, 0) + \sum_i^l \frac{ds}{d\alpha_i}(p, 0)\, \alpha_i + O(\|\alpha\|_2^2) \approx p + T_p \alpha,$$

where T_p is the matrix

$$T_p = \left(\frac{ds}{d\alpha_1} \quad \frac{ds}{d\alpha_2} \quad \cdots \quad \frac{ds}{d\alpha_l} \right),$$

and the derivatives are all evaluated in the point $(p, 0)$.

Thus the *tangent distance* between the points p and e is defined as the smallest possible residual in the least squares problem

$$\min_{\alpha_p, \alpha_e} \|p + T_p \alpha_p - e - T_e \alpha_e\|_2 = \min_{\alpha_p, \alpha_e} \left\| (p - e) - (-T_p \quad T_e) \begin{pmatrix} \alpha_p \\ \alpha_e \end{pmatrix} \right\|_2.$$

The least squares problem can be solved, *e.g.*, using the QR decomposition of $A = (-T_p \quad T_e)$. Note that we are in fact not interested in the solution itself but only in the norm of the residual. Write the least squares problem in the form

$$\min_{\alpha} \|b - A\alpha\|_2, \qquad b = p - e, \qquad \alpha = \begin{pmatrix} \alpha_p \\ \alpha_e \end{pmatrix}.$$

With the QR decomposition[18]

$$A = Q \begin{pmatrix} R \\ 0 \end{pmatrix} = (Q_1 \, Q_2) \begin{pmatrix} R \\ 0 \end{pmatrix} = Q_1 R,$$

the norm of the residual is given by

$$\min_{\alpha} \|b - A\alpha\|_2^2 = \min_{\alpha} \left\| \begin{pmatrix} Q_1^T b - R\alpha \\ Q_2^T b \end{pmatrix} \right\|^2$$

$$= \min_{\alpha} \{ \| (Q_1^T b - R\alpha) \|_2^2 + \|Q_2^T b\|_2^2 \} = \|Q_2^T b\|_2^2.$$

The case when the matrix A does not have full column rank is easily dealt with using the SVD. The probability that the columns of the tangent matrix are almost linearly dependent is high when the two patterns are close.

The most important property of this distance function is that it is *invariant under movements of the patterns on the tangent planes*. For instance, if we make a small translation in the x-direction of a pattern, then with this measure the distance it has been moved is equal to zero.

Simard *et al.* (1993) and (2001) considered the following transformation: *horizontal* and *vertical translation, rotation, scaling, parallel* and *diagonal hyperbolic transformation*, and *thickening*. If we consider the image pattern as a function of two variables, $p = p(x, y)$, then the derivative of each transformation can be expressed as a differentiation operator that is a linear combination of the derivatives $p_x = \frac{dp}{dx}$ and $p_y = \frac{dp}{dy}$. For instance, the rotation derivative is

$$y p_x - x p_y,$$

and the scaling derivative is

$$x p_x + y p_y.$$

The derivative of the diagonal hyperbolic transformation is

$$y p_x + x p_y,$$

and the 'thickening' derivative is

$$(p_x)^2 + (p_y)^2.$$

---

[18] A has dimension $256 \times 2l$; since the number of transformations is usually less than 10, the linear system is over-determined.

The algorithm is summarized as follows.

A tangent distance classification algorithm

Training. For each digit in the training set, compute its tangent matrix T_p.

Classification. For each test digit:

- Compute its tangent matrix.
- Compute the tangent distance to all training digits and classify it as the one with shortest distance.

This algorithm is quite good in terms of classification performance (96.9% correct classification for the US Postal Service data set (Savas 2002)), but it is very expensive, since each test digit is compared to all the training digits. In order to make it competitive it must be combined with some other algorithm that reduces the number of tangent distance comparisons to make.

We end this section by remarking that it is necessary to pre-process the digits in different ways in order to enhance the classification: see LeCun et al. (1998). For instance, performance is improved if the images are smoothed (convolved with a Gaussian kernel): see Simard et al. (2001). In Savas (2002) the derivatives p_x and p_y are computed numerically by finite differences.

6. Eigenvalue methods in data mining

When an Internet search is made using a search engine, there is first a traditional text processing part, where the aim is to find all the web pages containing the words of the query. Because of the massive size of the Web, the number of hits is likely to be much too large to be handled by the user. Therefore, some measure of quality is needed to sort out the pages that are likely to be most relevant to the particular query.

When one uses a web search engine, then typically the search phrase is under-specified.

Example 6.1. A Google search conducted on September 29, 2005, using the search phrase *university*, gave as a result links to the following well-known universities: *Harvard, Stanford, Cambridge, Yale, Cornell, Oxford*. The total number of web pages relevant to the search phrase was more than 2 billion.

Obviously Google uses an algorithm for ranking all the web pages that agrees rather well with a common-sense quality measure. Loosely speaking, Google assigns a high rank to a web page if it has inlinks from other pages that have a high rank. We will see that this 'self-referencing' statement can be formulated mathematically as an eigenvalue equation for a certain matrix.

In the context of automatic text summarization, similar self-referencing statements can be formulated mathematically as the defining equations of a singular value problem. We treat this application briefly in Section 6.3.

6.1. PageRank

It is of course impossible to define a generally valid measure of relevance that would be acceptable for a majority of users of a search engine. Google uses the concept of *PageRank* as a quality measure of web pages. It is based on the assumption that the number of links to and from a page give information about the importance of a page. We will give a description of PageRank based primarily on Page, Brin, Motwani and Winograd (1998). Concerning Google, see Brin and Page (1998).

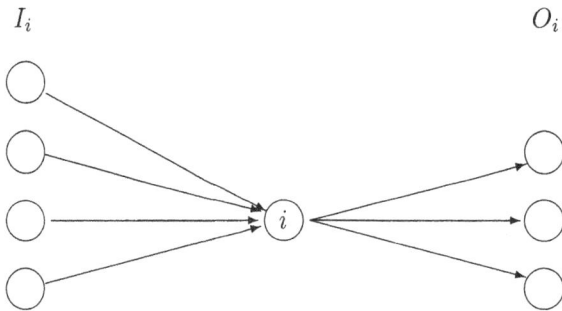

Figure 6.1. Inlinks and outlinks.

Let all web pages be ordered from 1 to n, and let i be a particular web page. Then O_i will denote the set of pages that i is linked to, the *outlinks*. The number of outlinks is denoted by $N_i = |O_i|$. The set of *inlinks*, denoted by I_i, are the pages that have an outlink to i: see Figure 6.1.

In general, a page i can be considered as more important the more inlinks it has. However, a ranking system based only on the number of inlinks is easy to manipulate.[19] When you design a web page i that (*e.g.*, for commercial reasons) you would like to be seen by as many as possible, you could simply create a large number of (information-less and unimportant) pages that have outlinks to i. In order to discourage this, one may define

---

[19] For an example of attempts to fool a search engine: see Totty and Mangalindan (2003).

the rank of i in such a way that if a highly ranked page j, has an outlink to i, this should add to the importance of i in the following way: the rank of page i is a weighted sum of the ranks of the pages that have outlinks to i. The weighting is such that the rank of a page j is divided evenly among its outlinks. The preliminary definition of PageRank is

$$r_i = \sum_{j \in I_i} \frac{r_j}{N_j}, \qquad i = 1, 2, \ldots, n. \tag{6.1}$$

The definition (6.1) is recursive, so PageRank cannot be computed directly. The equation can be reformulated as an eigenvalue problem for a matrix representing the graph of the Internet. Let Q be a square matrix of dimension n, and let

$$Q_{ij} = \begin{cases} 1/N_j, & \text{if there is a link from } j \text{ to } i, \\ 0, & \text{otherwise.} \end{cases}$$

This definition means that row i has nonzero elements in those positions that correspond to inlinks of i. Similarly, column j has nonzero elements equal to N_j in those positions that correspond to the outlinks of j, and, provided that the page has outlinks, the sum of all the elements in column j is equal to one. In the following symbolic picture of the matrix Q, nonzero elements are denoted $*$:

$$
\begin{array}{c}
 \quad\quad\quad\quad j \\
i \left(
\begin{array}{cccccc}
 & * & & & & \\
 & 0 & & & & \\
 & \vdots & & & & \\
0 & * & \cdots & * & * & \cdots \\
 & \vdots & & & & \\
 & 0 & & & & \\
 & * & & & & \\
\end{array}
\right) \quad \leftarrow \text{ inlinks} \\
\quad\quad\quad \uparrow \\
\quad\quad \text{outlinks}
\end{array}
$$

Obviously, (6.1) can be written as

$$\lambda r = Qr, \tag{6.2}$$

i.e., r is an eigenvector of Q with eigenvalue $\lambda = 1$. However, at this point it is not clear that PageRank is well defined, as we do not know if there exists an eigenvalue equal to 1.

It turns out that the theory of random walk and Markov chains gives an intuitive explanation of the concepts involved. Assume that a surfer visiting a web page always chooses the next page among the outlinks with equal probability. This random walk induces a Markov chain with transition matrix Q^T: see, *e.g.*, Meyer (2000) and Langville and Meyer (2005a).[20] A Markov chain is a random process for which the next state is determined completely by the present state; the process has no memory. The eigenvector r of the transition matrix with eigenvalue 1 corresponds to a stationary probability distribution for the Markov chain: The element in position i, r_i, is the asymptotic probability that the random walker is at web page i.

The random surfer should never get stuck. In other words, there should be no web pages without outlinks (such a page corresponds to a zero column in Q). Therefore the model is modified so that zero columns are replaced by a constant value in each position (equal probability to go to any other page in the net). Define the vectors

$$d_j = \begin{cases} 1, & \text{if } N_j = 0, \\ 0, & \text{otherwise,} \end{cases}$$

for $j = 1, \ldots, n$, and

$$e = \begin{pmatrix} 1 \\ 1 \\ \vdots \\ 1 \end{pmatrix} \in \mathbb{R}^n.$$

Then the modified matrix is defined by

$$P = Q + \frac{1}{n}ed^T. \tag{6.3}$$

Now P is a proper *column-stochastic matrix*: it has nonnegative elements $(P \geq 0)$, and

$$e^T P = e^T. \tag{6.4}$$

By analogy with (6.2), we would like to define the PageRank vector as a unique eigenvector of P with eigenvalue 1,

$$Pr = r.$$

However, uniqueness is still not guaranteed. To ensure this, the directed graph corresponding to the matrix must be *strongly connected*: given any two nodes (N_i, N_j), in the graph, there must exist a path leading from N_i

[20] Note that we use a slightly different notation to that common in the theory of stochastic processes.

to N_j. In matrix terms, P must be *irreducible*.[21] Equivalently, there must not exist any subgraph that has no outlinks.

The uniqueness of the eigenvalue is now guaranteed by the Perron–Frobenius theorem; we state it for the special case treated here.

Theorem 6.2. Let A be an irreducible column-stochastic matrix. Then the largest eigenvalue in magnitude is equal to 1. There is a unique corresponding eigenvector r satisfying $r > 0$, and $\|r\|_1 = 1$; this is the only eigenvector that is nonnegative. If $A > 0$, then $|\lambda_i| < 1$, $i = 2, 3, \ldots, n$.

Proof. Because A is column-stochastic we have $e^T A = e^T$, which means that 1 is an eigenvalue of A. The rest of the statement can be proved using Perron–Frobenius theory (Meyer 2000, Chapter 8). □

Given the size of the Internet and reasonable assumptions about its structure, it is highly probable that the link graph is *not* strongly connected, which means that the PageRank eigenvector of P is not well defined. To ensure connectedness, *i.e.*, to make it impossible for the random walker to get trapped in a subgraph, one can add, artificially, a link from every web page to all the other. In matrix terms, this can be made by taking a convex combination of P and a rank-one matrix,

$$A = \alpha P + (1 - \alpha)\frac{1}{n}ee^T, \tag{6.5}$$

for some α satisfying $0 \leq \alpha \leq 1$. Obviously A is irreducible (since $A > 0$) and column-stochastic:

$$e^T A = \alpha e^T P + (1 - \alpha)\frac{1}{n}e^T ee^T = \alpha e^T + (1 - \alpha)e^T = e^T.$$

The random walk interpretation of the additional rank-one term is that each time step a page is visited, the surfer will jump to any page in the whole web with probability $1 - \alpha$ (sometimes referred to as *teleportation*).

For the convergence of the numerical eigenvalue algorithm, it is essential to know how the eigenvalues of P are changed by the rank one modification (6.5).

Proposition 6.3. Given that the eigenvalues of the column-stochastic matrix P are $\{1, \lambda_2, \lambda_3 \ldots, \lambda_n\}$, the eigenvalues of $A = \alpha P + (1 - \alpha)\frac{1}{n}ee^T$ are $\{1, \alpha\lambda_2, \alpha\lambda_3, \ldots, \alpha\lambda_n\}$.

Several proofs of the proposition have been published (Haveliwala and Kamvar 2003b, Langville and Meyer 2005a). An elementary and simple variant (Eldén 2004a) is given here.

---

[21] A matrix P is *reducible* if there exist a permutation matrix Π such that $\Pi P \Pi^T = \begin{pmatrix} X & Y \\ 0 & Z \end{pmatrix}$, where both X and Z are square matrices.

Proof. Define $\hat{e}$ to be e normalized to Euclidean length 1, and let $U_1 \in \mathbb{R}^{n \times (n-1)}$ be such that $U = (\hat{e} \ \ U_1)$ is orthogonal. Then, since $\hat{e}^T P = \hat{e}^T$,

$$U^T P U = \begin{pmatrix} \hat{e}^T P \\ U_1^T P \end{pmatrix} (\hat{e} \ \ U_1) = \begin{pmatrix} \hat{e}^T \\ U_1^T P \end{pmatrix} (\hat{e} \ \ U_1)$$

$$= \begin{pmatrix} \hat{e}^T \hat{e} & \hat{e}^T U_1 \\ U_1^T P \hat{e} & U_1^T P^T U_1 \end{pmatrix} = \begin{pmatrix} 1 & 0 \\ w & T \end{pmatrix}, \qquad (6.6)$$

where $w = U_1^T P \hat{e}$, and $T = U_1^T P^T U_1$. Since we have made a similarity transformation, the matrix T has the eigenvalues $\lambda_2, \lambda_3, \ldots, \lambda_n$. We further have

$$U^T v = \begin{pmatrix} 1/\sqrt{n} \, e^T v \\ U_1^T v \end{pmatrix} = \begin{pmatrix} 1/\sqrt{n} \\ U_1^T v \end{pmatrix}.$$

Therefore,

$$U^T A U = U^T (\alpha P + (1 - \alpha) v e^T) U$$

$$= \alpha \begin{pmatrix} 1 & 0 \\ w & T \end{pmatrix} + (1 - \alpha) \begin{pmatrix} 1/\sqrt{n} \\ U_1^T v \end{pmatrix} (\sqrt{n} \ \ 0)$$

$$= \alpha \begin{pmatrix} 1 & 0 \\ w & T \end{pmatrix} + (1 - \alpha) \begin{pmatrix} 1 & 0 \\ \sqrt{n} \, U_1^T v & 0 \end{pmatrix} =: \begin{pmatrix} 1 & 0 \\ w_1 & \alpha T \end{pmatrix}.$$

The statement now follows immediately. □

This means that even if P has a multiple eigenvalue equal to 1, the second-largest eigenvalue in magnitude of A is always equal to α.

The vector e in (6.5) can be replaced by a nonnegative vector v with $\|v\|_1 = 1$ that can be chosen in order to make the search biased towards certain kinds of web pages. Therefore, it is referred to as a *personalization vector* (Page *et al.* 1998, Haveliwala and Kamvar 2003*a*). The vector v can also be used for avoiding manipulation by so-called link farms (Langville and Meyer 2005*a*). Proposition 6.3 also holds in this case.

We want to solve the eigenvalue problem

$$Ar = r,$$

where r is normalized, $\|r\|_1 = 1$. Because of the sparsity and the dimension of A it is not possible to use sparse eigenvalue algorithms that require the storage of more than a very few vectors. The only viable method so far for PageRank computations on the whole Web seems to be the *power method*.

It is well known (see, *e.g.*, Golub and Van Loan (1996, Section 7.3)) that the rate of convergence of the power method depends on the ratio of the second-largest and the largest eigenvalue in magnitude. Here we have

$$|\lambda^{(k)} - 1| = O(\alpha^k),$$

due to Proposition 6.3.

In view of the huge dimension of the Google matrix, it is nontrivial to compute the matrix-vector product $y = Az$, where $A = \alpha P + (1 - \alpha)\frac{1}{n}ee^T$. First, we see that if the vector z satisfies $\|z\|_1 = e^T z = 1$, then

$$\|y\|_1 = e^T y = e^T Az = e^T z = 1, \tag{6.7}$$

since A is column-stochastic ($e^T A = e^T$). Therefore normalization of the vectors produced in the power iteration is unnecessary.

Then recall that P was constructed from the actual link matrix Q as

$$P = Q + \frac{1}{n}ed^T,$$

where the row vector d has an element 1 in all those positions that correspond to web pages with no outlinks, see (6.3). This means that to represent P as a sparse matrix, we insert a large number of full vectors in Q, each of the same dimension as the total number of web pages. Consequently, one cannot afford to represent P explicitly. Let us look at the multiplication $y = Az$ in some more detail:

$$y = \alpha\left(Q + \frac{1}{n}ed^T\right)z + \frac{(1 - \alpha)}{n}e(e^T z) = \alpha Qz + \beta\frac{1}{n}e, \tag{6.8}$$

where

$$\beta = \alpha d^T z + (1 - \alpha)e^T z.$$

However, we do not need to compute β from this equation. Instead we can use (6.7) in combination with (6.8):

$$1 = e^T(\alpha Qz) + \beta e^T\left(\frac{1}{n}e\right) = e^T(\alpha Qz) + \beta.$$

Thus, we have $\beta = 1 - \|\alpha Qz\|_1$. An extra bonus is that we do not use the vector d, i.e., we do not need to know which pages lack outlinks.

The following Matlab code implements the matrix vector multiplication.

```
yhat=alpha*Q*z;
beta=1-norm(yhat,1);
y=yhat+beta*v;
residual=norm(y-z,1);
```

Here $v = (1/n)e$, or a personalized teleportation vector: see p. 372.

From Proposition 6.3 we know that the second eigenvalue of the Google matrix satisfies $\lambda_2 = \alpha$. A typical value of α is 0.85. Approximately $k = 57$ iterations are needed to make the factor 0.85^k equal to 10^{-4}. This is reported (Langville and Meyer 2005a) to be close the number of iterations used by Google.

In view of the fact that one PageRank calculation using the power method can take several days, several enhancements of the iteration procedure have been proposed. Kamvar, Haveliwala and Golub (2003a) describe an adaptive method that checks the convergence of the components of the PageRank vector and avoids performing the power iteration for those components. The block structure of the Web is used in Kamvar, Haveliwala, Manning and Golub (2003b), and speed-ups of a factor 2 have been reported. An acceleration method based on Aitken extrapolation is discussed in Kamvar, Haveliwala, Manning and Golub (2003c). Aggregation methods are discussed in several papers by Langville and Meyer and in Ipsen and Kirkland (2006).

If the PageRank is computed for a subset of the Internet, one particular domain, say, then the matrix A may be of sufficiently small dimension to use methods other than the power method: *e.g.*, the Arnoldi method (Golub and Greif 2004).

A variant of PageRank is proposed in Gyöngyi, Garcia-Molina and Pedersen (2004). Further properties of the PageRank matrix are given in Serra-Capizzano (2005).

6.2. HITS

Another method based on the link structure of the Web was introduced at the same time as PageRank (Kleinberg 1999). It is called HITS (hypertext induced topic search), and is based on the concepts of *authorities* and *hubs*. An authority is a web page with several inlinks and a hub has several outlinks. The basic idea is: *good hubs point to good authorities and good authorities are pointed to by good hubs*. Each web page is assigned both a hub score y and an authority score x.

Let L be the adjacency matrix of the directed web graph. Then two equations are given that mathematically define the relation between the two scores, based on the basic idea:

$$x = L^T y, \qquad y = Lx. \tag{6.9}$$

The algorithm for computing the scores is the power method, which converges to the left and right singular vectors corresponding to the largest singular value of L. In the implementation of HITS it is not the adjacency matrix of the whole web that is used, but of all the pages relevant to the query.

There is now an extensive literature on PageRank, HITS and other ranking methods. For overviews, see Langville and Meyer (2005b, 2005c) and Berkin (2005). A combination of HITS and PageRank has been proposed in Lempel and Moran (2001).

Obviously the ideas underlying PageRank and HITS are not restricted to web applications, but can be applied to other network analyses. For instance, a variant of the HITS method was recently used in a study of Supreme Court precedent (Fowler and Jeon 2005). A generalization of HITS is given in Blondel, Gajardo, Heymans, Senellart and Dooren (2004), which also treats synonym extraction.

6.3. Text summarization

Because of the explosion in the amount of textual information available, there is a need to develop automatic procedures for text summarization. One typical situation is when a web search engine presents a small amount of text from each document that matches a certain query. Another relevant area is the summarization of news articles.

Automatic text summarization is an active research field with connections to several other research areas such as information retrieval, natural language processing, and machine learning. Informally, the goal of text summarization is to *extract content from a text document, and present the most important content to the user in a condensed form and in a manner sensitive to the user's or application's need* (Mani 2001). In this section we will have a considerably less ambitious goal: we present a method (Zha 2002), related to HITS, for automatically extracting key words and key sentences from a text. There are connections to the vector space model in information retrieval, and to the concept of PageRank.

Consider a text from which we want to extract key words and key sentences. As one of the preprocessing steps, one should perform stemming and eliminate stop words. Similarly, if the text carries special symbols, *e.g.*, mathematics, or mark-up language tags (HTML, LaTeX), it may be necessary to remove those. Since we want to compare word frequencies in different sentences, we must consider each sentence as a separate document (in the terminology of information retrieval). After the preprocessing has been done, we parse the text, using the same type of parser as in information retrieval. This way a term-document matrix is prepared, which in this section we will refer to as a *term-sentence* matrix. Thus we have a matrix $A \in \mathbb{R}^{m \times n}$, where m denotes the number of different terms, and n the number of sentences. The element a_{ij} is defined as the frequency[22] of term i in document j.

The basis of the procedure in Zha (2002) is the simultaneous, but separate *ranking* of the terms and the sentences. Thus, term i is given a nonnegative *saliency score*, denoted u_i. The higher the saliency score, the more important the term. The saliency score of sentence j is denoted by v_j.

---

[22] Naturally, a term and document weighting scheme (see Berry and Browne (2005, Section 3.2.1)) should be used.

The assignment of saliency scores is made based on the *mutual reinforcement principle* (Zha 2002):

A term should have a high saliency score if it appears in many sentences with high saliency scores. A sentence should have a high saliency score if it contains many words with high saliency scores.

More precisely, we assert that the saliency score of term i is proportional to the sum of the scores of the sentences where it appears; in addition, each term is weighted by the corresponding matrix element,

$$u_i \propto \sum_{j=1}^{n} a_{ij} v_j, \qquad i = 1, 2, \ldots, m.$$

Similarly, the saliency score of sentence j is defined to be proportional to the scores of its words, weighted by the corresponding a_{ij},

$$v_j \propto \sum_{i=1}^{m} a_{ij} u_i, \qquad j = 1, 2, \ldots, n.$$

Collecting the saliency scores in two vectors, $u \in \mathbb{R}^m$, and $v \in \mathbb{R}^n$, these two equations can be written as

$$\sigma_u u = Av, \qquad\qquad\qquad\qquad (6.10)$$

$$\sigma_v v = A^T u, \qquad\qquad\qquad\qquad (6.11)$$

where σ_u and σ_v are proportionality constants. In fact, the constants must be equal: inserting one equation into the other, we get

$$\sigma_u u = \frac{1}{\sigma_v} A A^T u,$$

$$\sigma_v v = \frac{1}{\sigma_u} A^T A v,$$

which shows that u and v are singular vectors corresponding to the same singular value. If we choose the largest singular value, then we are guaranteed that the components of u and v are nonnegative.

Example 6.4. We created a term-sentence matrix using (a slightly earlier version of) the text in Section 4. Since the text is written using LaTeX, we first had to remove all LaTeX typesetting commands. This was done using a lexical scanner called **detex**.[23] Then the text was stemmed and stop words were removed. A term-sentence matrix A was constructed using a text parser: there turned out to be 435 terms in 218 sentences. The first singular vectors were computed in Matlab.

---

[23] http://www.cs.purdue.edu/homes/trinkle/detex/

By determining the ten largest components of u_1, and using the dictionary produced by the text parser, we found that the following ten words are the most important in the section.

document, term, matrix, approximation (approximate), query, vector, space, number, basis, cluster.

The three most important sentences are, in order, as follows.

(1) *Latent semantic indexing* (LSI) 'is based on the assumption that there is some underlying latent semantic structure in the data ... that is corrupted by the wide variety of words used ... ' (quoted from Park, Jeon and Rosen (2001)) and that this semantic structure can be enhanced by projecting the data (the term-document matrix and the queries) onto a lower-dimensional space using the singular value decomposition.

(2) In view of the large approximation error in the truncated SVD approximation of the term-document matrix, one may question whether the 'optimal' singular vectors constitute the best basis for representing the term-document matrix.

(3) For example, one can define the elements in A by

$$a_{ij} = f_{ij}\log(n/n_i),$$

where f_{ij} is the term frequency, the number of times term i appears in document j, and n_i is the number of documents that contain term i (inverse document frequency).

It is apparent that this method prefers long sentences. On the other hand, these sentences are undeniably key sentences for the text.

7. New directions

Multidimensional arrays (tensors) have been used for data analysis in psychometrics and chemometrics since the 1960s; for overviews see, *e.g.*, Kroonenberg (1992), Smilde, Bro and Geladi (2004) and the 'Three-Mode Company' web page.[24] In fact, 'three-mode analysis' appears to be a standard tool in those areas. Only in recent years has there been an increased interest among the numerical linear algebra community in tensor computations, especially for applications in signal processing and data mining. A particular generalization of the SVD, the *higher order SVD* (HOSVD), was studied in Lathauwer, Moor and Vandewalle (2000a).[25] This is a tensor decomposition

---

[24] http://three-mode.leidenuniv.nl/.
[25] However, related concepts had already been considered in Tucker (1964) and (1966), and are referred to as the *Tucker model* in psychometrics.

in terms of orthogonal matrices, which 'orders' the tensor in a way similar to that in which the singular values of the SVD are ordered, but which does not satisfy an Eckart–Young optimality property (Theorem 3.4); see Lathauwer, Moor and Vandewalle (2000b). Owing to the ordering property, this decomposition can be used for compression and dimensionality reduction, and it has been successfully applied to face recognition (Vasilescu and Terzopoulos 2002a, 2002b, 2003).

The SVD expansion (3.9) of a matrix, as a sum of rank-one matrices, has been generalized to tensors (Harshman 1970, Carroll and Chang 1970), and is called the *PARAFAC/CANDECOMP model*. For overviews, see Bro (1997) and Smilde *et al.* (2004). This does not give an exact decomposition of the tensor, and its theoretical properties are much more involved (*e.g.*, degeneracies occur – see Kruskal (1976, 1977), Bro (1997) and Sidiropoulos and Bro (2000)). A recent application of PARAFAC to network analysis is presented in Kolda, Bader and Kenny (2005a), where the hub and authority scores of the HITS method are complemented with topic scores for the anchor text of the web pages.

Recently several papers have appeared where standard techniques in data analysis and machine learning are generalized to tensors: see, *e.g.*, Yan, Xu, Yang, Zhang, Tang and Zhang (2005) and Cai, He and Han (2005).

It is not uncommon in the data mining/machine learning literature for data compression and rank reduction problems to be presented as matrix problems, while they can in fact be considered as tensor approximation problems: for examples see Tenenbaum and Freeman (2000) and Ye (2005).

Novel data mining applications, especially in link structure analysis, are presented and suggested in Kolda, Brown, Corones, Critchlow, Eliassi-Rad, Getoor, Hendrickson, Kumar, Lambert, Matarazzo, McCurley, Merrill, Samatova, Speck, Srikant, Thomas, Wertheimer and Wong (2005b).

REFERENCES

R. Baeza-Yates and B. Ribeiro-Neto (1999), *Modern Information Retrieval*, ACM Press, Addison-Wesley, New York.

B. Bergeron (2002), *Bioinformatics Computing*, Prentice-Hall.

P. Berkin (2005), 'A survey on PageRank computing', *Internet Mathematics* **2**, 73–120.

M. Berry, ed. (2001), *Computational Information Retrieval*, SIAM, Philadelphia, PA.

M. Berry and M. Browne (2005), *Understanding Search Engines: Mathematical Modeling and Text Retrieval*, 2nd edn, SIAM, Philadelphia, PA.

M. Berry and G. Linoff (2000), *Mastering Data Mining: The Art and Science of Customer Relationship Management*, Wiley, New York.

M. Berry, Z. Drmac and E. Jessup (1999), 'Matrices, vector spaces and information retrieval', *SIAM Review* **41**, 335–362.

M. Berry, S. Dumais and G. O'Brien (1995), 'Using linear algebra for intelligent information retrieval', *SIAM Review* **37**, 573–595.

Å. Björck (1996), *Numerical Methods for Least Squares Problems*, SIAM, Philadelphia, PA.

K. Blom and A. Ruhe (2005), 'A Krylov subspace method for information retrieval', *SIAM J. Matrix Anal. Appl.* **26**, 566–582.

V. D. Blondel, A. Gajardo, M. Heymans, P. Senellart and P. V. Dooren (2004), 'A measure of similarity between graph vertices: Applications to synonym extraction and web searching', *SIAM Review* **46**, 647–666.

S. Brin and L. Page (1998), 'The anatomy of a large-scale hypertextual web search engine', *Computer Networks and ISDN Systems* **30**, 107–117.

R. Bro (1997), 'PARAFAC: Tutorial and applications', *Chemometrics and Intelligent Laboratory Systems* **38**, 149–171.

M. Burl, L. Asker, P. Smyth, U. Fayyad, P. Perona, L. Crumpler and J. Aubele (1998), 'Learning to recognize volcanoes on Venus', *Machine Learning* **30**, 165–195.

D. Cai, X. He and J. Han (2005), Subspace learning based on tensor analysis. Technical Report UIUCDCS-R-2005-2572, UILU-ENG-2005-1767, Computer Science Department, University of Illinois, Urbana-Champaign.

J. D. Carroll and J. J. Chang (1970), 'Analysis of individual differences in multidimensional scaling via an N-way generalization of "Eckart–Young" decomposition', *Psychometrika* **35**, 283–319.

N. Christianini and J. Shawe-Taylor (2000), *An Introduction to Support Vector Machines*, Cambridge University Press.

M. Chu, R. Funderlic and G. Golub (1995), 'A rank-one reduction formula and its applications to matrix factorization', *SIAM Review* **37**, 512–530.

K. Cios, W. Pedrycz and R. Swiniarski (1998), *Data Mining: Methods for Knowledge Discovery*, Kluwer, Boston.

B. De Moor and P. Van Dooren (1992), 'Generalizations of the singular value and QR decompositions', *SIAM J. Matrix Anal. Appl.* **13**, 993–1014.

S. Deerwester, S. Dumais, G. Furnas, T. Landauer and R. Harsman (1990), 'Indexing by latent semantic analysis', *J. Amer. Soc. Information Science* **41**, 391–407.

I. Dhillon (2001), Co-clustering documents and words using bipartite spectral graph partitioning, in *Proc. 7th ACM–SIGKDD Conference*, pp. 269–274.

I. Dhillon and D. Modha (2001), 'Concept decompositions for large sparse text data using clustering', *Machine Learning* **42**, 143–175.

I. Dhillon, J. Fan and Y. Guan (2001), Efficient clustering of very large document collections, in *Data Mining For Scientific and Engineering Applications* (V. Grossman, C. Kamath and R. Namburu, eds), Kluwer.

I. Dhillon, Y. Guan and B. Kulis (2005), A unified view of kernel k-means, spectral clustering and graph partitioning. Technical Report UTCS TR-04-25, University of Texas at Austin, Department of Computer Sciences.

D. Di Ruscio (2000), 'A weighted view on the partial least-squares algorithm', *Automatica* **36**, 831–850.

R. Duda, P. Hart and D. Storck (2001), *Pattern Classification*, 2nd edn, Wiley-Interscience.

G. Eckart and G. Young (1936), 'The approximation of one matrix by another of lower rank', *Psychometrika* **1**, 211–218.

J. Einbeck, G. Tutz and L. Evers (2005), 'Local principal curves', *Statistics and Computing* **15**, 301–313.

L. Eldén (2004*a*), The eigenvalues of the Google matrix. Technical Report LiTH-MAT-R–04-01, Department of Mathematics, Linköping University.

L. Eldén (2004*b*), 'Partial least squares vs. Lanczos bidiagonalization I: Analysis of a projection method for multiple regression', *Comput. Statist. Data Anal.* **46**, 11–31.

U. Fayyad, G. Piatetsky-Shapiro, P. Smyth and R. Uthurusamy, eds (1996), *Advances in Knowledge Discovery and Data Mining*, AAAI Press/The MIT Press, Menlo Park, CA.

J. Fowler and S. Jeon (2005), The authority of supreme court precedent: A network analysis. Technical report, Department of Political Science, UC Davis.

I. Frank and J. Friedman (1993), 'A statistical view of some chemometrics regression tools', *Technometrics* **35**, 109–135.

J. Giles, L. Wo and M. Berry (2003), GTP (General Text Parser) software for text mining, in *Statistical Data Mining and Knowledge Discovery* (H. Bozdogan, ed.), CRC Press, Boca Raton, pp. 455–471.

N. Goharian, A. Jain and Q. Sun (2003), 'Comparative analysis of sparse matrix algorithms for information retrieval', *J. Systemics, Cybernetics and Informatics* **1**(1).

G. Golub and C. Greif (2004), Arnoldi-type algorithms for computing stationary distribution vectors, with application to PageRank. Technical Report SCCM-04-15, Department of Computer Science, Stanford University.

G. H. Golub and W. Kahan (1965), 'Calculating the singular values and pseudo-inverse of a matrix', *SIAM J. Numer. Anal. Ser. B* **2**, 205–224.

G. H. Golub and C. F. Van Loan (1996), *Matrix Computations*, 3rd edn, Johns Hopkins Press, Baltimore, MD.

G. Golub, K. Sølna and P. Van Dooren (2000), 'Computing the SVD of a general matrix product/quotient', *SIAM J. Matrix Anal. Appl.* **22**, 1–19.

D. Grossman and O. Frieder (1998), *Information Retrieval: Algorithms and Heuristics*, Kluwer.

L. Guttman (1957), 'A necessary and sufficient formula for matrix factoring', *Psychometrika* **22**, 79–81.

Z. Gyöngyi, H. Garcia-Molina and J. Pedersen (2004), Combating web spam with TrustRank, in *Proc. 30th International Conference on Very Large Databases*, Morgan Kaufmann, pp. 576–587.

J. Han and M. Kamber (2001), *Data Mining: Concepts and Techniques*, Morgan Kaufmann, San Francisco.

D. Hand, H. Mannila and P. Smyth (2001), *Principles of Data Mining*, MIT Press, Cambridge, MA.

R. A. Harshman (1970), 'Foundations of the PARAFAC procedure: Models and conditions for an "explanatory" multi-modal factor analysis', *UCLA Working Papers in Phonetics* **16**, 1–84.

T. Hastie (1984), Principal curves and surfaces. Technical report, Stanford University.

T. Hastie, R. Tibshirani and J. Friedman (2001), *The Elements of Statistical Learning: Data mining, Inference and Prediction*, Springer, New York.

T. Haveliwala and S. Kamvar (2003a), An analytical comparison of approaches to personalizing PageRank. Technical report, Computer Science Department, Stanford University.

T. Haveliwala and S. Kamvar (2003b), The second eigenvalue of the Google matrix. Technical report, Computer Science Department, Stanford University.

M. Hegland (2001), 'Data mining techniques', in *Acta Numerica*, Vol. 10, Cambridge University Press, pp. 313–355.

I. Helland (1988), 'On the structure of partial least squares regression', *Commun. Statist. Simulation* **17**, 581–607.

P. Howland and H. Park (2004), 'Generalizing discriminant analysis using the generalized singular value decomposition', *IEEE Trans. Pattern Anal. Machine Intelligence* **26**, 995– 1006.

P. Howland, M. Jeon and H. Park (2003), 'Structure preserving dimension reduction based on the generalized singular value decomposition', *SIAM J. Matrix Anal. Appl.* **25**, 165–179.

L. Hubert, J. Meulman and W. Heiser (2000), 'Two purposes for matrix factorization: A historical appraisal', *SIAM Review* **42**, 68–82.

I. C. Ipsen and S. Kirkland (2006), 'Convergence analysis of a Pagerank updating algorithm by Langville and Meyer', *SIAM J. Matrix Anal. Appl.* **27**, 952–967.

E. Jessup and J. Martin (2001), Taking a new look at the latent semantic analysis approach to information retrieval, in *Computational Information Retrieval* (M. Berry, ed.), SIAM, Philadelphia, PA, pp. 121–144.

I. Joliffe (1986), *Principal Component Analysis*, Springer, New York.

S. Kamvar, T. Haveliwala and G. Golub (2003a), 'Adaptive methods for the computation of PageRank', *Linear Algebra Appl.* **386**, 51–65.

S. Kamvar, T. Haveliwala, C. Manning and G. Golub (2003b), Exploiting the block structure of the Web for computing PageRank. Technical report, Computer Science Department, Stanford University.

S. Kamvar, T. Haveliwala, C. Manning and G. Golub (2003c), Extrapolation methods for accelerating PageRank computations, in *Proc. 12th International World Wide Web Conference, Budapest, May 2003*, pp. 261–270.

J. M. Kleinberg (1999), 'Authoritative sources in a hyperlinked environment', *J. Assoc. Comput. Mach.* **46**, 604–632.

T. Kolda, B. Bader and J. Kenny (2005a), Higher-order web link analysis using multilinear algebra, in *Proc. 5th IEEE International Conference on Data Mining, ICDM05*, IEEE Computer Society Press.

T. Kolda, D. Brown, J. Corones, T. Critchlow, T. Eliassi-Rad, L. Getoor, B. Hendrickson, V. Kumar, D. Lambert, C. Matarazzo, K. McCurley, M. Merrill, N. Samatova, D. Speck, R. Srikant, J. Thomas, M. Wertheimer and P. C. Wong (2005b), Data sciences technology for homeland security information management and knowledge discovery. Technical Report SAND2004-6648, Sandia National Laboratories.

R. Korfhage (1997), *Information Storage and Retrieval*, Wiley, New York.

P. M. Kroonenberg (1992), 'Three-mode component models: A survey of the literature', *Statistica Applicata* **4**, 619–633.

J. B. Kruskal (1976), 'More factors than subjects, tests and treatments: An indeterminacy theorem for canonical decomposition and individual differences scaling', *Psychometrika* **41**, 281–293.

J. B. Kruskal (1977), 'Three-way arrays: Rank and uniqueness of trilinear decompositions, with application to arithmetic complexity and statistics (Corrections, 17-1-1984; available from author)', *Linear Algebra Appl.* **18**, 95–138.

A. Langville and C. Meyer (2005*a*), 'Deeper inside PageRank', *Internet Mathematics* **1**, 335–380.

A. N. Langville and C. D. Meyer (2005*b*), 'A survey of eigenvector methods for web information retrieval', *SIAM Review* **47**, 135–161.

A. N. Langville and C. D. Meyer (2005*c*), *Understanding Web Search Engine Rankings: Google's PageRank, Teoma's HITS, and Other Ranking Algorithms*, Princeton University Press.

L. D. Lathauwer, B. D. Moor and J. Vandewalle (2000*a*), 'A multilinear singular value decomposition', *SIAM J. Matrix Anal. Appl.* **21**, 1253–1278.

L. D. Lathauwer, B. D. Moor and J. Vandewalle (2000*b*), 'On the best rank-1 and rank-$(R_1, R_2, \ldots, R_N)$ approximation of higher-order tensor', *SIAM J. Matrix Anal. Appl.* **21**, 1324–1342.

Y. LeCun, L. Bottou, Y. Bengio and P. Haffner (1998), 'Gradient-based learning applied to document recognition', *Proc. IEEE* **86**, 2278–2324.

R. Lempel and S. Moran (2001), 'Salsa: the stochastic approach for link-structure analysis', *ACM Trans. Inf. Syst.* **19**, 131–160.

I. Mani (2001), *Automatic Summarization*, John Benjamins.

W. Massy (1965), 'Principal components regression in exploratory statistical research', *J. Amer. Statist. Assoc.* **60**, 234–246.

J. Mena (1999), *Data Mining Your Website*, Digital Press, Boston.

C. Meyer (2000), *Matrix Analysis and Applied Linear Algebra*, SIAM, Philadelphia.

L. Mirsky (1960), 'Symmetric gauge functions and unitarily invariant norms', *Quart. J. Math. Oxford* **11**, 50–59.

C. Moler (2002), 'The world's largest matrix computation', *Matlab News and Notes*, October 2002, pp. 12–13.

L. Page, S. Brin, R. Motwani and T. Winograd (1998), 'The PageRank citation ranking: Bringing order to the Web', Stanford Digital Library Working Papers.

C. Paige and M. Saunders (1981), 'Towards a generalized singular value decomposition', *SIAM J. Numer. Anal.* **18**, 398–405.

C. Paige and M. Saunders (1982), 'LSQR: An algorithm for sparse linear equations and sparse least squares', *ACM Trans. Math. Software* **8**, 43–71.

H. Park and L. Eldén (2005), Matrix rank reduction for data analysis and feature extraction, in *Handbook of Parallel Computing and Statistics* (E. Kontoghiorghes, ed.), CRC Press, Boca Raton.

H. Park, M. Jeon and J. B. Rosen (2001), Lower dimensional representation of text data in vector space based information retrieval, in *Computational Information Retrieval* (M. Berry, ed.), SIAM, Philadelphia, PA, pp. 3–23.

H. Park, M. Jeon and J. B. Rosen (2003), 'Lower dimensional representation of text data based on centroids and least squares', *BIT* **43**, 427–448.

A. Phatak and F. de Hoog (2002), 'Exploiting the connection between PLS, Lanczos methods and conjugate gradients: Alternative proofs of some properties of PLS', *J. Chemometrics* **16**, 361–367.

Y. Saad (2003), *Iterative Methods for Sparse Linear Systems*, 2nd edn, SIAM.

G. Salton, C. Yang and A. Wong (1975), 'A vector-space model for automatic indexing', *Comm. Assoc. Comput. Mach.* **18**, 613–620.

B. Savas (2002), Analyses and test of handwritten digit algorithms. Master's thesis, Mathematics Department, Linköping University.

S. Serra-Capizzano (2005), 'Jordan canonical form of the Google matrix: A potential contribution to the Pagerank computation', *SIAM J. Matrix Anal. Appl.* **27**, 305–312.

N. D. Sidiropoulos and R. Bro (2000), 'On the uniqueness of multilinear decomposition of N-way arrays', *J. Chemometrics* **14**, 229–239.

P. Simard, Y. LeCun and J. Denker (1993), Efficient pattern recognition using a new transformation distance, in *Advances in Neural Information Processing Systems 5* (J. Cowan, S. Hanson and C. Giles, eds), Morgan Kaufmann, pp. 50–58.

P. Simard, Y. LeCun, J. Denker and B. Victorri (2001), 'Transformation invariance in pattern recognition: Tangent distance and tangent propagation', *Internat. J. Imaging System Techn.* **11**, 181–194.

H. D. Simon, A. Sohn and R. Biswas (1998), 'Harp: A dynamic spectral partitioner.', *J. Parallel Distrib. Comput.* **50**, 83–103.

M. Sjöström and S. Wold (1980), SIMCA: A pattern recognition method based on principal components models, in *Pattern Recognition in Practice* (E. S. Gelsema and L. N. Kanal, eds), North-Holland, pp. 351–359.

A. Smilde, R. Bro and P. Geladi (2004), *Multi-Way Analysis: Applications in the Chemical Sciences*, Wiley.

J. Tenenbaum and W. Freeman (2000), 'Separating style and content with bilinear models', *Neural Computation* **12**, 1247–1283.

M. Totty and M. Mangalindan (2003), 'As Google becomes Web's gatekeeper, sites fight to get in', *Wall Street Journal* **CCXLI**.

L. Tucker (1964), The extension of factor analysis to three-dimensional matrices, in *Contributions to Mathematical Psychology* (H. Gulliksen and N. Frederiksen, eds), Holt, Rinehart and Winston, New York, pp. 109–127.

L. Tucker (1966), 'Some mathematical notes on three-mode factor analysis', *Psychometrika* **31**, 279–311.

C. Van Loan (1976), 'Generalizing the singular value decomposition', *SIAM J. Numer. Anal.* **13**, 76–83.

M. Vasilescu and D. Terzopoulos (2002a), Multilinear analysis of image ensembles: Tensorfaces, in *Proc. 7th European Conference on Computer Vision (ECCV'02)*, Vol. 2350 of *Lecture Notes in Computer Science*, Springer, Copenhagen, Denmark, pp. 447–460.

M. Vasilescu and D. Terzopoulos (2002b), Multilinear image analysis for facial recognition, in *International Conference on Pattern Recognition, Quebec City, Canada (ICPR '02)*, IEEE Computer Society, pp. 511–514.

M. Vasilescu and D. Terzopoulos (2003), Multilinear subspace analysis of image
ensembles, in *IEEE Conference on Computer Vision and Pattern Recognition,
Madison WI* (CVPR'03), pp. 93–99.

J. Wedderburn (1934), *Lectures on Matrices*, Colloquium Publications, AMS, New
York.

I. Witten and E. Frank (2000), *Data Mining: Practical Machine Learning Tools and
Techniques with Java Implementations*, Morgan Kaufmann, San Francisco.

H. Wold (1975), Soft modeling by latent variables: The nonlinear iterative partial
least squares approach, in *Perspectives in Probability and Statistics: Papers
in Honour of M. S. Bartlett* (J. Gani, ed.), Academic Press, London.

S. Wold (1976), 'Pattern recognition by means of disjoint principal components
models', *Pattern Recognition* **8**, 127–139.

S. Wold, A. Ruhe, H. Wold and W. Dunn (1984), 'The collinearity problem in
linear regression: The partial least squares (PLS) approach to generalized
inverses', *SIAM J. Sci. Statist. Comput.* **5**, 735–743.

S. Wold, M. Sjöström and L. Eriksson (2001), 'PLS-regression: A basic tool of
chemometrics', *Chemometrics and Intell. Lab. Systems* **58**, 109–130.

S. Yan, D. Xu, Q. Yang, L. Zhang, X. Tang and H. Zhang (2005), Discriminant
analysis with tensor representation, in *Proc. 2005 IEEE Computer Society
Conference on Computer Vision and Pattern Recognition* (CVPR'05).

J. Ye (2005), 'Generalized low rank approximations of matrices', *Machine Learning*
61, 167–191.

D. Zeimpekis and E. Gallopoulos (2005), Design of a MATLAB toolbox for term-
document matrix generation, in *Proc. Workshop on Clustering High Dimensional Data and its Applications* (I. Dhillon, J. Kogan and J. Ghosh, eds),
Newport Beach, CA, pp. 38–48.

H. Zha (2002), Generic summarization and keyphrase extraction using mutual reinforcement principle and sentence clustering, in *Proc. 25th Annual International ACM–SIGIR Conference on Research and Development in Information
Retrieval*, Tampere, Finland, pp. 113–120.

H. Zha, C. Ding, M. Gu, X. He and H. Simon (2002), Spectral relaxation for
k-means clustering, in *Advances in Neural Information Processing Systems*
(T. Dietterich, S. Becker and Z. Ghahramani, eds), MIT Press, pp. 1057–
1064.

Acta Numerica (2006), pp. 385–470 © Cambridge University Press, 2006
doi: 10.1017/S0962492906250013

Numerical modelling of ocean circulation

Robert L. Higdon
Department of Mathematics,
Oregon State University,
Corvallis, Oregon 97331-4605, USA
E-mail: `higdon@math.oregonstate.edu`

Computational simulations of ocean circulation rely on the numerical solution of partial differential equations of fluid dynamics, as applied to a relatively thin layer of stratified fluid on a rotating globe. This paper describes some of the physical and mathematical properties of the solutions being sought, some of the issues that are encountered when the governing equations are solved numerically, and some of the numerical methods that are being used in this area.

CONTENTS

1. Introduction

The circulation of the world's oceans plays a major role in the global climate system. For example, the circulations of the atmosphere and ocean move large amounts of heat from tropical regions to higher latitudes, and this transport serves to moderate the temperature differences caused by unequal solar heating. In the case of the ocean, much of the transport is due to intense currents that are typically found along the western boundaries of ocean basins, and some is also due to turbulent mixing due to eddies.

An important part of the picture is the buoyancy-driven 'thermohaline circulation', in which a portion of the warm water in the Gulf Stream flows to the far northern Atlantic, becomes colder and saltier due to atmospheric forcing, sinks due to the increased density, and then moves slowly southward along the ocean's bottom to become part of a global circulation with a period of roughly a millennium.

An illustration of the circulation of the ocean is given in Figure 1.1. The figure shows the sea-surface temperature in the western Atlantic, at a fixed time, as computed in a numerical simulation using the Miami Isopycnic Coordinate Ocean Model (Bleck, Rooth, Hu and Smith 1992, Bleck 2002). The black regions represent land masses, and the various shades of grey indicate water temperatures. (Bright regions do not indicate the highest temperatures. Instead, this greyscale plot was reproduced from a colour original, and in some cases different colours were mapped to similar shades of grey.) The warmest waters are located in the southernmost region, and these are seen to move northward along the eastern coast of the United States and then into the interior of the North Atlantic. The figure suggests the role of the ocean in the Earth's climate, and it also illustrates how ocean currents exhibit small-scale meanders and eddies in addition to their larger-scale patterns.

In order to assess various scenarios for the future evolution of the Earth's climate, numerous groups have used numerical simulations based on coupled climate models that include models of the atmosphere, ocean, sea ice, and effects of the land surface. These models are run for centuries of model time and are very computationally intensive. Extensive information about the usage of such models is included in a recent report by the Intergovernmental Panel on Climate Change (2001).

More localized aspects of ocean circulation are also of interest, for scientific and societal reasons. For example, in certain coastal regions (such as along the northwestern United States) the prevailing winds sometimes cause the surface waters to shift offshore, so that the relatively cooler deep waters well upward to the surface. This upwelling brings nutrients that support the base of the food chain. These coastal upwelling regions occupy at most a few per cent of the area of the world's oceans, but in biological terms they are extraordinarily productive (Gill 1982).

There are multiple reasons for obtaining a better understanding of the circulation of the ocean, and numerical simulation is an important tool for gaining that understanding. The design of numerical algorithms for this purpose is heavily influenced by the physical properties of the flows being modelled and by the mathematical properties of the solutions of the governing equations. A description of some of these properties is a major emphasis of this review.

Figure 1.1. Sea-surface temperature in the western Atlantic Ocean, as computed in a numerical simulation. (Figure provided by Rainer Bleck, NASA Goddard Institute for Space Studies.)

Section 2 gives an overview of length, time, and velocity scales and some other physical properties of oceanic flows. Section 3 describes the problem of choosing a vertical coordinate for an ocean circulation model; the geometrical height z is a traditional choice, but there are thermodynamic and hybrid alternatives that are currently receiving serious attention. Section 4 summarizes some properties of solutions of the governing equations, including geostrophic balance, conservation of potential vorticity, Rossby waves, and the multiple time scales resulting from the contrast between fast external waves and the slower internal and advective motions. Section 5 discusses the numerical treatment of these multiple time scales, and it also describes some time-stepping schemes. Section 6 discusses grids and spatial discretizations, the numerical simulation of advection, and the numerical solution of the momentum equations.

Section 7 gives a detailed derivation of the partial differential equations that describe the conservation of mass, momentum, and tracers for a stratified and hydrostatic fluid that is in motion relative to a rotating spheroid. In this discussion the horizontal coordinates are arbitrary orthogonal curvilinear coordinates, and the vertical coordinate is a generalized coordinate that includes all of the cases discussed in Section 3. This derivation is placed at the end of the paper because of its length, and the preceding sections refer forward to it when needed.

2. Physical characteristics of large-scale oceanic flows

The physical motions of the ocean exhibit a wide range of space and time scales, and when viewed on a sufficiently large scale the circulation of the ocean is very nearly hydrostatic. In addition, the interior of the ocean is stratified, and over much of the ocean the relatively warm upper layers are nearly isolated from the relatively cold lower layers. These and other physical properties of the ocean are outlined in the present section.

2.1. Length scales and the hydrostatic assumption

Table 2.1 shows some features of oceanic flows that are resolved in simulations on a basin scale or global scale. The phrase 'horizontal gyres' refers to closed loops of circulation that are driven by the wind stress acting at the upper surface of the ocean. For example, a portion of the Gulf Stream returns laterally and southward to complete a closed loop, and the Kuroshio current plays an analogous role in the North Pacific.

The table indicates that the vertical length scale in the ocean is much smaller than the lateral dimensions of the various features listed there. The statement that the vertical length scale is small compared to the horizontal length scale is known as the shallow water condition, and it is one of the fundamental assumptions of large-scale ocean modelling. In colloquial terms,

Table 2.1. Approximate length scales in the ocean. All quantities
are horizontal dimensions, unless otherwise stated.

Ocean basins	$\sim$ 5000 to 10000 km
Horizontal gyres	basin scale
Western boundary currents	$\sim$ 100 km
Eddies	tens to hundreds of km
Depth of the ocean	$\sim$ 4 to 5 km, in the mid-ocean
Depth of surface currents	hundreds of metres

this condition can be phrased as follows: *The Pacific Ocean is shallow. Your coffee cup is deep.*

One consequence of the shallow water condition is that, for large-scale modelling, the ocean can be regarded as nearly hydrostatic, *i.e.*, the vertical acceleration of the fluid is negligible. This conclusion can be reached by a formal scaling analysis, as given, for example, by Holton (1992) or Pedlosky (1987). For a more informal argument, let L be a horizontal distance over which the horizontal velocity varies significantly. This variation in horizontal velocity can induce a horizontal divergence that causes the elevation of the free surface at the top of the fluid to rise or fall. However, this divergence acts over a vertical distance that is much smaller than L, and since the effect of this divergence is spread out over a large horizontal extent, the elevation of the free surface must change very slowly. For an analogy, suppose that a large swimming pool is being filled by a high-volume pump. An observer next to the pump may think that the water is flowing rapidly into the pool, but the water level changes slowly because the effect of the pumping is spread over a large horizontal area.

The hydrostatic assumption is not valid for motions whose horizontal scale is comparable to the vertical scale, or smaller. For example, in the case of the thermohaline circulation, the sinking of water in the far northern Atlantic takes place in fields of convective plumes; the fields are on the order of a hundred kilometres wide, and the width of each individual plume is on the order of a kilometre or less. The downdraft in such a plume is not hydrostatically balanced, and it also cannot be resolved in a climate model using present-day computers. Instead, this process must be represented by some kind of parametrization (Marshall and Schott 1999). In addition, limited-area models of near-shore processes can have grid spacings that are fine enough to resolve nonhydrostatic motions, and in that case the models must be configured accordingly. However, the emphasis in this review is on physical processes and numerical methods associated with large-scale circulation, and so the hydrostatic condition will assumed here except when stated otherwise.

2.2. Stratification

One simple model of oceanic motions is provided by the shallow water equations (Section 4.2), which describe a hydrostatic fluid having constant density. This system of equations is sufficient, for example, to enable highly accurate modelling of global tides (Bennett 2002). However, the assumption of constant density is only an approximation. Within a given water column, the density can vary by a few per cent from the top of the fluid to the bottom, and additional variation is found over the lateral extent of the ocean (Gill 1982). It turns out that these variations in density play a crucial role in major features of the ocean's dynamics, such as the thermohaline circulation, and so they must be incorporated into models of the general circulation of the ocean.

The density in the ocean is determined by temperature, salinity, and pressure, with most of the vertical variation within a water column being due to the pressure. However, the dependence on pressure is dynamically not very significant (Sun et al. 1999), and in physical oceanography it is common practice to use the concept of 'potential density', which is the density after adiabatic adjustment to a reference pressure. That is, imagine that the pressure acting on a fluid parcel is changed to some fixed reference pressure, without any heat flowing into or out of the parcel and without any changes in the salinity of the parcel. The reference pressure could be atmospheric pressure or the pressure at some specified depth. The density after this adjustment depends only on temperature and salinity, and it is referred to as the potential density. Similarly, the 'potential temperature' is defined to be the temperature of a fluid parcel after the same kind of adjustment to a reference pressure.

The stratification of the ocean is illustrated in Figure 2.1, which shows a contour plot of the zonal (east–west) average of potential density in the Pacific Ocean, with the potential density referenced to atmospheric pressure. This plot represents a climatological mean taken from the atlas of Levitus (1982). In this plot the vertical coordinate is the depth, in metres, and the horizontal coordinate is the latitude, in degrees, from 90°S to 90°N. The horizontal axis thus extends for approximately 20,000 kilometres, whereas the vertical axis covers 5.5 kilometres. One feature seen in the plot is that some of the surfaces of constant potential density outcrop to the surface, owing to the lateral variation in temperature, and some of the surfaces intersect the bottom topography. Another feature is that most of the variation of potential density is found near the upper surface, whereas the abyssal waters are relatively homogeneous.

One feature not shown in Figure 2.1 is the mixed layer at the top of the ocean. The uppermost part of the ocean tends to be vertically homogeneous, due in part to mixing caused by wind forcing. In circumstances where the

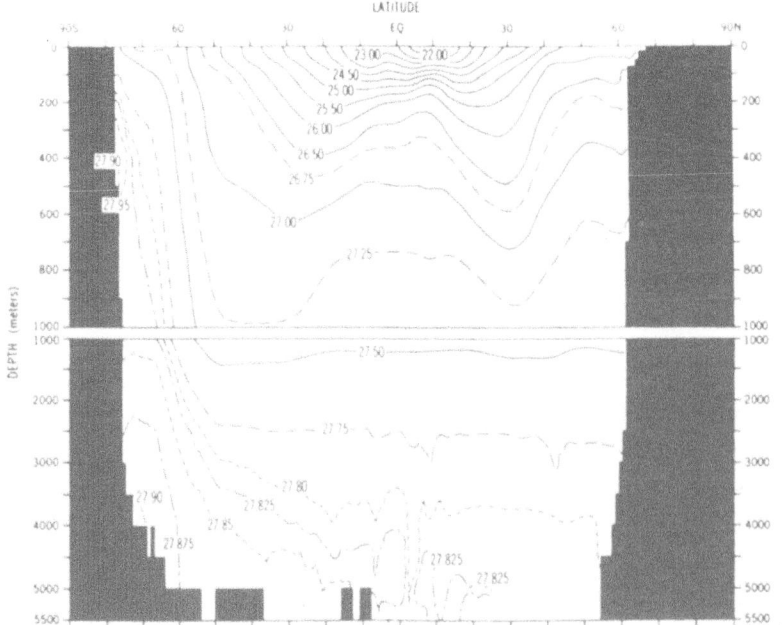

Figure 2.1. East–west average of potential density in the Pacific Ocean. The contour labels represent values of potential density minus 1000 kg/m^3; for example, the contour labelled 27.00 corresponds to potential density 1027 kg/m^3. (From Levitus (1982).)

atmosphere is colder than the ocean, mixing is also generated by convective overturning within the upper ocean. The thickness of the mixed layer is typically on the order of tens of metres to perhaps hundreds of metres. However, in deeply convective regions such as those in the far northern Atlantic, the mixed layer can sometimes be more than 2000 metres thick (Marshall and Schott 1999).

2.3. Time and velocity scales

Table 2.2 lists some approximate time and velocity scales for prominent motions in the ocean. The table includes the periods of some large-scale circulation patterns, and it also lists the velocities of currents and of two types of external waves and two types of internal waves: see, e.g., Gill (1982). In the case of an external wave, all fluid layers thicken or thin by approximately the same proportion at a given horizontal location and time, as illustrated in Figure 2.2. For such a wave, the behaviour of the free surface at the top of the fluid reveals the nature of the wave motion throughout

Table 2.2. Time and velocity scales.

Thermohaline circulation	$\sim$ millennium
Wind-driven gyres	decades
Currents	$\sim$ 1 m/sec for strong currents, elsewhere much less
External gravity waves	$\sim \sqrt{gH}$, e.g., $\sim$ 200 m/sec
Internal gravity waves	a few m/sec or less
External Rossby waves	$\sim$ 20 m/sec
Internal Rossby waves	$\sim$ 0.02 to $\sim$ 1 m/sec, depending on the latitude

the interior, and the horizontal velocity field is essentially independent of vertical position. On the other hand, an internal wave consists of undulations of surfaces of constant density within the fluid, and the free surface remains nearly level.

In the case of external gravity waves the restoring force is due to the density contrast between water and air, whereas for internal gravity waves the restoring force is due to variations of density within the fluid. The restoring force is much weaker in the latter case, so internal gravity waves move much more slowly than external gravity waves. In the case of Rossby waves the restoring mechanism is based on vorticity instead of gravity, and the existence of such waves depends on the variation of the Coriolis parameter with latitude and/or variations in the topography at the bottom of the fluid domain. Some properties of the above motions are derived mathematically in Section 4.

In general, fluids can also admit sound waves, which travel much more rapidly than the waves listed in Table 2.2. However, the circulation of the ocean is not affected by the propagation of sound waves within the ocean,

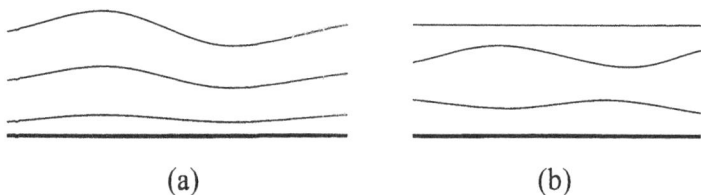

(a)   (b)

Figure 2.2. (a) External wave. (b) Internal wave. In each case, the uppermost curve represents the free surface at the top of the fluid, and the bottom line represents the bottom of the fluid domain. The intermediate curves represent surfaces of constant density within the fluid. The vertical displacements are exaggerated for the sake of visibility.

and these waves can be filtered from the equations of motion by using either the hydrostatic approximation or the Boussinesq approximation (*e.g.*, Gill (1982), Griffies (2004)). With the latter approximation, the density of the fluid is assumed constant in the momentum equations, except in the buoyancy term in the vertical momentum equation.

In a numerical model of the general circulation of the ocean, the fastest motions are typically the external gravity waves. These travel much more rapidly than any of the other motions that are present in the system, and their presence can cause serious problems with the efficiency of a numerical algorithm. In the ocean modelling community it has therefore become common practice to split the external motions into a separate subsystem that is solved by different techniques from the remainder of the system. This issue is discussed in Section 5.1.

One implication of the space and time scales of the large-scale circulation of the ocean is that the Coriolis term and lateral pressure gradient are typically in an approximate balance, known as the geostrophic balance. The same property also applies to the circulation of the atmosphere, and it is one of the most prominent features of geophysical fluid dynamics. When a system is in a geostrophically balanced state, the fluid flows along curves of constant pressure. The concept of geostrophic balance is developed in Section 4.3.

2.4. Subgrid-scale processes

If a numerical model of ocean circulation has a horizontal grid spacing of 0.1° latitude and longitude, for example, then the grid spacing near the equator is roughly 10 km. For a global ocean model this is presently considered high resolution, and many climate-scale studies are currently run with grid spacings that are much coarser, given the long time intervals that are involved (Intergovernmental Panel on Climate Change 2001). However, the dynamics of the ocean are influenced by the cumulative effects of turbulent processes that occur on scales that are too small to be resolved explicitly in a numerical model. For example, these scales can extend downward as far as centimetres or less (Gill 1982, Bleck 2005). It is then necessary to parametrize the large-scale effects of these subgrid-scale motions in terms of the dependent variables that are used in the model.

These processes affect both the transport of momentum and the transport of tracers. The term 'tracer' refers to a scalar-valued quantity that is transported with the flow, such as potential temperature, salinity, or the concentrations of various chemical or biological species. The first two of these affect the density of the fluid and thus its dynamics, whereas the last do not. These classes of tracers are labelled as active tracers and passive tracers, respectively.

On the scales of interest here, the mixing of momentum and tracers is highly anisotropic, and it occurs primarily along directions that are approximately horizontal. This phenomenon has been described in terms of directions of neutral buoyancy, which at any point in the fluid are defined locally as lying in the plane of constant buoyancy through that point; this plane is tangent to the surface of constant potential density through that point, where in this case the potential density is referenced to the local pressure (McDougall 1987). If a fluid parcel leaves such a neutral plane, then the buoyancy contrast with the surrounding fluid tends to force the parcel back towards that plane. On the other hand, adiabatic motions within neutral planes encounter no such impediment, so mixing within neutral planes is far more efficient than mixing across such planes. This picture is a local description, and due to a lack of integrability it can be extended globally to a concept of 'neutral surface' in only an approximate manner (Eden and Willebrand 1999). The preceding issues are closely related to the problem of choosing a thermodynamic variable that can be used as a vertical coordinate for an isopycnic model, which is discussed in Section 3.1.

Empirical observations indicate that, in the interior of the ocean, the rate of transport along the principal directions of mixing can be as much as 10^8 times the rate of mixing in orthogonal directions (Ledwell, Watson and Law 1993, Davis 1994). However, this discrepancy is not as large in boundary layers, especially in the mixed layer at the top of the ocean.

Proper choices of subgrid-scale parametrizations have long been an area of active research. Irreversible mixing processes tend to be modelled by some kind of downgradient diffusion. For another example, Gent and McWilliams (1990) and Gent, Willebrand, McDougall and McWilliams (1995) introduced a parametrization of unresolved eddies that can be used, for instance, to represent the transfer of potential energy into eddy kinetic energy when slanted density surfaces are flattened. Extensive reviews of the problem of subgrid-scale parametrization are given by Griffies (2004) and by Haidvogel and Beckmann (1999).

3. The choice of vertical coordinate

A crucial decision in the construction of an ocean model is the choice of vertical coordinate, and an obvious choice is the geometrical height z. This coordinate is used, for example, in the Bryan–Cox class of ocean models. The development of the original Bryan–Cox model began in the 1960s (Bryan 1969), and the various descendants of this model constitute what is currently the most widely used class of ocean circulation models; a genealogy of this class is given by Semtner (1997). Examples include the Modular Ocean Model (Griffies, Harrison, Pacanowski and Rosati 2004), the Parallel Ocean Program (Dukowicz and Smith 1994), and the Parallel

Ocean Climate Model (Semtner and Chervin 1992). Another example of a z-coordinate model is the MIT general circulation model (Marshall, Hill, Perelman and Adcroft 1997), which can represent both large-scale hydrostatic and small-scale nonhydrostatic processes.

Models that use z as the vertical coordinate are known as level models. However, other possibilities for the vertical coordinate are also used, and the purpose of this section is to describe and compare these alternatives.

3.1. Isopycnic coordinates

One alternative to z is isopycnic coordinates, for which the vertical coordinate is the reciprocal of potential density or some other related quantity. The term 'isopycnic' means 'equal density', which is an approximate statement about the surfaces of constant vertical coordinate in this setting. In the following discussion, the term 'isopycnal' refers to a surface of constant potential density (or related quantity), and 'diapycnal' refers to transport across such a surface.

In the setting of isopycnic coordinates, one seeks a quantity s that is (approximately) conserved along particle paths. In other words, one wants $Ds/Dt \approx 0$, where D/Dt denotes the material derivative (see Section 4.2). One also wants surfaces of constant s to be approximate neutral surfaces. For the moment, assume $Ds/Dt = 0$ exactly, everywhere in the fluid. In that event, a surface of constant s is a material surface; if a fluid parcel lies initially in such a surface, then as time evolves the fluid parcel will retain the same value of s and thus remain in the same surface. The elevation of such a surface can vary with horizontal position and time, but any two such surfaces will always enclose the same mass of fluid. If the fluid domain is discretized with respect to s, for purposes of numerical solution, then the fluid is divided into physical layers that do not mix. Water masses with distinct physical properties are then distinguished automatically by the choice of coordinate system. For this reason, isopycnic models are also referred to as 'layered' models.

If s is chosen to be the reciprocal of potential density or a related quantity, then the preceding statements are approximately true, except in the vertically homogeneous mixed layer at the top of the ocean. If such a quantity s is used as the vertical coordinate, then a graph such as Figure 2.1 provides a plot of the model's vertical coordinate system.

However, potential density, as such, is not necessarily an optimal choice for a vertical coordinate. For example, potential density is not exactly a neutral variable, and under certain conditions potential density can be non-monotonic as a function of z. In addition, if potential density is used as the vertical coordinate, then the form of the lateral pressure forcing in the momentum equations allows the possibility of numerical inaccuracy and even numerical instability. Several investigators have recently explored these issues and have developed variations on potential density that are more suitable for use in an ocean circulation model: see, *e.g.*, Section 7.14 and the work of Sun *et al.* (1999), de Szoeke (2000), de Szoeke, Springer and Oxilia (2000), and Hallberg (2005).

As noted in Section 7, the equation for conservation of mass in isopycnic coordinates amounts to a statement about the thicknesses of the coordinate layers. The vertical diffusion of heat and/or salt causes a vertical movement of isopycnals relative to the fluid, and in the context of isopycnic coordinates this movement can be viewed as an advection of fluid across a coordinate surface.

The development of isopycnic coordinate ocean models began in earnest in approximately the 1980s, in analogy to the usage of isentropic coordinates in atmospheric models (Hsu and Arakawa 1990). Examples of isopycnic ocean models are the Miami Isopycnic Coordinate Ocean Model (Bleck and Smith 1990, Bleck 2002) and the Hallberg Isopycnal Model (Hallberg 1997).

3.2. Comparison of z and isopycnic coordinates

Over most of the real ocean, the warm upper layers are nearly isolated from the colder deep layers, and the diapycnal transports between layers are typically subtle. However, it is important to represent these transports accurately. For example, the thermohaline circulation involves warm water flowing to the far northern Atlantic, becoming colder and saltier, and then sinking; the dynamics of this circulation could be misrepresented in a numerical model if the model allows inaccurate or spurious diffusion of heat or salt between the upper and lower regions.

In the case of isopycnic coordinates, the representation of the diapycnal transport is under the explicit control of the modeller. In contrast, a level (z-coordinate) model can implicitly allow spurious, non-physical diapycnal transports. The transports of heat and salt are typically represented by advection-diffusion equations, and computational algorithms for representing advection typically exhibit diffusion that is purely numerical. For the terms involving horizontal advection, this numerical diffusion acts along surfaces of constant z. However, surfaces of constant z can intersect isopycnals, as suggested by Figure 2.1, and this numerical diffusion can then cause artificial transport between different water masses (Griffies, Pacanowski and Hallberg 2000*b*).

The diffusion terms in transport equations are another potential source of spurious transport in level models. However, it is possible to rotate the diffusion operator to yield a component tangent to isopycnals and a component perpendicular to isopycnals, with the rate of diapycnal diffusion being orders of magnitude smaller than the rate of diffusion tangent to isopycnals. This process is reviewed in detail by Griffies (2004).

The above difficulties are avoided if isopycnic coordinates are used, but isopycnic coordinates have their own limitations. As seen in Figure 2.1, isopycnals can intersect the upper or lower boundaries of the fluid, and this implies a loss of vertical resolution, especially at the higher latitudes. A loss of resolution is especially pronounced in the mixed layer at the upper boundary of the ocean, as this layer is vertically homogeneous. In addition, when an isopycnal intersects the top (bottom) of the fluid domain, the layer above (below) the isopycnal approaches zero thickness. The vanishing of layers introduces algorithmic complications that are described in Section 6.

3.3. Sigma-coordinates

Another vertical coordinate is the terrain-following coordinate σ, which is chosen to vary linearly from $\sigma = 0$ at the upper boundary of the fluid to $\sigma = -1$ at the bottom boundary. This coordinate allows for an accurate representation of the topography of the ocean's bottom, and it facilitates the modelling of the bottom boundary layer.

However, σ-coordinate surfaces can intersect isopycnals, so the use of σ allows the possibility of spurious diapycnal transport, in analogy to the case of z-coordinates discussed above. Another limitation concerns the accurate computation of the lateral pressure gradient. When the pressure is written in σ-coordinates, the lateral pressure gradient $\nabla_\sigma p = (\partial p/\partial x, \partial p/\partial y)$ actually represents differentiation along surfaces of constant σ, not constant z. Conversion to a pressure gradient $\nabla_z p$ that is truly horizontal requires a correction term involving the hydrostatic balance and the slope of the σ-surface (Griffies 2004). In some regions this correction can have a magnitude comparable to that of $\nabla_\sigma p$. In such situations $\nabla_z p$ can be the difference of two large quantities of opposite signs, and errors in those quantities can dominate the result.

The use of σ as a vertical coordinate has been especially prominent in limited-area regional and coastal modelling, as in the Princeton Ocean Model (Blumberg and Mellor 1987) and the Regional Oceanic Modeling System (Shchepetkin and McWilliams 2005).

3.4. Hybrid coordinates

As noted in Section 3.2, isopycnic coordinates suffer the disadvantage of losing vertical resolution when isopycnals intersect the upper or lower boundaries of the fluid domain, and in addition they provide little resolution in vertically homogeneous regions such as the mixed layer at the top of the ocean. An attempt to overcome these limitations is represented by the class of hybrid coordinates, which combine features of isopycnic coordinates, z-coordinates, and σ-coordinates. Examples of hybrid coordinate models include the Hybrid Coordinate Ocean Model (Bleck 2002), which is derived from the Miami Isopycnic Coordinate Ocean Model; HYPOP (Dukowicz 2004); and Poseidon (Schopf and Loughe 1995).

In the general framework used in these models, each coordinate layer is assigned a target density. (For convenience, the present subsection uses the word 'density' to refer to a quantity related to potential density that would be used in an isopycnic framework, as discussed in Section 3.1.) If a given layer's target density lies within the range that is represented in the fluid at a given horizontal location and time, then the layer is assigned that density, and the layer at that location is isopycnic. However, if a given layer's target density is not present in the fluid, then in a purely isopycnic framework the layer would reduce to zero thickness; this would be the case, for example, for a layer of relatively low density at a high latitude. With a hybrid coordinate, such a layer is inflated in order to have positive thickness and thus contribute to the vertical resolution of the model. There is considerable freedom in the choice of methods for carrying out this process, but the basic idea is to impose a minimum thickness and then provide a recipe for the transition between the isopycnic state and the state where the layers are purely geometric (Bleck 2005). This use of a moving mesh resembles the Arbitrary Lagrangian–Eulerian (ALE) method of Hirt, Amsden and Cook (1974), although the ALE technique does not involve restoration to a target density.

Bleck (2002) recommends that any isopycnic layers that intersect bottom topography not be inflated, as the computation of horizontal pressure gradients in that situation could be inaccurate for the same reasons discussed in Section 3.3 for the case of σ-coordinates. Instead, the hybridization process should focus on layers near the top of the fluid. He also recommends that the minimum layer thickness be constant in the ocean's interior in order to avoid spurious dynamics resulting from the stretching of fluid columns within a layer. However, in relatively shallow regions near coastlines, the nonzero coordinate layers can be scaled with depth as with the terrain-following σ-coordinate, since vortex columns are likely to extend over the entire depth of the fluid in that case.

One difficulty with hybrid coordinates, as stated by Bleck (2002), is related to the seasonal cycle of the surface mixed layer. During winter, this layer is relatively thick, due to strong mixing caused by storms and by convective overturning. However, during the summer this forcing is much weaker, and consequently much of the fluid in this layer reverts to a stratified state. In a hybrid model, isopycnic layers are thus alternately destroyed and then re-created near the top of the fluid as part of the annual cycle. If the geometrically constrained, non-isopycnic layers are evenly distributed throughout a deep mixed layer during winter, for the sake of resolution, then they may need to migrate a long distance upward in order to reach their target densities in the spring or summer. This long migration implies large transports of fluid between coordinate layers; numerical errors in representing such transports can then cause a vertical dispersal of water properties, such as the concentrations of tracers in various layers. Such a dispersal could be alleviated by clustering the non-isopycnic layers near the top during the winter, but this would imply an uneven vertical resolution in that case.

The use of hybrid coordinates is relatively new and is a topic of current research. Issues currently debated include the details of transport and interpolation algorithms for carrying out the re-gridding from one state to the next, and whether the re-gridding should be performed at every time step or less frequently.

4. Analytical properties of solutions

We next describe some basic properties of solutions of the governing equations for the physical system considered here. This is done as a prelude to discussing methods for solving those equations numerically.

4.1. Governing equations

Section 7 gives a detailed derivation of partial differential equations that describe the conservation of mass, momentum, and tracers for a hydrostatic and stratified fluid that is in motion relative to a rotating spheroid such as the Earth. These equations are often called the 'primitive equations'; the term 'primitive' refers to the fact that the momentum equation is expressed in terms of velocity or momentum. In contrast, certain simplified models used for analytical or numerical studies employ dependent variables that are derived from velocity, such as vorticity, divergence, and streamfunction (e.g., Pedlosky (1996), Miller (2006)). The general circulation models quoted in this review largely employ the primitive equations, with the exception that nonhydrostatic effects are sometimes considered.

In Section 7 it is assumed that the horizontal position on the spheroid is represented with general orthogonal curvilinear coordinates and that the vertical coordinate is a generalized coordinate s that includes all of the cases discussed in Section 3. However, in Section 7 the representation of the diffusion of momentum and tracers assumes that the vertical coordinate is an isopycnic coordinate, and for other vertical coordinates the diffusion terms would have to be modified appropriately.

In Section 7 the equations are first derived for the continuous case. In anticipation of solving these equations numerically, the equations are also

integrated vertically between surfaces of constant s to yield equations involving mass-weighted vertical averages. Such equations can be used in
a vertically discrete system obtained by partitioning the fluid into coordinate layers.

The conservation of mass for the fluid is expressed in equations (7.32)–
(7.33) for the continuous case and in equation (7.38) for the case of a discrete
layer. In the latter case the dependent variable is essentially the mass per
unit horizontal area in the layer, and its evolution is determined by lateral
mass transport and by the vertical motions of layer interfaces relative to
the fluid. The conservation of mass for tracers in the fluid is described
by equations (7.106) and (7.107) for the continuous and vertically discrete
cases, respectively.

The conservation of horizontal momentum is expressed by the equations
(7.92) for the continuous case and by equations (7.101) and (7.104) for the
vertically discrete case. In the latter equations the dependent variables are
momentum density (velocity times the mass per unit horizontal area) and
the transport terms are written in flux form, which facilitates the usage of
(nearly) non-oscillatory advection schemes for those terms. The momentum and mass equations can also be combined to yield equations for which
the unknowns are components of velocity instead of momentum density.
Also needed for the momentum equation is the relation (7.105), which is a
discretization of the hydrostatic condition in generalized coordinates. This
relation provides a means for communicating pressure effects between layers.

The equations for mass, tracers, and momentum are supplemented with
a nonlinear equation of state that relates density, salinity, pressure, and
temperature (or potential temperature). Discussions of the equation of state
are given, for example, by Gill (1982) and Griffies (2004).

In the derivations in Section 7, the horizontal coordinates are denoted
by x_1 and x_2. These quantities could be latitude and longitude or some
other suitable parameters. Increments Δx_1 and Δx_2 in these coordinates
correspond to spatial displacements $m_1 \Delta x_1$ and $m_2 \Delta x_2$, respectively, where
m_1 and m_2 are metric coefficients developed in Section 7. Also appearing
in that section is a quantity $G = m_1 m_2$, which relates 'area' in the space
of the parameters x_1 and x_2 to physical area on the surface of the rotating
spheroid. The special case of Cartesian coordinates on a tangent plane is
then obtained by setting $m_1 = m_2 = G = 1$. This case is assumed in the
present section, for the sake of simplicity.

4.2. Shallow water equations

For the present discussion it is also useful to refer to the shallow water
equations, which describe the motions of a hydrostatic fluid of constant
density (*e.g.*, Pedlosky (1987)). If the system described in Section 7 is

discretized in the vertical by integrating over each of a set of coordinate layers, then the result can be regarded as a stack of shallow water models, with mechanisms for transferring mass, momentum, and pressure effects between adjacent layers.

In order to describe the shallow water system for a single-layer fluid, let x and y denote the horizontal coordinates; $u(x, y, t)$ and $v(x, y, t)$ denote the x- and y-components of velocity, respectively, defined relative to a rotating reference frame; $z_{\text{top}}(x, y, t)$ denote the elevation of the free surface at the top of the fluid; and $h(x, y, t) = z_{\text{top}}(x, y, t) - z_{\text{bot}}(x, y)$ denote the thickness of the fluid layer, where $z_{\text{bot}}(x, y)$ denotes the elevation of the bottom topography. Then

$$\frac{Du}{Dt} - fv = -g\frac{\partial z_{\text{top}}}{\partial x},$$

$$\frac{Dv}{Dt} + fu = -g\frac{\partial z_{\text{top}}}{\partial y}, \qquad (4.1)$$

$$\frac{\partial h}{\partial t} + \frac{\partial}{\partial x}(hu) + \frac{\partial}{\partial y}(hv) = 0,$$

where g is the magnitude of the acceleration due to gravity. The material derivative D/Dt is defined by $D/Dt = \partial/\partial t + u\partial/\partial x + v\partial/\partial y$, and it represents the time derivative as seen by an observer moving with the fluid. Since the velocity components u and v are defined relative to a rotating reference frame, the material derivatives Du/Dt and Dv/Dt do not represent acceleration relative to an inertial frame, and the Coriolis terms $-fv$ and fu are included to yield such an acceleration. In the case of a rotating spheroid, the Coriolis parameter f is given by equation (7.94), $f = 2\Omega \sin\theta$, where Ω is the angular rate of rotation and θ is the latitude.

The first two equations in (4.1) describe the conservation of momentum, and in particular they state that the lateral acceleration of the fluid is due to variations in the free-surface elevation, which correspond to lateral variations of pressure within the fluid. The third equation in (4.1) describes the conservation of mass, and for this constant-density fluid the equation relates time variations in the layer thickness h to lateral variations in the volume flux (hu, hv).

An alternate formulation of the pressure forcing is given by the following. Define the Montgomery potential M by

$$M(x, y, z, t) = \alpha p(x, y, z, t) + gz, \qquad (4.2)$$

where α is the specific volume (reciprocal of density ρ) of the fluid, and $p(x, y, z, t)$ is the pressure (Montgomery 1937). The hydrostatic condition $\partial p/\partial z = -\rho g$ (see (7.8)) implies that M is independent of z in a layer of constant density. Assuming that the pressure at the top of the fluid has a constant value, such as atmospheric pressure, then $\partial M/\partial x = g\partial z_{\text{top}}/\partial x$

and $\partial M/\partial y = g \partial z_{\text{top}}/\partial y$. The shallow water system (4.1) can then be written as

$$\frac{D\mathbf{u}}{Dt} + f\mathbf{u}^{\perp} = -\nabla M,$$

$$\frac{\partial h}{\partial t} + \nabla \cdot (h\mathbf{u}) = 0, \tag{4.3}$$

where $\mathbf{u} = (u, v)$, $\mathbf{u}^{\perp} = (-v, u)$, and $\nabla = (\partial/\partial x, \partial/\partial y)$. In addition, $-\nabla M = -\frac{1}{\rho}\nabla p$, which is a common form for representing the pressure term in fluid dynamics.

If one were to model a stratified fluid as a stack of constant-density immiscible shallow water models, then the equations (4.3) could be applied in each layer. At an interface between layers, the pressure p and elevation z are continuous, so equation (4.2) implies $\Delta M = p\Delta\alpha$ at the interface. Here, ΔM and $\Delta\alpha$ are the jumps in M and α, respectively, across that interface. The relation $\Delta M = p\Delta\alpha$ is an analogue of the discretization (7.105) of the hydrostatic condition in generalized coordinates.

If a continuously stratified fluid is modelled with a generalized vertical coordinate, then the Montgomery potential enables gradients along slanting coordinate surfaces to represent pressure forcing that is truly horizontal. This idea is developed in Section 7.11.

4.3. Geostrophic balance

A striking feature of large-scale flows in the ocean and atmosphere is the approximate balance between the Coriolis and pressure terms, which causes the fluid to flow along curves of constant pressure. Here this 'geostrophic balance' is derived, in the context of the shallow water equations, by using an approximate scale analysis.

Let T, L, and U denote numbers that represent typical scales for time, horizontal distance, and horizontal velocity, respectively. A representative scale for the nonlinear terms in the momentum equation in (4.3) (*i.e.*, $uu_x + vu_y$ and $uv_x + vv_y$) is then UU/L, and a representative value for the Coriolis term is fU. The ratio of these scales is the Rossby number $Ro = U/(fL)$.

As noted in Section 7.15, the Coriolis parameter is zero at the equator, and at the middle and high latitudes of the Earth this parameter is on the order of 10^{-4} sec^{-1}. A velocity scale $U = 1$ m/sec represents a strong current, and a length scale $L = 100$ km $= 10^5$ m represents the approximate width of a western boundary current. With these scales, and with a Coriolis parameter equal to 10^{-4} sec^{-1}, the Rossby number is 0.1. However, for basin-scale flows, U is typically smaller and L is larger, so the Rossby number is much smaller in that case. Similarly, the time derivative in the momentum equation in (4.1) has the scale U/T, and the ratio of this scale to the scale of the Coriolis term is $1/(fT)$. If the time, space, and length scales are related by $U = L/T$, then the parameter $1/(fT)$ is equal to the Rossby number. If a process has a time scale of a day (86400 seconds), then $1/(fT) \approx 0.1$ for the middle and high latitudes; for a time scale of 100 days, $1/(fT) \approx 10^{-3}$. It then follows that, for large-scale flows, the time derivative and nonlinear terms in the momentum equation are much smaller than the Coriolis term, except near the equator.

A remaining consideration is the size of a possible diffusion term. In geophysical fluid dynamics it is common practice to include some sort of diffusion term in the momentum equation to represent the large-scale effect of subgrid-scale mixing. For the system (4.3), such a term could take the form $A_H \nabla^2 \mathbf{u}$, where ∇^2 denotes the Laplace operator and A_H is an eddy viscosity coefficient. (Also see Section 7.12.) A scale for the diffusion term is then $A_H U/L^2$, and the ratio of the scale of the diffusion term to the scale of the Coriolis term is given by the Ekman number, $E = A_H/(fL^2)$. A physically appropriate choice of A_H involves considerable uncertainty; for example, values in the range 10 m^2/sec to 10^4 m^2/sec have been proposed (Pedlosky 1987). A horizontal length scale $L = 100$ km $= 10^5$ m and an eddy viscosity $A_H = 10^4$ m^2/sec leads to an Ekman number on the order of 10^{-2}. Larger length scales and/or smaller viscosities produce smaller Ekman numbers. For large-scale flows, the diffusion term would therefore be much smaller than the Coriolis term, except near the equator.

The only remaining term in the momentum equation is the pressure term $-g\nabla z_{\text{top}} = -\nabla M$. Since the other terms are much smaller than the Coriolis term, the pressure term and the Coriolis term must be approximately equal, i.e., the flow is approximately in a geostrophic balance. If this balance is exact, then $f\mathbf{u}^\perp = -\nabla M$, or

$$-fv = -g\frac{\partial z_{\text{top}}}{\partial x},$$
$$fu = -g\frac{\partial z_{\text{top}}}{\partial y}. \tag{4.4}$$

In this case the horizontal velocity vector $\mathbf{u} = (u, v)$ is orthogonal to ∇z_{top}. If the fluid is viewed from the top, then the fluid is seen to flow along contours of constant free-surface elevation z_{top}, which coincide with contours of constant pressure within the fluid. The geostrophic balance is also seen in weather maps, which show upper-level winds blowing along isobars. In general, a geophysical fluid flow is not in a state of exact balance, and the evolution of the flow is driven by small departures from a balanced state.

The geostrophic balance may seem counter-intuitive, as everyday experience suggests that fluid should flow from a region of high pressure to a region of low pressure, not along curves that separate the two regions. However, the key point is that the flows considered here are dominated by the effects of rotation. For a simple analogy, try walking in a straight line on a rotating merry-go-round; this 'straight' path is curved relative to the ground, and a force from the side is required to keep you on that straight/curved path.

4.4. Potential vorticity and Rossby waves

The vorticity of a fluid flow is the curl of the velocity field. If a fluid is stationary relative to a rotating spheroid, then the vorticity of the fluid relative to an inertial reference frame is 2Ω, where $\Omega = |\Omega|$ is the angular rate of rotation, the direction of the vector Ω aligns with the axis of rotation, and the direction of Ω and the direction of rotation are related by the right-hand rule. If a fluid is in motion relative to the spheroid, then the relative velocity gives rise to a vorticity relative to the rotating reference frame. In the case of the shallow water equations (4.1) and (4.3) on a tangent plane, the relative vorticity is $\zeta\mathbf{k}$, where $\mathbf{k}$ is the unit vector in the (local) upward direction, and

$$\zeta(x,y,t) = \frac{\partial v}{\partial x} - \frac{\partial u}{\partial y}.$$

However, as noted in Section 7.15, the Coriolis parameter f is the local vertical component of the planetary vorticity 2Ω. Therefore $\zeta + f$ is the local vertical component of the absolute (relative plus planetary) vorticity. In the present subsection we develop a conservation property associated with this quantity.

For the sake of simplicity, this property is developed in the context of the shallow water system (4.1), (4.3). The momentum equations in that system can be written in the form

$$\begin{aligned}
\frac{\partial u}{\partial t} + \frac{\partial}{\partial x}\left[\frac{1}{2}(u^2 + v^2)\right] - (\zeta + f)v &= -\frac{\partial M}{\partial x}, \\
\frac{\partial v}{\partial t} + \frac{\partial}{\partial y}\left[\frac{1}{2}(u^2 + v^2)\right] + (\zeta + f)u &= -\frac{\partial M}{\partial y};
\end{aligned} \tag{4.5}$$

that is, the nonlinear terms are written in terms of the relative vorticity ζ and the kinetic energy per unit mass, $\frac{1}{2}(u^2 + v^2)$. Now compute the x-derivative of the second equation minus the y-derivative of the first equation. The result is

$$\frac{\partial \zeta}{\partial t} + u\frac{\partial}{\partial x}(\zeta + f) + v\frac{\partial}{\partial y}(\zeta + f) + (\zeta + f)\left(\frac{\partial u}{\partial x} + \frac{\partial v}{\partial y}\right) = 0.$$

A comparison with the mass equation in (4.1), coupled with the fact that f is independent of t, yields

$$\frac{D}{Dt}(\zeta + f) - (\zeta + f)\frac{1}{h}\frac{Dh}{Dt} = 0,$$

or

$$\frac{D}{Dt}\left(\frac{\zeta + f}{h}\right) = 0. \qquad (4.6)$$

The quantity $(\zeta + f)/h$ is known as the potential vorticity, and it is the ratio of the absolute vorticity to the layer thickness.

Equation (4.6) states that the potential vorticity is constant, when seen by an observer moving with the fluid. For example, if a column of water moves laterally and experiences a change in h, then the absolute vorticity $\zeta + f$ changes, due to vortex stretching.

Of particular interest is the effect of variations in the Coriolis parameter on the relative vorticity. Suppose that ζ is initially zero, and consider a region where the bottom topography z_{bot} is constant, so that the layer thickness h is nearly constant. If a water column moves northward, for example, then the resulting increase in f causes a decrease in ζ, which corresponds to a clockwise rotation when seen from above. This local rotation causes water columns to the east and west to move southward and northward, respectively. The resulting changes in relative vorticity for those columns tend to force the first column to return towards its original position. The resulting wave motion is known as a Rossby wave, and for this wave the restoring mechanism is based on vorticity instead of gravity. Variations in bottom topography have an analogous effect, and a similar discussion of topographic Rossby waves is given by Pedlosky (1987). The existence of Rossby waves depends fundamentally on variations in bottom topography and/or the variation of the Coriolis parameter with latitude.

Detailed analyses of the dynamics of Rossby waves are given in several references, e.g., Gill (1982) and Pedlosky (1987, 2003). These waves propagate with low frequencies, and the resulting time derivatives in the system (4.1) are so small that a Rossby wave is in an approximate geostrophic balance; as the bumps and depressions in the free surface propagate, the velocity field is tangent to contours of constant elevation.

The phase velocities of Rossby waves are always westward. Relatively long Rossby waves are nearly nondispersive, but relatively short Rossby waves are strongly dispersive with eastward group velocity. Long Rossby waves then propagate westward from the eastern boundaries of ocean basins, and short Rossby waves propagate eastward (in the sense of group velocity) from western boundaries. The latter propagate more slowly, and the resulting accumulation of short waves near the western boundaries helps to explain the east–west asymmetry in ocean circulation patterns (Gill 1982).

4.5. External and internal modes

A stratified fluid can admit both external and internal wave motions. External waves and internal waves propagate at very different speeds, and the separation of time scales has a major impact on the process of solving the governing equations numerically. Internal waves cannot be modelled with the shallow water equations, since those equations apply to a fluid of constant density, so instead the present discussion refers to a simple linearized version of the governing equations that are developed in Section 7 for a stratified fluid. The following analysis is similar to one given by Higdon and Bennett (1996).

For the sake of simplicity, assume that the vertical coordinate s is the specific volume α (reciprocal of density) over the range $\alpha_{\text{bot}} \leq \alpha \leq \alpha_{\text{top}}$ and that $\dot{s} = \dot{\alpha} = 0$, $i.e.$, α is constant in time for each fluid parcel; an analysis of external and internal modes with a more general equation of state is given by Dukowicz (2006). In order to obtain a linearized model, assume that the bottom of the fluid domain is level and that the flow is a small perturbation of a static state having a level free surface at the top of the fluid and level surfaces of constant density within the fluid. Let $\tilde{p}(\alpha)$ denote the pressure in the fluid at the stationary state, and let $\tilde{M}(\alpha)$ denote the Montgomery potential at that state. Then let $p(x, y, \alpha, t)$ and $M(x, y, \alpha, t)$ denote the perturbations in those quantities from the static values. The pressure and Montgomery potential at an arbitrary state are then $\tilde{p}(\alpha) + p(x, y, \alpha, t)$ and $\tilde{M}(\alpha) + M(x, y, \alpha, t)$, respectively, with the present notation. Also assume that the viscosity and applied stresses are zero. Under these assumptions the governing equations from Section 7 for a continuously stratified fluid simplify to

$$
\begin{aligned}
u_t - fv &= -M_x, \\
v_t + fu &= -M_y, \\
p_{\alpha t} + \tilde{p}_\alpha(\alpha)(u_x + v_y) &= 0, \\
M_\alpha &= p.
\end{aligned}
\tag{4.7}
$$

The subscripts denote partial derivatives. The first two equations in (4.7) are the momentum equations (7.92) as applied to the present case, the third equation is the equation (7.32) for conservation of mass, and the fourth equation is the hydrostatic condition (7.65).

Upper and lower boundary conditions for the system (4.7) can be formulated as follows. At the top of the fluid the pressure is equal to the atmospheric pressure, which is assumed here to be a constant. The perturbation in pressure is then zero, so in the present notation

$$
M_\alpha = p = 0 \qquad \text{if } \alpha = \alpha_{\text{top}}.
\tag{4.8}
$$

At the bottom of the fluid the perturbation in elevation is zero, so the perturbation in Montgomery potential satisfies

$$
M = \alpha p = \alpha M_\alpha \qquad \text{if } \alpha = \alpha_{\text{bot}}.
\tag{4.9}
$$

Special solutions of the system (4.7)–(4.9) can be constructed by separating the vertical dependence from the time and horizontal dependences. Let $u(x, y, \alpha, t) = \hat{u}(x, y, t)\phi(\alpha)$, $v(x, y, \alpha, t) = \hat{v}(x, y, t)\phi(\alpha)$, and $M(x, y, \alpha, t) = \hat{M}(x, y, t)\phi(\alpha)$. Then $p(x, y, \alpha, t) = \hat{M}(x, y, t)\phi_\alpha(\alpha)$, and the system (4.7) becomes

$$\hat{u}_t - f\hat{v} - - \hat{M}_x,$$

$$\hat{v}_t + f\hat{u} = -\hat{M}_y, \tag{4.10}$$

$$\tilde{M}_t + (1/\lambda)(\hat{u}_x + \hat{v}_y) = 0,$$

where λ and ϕ satisfy

$$\phi_{\alpha\alpha} = \lambda\tilde{p}_\alpha(\alpha)\phi(\alpha) \quad \text{if } \alpha_{\text{bot}} < \alpha < \alpha_{\text{top}},$$

$$\phi_\alpha = 0 \qquad\qquad \text{if } \alpha = \alpha_{\text{top}}, \tag{4.11}$$

$$\phi = \alpha\phi_\alpha \qquad\qquad \text{if } \alpha = \alpha_{\text{bot}}.$$

The Sturm–Liouville problem (4.11) admits a countable set of eigenvalues and a complete set of eigenfunctions. The eigenvalues are all positive, since $\tilde{p}_\alpha < 0$, and they can be denoted by $0 < \lambda_0 < \lambda_1 < \lambda_2 < \ldots$. In the general solution of the system (4.7), u can be represented as

$$u(x, y, \alpha, t) = \sum_{j=1}^{\infty} \hat{u}^{(j)}(x, y, t)\phi^{(j)}(\alpha), \tag{4.12}$$

where $\phi^{(j)}$ is an eigenfunction corresponding to eigenvalue λ_j, and $\hat{u}^{(j)}(x, y, t)$ is obtained from the reduced system (4.10) with $\lambda = \lambda_j$. Orthogonality of the eigenfunctions with respect to the weight function $\tilde{p}_\alpha$ implies

$$\hat{u}^{(j)}(x, y, t) = \frac{\int_{\alpha_{\text{bot}}}^{\alpha_{\text{top}}} u(x, y, \alpha, t)\, \phi^{(j)}(\alpha)\, \tilde{p}_\alpha(\alpha)\, d\alpha}{\int_{\alpha_{\text{bot}}}^{\alpha_{\text{top}}} \phi^{(j)}(\alpha)^2\, \tilde{p}_\alpha(\alpha)\, d\alpha}. \tag{4.13}$$

Similar expansions apply to v, M, and $p = M_\alpha$.

The reduced system (4.10), which describes the dependence with respect to (x, y, t), has the form of the linearization of the shallow water equations (4.1). In the case where the Coriolis parameter f is constant, this system admits gravity wave solutions with speed $c = \sqrt{1/\lambda}$. In the case where f varies with latitude, the gravity waves are still admitted, and in addition Rossby waves can also be present (Gill 1982, Pedlosky 1987).

The vertical structures of the modal solutions are determined by the eigenfunctions $\phi^{(j)}$, which can be visualized as indicated in Figure 4.1. Assume that the α-axis is vertical and that the ϕ-axis is horizontal. The upper boundary condition $\phi_\alpha(\alpha_{\text{top}}) = 0$ states that the tangent line to the graph of ϕ is vertical when $\alpha = \alpha_{\text{top}}$. The lower boundary condition

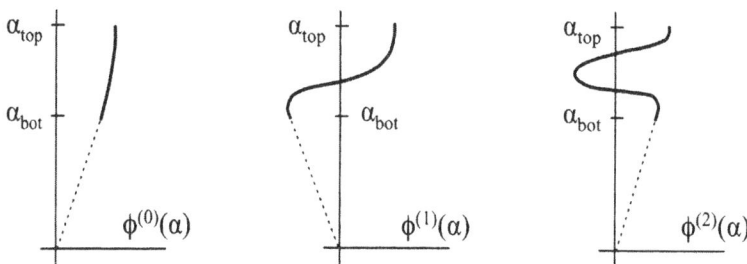

Figure 4.1. Graphs of eigenfunctions which give the vertical
dependences in modal solutions of the linearized system (4.7).
The eigenfunction $\phi^{(0)}$ corresponds to the external mode, and
$\phi^{(1)}$ and $\phi^{(2)}$ correspond to the first two internal modes.
The dotted lines illustrate the lower boundary condition
$\phi_\alpha(\alpha_{bot}) = \phi(\alpha_{bot})/\alpha_{bot}$. In these plots, the α-axes are not to
scale, as typically $(\alpha_{top} - \alpha_{bot})/\alpha_{bot} \approx 10^{-2}$. On the interval
$\alpha_{bot} \leq \alpha \leq \alpha_{top}$, the relative variation in $\phi^{(0)}$ is bounded by
the relative variation in α.

$\phi_\alpha(\alpha_{bot}) = \phi(\alpha_{bot})/\alpha_{bot}$ states that the tangent line to the graph of ϕ
at $\alpha = \alpha_{bot}$ passes through the origin. Without loss of generality, assume
$\phi(\alpha_{top}) = 1$, and consider the variation of $\phi(\alpha)$ as α varies from α_{top} downward towards α_{bot}. If $\lambda > 0$ then $\phi_{\alpha\alpha}$ and ϕ have opposite signs, since
$\tilde{p}_\alpha < 0$, so the graph of ϕ always bends so as to return toward the α-axis.
For a given value of $\phi(\alpha)$, a larger value of λ yields a larger value for the
concavity $\phi_{\alpha\alpha}$.

    In the case of the smallest eigenvalue λ_0, the corresponding eigenfunction
$\phi^{(0)}$ is monotone and maintains constant sign, and its derivative is largest
when $\alpha = \alpha_{bot}$. This condition on the derivative, coupled with the mean
value theorem, implies

$$\left| \phi^{(0)}(\alpha_{top}) - \phi^{(0)}(\alpha_{bot}) \right| < \frac{\phi^{(0)}(\alpha_{bot})}{\alpha_{bot}} \left| \alpha_{top} - \alpha_{bot} \right|. \qquad (4.14)$$

It follows that the relative variation of the eigenfunction $\phi^{(0)}$ over the inter-
val $\alpha_{bot} \leq \alpha \leq \alpha_{top}$ is less than the relative variation of α over that interval.
In the ocean, α typically varies by at most a few per cent over the vertical
extent of a water column. With such behaviour of α, the eigenfunction $\phi^{(0)}$
is nearly constant.

    For the eigenvalue λ_1, the corresponding eigenfunction $\phi^{(1)}$ changes sign
once; in general, the eigenfunction $\phi^{(j)}$, corresponding to the eigenvalue
λ_j, changes sign j times. The orthogonality of eigenfunctions with respect

to $\tilde{p}_\alpha$, coupled with the observation that $\phi^{(0)} \approx 1$, implies

$$\int_{\alpha_{\text{bot}}}^{\alpha_{\text{top}}} \phi^{(j)}(\alpha)\, \tilde{p}_\alpha(\alpha)\, d\alpha \approx \int_{\alpha_{\text{bot}}}^{\alpha_{\text{top}}} \phi^{(j)}(\alpha)\, \phi^{(0)}(\alpha)\, \tilde{p}_\alpha(\alpha)\, d\alpha = 0 \qquad (4.15)$$

for $j \geq 1$. The mass-weighted vertical average of $\phi^{(j)}$ is thus nearly zero, if $j \geq 1$.

A modal solution for $j = 0$ is an external mode, whereas the modes for $j \geq 1$ are internal modes. To see this, partition the interval $\alpha_{\text{bot}} \leq \alpha \leq \alpha_{\text{top}}$ into layers separated by surfaces of constant α. For the case $j = 0$, the horizontal velocity divergence $u_x + v_y = \phi^{(0)}(\alpha)(\hat{u}_x + \hat{v}_y)$ is nearly independent of depth. If the third equation in the linearized system (4.7) is integrated with respect to α over one of those layers, the result is that $\partial/\partial t(\Delta p/\Delta\tilde{p})$ is approximately independent of depth. Here, Δp and $\Delta\tilde{p}$ represent the vertical differences of p and $\tilde{p}$, respectively, across the layer. It follows that if $j = 0$ then all layers are thickened or thinned by approximately the same proportion, for given (x, y, t). The behaviour of the free surface at the top of the fluid then indicates the nature of the wave motion throughout the interior.

On the other hand, if $j \geq 1$ then the horizontal divergence $\phi^{(j)}(\alpha)(\hat{u}_x + \hat{v}_y)$ varies in sign with depth, so some layers are thickened and some are thinned, for given (x, y, t). The wave motion is then manifested by undulations of surfaces of constant density within the fluid. Furthermore, a vertical integration of the third equation in the linearized system (4.7) over the entire depth of the fluid, coupled with the zero-integral condition (4.15), implies that the elevation of the free surface remains nearly unperturbed if $j \geq 1$.

Estimates of wave speeds can be obtained as follows. Vertical integration of the equation $\phi_{\alpha\alpha}^{(0)} = \lambda_0 \tilde{p}_\alpha(\alpha)\phi^{(0)}(\alpha)$ over the interval $\alpha_{\text{bot}} \leq \alpha \leq \alpha_{\text{top}}$, combined with the boundary conditions in (4.11) and the condition $\phi^{(0)} \approx 1$, implies that the gravity-wave speed for the external mode is $c_0 = \sqrt{1/\lambda_0} \approx \sqrt{\alpha_{\text{bot}}(\tilde{p}_{\text{bot}} - \tilde{p}_{\text{top}})}$. However, the hydrostatic condition $\partial p/\partial z = -\rho g$ (see (7.8)) implies $\alpha_{\text{bot}}(\tilde{p}_{\text{bot}} - \tilde{p}_{\text{top}}) \approx gH$, where H is the total depth of the fluid. The speed of gravity waves as seen in the linearized shallow water equations is $\sqrt{gH}$, which is approximately the speed seen in the external mode. If, for example, $H = 4000$ m, then $c_0 \approx \sqrt{gH} \approx 200$ m/sec. On the other hand, the speeds of internal gravity waves in the ocean are typically at most a few metres per second (Gill 1982).

The external mode is essentially a two-dimensional phenomenon, in the sense that it can be approximately described by functions of (x, y, t), and the dynamics of this mode are very similar to those seen in the linearized shallow water equations. Furthermore, the velocity field for the external mode is obtained by the projection (4.13) for the case $j = 0$. Since $\phi^{(0)} \approx 1$, equation (4.13) implies

$$\hat{u}^{(0)}(x, y, t) \approx \frac{1}{\tilde{p}_{\text{bot}} - \tilde{p}_{\text{top}}} \int_{\alpha_{\text{bot}}}^{\alpha_{\text{top}}} u(x, y, \alpha, t) \left(-\tilde{p}_\alpha(\alpha)\right) d\alpha, \qquad (4.16)$$

and an analogous equation holds for $\hat{v}^{(0)}$. The velocity field in the external mode is thus approximately given by the mass-weighted vertical average of the horizontal velocity field for the three-dimensional system. These observations form the basis for techniques for isolating the fast dynamics into a relatively simple two-dimensional subsystem, as described in Section 5.1.

The preceding analysis assumes that the fluid is continuously stratified. One can also consider a vertical discretization of the system (4.7), or equivalently, a multi-layer stack of shallow water models. In that event, the system admits an external mode and finitely many internal modes. For example, Higdon (2005) developed a test problem involving linearized flow in a two-layer fluid in a straight channel having a level bottom, with the Coriolis parameter varying linearly in the cross-channel direction y. In this case the system admits both external and internal waves, and within each class are Rossby waves and gravity waves. The gravity waves include Poincaré waves and Kelvin waves; the former are approximately sinusoidal in y, whereas the latter decay exponentially from the one or another of the solid boundaries. The domain is discretized in space, and after a Fourier transform in time and in the along-channel direction x, the time frequencies and y-dependences are obtained via numerical solution of a matrix eigenvalue problem. The resulting modal solutions can then be used to test numerical methods for the time splittings that are discussed in Section 5.1. In particular, the external and internal modal solutions make it possible to test the fast and slow subsystems independently.

5. Time discretization

The present section outlines some aspects of time discretization for numerical models of ocean circulation. Included is a description of the process of splitting the fast and slow dynamics into separate subsystems and a discussion of some time-stepping methods.

5.1. Barotropic–baroclinic splitting

As noted in Section 2.3, the fastest motions in a numerical model of ocean circulation are typically the external gravity waves, and these travel much more rapidly than any of the other motions that are present in the system. If an explicit time discretization is used to solve a system of partial differential equations numerically, then the time step Δt is limited in terms of the fastest motions that can be present. However, in the present case the fastest motions are essentially two-dimensional, as demonstrated in Section 4.5. The numerical solution of a complex three-dimensional system is therefore severely constrained by a special set of two-dimensional motions.

It has therefore become common practice to split the dynamics of the system into two subsystems, one a relatively simple two-dimensional system that represents the fast motions, and the other a relatively complex three-dimensional system that represents the remaining (slow) processes. The latter system is solved explicitly with a relatively long time step that is appropriate for resolving the slow motions. The fast system is either solved implicitly with the same time step or explicitly with short steps. The resulting algorithm is much more efficient than solving the entire unsplit system explicitly with short steps.

Such a splitting is traditionally referred to as a barotropic–baroclinic splitting. In classical fluid dynamics, a flow is labelled barotropic if the density is a function of pressure only (Pedlosky 1987), and it is labelled baroclinic otherwise. In the case of a stratified fluid, a barotropic state implies that the surfaces of constant density coincide with the surfaces of constant pressure. If a fluid exhibits only an external wave, then all fluid layers thicken or thin by approximately the same proportion, at a given horizontal location and time (see Figure 2.2). The surfaces of constant density within the fluid thus have essentially the same shape, with the amplitudes of variation increasing from the bottom of the fluid domain to the top. Each such surface remains approximately the same distance below the free surface at the top of the fluid, at all horizontal locations and times, so the flow is approximately barotropic. On the other hand, this is not the case with an internal wave, and in that case the fluid is in a baroclinic state. In the following, the fast and slow subsystems are referred to as the barotropic and baroclinic equations, respectively.

The early versions of the z-coordinate Bryan–Cox class of ocean models used a rigid-lid boundary condition $w = 0$ at the top of the fluid, where w is the vertical component of fluid velocity. This assumption has the effect of replacing the fast speed of external gravity waves with an infinite speed, and a two-dimensional elliptic partial differential equation for this mode must then be solved at each (long) time step. More recent efforts have replaced the rigid lid with a free-surface boundary condition, which restores the finite speed of propagation for the barotropic subsystem. This system is obtained by a vertical integration of the three-dimensional momentum and mass equations, and it has a structure similar to that of the shallow water equations. Killworth, Stainforth, Webb and Paterson (1991) solved the barotropic equations explicitly with short steps, whereas Dukowicz and Smith (1994) solved the barotropic equations implicitly with same (long) time step as for the baroclinic equations. A hybrid-coordinate update of the latter model solves the barotropic equations explicitly with short steps (Dukowicz 2004). The current version of the Modular Ocean Model also solves the barotropic equations explicitly with short steps (Griffies *et al.* 2004).

A barotropic–baroclinic splitting for isopycnic models was developed by Bleck and Smith (1990). With that splitting, the barotropic velocity $\bar{\mathbf{u}}(x, y, t)$ is the mass-weighted vertical average of the horizontal velocity $\mathbf{u}(x, y, s, t) = \big(u(x, y, s, t), v(x, y, s, t)\big)$ appearing in the full three-dimensional system, in analogy to the approximate projection onto the external mode given in (4.16). A baroclinic velocity is then given by $\mathbf{u}'(x, y, s, t) = \mathbf{u}(x, y, s, t) - \bar{\mathbf{u}}(x, y, t)$, so that $\mathbf{u} = \bar{\mathbf{u}} + \mathbf{u}'$. Here, the mass-weighted vertical averages are the same as those used in Section 7.16, except that the averages in that section involve a single coordinate layer, whereas in the present case the averages are taken over the entire depth of the fluid. The splitting of the mass field is based on the idea that an external wave causes all layers to thicken or thin by approximately the same proportion. In particular, the pressure is represented as $p(x, y, s, t) = \big(1 + \eta(x, y, t)\big)p'(x, y, s, t)$, where p' represents the baroclinic component of the pressure, and η is a barotropic mass variable that represents the relative thickening of coordinate layers.

With this formulation, the splittings of the velocity and mass fields are inexact, even in the linearized case where one can discuss decompositions of the solution into external and internal modes. In addition, as formulated by Bleck and Smith (1990), the pressure forcing in the barotropic momentum equation is the same as in the shallow water equations for a fluid of constant density, and this introduces a further approximation that contributes to the inexactness of the splitting. This inexactness implies that the baroclinic equations can represent a (small) portion of the fast external motions, even though these equations are intended to represent the slow motions. If the baroclinic equations are solved explicitly with a long time step that is appropriate for resolving the slow motions, then the Courant–Friedrichs–Lewy condition is violated, strictly speaking. This CFL violation raises the possibility of numerical instability.

Higdon and Bennett (1996) showed that this instability can in fact occur with the splitting of Bleck and Smith (1990), as applied to the linearized case. However, Higdon and de Szoeke (1997) subsequently showed that the instability can be removed by modifying the barotropic momentum equation so as to replace the pressure-gradient term from the shallow water equations with the mass-weighted vertical average of $\nabla M = (\partial M/\partial x, \partial M/\partial y)$. In this case the splitting is still inexact, due to the inexactness in splitting the velocity and mass fields. However, the amount of 'fast' energy in the 'slow' baroclinic equations is low enough to obtain stability, even in the face of the formal violation of the CFL condition. In other words, the splitting of the external and internal modes does not need to be exact, but it does need to be sufficiently accurate.

The analysis of external and internal modes in Section 4.5 assumes that the bottom of the fluid domain is level. Without this assumption, the linearized system (4.7) is not separable, and the solutions do not decompose exactly into external and internal modes. In addition, realistic ocean circulation models include nonlinearity, and this provides another mechanism for interactions between fast and slow motions. Nonetheless, in the ocean modelling community it is common practice to split the dynamics of the governing equations based on the preceding ideas, even for nonlinear problems, and this approach has proved to be an effective method for gaining computational efficiency while maintaining numerical stability.

5.2. Leapfrog time-stepping

One time-stepping method that has been widely used in geophysical fluid dynamics is the leapfrog method. For example, this method is used to solve the baroclinic equations in many versions of the Bryan–Cox class of models (Griffies *et al.* 2000*a*) and in the Miami Isopycnic Coordinate Ocean Model (Bleck *et al.* 1992). The leapfrog method is a three-level scheme based on centred differencing about the middle time level, and for an equation of the form $u_t = F(t, u)$ it can be written in the form $u^{n+1} = u^{n-1} + 2\Delta t F(t_n, u^n)$. Here, u^n is an approximation to $u(t_n)$.

The leapfrog method is straightforward to implement, but it suffers the disadvantage of allowing a nonphysical computational mode consisting of sawtooth oscillations in t. For example, in the special case $F = 0$ the leapfrog method for $u_t = F(t, u)$ is simply $u^{n+1} = u^{n-1}$, and this scheme allows both constant solutions and sawtooth solutions of the form $u^n = c(-1)^n$. In a nonlinear model such a mode, once stimulated, can grow without bound unless measures are taken to suppress the oscillations. One widely used method for doing this is the Asselin filter (Asselin 1972), which uses a weighted average defined by $\bar{\phi}^n := \gamma \phi^{n+1} + (1 - 2\gamma)\phi^n + \gamma \bar{\phi}^{n-1}$. Here, γ is a positive constant, and $\bar{\phi}$ is a smoothed version of the quantity ϕ. After the quantity ϕ^{n+1} is computed, the filter is then used to smooth the solution at time t_n. The leapfrog method has order two, but using the Asselin filter reduces the method to first-order accuracy (Durran 1999). An alternative to this filter is given by Dukowicz and Smith (1994), who periodically interrupt the computation in order to average the solution between consecutive time levels. This is done for two consecutive pairs of time levels, and the leapfrog method is then restarted from the values obtained for the half-integer points.

If a filter succeeds in maintaining numerical stability but does not suppress the computational mode completely, then numerical difficulties can still persist. For example, Griffies (2004) describes some experience with the Modular Ocean Model using leapfrog time-stepping and an Asselin filter. In those experiments, grid noise remained and was especially prevalent near the equator, and it caused a complete reversal of the directions of strong zonal (east–west) currents in that region from one time step to the next.

Additional difficulties arise if a computational mode is present and if the state of a model is altered when certain variables reach threshold values (Rainer Bleck, personal communcation). For example, in a hybrid-coordinate model, a coordinate layer is in an isopycnic state if and only if its target density lies within the range of densities that actually exist in the fluid. If this state is updated at each time step, and if the target density lies near the boundary of that range but the computed density field displays sawtooth oscillations in time, then the layer may be in one state at the even time steps and in another state at the odd steps.

5.3. Two-level time-stepping

The preceding difficulties can be avoided if a model employs a time-stepping scheme that uses only two time levels, as such a scheme cannot support the sawtooth mode described above. This matter has been a subject of recent attention.

For example, two-level methods are now available as options, along with the leapfrog method, in the Modular Ocean Model (Griffies 2004). The two-level method for the baroclinic equations is staggered in time, with velocity defined at integer time steps, and tracers, pressure, and density defined at half-integer steps. The staggered time grid enables second-order centred differencing without allowing the computational mode associated with the leapfrog method. In the case of the barotropic equations, the two-level method is a predictor–corrector method.

Hallberg (1997) developed a two-level predictor–corrector scheme for layered models. This method involves a prediction and correction of all of the dependent variables in the baroclinic and barotropic subsystems, and various weighted averages are used during the correction steps. In addition, Shchepetkin and McWilliams (2005) give an extensive survey of a variety of two-level and multi-level time-stepping methods.

A two-level non-staggered method for layered models was developed by Higdon (2002, 2005). After an initial forward-Euler prediction of the baroclinic velocity from time t_n to time t_{n+1}, all time discretizations for the baroclinic equations involve centred differencing and unweighted averaging about the middle of the time interval $[t_n, t_{n+1}]$. The Coriolis terms are implemented implicitly; if the horizontal velocity components u and v are defined at different points, as on the C-grid (Section 6.1), then the Coriolis terms are implemented with a simple iteration. In a linearized analysis where the barotropic equations are assumed to be solved exactly in t, the algorithm is stable and essentially nondissipative. In some experiments with a nonlinear model and a standard dissipative advection scheme for mass and momentum, the algorithm behaves stably for very long times with zero explicit viscosity.

6. Spatial discretization and related issues

This section discusses some topics involving spatial discretization, the numerical simulation of advection, and the numerical solution of the momentum equations.

6.1. Horizontal grids: quadrilaterals

In existing numerical models of ocean circulation, it is commonplace to use finite difference and/or finite volume methods on quadrilateral spatial grids. This subsection focuses on such discretizations, and some alternatives are discussed in the following subsection.

A natural choice for a horizontal coordinate system for the Earth is the one based on latitude and longitude. However, this system admits a singularity at the North and South Poles, as the metric coefficient relating increments in longitude to increments in space tends to zero at those locations (see Section 7.18). This singularity is not an issue in the case of the South Pole, which lies on land, but it in the case of the North Pole the singularity lies within the fluid domain for a global model. If a spatial discretization is based on this coordinate system, then the convergence of grid lines causes substantial difficulties with explicit time-stepping methods, as the Courant–Friedrichs–Lewy condition implies that the time increment Δt must tend to zero as the spatial increment tends to zero.

This problem can be avoided with a coordinate system that does not place any singularities within the fluid domain. For example, Smith, Kortas and Meltz (1995) developed a class of orthogonal curvilinear grids with two coordinate poles that can be placed at arbitrary locations. In practice, the south coordinate pole can be located at the south geographical pole, and the latitude–longitude grid is deformed smoothly so that the north coordinate pole is located on a land mass.

With a coordinate system developed by Murray (1996), the coordinates are latitude and longitude south of a specified latitude. The remaining northern portion of the globe is covered with an orthogonal coordinate system that involves two poles that can be located on land. Models that have employed this grid include the Parallel Ocean Program, the Modular Ocean Model, and the Hybrid Coordinate Ocean Model. Figure 6.1 shows an example of such a grid.

Figure 6.1. Tripolar grid. To the south of a specified
latitude, the coordinate system uses latitude and longitude.
To the north of that latitude, each line of constant longitude
connects smoothly to a longitude line on the other side of
the pair of coordinate poles. (Figure provided by Philip
Jones, Los Alamos National Laboratory.)

A horizontal coordinate system developed by Rancic, Purser and Mesinger
(1996) employs a conformally expanded spherical cube. With this formulation, a cube is inscribed within a sphere, and a rectangular mesh on the
surface of the cube is then projected conformally onto the sphere. The
resulting mesh on the sphere is orthogonal, except at the singular points
corresponding to the corners of the cube. Adcroft, Campin, Hill and Marshall (2004) employ finite volume methods for the equations for mass and
momentum on this grid, and they demonstrate that the singularities cause
no difficulty in their setting. They also employ a re-scaling of the coordinates on the sphere so that the resulting grid is more uniform than the
grid of Rancic et al. (1996). A non-orthogonal expanded spherical cube was
used by Rossmanith (2006) to develop a finite volume method, based on the
framework of LeVeque (2002), for hyperbolic systems on a sphere.

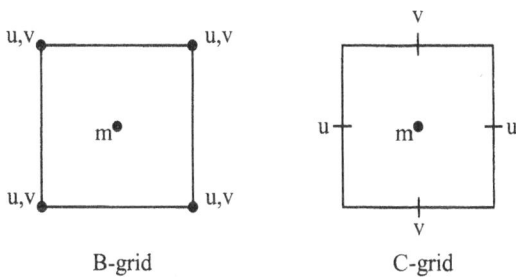

Figure 6.2. Arrangements of dependent variables on the
B-grid and C-grid. The symbol m refers to mass variables
such as density, layer thickness, and pressure; and u and v
are the x- and y-components of velocity, respectively.

Relative to a given grid, there are several options for specifying the points
where the various dependent variables are defined. For a rectangular grid,
Winninghoff (1968) and Arakawa and Lamb (1977) analysed five such arrangements, which they labelled A–E. These grids are also illustrated and
discussed in other references, such as Dukowicz (1995) and Haidvogel and
Beckmann (1999). With the A-grid, all dependent variables are defined at
the centres of grid cells. In the case of the B-grid, mass variables are defined
at the centres of grid cells, and the components of velocity are defined at the
corners. With the C-grid, the normal components of velocity are defined at
the centres of the edges of the mass cells. That is, if u and v denote the x-
and y-components of velocity, respectively, then the values of u are defined
at the centres of edges corresponding to minimal and maximal x, and the
values of v are defined at the centres of the edges corresponding to minimal
and maximal y: see Figure 6.2.

Among the Arakawa grids, the B-grid and the C-grid are the most commonly used in ocean models. For example, the B-grid is used throughout
the Bryan–Cox class of level models. Models that use the C-grid include the
Miami Isopycnic Coordinate Ocean Model, the Hybrid Coordinate Ocean
Model, and the Hallberg Isopycnal Model. A variation on the C-grid used
by the MIT model (Adcroft, Hill and Marshall 1999) is described below.

One approach to comparing grid arrangements is to compare the accuracy
of linear wave propagation as represented on those grids. For the case
of gravity waves modelled by the linearized shallow water equations for
a single-layer fluid, analyses of dispersion relations of numerical methods
indicate that the C-grid is more accurate than the B-grid for relatively well-
resolved waves, whereas the B-grid is more accurate in the case of relatively
coarser resolution (Arakawa and Lamb 1977, Dukowicz 1995). Here, the
resolution is defined in terms of the size of the grid spacing relative to the
Rossby radius c/f, where c is the speed of gravity waves and f is the Coriolis

parameter, and it is assumed that the spatial discretizations are based on second-order centred finite differences and averages that are natural for the grid in question. Comparisons based on gravity waves thus suggest that the C-grid is preferable for higher-resolution models and the B-grid is preferable for coarser-resolution runs.

However, in a similar analysis of Rossby waves, Dukowicz (1995) concluded that for Rossby waves the B-grid is more accurate at higher resolution. This is the opposite case to gravity waves. For coarser resolution the B-grid is generally more accurate for Rossby waves, but this can depend on the location in wavenumber space. Rossby waves play a fundamental role in the development of large-scale circulation systems, whereas gravity waves are primarily involved in the details of adjustments between states of geostrophic balance (Gill 1982).

One practical problem with the B-grid is the possibility of a chessboard pattern in the mass fields (see, e.g., Killworth et al. (1991).) On the B-grid, a given velocity point has four neighbouring mass cells, and a natural approximation to $\partial M/\partial x$ (equivalently, $\partial p/\partial x$) can be obtained by a difference in x of an average with respect to y. However, if the mass field displays a $+1/-1$ chessboard pattern then the computed pressure gradient is zero, and, more generally, a $+1/-1$ pattern can be added to an arbitrary mass field without affecting the computed pressure gradient. Grid noise of this nature can be introduced into a solution by grid-scale forcing, such as the forcing associated with variable bottom topography or variable coastlines. Spatial smoothing can be used to suppress this B-grid computational mode.

With the C-grid a given velocity point has only two neighbouring mass cells, and the above problem with the pressure gradient does not occur. However, there is an analogous problem with the Coriolis terms. The velocity components u and v are defined at different points, so some spatial averaging is required to implement the terms $-fv$ and fu in the u- and v-equations, respectively. Chessboard patterns in the velocity fields do not affect the computed values of those terms, and these patterns can persist once they are stimulated by grid-scale forcing.

Adcroft et al. (1999) state that the C-grid noise primarily occurs in low-resolution configurations, where again the resolution is defined in terms of the size of the grid spacing relative to the Rossby radius. For low-resolution simulations they obtain better results by implementing the Coriolis terms with a hybrid of the C-grid and the D-grid. With the D-grid, the *tangential* components of velocity are defined at the centres of the edges of mass cells; that is, the values of u and v are defined at the points where v and u, respectively, would be defined on the C-grid. In the method of Adcroft et al. (1999), the u-equation on the C-grid uses the value of $-fv$ from the D-grid. Equations are also solved in the D-grid, and for the u-equation on the D-grid, the value of $-fv$ is taken from the C-grid. The introduction of additional equations and unknowns leads to the existence of nonphysical computational modes, which in this case take the form of spatially independent inertial oscillations. These modes are damped by using a weighted time average when implementing the Coriolis terms.

An alternate approach is given by Nechaev and Yaremchuk (2004). At a given u-point on the C-grid, consider implementing the Coriolis term $-fv$ by using the value of v at *one* of the neighbouring v-points; similarly, at that v-point, implement the Coriolis term fu by referring to the u-point just mentioned. This scheme, by itself, would give an uncentred implementation of the Coriolis terms. Instead, Nechaev and Yaremchuk use all of the four possible pairings of u-points and v-points and average the results. This procedure gives a centred approximation that avoids the C-grid noise described above. In their analysis, these authors use the leapfrog time discretization, except that the Coriolis terms are represented with a weighted time average involving three consecutive time levels.

6.2. *Alternative spatial discretizations: spherical geodesic grids, spectral elements, and unstructured grids*

One alternative to a quadrilateral grid is a spherical geodesic grid. The construction of such a grid is described and illustrated, for example, by Lipscomb and Ringler (2005). First, inscribe within the sphere a regular icosahedron, which has 20 faces that are equilateral triangles. Subdivide each triangle into four triangles by connecting the midpoints of the edges, and then project each of the resulting vertices onto the sphere. Continue this process, yielding finer and finer triangulations of the sphere at each step. For any such triangulation, regard each vertex as the centre of a grid cell; such a cell is defined by including all points on the sphere that are closer to that vertex than to any other vertex. The resulting cells are all hexagons, except that the cells corresponding to the 12 vertices of the original icosahedron are pentagons. The resulting mesh of hexagons and pentagons is nearly uniform, and it avoids the problems with coordinate poles described in the preceding subsection.

Ringler and Randall (2002) developed numerical methods for solving the shallow water equations on such a geodesic grid. In their formulation, scalar quantities such as mass variables, divergence of velocity, and the vertical component of the curl of velocity are defined at the centres of grid cells. Velocity vectors are defined at the vertices of the grid cells, and the cell averages of the divergence and curl are obtained by computing line integrals around the boundaries of the grid cells. The pressure gradient at a given vertex of a grid cell is obtained by linearly interpolating the pressure at the centres of the three neighbouring cells and then computing the gradient of the linear interpolant.

Another type of spatial discretization is used in the Spectral Element Ocean Model (Haidvogel and Beckmann 1999, Iskandarani, Haidvogel and Levin 2003). This model employs a finite element discretization using high-degree polynomials on hexahedral elements, which are cubes with curved surfaces. Numerical convergence is obtained by refining the grid and/or increasing the order of the approximating polynomials. The number of elements in the vertical direction is independent of the horizontal position, and the three-dimensional grid amounts to a stack of two-dimensional unstructured grids. The elements are constructed so as to conform to bottom topography and coastlines; given the structured nature of the grid in the vertical direction, the discrete algorithm is essentially a terrain-following σ-coordinate model.

The consideration of unstructured grids is a relatively recent development in ocean modelling. The review of Pain et al. (2005) describes some recent work involving three-dimensional unstructured grids, with an emphasis on finite element methods. Issues discussed include adapting the grid resolution to evolving flow conditions, parametrizing subgrid-scale processes in the presence of variable resolution, maintaining hydrostatic and geostrophic balance, advection schemes, iterative methods, and parallelization. This paper is the lead article in a special issue of *Ocean Modelling* devoted entirely to unstructured grids.

6.3. Advection

Among the processes represented in a numerical model of ocean circulation is the advection of quantities such as density, layer thickness, and tracers. The purpose of the present subsection is to describe some aspects of advection that are of particular interest in ocean modelling, namely, positive definiteness and compatibility, and to describe briefly some of the advection schemes that are used in this field. The literature on advection schemes is vast, and there is no point in trying to summarize the field here; a comprehensive development of the subject is given, for example, in the recent text by LeVeque (2002).

Section 7 of the present paper includes derivations of partial differential equations that describe the conservation of mass of the fluid and the conservation of tracers. In those derivations it is assumed that the horizontal coordinates are arbitrary curvilinear coordinates on a spheroid and that the generalized vertical coordinate s can satisfy $\dot{s} \neq 0$, where $\dot{s} = Ds/Dt$, so that s can change with time following fluid parcels. For the sake of simplicity in the present discussion, assume that $\dot{s} = 0$ and that the diffusion of tracers is zero. Also assume that the horizontal coordinates are Cartesian coordinates on a tangent plane, and consider the portion of the fluid bounded below and above by two coordinate surfaces $s = s_0$ and $s = s_1$, respectively. Under these assumptions, equation (7.38) for the conservation of mass in such a coordinate layer is

$$\frac{\partial}{\partial t}(\Delta p) + \frac{\partial}{\partial x}(u\Delta p) + \frac{\partial}{\partial y}(v\Delta p) = \frac{\partial}{\partial t}(\Delta p) + \nabla \cdot (\mathbf{u}\Delta p) = 0, \qquad (6.1)$$

and equation (7.107) for the conservation of a tracer in that layer is

$$\frac{\partial}{\partial t}(q\Delta p) + \nabla \cdot (\mathbf{u}(q\Delta p)) = 0. \qquad (6.2)$$

Here, $\mathbf{u}(x, y, t) = (u(x, y, t), v(x, y, t))$ is the mass-weighted vertical average of the horizontal velocity in the layer; $q(x, y, t)$ is the mass-weighted vertical average of the tracer concentration, which is the quantity of tracer per unit mass of fluid; and $\Delta p(x, y, t)$ is the vertical pressure difference across the layer. In Section 7, overbars are used to indicate mass-weighted vertical averages, but these are deleted in the present notation.

Because of the hydrostatic assumption, Δp is the weight per unit horizontal area in the layer, and it is thus g times the mass per unit horizontal area when viewed from above. The quantity $q\Delta p$ is then equal to g times the quantity of tracer per unit horizontal area. In the case where s is the geometrical height z, the hydrostatic condition $\partial p/\partial z = -\rho g$ (see (7.8)) implies that Δp is equal to g times the vertical integral of the density ρ in the coordinate layer. In the case of an isopycnic model, the quantity Δp can be regarded as a measure of layer thickness.

The tracers used in a model of ocean circulation typically include potential temperature and salinity, and depending on the usage of the model they may also include the concentrations of various chemical and/or biological species. In a model of the dynamics of sea ice, Lipscomb and Hunke (2004) transport 46 different scalar fields, including area fractions and thermodynamic variables associated with each of five different ice thickness categories. They also anticipate that the number of transported fields will increase as ice models become more realistic.

Numerical advection schemes can be evaluated according to standard requirements such as stability, accuracy, efficiency, and avoiding spurious oscillations. However, some additional requirements are of special interest in the context of ocean modelling.

In the case of an isopycnic model, the layer thickness Δp can tend to zero, as indicated by Figure 2.1. When equation (6.1) is solved numerically, it is then necessary to maintain nonnegative values of Δp in the computed solution; that is, solutions that are initially nonnegative must remain nonnegative in the absence of forcing. A numerical method with this property can be labelled positive definite.

An example of a positive definite method is MPDATA, the Multidimensional Positive Definite Advection Transport Algorithm (Smolarkiewicz and

Margolin 1998). At each time step, this method employs multiple iterations. The first iteration uses the standard upwind (donor-cell) method, for which the mass flux at a cell edge is equal to the normal velocity at that edge times the density (such as Δp) in the upwind direction, *i.e.*, in the cell that is being drained. The upwind method is positive definite, assuming that a Courant–Friedrichs–Lewy condition is satisfied. Subsequent iterations are used to cancel the leading term in the truncation error, and these iterations are written in the form of upwind steps with non-physical pseudovelocities. The pseudovelocities are bounded by the physical velocities, so these subsequent iterations are also positive definite. The upwind method, by itself, is highly diffusive, and the subsequent iterations can be regarded as antidiffusive steps that cancel some of the numerical diffusion produced by the initial upwind step.

Another desirable property of an advection scheme is the synchronous transport of fluid and tracers. At a given time step, the mass equation (6.1) is used to update Δp, and the tracer equation (6.2) is used to update $q\Delta p$. The tracer concentration q is then obtained as a ratio of the computed $q\Delta p$ and the computed Δp. However, this ratio can display nonphysical oscillations and extrema, even if the fields $q\Delta p$ and Δp are individually well-behaved. This is especially a problem in cases where Δp tends to zero.

This problem can be characterized as follows. The mass equation (6.1) and the tracer equation (6.2) together imply

$$\frac{Dq}{Dt} = \frac{\partial q}{\partial t} + \mathbf{u} \cdot \nabla q = 0. \tag{6.3}$$

That is, the material derivative of q is zero, so q is constant along particle paths. The value of $q(x, y, t + \Delta t)$ is therefore bounded by values of q in a neighbourhood of (x, y) at time t. It is desirable that a numerical method produce solutions having an analogous property, namely, that the computed value of q at position (x_i, y_j) at time t_{n+1} be bounded by the values of q at time t_n at position (x_i, y_j) and the immediately adjacent grid points. Schär and Smolarkiewicz (1996) use the term 'compatible' to refer to a numerical method with this property, and they demonstrate that compability can be obtained through a suitable limiting of antidiffusive correction fluxes.

An alternate approach to compatibility, and to advection in general, is given by the method of incremental remapping developed by Dukowicz and Baumgardner (2000). This method was subsequently extended to the modelling of sea ice transport by Lipscomb and Hunke (2004). Lipscomb and Ringler (2005) developed a version of incremental remapping for use with geodesic grids, in contrast to the rectangular grids discussed by Dukowicz and Baumgardner.

With the method of incremental remapping, each fixed grid cell is regarded as an 'arrival cell' for the location of a portion of mass at time t_{n+1}.

The values of the velocity at the corners of that cell are used to trace the cell backward in time to yield an approximate 'departure cell' for the location of that mass at time t_n. The solution at time t_n is regarded as linear in each cell. In general, a departure cell contains portions of several of the fixed cells, and the contributions from these various portions are summed to yield the total mass in the departure cell at time t_n. This mass is then the mass in the arrival cell at time t_{n+1}. The cell masses at time t_{n+1} yield cell averages that are used to construct linear approximations in preparation for the next step.

This process resembles the 'reconstruct-evolve-average' (REA) approach to deriving advection schemes, in which piecewise fields are reconstructed from cell averages, propagated forward in time, and then averaged to yield new cell averages (LeVeque 2002). One difference between incremental remapping and the REA approach is that the former traces cells backward in time, not forward.

In the linear approximations used by Dukowicz and Baumgardner (2000), the slopes are determined by centred differences of the values of the solution in the adjacent cells. However, when necessary, these slopes are limited to ensure that the method is monotone, i.e., that the solution at all spatial positions at a fixed time is a monotone increasing function of the solution at all positions at the preceding time level. This condition implies that the method is positive definite. The method of incremental remapping is second-order accurate in space, except at locations where the limiting is applied.

If this method is applied to a pair of equations of the form (6.1) and (6.2), then the computed tracer concentration q in a given cell at time t_{n+1} is a weighted average of the values of q at time t_n in that cell and in neighbouring cells, and compatibility is therefore ensured. In addition, the method is also conservative; by construction, the total mass at time t_{n+1} is the sum of the masses in the departure cells at time t_n, and this is the total mass at time t_n.

With the method of incremental remapping, substantial computational cost is incurred when constructing the departure regions. However, if multiple quantities are transported with the same velocity field, then the geometrical constructions need to be performed only once per time step, as the same results are used for all of the transported quantities. This method is therefore well-suited for models that transport large numbers of tracers. In contrast, the pseudovelocities used in MPDATA are different for each transported quantity, so the marginal cost of introducing an additional tracer is greater with MPDATA than in the case of incremental remapping (Lipscomb and Hunke 2004).

The mass and tracer equations (6.1) and (6.2) are equivalent to stating that the masses of the fluid and tracers are conserved in Lagrangian volumes, i.e., in volumes that move with the fluid; this principle is used in

Section 7 to derive equations (7.38) and (7.107), of which (6.1) and (6.2) are special cases. The method of incremental remapping is a discrete, finite-volume, representation of this Lagrangian principle. This method is related to the class of semi-Lagrangian methods, for which grid *points* are traced backward from time t_{n+1} to departure points at time t_n. Values of the solution at departure points are obtained via interpolation from grid point values, and time differences represent approximations to material derivatives along particle paths. These methods are reviewed, for example, in the text by Durran (1999). In general, semi-Lagrangian methods do not conserve mass exactly. However, a conservative method by Lin and Rood (1996) has a semi-Lagrangian flavour, and it involves directional splitting and a composition of one-dimensional advection schemes that are essentially one-dimensional versions of the method of incremental remapping.

The preceding discussion of advection is not all-inclusive. For example, the review by Griffies *et al.* (2000*a*) includes a list of some advection schemes that were used in operational ocean models as of the date of that review. These include flux-corrected transport (Zalesak 1979), three different variations on the Quick scheme of Leonard (1979), an advection scheme of Easter (1993), various centred schemes, and MPDATA.

In addition, Griffies *et al.* (2005) describe a third-order upwind biased method, with flux limiting, that has been implemented in the Modular Ocean Model and in the MIT general circulation model. Iskandarani, Levin, Choi and Haidvogel (2005) compare four advection schemes for high-order finite element and finite volume methods, including the discontinuous Galerkin method and a spectral finite volume method with flux limiting. Hecht (2006) reviews several advection methods for ocean modelling, especially in the context of forward-in-time discretizations which involve only two time levels.

6.4. *Solution of the momentum equations*

We next give an overview of some different approaches that are taken to solving the momentum equations. In this subsection the horizontal coordinates are assumed to be Cartesian coordinates on a tangent plane, for the sake of notational simplicity.

Section 7 includes a derivation of partial differential equations that describe the conservation of momentum for a fluid that is in motion relative to a rotating spheroid. In the case of Cartesian coordinates, the momentum equations (7.92) can be written in the form

$$
\begin{aligned}
&\frac{\partial}{\partial t}(up_s) + \frac{\partial}{\partial x}\big(u(up_s)\big) + \frac{\partial}{\partial y}\big(v(up_s)\big) + \frac{\partial}{\partial s}\big(\dot{s}(up_s)\big) - fvp_s = F_u, \\
&\frac{\partial}{\partial t}(vp_s) + \frac{\partial}{\partial x}\big(u(vp_s)\big) + \frac{\partial}{\partial y}\big(v(vp_s)\big) + \frac{\partial}{\partial s}\big(\dot{s}(vp_s)\big) + fup_s = F_v.
\end{aligned}
\tag{6.4}
$$

Here, s is a generalized vertical coordinate, f is the Coriolis parameter, and F_u and F_v represent forcing due to pressure and viscosity.

A vertically discrete model can be obtained by partitioning the fluid domain into layers, each of which is bounded above and below by surfaces of constant s. The conservation of momentum in a coordinate layer bounded by the surfaces $s = s_0$ and $s = s_1$ is described by equations (7.101) and (7.104). In the case of Cartesian coordinates, the u-equation (7.101) can be written in the form

$$
\frac{\partial}{\partial t}(\bar{u}\Delta p) + \frac{\partial}{\partial x}\left(\bar{u}(\bar{u}\Delta p)\right) + \frac{\partial}{\partial y}\left(\bar{v}(\bar{u}\Delta p)\right)
$$
$$
+ \left[(\dot{s}p_s u)_{s=s_0} - (\dot{s}p_s u)_{s=s_1}\right] - f\bar{v}\Delta p = \bar{F}_u,
$$
(6.5)

and the v-equation (7.104) can be expressed similarly. The horizontal velocity components $\bar{u}(x, y, t)$ and $\bar{v}(x, y, t)$ are mass-weighted vertical averages within the layer. The quantities $\bar{u}\Delta p$ and $\bar{v}\Delta p$ are the components of momentum per unit horizontal area, times g, and so they will be regarded as components of momentum density. The second and third terms in (6.5) represent the lateral advection of momentum, and the terms in square brackets represent the transport of momentum between layers due to the movement of coordinate surfaces relative to the fluid. As noted in Section 7.8, the quantity $-\dot{s}p_s$ is the rate of flow of mass per unit horizontal area (times g) across a coordinate surface due to material changes in s. In the case of a hydrostatic z-coordinate model, $-\dot{s}p_s = -\dot{z}p_z = \rho g w$, where w is the vertical component of velocity. In the case of a hybrid coordinate model, the rate of mass flow depends on the actions of the grid generator that establishes and moves the coordinate surfaces (Bleck 2002). With an isopycnic model, the flow across coordinate surfaces is due to the diapycnal diffusion of heat and salt, which causes the surfaces to move relative to the fluid; the modelling of this transport is discussed, for example, by McDougall and Dewar (1998), Hallberg (2000), and de Szoeke and Springer (2003).

One approach to solving the momentum equations is to use an advection scheme to implement the advective terms in the layer u-equation (6.5) and in the analogous v-equation. The Coriolis terms and the effects of pressure and viscosity can be regarded as forcing terms that are implemented with standard differencing and/or averaging. This approach was used by Smolarkiewicz and Margolin (1998) in the context of the shallow water equations for a single layer. In their formulation, the forcing is implemented with a Strang splitting; starting with the computed solution at time t_n, add half the forcing at time t_n (times Δt), apply an advection scheme to the result, and then add half the forcing at time t_{n+1} (times Δt).

A different strategy for the momentum equations is used in the Bryan–Cox class of z-coordinate models. The initial model in this class is described by Bryan (1969). In that reference it is assumed that the Boussinesq approximation applies, *i.e.*, that the density is constant in the momentum equations except in the buoyancy term in the vertical component. In that case $p_s = p_z = -\rho g \approx -\rho_0 g$, and the u-equation in (6.4) has the form

$$\frac{\partial u}{\partial t} + \frac{\partial}{\partial x}(uu) + \frac{\partial}{\partial y}(vu) + \frac{\partial}{\partial z}(wu) - fv = -F_u/(\rho_0 g). \qquad (6.6)$$

In this equation the nonlinear terms on the left side represent an advection of the quantity u with advective velocity (u, v, w). To discretize these terms, consider grid cells centred at velocity points on the B-grid, and use spatial averages to obtain the necessary fluxes along the faces of those cells. The choice of discretization is guided by a desire to maintain energetic consistency in the model, namely, that the discrete nonlinear terms have no effect on the total kinetic energy in the solution and that the exchanges between kinetic and potential energy are represented correctly. These choices constrain the fluxes at cell faces to be calculated in terms of centred second-order averages.

In their description of the current version of the Modular Ocean Model, a direct descendant of the original Bryan–Cox model, Griffies *et al.* (2004) cite additional aspects of energetic consistency. One example is that the work done by the fluid against the pressure gradient can be converted into compressibility effects and/or work against gravity. The Modular Ocean Model no longer uses the Boussinesq approximation; one reason is that this approximation causes a model to preserve volume, which precludes the accurate modelling of sea-level rise due to thermal expansion. In order to employ the approach outlined above, Griffies *et al.* (2004) use a substitution $\rho_0 \tilde{\mathbf{u}} = \rho \mathbf{u}$ to convert the momentum equation into a form similar to (6.6). They give a detailed discussion of the averaging that they use to compute the necessary fluxes.

A third approach to solving the momentum equations is to convert the dependent variables to velocity and write the horizontal transport terms in terms of kinetic energy and vorticity. In the present case of Cartesian coordinates, the mass equation (7.32) is

$$\frac{\partial}{\partial t}(p_s) + \frac{\partial}{\partial x}(up_s) + \frac{\partial}{\partial y}(vp_s) + \frac{\partial}{\partial s}(\dot{s}p_s) = 0;$$

when this equation is combined with the u-momentum equation in (6.4), the result is

$$\frac{\partial u}{\partial t} + u\frac{\partial u}{\partial x} + v\frac{\partial u}{\partial y} + \dot{s}\frac{\partial u}{\partial s} - fv = F_u/p_s,$$

or

$$\frac{\partial u}{\partial t} + \frac{\partial}{\partial x}\left[\frac{1}{2}(u^2 + v^2)\right] - (\zeta + f)v + \dot{s}p_s\frac{\partial u}{\partial p} = F_u/p_s. \qquad (6.7)$$

The last equation is obtained by writing $uu_x + vu_y$ in terms of the relative vorticity $\zeta = v_x - u_y$ and the kinetic energy per unit mass, $\frac{1}{2}(u^2 + v^2)$. The same procedure was used in Section 4.4 in preparation for proving the conservation of potential vorticity for the shallow water equations. As noted in that section, $\zeta + f$ is the local vertical component of the absolute (relative plus planetary) vorticity.

The kinetic energy and vorticity terms can be approximated with centred finite differences; if the C-grid is used, some spatial averaging is also required. In the case of a vertically discrete layered model, the vorticity term can be written as

$$(\zeta + f)v = \left(\frac{\zeta + f}{\Delta p}\right)(v\Delta p). \tag{6.8}$$

The first factor on the right side of (6.8) is proportional to the potential vorticity in a coordinate layer, and the second factor is proportional to mass flux. Sadourny (1975) describes two methods for implementing (6.8) on a C-grid, each of which uses various kinds of spatial averages to compute the potential vorticity and mass flux. With one such discretization, the total energy is conserved when applied to the shallow water equations with exact integration in time. The other discretization yields the conservation of potential enstrophy in the same context; in the case of the shallow water equations, the potential enstrophy is the spatial integral of $\frac{1}{2}(\zeta + f)^2/h$. The conservation of this quantity is of interest because it inhibits the accumulation of energy at grid scales and thus promotes numerical stability. The potential-enstrophy-conserving method is used, for example, in the Miami Isopycnic Coordinate Ocean Model (Bleck 2002), whereas the energy-conserving method is used in the Hallberg Isopycnal Model (Hallberg 1997). Because of the particular nature of the spatial averaging used in the potential-enstrophy-conserving method, special procedures must be used in situations where Δp tends to zero. However, no special procedures are needed for the energy-conserving method.

Adcroft et al. (2004) use the formulation (6.7) when solving the momentum equations on the conformally expanded spherical cube (see Section 6.1). In their setting the vorticity is computed by a finite volume approach; a vorticity point is regarded as the centre of a grid cell, and the circulation around the boundary of that cell, divided by area, gives the vorticity. This formulation avoids problems associated with the singular points corresponding to the corners of the cube.

7. Governing equations in general coordinates

We now address the derivation of the governing equations that are to be solved numerically. In particular, the goal of the present section is to derive partial differential equations that describe the conservation of mass, momentum, and tracers for a hydrostatic and stratified fluid in a rotating reference frame. In this derivation the vertical coordinate is a generalized coordinate that includes all of the cases described in Section 3, and the horizontal coordinates are arbitrary orthogonal coordinates on a spheroid. Some aspects of the following discussion are also contained in various references, such as Bleck (2002), Gill (1982), Griffies (2004), Holton (1992), Miller (2006), and Pedlosky (1987).

7.1. Definition of level surfaces

The Earth's surface is not exactly a sphere, but instead can be described more accurately as an oblate spheroid (Gill 1982). In particular, the surface of constant gravitational potential at sea level is approximated by a spheroid with a polar radius approximately equal to 6357 km and an equatorial radius of approximately 6378 km. The maximum deviation from the spheroid of best fit is on the order of 100 metres.

In the following, we consider fluid motion relative to a spheroid that rotates with uniform angular velocity Ω. Assume that the axis of rotation is stationary in a rectangular coordinate system that serves as an inertial reference frame. As a first approximation, one can define directions tangent to the spheroid as horizontal, and directions normal to the spheroid as vertical. However, for later use it will be useful to incorporate the effects of centripetal acceleration into such a definition.

Suppose that a layer of fluid is stationary relative to the rotating spheroid, and assume that the only forces acting on the fluid are due to gravity and pressure. For a given fluid parcel and a given time, let $\mathbf{a}_p$ denote the net acceleration (force per unit mass) due to pressure forcing, and denote the acceleration due to gravity by $\mathbf{a}_g = -\nabla\phi$. Here, ϕ is the gravitational potential, ∇ is the three-dimensional spatial gradient, and all quantities are viewed in the inertial frame. Since the parcel rotates with uniform angular velocity Ω, the resultant force per unit mass must be the centripetal acceleration that is required to maintain the rotation. The centripetal acceleration is directed toward the axis of rotation and has magnitude $\Omega^2 r_\perp$, where $r_\perp$ denotes the distance from the parcel to the axis. This acceleration can also be written in terms of a potential function as $-\nabla\phi_c$, where $\phi_c = \frac{1}{2}\Omega^2 r_\perp^2$ (Pedlosky 1987). It then follows that the centripetal acceleration is $-\nabla\phi_c = \mathbf{a}_p + \mathbf{a}_g = \mathbf{a}_p - \nabla\phi$: see Figure 7.1.

With the Earth's rotation rate of approximately $2\pi + 2\pi/360.24$ radians in 24 hours (Griffies 2004), the centripetal acceleration at radius 6300 km has a magnitude of approximately 0.03 m/sec^2. In contrast, the acceleration due to gravity is approximately 9.8 m/sec^2. It is thus a slight imbalance between gravitational and pressure forces that provides the necessary centripetal force.

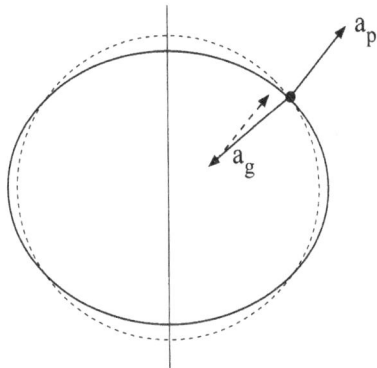

Figure 7.1. Combined effect of gravitational and pressure
accelerations. The vertical line segment is the axis of
rotation. The dashed ellipse is a cross-section of a spheroid
of constant gravitational potential ϕ; the gravitational
acceleration $\mathbf{a}_g = -\nabla\phi$ is orthogonal to this spheroid. The
solid ellipse is a cross-section of the spheroid of constant
gravitational–centripetal potential $\Phi = \phi - \phi_c$ that
intersects the first spheroid at the indicated point. The
pressure acceleration $\mathbf{a}_p$ for a relatively stationary fluid is
orthogonal to the surface of constant Φ. The dashed arrow
is a translation of $\mathbf{a}_p$, and the sum $\mathbf{a}_g + \mathbf{a}_p = -\nabla\phi + \mathbf{a}_p$ is
the centripetal acceleration that is required to maintain the
rotation. Locally horizontal and vertical directions are
defined relative to surfaces of constant Φ.

The effects of the gravitational and centripetal potentials can be combined
into a potential function $\Phi = \phi - \phi_c = \phi - \frac{1}{2}\Omega^2 r_\perp^2$. Surfaces of constant Φ
are also spheroids, and they bulge out slightly near the equator, compared
to surfaces of constant ϕ. For a layer of fluid that is stationary relative
to the rotating spheroid, $\mathbf{a}_p - \nabla\Phi = 0$, and $\mathbf{a}_p$ is then orthogonal to the
surfaces of constant Φ.

In the following discussions, directions that are tangent to surfaces of
constant Φ will be regarded as 'level', or locally 'horizontal', and directions
normal to such surfaces will be regarded as locally 'vertical'.

In the case of a fluid that is relatively stationary, the pressure acceleration
$\mathbf{a}_p$ points vertically, and it exactly cancels the effect of the gravitational–
centripetal potential Φ. More generally, for a moving fluid we will impose
the 'hydrostatic assumption' that the vertical component of the pressure
acceleration equals $\nabla\Phi$. The justification for this assumption follows from
the scaling of large-scale flows that is discussed in Section 2.1.

Assume that $\mathbf{a}_p$ is proportional to the pressure gradient ∇p. For a relatively stationary fluid, $\mathbf{a}_p$ is parallel to $\nabla\Phi$, and the surfaces of constant

pressure coincide with the surfaces of constant Φ. As noted above, the latter surfaces bulge out slightly near the equator, relative to the rotating spheroid. This bulging of the pressure surfaces provides the centripetal acceleration that enables the fluid to rotate with the spheroid without assistance from any forcing other than gravity and hydrostatic pressure.

The magnitude of the gravitational acceleration is nearly constant over the entire fluid domain. At latitude $45°$, this magnitude is approximately 9.806 m/sec^2, whereas the area average over the entire Earth is approximately 9.7976 m/sec^2 (Griffies 2004). In the following, it will be assumed that $|\nabla\Phi| = g$, where g is a constant.

7.2. Horizontal and vertical coordinates

Let Σ denote a surface of constant gravitational–centripetal potential Φ, such as at mean sea level. As before, assume that Σ rotates with uniform angular velocity Ω about an axis that is fixed in an inertial reference frame. Parametrize all or part of Σ with coordinates $\mathbf{x} = (x_1, x_2)$, and let z denote a distance perpendicular to Σ, with z increasing outward and $z = 0$ corresponding to points on Σ. The coordinate z has units of length, but x_1 and x_2 do not necessarily have those units.

The coordinates $(\mathbf{x}, z) = (x_1, x_2, z)$ define the position of a point relative to the rotating surface Σ. For any such point and any time t, let $\mathbf{r}(\mathbf{x}, z, t)$ denote the position of that point relative to the inertial reference frame at time t. That is, $\mathbf{r}(\mathbf{x}, z, t)$ is a vector of three numbers, with units of length, that gives the position of the point relative to the rectangular coordinate system in which the axis of rotation is stationary. The effects of rotation can be expressed as

$$\mathbf{r}(\mathbf{x}, z, t) = Q(\Omega t)\, \mathbf{r}(\mathbf{x}, z, 0), \tag{7.1}$$

where $Q(\Omega t)$ is an orthogonal matrix that represents a rotation through angle Ωt about the given axis. In (7.1), $\mathbf{r}(\mathbf{x}, z, t)$ and $\mathbf{r}(\mathbf{x}, z, 0)$ are regarded as column vectors.

Define the metric coefficients m_1 and m_2 by

$$m_i(\mathbf{x}, z) = \left| \frac{\partial \mathbf{r}}{\partial x_i}(\mathbf{x}, z, t) \right|, \qquad i = 1, 2. \tag{7.2}$$

Here, the Euclidean vector norm is used. A comparison with (7.1) shows that the right side of (7.2) is independent of t, so the notation on the left side indicates no dependence on t. Equation (7.2) implies that an increment Δx_i in the parameter x_i corresponds to a horizontal spatial displacement approximately equal to $m_i \Delta x_i$. In the case of an ocean model the metric coefficient m_i is nearly independent of z, as the ocean is a few kilometres deep, whereas the Earth has a radius of over 6300 km.

Next define the unit vectors $\mathbf{i}$ and $\mathbf{j}$ by

$$
\mathbf{i}(\mathbf{x}, t) = \frac{1}{m_1(\mathbf{x}, 0)} \frac{\partial \mathbf{r}}{\partial x_1}(\mathbf{x}, 0, t),
$$
$$
\mathbf{j}(\mathbf{x}, t) = \frac{1}{m_2(\mathbf{x}, 0)} \frac{\partial \mathbf{r}}{\partial x_2}(\mathbf{x}, 0, t).
$$
(7.3)

The partial derivatives in (7.3) are taken for fixed $z = 0$, so the vectors $\mathbf{i}$ and $\mathbf{j}$ are tangent to the surface Σ and thus locally horizontal. These vectors can be regarded as unit vectors pointing in the x_1- and x_2-directions, respectively, that are attached to Σ and rotate with Σ. Throughout the following discussion, the horizontal coordinate system is assumed to be orthogonal, in the sense that $\mathbf{i}$ and $\mathbf{j}$ are everywhere orthogonal. The unit vector in the upward vertical direction is

$$
\mathbf{k}(\mathbf{x}, t) = \frac{\partial \mathbf{r}}{\partial z}(\mathbf{x}, z, t).
$$
(7.4)

Since the distance z is taken orthogonal to the surface Σ, the vectors $\mathbf{i}$, $\mathbf{j}$, and $\mathbf{k}$ are mutually orthogonal. In Sections 7.12 and 7.15 it is assumed that Σ is parametrized so that the coordinate system is right-handed, in the sense that $\mathbf{i} \times \mathbf{j} = \mathbf{k}$, $\mathbf{k} \times \mathbf{j} = -\mathbf{i}$, and $\mathbf{i} \times \mathbf{k} = -\mathbf{j}$; however, this assumption is not used elsewhere.

In later discussions we use the approximations

$$
\frac{\partial \mathbf{r}}{\partial x_1}(\mathbf{x}, z, t) = m_1(\mathbf{x}, z) \, \mathbf{i}(\mathbf{x}, t),
$$
$$
\frac{\partial \mathbf{r}}{\partial x_2}(\mathbf{x}, z, t) = m_2(\mathbf{x}, z) \, \mathbf{j}(\mathbf{x}, t),
$$
(7.5)

which are exact when $z = 0$. In effect, these approximations assume that the tangent vectors to surfaces of constant z are parallel.

For an example of a coordinate system, assume that Σ is a sphere of radius a centred at the origin in the inertial frame. Also assume that the axis of rotation aligns with the third coordinate axis in that reference frame and that the rotation is anticlockwise when the sphere is viewed from the positive portion of that axis. Use spherical coordinates, with longitude $x_1 = \lambda$ and latitude $x_2 = \theta$, and assume that at time $t = 0$ the points with $\lambda = 0$ align with the first coordinate in the inertial frame. Then

$$
\mathbf{r}(\mathbf{x}, z, t) = \mathbf{r}(\lambda, \theta, z, t) = \left(r \cos\theta \cos(\lambda + \Omega t), \; r \cos\theta \sin(\lambda + \Omega t), \; r \sin\theta \right),
$$
(7.6)

with $r = a + z \approx a$. In this case the unit vectors $\mathbf{i}$ and $\mathbf{j}$ point eastward and northward, respectively; the metric coefficients are

$$m_\lambda = \left| \frac{\partial \mathbf{r}}{\partial \lambda} \right| = (a + z) \cos \theta \approx a \cos \theta,$$

$$m_\theta = \left| \frac{\partial \mathbf{r}}{\partial \theta} \right| = a + z \approx a;$$

(7.7)

and the relations (7.5) are exact.

Gill (1982) points out that the horizontal spheroidal coordinates for the Earth can be approximated accurately in terms of spherical coordinates. However, as noted in Section 6, for a numerical model of the global ocean circulation it is inadvisable to use spherical (or spheroidal) coordinates, because of the convergence of grid lines at the North Pole.

7.3. Generalized vertical coordinate

Here we develop a generalized vertical coordinate s for a hydrostatic, stratified fluid lying in a basin that rotates along with the equipotential surface Σ described in the preceding section. Possible choices for s include the level coordinate z, an isopycnic coordinate using the reciprocal of potential density (or a related quantity), the terrain-following coordinate σ, or some hybrid of the preceding.

Assume that, for fixed time t and horizontal coordinates $\mathbf{x} = (x_1, x_2)$, a quantity s is an increasing function of z. Equivalently, z is an increasing function of s. A surface of constant s can move upward and downward with time, and for fixed time the elevation of that surface can vary with horizontal position. For each $\mathbf{x}$ and t, let $z(\mathbf{x}, s, t)$ denote the elevation of such a surface.

The derivative $\partial z / \partial s$ relates increments in the parameter s to increments in space, and it can therefore serve as a metric coefficient for the vertical direction. For later use, it will be useful to relate this quantity to pressure and density.

Let $P(\mathbf{x}, z, t)$ denote the pressure in the fluid corresponding to horizontal position $\mathbf{x}$, elevation z, and time t. The hydrostatic assumption of Section 7.1 implies

$$\frac{\partial P}{\partial z} = -\rho g,$$

(7.8)

where ρ is the density of the fluid. In particular, the net vertical component of the pressure force acting on an element of fluid having horizontal area ΔA and height Δz is approximately $-\frac{\partial P}{\partial z} \Delta z \Delta A = -\frac{1}{\rho} \frac{\partial P}{\partial z} (\rho \Delta z \Delta A)$, so the pressure force per unit mass is approximately $-\frac{1}{\rho} \frac{\partial P}{\partial z}$. According to the hydrostatic assumption, this quantity equals $|\nabla \Phi|$, which was assumed to equal a constant value g at the end of Section 7.1.

Now let $p(\mathbf{x}, s, t)$ denote the pressure in terms of s, so that $p(\mathbf{x}, s, t) = P(\mathbf{x}, z(\mathbf{x}, s, t), t)$. Then

$$
\frac{\partial p}{\partial s}(\mathbf{x}, s, t) = \frac{\partial P}{\partial z}(\mathbf{x}, z(\mathbf{x}, s, t), t) \, \frac{\partial z}{\partial s}(\mathbf{x}, s, t)
$$

$$
= \frac{-g}{\alpha(\mathbf{x}, s, t)} \, \frac{\partial z}{\partial s}(\mathbf{x}, s, t).
$$

(7.9)

Here, $\alpha(\mathbf{x}, s, t) = 1/\rho(\mathbf{x}, s, t)$ denotes the specific volume (volume per unit mass). A metric coefficient for the coordinate s is then defined by

$$
m_s(\mathbf{x}, s, t) = z_s(\mathbf{x}, s, t) = -\frac{\alpha(\mathbf{x}, s, t) p_s(\mathbf{x}, s, t)}{g}.
$$

(7.10)

The subscripts on z and p denote partial derivatives. The coefficient m_s is positive, since $\partial p/\partial s < 0$.

An increment Δs in s corresponds to a vertical distance approximately equal to $m_s \Delta s$, in analogy to the horizontal displacement $m_i \Delta x_i$ corresponding to an increment Δx_i in the horizontal parameter x_i. Note that the lateral displacements are not measured along surfaces of constant s, but instead are measured in terms of projections onto the horizontal. This formulation is also used, for example, by Bleck (1978, 2002).

7.4. Integration in parameter space

The partial differential equations for conservation of mass and momentum will be derived so that the independent spatial variables are the parameters $(\mathbf{x}, s) = (x_1, x_2, s)$. However, quantities such as mass per unit volume and momentum per unit volume are naturally expressed in terms of rectangular coordinates having units of length. In the following, the term 'parameter space' refers to the coordinates $(\mathbf{x}, s)$ attached to the rotating spheroid Σ, and the term 'rectangular coordinates' refers to the rectangular coordinate system in which the axis of rotation is stationary. The inertial nature of this frame is immaterial to the formulation of conservation of mass, but it is essential to the formulation of conservation of momentum. The purpose of the present subsection is to develop the necessary change of variables between integrals in these two coordinate systems.

Let $A(t)$ be a region in parameter space, and let $B(t)$ denote the corresponding region in the rectangular coordinate system in the inertial reference frame at time t. ($A(t)$ and $B(t)$ provide different mathematical descriptions of the same physical entity, and in the following they are regarded as different mathematical objects. The dependence on t is immaterial to the present subsection, but it is included in the notation for the sake of further developments.) The coordinate mapping from $A(t)$ to $B(t)$ can be represented as

$$
\tilde{\mathbf{r}}(\mathbf{x}, s, t) = \mathbf{r}(\mathbf{x}, z(\mathbf{x}, s, t), t)
$$

(7.11)

for all $(\mathbf{x}, s) \in A(t)$. Let $\tilde{\psi}(\cdot, t)$ denote an integrable function defined on $B(t)$, and let $\psi(\cdot, t)$ be the corresponding function on $A(t)$ defined by $\psi(\mathbf{x}, s, t) = \tilde{\psi}\big(\tilde{\mathbf{r}}(\mathbf{x}, s, t), t\big)$. Then

$$
\int_{B(t)} \tilde{\psi}(\cdot, t) \, dV_{B(t)} = \int_{A(t)} \tilde{\psi}\big(\tilde{\mathbf{r}}(\mathbf{x}, s, t), t\big) \, J(\mathbf{x}, s, t) \, dx_1 \, dx_2 \, ds
$$
$$
= \int_{A(t)} \psi(\mathbf{x}, s, t) \, J(\mathbf{x}, s, t) \, d\mathbf{x} \, ds,
$$
(7.12)

where $dV_{B(t)}$ denotes integration on $B(t)$, and $J(\mathbf{x}, s, t)$ is the absolute value of the Jacobian,

$$
J(\mathbf{x}, s, t) = \left| \det\left[\frac{\partial \tilde{\mathbf{r}}}{\partial x_1}(\mathbf{x}, s, t), \ \frac{\partial \tilde{\mathbf{r}}}{\partial x_2}(\mathbf{x}, s, t), \ \frac{\partial \tilde{\mathbf{r}}}{\partial s}(\mathbf{x}, s, t) \right] \right|.
$$
(7.13)

Here, $\tilde{\mathbf{r}}$ is interpreted as a column vector, and the quantity in square brackets is a 3×3 matrix. Insert the definition (7.11) of $\tilde{\mathbf{r}}$ into (7.13) to obtain

$$
J(\mathbf{x}, s, t) = \left| \det\left[\frac{\partial \mathbf{r}}{\partial x_1} + \frac{\partial \mathbf{r}}{\partial z}\frac{\partial z}{\partial x_1}, \ \frac{\partial \mathbf{r}}{\partial x_2} + \frac{\partial \mathbf{r}}{\partial z}\frac{\partial z}{\partial x_2}, \ \frac{\partial \mathbf{r}}{\partial z}\frac{\partial z}{\partial s} \right] \right|
$$
(7.14)

$$
= \left| \det\left[\frac{\partial \mathbf{r}}{\partial x_1}, \ \frac{\partial \mathbf{r}}{\partial x_2}, \ \frac{\partial \mathbf{r}}{\partial z}\frac{\partial z}{\partial s} \right] \right|
$$
(7.15)

$$
= m_1 m_2 m_s \left| \det\left[\mathbf{i}(\mathbf{x}, t), \mathbf{j}(\mathbf{x}, t), \mathbf{k}(\mathbf{x}, t) \right] \right|.
$$
(7.16)

In (7.14)–(7.15), the partial derivatives of $\mathbf{r}$ are evaluated at $\big(\mathbf{x}, z(\mathbf{x}, s, t), t\big)$, and the partial derivatives of z are evaluated at $(\mathbf{x}, s, t)$. The representation (7.15) arises from expanding the sums in (7.14) and using the fact that a determinant is zero if one column is a multiple of another. The representation (7.16) relies on the relations (7.5); the metric coefficients m_1 and m_2 are evaluated at $\big(\mathbf{x}, z(\mathbf{x}, s, t)\big)$; and m_s is evaluated at $(\mathbf{x}, s, t)$. The unit vectors $\mathbf{i}$, $\mathbf{j}$, and $\mathbf{k}$ are mutually orthogonal, so the determinant in (7.16) has absolute value equal to 1. It then follows that

$$
J(\mathbf{x}, s, t) = m_1\big(\mathbf{x}, z(\mathbf{x}, s, t)\big) \, m_2\big(\mathbf{x}, z(\mathbf{x}, s, t)\big) \, m_s(\mathbf{x}, s, t)
$$
(7.17)

and thus

$$
\int_{B(t)} \tilde{\psi}(\cdot, t) \, dV_{B(t)} = \int_{A(t)} \psi(\mathbf{x}, s, t) \, m_1 m_2 m_s \, dx_1 \, dx_2 \, ds.
$$
(7.18)

For an interpretation of the preceding, consider a rectangular solid having sides Δx_1, Δx_2, Δs in the region $A(t)$ in parameter space. The corresponding subset of $B(t)$ is approximately a parallelepiped generated by the vectors

$$
\frac{\partial \tilde{\mathbf{r}}}{\partial x_1}\Delta x_1, \quad \frac{\partial \tilde{\mathbf{r}}}{\partial x_2}\Delta x_2, \quad \frac{\partial \tilde{\mathbf{r}}}{\partial s}\Delta s.
$$

The first two of these vectors are tangent to a surface of constant s, and the expressions in (7.14) amount to resolving these vectors into horizontal and vertical components. The subsequent expressions (7.15)–(7.16) give the volume of a rectangular solid having the same volume as the parallelepiped. The appropriate volume element is then $m_1 m_2 m_s \, dx_1 \, dx_2 \, ds$.

7.5. Integration on a material volume

Now consider integration on a time-dependent region that moves with a fluid flow. In particular, suppose that a volume of fluid occupies a region $A(t)$ in parameter space at time t, and let $B(t)$ denote the corresponding region in rectangular coordinates in the inertial reference frame. Assume that a function $\tilde{\psi}(\cdot, t)$ is defined and integrable on $B(t)$ for each t, and let $\psi(\cdot, t)$ be the corresponding function on $A(t)$ defined by $\psi(\mathbf{x}, s, t) = \tilde{\psi}\big(\tilde{\mathbf{r}}(\mathbf{x}, s, t), t\big)$ for all $(\mathbf{x}, s) \in A(t)$.

In the following usage, $\tilde{\psi}$ represents a fluid property such as mass density or a component of momentum density, and an integral on $B(t)$ is used when representing a conservation principle for that mass of fluid. In particular, it will be necessary to compute a time derivative of such an integral, and for that purpose it is useful to relate an integral on $B(t)$ to an integral on a fixed region in parameter space at time $t = 0$.

Let $(\mathbf{X}, S) \in A(0)$ denote the position of a fluid parcel in parameter space at time 0, and denote the position of that parcel in parameter space at any time t by $\big(\mathbf{x}(\mathbf{X}, S, t), s(\mathbf{X}, S, t)\big)$. That is, $\mathbf{X}$ and S are Lagrangian variables, and $\mathbf{x}$ and s are Eulerian variables. Then

$$\int_{B(t)} \tilde{\psi}(\cdot, t) \, dV_{B(t)} = \int_{A(t)} \psi(\mathbf{x}, s, t) \, J(\mathbf{x}, s, t) \, d\mathbf{x} \, ds$$

$$= \int_{A(0)} \psi(\mathbf{x}, s, t) \, J(\mathbf{x}, s, t) \, H(\mathbf{X}, S, t) \, d\mathbf{X} \, ds. \tag{7.19}$$

Here, $\mathbf{x}$ and s are functions of $(\mathbf{X}, S, t)$, and $H(\mathbf{X}, S, t)$ is the Jacobian of the transformation from $(\mathbf{X}, S)$ to $(\mathbf{x}, s)$, i.e.,

$$H(\mathbf{X}, S, t) = \begin{pmatrix} \frac{\partial x_1}{\partial X_1} & \frac{\partial x_1}{\partial X_2} & \frac{\partial x_1}{\partial S} \\ \frac{\partial x_2}{\partial X_1} & \frac{\partial x_2}{\partial X_2} & \frac{\partial x_2}{\partial S} \\ \frac{\partial s}{\partial X_1} & \frac{\partial s}{\partial X_2} & \frac{\partial s}{\partial S} \end{pmatrix}. \tag{7.20}$$

The last integral in (7.19) uses H instead of $|H|$, as $H(\mathbf{X}, S, t) > 0$ for all $(\mathbf{X}, S, t)$. This can be demonstrated as follows. Impose the physically realistic assumption that this Jacobian is everywhere nonzero for all t, so that positive volumes are mapped to positive volumes, and also assume that the flow is smooth enough that H is a continuous function. At time $t = 0$ the mapping $(\mathbf{X}, S) \mapsto (\mathbf{x}, s)$ is the identity mapping. In that case H is the determinant of the identity matrix, so $H(\mathbf{X}, S, 0) = 1$ for all $(\mathbf{X}, S)$.

If $H(\mathbf{X}, S, t) < 0$ for some $(\mathbf{X}, S, t)$, then continuity in time implies that $H(\mathbf{X}, S, t_0) = 0$ for some t_0, which is a contradiction.

7.6. Time derivative of a material integral

Next consider the time derivative of the integral (7.19) over a material volume. For that purpose, it will be useful to define the following quantities.

For any point $(\mathbf{x}, s) \in A(t)$, let $\dot{\mathbf{x}}(\mathbf{x}, s, t)$ and $\dot{s}(\mathbf{x}, s, t)$ denote the time derivatives of $\mathbf{x}$ and s, respectively, as seen by the fluid parcel that is located at position $(\mathbf{x}, s)$ in parameter space at time t. More precisely,

$$\frac{\partial \mathbf{x}}{\partial t}(\mathbf{X}, S, t) = \dot{\mathbf{x}}\big(\mathbf{x}(\mathbf{X}, S, t), s(\mathbf{X}, S, t), t\big),$$
$$\frac{\partial s}{\partial t}(\mathbf{X}, S, t) = \dot{s}\big(\mathbf{x}(\mathbf{X}, S, t), s(\mathbf{X}, S, t), t\big) \tag{7.21}$$

for all $(\mathbf{X}, S) \in A(0)$ and all t.

In general, the quantities $\dot{x}_1$, $\dot{x}_2$, and $\dot{s}$ are not components of linear velocity. For example, if spherical coordinates are used for the horizontal parameters, then $\dot{x}_1 = \dot{\lambda}$ and $\dot{x}_2 = \dot{\theta}$ represent the time derivatives of longitude and latitude, respectively, as seen by an observer moving with the fluid. Corresponding components of linear velocity are obtained by using the appropriate metric coefficients, as described in Section 7.9. In addition, the values of $\dot{s}$ need not be related to any geometrical motion at all; for example, if s is the reciprocal of potential density, then nonzero values of $\dot{s}$ could result entirely from a warming or cooling of the fluid.

Now define a material derivative of a quantity in terms of the coordinates used here. Assume that $F(\mathbf{x}, s, t)$ is defined for all $(\mathbf{x}, s) \in A(t)$, and let $\hat{F}(\mathbf{X}, S, t) = F\big(\mathbf{x}(\mathbf{X}, S, t), s(\mathbf{X}, S, t), t\big)$ for all $(\mathbf{X}, S) \in A(0)$. Then

$$\frac{\partial \hat{F}}{\partial t}(\mathbf{X}, S, t) = \frac{\partial F}{\partial t} + \frac{\partial \mathbf{x}}{\partial t}(\mathbf{X}, S, t) \cdot \nabla F + \frac{\partial s}{\partial t}(\mathbf{X}, S, t)\frac{\partial F}{\partial s}$$
$$= \frac{\partial F}{\partial t} + \dot{\mathbf{x}} \cdot \nabla F + \dot{s}\frac{\partial F}{\partial s}. \tag{7.22}$$

Here, $\nabla F(\mathbf{x}, s, t) = \big(\frac{\partial F}{\partial x_1}, \frac{\partial F}{\partial x_2}\big)$ for all $(\mathbf{x}, s) \in A(t)$ and all t. The notation ∇ denotes the gradient with respect to (x_1, x_2) for fixed s; the notation ∇_s is also used in this context, but a subscript s is unnecessary here, given the definition of F. In (7.22), $\dot{\mathbf{x}}$, $\dot{s}$, and all derivatives of F are evaluated at $\big(\mathbf{x}(\mathbf{X}, S, t), s(\mathbf{X}, S, t), t\big)$. Now define the material derivative

$$\frac{DF}{Dt}(\mathbf{x}, s, t) = \frac{\partial F}{\partial t} + \dot{\mathbf{x}} \cdot \nabla F + \dot{s}F_s \tag{7.23}$$

for all $(\mathbf{x}, s) \in A(t)$ and all t, where all quantities on the right side are evaluated at $(\mathbf{x}, s, t)$, and $F_s = \partial F/\partial s$. The quantity DF/Dt is the time derivative of F following fluid parcels, as opposed to the partial derivative $\partial F/\partial t$ for fixed position, and $DF/Dt(\mathbf{x}, s, t)$ refers to the parcel that is located at position $(\mathbf{x}, s)$ at time t.

The time derivative of the integral (7.19) includes the time derivative of the Jacobian H in (7.20). The calculation of this integral is an analogue of the calculation done for standard Cartesian coordinates (*e.g.*, Chorin and Marsden (1990)), and it will not be given here. Instead, we simply state the result, which is

$$\frac{\partial H}{\partial t}(\mathbf{X}, S, t) = \left(\frac{\partial \dot{x}_1}{\partial x_1} + \frac{\partial \dot{x}_2}{\partial x_2} + \frac{\partial \dot{s}}{\partial s} \right) H(\mathbf{X}, S, t). \tag{7.24}$$

Each of the terms in the parentheses is evaluated at $(\mathbf{x}(\mathbf{X}, S, t), s(\mathbf{X}, S, t), t)$.

The result (7.24) can be interpreted as follows. The Jacobian H represents the 'volume' of a fluid parcel in parameter space at time t, relative to its volume at time 0. If $\partial \dot{x}_1/\partial x_1 > 0$ then the rate of fluid flow in the x_1-direction increases with x_1, so the fluid expands in that direction. The quantity in parentheses in (7.24) represents the net effect of such quantities over the three coordinate directions, and it is a generalization of the usual divergence of velocity. If the generalized divergence is positive, then $\partial H/\partial t > 0$, and the fluid parcel expands as t increases, as expected.

Now let $E(t)$ denote the time-dependent integral (7.19), and compute $E'(t)$. The quantities ψ and J in (7.19) are evaluated at the point $(\mathbf{x}(\mathbf{X}, S, t), s(\mathbf{X}, S, t), t)$, so the time derivative of $J\psi$ in (7.19) is the material derivative. Then

$$\begin{aligned} E'(t) &= \int_{A(0)} \left[\frac{D(J\psi)}{Dt} + J\psi \left(\frac{\partial \dot{x}_1}{\partial x_1} + \frac{\partial \dot{x}_2}{\partial x_2} + \frac{\partial \dot{s}}{\partial s} \right) \right] H(\mathbf{X}, S, t) \, \mathrm{d}\mathbf{X} \, \mathrm{d}S \\ &= \int_{A(t)} \left[\frac{D(J\psi)}{Dt} + J\psi \left(\frac{\partial \dot{x}_1}{\partial x_1} + \frac{\partial \dot{x}_2}{\partial x_2} + \frac{\partial \dot{s}}{\partial s} \right) \right] \mathrm{d}\mathbf{x} \, \mathrm{d}s. \end{aligned} \tag{7.25}$$

In the first integral, the terms in the square brackets are evaluated at $(\mathbf{x}(\mathbf{X}, S, t), s(\mathbf{X}, S, t), t)$, and in the second integral they are evaluated at $(\mathbf{x}, s, t)$. A comparison with the expression (7.23) for the material derivative yields

$$E'(t) = \int_{A(t)} D_f(J\psi) \, \mathrm{d}\mathbf{x} \, \mathrm{d}s, \tag{7.26}$$

where

$$D_f(J\psi) = \frac{\partial}{\partial t}(J\psi) + \frac{\partial}{\partial x_1}(\dot{x}_1 J\psi) + \frac{\partial}{\partial x_2}(\dot{x}_2 J\psi) + \frac{\partial}{\partial s}(\dot{s} J\psi). \tag{7.27}$$

The quantity $D_f(J\psi)$ will be termed a 'flux derivative' of $J\psi$, as the spatial terms are expressed in terms of derivatives of fluxes.

7.7. *Conservation of mass*

Now suppose that $\tilde{\psi}(\cdot, t) = \tilde{\rho}(\cdot, t)$ is the density function (mass per unit volume) at time t in some fluid domain, as seen in rectangular coordinates in the inertial frame. The corresponding function on parameter space is then defined by $\psi(\mathbf{x}, s, t) = \rho(\mathbf{x}, s, t) = \tilde{\psi}(\tilde{\mathbf{r}}(\mathbf{x}, s, t), t)$. Consider a volume of fluid within this domain, and let $A(t)$ and $B(t)$ denote the regions in parameter space and rectangular coordinates, respectively, that are occupied by this volume at time t. The total mass of this volume of fluid at time t is

$$\int_{B(t)} \tilde{\psi}(\cdot, t) \, dV_{B(t)} = \int_{B(t)} \tilde{\rho}(\cdot, t) \, dV_{B(t)}. \tag{7.28}$$

Since the region $B(t)$ follows the flow, the mass (7.28) must remain constant in time. Equation (7.26) then implies

$$\int_{A(t)} D_f(J\rho) \, d\mathbf{x} \, ds = 0$$

for all t. This relation holds for all regions $A(t)$ within the fluid domain. If the function $D_f(J\rho)$ is continuous everywhere, then $D_f(J\rho) = 0$ at all positions in space at all times, *i.e.*,

$$\frac{\partial}{\partial t}(J\rho) + \frac{\partial}{\partial x_1}(\dot{x}_1 J\rho) + \frac{\partial}{\partial x_2}(\dot{x}_2 J\rho) + \frac{\partial}{\partial s}(\dot{s}J\rho) = 0. \tag{7.29}$$

All quantities in (7.29) are functions of $(\mathbf{x}, s, t)$.

   Equation (7.29) is the basic statement of conservation of mass, but it is useful to re-write it as follows. The definition (7.10) of the metric coefficient m_s implies

$$\rho(\mathbf{x}, s, t) = \frac{1}{\alpha(\mathbf{x}, s, t)} = -\frac{p_s(\mathbf{x}, s, t)}{g \, m_s(\mathbf{x}, s, t)}. \tag{7.30}$$

The definition $J = m_1 m_2 m_s$ from (7.17) implies

$$J\rho = -m_1 m_2 p_s g^{-1}, \tag{7.31}$$

and equation (7.29) becomes

$$\frac{\partial}{\partial t}(Gp_s) + \frac{\partial}{\partial x_1}(\dot{x}_1 Gp_s) + \frac{\partial}{\partial x_2}(\dot{x}_2 Gp_s) + \frac{\partial}{\partial s}(\dot{s}Gp_s) = 0, \tag{7.32}$$

or equivalently,

$$\frac{\partial}{\partial t}(Gp_s) + \nabla \cdot (\dot{\mathbf{x}}Gp_s) + \frac{\partial}{\partial s}(\dot{s}Gp_s) = 0. \tag{7.33}$$

Here,

$$\begin{aligned} G(\mathbf{x}, s, t) &= m_1\big(\mathbf{x}, z(\mathbf{x}, s, t)\big) \, m_2\big(\mathbf{x}, z(\mathbf{x}, s, t)\big) \\ &\approx m_1(\mathbf{x})m_2(\mathbf{x}), \end{aligned} \tag{7.34}$$

and $p_s = \partial p / \partial s$. As noted in Section 7.2, the metric coefficients m_1 and m_2 are nearly independent of z, so to good approximation one can use $G(\mathbf{x}) = m_1(\mathbf{x}) m_2(\mathbf{x})$. In equations (7.32)–(7.33), the operations $\partial / \partial x_i$ and ∇ are taken for fixed s, as the functions involved depend on $(\mathbf{x}, s, t)$.

In the mass conservation equation (7.32)–(7.33), the dependent variable for mass is p_s. This quantity indicates the change in hydrostatic pressure, and thus the amount of mass, over the vertical distance between nearby surfaces of constant s. In the special case where s is the elevation z, $p_s = p_z = -\rho g$, and in that case the density ρ can be used in place of p_s in (7.32)–(7.33).

7.8. Conservation of mass in layers

The role of $p_s = \partial p / \partial s$ as the mass variable can be illuminated by examining the conservation of mass in a layer bounded above and below by surfaces of constant s. The following analysis also anticipates the vertical discretization that is needed for solving the governing equations numerically.

Consider the fluid lying between the coordinate surfaces $s = s_0$ and $s = s_1$, where $s_0 < s_1$. For any $\mathbf{x}$ and t, let

$$\Delta p(\mathbf{x}, t) = p(\mathbf{x}, s_0, t) - p(\mathbf{x}, s_1, t) = \int_{s_0}^{s_1} (-p_s) \, ds > 0. \qquad (7.35)$$

Due to the hydrostatic assumption, $\Delta p(\mathbf{x}, t)$ represents the weight per unit horizontal area in this layer, at horizontal position $\mathbf{x}$ at time t. Therefore Δp equals the mass per unit horizontal area in the layer, times g, so Δp can be regarded as a two dimensional density when the layer is viewed from above. In an isopycnic setting this quantity indicates the thickness of the layer, and Δp is often labelled informally as the 'layer thickness'.

Also define the mass-weighted vertical average of $\dot{\mathbf{x}}$ over the layer by

$$\overline{\dot{\mathbf{x}}}(\mathbf{x}, t) = \frac{1}{\Delta p} \int_{s_0}^{s_1} \dot{\mathbf{x}}(\mathbf{x}, s, t)(-p_s) \, ds, \qquad (7.36)$$

and use the approximation $G(\mathbf{x}) = m_1(\mathbf{x}) m_2(\mathbf{x})$. Now multiply the mass equation (7.33) by -1 and integrate over s from s_0 to s_1 to obtain

$$\frac{\partial}{\partial t}(G \Delta p) + \nabla \cdot (G \overline{\dot{\mathbf{x}}} \Delta p) + G \left[(\dot{s} p_s)_{s=s_0} - (\dot{s} p_s)_{s=s_1} \right] = 0. \qquad (7.37)$$

The independent variables in (7.37) are the horizontal coordinates x_1 and x_2 and the time t.

Equation (7.37) can be expressed in terms of linear velocity as follows. In Section 7.9 it is shown that the horizontal components of linear velocity in the x_1- and x_2-directions, relative to the rotating spheroid Σ, are $u(\mathbf{x}, s, t) = m_1 \dot{x}_1(\mathbf{x}, s, t)$ and $v(\mathbf{x}, s, t) = m_2 \dot{x}_2(\mathbf{x}, s, t)$, respectively. Let $\bar{u}(\mathbf{x}, t)$ and

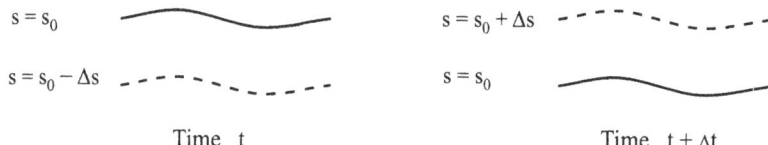

$s = s_0$

$s = s_0 + \Delta s$

$s = s_0 - \Delta s$

$s = s_0$

Time t

Time t + Δt

Figure 7.2. Illustration of mass transport across coordinate
surfaces due to material changes in the vertical coordinate s.
If s is a thermodynamic variable, then a warming or cooling
of the fluid could cause changes in s for given fluid parcels.
Fluid can then cross coordinate surfaces, even if the fluid is
motionless in space. Similarly, if s is a hybrid coordinate,
then such transport could occur during a conversion between
isopycnic and geometric coordinates.

$\bar{v}(\mathbf{x}, t)$ denote the mass-weighted vertical averages of u and v, respectively,
over the layer. Then $\bar{u}(\mathbf{x}, t) = m_1 \dot{x}_1$ and $\bar{v}(\mathbf{x}, t) = m_2 \dot{x}_2$, and equation
(7.37) can be written as

$$\frac{\partial}{\partial t}(G\Delta p) + \frac{\partial}{\partial x_1}(m_2 \bar{u} \Delta p) + \frac{\partial}{\partial x_2}(m_1 \bar{v} \Delta p)$$

$$+ G\left[(\dot{s} p_s)_{s=s_0} - (\dot{s} p_s)_{s=s_1} \right] = 0. \tag{7.38}$$

Equation (7.38) can be interpreted as follows. Integrate (7.38) with re-
spect to x_1 and x_2 on a rectangle R (in the $\mathbf{x}$ domain) having sides Δx_1
and Δx_2. The quantities $m_1 \Delta x_1$ and $m_2 \Delta x_2$ give the (approximate) linear dimensions of the corresponding rectangular region $\tilde{R}$ on the surface of
the rotating spheroid Σ. The area of $\tilde{R}$ is approximately $m_1 m_2 \Delta x_1 \Delta x_2 =$
$G\Delta x_1 \Delta x_2$, so G is the factor that relates horizontal 'area' in parameter
space to physical area on the surface of Σ. The quantity $G\Delta x_1 \Delta x_2 \Delta p$ is
the mass in the given layer in $\tilde{R}$, times g, and the integrals on R of the
terms in (7.38) involving $\partial/\partial x_1$ and $\partial/\partial x_2$ can be interpreted in terms of
lateral mass transport across the edges of $\tilde{R}$.

The terms in (7.38) involving $\dot{s} p_s$ are less standard and represent transport
between layers due to material changes in the coordinate s. Consider the
coordinate surface $s = s_0$, and for definiteness assume that $\dot{s} > 0$ on that
surface. The quantity $\dot{s}$ is the time derivative of s as seen by an observer
that is fixed relative to the fluid, and nonzero values of $\dot{s}$ may or may not
be due to any motion of the fluid. For example, if s is the reciprocal of
potential density or a related quantity, then a warming of the fluid could
cause $\dot{s} > 0$, even if the fluid is motionless relative to Σ. Over a time

increment Δt, the value of s seen by a fluid parcel changes by approximately the amount $\Delta s = \dot{s}\Delta t$. Now consider the two coordinate surfaces $s = s_0$ and $s = s_0 - \Delta s$ at time t. From time t to time $t + \Delta t$, the fluid parcels on these surfaces experience a change from s_0 and $s_0 - \Delta s$ to $s_0 + \Delta s$ and s_0, respectively (see Figure 7.2). The coordinate surface $s = s_0$ thus moves downward relative to the fluid, but an observer on that surface sees fluid crossing the surface in the upward direction. For the fluid that crosses the surface $s = s_0$ during the time increment Δt, the weight per unit area is approximately $\Delta s(-\partial p/\partial s) = \dot{s}\Delta t(-\partial p/\partial s)$, so the rate of mass flow per unit time per unit horizontal area (times g) is $-\dot{s}p_s$. Here, a positive rate of flow indicates an upward movement of fluid relative to the coordinate surface. The last two terms in equation (7.38) represent the net effect of transport into or out of the layer by this mechanism.

7.9. Velocity

In preparation for deriving the equation for conservation of momentum, the present subsection develops a representation for fluid velocity. This velocity is expressed as a motion relative to the rotating spheroid Σ, plus a rigid-body rotation of points attached to Σ.

As in Section 7.5, let $(\mathbf{X}, S)$ denote the position of a fluid parcel in parameter space at time 0, and denote the position of that parcel in parameter space at any time t by $\big(\mathbf{x}(\mathbf{X}, S, t), s(\mathbf{X}, S, t)\big)$. The position of that parcel, relative to the rectangular coordinate system in the inertial reference frame, can then be written in the form

$$\mathbf{R}(\mathbf{X}, S, t) = \mathbf{r}\big(\mathbf{x}(\mathbf{X}, S, t), \hat{z}(\mathbf{X}, S, t), t\big).$$

Here, $\mathbf{r}(\mathbf{x}, z, t)$ denotes the position in rectangular coordinates of the point with coordinates $(\mathbf{x}, z)$ relative to Σ, as defined in Section 7.2, and $\hat{z}(\mathbf{X}, S, t)$ denotes the elevation of the fluid parcel in question. More precisely,

$$\hat{z}(\mathbf{X}, S, t) = z\big(\mathbf{x}(\mathbf{X}, S, t), s(\mathbf{X}, S, t), t\big),$$

where $z(\mathbf{x}, s, t)$ is the elevation of the coordinate surface as defined in Section 7.3.

The velocity of a fluid parcel, as seen in rectangular coordinates in the inertial frame, is then

$$\frac{\partial \mathbf{R}}{\partial t}(\mathbf{X}, S, t) = \frac{\partial \mathbf{r}}{\partial x_1}\frac{\partial x_1}{\partial t} + \frac{\partial \mathbf{r}}{\partial x_2}\frac{\partial x_2}{\partial t} + \frac{\partial \mathbf{r}}{\partial z}\frac{\partial \hat{z}}{\partial t} + \frac{\partial \mathbf{r}}{\partial t}. \qquad (7.39)$$

Here, the partial derivatives of $\mathbf{r}$ are evaluated at $\big(\mathbf{x}(\mathbf{X}, S, t), \hat{z}(\mathbf{X}, S, t), t\big)$, and the partial derivatives of x_1, x_2, and $\hat{z}$ are evaluated at $(\mathbf{X}, S, t)$. The relations (7.4) and (7.5) involving $\mathbf{i}$, $\mathbf{j}$, and $\mathbf{k}$, and the definitions (7.21) of $\dot{\mathbf{x}}$ and $\dot{s}$, imply

$$\frac{\partial \mathbf{R}}{\partial t}(\mathbf{X}, S, t) = m_1 \dot{x}_1 \mathbf{i} + m_2 \dot{x}_2 \mathbf{j} + \frac{\partial \hat{z}}{\partial t} \mathbf{k} + \frac{\partial \mathbf{r}}{\partial t}. \tag{7.40}$$

In (7.40), m_1 and m_2 are evaluated at the point $\big(\mathbf{x}(\mathbf{X}, S, t), \hat{z}(\mathbf{X}, S, t)\big)$; $\mathbf{i}$, $\mathbf{j}$, and $\mathbf{k}$ are evaluated at $\big(\mathbf{x}(\mathbf{X}, S, t), t\big)$; and $\dot{x}_1$ and $\dot{x}_2$ are evaluated at $\big(\mathbf{x}(\mathbf{X}, S, t), s(\mathbf{X}, S, t), t\big)$.

In equations (7.39) and (7.40), the partial derivative $\partial \mathbf{r}/\partial t$ is taken for fixed position $(\mathbf{x}, z)$ relative to the rotating surface Σ, and it therefore represents the velocity associated with a rigid-body rotation about the given axis. The remaining terms in (7.39) and (7.40) involve derivatives of $\mathbf{r}$ with respect to position on Σ. They thus represent motion relative to Σ, and their sum is termed the 'relative velocity'.

For the horizontal components of the relative velocity, define

$$\begin{aligned} u(\mathbf{x}, s, t) &= m_1\big(\mathbf{x}, z(\mathbf{x}, s, t)\big)\, \dot{x}_1(\mathbf{x}, s, t), \\ v(\mathbf{x}, s, t) &= m_2\big(\mathbf{x}, z(\mathbf{x}, s, t)\big)\, \dot{x}_2(\mathbf{x}, s, t) \end{aligned} \tag{7.41}$$

and define a vertical component $w(\mathbf{x}, s, t)$ by

$$\frac{\partial \tilde{z}}{\partial t}(\mathbf{X}, S, t) = w\big(\mathbf{x}(\mathbf{X}, S, t), s(\mathbf{X}, S, t), t\big). \tag{7.42}$$

The particle velocity (7.40) can then be written as

$$\frac{\partial \mathbf{R}}{\partial t}(\mathbf{X}, S, t) = u\mathbf{i} + v\mathbf{j} + w\mathbf{k} + \frac{\partial \mathbf{r}}{\partial t}, \tag{7.43}$$

where the components u, v, and w are evaluated at $\big(\mathbf{x}(\mathbf{X}, S, t), s(\mathbf{X}, S, t), t\big)$; the unit vectors $\mathbf{i}$, $\mathbf{j}$, and $\mathbf{k}$ are evaluated at $\big(\mathbf{x}(\mathbf{X}, S, t), t\big)$; and the rigid-body velocity $\partial \mathbf{r}/\partial t$ is evaluated at $\big(\mathbf{x}(\mathbf{X}, S, t), \hat{z}(\mathbf{X}, S, t), t\big)$.

The terms $u\mathbf{i}$ and $v\mathbf{j}$ are defined in terms of $\partial \mathbf{r}/\partial x_1$ and $\partial \mathbf{r}/\partial x_2$, via (7.39). Given the definition of the position vector $\mathbf{r}$, these partial derivatives are taken for fixed z, so the vectors $\mathbf{i}$ and $\mathbf{j}$ are horizontal, $i.e.$, tangent to Σ. The quantities u and v thus represent components of velocity that are truly horizontal, not components along surfaces of constant s. This is consistent with the observation, made at the end of Section 7.3, that lateral displacements are measured in terms of projections onto the horizontal instead of along s-coordinate surfaces.

7.10. Momentum

In the present subsection we formulate the momentum of a volume of fluid and calculate its time derivative.

Assume that a volume of fluid occupies a region $A(t)$ in parameter space at time t, and let $B(t)$ denote the corresponding region in rectangular coordinates in the inertial reference frame. Also let $\mathbf{Q}(t)$ denote the total momentum of this volume of fluid, as seen in the inertial frame. Then $\mathbf{Q}(t)$ is the integral on $B(t)$ of the velocity times mass density; the change of variables given in (7.19) implies

$$\mathbf{Q}(t) = \int_{A(t)} \rho(\mathbf{x}, s, t)\left[u\mathbf{i} + v\mathbf{j} + w\mathbf{k} + \frac{\partial \mathbf{r}}{\partial t}\right] J(\mathbf{x}, s, t)\, d\mathbf{x}\, ds \qquad (7.44)$$

$$= \int_{A(0)} \rho(\mathbf{x}, s, t)\left[u\mathbf{i} + v\mathbf{j} + w\mathbf{k} + \frac{\partial \mathbf{r}}{\partial t}\right] J(\mathbf{x}, s, t)\, H(\mathbf{X}, S, t)\, d\mathbf{X}\, dS.$$

Equation (7.30) and subsequent discussions imply $\rho J = -m_1 m_2 p_s g^{-1} = -Gp_s g^{-1}$, so

$$\mathbf{Q}(t) = -\frac{1}{g}\int_{A(0)}\left[(Gup_s)\mathbf{i} + (Gvp_s)\mathbf{j} + (Gwp_s)\mathbf{k} + Gp_s\frac{\partial \mathbf{r}}{\partial t}\right] H(\mathbf{X}, S, t)\, d\mathbf{X}\, dS.$$
$$(7.45)$$

Here, G, u, v, w, p_s are evaluated at $(\mathbf{x}(\mathbf{X}, S, t), s(\mathbf{X}, S, t), t)$; $\mathbf{i}$, $\mathbf{j}$, and $\mathbf{k}$ are evaluated at $(\mathbf{x}(\mathbf{X}, S, t), t)$; $\partial \mathbf{r}/\partial t$ is evaluated at $(\mathbf{x}(\mathbf{X}, S, t), \hat{z}(\mathbf{X}, S, t), t)$; and $H(\mathbf{X}, S, t)$ is the Jacobian from $(\mathbf{X}, S)$ to $(\mathbf{x}, s)$ defined in (7.20).

The next main task is to calculate the time derivative of (7.45). Several terms in the integrand have the form

$$\hat{F}(\mathbf{X}, S, t) = F\big(\mathbf{x}(\mathbf{X}, S, t), s(\mathbf{X}, S, t), t\big)\, H(\mathbf{X}, S, t),$$

and some calculations similar to those in (7.25)–(7.27) imply

$$\frac{\partial \hat{F}}{\partial t}(\mathbf{X}, S, t) = D_f(F)\, H(\mathbf{X}, S, t),$$

where the flux derivative

$$D_f(F) = \frac{\partial F}{\partial t} + \frac{\partial}{\partial x_1}(\dot{x}_1 F) + \frac{\partial}{\partial x_2}(\dot{x}_2 F) + \frac{\partial}{\partial s}(\dot{s}F) \qquad (7.46)$$

$$= \frac{\partial F}{\partial t} + \nabla \cdot (\dot{\mathbf{x}}F) + \frac{\partial}{\partial s}(\dot{s}F)$$

is evaluated at $(\mathbf{x}(\mathbf{X}, S, t), s(\mathbf{X}, S, t), t)$. It then follows that

$$\mathbf{Q}'(t) = -\frac{1}{g}\int_{A(0)}\bigg\{D_f(Gup_s)\mathbf{i} + D_f(Gvp_s)\mathbf{j} + D_f(Gwp_s)\mathbf{k}$$

$$+ (Gup_s)\frac{D\mathbf{i}}{Dt} + (Gvp_s)\frac{D\mathbf{j}}{Dt} + (Gwp_s)\frac{D\mathbf{k}}{Dt} \qquad (7.47)$$

$$+ Gp_s\frac{\partial}{\partial t}\left[\frac{\partial \mathbf{r}}{\partial t}(\mathbf{x}(\mathbf{X}, S, t), \hat{z}(\mathbf{X}, S, t), t)\right]\bigg\}$$

$$H(\mathbf{X}, S, t)\, d\mathbf{X}\, dS.$$

The derivation of (7.47) uses the relation $D_f(Gp_s) = 0$, which is a way of stating the equation (7.32)–(7.33) of conservation of mass. In (7.47), the notation Di/Dt refers to the time derivative of $\mathbf{i}(\mathbf{x}(\mathbf{X}, S, t), t)$, so

$$\frac{D\mathbf{i}}{Dt} = \frac{\partial \mathbf{i}}{\partial t} + \dot{x}_1 \frac{\partial \mathbf{i}}{\partial x_1} + \dot{x}_2 \frac{\partial \mathbf{i}}{\partial x_2}, \tag{7.48}$$

where $\dot{x}_1$ and $\dot{x}_2$ are evaluated at $(\mathbf{x}(\mathbf{X}, S, t), s(\mathbf{X}, S, t), t)$. The terms $D\mathbf{j}/Dt$ and $D\mathbf{k}/Dt$ are analogous.

The time derivative in the third line of (7.47) is

$$\frac{\partial^2 \mathbf{r}}{\partial t \partial x_1} \dot{x}_1 + \frac{\partial^2 \mathbf{r}}{\partial t \partial x_2} \dot{x}_2 + \frac{\partial^2 \mathbf{r}}{\partial t \partial z} \frac{\partial \hat{z}}{\partial t} + \frac{\partial^2 \mathbf{r}}{\partial t^2}, \tag{7.49}$$

where the partial derivatives of $\mathbf{r}$ are evaluated at $(\mathbf{x}(\mathbf{X}, S, t), \hat{z}(\mathbf{X}, S, t), t)$, $\dot{x}_1$ and $\dot{x}_2$ are evaluated at $(\mathbf{x}(\mathbf{X}, S, t), s(\mathbf{X}, S, t), t)$, and $\partial \hat{z}/\partial t$ is evaluated at $(\mathbf{X}, S, t)$. According to the approximation introduced in (7.5),

$$\begin{aligned} \frac{\partial^2 \mathbf{r}}{\partial x_1 \partial t}(\mathbf{x}, z, t) &= m_1(\mathbf{x}, z) \frac{\partial \mathbf{i}}{\partial t}(\mathbf{x}, t), \\ \frac{\partial^2 \mathbf{r}}{\partial x_2 \partial t}(\mathbf{x}, z, t) &= m_2(\mathbf{x}, z) \frac{\partial \mathbf{j}}{\partial t}(\mathbf{x}, t), \end{aligned} \tag{7.50}$$

and the definition (7.4) of $\mathbf{k}$ implies

$$\frac{\partial^2 \mathbf{r}}{\partial z \partial t}(\mathbf{x}, z, t) = \frac{\partial \mathbf{k}}{\partial t}(\mathbf{x}, t). \tag{7.51}$$

The definitions (7.41) and (7.42) of the velocity components u, v, and w imply that the expression (7.49) is equal to

$$u \frac{\partial \mathbf{i}}{\partial t} + v \frac{\partial \mathbf{j}}{\partial t} + w \frac{\partial \mathbf{k}}{\partial t} + \mathbf{a}_c, \tag{7.52}$$

where $\mathbf{a}_c = \partial^2 \mathbf{r}/\partial t^2$; u, v, and w are evaluated at $(\mathbf{x}(\mathbf{X}, S, t), s(\mathbf{X}, S, t), t)$; and the derivatives of $\mathbf{i}$, $\mathbf{j}$, and $\mathbf{k}$ are evaluated at $(\mathbf{x}(\mathbf{X}, S, t), t)$.

The quantity $\partial^2 \mathbf{r}/\partial t^2(\mathbf{x}, z, t)$ is the second-order time derivative of the position vector $\mathbf{r}$, for a fixed position $(\mathbf{x}, z)$ relative to the rotating spheroid Σ. The vector $\mathbf{a}_c$ is thus the centripetal acceleration associated with a rigid-body rotation. For present purposes, $\mathbf{a}_c$ will be regarded as a function of $(\mathbf{x}, s, t)$, where $\mathbf{x}$ and s are evaluated at $(\mathbf{X}, S, t)$ in (7.52).

Equation (7.47) can now be written as

$$\mathbf{Q}'(t) = -\frac{1}{g} \int_{A(t)} \Big[D_f(Gup_s)\mathbf{i} + D_f(Gvp_s)\mathbf{j} + D_f(Gwp_s)\mathbf{k}$$

$$+ Gp_s\mathbf{a}_c + \boldsymbol{\Psi} \Big] \, d\mathbf{x} \, ds \tag{7.53}$$

where

$$\boldsymbol{\Psi} = 2(Gup_s)\frac{\partial \mathbf{i}}{\partial t} + 2(Gvp_s)\frac{\partial \mathbf{j}}{\partial t} + 2(Gwp_s)\frac{\partial \mathbf{k}}{\partial t}$$

$$+ (Gup_s)\left(\dot{x}_1 \frac{\partial \mathbf{i}}{\partial x_1} + \dot{x}_2 \frac{\partial \mathbf{i}}{\partial x_2} \right) \tag{7.54}$$

$$+ (Gvp_s)\left(\dot{x}_1 \frac{\partial \mathbf{j}}{\partial x_1} + \dot{x}_2 \frac{\partial \mathbf{j}}{\partial x_2} \right)$$

$$+ (Gwp_s)\left(\dot{x}_1 \frac{\partial \mathbf{k}}{\partial x_1} + \dot{x}_2 \frac{\partial \mathbf{k}}{\partial x_2} \right).$$

In equations (7.53) and (7.54), $\mathbf{i}$, $\mathbf{j}$, $\mathbf{k}$ and their derivatives depend on $(\mathbf{x}, t)$, and all other quantities depend on $(\mathbf{x}, s, t)$.

The time derivatives $\partial \mathbf{i}/\partial t$, $\partial \mathbf{j}/\partial t$, and $\partial \mathbf{k}/\partial t$ in (7.54) are taken with the position fixed relative to Σ. The terms containing these derivatives are therefore due to the rotation of Σ. The remaining terms in (7.54) involve the variation of the unit vectors $\mathbf{i}$, $\mathbf{j}$, and $\mathbf{k}$ with respect to position on Σ, so these terms are due to the curvature of Σ and/or properties of the parametrization of Σ in terms of $\mathbf{x} = (x_1, x_2)$.

According to the principle of conservation of momentum, $\mathbf{Q}'(t)$ is equal to the sum of the various forces acting on the fluid volume. These forces are due to pressure, gravity, stresses, and viscosity, and they will be discussed in the next subsections.

7.11. Pressure and the Montgomery potential

As in the preceding subsection, assume that a volume of fluid occupies a region $A(t)$ in parameter space at time t, and let $B(t)$ denote the corresponding region in rectangular coordinates in the inertial reference frame. Also, as in Section 7.3, let $P(\mathbf{x}, z, t)$ denote the pressure in the fluid corresponding to horizontal position $\mathbf{x}$, elevation z, and time t. Now let $\tilde{P}(\mathbf{r}, t)$ denote the pressure as seen in rectangular coordinates in the inertial frame, i.e., $P(\mathbf{x}, z, t) = \tilde{P}(\mathbf{r}(\mathbf{x}, z, t), t)$.

Let $\partial B(t)$ denote the boundary of the region $B(t)$ in rectangular coordinates, and let $\mathbf{n}(\mathbf{r}, t)$ denote the unit outward normal vector to $\partial B(t)$. The net pressure force acting on the volume of fluid is then

$$\int_{\partial B(t)} -\tilde{P}\mathbf{n}\, dA = -\int_{\partial B(t)} (\tilde{P}n_1, \tilde{P}n_2, \tilde{P}n_3)\, dA$$

$$= -\int_{\partial B(t)} \left((\tilde{P},0,0)\cdot\mathbf{n},\ (0,\tilde{P},0)\cdot\mathbf{n},\ (0,0,\tilde{P})\cdot\mathbf{n}\right) dA$$

$$= -\int_{B(t)} \left(\mathrm{div}(\tilde{P},0,0),\ \mathrm{div}(0,\tilde{P},0),\ \mathrm{div}(0,0,\tilde{P})\right) dV_{B(t)}$$

$$= -\int_{B(t)} \left(\frac{\partial\tilde{P}}{\partial r_1}, \frac{\partial\tilde{P}}{\partial r_2}, \frac{\partial\tilde{P}}{\partial r_3}\right) dV_{B(t)}$$

$$= -\int_{A(t)} \nabla\tilde{P}\big(\mathbf{r}(\mathbf{x}, z(\mathbf{x}, s, t), t), t\big) J(\mathbf{x}, s, t)\, d\mathbf{x}\, ds. \qquad (7.55)$$

Here, the integration is performed component-by-component, the divergence theorem is used on each component to obtain the third line, and the change of variables described in Section 7.4 is used to obtain the last line. In the last line, the notation $\nabla\tilde{P}\big(\mathbf{r}(\mathbf{x}, z(\mathbf{x}, s, t), t), t\big)$ refers to the value of $\nabla\tilde{P}$ at the indicated point; it does not indicate the derivative of a composite function.

The integral (7.55) can also be be expressed as

$$\int_{\partial B(t)} -\tilde{P}\mathbf{n}\, dA = -\int_{A(t)} \left((\nabla\tilde{P}\cdot\mathbf{i})\mathbf{i} + (\nabla\tilde{P}\cdot\mathbf{j})\mathbf{j} + (\nabla\tilde{P}\cdot\mathbf{k})\mathbf{k}\right) J(\mathbf{x}, s, t)\, d\mathbf{x}\, ds,$$

$$(7.56)$$

where $\mathbf{i}$, $\mathbf{j}$, and $\mathbf{k}$ are evaluated at $(\mathbf{x}, t)$. The relation (7.5) implies

$$\nabla\tilde{P}\cdot\mathbf{i} = \frac{1}{m_1}\nabla\tilde{P}\big(\mathbf{r}(\mathbf{x}, z(\mathbf{x}, s, t), t), t\big)\cdot\frac{\partial\mathbf{r}}{\partial x_1}(\mathbf{x}, z(\mathbf{x}, s, t), t)$$

$$= \frac{1}{m_1}\frac{\partial P}{\partial x_1}(\mathbf{x}, z(\mathbf{x}, s, t), t), \qquad (7.57)$$

where m_1 is evaluated at $(\mathbf{x}, z(\mathbf{x}, s, t))$. Given the definition of the function P, the partial derivative $\partial P/\partial x_1$ is taken for fixed z and then evaluated at $(\mathbf{x}, z(\mathbf{x}, s, t), t)$. Similarly,

$$\nabla\tilde{P}\cdot\mathbf{j} = \frac{1}{m_2}\frac{\partial P}{\partial x_2}(\mathbf{x}, z(\mathbf{x}, s, t), t), \qquad (7.58)$$

and the definition (7.4) of $\mathbf{k}$ implies

$$\nabla\tilde{P}\cdot\mathbf{k} = \frac{\partial P}{\partial z}(\mathbf{x}, z(\mathbf{x}, s, t), t). \qquad (7.59)$$

Now express the preceding in terms of functions of $(\mathbf{x}, s, t)$, as these will be the independent variables in the final form of the governing equations.

As in Section 7.3, let $p(\mathbf{x}, s, t)$ denote the pressure in terms of s, so that $p(\mathbf{x}, s, t) = P(\mathbf{x}, z(\mathbf{x}, s, t), t)$. Then, for $i = 1$ and $i = 2$,

$$
\begin{aligned}
\frac{\partial p}{\partial x_i}(\mathbf{x}, s, t) &= \frac{\partial P}{\partial x_i} + \frac{\partial P}{\partial z}\frac{\partial z}{\partial x_i}(\mathbf{x}, s, t) \\
&= \frac{\partial P}{\partial x_i} - \frac{g}{\alpha(\mathbf{x}, s, t)}\frac{\partial z}{\partial x_i}(\mathbf{x}, s, t),
\end{aligned}
\tag{7.60}
$$

where $\alpha(\mathbf{x}, s, t)$ is the specific volume (reciprocal of density). The second line is obtained by using the hydrostatic condition (7.8). Equation (7.60) then implies

$$
\begin{aligned}
\alpha(\mathbf{x}, s, t)\frac{\partial P}{\partial x_i}(\mathbf{x}, z(\mathbf{x}, s, t), t) &= \alpha\frac{\partial p}{\partial x_i} + g\frac{\partial z}{\partial x_i} \\
&= \frac{\partial M}{\partial x_i} - p\frac{\partial \alpha}{\partial x_i},
\end{aligned}
\tag{7.61}
$$

where

$$
M(\mathbf{x}, s, t) = \alpha(\mathbf{x}, s, t)p(\mathbf{x}, s, t) + gz(\mathbf{x}, s, t)
\tag{7.62}
$$

is the Montgomery potential (Montgomery 1937). The hydrostatic condition (7.8) also implies

$$
\alpha(\mathbf{x}, s, t)\frac{\partial P}{\partial z}(\mathbf{x}, z(\mathbf{x}, s, t), t) = -g.
$$

The pressure force (7.56) can then be expressed in the form

$$
\begin{aligned}
\int_{\partial B(t)} -\tilde{P}\mathbf{n}\, dA = &-\int_{A(t)} \left[\frac{1}{m_1}\left(\frac{\partial M}{\partial x_1} - p\frac{\partial \alpha}{\partial x_1}\right)\mathbf{i} \right. \\
&\left. + \frac{1}{m_2}\left(\frac{\partial M}{\partial x_2} - p\frac{\partial \alpha}{\partial x_2}\right)\mathbf{j} - g\mathbf{k}\right] \frac{J(\mathbf{x}, s, t)}{\alpha(\mathbf{x}, s, t)}\, dx\, ds,
\end{aligned}
\tag{7.63}
$$

where m_1 and m_2 are evaluated at $(\mathbf{x}, z(\mathbf{x}, s, t))$. According to equations (7.31) and (7.34), $J/\alpha = J\rho = -m_1 m_2 p_s g^{-1} = -Gp_s g^{-1}$, and (7.63) becomes

$$
\begin{aligned}
\int_{\partial B(t)} -\tilde{P}\mathbf{n}\, dA = &\frac{1}{g}\int_{A(t)} \left[\frac{1}{m_1}\left(\frac{\partial M}{\partial x_1} - p\frac{\partial \alpha}{\partial x_1}\right)\mathbf{i} \right. \\
&\left. + \frac{1}{m_2}\left(\frac{\partial M}{\partial x_2} - p\frac{\partial \alpha}{\partial x_2}\right)\mathbf{j} - g\mathbf{k}\right] Gp_s\, dx\, ds.
\end{aligned}
\tag{7.64}
$$

In the earlier expression (7.56) for the pressure force, the coefficients of $\mathbf{i}$ and $\mathbf{j}$ represent the components that are horizontal, i.e., in directions tangent to the rotating spheroid Σ, and the derivatives in (7.57) and (7.58) are taken for fixed z. However, when the independent variables are $(\mathbf{x}, s, t)$, derivatives with respect to x_1 and x_2 are taken for fixed s, and these directions need not be horizontal. The usage of the Montgomery potential (7.62) enables derivatives for fixed s to yield components of the pressure force that are in the directions of the horizontal vectors $\mathbf{i}$ and $\mathbf{j}$.

The vertical variation of the Montgomery potential can be determined by observing

$$\frac{\partial M}{\partial s}(\mathbf{x}, s, t) = p\frac{\partial \alpha}{\partial s} + \alpha\frac{\partial p}{\partial s} + g\frac{\partial z}{\partial s}$$
$$= p\frac{\partial \alpha}{\partial s}. \tag{7.65}$$

The condition $\alpha p_s + g z_s = 0$ follows from equation (7.9), which in turn follows from the hydrostatic condition (7.8). Equation (7.65) can be regarded as a statement of the hydrostatic condition in a generalized vertical coordinate.

7.12. Diffusion of momentum

One mechanism for the transport of momentum within a fluid is the mixing caused by small-scale turbulence and molecular diffusion. In the ocean the effects of molecular diffusion are negligible compared to the effects of turbulence, so molecular diffusion will not be considered here. In numerical models of ocean circulation it is generally not possible to represent the details of turbulent motions, owing to insufficient grid resolution, so instead it is necessary to parametrize the large-scale effects of these subgrid-scale motions in terms of the dependent variables that are used in the model. This problem has long been an active area of research; see, e.g., Griffies (2004) or Pedlosky (1987). In present-day practice, operational ocean models often use a parametrization that involves some sort of Laplacian or biharmonic diffusion.

As noted in Section 2.4, mixing in the ocean's interior is highly anisotropic, with the principle directions of mixing being approximately horizontal. In the present subsection, we assume that the vertical coordinate s is an isopycnic coordinate and that the mixing occurs primarily along surfaces of constant s. In addition, we represent the subgrid-scale processes in terms of a diffusion that is proportional to the gradient of velocity, in analogy to molecular diffusion, and the diffusion is separated into a component tangent to coordinate surfaces and a component normal to such surfaces. The following analysis produces a parametrization similar to the one used, for example, by Bleck and Smith (1990) and Bleck (2002) for isopycnic models.

As in the preceding two subsections, assume that a volume of fluid occupies a region $A(t)$ in parameter space at time t, and denote by $B(t)$ the corresponding region in rectangular coordinates in the inertial reference frame. For each $(\mathbf{x}, s) \in A(t)$, let $\tilde{\mathbf{r}}(\mathbf{x}, s, t) = \mathbf{r}(\mathbf{x}, z(\mathbf{x}, s, t), t)$ denote the corresponding point in $B(t)$, as in equation (7.11).

In order to formulate the highly anisotropic diffusion considered here, it is useful to define a local coordinate system, at each point in the fluid, that involves a plane tangent to a surface of constant s and a vector normal to that surface. A starting point is provided by the tangent vectors $\partial \tilde{\mathbf{r}}/\partial x_1$ and $\partial \tilde{\mathbf{r}}/\partial x_2$. Equations (7.11), (7.4), and (7.5) imply

$$\frac{\partial \tilde{\mathbf{r}}}{\partial x_1}(\mathbf{x}, s, t) = \frac{\partial \mathbf{r}}{\partial x_1} + \frac{\partial \mathbf{r}}{\partial z}\frac{\partial z}{\partial x_1}$$

$$= m_1 \mathbf{i} + \frac{\partial z}{\partial x_1}\mathbf{k}$$

$$= m_1 (\mathbf{i} + \delta_1 \mathbf{k}),$$

where $\delta_1(\mathbf{x}, s, t)$ is the slope of a surface of constant s in the x_1-direction, m_1 is evaluated at $(\mathbf{x}, z(\mathbf{x}, s, t))$, and $\mathbf{i}$ and $\mathbf{k}$ are evaluated at $(\mathbf{x}, t)$. Similarly,

$$\frac{\partial \tilde{\mathbf{r}}}{\partial x_2}(\mathbf{x}, s, t) = m_2 (\mathbf{j} + \delta_2 \mathbf{k}),$$

where $\delta_2(\mathbf{x}, s, t)$ is the slope of a surface of constant s in the x_2-direction. Unit tangent vectors $\tilde{\mathbf{i}}$ and $\tilde{\mathbf{j}}$ to a coordinate surface are then defined by

$$\frac{\partial \tilde{\mathbf{r}}}{\partial x_1}(\mathbf{x}, s, t) = \tilde{m}_1(\mathbf{x}, s, t)\frac{\mathbf{i} + \delta_1 \mathbf{k}}{\sqrt{1 + \delta_1^2}} \equiv \tilde{m}_1(\mathbf{x}, s, t)\,\tilde{\mathbf{i}}(\mathbf{x}, s, t),$$

$$\frac{\partial \tilde{\mathbf{r}}}{\partial x_2}(\mathbf{x}, s, t) = \tilde{m}_2(\mathbf{x}, s, t)\frac{\mathbf{j} + \delta_2 \mathbf{k}}{\sqrt{1 + \delta_2^2}} \equiv \tilde{m}_2(\mathbf{x}, s, t)\,\tilde{\mathbf{j}}(\mathbf{x}, s, t),$$

where

$$\tilde{m}_i(\mathbf{x}, s, t) = \left| \frac{\partial \tilde{\mathbf{r}}}{\partial x_i}(\mathbf{x}, s, t) \right| = m_i\sqrt{1 + \delta_i^2} = m_i(1 + O(\delta^2)) \qquad (7.66)$$

for $i = 1, 2$. Here, δ is an upper bound for $|\delta_1|$ and $|\delta_2|$. In the interior of the ocean, the slopes of isopycnals generally do not reach much above 10^{-2} (Griffies et al. 2000a), so typically $\delta \approx 10^{-2}$. The quantity $\tilde{m}_i$ can be regarded as a metric coefficient that relates increments in the parameter x_i to spatial increments along s-coordinate surfaces, whereas the metric coefficient m_i relates increments in x_i to spatial increments that are horizontal.

For any $(\mathbf{x}, s, t)$, the vectors $\tilde{\mathbf{i}}(\mathbf{x}, s, t)$ and $\tilde{\mathbf{j}}(\mathbf{x}, s, t)$ provide a basis for the tangent plane to the coordinate surface at that position and time. However, these vectors are not exactly orthogonal, as $\tilde{\mathbf{i}} \cdot \tilde{\mathbf{j}} = O(\delta^2)$. Orthogonal unit vectors can be obtained by rotating $\tilde{\mathbf{i}}$ and/or $\tilde{\mathbf{j}}$ through an angle of magnitude $O(\delta^2)$ in that plane. This process is not uniquely determined, but assume that orthogonal unit vectors $\hat{\mathbf{i}}$ and $\hat{\mathbf{j}}$ in that plane have been obtained so that

$$\hat{\mathbf{i}}(\mathbf{x}, s, t) = \tilde{\mathbf{i}} + \mathbf{O}(\delta^2) = \frac{1}{\tilde{m}_1(\mathbf{x}, s, t)} \frac{\partial \tilde{\mathbf{r}}}{\partial x_1}(\mathbf{x}, s, t) + \mathbf{O}(\delta^2),$$

$$\hat{\mathbf{j}}(\mathbf{x}, s, t) = \tilde{\mathbf{j}} + \mathbf{O}(\delta^2) = \frac{1}{\tilde{m}_2(\mathbf{x}, s, t)} \frac{\partial \tilde{\mathbf{r}}}{\partial x_2}(\mathbf{x}, s, t) + \mathbf{O}(\delta^2).$$

(7.67)

The boldface font in the order terms is used to indicate vector quantities.

Now define a unit normal vector $\hat{\mathbf{k}}$ by

$$\hat{\mathbf{k}}(\mathbf{x}, s, t) = \hat{\mathbf{i}} \times \hat{\mathbf{j}}$$

$$= \left[\frac{\mathbf{i} + \delta_1 \mathbf{k}}{\sqrt{1 + \delta_1^2}} + \mathbf{O}(\delta^2) \right] \times \left[\frac{\mathbf{j} + \delta_2 \mathbf{k}}{\sqrt{1 + \delta_2^2}} + \mathbf{O}(\delta^2) \right]$$

(7.68)

$$= \mathbf{k} - \delta_1 \mathbf{i} - \delta_2 \mathbf{j} + \mathbf{O}(\delta^2).$$

Here, it is assumed that the coordinate system on the rotating spheroid Σ is right-handed, in the sense that $\mathbf{i} \times \mathbf{j} = \mathbf{k}$, $\mathbf{k} \times \mathbf{j} = -\mathbf{i}$, and $\mathbf{i} \times \mathbf{k} = -\mathbf{j}$. Equation (7.11), $\tilde{\mathbf{r}}(\mathbf{x}, s, t) = \mathbf{r}\big(\mathbf{x}, z(\mathbf{x}, s, t), t\big)$, together with equations (7.4) and (7.10), imply

$$\frac{\partial \tilde{\mathbf{r}}}{\partial s}(\mathbf{x}, s, t) = \frac{\partial \mathbf{r}}{\partial z} \frac{\partial z}{\partial s} = m_s(\mathbf{x}, s, t)\, \mathbf{k}(\mathbf{x}, t).$$

It then follows from (7.68) that

$$\hat{\mathbf{k}}(\mathbf{x}, s, t) = \frac{1}{m_s(\mathbf{x}, s, t)} \frac{\partial \tilde{\mathbf{r}}}{\partial s}(\mathbf{x}, s, t) + \mathbf{O}(\delta).$$

(7.69)

The representation of diffusion will initially be formulated in the region $B(t)$ in rectangular coordinates in the inertial reference frame. For that purpose, define the unit vectors $\hat{\mathbf{I}}$, $\hat{\mathbf{J}}$, and $\hat{\mathbf{K}}$ on $B(t)$ by

$$\hat{\mathbf{i}}(\mathbf{x}, s, t) = \hat{\mathbf{I}}\big(\tilde{\mathbf{r}}(\mathbf{x}, s, t), t\big),$$

$$\hat{\mathbf{j}}(\mathbf{x}, s, t) = \hat{\mathbf{J}}\big(\tilde{\mathbf{r}}(\mathbf{x}, s, t), t\big),$$

$$\hat{\mathbf{k}}(\mathbf{x}, s, t) = \hat{\mathbf{K}}\big(\tilde{\mathbf{r}}(\mathbf{x}, s, t), t\big),$$

for all $(\mathbf{x}, s) \in A(t)$. Also define components U and V of horizontal velocity, with the independent spatial variables in $B(t)$, by

$$u(\mathbf{x}, s, t) = U\big(\tilde{\mathbf{r}}(\mathbf{x}, s, t), t\big),$$

$$v(\mathbf{x}, s, t) = V\big(\tilde{\mathbf{r}}(\mathbf{x}, s, t), t\big),$$

(7.70)

for all $(\mathbf{x}, s) \in A(t)$.

Now consider a parametrization of the diffusion of the component u. This will provide a term in the coefficient of $\mathbf{i}$ in equation (7.53), which gives the derivative $\mathbf{Q}'(t)$ of the momentum of the volume of fluid considered here. The diffusion of the component v is represented similarly. A component in the direction of $\mathbf{k}$ will not be considered here, as the vertical component of momentum conservation will be represented with the hydrostatic balance.

Assume that the momentum flux associated with u, $i.e.$, the rate of transport of u-momentum per unit cross-sectional area per unit time, is given by

$$-\rho A_H(\nabla U \cdot \hat{\mathbf{I}})\hat{\mathbf{I}} - \rho A_H(\nabla U \cdot \hat{\mathbf{J}})\hat{\mathbf{J}} - \rho A_D(\nabla U \cdot \hat{\mathbf{K}})\hat{\mathbf{K}}, \qquad (7.71)$$

where ρ is the density of the fluid, A_H is a kinematic viscosity coefficient for diffusion in directions that are tangent to s-coordinate surfaces, and A_D is a kinematic viscosity for diapycnal diffusion normal to such surfaces. All quantities in (7.71) are evaluated at points $(\mathbf{r}, t)$ for $\mathbf{r} \in B(t)$, except that A_H and A_D could perhaps be taken as constant, and $\nabla U = (\partial U/\partial r_1, \partial U/\partial r_2, \partial U/\partial r_3)$.

The net rate of diffusion of momentum into the region $B(t)$ is given by the integral of the negative of the outward normal component of (7.71) over $\partial B(t)$. This integral needs to be represented in terms of an integral over the region $A(t)$, as the quantity $\mathbf{Q}'(t)$ in (7.53) also involves an integral over that region. Toward that end, represent the boundary $\partial A(t)$ of $A(t)$ in terms of parameters $\sigma = (\sigma_1, \sigma_2)$ by

$$\mathbf{b}(\sigma, t) = \mathbf{b}(\sigma_1, \sigma_2, t) = (\mathbf{x}(\sigma, t), s(\sigma, t)) \qquad (7.72)$$

for all σ in a parameter region Γ. The boundary $\partial B(t)$ of $B(t)$ is then represented by

$$\mathbf{B}(\sigma, t) = \tilde{\mathbf{r}}(\mathbf{b}(\sigma_1, \sigma_2, t), t) = \tilde{\mathbf{r}}(\mathbf{x}(\sigma, t), s(\sigma, t), t) \qquad (7.73)$$

for $\sigma \in \Gamma$. The net rate of diffusion of momentum into the region $B(t)$ in rectangular coordinates in the inertial frame is then

$$\int_\Gamma \left[\rho A_H(\nabla U \cdot \hat{\mathbf{I}})\hat{\mathbf{I}} + \rho A_H(\nabla U \cdot \hat{\mathbf{J}})\hat{\mathbf{J}} + \rho A_D(\nabla U \cdot \hat{\mathbf{K}})\hat{\mathbf{K}}\right] \cdot \left(\frac{\partial \mathbf{B}}{\partial \sigma_1} \times \frac{\partial \mathbf{B}}{\partial \sigma_2}\right) d\sigma_1 \, d\sigma_2.$$

$$(7.74)$$

Here, it is assumed that the parameters $\sigma = (\sigma_1, \sigma_2)$ are chosen so that the cross product in (7.74) points out of the region $B(t)$. The quantities in the square brackets are evaluated at $\mathbf{B}(\sigma, t)$.

Equations (7.73), (7.67), and (7.69) imply

$$\frac{\partial \mathbf{B}}{\partial \sigma_i} = \frac{\partial \tilde{\mathbf{r}}}{\partial x_1}\frac{\partial x_1}{\partial \sigma_i} + \frac{\partial \tilde{\mathbf{r}}}{\partial x_2}\frac{\partial x_2}{\partial \sigma_i} + \frac{\partial \tilde{\mathbf{r}}}{\partial s}\frac{\partial s}{\partial \sigma_i}$$

$$= \tilde{m}_1(\hat{\mathbf{i}} + \mathbf{O}(\delta^2))\frac{\partial x_1}{\partial \sigma_i}$$

$$+ \tilde{m}_2(\hat{\mathbf{j}} + \mathbf{O}(\delta^2))\frac{\partial x_2}{\partial \sigma_i} \qquad (7.75)$$

$$+ m_s(\hat{\mathbf{k}} + \mathbf{O}(\delta))\frac{\partial s}{\partial \sigma_i}$$

$$= D(\sigma, t)\mathbf{q}_i(\sigma, t) + \mathbf{e}_i,$$

where

$$D(\sigma, t) = [\hat{\mathbf{i}}, \ \hat{\mathbf{j}}, \ \hat{\mathbf{k}}]$$

$$\mathbf{q}_i(\sigma, t) = \left(\tilde{m}_1 \frac{\partial x_1}{\partial \sigma_i}, \ \tilde{m}_2 \frac{\partial x_2}{\partial \sigma_i}, \ m_s \frac{\partial s}{\partial \sigma_i} \right)^T, \tag{7.76}$$

where e_i is an error vector involving $\delta^2(\tilde{m}_1 \partial x_1/\partial \sigma_i)$, $\delta^2(\tilde{m}_2 \partial x_2/\partial \sigma_i)$, and $\delta(m_s \partial s/\partial \sigma_i)$. The notation for $D(\sigma, t)$ specifies a 3×3 matrix having the indicated columns, and $\mathbf{q}_i$ is a column vector. In (7.75) and (7.76), the quantities $\tilde{m}_1$, $\tilde{m}_2$, m_s, $\hat{\mathbf{i}}$, $\hat{\mathbf{j}}$, and $\hat{\mathbf{k}}$ are evaluated at $(\mathbf{x}(\sigma, t), s(\sigma, t), t)$; and the partial derivatives of x_1, x_2, and s are evaluated at (σ, t). Since D is a unitary matrix, it follows that

$$\frac{\partial \mathbf{B}}{\partial \sigma_1} \times \frac{\partial \mathbf{B}}{\partial \sigma_2} = D(\sigma, t)(\mathbf{q}_1 \times \mathbf{q}_2) + \mathbf{O}(\mathbf{e} \times \mathbf{q}),$$

where $\mathbf{q}_1 \times \mathbf{q}_2$ is regarded as a column vector, and the error term $\mathbf{O}(\mathbf{e} \times \mathbf{q})$ involves δ, δ^2, and higher powers.

The net rate of diffusion (7.74), after neglecting the error terms involving δ, can then be written in the form

$$\int_\Gamma [F_1\hat{\mathbf{i}} + F_2\hat{\mathbf{j}} + F_3\hat{\mathbf{k}}] \cdot [D(\sigma, t)(\mathbf{q}_1 \times \mathbf{q}_2)] \, d\sigma_1 \, d\sigma_2, \tag{7.77}$$

where $F_1\hat{\mathbf{i}} + F_2\hat{\mathbf{j}} + F_3\hat{\mathbf{k}}$ is evaluated at $(\mathbf{x}(\sigma, t), s(\sigma, t), t)$, and the terms F_1, F_2, and F_3 are defined by comparing with (7.74). The dot product in (7.77) can be represented as a row vector times a column vector, so (7.77) is equal to

$$\int_\Gamma (F_1, F_2, F_3) D^T D (\mathbf{q}_1 \times \mathbf{q}_2) \, d\sigma_1 \, d\sigma_2 \tag{7.78}$$

$$= \int_\Gamma (F_1, F_2, F_3) \mathbf{q}_1 \times \mathbf{q}_2 \, d\sigma_1 \, d\sigma_2$$

$$= \int_\Gamma (\tilde{m}_2 m_s F_1, \ \tilde{m}_1 m_s F_2, \ \tilde{m}_1 \tilde{m}_2 F_3) \frac{\partial \mathbf{b}}{\partial \sigma_1} \times \frac{\partial \mathbf{b}}{\partial \sigma_2} \, d\sigma_1 \, d\sigma_2$$

$$= \int_{\partial A(t)} (\tilde{m}_2 m_s F_1, \ \tilde{m}_1 m_s F_2, \ \tilde{m}_1 \tilde{m}_2 F_3) \cdot \mathbf{n}_{\partial A(t)} \, dA_{\partial A(t)}$$

$$= \int_{A(t)} \left[\frac{\partial}{\partial x_1} (\tilde{m}_2 m_s F_1) + \frac{\partial}{\partial x_2} (\tilde{m}_1 m_s F_2) + \frac{\partial}{\partial s} (\tilde{m}_1 \tilde{m}_2 F_3) \right] dx_1 \, dx_2 \, ds.$$

The third line in (7.78) arises from examining the roles of the factors $\tilde{m}_1$, $\tilde{m}_2$, and m_s in $\mathbf{q}_i$ on the cross product and observing that the components of $\mathbf{b}(\sigma, t)$ are $x_1(\sigma, t)$, $x_2(\sigma, t)$, and $s(\sigma, t)$. The integrands in the second and third lines have the form of a row vector times a column vector. The fourth line is an integral on the boundary $\partial A(t)$ of the region $A(t)$ in parameter space, and $\mathbf{n}_{\partial A(t)}(\mathbf{x}, s, t)$ is the unit outward normal to that boundary. The last line is obtained by using the divergence theorem in parameter space, with independent variables (x_1, x_2, s).

Equations (7.74), (7.77), (7.67), and (7.70) imply

$$F_1(\mathbf{x}, s, t) = \rho A_H \nabla U \left(\tilde{\mathbf{r}}(\mathbf{x}, s, t), t \right) \cdot \hat{\mathbf{i}}(\mathbf{x}, s, t)$$

$$= \rho A_H \nabla U \cdot \left[\frac{1}{\tilde{m}_1(\mathbf{x}, s, t)} \frac{\partial \tilde{\mathbf{r}}}{\partial x_1}(\mathbf{x}, s, t) + \mathbf{O}(\delta^2) \right]$$

$$= \rho A_H \left[\frac{1}{\tilde{m}_1(\mathbf{x}, s, t)} \frac{\partial u}{\partial x_1}(\mathbf{x}, s, t) + O(\delta^2) \right].$$

Similarly,

$$F_2(\mathbf{x}, s, t) = \rho A_H \left[\frac{1}{\tilde{m}_2(\mathbf{x}, s, t)} \frac{\partial u}{\partial x_2}(\mathbf{x}, s, t) + O(\delta^2) \right],$$

$$F_3(\mathbf{x}, s, t) = \rho A_D \left[\frac{1}{m_s(\mathbf{x}, s, t)} \frac{\partial u}{\partial s}(\mathbf{x}, s, t) + O(\delta) \right].$$

According to equation (7.66), $\tilde{m}_i$ can be replaced by m_i for $i = 1$ and $i = 2$, to order δ^2. If the error terms involving δ and higher powers are neglected, then (7.78) can be written as

$$\int_{A(t)} \left[\frac{\partial}{\partial x_1} \left(\rho A_H \left(\frac{1}{m_1} \frac{\partial u}{\partial x_1} \right) m_2 m_s \right) + \frac{\partial}{\partial x_2} \left(\rho A_H \left(\frac{1}{m_2} \frac{\partial u}{\partial x_2} \right) m_1 m_s \right) \right.$$

$$\left. + \frac{\partial}{\partial s} \left(\rho A_D \left(\frac{1}{m_s} \frac{\partial u}{\partial s} \right) m_1 m_2 \right) \right] dx_1 \, dx_2 \, ds. \tag{7.79}$$

The integral (7.79) represents the net rate of diffusion of the u-component of momentum into the volume of fluid that occupies the region $A(t)$ in parameter space and the region $B(t)$ in rectangular coordinates in the inertial reference frame. The rate of diffusion of the v-component is obtained by replacing u with v in (7.79).

7.13. Applied stresses plus diffusion

The ocean is subjected to forces due to wind stress at the top of the fluid and frictional stress along the bottom. In the present subsection we develop a representation of these forces and then combine that result with the representation (7.79) of diffusion.

Here, it is assumed that each of these stresses acts as a body force near the top or bottom boundary, with the amplitude of the stress decaying with distance from the boundary. In the case of an isopycnic-coordinate ocean model, layer thicknesses can tend to zero due to interfaces intersecting the top or bottom of the fluid domain. If the wind or bottom stress were applied strictly as a surface force, then nonzero forcing could be applied to layers having essentially zero thickness. This would have adverse algorithmic consequences, whereas the assumption of a decaying body force enables the forcing to be distributed proportionately when thin layers are found near the boundary.

Assume that the wind and bottom stresses act horizontally, and denote the sum of the stresses by $\boldsymbol{\tau}^{wb} = \tau_u^{wb}\mathbf{i} + \tau_v^{wb}\mathbf{j}$. Near the upper boundary $\boldsymbol{\tau}^{wb}$ is due to the wind, near the bottom boundary $\boldsymbol{\tau}^{wb}$ is due to bottom friction, and in the ocean's interior $\boldsymbol{\tau}^{wb}$ is zero. Along any horizontal plane, $\boldsymbol{\tau}^{wb}$ represents a horizontal force, per unit horizontal area, that is exerted by the upper region on the lower region. The net force, due to $\boldsymbol{\tau}^{wb}$, acting on the volume of fluid contained in the region $B(t)$ in rectangular coordinates is then given by

$$\int_{\partial B(t)} \boldsymbol{\tau}^{wb}(\mathbf{k} \cdot \mathbf{n})\,\mathrm{d}A, \tag{7.80}$$

where $\mathbf{n}(\mathbf{r}, t)$ denotes the unit outward normal to the boundary $\partial B(t)$.

The statement (7.80) can be justified as follows. The quantity $\mathbf{k} \cdot \mathbf{n}$ is the vertical component of the unit outward normal. If an element ΔA of $\partial B(t)$ is horizontal and lies on the upper part of $\partial B(t)$, then $\mathbf{k} \cdot \mathbf{n} = 1$, and the corresponding contribution to (7.80) is $\boldsymbol{\tau}^{wb}\Delta A$. This is the product of the area and the force per unit area, and it is consistent with the sign convention that a stress represents the force exerted by an upper region on a lower region. On the other hand, if ΔA is horizontal and lies on the lower part of $\partial B(t)$, then $\mathbf{k} \cdot \mathbf{n} = -1$, and the corresponding contribution to (7.80) is $-\boldsymbol{\tau}^{wb}\Delta A$. The minus sign indicates the effect of a lower region on an upper region.

For a general element of area ΔA on $\partial B(t)$, the quantity $(\mathbf{k} \cdot \mathbf{n})\Delta A$ is a signed projection of that area onto a horizontal plane. Consider the element of fluid bounded by ΔA, the horizontal projection, and a vertical surface. The quantity $\mathbf{F}_H = -\boldsymbol{\tau}^{wb}(\mathbf{k} \cdot \mathbf{n})\Delta A$ represents the force exerted on the element of fluid along that horizontal projection. If $\mathbf{F}_{\Delta A}$ represents the force exerted along the area element ΔA, then the net force acting on the fluid element, due to the horizontal stress $\boldsymbol{\tau}^{wb}$, is $\mathbf{F}_{\Delta A} + \mathbf{F}_H$. This net force is proportional to the volume of the fluid element and thus varies as the cube of distance. However, $\mathbf{F}_{\Delta A}$ and $\mathbf{F}_H$ vary as the square of distance, so for sufficiently small elements $\mathbf{F}_{\Delta A} \approx -\mathbf{F}_H = \boldsymbol{\tau}^{wb}(\mathbf{k} \cdot \mathbf{n})\Delta A$.

The force (7.80) can be represented as an integral on the region $A(t)$ in parameter space by using techniques similar to those used in the preceding subsection. In the present case, parametrize $\partial B(t)$ by

$$\mathbf{B}^z(\sigma, t) = \mathbf{r}\big(\mathbf{x}(\sigma, t), z(\sigma, t), t\big), \tag{7.81}$$

so that

$$\frac{\partial \mathbf{B}^z}{\partial \sigma_i} = \frac{\partial \mathbf{r}}{\partial x_1} \frac{\partial x_1}{\partial \sigma_i} + \frac{\partial \mathbf{r}}{\partial x_2} \frac{\partial x_2}{\partial \sigma_i} + \frac{\partial \mathbf{r}}{\partial z} \frac{\partial z}{\partial \sigma_i}$$

$$= m_1 \frac{\partial x_1}{\partial \sigma_i} \mathbf{i} + m_2 \frac{\partial x_2}{\partial \sigma_i} \mathbf{j} + m_s \frac{\partial s}{\partial \sigma_i} \mathbf{k}.$$

(7.82)

Calculations similar to those in (7.74)–(7.78) imply that the force (7.80) is equal to

$$\int_{A(t)} \left[\frac{\partial}{\partial s} (m_1 m_2 \tau_u^{wb}) \mathbf{i} + \frac{\partial}{\partial s} (m_1 m_2 \tau_v^{wb}) \mathbf{j} \right] dx_1 \, dx_2 \, ds.$$

(7.83)

During these calculations, $\tilde{\mathbf{i}}$, $\tilde{\mathbf{j}}$, $\tilde{\mathbf{k}}$, $\tilde{m}_1$, and $\tilde{m}_2$ are replaced by $\mathbf{i}$, $\mathbf{j}$, $\mathbf{k}$, m_1, and m_2, respectively, and there are no error terms involving δ.

Now combine the representation (7.83) of the applied stresses with the representation (7.79) for the diffusion of the u-component of momentum and the analogous formula for the v-component. According to (7.10), the hydrostatic condition implies $\rho m_s = -p_s/g$, and this relation can be inserted into the first two terms in (7.79). For the u-component, the combined effect of diffusion and applied stress is

$$-\frac{1}{g} \int_{A(t)} \left[\frac{\partial}{\partial x_1} \left(A_H \left(\frac{1}{m_1} \frac{\partial u}{\partial x_1} \right) m_2 p_s \right) + \frac{\partial}{\partial x_2} \left(A_H \left(\frac{1}{m_2} \frac{\partial u}{\partial x_2} \right) m_1 p_s \right) \right.$$

$$\left. - \frac{\partial}{\partial s} (g \tau_u G) \right] dx_1 \, dx_2 \, ds,$$

(7.84)

where $G = m_1 m_2$ and

$$\tau_u = \tau_u^{wb} + \rho A_D \left(\frac{1}{m_s} \frac{\partial u}{\partial s} \right).$$

(7.85)

The corresponding formulas for the v-component are obtained by replacing u with v in (7.84) and (7.85).

The second term in (7.85) can also be regarded as $\rho A_D \partial u/\partial z$, and it represents the internal friction due to vertical variations in horizontal velocity. This term represents a form of shear stress, and in (7.85) it is combined with the shear stress τ_u^{wb} due to wind forcing and bottom friction to yield the total shear stress τ_u in the u-direction. Analogous remarks apply to the total shear stress τ_v in the v-direction.

7.14. Conservation of momentum

Equation (7.53) gives the derivative $\mathbf{Q}'(t)$ of the momentum of the volume of fluid that occupies the region $A(t)$ in parameter space and the region $B(t)$ in rectangular coordinates in the inertial reference frame, at time t. According to the principle of conservation of momentum, $\mathbf{Q}'(t)$ is equal to the sum of the forces acting on that volume of fluid due to pressure, gravity, applied stresses, and diffusion of momentum. In the present subsection the results of the preceding subsections are combined to express this principle in terms of partial differential equations.

The formula for $\mathbf{Q}'(t)$ in (7.53) includes a term involving the centripetal acceleration $\mathbf{a}_c$. According to the hydrostatic assumption stated in Section 7.1, $\mathbf{a}_c$ is equal to the gravitational acceleration plus the vertical component of the pressure force per unit mass. It then follows that the term involving centripetal acceleration in the formula for $\mathbf{Q}'(t)$ is balanced by terms involving gravity and the vertical component of the pressure force. These terms will then be deleted from the following discussion. The vertical velocity w will also be neglected, due to the hydrostatic assumption, and from now on the discussion focuses on the horizontal component of $\mathbf{Q}'(t)$.

This component can be expressed as

$$\mathbf{Q}'_H(t) = -\frac{1}{g} \int_{A(t)} \left[(D_f(Gup_s) + \mathbf{\Psi} \cdot \mathbf{i})\mathbf{i} + (D_f(Gvp_s) + \mathbf{\Psi} \cdot \mathbf{j})\mathbf{j} \right] dx\, ds, \quad (7.86)$$

where $\mathbf{\Psi}$ is defined in (7.54). When $\mathbf{\Psi} \cdot \mathbf{i}$ and $\mathbf{\Psi} \cdot \mathbf{j}$ are calculated, the terms in $\mathbf{\Psi}$ involving $\mathbf{k}$ can be neglected, as these involve the vertical velocity w. In addition, $\mathbf{i}(\mathbf{x}, t) \cdot \mathbf{i}(\mathbf{x}, t) = 1$, $\mathbf{j}(\mathbf{x}, t) \cdot \mathbf{j}(\mathbf{x}, t) = 1$, and $\mathbf{i}(\mathbf{x}, t) \cdot \mathbf{j}(\mathbf{x}, t) = 0$ for all $(\mathbf{x}, t)$, so

$$\mathbf{i} \cdot \frac{\partial \mathbf{i}}{\partial t} = \mathbf{i} \cdot \frac{\partial \mathbf{i}}{\partial x_i} = \mathbf{j} \cdot \frac{\partial \mathbf{j}}{\partial t} = \mathbf{j} \cdot \frac{\partial \mathbf{j}}{\partial x_i} = 0,$$

$$\mathbf{i} \cdot \frac{\partial \mathbf{j}}{\partial t} = -\mathbf{j} \cdot \frac{\partial \mathbf{i}}{\partial t}, \quad (7.87)$$

$$\mathbf{i} \cdot \frac{\partial \mathbf{j}}{\partial x_i} = -\mathbf{j} \cdot \frac{\partial \mathbf{i}}{\partial x_i}.$$

It then follows from (7.54) that

$$\mathbf{\Psi} \cdot \mathbf{i} = (Gvp_s)\left[2\,\mathbf{i} \cdot \frac{\partial \mathbf{j}}{\partial t} + \mathbf{i} \cdot \left(\dot{x}_1 \frac{\partial \mathbf{j}}{\partial x_1} + \dot{x}_2 \frac{\partial \mathbf{j}}{\partial x_2} \right) \right],$$

$$\mathbf{\Psi} \cdot \mathbf{j} = (Gup_s)\left[2\,\mathbf{j} \cdot \frac{\partial \mathbf{i}}{\partial t} + \mathbf{j} \cdot \left(\dot{x}_1 \frac{\partial \mathbf{i}}{\partial x_1} + \dot{x}_2 \frac{\partial \mathbf{i}}{\partial x_2} \right) \right]. \quad (7.88)$$

The last two lines in (7.87) imply that the bracketed quantities in (7.88) are negatives of each other.

Define the Coriolis parameter f by

$$f(\mathbf{x}) = 2\,\mathbf{j} \cdot \frac{\partial \mathbf{i}}{\partial t} \quad (7.89)$$

for all $\mathbf{x}$. The partial derivative $\partial \mathbf{i}/\partial t$ is taken for a fixed position on the rotating spheroid Σ, so the presence of f is due to the rotation of Σ. The right side of (7.89) appears to depend on t; however, the alternate representation developed in Section 7.15 shows that f is independent of t and is also independent of the choice of coordinate system on Σ.

The remaining terms in the square brackets in (7.88) involve the variation of $\mathbf{i}$ and $\mathbf{j}$ with respect to position on Σ, for fixed t. These terms are therefore due to the curvature of Σ and/or the properties of the parametrization of Σ in terms of $\mathbf{x}$. The quantities $\dot{x}_1$ and $\dot{x}_2$ can depend on $(\mathbf{x}, s, t)$, but the dot products of the quantities involving $\mathbf{i}$ and $\mathbf{j}$ depend only on $\mathbf{x}$.

Let

$$\tilde{f}(\mathbf{x}, s, t) = f(\mathbf{x}) + \mathbf{j} \cdot \left(\dot{x}_1 \frac{\partial \mathbf{i}}{\partial x_1} + \dot{x}_2 \frac{\partial \mathbf{i}}{\partial x_2} \right) \tag{7.90}$$

denote the sum of the Coriolis and geometric parameters. The derivative (7.86) of the horizontal component of momentum then becomes

$$\mathbf{Q}'_H(t) = -\frac{1}{g} \int_{A(t)} \Big[\big(D_f(Gup_s) - \tilde{f}Gvp_s \big) \mathbf{i}$$
$$+ \big(D_f(Gvp_s) + \tilde{f}Gup_s \big) \mathbf{j} \Big] \, d\mathbf{x}\, ds. \tag{7.91}$$

Now compare to the pressure force (7.64) and the combined effect (7.84) of the diffusion and applied stresses to obtain the momentum equations

$$D_f(Gup_s) - \tilde{f}Gvp_s = -\frac{1}{m_1} \left(\frac{\partial M}{\partial x_1} - p\frac{\partial \alpha}{\partial x_1} \right) Gp_s - \frac{\partial}{\partial s}(g\tau_u G) + \psi_u,$$
$$D_f(Gvp_s) + \tilde{f}Gup_s = -\frac{1}{m_2} \left(\frac{\partial M}{\partial x_2} - p\frac{\partial \alpha}{\partial x_2} \right) Gp_s - \frac{\partial}{\partial s}(g\tau_v G) + \psi_v. \tag{7.92}$$

The flux derivative D_f is defined in (7.46), $G = m_1 m_2$, the Montgomery potential $M = \alpha p + gz$ is defined in (7.62), the total shear stress τ_u is defined in (7.85), and

$$\psi_u = \frac{\partial}{\partial x_1}\left(A_H \left(\frac{1}{m_1}\frac{\partial u}{\partial x_1} \right) m_2 p_s \right) + \frac{\partial}{\partial x_2}\left(A_H \left(\frac{1}{m_2}\frac{\partial u}{\partial x_2} \right) m_1 p_s \right) \tag{7.93}$$

represents the diffusion of u along the s-coordinate surfaces. The quantities τ_v and ψ_v are analogous. The terms ψ_u and ψ_v, and the terms involving A_D in τ_u and τ_v, depend on the particular parametrization of subgrid-scale mixing developed in Section 7.12.

The pressure forcing on the right sides in (7.92) can be expressed in the form $-(\nabla M - p\nabla \alpha)Gp_s$. Depending on the exact choice of the vertical coordinate s, it is possible for the terms ∇M and $p\nabla \alpha$ to be far larger than their difference, under certain conditions. This situation can lead to numerical inaccuracy and even numerical instability. However, these problems can be remedied by proper choices of the details of the vertical coordinate and a proper implementation of the pressure term: see, e.g., Sun et al. (1999), Bleck (2002), and Hallberg (2005).

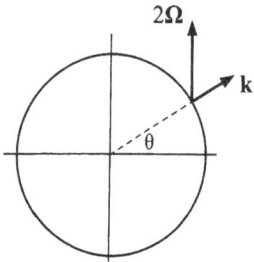

Figure 7.3. The Coriolis parameter $f = 2\boldsymbol{\Omega} \cdot \mathbf{k} = 2\Omega \sin \theta$ is
the local vertical component of the planetary vorticity $2\boldsymbol{\Omega}$.
This parameter is zero at the equator and has maximum
magnitude at the poles.

7.15. The Coriolis parameter

Here we develop a more transparent representation of the Coriolis parameter
$f(\mathbf{x}) = 2\mathbf{j} \cdot \frac{\partial \mathbf{i}}{\partial t}$ given in (7.89). Equations (7.1) and (7.3) imply

$$\mathbf{i}(\mathbf{x}, t) = Q(\Omega t)\mathbf{i}(\mathbf{x}, 0) \quad \text{and} \quad \mathbf{j}(\mathbf{x}, t) = Q(\Omega t)\mathbf{j}(\mathbf{x}, 0),$$

where $Q(\Omega t)$ is a rotation matrix. Then

$$\frac{\partial \mathbf{i}}{\partial t}(\mathbf{x}, t) = \Omega Q'(\Omega t)\mathbf{i}(\mathbf{x}, 0) = \Omega Q'(\Omega t)Q^T(\Omega t)\mathbf{i}(\mathbf{x}, t),$$

where the superscript T denotes the transpose. A calculation shows that
$\frac{\partial \mathbf{i}}{\partial t}(\mathbf{x}, t) = \boldsymbol{\Omega} \times \mathbf{i}(\mathbf{x}, t)$, which is the velocity associated with a rigid-body
rotation. Here, $\boldsymbol{\Omega}$ is a vector with length Ω that is aligned with the axis of
rotation, and the direction of $\boldsymbol{\Omega}$ and the direction of rotation are related by
the right-hand rule. It then follows that $f = 2\mathbf{j} \cdot (\boldsymbol{\Omega} \times \mathbf{i}) = 2\boldsymbol{\Omega} \cdot (\mathbf{i} \times \mathbf{j})$. Now
assume that the local coordinate system on Σ is right-handed, in the sense
that $\mathbf{i} \times \mathbf{j} = \mathbf{k}$. The Coriolis parameter can then be written as

$$f(\mathbf{x}) = 2\boldsymbol{\Omega} \cdot \mathbf{k} = 2\Omega \sin \theta, \tag{7.94}$$

where θ is the latitude. The second equality neglects the slight departure
of Σ from being a perfect sphere.

A calculation shows that if the fluid is motionless relative to the rotating
spheroid Σ, then the vorticity (curl of velocity) of that fluid relative to the
inertial reference frame is $2\boldsymbol{\Omega}$. The term 'planetary vorticity' is commonly
applied to this vorticity. At any point on the spheroid, the value of the
Coriolis parameter is then the local vertical component of the planetary
vorticity, as illustrated in Figure 7.3.

The Coriolis parameter is zero at the equator, positive in the northern hemisphere, and negative in the southern hemisphere. The Earth rotates at a rate of approximately $2\pi + 2\pi/360.24$ radians in 24 hours, and for the Earth the maximum value of $|f|$ is $2\Omega \approx 1.46 \times 10^{-4}$ sec^{-1}.

7.16. Conservation of momentum in layers

The partial differential equations (7.92) describe the conservation of momentum in the x_1- and x_2-directions, respectively. In anticipation of solving these equations numerically, we now develop the vertically discrete equations obtained by integrating the equations (7.92) with respect to s between the coordinate surfaces $s = s_0$ and $s = s_1$, with $s_0 < s_1$. The present discussion is an analogue of the discussion of conservation of mass in layers given in Section 7.8.

As in Section 7.8, let

$$\Delta p(\mathbf{x}, t) = p(\mathbf{x}, s_0, t) - p(\mathbf{x}, s_1, t) = \int_{s_0}^{s_1} (-p_s) \, ds > 0, \tag{7.95}$$

and define mass-weighted vertical averages of the velocity components u and v by

$$
\begin{aligned}
\bar{u}(\mathbf{x}, t) &= \frac{1}{\Delta p} \int_{s_0}^{s_1} u(\mathbf{x}, s, t)(-p_s) \, ds, \\
\bar{v}(\mathbf{x}, t) &= \frac{1}{\Delta p} \int_{s_0}^{s_1} v(\mathbf{x}, s, t)(-p_s) \, ds.
\end{aligned}
\tag{7.96}
$$

Denote the deviations from these averages by δ_u and δ_v, respectively, so that $u(\mathbf{x}, s, t) = \bar{u}(\mathbf{x}, t) + \delta_u(\mathbf{x}, s, t)$ and $v(\mathbf{x}, s, t) = \bar{v}(\mathbf{x}, t) + \delta_v(\mathbf{x}, s, t)$. It then follows that

$$\int_{s_0}^{s_1} \delta_u(\mathbf{x}, s, t)(-p_s) \, ds = \int_{s_0}^{s_1} \delta_v(\mathbf{x}, s, t)(-p_s) \, ds = 0 \tag{7.97}$$

and $\delta_u = O(\Delta s)$ and $\delta_v = O(\Delta s)$, where $\Delta s = s_1 - s_0$.

Now consider the vertical integral of the u-equation in (7.92). The flux derivative in the u-equation is

$$
\begin{aligned}
D_f(Gup_s) &= \frac{\partial}{\partial t}(Gup_s) + \frac{\partial}{\partial x_1}(\dot{x}_1 Gup_s) + \frac{\partial}{\partial x_2}(\dot{x}_2 Gup_s) + \frac{\partial}{\partial s}(\dot{s} Gup_s) \\
&= \frac{\partial}{\partial t}(Gup_s) + \frac{\partial}{\partial x_1}(uup_s m_2) + \frac{\partial}{\partial x_2}(vup_s m_1) + \frac{\partial}{\partial s}(\dot{s} Gup_s).
\end{aligned}
\tag{7.98}
$$

The second equality uses the relations $u = m_1 \dot{x}_1$, $v = m_2 \dot{x}_2$, and $G = m_1 m_2$. From now on, use the approximation $G(\mathbf{x}) = m_1(\mathbf{x}) m_2(\mathbf{x})$, as discussed in

Section 7.7. Multiply (7.98) by -1 and integrate over s to obtain

$$-\int_{s_0}^{s_1} D_f(Gup_s)\,ds = \frac{\partial}{\partial t}(G\bar{u}\Delta p) + \frac{\partial}{\partial x_1}\left(m_2\int_{s_0}^{s_1} uu(-p_s)\,ds\right)$$
$$+ \frac{\partial}{\partial x_2}\left(m_1\int_{s_0}^{s_1} vu(-p_s)\,ds\right) \tag{7.99}$$
$$+ G\left[(\dot{s}p_su)_{s=s_0} - (\dot{s}p_su)_{s=s_1}\right].$$

The relations (7.95) and (7.97) imply

$$\int_{s_0}^{s_1} uu(-p_s)\,ds = \int_{s_0}^{s_1}(\bar{u}+\delta_u)(\bar{u}+\delta_u)(-p_s)\,ds$$
$$= \bar{u}\bar{u}\Delta p + O(\Delta s)^3,$$

so equation (7.99) becomes

$$-\int_{s_0}^{s_1} D_f(Gup_s)\,ds = \frac{\partial}{\partial t}(G(\bar{u}\Delta p)) + \frac{\partial}{\partial x_1}(m_2\bar{u}(\bar{u}\Delta p))$$
$$+ \frac{\partial}{\partial x_2}(m_1\bar{v}(\bar{u}\Delta p)) + O(\Delta s)^3 \tag{7.100}$$
$$+ G\left[(\dot{s}p_su)_{s=s_0} - (\dot{s}p_su)_{s=s_1}\right].$$

The remaining terms in the u-equation in (7.92) can be handled in a similar manner. The resulting vertically integrated u-equation, after the deletion of error terms, is

$$\frac{\partial}{\partial t}(G(\bar{u}\Delta p)) + \frac{\partial}{\partial x_1}(m_2\bar{u}(\bar{u}\Delta p)) + \frac{\partial}{\partial x_2}(m_1\bar{v}(\bar{u}\Delta p))$$
$$+ G\left[(\dot{s}p_su)_{s=s_0} - (\dot{s}p_su)_{s=s_1}\right]$$
$$- \bar{f}G(\bar{v}\Delta p) \tag{7.101}$$
$$= - G\Delta p\,\frac{1}{m_1}\left(\frac{\partial\overline{M}}{\partial x_1} - \bar{p}\frac{\partial\bar{\alpha}}{\partial x_1}\right) + Gg\Delta\tau_u + \bar{\psi}_u$$

where $\Delta\tau_u = (\tau_u)_{s=s_1} - (\tau_u)_{s=s_0}$ denotes the vertical difference of the total shear stress defined in (7.85); $\bar{f}(\mathbf{x},t)$ is the mass-weighted vertical average of the Coriolis/geometric parameter (7.90), *i.e.*,

$$\bar{f}(\mathbf{x},t) = f(\mathbf{x}) + \mathbf{j}\cdot\left(\bar{u}\frac{1}{m_1}\frac{\partial\mathbf{i}}{\partial x_1} + \bar{v}\frac{1}{m_2}\frac{\partial\mathbf{i}}{\partial x_2}\right); \tag{7.102}$$

$\overline{M}$, $\bar{\alpha}$, and $\bar{p}$, respectively, are the mass-weighted vertical averages of M, α, and p; and $\bar{\psi}_u$ is the vertical integral of the lateral diffusion term (7.93), *i.e.*,

$$\bar{\psi}_u = \frac{\partial}{\partial x_1}\left(A_H\left(\frac{1}{m_1}\frac{\partial\bar{u}}{\partial x_1}\right)m_2\Delta p\right) + \frac{\partial}{\partial x_2}\left(A_H\left(\frac{1}{m_2}\frac{\partial\bar{u}}{\partial x_2}\right)m_1\Delta p\right). \tag{7.103}$$

The u-momentum equation (7.101) can be interpreted as follows. As noted in Section 7.8, $\Delta p(\mathbf{x}, t)$ is the weight per unit horizontal area in the fluid layer lying between the coordinate surfaces $s = s_0$ and $s = s_1$. The quantities $\bar{u}\Delta p$ and $\bar{v}\Delta p$ then represent components of momentum density (momentum per unit horizontal area), times g. In analogy to the discussion of the mass equation given in Section 7.8, integrate the u-momentum equation (7.101) on a rectangle R in the $\mathbf{x}$ domain having sides Δx_1 and Δx_2. The quantities $m_1 \Delta x_1$ and $m_2 \Delta x_2$ give the (approximate) linear dimensions of the corresponding rectangular region $\tilde{R}$ on the surface of the rotating spheroid Σ. The quantity $G\Delta x_1 \Delta x_2 (\bar{u}\Delta p) = m_1 \Delta x_1 m_2 \Delta x_2 (\bar{u}\Delta p)$ is then the u-component of momentum in the given layer in $\tilde{R}$, times g. The integrals of the second and third terms in (7.101) (involving $\partial/\partial x_1$ and $\partial/\partial x_2$) can be interpreted in terms of the lateral advection of momentum across the edges of $\tilde{R}$.

As described in Section 7.8, the quantity $-\dot{s}p_s$ is the rate of flow of mass per unit area (times g) across a coordinate surface due to material changes in s. The terms involving $\dot{s}p_s u$ then represent the rate of transport of u-momentum across coordinate surfaces due to the movement of such surfaces relative to the fluid.

The vertical integral of the v-equation in (7.92) is derived in a manner similar to that of (7.101), and the result is

$$
\frac{\partial}{\partial t}\big(G(\bar{v}\Delta p)\big) + \frac{\partial}{\partial x_1}\big(m_2\bar{u}(\bar{v}\Delta p)\big) + \frac{\partial}{\partial x_2}\big(m_1\bar{v}(\bar{v}\Delta p)\big)
$$

$$
+ G\Big[\big(\dot{s}p_s v\big)_{s=s_0} - \big(\dot{s}p_s v\big)_{s=s_1}\Big]
$$

$$
+ \bar{f}G(\bar{u}\Delta p)
$$

$$
= -G\Delta p\,\frac{1}{m_2}\Big(\frac{\partial \overline{M}}{\partial x_2} - \bar{p}\frac{\partial \bar{\alpha}}{\partial x_2}\Big) + Gg\Delta \tau_v + \bar{\psi}_v,
$$

(7.104)

where $\bar{\psi}_v$ is obtained by replacing $\bar{u}$ with $\bar{v}$ in (7.103).

Equations (7.101) and (7.104) describe the conservation of horizontal momentum in a coordinate layer. In these equations the dependent variables are the components of momentum density, and the second and third terms in each equation are written in terms of the horizontal flux of that density. This formulation facilitates the usage of (nearly) nonoscillatory advection schemes for solving these equations numerically. However, it has been commonplace in the ocean modelling community to use the components of velocity as the dependent variables, and the corresponding form of the momentum equations can be obtained by combining equations (7.101) and (7.104) with the layer mass equation (7.38). A derivation for the case of Cartesian coordinates on a tangent plane is given in Section 6.4.

Also needed for solving the momentum equations is a vertically discrete analogue of equation (7.65), $\partial M/\partial s = p\partial\alpha/\partial s$, which is essentially a statement of the hydrostatic condition in a generalized vertical coordinate. This relation can also be expressed in the form $\partial M/\partial\alpha = p$, and this latter form provides a guide for the following discretization.

Define $\widetilde{M}(\mathbf{x}, \alpha, t)$ by $M(\mathbf{x}, s, t) = \widetilde{M}(\mathbf{x}, \alpha(\mathbf{x}, s, t), t)$ for all $\mathbf{x}$ and t. Then $p\partial\alpha/\partial s = \partial M/\partial s = (\partial\widetilde{M}/\partial\alpha)(\partial\alpha/\partial s)$, so

$$\frac{\partial\widetilde{M}}{\partial\alpha}(\mathbf{x}, \alpha(\mathbf{x}, s, t), t) = p(\mathbf{x}, s, t).$$

Consider two adjacent coordinate layers bounded by the surfaces $s = s_0$, $s = s_1$, and $s = s_2$, with $s_0 < s_1 < s_2$. The interface between the two layers is then defined by $s = s_1$. Denote the values of M, α, and p on that interface by $M_{\text{int}}(\mathbf{x}, t)$, $\alpha_{\text{int}}(\mathbf{x}, t)$, and $p_{\text{int}}(\mathbf{x}, t)$, respectively. Then

$$M(\mathbf{x}, s, t) = \widetilde{M}(\mathbf{x}, \alpha(\mathbf{x}, s, t), t)$$
$$= M_{\text{int}}(\mathbf{x}, t) + p_{\text{int}}(\mathbf{x}, t)\big(\alpha(\mathbf{x}, s, t) - \alpha_{\text{int}}(\mathbf{x}, t)\big) + O\big(\alpha - \alpha_{\text{int}}\big)^2$$

for all $(\mathbf{x}, s)$ in each of the two layers. Denote the mass-weighted vertical averages of M and α on the upper layer $s_1 < s < s_2$ by $\overline{M}_{\text{upper}}(\mathbf{x}, t)$ and $\bar{\alpha}_{\text{upper}}(\mathbf{x}, t)$, respectively; and denote their averages on the lower layer $s_0 < s < s_1$ by $\overline{M}_{\text{lower}}(\mathbf{x}, t)$ and $\bar{\alpha}_{\text{lower}}(\mathbf{x}, t)$, respectively. Then

$$\overline{M}_{\text{upper}}(\mathbf{x}, t) = M_{\text{int}}(\mathbf{x}, t) + p_{\text{int}}(\mathbf{x}, t)\big(\bar{\alpha}_{\text{upper}}(\mathbf{x}, t) - \alpha_{\text{int}}(\mathbf{x}, t)\big) + O(\Delta s)^3,$$
$$\overline{M}_{\text{lower}}(\mathbf{x}, t) = M_{\text{int}}(\mathbf{x}, t) + p_{\text{int}}(\mathbf{x}, t)\big(\bar{\alpha}_{\text{lower}}(\mathbf{x}, t) - \alpha_{\text{int}}(\mathbf{x}, t)\big) + O(\Delta s)^3,$$

and thus $\overline{M}_{\text{upper}} - \overline{M}_{\text{lower}} = p_{\text{int}}(\bar{\alpha}_{\text{upper}} - \bar{\alpha}_{\text{lower}}) + O(\Delta s)^3$. A second-order approximation to the hydrostatic condition (7.65) is then

$$\frac{\Delta\overline{M}}{\Delta\bar{\alpha}} = \frac{\overline{M}_{\text{upper}} - \overline{M}_{\text{lower}}}{\bar{\alpha}_{\text{upper}} - \bar{\alpha}_{\text{lower}}} = p_{\text{int}}. \qquad (7.105)$$

The relation (7.105) provides a means for communicating pressure effects between layers.

7.17. Transport of tracers

Next consider the transport by the fluid of a tracer such as heat, salt, or a chemical or biological component. Let $q(\mathbf{x}, s, t)$ denote the quantity of tracer per unit mass of the fluid. Then the quantity of tracer per unit volume is $\rho(\mathbf{x}, s, t)q(\mathbf{x}, s, t)$, where ρ is the density (mass per unit volume) of the fluid itself.

In the discussion of conservation of mass of the fluid given in Section 7.7, $A(t)$ denotes the region of parameter space occupied by a volume of fluid, and $B(t)$ denotes the corresponding region in rectangular coordinates in the inertial reference frame. The total mass of the fluid occupying those regions remains constant in time, since the regions follow the fluid. In the case of a tracer, it is assumed here that the amount of tracer in the volume of fluid can vary due to diffusion across the boundary. It is also assumed that the diffusion is due to the gradient of q and occurs predominantly along surfaces of constant s, as in the discussion of diffusion of momentum in Section 7.12.

In analogy to the discussions in Sections 7.5–7.7, the total quantity of tracer in the region $A(t)$ in parameter space at time t is

$$\int_{A(t)} \rho(\mathbf{x}, s, t) q(\mathbf{x}, s, t) \, J(\mathbf{x}, s, t) \, d\mathbf{x} \, ds,$$

where $J = m_1 m_2 m_s$ is the Jacobian defined in (7.17). The time derivative of this quantity is

$$\int_{A(t)} D_f(J\rho q) \, d\mathbf{x} \, ds,$$

where D_f is the flux derivative given in (7.27) and (7.46). The diffusion of the tracer can be modelled by the same form (7.71) as used for the diffusion of momentum, with q replacing U in that formula. Calculations analogous to those in Sections 7.12 and 7.14 imply

$$D_f(Gqp_s) = \psi_q + \frac{\partial}{\partial s}(GgF_D). \tag{7.106}$$

Here,

$$F_D = -\rho A_D \left(\frac{1}{m_s} \frac{\partial q}{\partial s} \right)$$

denotes the rate of diapycnal diffusion of q per unit cross-sectional area, and

$$\psi_q = \frac{\partial}{\partial x_1} \left(A_H \left(\frac{1}{m_1} \frac{\partial q}{\partial x_1} \right) m_2 p_s \right) + \frac{\partial}{\partial x_2} \left(A_H \left(\frac{1}{m_2} \frac{\partial q}{\partial x_2} \right) m_1 p_s \right).$$

Integration over a layer between two coordinate surfaces $s = s_0$ and $s = s_1$, analogous to the calculations in Section 7.16, yields

$$\frac{\partial}{\partial t}(G(\bar{q}\Delta p)) + \frac{\partial}{\partial x_1}(m_2 \bar{u}(\bar{q}\Delta p)) + \frac{\partial}{\partial x_2}(m_1 \bar{v}(\bar{q}\Delta p))$$
$$+ G\left[(\dot{s}p_s q)_{s=s_0} - (\dot{s}p_s q)_{s=s_1} \right] \tag{7.107}$$
$$= \bar{\psi}_q + Gg\left[(F_D)_{s=s_0} - (F_D)_{s=s_1} \right],$$

where $\bar{q}(\mathbf{x}, t)$ is the mass-weighted vertical average of q over the layer, and

$$\bar{\psi}_q = \frac{\partial}{\partial x_1} \left(A_H \left(\frac{1}{m_1} \frac{\partial \bar{q}}{\partial x_1} \right) m_2 \Delta p \right) + \frac{\partial}{\partial x_2} \left(A_H \left(\frac{1}{m_2} \frac{\partial \bar{q}}{\partial x_2} \right) m_1 \Delta p \right). \tag{7.108}$$

The term $\bar{q}\Delta p$ is the quantity of tracer per unit horizontal area, times g. The terms involving $\dot{s}p_s q$ represent the rate of transport of tracer across coordinate surfaces due to material changes in s, $i.e.$, due to the movement of coordinate surfaces relative to the fluid. The velocity components $\bar{u}$ and $\bar{v}$ in the momentum equations (7.101) and (7.104) are components of momentum per unit mass, so these quantities can be regarded as examples of the quantity $\bar{q}$, except for the forcing due to the Coriolis, pressure, and applied stress terms. If those terms are deleted from equations (7.101) and (7.104), then the resulting equations have the same form as (7.107).

7.18. Planar and spherical coordinates

The layer mass equation (7.38), the layer momentum equations (7.101) and (7.104), and the layer tracer equation (7.107) are derived under the assumption that the horizontal coordinates are general orthogonal curvilinear coordinates $\mathbf{x} = (x_1, x_2)$. The present subsection discusses the forms of these equations in the special cases of planar and spherical coordinates.

For the case of planar coordinates, $i.e.$, Cartesian coordinates on a tangent plane, x_1 and x_2 can be taken as literal measures of distance. The metric coefficients (7.2) are then given by $m_1 = 1$ and $m_2 = 1$, and in addition $G = m_1 m_2 = 1$. The factors m_1, m_2, and G can then be deleted from the partial differential equations (7.38), (7.101), (7.104), and (7.107). In the Coriolis/geometric parameter (7.102), the geometric terms (involving $\partial \mathbf{i}/\partial x_1$ and $\partial \mathbf{i}/\partial x_2$) are equal to zero, and the values of the Coriolis parameter can be taken from the formula $f = 2\Omega \sin\theta$ given in (7.94).

Spherical coordinates are defined in equation (7.6). In that case the horizontal coordinates are the longitude $x_1 = \lambda$ and the latitude $x_2 = \theta$. The metric coefficients (7.7) are $m_1 = m_\lambda = r\cos\theta \approx a\cos\theta$ and $m_2 = m_\theta = r \approx a$, where a is the radius of the sphere, and thus $G = m_\lambda m_\theta = r^2 \cos\theta \approx a^2 \cos\theta$. It follows from (7.6) and (7.3) that

$$\mathbf{i}(\lambda, \theta, t) = \frac{1}{m_\lambda} \frac{\partial \mathbf{r}}{\partial \lambda} = \big(-\sin(\lambda + \Omega t), \ \cos(\lambda + \Omega t), \ 0\big), \qquad (7.109)$$

$$\mathbf{j}(\lambda, \theta, t) = \frac{1}{m_\theta} \frac{\partial \mathbf{r}}{\partial \theta} = \big(-\sin\theta\cos(\lambda + \Omega t), \ -\sin\theta\sin(\lambda + \Omega t), \ \cos\theta\big).$$

A calculation reproduces the formula (7.94), $f = 2\,\mathbf{j}\cdot\frac{\partial \mathbf{i}}{\partial t} = 2\Omega\sin\theta$. The geometric parameter in (7.102) is

$$\mathbf{j}\cdot\left(\bar{u}\frac{1}{m_\lambda}\frac{\partial \mathbf{i}}{\partial \lambda} + \bar{v}\frac{1}{m_\theta}\frac{\partial \mathbf{i}}{\partial \theta}\right) = \mathbf{j}\cdot\left[\frac{\bar{u}}{r\cos\theta}\big(-\cos(\lambda + \Omega t), \ -\sin(\lambda + \Omega t), \ 0\big)\right]$$

$$= \frac{\bar{u}\tan\theta}{r},$$

so the combined Coriolis/geometric parameter (7.102) is

$$\bar{f} = 2\Omega \sin\theta + \frac{\bar{u}\tan\theta}{r}. \tag{7.110}$$

As noted in Section 7.15, $2\Omega \approx 1.46 \times 10^{-4}$ sec^{-1} in the case of the rotating Earth. For a fluid velocity $\bar{u} = 1$ m/sec, the factor $\bar{u}/r$ in the geometric parameter in (7.110) is approximately 1.6×10^{-7} sec^{-1}. In the low and middle latitudes, the Coriolis parameter thus greatly exceeds the geometric parameter for spherical coordinates. However, as the latitude θ approaches 90°, $\tan\theta$ increases without bound, owing to the singularity in the spherical coordinate system at that point. As pointed out in Section 6.1, this singularity corresponds to a convergence of grid lines at the poles, and for a global ocean model it is better to use a coordinate system that does not produce a convergence of grid lines within the fluid domain.

8. Summary

The goal of this paper is to provide an introduction to the mathematical and computational modelling of ocean circulation. The paper includes a detailed derivation of partial differential equations that describe the conservation of mass, momentum, and tracers for a hydrostatic and stratified fluid on a rotating spheroid. Also included are a description of some physical properties of oceanic flows, a discussion of some of the issues that are encountered when the governing equations are solved numerically, and a summary of some of the numerical methods that are used in this field.

In the derivation of the governing equations, it is assumed that the vertical coordinate is a generalized coordinate that includes the cases of level, isopycnic, sigma, and hybrid coordinates. Each of these coordinates presents advantages and disadvantages that are discussed here. Among the considerations in the choice and implementation of a vertical coordinate is the need to represent accurately the vertical exchanges between water masses that are often subtle but nevertheless important for long-term integrations. It is also assumed that the horizontal coordinates are general orthogonal curvilinear coordinates instead of spherical coordinates, in anticipation of using a grid that is suitable for a global model.

Several numerical issues are discussed here. These include time-stepping methods and the numerical treatment of multiple time scales; spatial grids and spatial discretization; the solution of the nonlinear equations for the conservation of momentum; and the numerical simulation of advection, especially in the context of maintaining nonnegative solutions and transporting the multiple tracers that are typically included in numerical models of ocean circulation.

Acknowledgements

Over a period of several years I have benefited greatly from conversations with numerous ocean modellers, especially members of the physical oceanography group at Oregon State University, members of the ocean and climate modelling group at Los Alamos National Laboratory, and attendees at the annual Layered Ocean Model Users' Workshop at the University of Miami. Special thanks are extended to Rainer Bleck, Andrew Bennett, Mats Bentsen, Roland de Szoeke, John Dukowicz, Roger Samelson, and Scott Springer. This work was supported by National Science Foundation grant DMS-0511782.

REFERENCES

A. J. Adcroft, C. N. Hill and J. C. Marshall (1999), A new treatment of the Coriolis terms in C-grid models at both high and low resolutions, *Monthly Weather Review* **127**, 1928–1936.

A. Adcroft, J.-M. Campin, C. Hill and J. Marshall (2004), Implementation of an atmosphere-ocean general circulation model on the expanded spherical cube, *Monthly Weather Review* **132**, 2845–2863.

A. Arakawa and V. R. Lamb (1977), Computational design of the basic dynamical processes of the UCLA general circulation model, *Methods in Computational Physics* **17**, 173–265.

R. Asselin (1972), Frequency filter for time integrations, *Monthly Weather Review* **100**, 487–490.

A. F. Bennett (2002), *Inverse Modeling of the Ocean and Atmosphere*, Cambridge University Press, Cambridge.

R. Bleck (1978), Finite difference equations in generalized vertical coordinates, Part I: Total energy conservation, *Contributions to Atmospheric Physics* **51**, 360–372.

R. Bleck (2002), An oceanic general circulation model framed in hybrid isopycnic-cartesian coordinates, *Ocean Modelling* **4**, 55–88.

R. Bleck (2005), On the use of hybrid vertical coordinates in ocean circulation modeling, in *An Integrated View of Oceanography: Ocean Weather Forecasting in the 21st Century* (E. Chassignet and J. Verron, eds), Kluwer, Dordrecht, pp. 109–126.

R. Bleck and L. T. Smith (1990), A wind-driven isopycnic coordinate model of the north and equatorial Atlantic Ocean 1: Model development and supporting experiments, *J. Geophysical Research* **95C**, 3273–3285.

R. Bleck, C. Rooth, D. Hu and L. T. Smith (1992), Salinity-driven thermocline transients in a wind- and thermohaline-forced isopycnic coordinate model of the North Atlantic, *J. Physical Oceanography* **22**, 1486–1505.

A. F. Blumberg and G. L. Mellor (1987), A description of a three-dimensional coastal ocean circulation model, in *Three-dimensional Coastal Ocean Models* (N. Heaps, ed.), American Geophysical Union, Washington, DC.

K. Bryan (1969), A numerical method for the study of the circulation of the world ocean, *J. Comput. Phys.* **4**, 347–376.

A. J. Chorin and J. E. Marsden (1990), *A Mathematical Introduction to Fluid Mechanics*, 2nd edn, Springer, New York.

R. E. Davis (1994), Diapycnal mixing in the ocean: equations for large-scale budgets, *J. Physical Oceanography* **24**, 777–800.

R. A. de Szoeke (2000), Equations of motion using thermodynamic coordinates, *J. Physical Oceanography* **30**, 2814–2829.

R. A. de Szoeke and S. R. Springer (2003), A diapycnal diffusion algorithm for isopycnal ocean circulation models with special application to mixed layers, *Ocean Modelling* **5**, 297–323.

R. A. de Szoeke, S. R. Springer and D. M. Oxilia (2000), Orthobaric density: A thermodynamic variable for ocean circulation studies, *J. Physical Oceanography* **30**, 2830–2852.

J. K. Dukowicz (1995), Mesh effects for Rossby waves, *J. Comput. Phys.* **119**, 188–194.

J. K. Dukowicz (2004), HYPOP Summary (Hybrid vertical coordinate version of POP) Current C-Grid Version. Technical Report LA-UR-04-8586, Los Alamos National Laboratory.

J. K. Dukowicz (2006), Structure of the barotropic mode in layered ocean models, *Ocean Modelling* **11**, 49–68.

J. K. Dukowicz and J. R. Baumgardner (2000), Incremental remapping as a transport/advection algorithm, *J. Comput. Phys.* **160**, 318–335.

J. K. Dukowicz and R. D. Smith (1994), Implicit free-surface method for the Bryan–Cox–Semtner ocean model, *J. Geophysical Research* **99**, 7991–8014.

D. R. Durran (1999), *Numerical Methods for Wave Equations in Geophysical Fluid Dynamics*, Springer, New York.

R. C. Easter (1993), Two modified versions of Bott's positive-definite numerical advection scheme, *Monthly Weather Review* **121**, 297–304.

C. Eden and J. Willebrand (1999), Neutral density revisited, *Deep-Sea Research Part II* **46**, 33–54.

P. R. Gent and J. C. McWilliams (1990), Isopycnal mixing in ocean circulation models, *J. Physical Oceanography* **20**, 150–155.

P. R. Gent, J. Willebrand, T. J. McDougall and J. C. McWilliams (1995), Parameterizing eddy-induced tracer transports in ocean circulation models, *J. Physical Oceanography* **25**, 463–474.

A. E. Gill (1982), *Atmosphere-Ocean Dynamics*, Academic Press, San Diego.

S. M. Griffies (2004), *Fundamentals of Ocean Climate Models*, Princeton University Press, Princeton, NJ.

S. M. Griffies, C. Bönig, F. O. Bryan, E. P. Chassignet, R. Gerdes, H. Hasumi, A. Hirst, A.-M. Treguier and D. Webb (2000*a*), Developments in ocean climate modeling, *Ocean Modelling* **2**, 123–192.

S. M. Griffies, R. C. Pacanowski and R. W. Hallberg (2000*b*). Spurious diapycnal mixing associated with advection in a *z*-coordinate ocean model, *Monthly Weather Review* **128**, 538–564.

S. M. Griffies, M. J. Harrison, R. C. Pacanowski and A. Rosati (2004), A technical guide to MOM4. NOAA/Geophysical Fluid Dynamics Laboratory, Princeton, NJ, available on-line.

S. M. Griffies, A. Gnanadesikan, K. W. Dixon, J. P. Dunne, R. Gerdes, M. J. Harrison, A. Rosati, J. L. Russell, B. L. Samuels, M. J. Spelman, M. Winton, and R. Zhang (2005), Formulation of an ocean model for global climate simulations, *Ocean Science* **1**, 45–79.

D. B. Haidvogel and A. Beckmann (1999), *Numerical Ocean Circulation Modeling*, Imperial College Press, London.

R. Hallberg (1997), Stable split time stepping schemes for large-scale ocean modeling, *J. Comput. Phys.* **135**, 54–65.

R. Hallberg (2000), Time integration of diapycnal diffusion and Richardson number-dependent mixing in isopycnal coordinate ocean models, *Monthly Weather Review* **128**, 1402–1419.

R. Hallberg (2005), A thermobaric instability of Lagrangian vertical coordinate ocean models, *Ocean Modelling* **8**, 279–300.

M. W. Hecht (2006), Forward-in-time upwind-weighted methods in ocean modelling, *Int. J. Numer. Meth. Fluids* **50**, 1159–1173.

R. L. Higdon (2002), A two-level time-stepping method for layered ocean circulation models, *J. Comput. Phys.* **177**, 59–94.

R. L. Higdon (2005), A two-level time-stepping method for layered ocean circulation models: Further development and testing, *J. Comput. Phys.* **206**, 463–504.

R. L. Higdon and A. F. Bennett (1996), Stability analysis of operator splitting for large-scale ocean modeling, *J. Comput. Phys.* **123**, 311–329.

R. L. Higdon and R. A. de Szoeke (1997), Barotropic–baroclinic time splitting for ocean circulation modeling, *J. Comput. Phys.* **135**, 30–53.

C. W. Hirt, A. A. Amsden and J. L. Cook (1974), An arbitrary Lagrangian–Eulerian computing method for all flow speeds, *J. Comput. Phys.* **14**, 227–253.

J. R. Holton (1992), *An Introduction to Dynamic Meteorology*, 3rd edn, Academic Press, San Diego.

Y.-J. Hsu and A. Arakawa (1990), Numerical modeling of the atmosphere with an isentropic vertical coordinate, *Monthly Weather Review* **118**, 1933–1959.

Intergovernmental Panel on Climate Change (2001), *IPCC Third Assessment Report: Climate Change 2001*, on-line, http://www.ipcc.ch/

M. Iskandarani, D. B. Haidvogel and J. C. Levin (2003), A three-dimensional spectral element model for the solution of the hydrostatic primitive equations, *J. Comput. Phys.* **186**, 397–425.

M. Iskandarani, J. C. Levin, B.-J. Choi and D. B. Haidvogel (2005), Comparison of advection schemes for high-order h-p finite element and finite volume methods, *Ocean Modelling* **10**, 233–252.

P. D. Killworth, D. Stainforth, D. J. Webb and S. M. Paterson (1991), The development of a free-surface Bryan–Cox–Semtner ocean model, *J. Physical Oceanography* **21**, 1333–1348.

J. R. Ledwell, A. J. Watson and C. S. Law (1993), Evidence for slow mixing across the pycnocline from an open-ocean tracer-release experiment, *Nature* **364**, 701–703.

B. P. Leonard (1979), A stable and accurate convective modeling procedure based on quadratic upstream interpolation, *Comput. Methods Appl. Mech. Engrg.* **19**, 59–98.

R. J. LeVeque (2003), *Finite Volume Methods for Hyperbolic Problems*, Cambridge University Press, Cambridge.

S. Levitus (1982), *Climatological Atlas of the World Ocean*, US Department of Commerce, National Oceanic and Atmospheric Administration, Rockville, MD.

S.-J. Lin and R. R. Rood (1996), Multidimensional flux-form semi-Lagrangian transport schemes, *Monthly Weather Review* **124**, 2046–2070.

W. H. Lipscomb and E. C. Hunke (2004), Modeling sea ice transport using incremental remapping, *Monthly Weather Review* **132**, 1341–1354.

W. H. Lipscomb and T. D. Ringler (2005), An incremental remapping transport scheme on a spherical geodesic grid, *Monthly Weather Review* **133**, 2335–2350.

T. J. McDougall (1987), Neutral surfaces, *J. Physical Oceanography* **17**, 1950–1967.

T. J. McDougall and W. K. Dewar (1998), Vertical mixing and cabbeling in layered models, *J. Physical Oceanography* **28**, 1458–1480.

J. Marshall and F. Schott (1999), Open-ocean convection: Observations, theory, and models, *Reviews of Geophysics* **37**, 1–64.

J. Marshall, C. Hill, L. Perelman and A. Adcroft (1997), Hydrostatic, quasi-hydrostatic, and nonhydrostatic ocean modeling, *J. Geophysical Research* **102**(C3), 5733–5752.

R. N. Miller (2006), *Numerical Modeling of Ocean Circulation*, Cambridge University Press, to appear.

R. B. Montgomery (1937), A suggested method for representing gradient flow in isentropic surfaces, *Bull. Amer. Meteorological Soc.* **18**, 210–212.

R. J. Murray (1996) Explicit generation of orthogonal grids for ocean models, *J. Comput. Phys.* **126**, 251–273.

D. Nechaev and M. Yaremchuk (2004), On the approximation of the Coriolis terms in C-grid models, *Monthly Weather Review* **132**, 2283–2289.

C. C. Pain, M. D. Piggott, A. J. H. Goddard, F. Fang, G. J. Gorman, D. P. Marshall, M. D. Eaton, P. W. Power and C. R. E. de Oliveira (2005), Three-dimensional unstructured mesh ocean modelling, *Ocean Modelling* **10**, 5–33.

J. Pedlosky (1987), *Geophysical Fluid Dynamics*, 2nd edn, Springer, New York.

J. Pedlosky (1996), *Ocean Circulation Theory*, Springer, Berlin.

J. Pedlosky (2003), *Waves in the Ocean and Atmosphere*, Springer, Berlin.

M. Rancic, R. J. Purser and F. Mesinger (1996), A global shallow-water model using an expanded spherical cube: Gnomonic versus conformal coordinates, *Quarterly J. Royal Meteorological Soc.* **122**, 959–982.

T. D. Ringler and D. A. Randall (2002), A potential enstrophy and energy conserving numerical scheme for solution of the shallow-water equations on a geodesic grid, *Monthly Weather Review* **130**, 1397–1410.

J. A. Rossmanith (2006), A wave propagation method for hyperbolic systems on the sphere, *J. Comput. Phys.* **213**, 629–658.

R. Sadourny (1975), The dynamics of finite-difference models of the shallow-water equations, *J. Atmospheric Sci.* **32**, 680–689.

C. Schär and P. K. Smolarkiewicz (1996), A synchronous and iterative flux-correction formalism for coupled transport equations, *J. Comput. Phys.* **128**, 101–120.

P. Schopf and A. Loughe (1995), A reduced gravity isopycnal ocean model, *Monthly Weather Review* **123**, 2839–2863.

A. J. Semtner (1997), Introduction to 'A numerical method for the study of the circulation of the world ocean', *J. Comput. Phys.* **135**, 149–153.

A. J. Semtner and R. M. Chervin (1992), Ocean general circulation from a global eddy-resolving model, *J. Geophysical Research* **97**, 5493–5550.

A. F. Shchepetkin and J. C. McWilliams (2005), The regional oceanic modeling system (ROMS): A split-explicit, free-surface, topography-following-coordinate oceanic model, *Ocean Modelling* **9**, 347–404.

R. D. Smith, S. Kortas and B. Meltz (1995), Curvilinear coordinates for global ocean models. Technical Report LA-UR-95-1146, Los Alamos National Laboratory.

P. K. Smolarkiewicz and L. G. Margolin (1998), MPDATA: A finite-difference solver for geophysical flows, *J. Comput. Phys.* **140**, 459–480.

S. Sun, R. Bleck, C. Rooth, J. Dukowicz, E. Chassignet and P. Killworth (1999), Inclusion of thermobaricity in isopycnic-coordinate ocean models, *J. Physical Oceanography* **29**, 2719–2729.

F. J. Winninghoff (1968), On the adjustment toward a geostrophic balance in a simple primitive equation model with application to the problems of initialization and objective analysis, PhD thesis, Department of Meteorology, University of California, Los Angeles.

S. T. Zalesak (1979), Fully multidimensional flux-corrected transport algorithms for fluids, *J. Comput. Phys.* **31**, 335–362.

Acta Numerica (2006), pp. 471–542
doi: 10.1017/S096249290626001X

© Cambridge University Press, 2006

The Lanczos and conjugate gradient algorithms in finite precision arithmetic

Gérard Meurant
CEA/DIF,
BP 12,
91680, Bruyères le Chatel, France
E-mail: `gerard.meurant@cea.fr`

Zdeněk Strakoš*
Institute of Computer Science,
Academy of Sciences of the Czech Republic,
Pod Vodárenskou věži 2,
182 07 Praha 8, Czech Republic
E-mail: `strakos@cs.cas.cz`

Dedicated to Chris Paige for his fundamental contributions
to the rounding error analysis of the Lanczos algorithm

The Lanczos and conjugate gradient algorithms were introduced more than five decades ago as tools for numerical computation of dominant eigenvalues of symmetric matrices and for solving linear algebraic systems with symmetric positive definite matrices, respectively. Because of their fundamental relationship with the theory of orthogonal polynomials and Gauss quadrature of the Riemann–Stieltjes integral, the Lanczos and conjugate gradient algorithms represent very interesting general mathematical objects, with highly nonlinear properties which can be conveniently translated from algebraic language into the language of mathematical analysis, and *vice versa*. The algorithms are also very interesting numerically, since their numerical behaviour can be explained by an elegant mathematical theory, and the interplay between analysis and algebra is useful there too.

Motivated by this view, the present contribution wishes to pay a tribute to those who have made an understanding of the Lanczos and conjugate gradient algorithms possible through their pioneering work, and to review recent solutions of several open problems that have also contributed to knowledge of the subject.

* Supported by the National Program of Research 'Information Society' under project 1ET400300415 and by the Institutional Research Plan AVOZ10300504.

CONTENTS

1. Introduction

The Lanczos algorithm is one of the most frequently used tools for computing a few dominant eigenvalues (and eventually eigenvectors) of a large sparse symmetric n by n matrix A. More specifically, if for instance the extreme eigenvalues of A are well separated, the Lanczos algorithm obtains good approximations to these eigenvalues in only a few iterations. Moreover, the matrix A need not be explicitly available. The Lanczos algorithm only needs a procedure performing the matrix-vector product Av for a given vector v. Hence, it can even be used in some applications for which the matrix cannot be stored as long as one is able to produce the result of the operation matrix times a given vector. Another interesting property is that when one just needs the eigenvalues, the Lanczos algorithm only requires a very small storage of a few vectors (besides storing the matrix where applicable), since a new basis vector is computed using only the two previous ones.

The Lanczos algorithm constructs a basis of Krylov subspaces which are defined for a square matrix A of order n and a vector v by

$$\mathcal{K}_k(v, A) = \mathrm{span}\{v, Av, \ldots, A^{k-1}v\}, \quad k = 1, 2, \ldots.$$

Since the natural basis $v, Av, \ldots, A^{k-1}v$ is badly conditioned, the algorithm constructs an orthonormal basis of $\mathcal{K}_k(v, A)$. The vectors in the natural basis can even become numerically dependent (within the accuracy of the floating point calculations) for a small value of k. In fact, computing successively $A^k v$ for a given vector v is, with a proper normalization, the basis of the power method. Unlike the power method, which focuses at the kth step only on the local information present in $A^{k-1}v$, and aims to converge to the eigenvector corresponding to the eigenvalue of largest modulus, the Lanczos algorithm exploits simultaneously all vector information accumulated in previous steps. Building an orthonormal basis of $\mathcal{K}_k(v, A)$ can therefore be seen as an effective numerical tool for extracting information from the sequence $v, Av, \ldots, A^{k-1}v$ while preventing any possible loss which could be caused by effects of existing dominance.

The orthonormal basis vectors v^j, $j = 1, \ldots, k$ are constructed recursively one at a time and can be considered columns of a matrix $V_k = (v^1, \ldots, v^k)$. The method also constructs at iteration k an unreduced symmetric tridiagonal k by k matrix T_k (which is obtained from T_{k-1} by adding one row and one column) having positive subdiagonal entries, whose eigenvalues are approximations to the eigenvalues of A: see, for instance, Lanczos (1950), Wilkinson (1965) and Parlett (1980). Moreover, in exact arithmetic $AV_m = V_m T_m$ for some $m \leq n$, n being the dimension of the problem. It means that the columns of V_m span an invariant subspace of the operator represented by A, and the eigenvalues of T_m are also eigenvalues of A.

All these properties are quite nice. However, it has been known since the introduction of the method by Cornelius Lanczos (1950) that, when used in finite precision arithmetic, this algorithm does not fulfil its theoretical properties. In particular, the computed basis vectors lose their orthogonality as the iteration number k increases. Moreover, as a consequence of the loss of orthogonality, in finite precision computations multiple approximations of the original eigenvalues appear within the set of computed approximate eigenvalues if we do a sufficiently large number of iterations. This phenomenon leads to a delay in the computation of some other eigenvalues. Sometimes it is also difficult to determine whether some computed approximations are additional copies caused by rounding error effects and the loss of orthogonality, or genuine close eigenvalues.

The finite precision behaviour of the Lanczos algorithm was analysed in great depth by Chris Paige in his pioneering PhD thesis, Paige (1971); see also Paige (1972, 1976, 1980). With no exaggeration, Paige's work was revolutionary. He showed that the effects of rounding errors in the Lanczos algorithm can be described by a rigorous and elegant mathematical theory. In the spirit of Wilkinson, the theory built by Paige reveals the mechanics of the finite precision Lanczos algorithm behaviour. It starts with bounds on the elementary round-off errors at each iteration, and ends up with elegant mathematical theorems which link convergence of the computed eigenvalue approximations to the loss of orthogonality. Following Paige, the theory was further developed and applied by Parlett and Scott (1979), Scott (1979), Parlett (1980) and Simon (1982, 1984a, 1984b). A forward error analysis was attempted by Grcar (1981).

Another fundamental step forward, similar in significance to that of Paige, was made by Anne Greenbaum (1989). On the foundations laid by Paige she developed a backward-like analysis of the Lanczos algorithm (and also of the closely related conjugate gradient algorithm). Her ideas, combined with thoughts of several other authors, stimulated further developments: see, *e.g.*, Druskin and Knizhnerman (1991), Strakoš (1991), Greenbaum and Strakoš (1992), Strakoš and Greenbaum (1992), Knizhnerman (1995a) and Druskin, Greenbaum and Knizhnerman (1998). Recently, new analysis of the open problems formulated in the literature has led to work by Zemke (2003), Wülling (2005, 2006) and Meurant (2006).

The Lanczos algorithm has been implemented and applied in two ways. In the first way it is applied without any additional measures designed to limit round-off. The number of iterations is not limited by strict rules and the 'good' eigenvalue approximations are identified by some convergence tests. In particular, this was advocated by Cullum and Willoughby (1985). The second way limits the unwanted effects of rounding errors by some form of reorthogonalization, of varied sophistication. Proposals in this direction were made by Parlett and Scott (1979), Grcar (1981), Simon (1982) and Parlett (1992). Here the theory developed by Paige almost immediately led to successful software implementations. The way the Lanczos algorithm is used in a particular application depends on a particular goal.

After a period of intensive discussions, particularly concentrated at the Institute of Numerical Analysis at UCLA (see Hestenes and Todd (1991) and Golub and O'Leary (1989)), the conjugate gradient (CG) algorithm, independently introduced by Magnus Hestenes and Eduard Stiefel, was thoroughly described in their seminal paper, Hestenes and Stiefel (1952). Intended for solving symmetric positive definite linear systems, it is closely linked to the Lanczos algorithm. Lanczos used his algorithm to solve linear systems in Lanczos (1952) but it was already clear in Lanczos (1950) that it can be used for that purpose. In fact, even though it was not introduced in this way, one can obtain the Hestenes–Stiefel CG from the Lanczos algorithm by doing an LU factorization (with L lower triangular and U upper triangular) of the positive definite matrix T_k given by the Lanczos coefficients (by introducing some intermediate variables). In exact arithmetic the CG residual vectors are proportional to the Lanczos vectors. In finite precision, the residual vectors lose their orthogonality just as the Lanczos vectors do.

The Lanczos algorithm, respectively CG, builds up (in exact arithmetic) orthogonal bases of Krylov subspaces $\mathcal{K}_k(v, A)$, $k = 1, 2, \dots$. and the basis vectors can be expressed in terms of polynomials in the matrix A applied to the initial vector v. Using the spectral decomposition of the symmetric (respectively the symmetric positive definite) matrix A, it is easy to see that the corresponding polynomials are orthogonal with respect to a Riemann–Stieltjes integral. Its piecewise constant distribution function is defined by the points of increase equal to the eigenvalues of A and by the sizes of the discontinuities equal to the squared components of v in the corresponding invariant subspaces. In this way, the Lanczos algorithm and CG are intimately related to orthogonal polynomials: see Hestenes and Stiefel (1952) and Fischer (1996). This fact has been emphasized for decades in the work of Gene Golub, who substantially contributed to the whole field by his deep understanding of the interconnections between different mathematical areas and by sharing his ideas with many collaborators: see, *e.g.*, Gautschi (2002). The Lanczos algorithm can be viewed as a matrix formulation of the discretized Stieltjes procedure (see, *e.g.*, Gautschi (1982)), and its roots can therefore be linked to the works of Stieltjes (1884), Christoffel (1877) and Darboux (1878). Such interconnections are fundamental to the understanding of the Lanczos algorithm and CG behaviour in both exact and finite precision arithmetic. In particular, in exact arithmetic the A-norm of the CG error can be written using the Gauss quadrature formula, and this clearly shows that the convergence rate depends, in a rather complicated way, on how well the eigenvalues of A are approximated by the eigenvalues of T_k. This also indicates possible differences in the effect of eigenvalues from different parts of the spectrum of A on the convergence behaviour. In finite precision arithmetic the Gauss quadrature formula is also verified, up to small terms involving the machine precision. However, the appearance of multiple approximations of the original eigenvalues leads to a delay in CG convergence.

The concept of delay is essential to analysis of the CG finite precision behaviour. In short, delay of convergence in a CG finite precision computation is determined by the rank-deficiencies of the computed Krylov subspaces. This understanding emerged from the work of Greenbaum (Greenbaum 1989, Greenbaum and Strakoš 1992) and Notay (1993), and it was strongly advocated by Paige and Strakoš (1999). Analysis and discussion of the Gauss quadrature relationship in finite precision arithmetic can be found in Golub and Strakoš (1994), Strakoš and Tichý (2002) and Meurant (2006).

A finite precision computation does not give the approximate solution with an arbitrarily small error. The error is not reduced below some level, called the maximal attainable accuracy. This is not so important for the Lanczos algorithm, as Paige (1971) shows, but it can become important in solving highly ill-conditioned linear systems and, in particular, in some inner iterations within nonlinear optimization algorithms. Maximal attainable accuracy of CG has been studied for a long time. The early results (see, *e.g.*, Wozniakowski (1978, 1980) and Bollen (1984), with a thorough survey given in Chapter 17 of Higham (2002)) were, however, not applicable to practical implementations. These were analysed more recently by Greenbaum (1997*a*, 1994), Sleijpen, van der Vorst and Fokkema (1994), Sleijpen, van der Vorst and Modersitzki (2001), Björck, Elfving and Strakoš (1998) and Gutknecht and Strakoš (2000). It turns out that a deterioration of the maximal attainable accuracy can be caused at a very early stage of the computation and that CG is unable to correct such a situation in later iterations.

The authors have previously published some surveys of the Lanczos and CG algorithms in exact and finite precision arithmetic as parts of more widely based publications: see Meurant (1999*b*), Strakoš (1998) and Strakoš and Liesen (2005). Following these works, this paper first recalls in Sections 2 and 3 the basic facts on the Lanczos and CG algorithms in exact arithmetic. Then we turn to our main goal – to review the main results on the behaviour of the Lanczos and CG algorithms in finite precision arithmetic, and to present some recent developments related in particular to the appearance of multiple computed approximations of simple original eigenvalues. Section 4 is devoted to the Lanczos algorithm and Section 5 to CG.

For simplicity of exposition we adopt in this paper several restrictions. We will consider real symmetric resp. symmetric positive definite problems. Restriction to real problems is not substantial; we use it for convenience of notation. We will not consider nonsymmetric problems, since this extension would necessarily bring into consideration fundamental issues not present in the symmetric case, some of them still not fully understood. This would, in our opinion, distract from the focus of this paper. In particular, for CG we will assume that the symmetric positive definite matrix A is not close to being singular. Solving near-singular problems (as in singular problems) needs specific approaches. Their presentation and the analysis of their behaviour in finite precision arithmetic is beyond the scope of this paper. We will consider problems with single right-hand sides only. In particular, we will not include the block Lanczos algorithm since that would require significant additional space. Although we understand that preconditioning represents an unavoidable and fundamental part of practical computations, we concentrate here on analysis of basic unpreconditioned algorithms. Most of the results can be extended to preconditioned algorithms: see Strakoš and Tichý (2005) and Meurant (2006).

Unless we need to relate the exact arithmetic quantities to the corresponding results of finite precision computations, we do not use any specific notation for the latter; the meaning will be clear from the context. When helpful, we will emphasize the distinction by using the word 'ideally' to refer to a result using exact arithmetic, and 'computationally' or 'numerically' to refer to a result of a finite precision computation.

2. The Lanczos algorithm

This section briefly describes the Lanczos algorithm in exact arithmetic and presents bounds for the convergence of the eigenvalue approximations. For an extensive and thorough description we refer to Parlett (1980).

Strictly speaking, we should not use the term 'convergence' since (with a proper initial vector) the algorithm ideally finds all distinct eigenvalues of A in less than (or equal to) n iterations. Similarly, the term 'convergence of CG' used throughout the paper must be understood differently from the classical asymptotic approach: see, *e.g.*, Hackbusch (1994, p. 270), Beckermann and Kuijlaars (2002) and Kuijlaars (2006). Here we must analyse the behaviour from the start since there is no transient phase which can be skipped, just as there is no asymptotic phase which eventually describes convergence.

2.1. Basic properties of the Lanczos algorithm

Let A be a real n by n nonsingular symmetric matrix and v be a given n-dimensional vector of Euclidean norm 1. The kth Krylov subspace is defined by

$$\mathcal{K}_k(v, A) = \mathrm{span}\{v, Av, \dots, A^{k-1}v\}.$$

Ideally, as long as k is less than or equal to the order of the minimal polynomial of v with respect to A (see Chapter VII, §1 and §2 in Gantmacher (1959)), the subspace $\mathcal{K}_k(v, A)$ is of dimension k and the vectors $A^j v, j = 0, \dots, k-1$ are linearly independent. Clearly, for any v the degree of the minimal polynomial of v with respect to A is always less than or equal to the degree of the minimal polynomial of A; there always exists a vector v such that the latter is reached.

Our goal is to construct an orthonormal basis of the Krylov subspace. Although historically things did not proceed in this way, let us consider what is now called the Arnoldi algorithm (Arnoldi 1951). This is a variant of the Gram–Schmidt orthogonalization process applied to the Krylov basis without assuming A to be symmetric. Starting from $v^1 = v$, the algorithm for computing the $(j+1)$st vector of the basis using the previous ones is

$$h_{i,j} = (Av^j, v^i), \quad i = 1, \dots, j,$$

$$\hat{v}^j = Av^j - \sum_{i=1}^{j} h_{i,j} v^i,$$

$$h_{j+1,j} = \|\hat{v}^j\|, \quad \text{if } h_{j+1,j} = 0 \text{ then stop,}$$

$$v^{j+1} = \frac{\hat{v}^j}{h_{j+1,j}}.$$

It is easy to verify that the vectors v^j span the Krylov subspace and that they are orthonormal. Collecting the basis vectors up to iteration k in an n by k matrix V_k, the relations defining the vector v^{k+1} can be written in a matrix form as

$$AV_k = V_k H_k + h_{k+1,k} v^{k+1} (e^k)^T,$$

where H_k is an unreduced upper Hessenberg matrix with elements $h_{i,j}$, which means that its elements are nonzero in the upper triangle and on the first subdiagonal, and zero below this. The vector e^k is the kth column of the k by k identity matrix (throughout this paper, e^j denotes the jth column of an identity matrix of the size determined by the context). From the orthogonality of the basis vectors,

$$V_k^T A V_k = H_k.$$

If we suppose that the matrix A is symmetric, then because of the last relation, H_k is also symmetric, and therefore tridiagonal. Consequently $\hat{v}^k$ and hence v^{k+1} can be computed using only the two previous vectors v^k and v^{k-1}, and this gives the elegant Lanczos algorithm. Starting from a vector $v^1 = v$, $\|v\| = 1$, $v^0 = 0$, $\eta_1 = 0$, the iterations are:

for $k = 1, 2, \ldots$

$$\alpha_k = (Av^k, v^k) = (v^k)^T A v^k,$$
$$\hat{v}^{k+1} = Av^k - \alpha_k v^k - \eta_k v^{k-1},$$
$$\eta_{k+1} = \|\hat{v}^{k+1}\|, \quad \text{if } \eta_{k+1} = 0 \text{ then stop},$$
$$v^{k+1} = \frac{\hat{v}^{k+1}}{\eta_{k+1}}.$$

We point out that the orthogonalization of the newly computed Av^k against the previously computed vectors in the Arnoldi algorithm and in the Lanczos algorithm described above corresponds to the classical version of the Gram–Schmidt orthogonalization. Here the individual orthogonalization coefficients are computed independently of each other. If a mathematically equivalent modified Gram–Schmidt orthogonalization is used, then the orthogonalization coefficients are computed and the orthogonalization is performed recursively, which in the case of the Lanczos algorithm gives the following implementation. Starting from $v^1 = v$, $\|v\| = 1$, $v^0 = 0$, $\eta_1 = 0$:

for $k = 1, 2, \ldots$

$$u^k = Av^k - \eta_k v^{k-1},$$
$$\alpha_k = (u^k, v^k),$$
$$\hat{v}^{k+1} = u^k - \alpha_k v^k, \tag{2.1}$$
$$\eta_{k+1} = \|\hat{v}^{k+1}\|, \quad \text{if } \eta_{k+1} = 0 \text{ then stop},$$
$$v^{k+1} = \frac{\hat{v}^{k+1}}{\eta_{k+1}}.$$

Clearly, this version can be implemented by storing two vectors instead of three. Although mathematically equivalent to the previous version, the last one advocated by Paige (1976, 1980) and Lewis (1977) can, because of the relationship between classical and modified Gram–Schmidt orthogonalization, be expected to be slightly numerically superior.

In matrix notation the Lanczos algorithm can be expressed as follows:

$$AV_k = V_k T_k + \eta_{k+1} v^{k+1} (e^k)^T,$$

where

$$T_k = \begin{pmatrix} \alpha_1 & \eta_2 & & & & \\ \eta_2 & \alpha_2 & \eta_3 & & & \\ & \ddots & \ddots & \ddots & & \\ & & \eta_{k-1} & \alpha_{k-1} & \eta_k & \\ & & & \eta_k & \alpha_k \end{pmatrix}$$

is an unreduced symmetric tridiagonal matrix with positive subdiagonal entries storing coefficients of the Lanczos recurrence.

We note that since $\|v^k\| = 1$, α_k is a so-called Rayleigh quotient. This implies that

$$\lambda_{\min}(A) \le \alpha_k \le \lambda_{\max}(A).$$

We denote the eigenvalues of A (which are real) by

$$\lambda_{\min}(A) = \lambda_1 \le \lambda_2 \le \cdots \le \lambda_n = \lambda_{\max}(A),$$

and the corresponding orthonormal eigenvectors $q^1, \ldots, q^n$, $Q \equiv (q^1, \ldots, q^n)$.

If $\eta_j \ne 0$ for $j = 2, \ldots, n$, then $AV_n = V_n T_n$ ($\hat{v}^{n+1}$ must be orthogonal to a set of n orthonormal vectors in a space of dimension n and must therefore vanish). Otherwise there exists an $m + 1 < n$ for which $\eta_{m+1} = 0$, $AV_m = V_m T_m$, and we have found an invariant subspace of A, the eigenvalues of T_m being a subset of the eigenvalues of A. When the Lanczos algorithm does not stop before $m = n$, the eigenvalues of A are simple since A is similar to the unreduced symmetric tridiagonal matrix T_n. On the other hand, if A has some multiple eigenvalues, then $\eta_{m+1} = 0$ for some $m + 1 < n$. Ideally, the Lanczos algorithm cannot detect the multiplicity of the individual eigenvalues. In exact arithmetic an eigenvalue of A is found as an eigenvalue of T_m only once.

Let

$$\theta_1^{(k)} < \theta_2^{(k)} < \cdots < \theta_k^{(k)}$$

be the eigenvalues of T_k with the corresponding normalized eigenvectors $z_{(k)}^j \equiv (\zeta_{1,j}^{(k)}, \ldots, \zeta_{k,j}^{(k)})^T$, $j = 1, \ldots, k$, $Z_k \equiv (z_{(k)}^1, \ldots, z_{(k)}^k)$. Since the Lanczos algorithm can be considered as a Rayleigh–Ritz procedure, the eigenvalues $\theta_j^{(k)}$ are called Ritz values and the associated vectors $x_{(k)}^j = V_k z_{(k)}^j$ are known as the Ritz vectors. They are the approximations to the eigenvectors of A given by the algorithm. The residual associated with an eigenpair $(\theta_j^{(k)}, x_{(k)}^j)$

obtained from T_k is

$$
\begin{aligned}
r_{(k)}^j &= A x_{(k)}^j - \theta_j^{(k)} x_{(k)}^j = (AV_k - V_k T_k) z_{(k)}^j \\
&= \eta_{k+1}(e^k)^T z_{(k)}^j\, v^{k+1} \\
&= \eta_{k+1} \zeta_{k,j}^{(k)}\, v^{k+1}.
\end{aligned}
$$

Therefore

$$
\| r_{(k)}^j \| = \eta_{k+1} |\zeta_{k,j}^{(k)}|.
$$

We see that for a given k all residual vectors are proportional to v^{k+1}. When the product of the coefficient η_{k+1} with the absolute value of the bottom element of $z_{(k)}^j$ is small, we have a small residual norm. Moreover, using the spectral decomposition of A and the fact that (in exact arithmetic!) $\| x_{(k)}^j \| = 1$,

$$
\min_i |\lambda_i - \theta_j^{(k)}| \le \| r_{(k)}^j \| = \eta_{k+1} |\zeta_{k,j}^{(k)}|.
$$

Consequently a small residual norm $\| r_{(k)}^j \|$ means convergence of $\theta_j^{(k)}$ to some eigenvalue of A.

2.2. Relationship to orthogonal polynomials

By using the three-term recurrence, the Lanczos basis vectors $v^2, v^3, \ldots$ can be expressed in terms of polynomials in the matrix A acting on the initial vector v^1. From (2.1) we see that

$$
v^{k+1} = p_{k+1}(A) v^1, \quad k = 0, 1, \ldots, \tag{2.2}
$$

where the polynomials p_k satisfy the three-term recurrence (with $p_0 \equiv 0$)

$$
p_1(\lambda) = 1; \quad \eta_{k+1} p_{k+1}(\lambda) = (\lambda - \alpha_k) p_k(\lambda) - \eta_k p_{k-1}(\lambda), \quad k = 1, 2, \ldots. \tag{2.3}
$$

Let $\chi_{1,k}(\lambda)$ (or, where appropriate, simply $\chi_k(\lambda)$) be the characteristic polynomial of T_k (determinant of $T_k - \lambda I$), so that $\chi_0(\lambda) = 1$, $\chi_1(\lambda) = (\alpha_1 - \lambda)$, $\chi_k(\lambda) = (\alpha_k - \lambda)\chi_{k-1}(\lambda) - \eta_k^2 \chi_{k-2}(\lambda)$; then for the degree k polynomial

$$
p_{k+1}(\lambda) = (-1)^k \frac{\chi_{1,k}(\lambda)}{\eta_2 \cdots \eta_{k+1}}.
$$

Using the orthogonality of the vectors $v^1, v^2, \ldots$ and the spectral decomposition of A, the normalized Lanczos polynomials $p_1(\lambda) = 1$, $p_2(\lambda)$, $p_3(\lambda), \ldots$ are orthonormal polynomials with respect to a scalar product defined by the Riemann–Stieltjes integral

$$
(p, q) = \int_{\lambda_1}^{\lambda_n} p(\lambda) q(\lambda)\, d\omega(\lambda) = \sum_{l=1}^n \omega_l\, p(\lambda_l) q(\lambda_l), \tag{2.4}
$$

where the distribution function ω is a non-decreasing piecewise constant function with at most n points of increase $\lambda_1, \ldots, \lambda_n$. For simplicity of exposition, suppose that

$$\lambda_1 < \lambda_2 < \cdots < \lambda_n,$$

i.e., all eigenvalues of A are distinct. Then

$$\omega(\lambda) = \begin{cases} 0 & \text{if } \lambda < \lambda_1, \\ \sum_{l=1}^{i} \omega_l & \text{if } \lambda_i \leq \lambda < \lambda_{i+1}, \\ \sum_{l=1}^{n} \omega_l = 1 & \text{if } \lambda_n \leq \lambda, \end{cases}$$

where $\omega_l = |(v^1, q^l)|^2$ is the squared component of the starting vector v^1 in the direction of the lth invariant subspace of A.

Writing $P_k(\lambda) = (p_1(\lambda), \ldots, p_k(\lambda))^T$, the recurrence for the orthonormal polynomials can be written in the matrix form

$$\lambda P_k(\lambda) = T_k P_k(\lambda) + \eta_{k+1} p_{k+1}(\lambda) e^k.$$

Since p_{k+1} is proportional to the characteristic polynomial of T_k, its roots are the eigenvalues of T_k, that is, the Ritz values $\theta_j^{(k)}$, $j = 1, \ldots, k$.

Since $\chi_{1,k}(\lambda)$ is (apart from multiplication by $(-1)^k$) a monic polynomial orthogonal with respect to the inner product defined by (2.4) to any polynomial of degree $k - 1$ or less, it must resolve the following minimization problem:

$$(-1)^k \chi_{1,k}(\lambda) = \arg \min_{\psi \in \mathcal{M}_k} \int_{\lambda_1}^{\lambda_n} \psi^2(\lambda) \, d\omega(\lambda), \quad k = 1, 2, \ldots, n,$$

where $\mathcal{M}_k$ denotes the set of all monic polynomials of degree less than or equal to k.

Consider the unreduced symmetric tridiagonal matrix T_k defined above. It stores the coefficients of the first k steps of the Lanczos algorithm applied to A with an initial vector v^1. The same T_k can be seen as a result of the Lanczos algorithm applied to T_k with the (k-dimensional) initial vector e^1. Consequently the polynomials $p_1 = 1$, $p_2, \ldots, p_{k+1}$ form a set of orthonormal polynomials with respect to a scalar product defined by the Riemann–Stieltjes integral

$$(p, q)_k = \int_{\lambda_1}^{\lambda_n} p(\lambda) q(\lambda) \, d\omega^{(k)}(\lambda) = \sum_{l=1}^{k} \omega_l^{(k)} p(\theta_l^{(k)}) q(\theta_l^{(k)}), \quad (2.5)$$

where the distribution function $\omega^{(k)}$ is a non-decreasing piecewise constant function with k points of increase $\theta_1^{(k)}, \ldots, \theta_k^{(k)}$,

$$
w^{(k)}(\lambda) = \begin{cases} 0 & \text{if } \lambda < \theta_1^{(k)}, \\ \sum_{l=1}^{i} w_l^{(k)} & \text{if } \theta_i^{(k)} \leq \lambda < \theta_{i+1}^{(k)}, \\ \sum_{l=1}^{k} w_l^{(k)} = 1 & \text{if } \theta_k^{(k)} \leq \lambda, \end{cases}
$$

and $w_l^{(k)} = |(z_{(k)}^l, e^1)|^2$. We see that the first components of the normalized eigenvectors of T_k determine the weights in the Riemann–Stieltjes integral (2.5). Here p_{k+1} represents the $(k+1)$st orthogonal polynomial in the sequence defined by (2.4), and, at the same time, the final polynomial with roots $\theta_1^{(k)}, \ldots, \theta_k^{(k)}$ in the same sequence of orthonormal polynomials defined by (2.5). This fact has the following fundamental consequence, formulated as a theorem.

Theorem 2.1. Using the previous notation, (2.5) represents the kth Gauss quadrature approximation to the Riemann–Stieltjes integral (2.4).

Proof. Consider a polynomial $\Phi(\lambda)$ of degree at most $2k-1$. Then we can write

$$
\Phi(\lambda) = p_{k+1}(\lambda)\Phi_1(\lambda) + \Phi_2(\lambda) = p_{k+1}(\lambda)\Phi_1(\lambda) + \sum_{l=2}^{k} \nu_l p_l(\lambda) + \nu_1,
$$

where $\Phi_1(\lambda)$, $\Phi_2(\lambda)$ are of degree at most $k-1$ and $\nu_1, \ldots, \nu_k$ are some scalar coefficients. From the orthogonality of $1, p_2(\lambda), \ldots, p_k(\lambda)$ with respect to both (2.4) and (2.5) it immediately follows that

$$
\int_{\lambda_1}^{\lambda_n} \Phi(\lambda)\, d\omega(\lambda) = \int_{\lambda_1}^{\lambda_n} \nu_1\, d\omega(\lambda) = \nu_1 = \int_{\lambda_1}^{\lambda_n} \nu_1\, d\omega^{(k)}(\lambda) = \int_{\lambda_1}^{\lambda_n} \Phi(\lambda)\, d\omega^{(k)}(\lambda).
$$

$\square$

Since $\chi_{k-1}(\lambda) = -\chi_k(\lambda)/(\lambda - \theta_l^{(k)}) +$ a polynomial of degree at most $k-2$,

$$
\int_{\lambda_1}^{\lambda_n} \chi_{k-1}^2(\lambda)\, d\omega(\lambda) = -\int_{\lambda_1}^{\lambda_n} \chi_{k-1}(\lambda)\frac{\chi_k(\lambda)}{(\lambda - \theta_l^{(k)})}\, d\omega(\lambda)
$$

$$
= -\int_{\lambda_1}^{\lambda_n} \chi_{k-1}(\lambda)\frac{\chi_k(\lambda)}{(\lambda - \theta_l^{(k)})}\, d\omega^{(k)}(\lambda)
$$

$$
= -\sum_{i=1}^{k} w_i^{(k)}\left[\chi_{k-1}(\lambda)\frac{\chi_k(\lambda)}{(\lambda - \theta_l^{(k)})}\right]_{\lambda=\theta_i^{(k)}}
$$

$$
= -w_l^{(k)}\chi_{k-1}(\theta_l^{(k)})\chi_k'(\theta_l^{(k)}).
$$

Consequently

$$w_l^{(k)} = |(z_{(k)}^l, e^1)|^2 = -\frac{\int_{\lambda_1}^{\lambda_n} \chi_{k-1}^2(\lambda)\,d\omega(\lambda)}{\chi_{k-1}(\theta_l^{(k)})\chi_k'(\theta_l^{(k)})} = -\frac{\eta_2^2 \eta_3^2 \cdots \eta_k^2}{\chi_{k-1}(\theta_l^{(k)})\chi_k'(\theta_l^{(k)})} \qquad (2.6)$$

gives for $l = 1, \ldots, k$ the weights $w_l^{(k)}$ of the kth Gauss quadrature applied to (2.4). It is worth noticing that this identity gives squares of the first elements of eigenvectors of any unreduced symmetric tridiagonal matrix T_k in terms of the values of the derivative $\chi_k'(\lambda)$ of its characteristic polynomial and of the values of the characteristic polynomial $\chi_{k-1}(\lambda)$ of the reduced matrix with the last row and column omitted. Here (and in several other places below) we do not use the positiveness of the subdiagonal entries of the coefficient matrices in the Lanczos algorithm, since in the theory of unreduced symmetric tridiagonal matrices the positiveness of the subdiagonal entries is insignificant: see Parlett (1980, Lemma 7.2.1).

Clearly we can consider the Lanczos algorithm applied to T_k with the initial vector e^k, leading to the Riemann–Stieltjes integral analogous to (2.5) but with the weights $|(z_{(k)}^l, e^k)|^2$. Then, analogously to (2.6),

$$|(z_{(k)}^l, e^k)|^2 = -\frac{\eta_2^2 \eta_3^2 \cdots \eta_k^2}{\chi_{2,k}(\theta_l^{(k)})\chi_k'(\theta_l^{(k)})}, \qquad (2.7)$$

where $\chi_{2,k}(\lambda)$ is the characteristic polynomial of the reduced matrix with the first row and column omitted. It is useful to exploit the knowledge about unreduced symmetric tridiagonal matrices: see, *e.g.*, Wilkinson (1965), Thompson and McEnteggert (1968), Golub (1973), Paige (1971, 1980), Parlett (1980), Elhay, Gladwell, Golub and Ram (1999) and also Strakoš and Greenbaum (1992). For other equivalent expressions for the components of the eigenvectors: see Meurant (2006). In particular,

$$\chi_{2,k}(\theta_l^{(k)})\chi_{k-1}(\theta_l^{(k)}) = \eta_2^2 \eta_3^2 \cdots \eta_k^2,$$

and

$$|(z_{(k)}^l, e^1)|^2 = -\frac{\chi_{2,k}(\theta_l^{(k)})}{\chi_k'(\theta_l^{(k)})}, \qquad |(z_{(k)}^l, e^k)|^2 = -\frac{\chi_{k-1}(\theta_l^{(k)})}{\chi_k'(\theta_l^{(k)})}. \qquad (2.8)$$

One of the most beautiful and most powerful features of mathematics is the translation of a given problem into appropriate language where the problem can easily be resolved. The Lanczos algorithm and related mathematical structures offer an excellent example.

- Given A and v^1, the Lanczos algorithm is usually formulated in n-dimensional vector space, and computes the orthonormal basis vectors $v^1, v^2, \ldots$ of the Krylov subspaces $\mathcal{K}_k(v^1, A)$, $k = 1, 2, \ldots$.

- The Lanczos algorithm can be formulated in terms of the unreduced symmetric tridiagonal matrices T_k, $k = 1, 2, \ldots$, with positive next to diagonal elements, where T_k is appended by a row and a column at each Lanczos step.

- The Lanczos algorithm can be formulated as a Stieltjes procedure in terms of polynomials $p_1(\lambda) = 1$, $p_2(\lambda)$, $p_3(\lambda), \ldots$ orthonormal with respect to the Riemann–Stieltjes integral (2.4).

- The Lanczos algorithm can be formulated in terms of Gauss quadrature approximations (2.5) to the original Riemann–Stieltjes integral (2.4).

Thus the purely algebraic formulation of the problem can be translated to a problem in the classical theory of orthogonal polynomials, and *vice versa*. Similarly, classical tools such as moments, continued fractions and interpolatory quadratures can be directly related to the algebraic tools, developed a century, or many decades, later. These connections are fundamental. They were promoted in modern numerical linear algebra by many distinguished mathematicians. Of these, particular recognition should be given to Gene Golub: see, *e.g.*, Gautschi (2002). For a comprehensive text on generating orthogonal polynomials, Stieltjes procedure and its computational aspects we refer to Gautschi (1982) and the book Gautschi (2004). Other useful information can be found, *e.g.*, in Strakoš and Tichý (2002) and Fischer (1996).

2.3. Approximation from subspaces and the persistence theorem

Approximation results for eigenvalues can be obtained by using the general theory of Rayleigh–Ritz approximations. Good expositions of the theory are given by Stewart (2001) or Parlett (1980). Here is an example of such a result for an eigenpair (λ_i, q^i) of A that we quote from Stewart (2001, p. 285).

Theorem 2.2. Let U be an orthonormal matrix, let $B = U^T A U$ be the matrix Rayleigh quotient, and let θ be the angle between the eigenvector q_i we want to approximate and the range of U, where $A q^i = \lambda_i q^i$. Then there exists a matrix E satisfying

$$\|E\| \leq \frac{\sin \theta}{\sqrt{1 - \sin^2 \theta}} \|A\|$$

such that λ_i is an eigenvalue of $B + E$.

Then one can apply a general theorem on eigenvalues of perturbed matrices: see Stewart (2001, pp. 285–286).

Corollary 2.3. With the notation of Theorem 2.2, there exists an eigenvalue μ of B such that

$$|\mu - \lambda_i| \leq \|E\|.$$

This shows that if, as a result of an iterative algorithm, we get a small angle θ between the wanted eigenvector q^i and U, then we get an approximate eigenvalue of $B = U^T A U$ converging towards the eigenvalue λ_i of A.

In the case of the Lanczos algorithm we build up a sequence $U = V_k$ and $B = T_k$, $k = 1, 2, \ldots$. We will not further describe the *a priori* error bounds which can be found elsewhere. We will rather concentrate on *a posteriori* bounds and properties important for analysis of the finite precision behaviour.

Let $A = Q\Lambda Q^T$ and $T_k = Z_k \Theta_k Z_k^T$ be the spectral decompositions of A and T_k respectively, $\Lambda = \text{diag}(\lambda_i)$, $\Theta_k = \text{diag}(\theta_j^{(k)})$. Denote by $\bar{V}_k \equiv Q^T V_k$ the matrix whose columns are composed of the projections of the Lanczos vectors on the eigenvectors of A. Since $T_k = V_k^T A V_k = \bar{V}_k^T \Lambda \bar{V}_k$, we have the following relationship between the Ritz values and the eigenvalues of A.

Proposition 2.4. Let $W_k = (w_{(k)}^1, \ldots, w_{(k)}^k) \equiv Q^T V_k Z_k = \bar{V}_k Z_k$, $w_{(k)}^j \equiv (\xi_{1,j}^{(k)}, \ldots, \xi_{n,j}^{(k)})^T$. Then,

$$\Theta_k = W_k^T \Lambda W_k,$$

$$\theta_j^{(k)} = \sum_{l=1}^{n} (\xi_{l,j}^{(k)})^2 \lambda_l \quad \text{and} \quad \sum_{l=1}^{n} (\xi_{l,j}^{(k)})^2 = 1.$$

Proof. The result follows from $T_k = V_k^T A V_k$ and the eigendecompositions of A and T_k, $W_k^T W_k = Z_k^T V_k^T Q Q^T V_k Z_k = I$. ☐

Clearly the Ritz values are convex combinations of the eigenvalues. We have seen above that a small residual norm $\|r_{(k)}^j\| = \eta_{k+1} |\zeta_{k,j}^{(k)}|$ means that $\theta_j^{(k)}$ is close to some eigenvalue λ_i of A. The following fundamental result proved by Paige, which we formulate for its importance as a theorem, shows that once an eigenvalue λ_i of A is at step t approximated by some Ritz value $\theta_s^{(t)}$ with a small residual norm, it must be approximated to a comparable accuracy by some Ritz value at all subsequent Lanczos steps.

Theorem 2.5. (Persistence Theorem) Let $t < k$. Then,

$$\min_j |\theta_s^{(t)} - \theta_j^{(k)}| \leq \eta_{t+1} |\zeta_{t,s}^{(t)}|.$$

Proof. A proof was given by Paige (1971) using the result in Wilkinson (1965, p. 171); see (3.9) on p. 241 of Paige (1980). ☐

Theorem 2.5 implies that for every $k > t$ and for any unreduced symmetric tridiagonal extension T_k of T_t there is an eigenvalue $\theta_j^{(k)}$ of T_k within $\eta_{t+1}|\zeta_{t,s}^{(t)}|$ of $\theta_s^{(t)}$. The situation deserves a formal definition. In order to avoid possible subtle ambiguities in the exposition, we slightly modify Definition 1 of Paige (1980).

Definition 2.6. We call an eigenvalue $\theta_s^{(t)}$ of the t by t unreduced symmetric tridiagonal matrix T_t *stabilized* to within $\delta \equiv \eta_{t+1}|\zeta_{t,s}^{(t)}|$. In short, if $\eta_{t+1}|\zeta_{t,s}^{(t)}|$ is small, we call $\theta_s^{(t)}$ stabilized to within small δ.

We will see in Section 4 that, for some Ritz value $\theta_s^{(t)}$ at step t of the Lanczos algorithm, it can happen that for any unreduced symmetric tridiagonal extension T_k of T_t there is an eigenvalue $\theta_j^{(k)}$ very close to $\theta_s^{(t)}$, even though $\delta = \eta_{t+1}|\zeta_{t,s}^{(t)}|$ is not small. However the subsequent Theorem 2.7 and results in Section 4 will also show that in such a case $\theta_s^{(t)}$ must be a close approximation to some eigenvalue of A.

Another useful result in Paige (1971) relates the difference $\theta_s^{(t)} - \theta_j^{(k)}$ to the scalar products of the corresponding eigenvectors.

Theorem 2.7. Using the same notation as in Theorem 2.5,

$$(\theta_s^{(t)} - \theta_j^{(k)})(z_{(k)}^j)^T \begin{bmatrix} z_{(t)}^s \\ 0 \end{bmatrix} = \eta_{t+1}\zeta_{t,s}^{(t)}\zeta_{t+1,j}^{(k)}.$$

Proof.   See Paige (1980, p. 241).   □

Using this theorem, it is interesting to compare Ritz values on successive steps of the Lanczos algorithm, *i.e.*, take $k = t + 1$. Then, because of the interlacing property of Ritz values it is enough to consider $j = s$ or $j = s+1$,

$$(\theta_s^{(t)} - \theta_j^{(t+1)}) \sum_{l=1}^{t} \zeta_{l,s}^{(t)}\zeta_{l,j}^{(t+1)} = \eta_{t+1}\zeta_{t,s}^{(t)}\zeta_{t+1,j}^{(t+1)}.$$

In particular this leads to

$$\eta_{t+1}|\zeta_{t,s}^{(t)}\zeta_{t+1,j}^{(t+1)}| \le |\theta_s^{(t)} - \theta_j^{(t+1)}|,$$

for $j = s$ or $j = s + 1$. Assuming that η_{t+1} is not small (a small η_{t+1} would mean the lucky event indicating closeness to an invariant subspace and convergence of all Ritz values), this shows that if the difference between the Ritz values $|\theta_s^{(t)} - \theta_j^{(t+1)}|$ from two successive steps is small, then the product of the last elements of the corresponding eigenvectors is small. This suggests that Ritz values in two successive steps which are close to each other indicate convergence to some eigenvalue of A. This question has been further investigated by Wülling (2005) following some earlier thoughts in Strakoš and Greenbaum (1992). We will discuss the related results in more detail in Section 4.

3. The conjugate gradient algorithm

The conjugate gradient (CG) algorithm was developed independently by Magnus Hestenes in the US and by Eduard Stiefel in Switzerland, at the beginning of the 1950s. Then they met during a conference in 1951 and wrote a famous joint paper, Hestenes and Stiefel (1952). The algorithm was derived using conjugacy and minimization of functionals. However, it turns out that it is very closely related to the Lanczos algorithm, which can easily be applied for solving linear algebraic systems (Lanczos 1950, 1952).

Consider a symmetric positive definite matrix A, right-hand side b, and the problem $Ax = b$. With an initial vector x^0 and the corresponding residual $r^0 = b - Ax^0$, we can seek an approximate solution to the given linear system in the form $x^k = x^0 + V_k y^k$, where V_k is the matrix of the orthonormal basis vectors of the Krylov subspace $\mathcal{K}_k(v^1, A)$ generated by the Lanczos algorithm with $v^1 = r^0 / \|r^0\|$. If we ensure that the residual $r^k = b - Ax^k$ is orthogonal to V_k, then $r^{n+1} = 0$. The resulting method will give (in exact arithmetic) the exact solution in at most n steps and therefore will represent a direct method. Since $r^k = r^0 - AV_k y^k$, this will give

$$0 = V_k^T r^k = V_k^T r^0 - T_k y^k,$$

implying that the coordinates of the approximate solution in V_k are given by the solution of the k by k system with matrix T_k. With the background of the Lanczos algorithm, the whole method can be formulated as

$$T_k y^k = \|r^0\| e^1, \quad x^k = x^0 + V_k y^k. \tag{3.1}$$

The residual r^k is proportional to v^{k+1}, since from the matrix form of (2.1)

$$r^k = r^0 - AV_k y^k = r^0 - (V_k T_k + \eta_{k+1} v^{k+1}(e^k)^T) y^k$$
$$= -\eta_{k+1}(y^k, e^k) v^{k+1} = (-1)^k v^{k+1} \|r^0\| \eta_2 \cdots \eta_{k+1} / \det(T_k),$$

using the adjugate of T_k. We next show that ideally (3.1) is equivalent to the CG algorithm of Hestenes and Stiefel (1952).

3.1. Relationship between the formulation of the CG and Lanczos algorithms

In our notation, the Hestenes–Stiefel formulation of the CG algorithm for solving $Ax = b$ with a symmetric positive definite matrix A given in Hestenes and Stiefel (1952) is as follows. Given x^0, $r^0 = b - Ax^0$, $p^0 = r^0$, the subsequent approximate solutions x^k and the corresponding residual vectors $r^k = b - Ax^k$ are computed by:

for $k = 1, 2, \ldots$

$$\gamma_{k-1} = \frac{\|r^{k-1}\|^2}{(p^{k-1}, Ap^{k-1})},$$

$$x^k = x^{k-1} + \gamma_{k-1}p^{k-1},$$

$$r^k = r^{k-1} - \gamma_{k-1}Ap^{k-1}, \qquad (3.2)$$

$$\beta_k = \frac{\|r^k\|^2}{\|r^{k-1}\|^2},$$

$$p^k = r^k + \beta_k p^{k-1}.$$

With $v^1 = r^0/\|r^0\|$ it can be seen, for example by induction, that

$$\mathcal{K}_k(v^1, A) = \mathrm{span}\{r^0, \ldots, r^{k-1}\} = \mathrm{span}\{p^0, \ldots, p^{k-1}\}.$$

Another straightforward induction (see Hestenes and Stiefel (1952)) gives

$$(r^i, r^j) = 0 \quad \text{and} \quad (p^i, Ap^j) = 0 \quad \text{for } i \neq j.$$

This immediately implies $r^k \perp \mathcal{K}_k(v^1, A)$ and therefore proves the equivalence (up to signs) with the Lanczos algorithm-based formulation described above. Eliminating p^{k-1} from the recurrence for the CG residual, we get, after a simple manipulation,

$$-\frac{1}{\gamma_{k-1}}r^k = Ar^{k-1} - \left(\frac{1}{\gamma_{k-1}} + \frac{\beta_{k-1}}{\gamma_{k-2}}\right)r^{k-1} + \frac{\beta_{k-1}}{\gamma_{k-2}}r^{k-2}. \qquad (3.3)$$

Comparing (3.2) with the Lanczos recurrence (2.1), or more easily with the 3-term recurrence (3.3) for r^k, shows that

$$v^{k+1} = (-1)^k \frac{r^k}{\|r^k\|}. \qquad (3.4)$$

If $\hat{v}^{m+1} = 0$ in (2.1), i.e., $\eta_{m+1} = 0$ and the Lanczos algorithm stops, then $r^0, \ldots, A^{m-1}r^0$ are linearly independent while $r^0 \in A\mathcal{K}_m(v^1, A)$, i.e., $r^m = b - Ax^m = r^0 - Au^m = 0$, $u^m \in \mathcal{K}_m(v^1, A)$. Consequently, termination of the Lanczos algorithm implies convergence of CG to the exact solution.

The Lanczos coefficients α_k, η_{k+1} can be determined from the CG coefficients γ_{k-1}, β_k in the following way. Using (3.4) in (3.3) and $\beta_k = \|r^k\|^2/\|r^{k-1}\|^2$, we obtain

$$\frac{\sqrt{\beta_k}}{\gamma_{k-1}}v^{k+1} = Av^k - \left(\frac{1}{\gamma_{k-1}} + \frac{\beta_{k-1}}{\gamma_{k-2}}\right)v^k - \frac{\sqrt{\beta_{k-1}}}{\gamma_{k-2}}v^{k-1},$$

and therefore we have for $k = 1, 2, \ldots$ the following relations between the coefficients:

$$\alpha_k = \frac{1}{\gamma_{k-1}} + \frac{\beta_{k-1}}{\gamma_{k-2}}, \qquad \beta_0 = 0, \qquad \gamma_{-1} = 1,$$

$$\eta_{k+1} = \frac{\sqrt{\beta_k}}{\gamma_{k-1}}.$$

On the other hand, the CG algorithm (3.2) can be derived from the Lanczos algorithm by the LDL^T decomposition (a variant of the Cholesky decomposition where L is a lower triangular, here lower bidiagonal, factor with ones on the diagonal, and D is a diagonal matrix) of the matrix T_k. This idea, presented in Section 5.7 of Householder (1964), and thoroughly exploited by Paige and Saunders (1975) (see also Stoer (1983)), offers a very insightful explanation of the behaviour of CG when A is indefinite. Since the CG approximate solution satisfies (see (3.1))

$$x^k = x^0 + \|r^0\| V_k T_k^{-1} e^1, \quad k = 1, 2, \ldots,$$

it does not exist whenever T_k is singular. When the Lanczos algorithm terminates, the matrix T_m has all its eigenvalues equal to some eigenvalues of the (symmetric and nonsingular) matrix A. Clearly T_m must also be nonsingular. An easy exercise shows that, whenever T_k and T_{k+1} for any $1 \le k < m - 1$ are simultaneously singular, then $T_{k+2}, \ldots, T_m$ must also be singular, a contradiction. Consequently, at least every second T_k in the sequence $T_1, \ldots, T_m$ must be nonsingular, which means that the CG approximation exists at least at every second step. It cannot, in general, be computed via the formulas (3.2), since the Cholesky decomposition of the singular T_k does not exist and the implementation (3.2) in such a case breaks down. If T_k is close to singular, the Cholesky decomposition is poorly determined numerically for all $j > k$, and so is the recurrence (3.2).

Paige and Saunders showed in a very instructive way how to compute the CG approximation x^k when it exists, and how to avoid numerical instabilities when T_k is close to singular. Their approach is based on exploiting the Lanczos algorithm, but it does not require storing the Lanczos basis V_k. The CG approximations x^k are computed recursively with the help of auxiliary approximations to the solution which exist at every step and which define the method called SYMMLQ. They also suggested an effective implementation of the Krylov subspace method MINRES, which minimizes residual norms and is used for symmetric indefinite problems. Paige and Saunders (1975) resolved open problems that had arisen from the earlier work by Fridman (1963) and Luenberger (1969, 1970). The relationship between different implementations was further studied by Fletcher (1976), and a numerically stable variant of the OD algorithm of Fridman (1963) called STOD was suggested by Stoer and Freund (1982); see the overview in Stoer (1983).

3.2. Orthogonality and optimality properties

In some important applications leading to systems with symmetric positive definite matrices it is natural to measure the error in the A-norm,

$$\|x - u\|_A = (x - u, A(x - u))^{\frac{1}{2}},$$

since the A-norm can be interpreted as the discretized measure of energy which is to be minimized: see, *e.g.*, Arioli (2004) and Arioli, Noulard and Russo (2001). The CG algorithm is, from this point of view, best suited to solving such problems, since it minimizes the A-norm of the error among all possible approximations from the same Krylov subspaces,

$$\|x - x^k\|_A = \min_{u \in x^0 + \mathcal{K}_k(v^1, A)} \|x - u\|_A. \tag{3.5}$$

Indeed, in order to reach the minimum (3.5), $x - x^k$ must be orthogonal with respect to the inner product defined by the matrix A to the Krylov subspace $\mathcal{K}_k(v^1, A)$, *i.e.*,

$$0 = (r^j, A(x - x^k)) = (r^j, r^k) = (v^{j+1}, r^k) \quad \text{for } j = 0, \ldots, k - 1,$$

which uniquely determines the approximate solution x^k generated by the CG algorithm described above.

Using the A-orthogonality of the direction vectors $p^0, p^1, p^2, \ldots$, it can be seen from (3.2) that the kth error, assuming that CG terminates at step m with $x^m = x$, can conveniently be written

$$x - x^0 = \sum_{l=1}^{k} \gamma_{l-1} p^{l-1} + x - x^k = \sum_{l=1}^{m} \gamma_{l-1} p^{l-1},$$

$$x - x^k = \sum_{l=k+1}^{m} \gamma_{l-1} p^{l-1},$$

$$\|x - x^0\|_A^2 = \sum_{l=1}^{m} \gamma_{l-1}^2 (p^{l-1}, A p^{l-1}) = \sum_{l=1}^{m} \gamma_{l-1} \|r^{l-1}\|^2,$$

$$\|x - x^k\|_A^2 = \sum_{l=k+1}^{m} \gamma_{l-1} \|r^{l-1}\|^2,$$

and, finally,

$$\|x - x^0\|_A^2 = \sum_{l=1}^{k} \gamma_{l-1} \|r^{l-1}\|^2 + \|x - x^k\|_A^2, \quad k = 1, 2, \ldots, m. \qquad (3.6)$$

The last identity reflects the mathematical elegance of the CG algorithm, but it also demonstrates complications which have to be dealt with in finite precision arithmetic computations. The derivation of (3.6) presented above relies upon the global A-orthogonality of all vectors $p^0, \ldots, p^{m-1}$,

$$(p^i, Ap^j) = 0 \quad \text{for } i \neq j,$$

which holds ideally, but which is not preserved numerically. Unless (3.6) is supported by arguments that also hold numerically, it cannot be used for the results of finite precision computations. This point is of crucial importance. A patient reader will, however, see in Section 5 that (3.6) indeed holds, up to a small insignificant inaccuracy, also numerically: see Strakoš and Tichý (2002).

As the relationship of CG with the Lanczos algorithm suggests, there is of course a three-term recurrence formulation ideally equivalent to (3.2): see, e.g., Rutishauser (1959) and Hageman and Young (1981). The three-term recurrence is reputed to have some disadvantages concerning the maximal attainable accuracy: see Section 5 and Gutknecht and Strakoš (2000). However, it is of some interest for parallel computation.

Convergence bounds for CG are typically derived from its polynomial formulation, which follows from (3.2) (see, e.g., the 3-term recurrence (3.3) for r^k):

$$r^k = \varphi_k(A)r^0, \qquad \varphi_k(0) = 1,$$
$$x - x^k = \varphi_k(A)(x - x^0),$$

where $\varphi_k(0) = 1$ follows, e.g., from induction on the 3-term recurrence for r^k, while the $x - x^k$ expression follows since A is nonsingular. The 3-term recurrences and the definition of the Lanczos polynomials p_k in (2.3) lead to

$$\varphi_k(\lambda) = \frac{p_{k+1}(\lambda)}{p_{k+1}(0)}.$$

Here the assumption that A is symmetric positive definite guarantees that all roots of p_k are no less than $\lambda_1 > 0$, and therefore $p_k(0) \neq 0$. With the spectral decomposition of A we can easily obtain the following theorem.

Theorem 3.1.

$$\|r^k\|^2 = \|r^0\|^2 \sum_{i=1}^{n} \prod_{l=1}^{k} \left(1 - \frac{\lambda_i}{\theta_l^{(k)}}\right)^2 \omega_i,$$

$$\|x - x^k\|^2 = \|r^0\|^2 \sum_{i=1}^{n} \prod_{l=1}^{k} \left(\frac{1}{\lambda_i} - \frac{1}{\theta_l^{(k)}}\right)^2 \omega_i,$$

$$\|x - x^k\|_A^2 = \|r^0\|^2 \sum_{i=1}^{n} \prod_{l=1}^{k} \left(\frac{1}{\sqrt{\lambda_i}} - \frac{\sqrt{\lambda_i}}{\theta_l^{(k)}}\right)^2 \omega_i,$$

where, as in (2.4), $\omega_i = |(v^1, q^i)|^2$, $v^1 = r^0/\|r^0\|$.

Proof. We remark that $x - x^0 = A^{-1}r^0$ and

$$\frac{p_{k+1}(\lambda_i)^2}{p_{k+1}(0)^2} = \prod_{l=1}^{k} \left(1 - \frac{\lambda_i}{\theta_l^{(k)}}\right)^2.$$

By using this, the proofs become straightforward.   □

Since $\|x - x^k\|_A \le \|\varphi_k(A)\| \, \|x - x^0\|_A$, we have the bound

$$\|x - x^k\|_A \le \min_{\varphi \in \Pi_k} \max_i |\varphi(\lambda_i)| \, \|x - x^0\|_A, \tag{3.7}$$

where Π_k denotes the set of all polynomials of degree at most k with the constant term equal to one (value one at zero). Any bound which is based on (3.7) holds for *any* initial error (initial residual) and therefore represents a worst case bound. Therefore, even analytic knowledge of the value

$$\min_{\varphi \in \Pi_k} \max_i |\varphi(\lambda_i)| = \left| \sum_{l=1}^{k+1} (-1)^{l-1} \prod_{j=1, j \ne l}^{k+1} \frac{\mu_j}{\mu_j - \mu_l} \right|^{-1}, \tag{3.8}$$

where $\{\mu_1, \dots, \mu_{k+1}\}$ is some properly chosen subset of the distinct eigenvalues of A (on which the kth minimax polynomial assumes its maximum absolute value – see Greenbaum (1979) and Liesen and Tichý (2005)) does not help in describing possible differences in the behaviour of CG for different initial residuals (right-hand sides): *cf.* Beckermann and Kuijlaars (2002) and Strakoš and Tichý (2005). The error bound (3.7) with (3.8) is sharp, *i.e.*, at any given step k it can be attained with a certain initial vector (which depends on k).

The generally known bound is derived from using the kth degree Chebyshev polynomial on the spectral interval $[\lambda_1, \lambda_n]$, which gives

$$\frac{\|x - x^k\|_A}{\|x - x^0\|_A} \le 2\left[\left(\frac{\sqrt{\kappa} - 1}{\sqrt{\kappa} + 1}\right)^k + \left(\frac{\sqrt{\kappa} + 1}{\sqrt{\kappa} - 1}\right)^k\right]^{-1} \le 2\left(\frac{\sqrt{\kappa} - 1}{\sqrt{\kappa} + 1}\right)^k, \tag{3.9}$$

where $\kappa \equiv \kappa(A) \equiv \lambda_n/\lambda_1$ denotes the condition number of A. This bound is frequently attributed to Kaniel (1966) or Daniel (1967), but it appeared even earlier in the paper by Meinardus (1963); see Li (2005). Though it is useful in the analysis of many model problems, it cannot be identified, except for some specific cases, with convergence behaviour of CG. The bound (3.9) describes linear convergence; it shows that the closer the condition number is to 1, the faster is the convergence of CG when measured in the A-norm. However, we have seen in Theorem 3.1, and we are going to see again in the next section, that CG convergence depends on the distribution of all the eigenvalues of A and not just on the condition number. If the distribution of eigenvalues is favourable, then the convergence of CG significantly accelerates as k increases. For an early investigation of convergence behaviour in relation to the spectrum see, e.g., Axelsson and Linskog (1986) and van der Vorst (1982).

3.3. Estimating quadratic forms and identities for the error norms in CG

We have seen that given A and v^1 respectively r^0, $v^1 = r^0/\|r^0\|$, the Lanczos algorithm and CG can be formulated in terms of the orthogonal polynomials $1, p_1(\lambda), p_2(\lambda), \dots$, and therefore in terms of the Gauss quadrature of the Riemann–Stieltjes integral determined by A, v^1. In this way, the Lanczos algorithm and CG can be viewed as matrix representations of Gauss quadrature. That explains the subtle character of problems related to the Lanczos and CG convergence behaviour. Here we will go a step forward to show how the A-norm of the error and the Euclidean norm of the error in CG can be computed using Gauss quadrature and how they can be bounded using some of its modifications.

Computing the A-norm of the error $\epsilon^k \equiv x - x^k$ is closely related to approximating quadratic forms. This has been studied extensively by Gene Golub with many collaborators during the last thirty-five years. The relationship to Gauss quadrature was summarized in Golub and Meurant (1994); see also Golub and Meurant (1997), Golub and Strakoš (1994), Fischer (1996), Golub and von Matt (1991) and Calvetti, Morigi, Reichel and Sgallari (2000). With A symmetric positive definite, the problem considered by Golub and Meurant (1994) was to find upper and lower bounds (or approximations) for the entries of a function of a matrix. This problem leads to the quadratic form

$$u^T f(A) u,$$

where u is a given vector and f is a smooth (possibly C^∞) function on a given interval of the real line. The more general case $u^T f(A) v$ can easily be converted into the previous one using the well-known identity

$$u^T f(A) v = \frac{1}{2} \left(u^T f(A) u + v^T f(A) v - (u-v)^T f(A)(u-v) \right).$$

This problem is of great importance in computational sciences such as computational quantum chemistry or solid state physics. The example we are interested in for CG is $f(\lambda) = 1/\lambda$. This is related to the problem of computing the A-norm of the error, since the error ϵ^k is related to the residual r^k by the equation $A\epsilon^k = r^k$. Therefore,

$$\|\epsilon^k\|_A^2 = (A\epsilon^k, \epsilon^k) = (A^{-1}r^k, r^k) = (r^k)^T A^{-1} r^k.$$

Using the spectral decomposition of A (as above, for simplicity we assume that the eigenvalues of A are distinct and ordered, $\lambda_1 < \lambda_2 < \cdots < \lambda_n$)

$$f(A) = Qf(\Lambda)Q^T.$$

Therefore,

$$
\begin{aligned}
u^T f(A)u &= u^T Q f(\Lambda) Q^T u \\
&= y^T f(\Lambda) y, \\
&= \sum_{i=1}^{n} \omega_i f(\lambda_i),\ y \equiv Q^T u,\ \omega_i \equiv |(u, q^i)|^2.
\end{aligned}
$$

We assume, without loss of generality, that $\|u\| = 1$. Clearly, as in Section 2, the last sum is a Riemann–Stieltjes integral, namely

$$I[f] = u^T f(A)u = \int_{\lambda_1}^{\lambda_n} f(\lambda)\, d\omega(\lambda),$$

where, as above, the distribution function ω is the non-decreasing piecewise constant function, with points of increase at the eigenvalues of A, and discontinuities of sizes $\omega_1, \ldots, \omega_n$.

We are looking for upper and lower bounds $L[f]$ and $U[f]$ for $I[f]$,

$$L[f] \leq I[f] \leq U[f].$$

They can be obtained, among other techniques, by using Gauss, Gauss–Radau and Gauss–Lobatto quadrature formulas; for the pioneering work see, in particular, Dahlquist, Eisenstat and Golub (1972) and Dahlquist, Golub and Nash (1978). We shall use the general formula

$$\int_{\lambda_1}^{\lambda_n} f(\lambda)\, d\omega(\lambda) = \sum_{j=1}^{k} \omega_j^{(k)} f(\tau_j^{(k)}) + \sum_{l=1}^{M} \vartheta_l^{(M)} f(\sigma_l^{(M)}) + R^{k,M}[f],$$

where the weights $\omega_j^{(k)}, j = 1, \ldots, k$, $\vartheta_l^{(M)}, l = 1, \ldots, M$, and the nodes $\tau_j^{(k)}, j = 1, \ldots, k$ are to be determined, while the nodes $\sigma_l^{(M)}, l = 1, \ldots, M$ are prescribed; see Davis and Rabinowitz (1984), Gautschi (1968, 1985) and Golub and Welsch (1969). It is well known (see the excellent survey by Gautschi (1981)) that

$$R^{k,M}[f] = \frac{f^{(2k+M)}(\eta)}{(2k+M)!} \int_{\lambda_1}^{\lambda_n} \prod_{l=1}^{M}(\lambda - \sigma_l^{(M)}) \left[\prod_{j=1}^{k}(\lambda - \tau_j^{(k)})\right]^2 d\omega(\lambda),$$

where $\lambda_1 < \eta < \lambda_n$. If $M = 0$, this leads to the Gauss rule with no prescribed nodes. If $M = 1$ and we fix the node at one of the end points, $\sigma_1^{(1)} = \lambda_1$ or $\sigma_1^{(1)} = \lambda_n$, we have the Gauss–Radau formula. If $M = 2$ and $\sigma_1^{(2)} = \lambda_1$, $\sigma_2^{(2)} = \lambda_n$, this is the Gauss–Lobatto formula.

As presented above, the nodes and weights in the Gauss rule are given by the eigenvalues of T_k (the Ritz values $\theta_j^{(k)}$) and the squared first elements of the normalized eigenvectors of T_k respectively (cf. Golub and Welsch (1969)), where T_k is the tridiagonal matrix of the recurrence coefficients generated by the Lanczos algorithm for A and the starting vector $v^1 = u$. For the Gauss quadrature rule, we have

$$\int_{\lambda_1}^{\lambda_n} f(\lambda) \, d\omega(\lambda) \equiv L_G^{(k)}[f] + R_G^{(k)}[f],$$

with

$$L_G^{(k)}[f] = \sum_{j=1}^{k} \omega_j^{(k)} f(\theta_j^{(k)}) = (e^1)^T f(T_k) \, e^1,$$

$$R_G^{(k)}[f] = \frac{f^{(2k)}(\eta)}{(2k)!} \int_{\lambda_1}^{\lambda_n} \left[\prod_{j=1}^{k}(\lambda - \theta_j^{(k)})\right]^2 d\omega(\lambda).$$

Consequently, in order to compute the value of the quadrature, we do not need to determine its nodes and weights. Suppose f is such that $f^{(2k)}(\xi) > 0$, $\forall k$, $\forall \xi$, $\lambda_1 < \xi < \lambda_n$. Then

$$L_G[f] \leq I[f], \quad k = 1, 2, \ldots$$

and the Gauss rule provides in this case a lower bound for the quadratic form. Note that this applies for $f(\lambda) = 1/\lambda$.

To summarize, for estimating the A-norm of the error in CG, we obtain

$$\|\epsilon^0\|_A^2 = (A^{-1}r^0, r^0) = \|r^0\|^2(T_n^{-1}e^1, e^1),$$

$$L_G^{(k)}\begin{bmatrix}1\\\lambda\end{bmatrix} = (T_k^{-1}e^1, e^1),$$

$$\|r^0\|^2\left[(T_n^{-1}e^1, e^1) - (T_k^{-1}e^1, e^1)\right] = \|r^0\|^2 \, R_G^{(k)}\begin{bmatrix}1\\\lambda\end{bmatrix} \geq 0.$$

and we formulate the key point as a theorem.

Theorem 3.2.   Using the previous notation, we get the following identities for the A-norm of the error in CG:

$$\|\epsilon^k\|_A^2 = \|r^0\|^2\, R_G^{(k)}\left[\frac{1}{\lambda}\right] = \|r^0\|^2[(T_n^{-1}e^1, e^1) - (T_k^{-1}e^1, e^1)],$$

i.e.,

$$\|\epsilon^k\|_A^2 = \|r^0\|^2\left[\sum_{j=1}^n \frac{(z_{(n)}^j, e^1)^2}{\lambda_j} - \sum_{j=1}^k \frac{(z_{(k)}^j, e^1)^2}{\theta_j^{(k)}}\right].$$

Proof.   This result is known: see Dahlquist, Golub and Nash (1978). The proof given here is, however, different from the original one.

By using the definition of the A-norm and $A\epsilon^k = r^k = r^0 - AV_ky^k$, we have

$$\|\epsilon^k\|_A^2 = (A\epsilon^k, \epsilon^k) = (A^{-1}r^0, r^0) - 2(r^0, V_ky^k) + (AV_ky^k, V_ky^k).$$

Since $T_k y^k = \|r^0\|e^1$,

$$(r^0, V_ky^k) = \|r^0\|^2(T_k^{-1}e^1, e^1),$$

and

$$(AV_ky^k, V_ky^k) = (V_k^T AV_ky^k, y^k) = (T_ky^k, y^k) = \|r^0\|^2(T_k^{-1}e^1, e^1),$$

the first identity is proved. The rest follows from the spectral decomposition of T_n and T_k.   □

We can conclude that the square of the A-norm of the CG error at the kth step divided by $\|r^0\|^2$ represents the remainder of the kth Gauss quadrature approximation of the corresponding Riemann–Stieltjes integral determined by A and $u = r^0/\|r^0\|$. Therefore the Gauss quadrature (here represented fully in the matrix form) gives lower bounds for the A-norm of the CG error. Upper bounds can be obtained with the Gauss–Radau rule if we have estimates of λ_1: see Golub and Meurant (1994). The second identity reflects the complicated relationship between the CG rate of convergence and the convergence of the Ritz values towards the eigenvalues of A. Another point on this is given by the following theorem.

Theorem 3.3.   For all k, there exists ϑ_k, $\lambda_1 \le \vartheta_k \le \lambda_n$ such that the A-norm of the error is given by

$$\|\epsilon^k\|_A^2 = \frac{\|r^0\|^2}{\vartheta_k^{2k+1}} \sum_{i=1}^n \left[\prod_{j=1}^k (\lambda_i - \theta_j^{(k)})^2\right]\omega_i,$$

where $\omega_i = |(v^1, q^i)|^2$.

Proof. The remainder of approximation $\int_{\lambda_1}^{\lambda_n} f(\lambda)\,d\omega(\lambda)$ with the Gauss quadrature is

$$\frac{f^{(2k)}(\vartheta)}{(2k!)}\int_{\lambda_1}^{\lambda_n}\prod_{j=1}^{k}(\lambda-\theta_j^{(k)})^2\,d\omega(\lambda),$$

with $\lambda_1 \le \vartheta \le \lambda_n$. Using $f = 1/\lambda$, this gives the statement of the theorem. □

This shows that, in exact arithmetic, when a Ritz value has converged to an eigenvalue of A, we have eliminated the component of the initial residual in the direction of the corresponding eigenvector of A. For related results on this subject we refer in particular to Axelsson and Linskog (1986) and van der Sluis and van der Vorst (1986).

The statement from Theorem 3.2 can be written as

$$\|\epsilon^0\|_A^2 = \|r^0\|^2(T_k^{-1}e^1, e^1) + \|\epsilon^k\|_A^2.$$

This recalls (3.6); restated as a theorem it reads as follows.

Theorem 3.4.

$$\|\epsilon^0\|_A^2 = \sum_{l=1}^{k}\gamma_{l-1}\|r^{l-1}\|^2 + \|\epsilon^k\|_A^2.$$

This means that the Gauss quadrature approximation $(T_k^{-1}e^1, e^1)$ can easily be computed as

$$L_G^{(k)}\left[\frac{1}{\lambda}\right] = \sum_{l=1}^{k}\gamma_l\,{}_1\frac{\|r^{l-1}\|^2}{\|r^0\|^2}.$$

Theorem 3.4 is in fact proved in Theorem 6:1 of Hestenes and Stiefel (1952, p. 416). The result was later derived and used, independently of the original paper, by many authors: see, *e.g.*, Deuflhard (1994), Axelsson and Kaporin (2001), Greenbaum (1997a) and Arioli (2004). It was used without being explicitly stated in Golub and Meurant (1997). In some of the given references the motivation is estimation of the error in CG.

The importance of the formula in Theorem 3.4 was emphasized by Strakoš and Tichý (2002, 2005). The first paper points to the original reference Hestenes and Stiefel (1952), proves the equivalence with the Gauss quadrature and gives an elementary proof which does not use the global orthogonality of the residuals or the global A-orthogonality of the direction vectors

$$\|\epsilon^k\|_A^2 - \|\epsilon^{k+1}\|_A^2 = \|x - x^{k+1} + x^{k+1} - x^k\|_A^2 - \|\epsilon^{k+1}\|_A^2$$

$$= \|x^{k+1} - x^k\|_A^2 + 2(x - x^{k+1})^T A(x^{k+1} - x^k)$$

$$= \gamma_k^2(p^k, Ap^k) + 2(r^{k+1}, x^{k+1} - x^k) = \gamma_k\|r^k\|^2.$$

The independence of the result on the global orthogonality is fundamental: it allows one to perform a detailed rounding error analysis, and to build up a mathematically rigorous argument that justifies validity of the given identity in finite precision CG computations. Without such an analysis, results derived using assumptions violated because of rounding errors are numerically useless, since they can give misleading information. Estimating errors in CG will be reviewed in Section 5.

Regarding the Euclidean norm, Theorem 6:3 of Hestenes and Stiefel (1952, pp. 416–417) gives the following result.

Theorem 3.5.

$$\|\epsilon^k\|^2 - \|\epsilon^{k+1}\|^2 = \frac{\|\epsilon^k\|_A^2 + \|\epsilon^{k+1}\|_A^2}{\mu(p^k)},$$

with

$$\mu(p^k) = \frac{(p^k, Ap^k)}{\|p^k\|^2}.$$

Hence, the Euclidean norm of the error is monotonically decreasing.

There is another expression for the Euclidean norm of the error derived by Meurant (2005).

Theorem 3.6.

$$\|\epsilon^k\|^2 = \|r^0\|^2[(e^1, T_n^{-2}e^1) - (e^1, T_k^{-2}e^1)] - 2\frac{(e^k, T_k^{-2}e^1)}{(e^k, T_k^{-1}e^1)}\|\epsilon^k\|_A^2.$$

This result (which is, of course, equivalent to the expression obtained by Hestenes and Stiefel) allows us, by using the spectral decomposition of T_n and T_k, to relate the norm of the error to the eigenvalues of A and to the Ritz values.

4. The Lanczos algorithm in finite precision

As an example, we consider a matrix that was introduced by Strakoš (1991) and used by Strakoš and Greenbaum (1992). The matrix of dimension n is diagonal, with the eigenvalues

$$\lambda_i = \lambda_1 + \left(\frac{i-1}{n-1}\right)(\lambda_1 - \lambda_n)\rho^{n-i}, \quad i = 1, \ldots, n.$$

The parameter ρ controls the distribution of the eigenvalues within the interval $[\lambda_1, \lambda_n]$. We shall use $n = 30$, $\lambda_1 = 0.1$, $\lambda_n = 100$ and $\rho = 0.8$, which gives well-separated large eigenvalues, and call this matrix D30. Figure 4.1 shows $\log_{10}$ of the elements of $|V_{30}^T V_{30}|$, each plotted against its index pair i, j, for the Lanczos algorithm applied to $A = $ D30 with the initial vector v^1 having equal components. Ideally $V_{30}^T V_{30}$ should be the identity matrix.

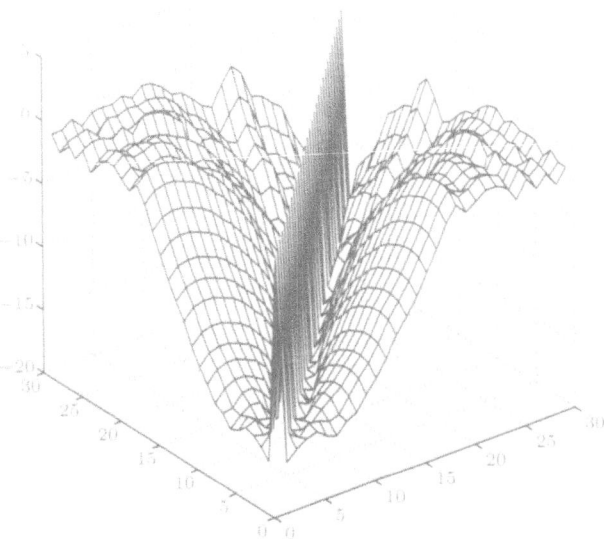

Figure 4.1. Matrix D30, $\log_{10}$ of $|V_{30}^T V_{30}|$.

But numerically this matrix is far from the identity. The magnitude of most nondiagonal entries is much larger than the level of elementary round-off. On the contrary, the magnitude of many of them is $\mathcal{O}(1)$.

It has been known since Lanczos (1950) that the behaviour of the algorithm in finite precision arithmetic is far from ideal. The Lanczos vectors v^k do not stay orthogonal as they ideally should. This also means that $V_k^T A V_k$ is no longer a tridiagonal matrix and the computed tridiagonal matrix T_k is not the projection of A on the computed Krylov subspace. Therefore the computed T_n is not similar to A, and the algorithm typically does not deliver sufficiently accurate numerical approximations to all eigenvalues of A in n iterations. Moreover, some eigenvalues of A can be numerically approximated by sets of very close Ritz values (called multiple copies) and it is difficult to decide whether such Ritz values are good approximations of the genuine close eigenvalues of A or just artifacts caused by rounding errors.

All these troubles are easily observable from numerical experiments; they are pointed out in practically all textbook expositions of the effect of rounding errors in the Lanczos algorithm. The same attention is, however, not paid to the analysis and resolution of the worst consequence of rounding errors, which every practical user of the Lanczos algorithm is inevitably faced with. Since numerically the Lanczos vectors are not orthogonal and they can very soon become linearly dependent, there is no guarantee whatsoever that the norm of the Ritz vector

$$x^j_{(k)} = V_k z^j_{(k)}$$

is close to one, where we assume for simplicity that the eigenvector $z^j_{(k)}$ of the computed T_k corresponding to the Ritz value $\theta^{(k)}_j$ is determined exactly. It can even numerically vanish, with norm close to the machine precision. In such a case it is absolutely unclear whether a small value of the convergence criterion $\eta_{k+1}|\zeta^{(k)}_{k,j}|$ described in Section 2.1 means convergence of $\theta^{(k)}_j$ to any eigenvalue λ_i of A using the finite precision Lanczos algorithm. Please notice that the trouble is not in computing $\eta_{k+1}|\zeta^{(k)}_{kj}|$ more or less accurately from T_k. We will assume (with negligible and easily quantifiable inaccuracy) that the exact spectral decomposition of T_k is known and that the quantity in question is computed exactly. The trouble consists in the fact that the derivation of the bound for $\min_i |\lambda_i - \theta^{(k)}_j|$ is based on the assumption that $V_k^T V_k = I$, which is usually drastically violated in finite precision computations.

It may seem that all the mathematical theory behind the Lanczos algorithm is lost because of rounding errors and the loss of orthogonality. Without a proper rounding error analysis we can not even interpret any Ritz value as a close approximation to an eigenvalue of A.

The first, and at the same time most original and most significant step in explaining the behaviour of the Lanczos algorithm in finite precision arithmetic was made by Chris Paige in his PhD thesis (Paige 1971). He proved the fundamental result that loss of orthogonality goes hand in hand with convergence of Ritz values, and developed a theory which formed a basis for practically all further progress in this area (except, perhaps, investigations of the maximal attainable accuracy of Krylov subspace linear algebraic solvers). His results were published in a series of papers, Paige (1972, 1976, 1980). Most of them are also included, together with some subsequent developments, in the classical monograph by Beresford Parlett (1980).

In this paper we would like to (partially) address the following questions.

- What theoretical properties of the Lanczos algorithm remain (with an insignificant inaccuracy) true in finite precision arithmetic?

- How can we describe the mechanics of the loss of orthogonality among the Lanczos vectors?

- What happens numerically to the equivalence of the Lanczos and CG algorithms as well as to the equivalence with orthogonal polynomials and Gauss quadrature?

- How do we evaluate convergence of CG in finite precision arithmetic?

The length of this paper is limited. Unless there is a good expository reason for presenting a part or the whole proof, in the following review we present theorems and statements without proofs. The reader interested in proofs or further details is referred to the original works.

4.1. Finite precision arithmetic

We use the standard model for floating point computations: see, *e.g.*, Higham (2002). Where needed we will denote by $fl(X)$ the result of the computation of X or denote the computed quantities by ˜. For any of the four basic operations $(+, -, *, /)$ denoted by op, we have

$$fl(x \text{ op } y) = (x \text{ op } y)(1 + \delta), \quad |\delta| \leq u_M$$

u_M being the unit round-off which is $(1/2)\beta^{1-t}$, where β is the base and t is the number of digits in the mantissa. This bound is obtained using rounding to the nearest floating point number, but this is generally the case. In IEEE standard 754, double precision, $(\beta = 2, t = 53)$ and

$$u_M = 1.110223024625157 \times 10^{-16},$$

which is half of the machine precision unit (machine epsilon) $\varepsilon_M = \beta^{1-t}$ representing the distance from 1 to the next-larger floating point number.

4.2. Paige's theory, loss of orthogonality, stabilization and convergence

The fundamental work of Chris Paige started at the end of the 1960s with some technical reports and papers, Paige (1969*a*), (1969*b*), (1970*a*) and (1970*b*), whose results led to his PhD thesis, Paige (1971), which clearly stated and proved, contrary to the common wisdom of the time, that even though the Lanczos algorithm in finite precision arithmetic does not keep its theoretical properties, it nevertheless works well as a reliable and highly efficient numerical tool for computing highly accurate approximations of dominant, and often other, eigenvalues of large sparse matrices.

The main contributions of Chris Paige presented in his thesis, or further developed from it, can be described as follows. He derived bounds for the local rounding errors in the Lanczos algorithm. He showed that the last elements of the eigenvectors of the computed tridiagonal matrix T_k indeed reliably tell us how well the eigenvalues of A are approximated by Ritz values, and how we can always obtain useful intervals containing eigenvalues of A. The computed Ritz values always lie between the extreme eigenvalues of A to within a small multiple of the machine precision. Moreover, at least one small interval containing an eigenvalue of A is found by the nth iteration. The algorithm behaves numerically like the Lanczos algorithm with full reorthogonalization until a very close eigenvalue approximation is found. Of course, the most (and rightly) celebrated of the results from

Paige's thesis is his proof that loss of orthogonality follows a rigorous pattern and implies that some Ritz values have converged.

Paige used a handy notation to bound combinations of rounding errors. Given $\varepsilon_1, \ldots, \varepsilon_p$ with each $|\varepsilon_i| \leq u_M$, then there exists a value α such that

$$\prod_{i=1}^{p}(1 + \varepsilon_i) = \alpha^p, \qquad |\alpha - 1| \leq u_M.$$

Let $D(\alpha)$ represent a diagonal matrix with elements not necessarily equal but satisfying the above bounds. Rules for manipulating such quantities are

$$\alpha^p \alpha^q = \alpha^{p+q},$$

$$x = \alpha(y + z) \Rightarrow x = \alpha y + \alpha z,$$

$$x = \left(\frac{\alpha^p}{\alpha^q}\right) y \text{ or } x = \alpha^p \alpha^q y \Rightarrow x = [1 + (p + q)\varepsilon]y,$$

where $|\varepsilon| \leq 1.01 u_M$. Using this notation, for the inner product we have, neglecting higher order terms in ε,

$$fl(x^T y) = x^T D(\alpha^n) y = x^T y + n\varepsilon |x^T| |y|,$$

and for the computation of the Euclidean norm,

$$fl(x^T x) = \alpha^n x^T x.$$

For the matrix vector product Paige used

$$fl(Ax) = (A + \delta A) x, \quad |\delta A| \leq m_A \varepsilon |A|,$$

where m_A is the maximum number of nonzero elements per row. This leads to

$$\|\delta A\| \leq m_A \varepsilon \| |A| \|.$$

Let β be such that $\| |A| \| = \beta \|A\|$. Then

$$\|\delta A\| \leq m_A \varepsilon \beta \|A\|.$$

Then Paige's thesis analysed various implementations of the Lanczos algorithm. This part of his work was published and complemented in Paige (1972), which justifies (2.1) as the preferable variant of the Lanczos algorithm.

In the subsequent part of the thesis, published in Paige (1976), the implementation (2.1) was studied further. The results were gathered in a theorem; see also Paige (1980, (2.10)–(2.16)).

Theorem 4.1. Let $\varepsilon_0 = 2(n + 4)\varepsilon_M < \frac{1}{12}$, $\varepsilon_1 = 2(7 + m_A\beta)\varepsilon_M$. Then the computed results of the Lanczos algorithm in finite precision arithmetic satisfy the matrix identity

$$AV_k = V_k T_k + \eta_{k+1} v^{k+1} (e^k)^T + \delta V_k,$$

where, for $j = 1, 2, \ldots, k$,

$$|(v^{j+1})^T v^{j+1} - 1| \le \varepsilon_0,$$

$$\|\delta v^j\| \le \varepsilon_1 \|A\|,$$

$$\eta_{j+1} |(v^j)^T v^{j+1}| \le 2\varepsilon_0 \|A\|,$$

$$|\eta_j^2 + \alpha_j^2 + \eta_{j+1}^2 - \|Av^j\|^2| \le 4j(3\varepsilon_0 + \varepsilon_1)\|A\|^2.$$

Since the local errors collected in δV_k are minor, the computed quantities satisfy the identity which formally looks very close to its exact precision counterpart. The presence of the extra term δV_k has, however, significant consequences. As an immediate one we get the following theorem.

Theorem 4.2. (Paige 1971, 1976 (21)–(23)) If R_k is the strictly upper triangular part of $V_k^T V_k$ such that

$$V_k^T V_k = R_k^T + \mathrm{diag}((v^j)^T v^j) + R_k,$$

then

$$T_k R_k - R_k T_k = \eta_{k+1} V_k^T v^{k+1}(e^k)^T + \delta R_k,$$

where δR_k is upper triangular with elements such that $|(\delta R_k)_{1,1}| \le 2\varepsilon_0 \|A\|$, and for $j = 2, 3, \ldots, k$

$$|(\delta R_k)_{j,j}| \le 4\varepsilon_0 \|A\|, \qquad |(\delta R_k)_{j-1,j}| \le 2(\varepsilon_0 + \varepsilon_1)\|A\|,$$
$$|(\delta R_k)_{i,j}| \le 2\varepsilon_1 \|A\|, \quad i = 1, 2, \ldots, j - 2.$$

This shows how the loss of orthogonality propagates through the algorithm.

A paper which finalizes publication of many of the results presented in Paige's thesis was published in 1980 in *Linear Algebra and its Applications* (Paige 1980). This paper is truly seminal; in this time of malign overemphasis on quantity of publications it should serve as a textbook example of a paper which could easily be split, although not for good reasons, into several publishable papers. The effect would have been similar to cutting a large diamond of superb quality into several pieces of more common size. The resulting pieces would still be easy to sell, but would be reduced to average quality. As a single brilliant piece, the paper Paige (1980) will continue to be read decades after its publication.

The paper starts by recalling the theorems of Paige (1976) quoted above (in Paige (1980) and here too the values of ε_0 and ε_1 are twice those of Paige (1976)). The matrix δR_k is bounded by

$$\|\delta R_k\|_F^2 \le 2[2(5k - 4)\varepsilon_0^2 + 4(k - 1)\varepsilon_0\varepsilon_1 + k(k - 1)\varepsilon_1^2] \|A\|^2,$$

where $\| \cdot \|_F$ denotes the Frobenius norm. If we denote $\varepsilon_2 = \sqrt{2} \max(6\varepsilon_0, \varepsilon_1)$ then

$$\| \delta R_k \|_F \le k\varepsilon_2 \| A \|.$$

The fundamental result relating the loss of orthogonality to eigenvalue convergence is given in the following theorem. We present the proof for its elegance and instructiveness.

Theorem 4.3. Let $z_{(k)}^j = (\zeta_{1,j}^{(k)}, \ldots, \zeta_{k,j}^{(k)})^T$ be the eigenvector of T_k corresponding to the Ritz value $\theta_j^{(k)}$ and $x_{(k)}^j = V_k z_{(k)}^j$ the corresponding Ritz vector, $j = 1, \ldots, k$. Let $\epsilon_{l,j}^{(k)} = (z_{(k)}^l)^T \delta R_k z_{(k)}^j$. Then,

$$|\epsilon_{l,j}^{(k)}| \le k\varepsilon_2 \| A \|,$$

and

$$(x_{(k)}^j)^T v^{k+1} = -\frac{\epsilon_{j,j}^{(k)}}{\eta_{k+1} |\zeta_{k,j}^{(k)}|}.$$

Proof. Multiplying the identity from Theorem 4.2 on both sides with a different eigenvector of T_k, we have

$$(z_{(k)}^l)^T (T_k R_k - R_k T_k) z_{(k)}^j = \eta_{k+1} (x_{(k)}^l)^T v^{k+1} \zeta_{k,j}^{(k)} + \epsilon_{l,j}^{(k)}.$$

Therefore,

$$(\theta_l^{(k)} - \theta_j^{(k)})(z_{(k)}^l)^T R_k z_{(k)}^j = \eta_{k+1} (x_{(k)}^l)^T v^{k+1} \zeta_{k,j}^{(k)} + \epsilon_{l,j}^{(k)}.$$

If we take $l = j$, we obtain the result. The bound on $\epsilon_{l,j}^{(k)}$ is a consequence of the bound on the norm of δR_k. $\qquad \square$

Hence, until $\eta_{k+1} |\zeta_{k,j}^{(k)}|$ is very small (at least proportional to $k\varepsilon_2 \| A \|$), the scalar product of the Ritz vector $x_{(k)}^j$ and v^{k+1} is small.

We point out that here and elsewhere in this expository paper the actual values of the upper bounds for the quantities which are small are not at all tight for realistic problems. They do not represent indicators of the maximal attainable accuracy using the Lanczos algorithm. Most of the known worst case bound techniques inevitably produce values of the bounds which are largely oversized. But this has little effect, if any, to the value of the results obtained by the worst case rounding error analysis. Their importance and strength is in the *insight*, not in values of the bounds.

Now we come to the point. Ideally, small $\eta_{k+1} |\zeta_{k,j}^{(k)}|$ means convergence of $\theta_j^{(k)}$ to some eigenvalue λ_i of A. Numerically, however, we must take into account that $\|x^j_{(k)}\|$ can be significantly smaller than unity, and, as given in Paige (1980, relation (3.15)),

$$\min_i |\lambda_i - \theta^{(k)}_j| \leq \frac{\eta_{k+1}|\zeta^{(k)}_{k,j}|(1 + \varepsilon_0) + \sqrt{k}\varepsilon_1 \|A\|}{\|x^j_{(k)}\|}.$$

A bound for the accuracy of the Ritz vector is then (see Paige (1971) and also Strakoš and Greenbaum (1992, Lemma 3.4))

$$\|x^j_{(k)} - (x^j_{(k)}, q^i)q^i\| \leq \frac{\eta_{k+1}|\zeta^{(k)}_{k,j}| + \sqrt{k}\varepsilon_1 \|A\|}{\min_{l \neq i} |\lambda_l - \theta^{(k)}_j|}.$$

Up to now, the analysis has been relatively simple and straightforward. This is no longer true for the remainder. In order to prove convergence of $\theta^{(k)}_j$ for $\|x^j_{(k)}\|$ significantly different from unity, Paige has ingeniously exploited properties of unreduced symmetric tridiagonal matrices. In particular, his concept of stabilized eigenvalues of T_k (see Section 2) plays the main role here. Paige has proved that if $\|x^j_{(k)}\|$ is significantly different from unity, then for some step $t < k$ there must be an eigenvalue of the left principal submatrix T_t of T_k which has stabilized to within a small δ and is close to $\theta^{(k)}_j$. This has further been used to prove that if $\eta_{k+1}|\zeta^{(k)}_{k,j}|$ is small, i.e., if $\theta^{(k)}_j$ has stabilized to within a small δ, then it is always close to some eigenvalue λ_i of A, regardless the size of $\|x^j_{(k)}\|$. Consequently, although the Lanczos algorithm can produce multiple Ritz approximations of single original eigenvalues, it can never produce any 'spurious' eigenvalues, i.e., Ritz values for which the convergence test $\eta_{k+1}|\zeta^{(k)}_{k,j}|$ is small and $\theta^{(k)}_j$ does not correspond to any eigenvalue λ_i of A. We summarize the result of Paige (1980, pp. 241–249) in the following theorem.

Theorem 4.4. Using the previous notation, for an eigenvalue $\theta^{(k)}_j$ of the matrix T_k computed via the Lanczos algorithm in finite precision arithmetic, we have

$$\min_i |\lambda_i - \theta^{(k)}_j| \leq \max\{2.5(\eta_{k+1}|\zeta^{(k)}_{k,j}| + \sqrt{k}\|A\|\varepsilon_1), [(k+1)^3 + \sqrt{3}n^2]\|A\|\varepsilon_2\}.$$

In the particular case when $\eta_{k+1}|\zeta^{(k)}_{k,j}|$ is small, the statement can be strengthened.

Theorem 4.5. If

$$\eta_{k+1}|\zeta^{(k)}_{k,j}| \leq \sqrt{3}k^2\|A\|\varepsilon_2,$$

then there exists a step $1 \leq t \leq k$ and an index $1 \leq s \leq t$ such that

$$\eta_{t+1}|\zeta_{s,t}^{(t)}| \leq \sqrt{3}t^2\|A\|\varepsilon_2 \quad \text{and} \quad \|x_{(t)}^s\| \geq \frac{1}{2},$$
$$\min_i |\lambda_i - \theta_s^{(t)}| \leq 5t^2\|A\|\varepsilon_2,$$

and $\theta_s^{(t)}, x_{(t)}^s$ is an exact eigenpair for a matrix within $5t^2\|A\|\varepsilon_2$ of A.

Please note that we are unable to prove a similar result for the given $\theta_j^{(k)}$. The difficulty is related to the possible existence of other Ritz values $\theta_l^{(k)}$ close to $\theta_j^{(k)}$. Using (2.8), Theorems 2.5, 2.7 and 4.3, Paige has proved that if $\theta_j^{(k)}$ is well separated from the other Ritz values at the same step, then $\|x_{(k)}^j\|$ cannot be significantly different from unity; see Paige (1980, (3.21), p. 243). In particular, if

$$\min_{l \neq j} |\theta_j^{(k)} - \theta_l^{(k)}| \geq k^{5/2}\|A\|\varepsilon_2,$$

we have

$$0.42 < \|x_{(k)}^j\| < 1.4.$$

Then, the result proved for $\theta_s^{(t)}$ will also hold for $\theta_j^{(k)}$. The strength of Theorem 4.4 is in the fact that the statement holds for $\theta_j^{(k)}$ no matter how many other eigenvalues of T_k are close to it.

The following theorem (see Paige (1980, Theorem 4.1)) shows that at least one eigenvalue of T_n (please recall that in exact arithmetic n represents the maximal number of steps of the Lanczos algorithm applied to A with any initial vector) must approximate some eigenvalue of A.

Theorem 4.6. If $n(3\varepsilon_0 + \varepsilon_1) \leq 1$, then at least one eigenvalue $\theta_j^{(n)}$ of T_n must be within $(n+1)^3\|A\|\varepsilon_2$ of an eigenvalue λ_i of the (n by n) matrix A. Moreover, there exist $1 \leq s \leq t \leq n$ such that

$$\eta_{t+1}|\zeta_{t,s}^{(t)}| \leq 5t^2\|A\|\varepsilon_2,$$

i.e., $\theta_s^{(t)}$ is within $5t^2\|A\|\varepsilon_2$, of λ_i.

One may question whether an analogous result can be proved for some Lanczos step $k < n$. The answer is negative, as follows from the beautiful result published in Scott (1979), which we now recall. Please note that the previous theory must hold for any initial vector v^1, $\|v^1\| = 1$. Scott's suggestion is to find, using the idea of reconstructing the unreduced symmetric tridiagonal matrix from the spectral data (for the history of this classical problem see Strakoš and Greenbaum (1992, p. 8)) a particular initial vector constructed in the following way.

Consider the diagonal matrix $A = \text{diag}(\lambda_i)$. Then (remember the assumption that the eigenvalues of A are distinct) the weights in the corresponding Riemann–Stieltjes integral (2.4) are determined by $\omega_l = |(v^1, e^l)|^2$. Using (2.6) for the last step of the ideal Lanczos algorithm, the same weights are given by

$$\omega_l = |(v^1, e^l)|^2 = -\frac{\hat{\eta}}{\chi_{n-1}(\lambda_l)\chi_n'(\lambda_l)}, \quad l = 1, \ldots, n, \qquad (4.1)$$

where $\hat{\eta}$ is a proper normalization constant chosen such that the constructed vector v^1 will have $\|v^1\| = 1$. Clearly, prescribing the eigenvalues of T_{n-1} (polynomial χ_{n-1}), (4.1) allows us to construct

$$v^1 \equiv (\sqrt{\omega_1}, \ldots, \sqrt{\omega_n})^T$$

such that the ideal Lanczos algorithm applied to $A = \text{diag}(\lambda_i)$ with this v^1 gives T_n in the last step (and T_{n-1} in the step $n-1$). The point is that the eigenvalues T_{n-1} (Ritz values $\theta_l^{(n-1)}$) can be chosen, e.g., as the midpoints of the intervals determined by the (distinct) eigenvalues of T_n (and A). Then no Ritz value $\theta_l^{(n-1)}$ at the step $n-1$ of the ideal Lanczos algorithm applied to A, with v^1 constructed as above, approximates an eigenvalue of A and no $\eta_n|\zeta_{n-1,l}^{(n-1)}|$ is small, $l = 1, \ldots, n-1$. But this means, by Theorem 2.5 (the Persistence Theorem), that no $\eta_{t+1}|\zeta_{t,s}^{(t)}|$ can be small for any choice $1 \le s \le t \le n-1$. Consequently, for this A and v^1 no Ritz value converges until step n.

A variant of the above construction of v^1 works for any given symmetric matrix A. Moreover, using a clever argument, Scott quantified the result in the following theorem; see Scott (1979, Section 4, Theorem 4.3).

Theorem 4.7. Let A be a symmetric n by n matrix with eigenvalues $\lambda_1 < \lambda_2 < \cdots < \lambda_n$, $\delta_A \equiv \min_{l \neq i} |\lambda_i - \lambda_l|$. Then there exists a starting vector v^1 such that, for the exact Lanczos algorithm applied to A with v^1, at any step $j < n$ the residual norm

$$\|Ax_{(j)} - \theta^{(j)} x_{(j)}\|$$

of any Ritz pair $\theta^{(j)}, x_{(j)}$ will be larger than $\delta_A/4$.

It should be emphasized that this result does not imply that no Ritz value can be close to an eigenvalue of A before step n. Under some lucky circumstances this can happen. Theorem 4.7 proves that such a situation cannot be revealed by the residual norm $\|Ax_{(j)} - \theta^{(j)} x_{(j)}\|$ or by the value $\eta_{j+1}|\zeta_{j,l}^{(j)}|$.

The previous results have remarkable consequences.

- First, since small $\eta_{k+1}|\zeta_{k,j}^{(k)}|$ means convergence of $\theta_j^{(k)}$ to some eigen-value λ_i of A, Theorem 4.3 may be restated:

  Orthogonality can be lost only in the directions of converged Ritz vectors.

  In contrast to this, we do not have a *proof* that convergence of a Ritz value is necessarily accompanied by the loss of orthogonality of v^{k+1} in the direction of the corresponding Ritz vector, since $\epsilon_{j,j}^{(k)}$ in the numerator in the statement of Theorem 4.3 can vanish. We have, however, not seen an example of such behaviour.

- Second, in the example constructed by Scott there is ideally no con-vergence of Ritz values until the final step. If this also remains true numerically, then for the particular A, v^1 given by Scott there is no significant loss of orthogonality among the computed Lanczos vectors $v^1, \ldots, v^n$! This means that loss of orthogonality in the finite precision Lanczos algorithm significantly depends for a given A on the choice of v^1. It should be admitted, though, that the particular initial vec-tors for which the loss of orthogonality is suppressed typically have rather weird components varying by many orders of magnitude. In-terested readers can check the validity of the above statements and the illustrative properties of the initial vectors suggested by Scott by numerical experiments.

An argument derived from the investigation of the accuracy of $\theta_j^{(k)}$ as the Rayleigh quotient shows (Paige 1980, (3.48))

$$\lambda_{\min}(A) - k^{\frac{5}{2}}\|A\|\varepsilon_2 \leq \theta_j^{(k)} \leq \lambda_{\max}(A) + k^{\frac{5}{2}}\|A\|\varepsilon_2,$$

which is true whether or not $\theta_j^{(k)}$ has stabilized to within a small δ.

We have seen that until some Ritz value stabilizes to within small δ, the orthogonality of numerically computed Lanczos vectors cannot be lost. This poses the question as to how closely the Lanczos algorithm in finite precision arithmetic can resemble the ideal one. Paige gives an elegant answer in terms of the backward error. In fact, if at step k

$$\eta_{l+1}|\zeta_{l,j}^{(l)}| \geq \sqrt{3}k^2\|A\|\varepsilon_2, \quad 1 \leq j \leq l \leq k, \tag{4.2}$$

then $\|R_k\|_F^2 < 1/12$ and all singular values of V_k lie in the open interval $(0.41, 1.6)$: see Paige (1980, p. 250). If the Lanczos algorithm is applied with full reorthogonalization at every step, implemented via the modified Gram–Schmidt algorithm, then under a mild restriction the computed columns V_k span the exact Krylov subspace of $A + \delta A$ (starting with the same v^1), where $\|\delta A\|$ is a multiple of $\|A\|\varepsilon_M$ (Paige 1970a). The following theorem (see Paige (1980, Theorem 4.2)) completes the argument.

Theorem 4.8. Using the previous notation, let (4.2) hold at step k of the finite precision Lanczos algorithm applied to A with v^1. Then there exists a matrix

$$A'(k) \quad \text{within} \quad (3k)^{1/2}\|A\|\varepsilon_2 \quad \text{of} \quad A$$

such that, for all $l = 1, \ldots, k+1$, the Lanczos vectors $v^1, \ldots, v^l$ span the Krylov subspaces of $A'(k)$ with the initial vector v^1.

Consequently, until the computed Krylov subspace contains an exact eigenvector of a matrix to within $5k^2\|A\|\varepsilon_2$ of the original matrix A (see Theorem 4.5), this subspace is the same as the Krylov subspace generated by a slightly perturbed matrix, *i.e.*, it is numerically stable in the backward error sense. It should, however, be noted that generally Krylov subspaces can be very sensitive to small changes in the matrix A.

We conclude the journey through the fascinating paper of Paige (1980) with the following comment. Until a Ritz value in steps 1 to k has stabilized to within $\sqrt{3}k^2\|A\|\varepsilon_2$, the Lanczos algorithm behaves numerically like the algorithm with full modified Gram–Schmidt reorthogonalization.

4.3. Backward-like analysis of Greenbaum and subsequent results

Consider a fixed step k of the finite precision Lanczos algorithm applied to A and v^1. We ask whether the results computed in steps 1 to k can be interpreted in some sense as results of the ideal Lanczos algorithm applied to some matrix B with an initial vector v_B^1. Indeed, as we have seen in Section 2, the numerically computed matrix T_k storing the Lanczos recurrence coefficients can be obtained in k steps of the k-dimensional ideal Lanczos algorithm applied to T_k with the initial vector e^1. Components of e^1 in the basis of the (orthonormal) eigenvectors $Z_k = (z_{(k)}^1, \ldots, z_{(k)}^k)$ of T_k are equal to the elements of the first row of the matrix Z_k; their squares representing weights in the corresponding k-dimensional Riemann–Stieltjes integral: see (2.6). Consequently, the matrix T_k can be obtained as a result of the exact Lanczos algorithm applied to any k by k matrix B having the same eigenvalues as T_k with the initial vector v_B^1 having the components in the corresponding eigenspaces of B equal to the elements of $Z_k^T e^1$.

This relationship, although interesting, does not tell us much, since T_k (or B) can have some eigenvalues close to the eigenvalues of A, but others can be very different from the eigenvalues of A. As we have seen, the finite precision Lanczos algorithm may form multiple copies of several eigenvalues of A, with the multiplicities growing as the number of iteration steps increases. But the algorithm will never give a Ritz value stabilized to within a small δ that does not approximate any eigenvalue of A. It therefore seems necessary to impose additional conditions on B and v_B^1.

Given T_k computed in k steps of the finite precision Lanczos algorithm applied to A with v^1, we look for B and v_B^1 such that *all* eigenvalues of B lie close to the eigenvalues of A. In addition to that, the sum of squares of the components of v_B^1 in the invariant subspaces corresponding to close approximations of some eigenvalue λ_i of A is required to be equal to the squared component of v^1 in the direction of the original eigenvector q^i. Finally, the point is that we require T_k to be determined in the first k steps of the *exact* Lanczos algorithm applied to B with v_B^1.

When Anne Greenbaum developed her highly original and deeply thought theory on the foundations laid by Paige, she supported the previous intuitive argument with a rigorous mathematical derivation. She showed that the exact Lanczos recurrence for a matrix whose eigenvalues are clustered in small intervals can be thought of as a slightly perturbed recurrence, analogous to that of Theorem 4.1, for a *new problem*. This new problem has, for each original cluster interval, one eigenvalue from this interval representing the whole cluster. The sum of the weights of the original eigenvalues in each cluster is equal to the weight of its chosen representing eigenvalue: see Greenbaum (1989). From that she set the goal of proving that *every* slightly perturbed Lanczos recurrence, including the finite precision Lanczos algorithm described in Theorem 4.1, is in the sense described above equivalent to an exact Lanczos recurrence for a matrix whose eigenvalues lie in small intervals about the eigenvalues of the given original matrix.

While the details of the theorems and proofs of Greenbaum (1989) are quite involved and have probably not been read carefully by many people, the basic ideas are ingenious, and the paper should be considered obligatory classical reading together with Paige (1980). We will try to recall the main points in order to reveal, within our abilities, the beauty of the construction given by Greenbaum.

To show that the matrix T_k generated at step k of the finite precision Lanczos recurrence applied to A with v^1 is the same as that given by the exact Lanczos algorithm applied to some B with v_B^1, where all eigenvalues of B are close to those of A, it is sufficient and also necessary to show that T_k can be extended to a larger unreduced symmetric tridiagonal matrix (having positive subdiagonal entries)

$$T_{k+K} = \begin{pmatrix} T_k & \eta_{k+1} & & & & & \\ \eta_{k+1} & \alpha_{k+1} & \eta_{k+2} & & & & \\ & \eta_{k+2} & \alpha_{k+2} & \eta_{k+3} & & & \\ & & \ddots & \ddots & \ddots & & \\ & & & \eta_{k+K-1} & \alpha_{k+K-1} & \eta_{k+K} \\ & & & & \eta_{k+K} & \alpha_{k+K} \end{pmatrix}$$

whose eigenvalues are all close to those of A. Then we can simply take $B \equiv T_{k+K}$, $v_B^1 \equiv e^1$.

Greenbaum has constructed T_{k+K} by a hypothetical continuation of the first k steps of the finite precision Lanczos algorithm applied to A with v^1. The needed situation $\eta_{k+K+1} = 0$ for some K can be reached in the following way. Based on the theory of Paige describing the loss of orthogonality among the Lanczos and Ritz vectors, Greenbaum has identified a set of $k - m_k$ vectors in the subspace generated by the computed $\{v^1, \ldots, v^k\}$ such that the chosen vectors are mutually (exactly) orthogonal and normalized, and the newly computed v^{k+1} is also approximately orthogonal to all of them. Let these vectors be stored as the columns of the n by $(k - m_k)$ matrix Y_{k-m_k}. Exact orthogonalization of v^{k+1} against them adds a small additional contribution into the error term. Then the Lanczos recurrence can hypothetically be continued with the exact orthogonalization of the newly generated Lanczos vectors against each other and with exact orthogonalization of them against Y_{k-m_k}. From the exact orthogonalization we must get $\eta_{k+K+1} = 0$ since $(Y_{k-m_k}, v^{k+1}, \ldots, v^{k+K})$, where $K = n + m_k - k$, represents a set of n orthogonal vectors in the n-dimensional space. Summarizing, we get

$$AV_{k+K} = V_{k+K}T_{k+K} + F_{k+K},$$

where in $F_{k+K} = \left(f^1, \ldots, f^{k-1}, f^k, \ldots, f^{k+K}\right)$ the first $k - 1$ columns are the perturbations in the steps of the original finite precision Lanczos algorithm and the other columns $f^k, \ldots, f^{k+K}$ are perturbations arising from reorthogonalizations in Greenbaum's construction. The way this is done cannot be described here since it involves many details which cannot be included in this expository paper. The key point is in the choice of the orthonormal vectors Y_{k-m_k}; they cannot contain, e.g., any vector in the subspace of the converged Ritz vectors corresponding to well-separated Ritz values, since these represent well-defined directions in which the orthogonality is definitely lost. More substantively, the clever choice of Y_{k-m_k} described in Greenbaum (1989) allows her to prove that the perturbation vectors $f^k, \ldots, f^{k+K}$, introduced in the hypothetical continuation of the finite precision Lanczos algorithm, are *small*. Paige's results summarized in Theorem 4.4 can then be applied to the $k + K = n + m_k$ steps of the hypothetically extended finite precision Lanczos recurrence described above with $\eta_{n+m_k+1} = 0$, where from the proofs in Paige (1980) it follows that the size of the errors corresponding to the perturbations $f^k, \ldots, f^{k+K}$ can be expressed in term of their norms.

Theorem 4.9. The matrix T_k generated at step k of the finite precision Lanczos algorithm applied to A with v^1 is equal to that generated by an exact Lanczos recurrence applied to an $(n + m_k)$ by $(n + m_k)$ matrix B whose eigenvalues lie within

$$\mathcal{O}(n + m_k)^3 \max\{\varepsilon_M \|A\|, \|f^k\|, \ldots, \|f^{n+m_k}\|\}$$

of some of the eigenvalues of A, where $f^k, \ldots, f^{n+m_k}$ are the smallest perturbations that will cause a coefficient η_{j+1} to be zero at or before step $n + m_k$.

The particular $f^k, \ldots, f^{n+m_k}$ given via the construction of Greenbaum (1989) are perhaps not the optimal ones, but they are small enough to justify this approach. Finally, Theorem 4.2 of Strakoš (1991) proves the intuitively expected fact that any matrix B with the property of Theorem 4.9 must have at least one eigenvalue close to each eigenvalue of the original matrix A for which the component of the initial vector v^1 in the corresponding invariant subspace is nonzero.

We remark that the value m_k and the matrix B depend on k. The matrix B with the required property described above is not unique; there might be other constructions giving similar results with matrices of different sizes. If we limit the number of steps in the application of the Lanczos algorithm to some reasonable number N, say, much smaller than $(n\varepsilon_M\|A\|)^{-1}$, then it is legitimate to ask whether one can take a matrix B with v_B^1 such that the exact Lanczos algorithm applied to B with v_B^1 will give in steps 1 to N not necessarily identical, but *very close* Ritz values, to those provided by the finite precision Lanczos algorithm applied to A, v^1. Here we do not mean determining B (and v_B^1) *a posteriori* for the step N, but *a priori* using the spectral decomposition of A and the components of v^1 in the individual invariant subspaces. This idea was thoroughly illustrated in Greenbaum and Strakoš (1992), where B was constructed by spreading sufficiently many eigenvalues in tiny intervals around each eigenvalue of A. Numerical experiments suggest that the size of such intervals is much smaller than the technically complicated bounds from Greenbaum (1989) would suggest. A rigorous mathematical quantification of this approach is still incomplete. When completed, it would also lead to a possibly very elegant matrix-free description of the Lanczos algorithm behaviour in finite precision arithmetic in terms of the Gauss quadratures of a Riemann–Stieltjes integral with a slightly blurred distribution function (see Section 5 of Golub and Strakoš (1994), Section 4.5 of Greenbaum (1997a), and Section 5 of Strakoš and Tichý (2002)). This must, however, include a sensitivity analysis of Gauss quadrature to small perturbations of the Riemann–Stieltjes integral, which appears to be a rather difficult problem (O'Leary and Strakoš 2004). A different but somewhat related problem concerning sensitivity of the Lanczos coefficients to perturbations of the distribution function in the Riemann–Stieltjes integral is investigated in Kautsky and Golub (1983); see also Gragg and Harrod (1984), Laurie (1999, 2001) and Druskin, Borcea and Knizhnerman (2005).

A frequently asked question is whether the finite precision Lanczos algorithm can simply miss an eigenvalue because it is constantly forming copies of others. This is known as the Lanczos phenomenon (see Cullum and Willoughby (1985)) and it can be considered resolved by the series of works by Druskin and Knizhnerman (Druskin and Knizhnerman 1991, Knizhnerman 1995*a*, 1995*b*, 1996); see also Druskin, Greenbaum and Knizhnerman (1998) and Greenbaum (1994). Using some technical assumptions, it is proved that each eigenvalue of A will indeed eventually be approximated by a Ritz value. The proven statement is, however, more of theoretical than practical interest. A considerable part of these papers is also devoted to approximation of matrix functions.

Existence of tight clusters of Ritz values is linked to most of the technical difficulties that complicate the bounds and proofs of Paige (1980) and Greenbaum (1989). We know that a Ritz value can stabilize to within a small δ only close to an eigenvalue of A. If the stabilized Ritz value is well separated, then the norm of the Ritz vector cannot significantly differ from unity, and the Ritz vector closely approximates the corresponding eigenvector of A. When a Ritz value is a part of a tight cluster, then some or *all* Ritz pairs corresponding to the cluster can have weird properties.

In Strakoš and Greenbaum (1992) several conjectures have been formulated, but not proved (except for some simple cases). In particular, it is important to ask the following questions.

C1 (Stabilization of clusters.) Does any tight well-separated cluster consisting of at least two Ritz values approximate an eigenvalue of A?

C2 (Stabilization of Ritz values in a cluster.) Is any Ritz value in a tight well-separated cluster stabilized to within a small δ? In particular, Strakoš and Greenbaum (1992) conjectured that the answer is positive, and that δ is proportional to the square root of the size of the cluster interval divided by the square root of the separation of the cluster from the other Ritz values.

C3 (Stabilization of weights.) Let Ritz values in a tight well-separated cluster, which may consist of one or more Ritz values, closely approximate some eigenvalue λ_i of A. Does the sum of weights of these Ritz values in the corresponding Riemann–Stieltjes integral closely approximate the weight of the original eigenvalue λ_i?

Similar questions can be formulated solely in terms of unreduced symmetric tridiagonal matrices, and they are therefore not specific to the finite precision Lanczos algorithm. In the latter case they are, however, of particular importance. We will not specify the intuitive meaning of the terms 'tight cluster', 'size of the cluster interval' and 'separation of the cluster' since that would need detailed notation which we cannot afford, because of lack of space. The intuitive meaning is clear; a technical quantification can be found in the papers by Wülling, which we are now going to recall.

The conjectures were investigated in Wülling (2005) and (2006) with the following outcome.

- Every tight well-separated cluster of at least two Ritz values must stabilize, *i.e.*, the answer to **C1** is positive.

- There are tight well-separated clusters of Ritz values (which, according to the previous point, must approximate an eigenvalue of A) in which none of the Ritz values is stabilized to within a small δ, *i.e.*, the answer to **C2** is negative.

- The weights in the Riemann–Stieltjes integral corresponding to the kth Gauss quadrature of the original Riemann–Stieltjes integral determined by A and v^1 must stabilize, *i.e.*, the answer to **C3** is positive. This is not proved directly in Wülling (2005), but it can be obtained by a combination of Wülling (2005) with the inequality (8.21) in Greenbaum (1989): see Wülling (2005, Section 5).

In contrast with Strakoš and Greenbaum (1992), where the results are based on relatively simple algebraic manipulations of the known formulas for eigenvalues and eigenvector elements of unreduced symmetric tridiagonal matrices, Wülling (2005) and (2006) are based on the following very clever observation. The bottom and top elements of the eigenvectors of T_k, which determine the stabilization criterion and the weights respectively, are expressed in terms of the values of polynomials $\chi_{k-1}(\theta)$ and $\chi'_k(\theta)$: see (2.6)–(2.8). Moreover, $\chi_{k-1}(\theta)$ and $\chi_k(\theta)$ have simple roots in the corresponding Ritz values. Therefore, using the residue theorem from complex analysis, the sum of squares of the bottom elements of the (normalized) eigenvectors of T_k, which correspond to the Ritz values in a cluster C, can be viewed as the result of the line integral

$$\sum_C (\zeta_{k,l}^{(k)})^2 = -\sum_C \frac{\chi_{k-1}(\theta_l^{(k)})}{\chi'_k(\theta_l^{(k)})} = \frac{1}{2\pi} \left| \int_{\partial D_C} \frac{\chi_{k-1}(z)}{\chi_k(z)} \, \mathrm{d}z \right|, \qquad (4.3)$$

where ∂D_C is the circle which contains all Ritz values belonging to C in its interior and all other eigenvalues of T_k in its exterior: see Wülling (2006). Similarly, omitting technicalities, the changes in the weights can be investigated using the line integral

$$\frac{1}{2\pi} \left| \int_{\partial D_C} \frac{\eta_2^2 \eta_3^2 \cdots \eta_k^2}{\chi_{k-1}(z)\chi_k(z)} \, \mathrm{d}z \right|; \qquad (4.4)$$

see Wülling (2005, (4.5)). The results are then obtained by bounding the line integrals (4.3) and (4.4), which represent an example of nontrivial technical work. We also point out that, concerning **C1** and **C2**, the results of Wülling (2006) are stronger than the formulations of the conjectures in Strakoš and Greenbaum (1992) have assumed.

The analysis of Wülling gives another example of interplay between analysis (here complex analysis, which is used to obtain bounds for algebraic expressions formulated in terms of values of orthogonal polynomials) and algebra, often observable while dealing with the Lanczos algorithm.

4.4. Intermediate quantities and the accuracy of Ritz approximations

As we have already seen, the finite precision Lanczos algorithm serves as an instructive example illustrating several fundamental principles. Its rounding error analysis is perhaps complicated, lengthy and full of unpleasant technical details, bounds and formulas. However, it reveals the pattern rigorously, and the conclusions can be formulated clearly, simply and in an elegant way.

In addition, the whole rounding error analysis reveals the following principal fact of 'philosophical' importance. The Ritz values as approximations to eigenvalues of the original matrix A can be computed to high accuracy despite the fact that the intermediate quantities, $i.e.$, the computed Lanczos coefficients stored in the matrix T_k, $k = 1, 2, \ldots$, can have from some (typically rather modest) value of k not a single digit of accuracy. In other words, the number of correct digits in the computed entries of T_k (in comparison with their ideal counterparts) is absolutely irrelevant for the obtainable accuracy of the approximations to the eigenvalues of A determined from T_k. Here we see the power of the backward-like analysis (cf. Parlett (1990, pp. 22 and 24)), and the limitations of the mechanically applied forward error analysis, when it considers comparison of all computed and ideal quantities.

4.5. Reorthogonalization strategies and rewards for maintaining semiorthogonality

Although the inaccuracy of T_k does not prevent accurate approximation of eigenvalues of A by Ritz values, it has rather unpleasant effects: multiple approximations of some eigenvalues of A, and delays in the approximation of another ones. The way to suppress these side effects, which is sometimes desirable, is to apply a correction procedure which preserves maximally, or to some suitable level, the mutual orthogonality of the computed Lanczos vectors. Reorthogonalization strategies and the rewards for maintaining a proper level of mutual orthogonality are thoroughly described in Scott (1978), Parlett and Scott (1979), Scott (1981), Simon (1982, 1984a) and Parlett (1994), and excellently summarized in Simon (1984b), Parlett (1992). Here we will briefly recall some main ideas. An extended exposition can be found in the last two papers.

We start with the PhD thesis of Grcar (1981), which, to our knowledge, was not published. In contrast to other researchers, his considerations are based on the forward error of the computed Lanczos vectors. Grcar's results suggest, though the formal proofs have not been completed, that until the above-mentioned forward error exceeds the level proportional to $\sqrt{\varepsilon_M}$, the computed Krylov subspace is correct to the level proportional to ε_M (the error stays largely within the ideal Krylov subspace). In order to maintain this so-called projection property, Grcar suggested periodic reorthogonalization. The forward approach of Grcar (1981) has to deal with some theoretical and practical difficulties. The way Grcar uses nonhomogeneous three-term recurrences inspired later solutions of other problems: see Gutknecht and Strakoš (2000) and Meurant (2006).

Beresford Parlett and his PhD students played the instrumental role in the other reorthogonalization strategies, which have been conveniently based on Paige's results and backward error analysis. It was discovered that, in order to largely suppress the unpleasant effects of round-off on the approximation of the eigenvalues of A, full reorthogonalization of the Lanczos vectors (in order to maintain their mutual orthogonality close to ε_M) is not necessary. It suffices to maintain some 'strong linear independence' of the computed Lanczos vectors. Scott has shown (see Parlett and Scott (1979), Scott (1978, 1981)) that it is beneficial to maintain *semi-orthogonality* of the numerically computed Lanczos vectors, *i.e.*, to satisfy

$$\|V_k^T v^{k+1}\| \leq \sqrt{\varepsilon_M}, \quad k = 1, 2, \ldots. \tag{4.5}$$

Since Theorem 4.3 proved by Paige shows that orthogonality can be lost only in the direction of converged Ritz vectors, one suggestion is to maintain semi-orthogonality by reorthogonalizing at each step k the newly computed Lanczos vector against all Ritz vectors for which

$$\eta_{k+1}|\zeta_{k,l}^{(k)}| < k\sqrt{\varepsilon_M}\|A\|;$$

cf. Simon (1984b, Theorem 6, p. 126). This strategy, called *selective reorthogonalization* (SO) requires computing Ritz vectors. That is avoided in the *partial reorthogonalization strategy* (PRO) of Simon. Based on the underlying rigorous analysis of Paige, Simon has suggested and justified a simplified model of finite precision behaviour of the Lanczos algorithm. His strategy is based on monitoring the loss of orthogonality among the Lanczos vectors via a three-term recurrence: see Simon (1984b, Theorem 1, p. 107). It reorthogonalizes the newly computed Lanczos vector at step k against those previously computed Lanczos vectors related through some heuristic to the threshold criteria for the loss of orthogonality proportional to $\sqrt{\varepsilon_M}/k$. Simon then proved the following theorem.

Theorem 4.10. Let T_k be the unreduced symmetric tridiagonal matrix computed by the Lanczos algorithm applied to A with v^1 that uses some reorthogonalization in order to maintain semi-orthogonality among the computed Lanczos vectors. Then, up to a (full) perturbation matrix having norm proportional to $\varepsilon\|A\|$, T_k is the orthogonal projection of A onto the subspace spanned by the computed Lanczos vectors.

This means (see also Simon (1984b, pp. 119–122), Parlett (1992, pp. 255–257)) that in the above sense semi-orthogonality is as good as orthogonality maintained proportional to full machine precision. Finally, Theorem 4.4 of Parlett (1992) proves that an additional full reorthonalization at a step k guarantees an improvement of the mutual orthogonality only if semi-orthogonality is maintained in steps 1 to k.

As mentioned above, in our exposition we assume that the exact spectral decomposition of the unreduced symmetric tridiagonal matrix T_k is known. Here such an assumption is reasonable, since an investigation of further issues related to computing this spectral decomposition is out of the scope of this review. Nevertheless, since T_k can have tight clusters of eigenvalues, we wish at least to point out several publications devoted to interesting issues arising from this problem; see Ye (1995), Parlett (1996), Parlett and Dhillon (2000) and Dhillon and Parlett (2003, 2004). .

4.6. Recent results on the loss of orthogonality and multiple approximation of eigenvalues

As we have said before, an attempt at forward error analysis of the Lanczos algorithm was given by Grcar (1981). Grcar obtained expressions for the computed Lanczos vectors in terms of the exact Lanczos vectors. In this section, we will summarize works (Zemke 2003, Meurant 2006) interested in the components of the Lanczos vectors in the directions of the eigenvectors of A. The goal of these works is to understand the behaviour of the projections of the Lanczos vectors, their relation to the loss of orthogonality, and the appearance of multiple copies of the eigenvalues. This problem leads to investigation of perturbed three-term scalar recurrences. There are different ways to write the solution of these recurrences (for instance, using polynomials or using the Lanczos matrix T_k). They show what equation (2.2), giving Lanczos vectors as polynomials in A applied to the initial vector, becomes in finite precision arithmetic.

Let us start by considering the D30 example. We look at components of the Lanczos vectors in the directions of the eigenvectors of A. Since the matrix D30 is diagonal, we simply consider the components of the Lanczos vectors. The initial vector has all its components equal. The eigenvalue which is first approximated by a Ritz value is the largest one, $\lambda_{30} = 100$. In Figure 4.2 the solid line is $\log_{10}(|v_{30}^k|)$ as a function of k, computed by the Lanczos algorithm using full reorthogonalization of the newly computed

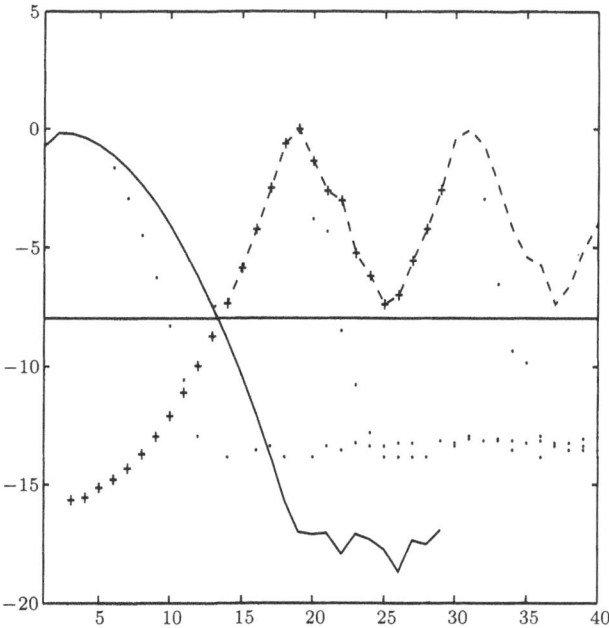

Figure 4.2. D30, $\log_{10}$ of the absolute value of
the last component of the Lanczos vectors.

Lanczos vector against the previously computed Lanczos vectors, with the
reorthogonalization done twice (which we call double reorthogonalization).
Before the component of interest v_{30}^k reaches the square root of machine
precision, the computed results of this example can be considered close approximations to the exact precision ones. As predicted by theory, the last
component of the Lanczos vector (with double reorthogonalization) converges to machine precision. The dashed line (which is hidden behind the
solid line until it is nearly at the horizontal line) is $\log_{10}(|\tilde{v}_{30}^k|)$ computed
by the standard finite precision Lanczos algorithm. The + signs represent
$\log_{10}$ of the absolute value of the differences $v_{30}^k - \tilde{v}_{30}^k$. The horizontal line
is $\log_{10}(\sqrt{\varepsilon_M})$. The dots give the distances of the Ritz values to λ_{30} after
they become smaller than a threshold of 0.1.

The computed component is almost equal to the ideal result down to $\sqrt{\varepsilon_M}$
but then, instead of continuing to go down, it starts going back up to $\mathcal{O}(1)$.
The difference is increasing almost from the beginning of the iterations up
to iteration 18. After that there is an almost periodic behaviour. Each
time the last component reaches $\mathcal{O}(1)$, a new copy of the largest eigenvalue
appears. This simple example shows there is an interesting structure in the
components of the Lanczos vectors in the directions of the eigenvectors of A.
Similar pictures with different examples are given and analysed in Zemke
(2003, pp. 210–217).

In exact arithmetic we have the relation

$$\eta_{k+1}v^{k+1} = Av^k - \alpha_k v^k - \eta_k v^{k-1}.$$

In finite precision computations, this relation becomes

$$\tilde{\eta}_{k+1}\tilde{v}^{k+1} = A\tilde{v}^k - \tilde{\alpha}_k \tilde{v}^k - \tilde{\eta}_k \tilde{v}^{k-1} + f^k, \tag{4.6}$$

where f^k represents the rounding errors that occurred while computing step $k+1$. Of course, the coefficients $\tilde{\alpha}_k$ and $\tilde{\eta}_k$ are different from those in exact arithmetic since they are determined (numerically) using the computed Lanczos vectors. This is what makes a forward analysis of the finite precision Lanczos algorithm difficult. Let $\bar{v}^k = Q^T \tilde{v}^k$ be the vector of the projections of the computed Lanczos vector on the eigenvectors of A. We have

$$\tilde{\eta}_{k+1}\bar{v}_i^{k+1} = \lambda_i \bar{v}_i^k - \tilde{\alpha}_k \bar{v}_i^k - \tilde{\eta}_k \bar{v}_i^{k-1} + \bar{f}_i^k, \tag{4.7}$$

where $\bar{f}^k = Q^T f^k$. Solutions of such three-term recurrences are studied in Meurant (2006) where the following result is proved.

Theorem 4.11. Let j be given and let $p_{j,k}$ be the polynomials determined by

$$p_{j,j-1}(\lambda) \equiv 0, \qquad p_{j,j}(\lambda) \equiv 1,$$
$$\zeta_{k+1}p_{j,k+1}(\lambda) = (\lambda - \tau_k)p_{j,k}(\lambda) - \zeta_k p_{j,k-1}(\lambda), \ \ k = j, j+1, \ldots.$$

The solution of the perturbed scalar recurrence

$$\zeta_{k+1}s_{k+1} = (\lambda - \tau_k)s_k - \zeta_k s_{k-1} + f_k, \tag{4.8}$$

starting from $s_0 = 0$ and s_1 is given by

$$s_{k+1} = p_{1,k+1}(\lambda)s_1 + \sum_{l=1}^{k} p_{l+1,k+1}(\lambda)\frac{f_l}{\zeta_{l+1}}.$$

The polynomials $p_{j,k}$, $j > 1$ are usually called the associated polynomials. They are orthogonal with respect to a Riemann–Stieltjes integral with a distribution function that depends on j. When applying this to the Lanczos algorithm, we use the following result.

Lemma 4.12. The polynomial $p_{j,k}$, $k \geq j$ is given by

$$p_{j,k}(\lambda) = (-1)^{k-j} \frac{\chi_{j,k-1}(\lambda)}{\tilde{\eta}_{j+1} \cdots \tilde{\eta}_k},$$

where $\chi_{j,k}(\lambda)$ is the determinant of $\tilde{T}_{j,k} - \lambda I$, where $\tilde{T}_{j,k}$ is the tridiagonal matrix obtained from the computed Lanczos matrix $\tilde{T}_k$ by deleting the first $j-1$ rows and columns.

The possible growth of the local round-off perturbations is therefore linked to the eigenvalues of the matrices $\tilde{T}_{j,k}$ for all $j \leq k$. A similar technique has also been used by Gutknecht and Strakoš (2000) in the investigation of the maximal attainable accuracy.

Applying these results to the finite precision Lanczos algorithm, that is, to (4.7), we obtain the following result.

Theorem 4.13. Let j be given and $\tilde{p}_{j,k}$ be the polynomials given by

$$\tilde{p}_{j,j-1}(\lambda) \equiv 0, \qquad \tilde{p}_{j,j}(\lambda) \equiv 1,$$
$$\tilde{\eta}_{k+1}\tilde{p}_{j,k+1}(\lambda) = (\lambda - \tilde{\alpha}_k)\tilde{p}_{j,k}(\lambda) - \tilde{\eta}_k\tilde{p}_{j,k-1}(\lambda), \; k = j, j+1, \ldots.$$

Then, the computed Lanczos vector at iteration $k+1$ is

$$\tilde{v}^{k+1} = \tilde{p}_{1,k+1}(A)v^1 + \sum_{l=1}^{k} \tilde{p}_{l+1,k+1}(A)\frac{f^l}{\tilde{\eta}_{l+1}}. \tag{4.9}$$

This is to be compared with (2.2) which gives the result in exact arithmetic. We note that the first term $\tilde{p}_{1,k+1}(A)v^1$ is different from what we have in exact arithmetic since the coefficients of the recurrence are different. If we want to pursue the forward analysis and consider the difference between ideal and computed Lanczos vectors, we have to link $\tilde{v}^{k+1}$ to v^{k+1}. Looking at the three-term recurrences for the ideal and computed polynomials we have

$$\eta_{k+1}p_{1,k+1}(\lambda) = (\lambda - \alpha_k)p_{1,k}(\lambda) - \eta_k p_{1,k-1}(\lambda),$$

and

$$\tilde{\eta}_{k+1}\tilde{p}_{1,k+1}(\lambda) = (\lambda - \tilde{\alpha}_k)\tilde{p}_{1,k}(\lambda) - \tilde{\eta}_k\tilde{p}_{1,k-1}(\lambda).$$

Setting $\Delta p_k(\lambda) = p_{1,k}(\lambda) - \tilde{p}_{1,k}(\lambda)$, this difference satisfies a three-term recurrence relation,

$$\tilde{\eta}_{k+1}\Delta p_{k+1}(\lambda) = (\lambda - \tilde{\alpha}_k)\Delta p_k(\lambda) - \tilde{\eta}_k\Delta p_{k-1}(\lambda) + g_k(\lambda), \tag{4.10}$$

with

$$g_k(\lambda) = (\tilde{\eta}_{k+1} - \eta_{k+1})p_{1,k+1}(\lambda) + (\tilde{\alpha}_k - \alpha_k)p_{1,k}(\lambda) + (\tilde{\eta}_k - \eta_k)p_{1,k-1}(\lambda).$$

From Theorem 4.11 we can obtain the solution of (4.10) and then derive an expression for the difference between the ideal and computed Lanczos vectors: see Meurant (2006).

Theorem 4.14. As long as $k < n$,

$$\tilde{v}^{k+1} = v^{k+1} + \sum_{l=1}^{k} \tilde{p}_{l+1,k+1}(A)g_l(A)\frac{v^1}{\tilde{\eta}_{l+1}} + \sum_{l=1}^{k} \tilde{p}_{l+1,k+1}(A)\frac{f^l}{\tilde{\eta}_{l+1}}. \tag{4.11}$$

Theorem 4.14 shows that the difference between the ideal and the computed Lanczos vectors arises from two sources: the local rounding errors f^l and the differences of the coefficients (which, of course, come from the differences of the previous Lanczos vectors). From the D30 example, we have seen that it is interesting to consider the behaviour of $(\bar{v}^{k+1})_i = (Q^T \tilde{v}^{k+1})_i$. This is given by

$$(Q^T \tilde{v}^{k+1})_i = \tilde{p}_{1,k+1}(\lambda_i)(Q^T v^1)_i + \sum_{l=1}^{k} \tilde{p}_{l+1,k+1}(\lambda_i) \frac{(Q^T f^l)_i}{\tilde{\eta}_{l+1}}. \qquad (4.12)$$

It is difficult to study the behaviour of the sum in (4.12). This shows again the limitations of a forward analysis. However, in order to get some insight, one can look at each term individually.

What can be shown is the fact that, for a given λ_i towards which a Ritz value is converging, the absolute value of the polynomials $|\tilde{p}_{1,k}(\lambda_i)|$, as a function of k, first decreases to the level $\sqrt{\varepsilon_M}$, and then increases back to $\mathcal{O}(1)$. The values $|\tilde{p}_{j,k}(\lambda_i)|$ for $j > 1$ increase as a function of k up to a maximum of $\mathcal{O}(1)$, and then decrease down to $\sqrt{\varepsilon_M}$. This can be proved rigorously for the beginning of the process until the first Ritz value has converged and $|\tilde{p}_{1,k}(\lambda_i)|$ is back to $\mathcal{O}(1)$. This is done by investigating the product $|\tilde{p}_{1,k}(\lambda)\tilde{p}_{j,k}(\lambda)|$ for $k > j > 1$: see Meurant (2006).

The approach using polynomials offers some insight into the numerical behaviour of the Lanczos algorithm. In the beginning, the growth of the individual terms in the sum representing the influence of the round-off on the components of the Lanczos vectors in the directions of the eigenvectors of A goes hand in hand with the decrease of the original component. But, the argument is incomplete since we cannot analyse the whole sums defining a component of $\bar{v}^k$.

One can also consider other ways to write the solution of a three-term nonhomogeneous recurrence: see Meurant (2006). We consider once again the recurrence (4.8) with s_1 given and $\zeta_2 s_2 = (\lambda - \tau_1)s_1 + f_1$. For simplicity we take $\lambda = 0$ and let

$$L_{k+1} = \begin{pmatrix} 1 & 0 & \cdots & \cdots & 0 & 0 \\ \tau_1 & \zeta_2 & 0 & \cdots & 0 & 0 \\ \zeta_2 & \tau_2 & \zeta_3 & & \vdots & \vdots \\ & \ddots & \ddots & \ddots & 0 & \vdots \\ & & \zeta_{k-1} & \tau_{k-1} & \zeta_k & 0 \\ & & & \zeta_k & \tau_k & \zeta_{k+1} \end{pmatrix}.$$

This matrix is written as

$$L_{k+1} = \begin{pmatrix} (e^1)^T & 0 \\ T_k & \zeta_{k+1} e^k \end{pmatrix},$$

where T_k is the tridiagonal matrix of the recurrence coefficients. Let $s^{k+1} = (s_1, \ldots, s_{k+1})^T$ and $g = s_1$, $h = (f_1, \ldots, f_k)^T$ then the non homogeneous recurrence (4.8) can be written as

$$L_{k+1} s^{k+1} = \begin{pmatrix} g \\ h \end{pmatrix}.$$

In the following we shall use this for the Lanczos algorithm with $\tilde{T}_k - \lambda_i I$ instead of T_k. To obtain the solution of the recurrence, the first step is to find an expression for the inverse of L_{k+1} involving T_k. This is given in the next theorem in which we only give the entries we are interested in, and with the proof left to Meurant (2006).

Theorem 4.15.

$$(L_{k+1}^{-1})_{(1:k,1)} = \frac{1}{(T_k^{-1})_{1,k}} T_k^{-1} e^k,$$

$$(L_{k+1}^{-1})_{(1:k,2:k+1)} = T_k^{-1} - \frac{1}{(T_k^{-1})_{1,k}} T_k^{-1} e^k (e^1)^T T_k^{-1}.$$

From Theorem 4.15 we have a characterization of the solution of the three-term recurrence (4.8) involving the inverse of T_k.

Theorem 4.16. The k first elements of the solution of the three-term recurrence (4.8) are given by

$$s^k = \left(L_{k+1}^{-1} \begin{pmatrix} s_1 \\ h \end{pmatrix} \right)_{1:k} = \frac{s_1}{(T_k^{-1})_{1,k}} T_k^{-1} e^k + \left[I - \frac{1}{(T_k^{-1})_{1,k}} T_k^{-1} e^k (e^1)^T \right] T_k^{-1} h. \tag{4.13}$$

Moreover, the last element is

$$s_k = (T_k^{-1} h)_k - \frac{(T_k^{-1})_{k,k}}{(T_k^{-1})_{1,k}} (T_k^{-1} h)_1 + \frac{(T_k^{-1})_{k,k}}{(T_k^{-1})_{1,k}} s_1.$$

The solution can also be written as

$$s_k = \frac{(T_k^{-1})_{k,k}}{(T_k^{-1})_{1,k}} s_1 + \frac{1}{\zeta_k (T_{k-1}^{-1})_{1,k-1}} \sum_{j=1}^{k-1} (T_{k-1}^{-1})_{j,1} f_j.$$

For the components of the Lanczos vectors in the directions of the eigenvectors of A we apply Theorem 4.16 with $\check{T}_k = \tilde{T}_k - \lambda_i I$ (which is nonsingular) instead of T_k, where $\tilde{T}_k$ is the computed Lanczos matrix. This gives

$$\bar{v}_i^{k+1} = \frac{(\check{T}_{k+1}^{-1})_{k+1,k+1}}{(\check{T}_{k+1}^{-1})_{1,k+1}} \bar{v}_i^1 + \frac{1}{\eta_{k+1} (\check{T}_k^{-1})_{1,k}} \sum_{j=1}^{k} (\check{T}_k^{-1})_{j,1} \bar{f}_i^j.$$

It can be shown that the first term on the right-hand side of the last identity is

$$\frac{(\check{T}_{k+1}^{-1})_{k+1,k+1}}{(\check{T}_{k+1}^{-1})_{1,k+1}}\bar{v}_i^1 = \tilde{p}_{1,k+1}(\lambda_i)\bar{v}_i^1,$$

where $\tilde{p}_{1,k}$ is the polynomial defined in Theorem 4.13.

We will finish this section by showing that the previous results are useful when bounding perturbation terms. Going back to (4.13) and denoting

$$U_k = I - \frac{1}{(\check{T}_k^{-1})_{1,k}}\check{T}_k^{-1}e^k(e^1)^T$$

and $h^{(i)} = (\bar{f}_i^1 \cdots \bar{f}_i^k)^T$, we can bound the perturbation term $U_k\check{T}_k^{-1}h^{(i)}$ by

$$\|U_k\check{T}_k^{-1}h^{(i)}\| \le \|U_k\|\,\|\check{T}_k^{-1}\|\,\|h^{(i)}\|.$$

It can be shown (see Meurant (2006)) that $\|U_k\|$ is bounded by $C\sqrt{k}/|\bar{v}_i^1|$, where C is a constant independent of k, when the component of the initial vector in the direction of the ith eigenvector $|\bar{v}_i^1| = (q^i, v^1)$ is different from zero. This result seems not to be optimal since, when $|\bar{v}_i^1|$ is small, the bound can be large. This can possibly reflect the fact that, in this case, $\tilde{T}_k - \lambda_i I$ can be close to singular. Using this bound, we have the following result.

Theorem 4.17. Using the previous notation and supposing $|\bar{v}_i^1| \neq 0$, the perturbation term in (4.13) is bounded by

$$\|U_k\check{T}_k^{-1}h^{(i)}\| \le \frac{C\sqrt{k}}{|\bar{v}_i^1|}\frac{\|h^{(i)}\|}{\min_j(\theta_j^{(k)} - \lambda_i)}.$$

We note that

$$\|h^{(i)}\|^2 = \sum_{j=1}^{k}(q^i, f^j)^2 \le \sum_{j=1}^{k}\|f^j\|^2.$$

Theorem 4.17 shows that if $\min_j(\theta_j^{(k)} - \lambda_i)$ is large (no Ritz value is close to λ_i), the perturbation term for the ith component $(Q^T\tilde{v}^{k+1})_i$ of the projection of the finite precision Lanczos vector stays bounded and small, as long as $|\bar{v}_i^1|$ is not too small.

This represents a different point of view to the behaviour of the finite precision Lanczos algorithm, which also helps in understanding some properties of CG convergence in presence of round-off errors. However, the approach here does not allow us to study how $|\bar{v}_i^k|$ varies, since $(\check{T}_k^{-1})_{j,1}$ seems to be difficult to analyse.

5. The conjugate gradient algorithm in finite precision

Let us start with an example. Figure 5.1 depicts the Euclidean norm of the residual when the conjugate gradient algorithm is applied to a linear system with the matrix D30, a right-hand side of all ones and starting vector equal to zero. The solid line corresponds to the finite precision CG computation and the dashed line to CG with full reorthogonalization of the iteratively computed residual vectors at each step. As expected, in the latter case the residual vanishes at iteration 30. However, in finite precision arithmetic it takes many more iterations to get a small residual. Notice that even to reach a modest decrease, the number of iterations is considerably larger than the order of the matrix.

In finite precision arithmetic CG exhibits similar problems to the Lanczos algorithm: the residual vectors lose their orthogonality. Moreover in comparison to what happens in exact arithmetic or with reorthogonalization, convergence of the CG approximate solution is delayed. Intuitively, this observed fact is closely related to convergence of Ritz values. In CG the tridiagonal matrix T_k and the Ritz values do not appear explicitly, therefore the appearance of multiple Ritz approximations to single original eigenvalues is hard to notice for a practical user of the algorithm. Since we know that ideally CG behaviour depends on convergence of the Ritz values to eigenvalues (see Section 3), we may also expect the same numerically. An appearance

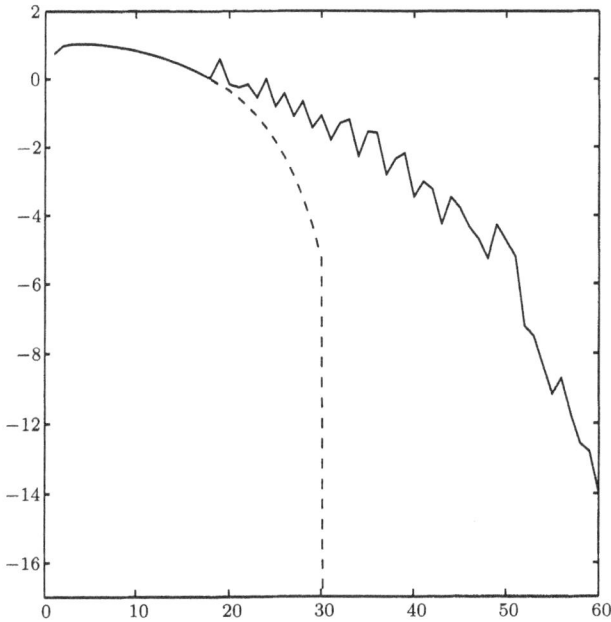

Figure 5.1. D30, $\log_{10}$ of the norms of residuals.

of multiple Ritz approximations to some eigenvalues delays convergence of Ritz values to other eigenvalues. Consequently it also delays convergence of the approximate solutions in the finite precision CG algorithm.

We first recall the relationships between the Lanczos and CG algorithms in finite precision arithmetic. For the finite precision CG algorithm we present, based on the existing literature, the corresponding CG–Lanczos recurrence which resembles, apart from the different perturbation error terms, the finite precision Lanczos algorithm from Section 4. Using the established correspondence, we then use the knowledge about the finite precision Lanczos algorithm in order to understand the finite precision CG behaviour.

In practical applications it is important to estimate the errors of computed approximate solutions. We recall the state-of-the-art error estimates and explain the instrumental role of rounding error analysis in convergence evaluation and in formulation of a meaningful stopping criteria. Finally, in addition to delaying convergence, rounding errors also limit the maximal attainable accuracy of the computed approximate solutions. We address this issue and end the section by pointing out some recent developments.

5.1. Local rounding errors and the CG–Lanczos recurrence

In analogy to the finite precision Lanczos algorithm, the recurrences for the CG quantities (*cf.* (3.2)) computed in finite precision arithmetic can be written in the form

$$\overline{\gamma}_{k-1} \equiv \gamma_{k-1} + \delta_\gamma^{k-1} \equiv \frac{\|r^{k-1}\|^2}{(p^{k-1}, Ap^{k-1})} + \delta_\gamma^{k-1},$$

$$x^k = x^{k-1} + \gamma_{k-1}p^{k-1} + \delta_x^k,$$

$$r^k = r^{k-1} - \gamma_{k-1}Ap^{k-1} + \delta_r^k, \qquad\qquad (5.1)$$

$$\overline{\beta}_k \equiv \beta_k + \delta_\beta^k \equiv \frac{\|r^k\|^2}{\|r^{k-1}\|^2} + \delta_\beta^k,$$

$$p^k = r^k + \beta_k p^{k-1} + \delta_p^k,$$

where the perturbation terms also depend, in addition to ε_M, n and $\|A\|$, on the norms and absolute values of the computed vector and scalar quantities respectively. The detailed bounds for the perturbation terms can be found in relations (7.9)–(7.14) of Strakoš and Tichý (2002, p. 71); see also Meurant (2006). The local orthogonality between the vectors r^{k+1} and r^k, r^{k+1} and p^k, p^{k+1} and Ap^k can also be bounded analogously to the local orthogonality among the computed subsequent Lanczos vectors in Theorem 4.1, but the bounds (and the proofs) are considerably more complicated. They depend, in addition to ε_M, n and $\|A\|$ also on $\kappa(A)$ and $\|r^k\|^2$: see Strakoš and Tichý (2002, Section 9).

As in the ideal CG algorithm in Section 3.1 we can write a three-term recurrence for the computed residuals:

$$r^k = -\gamma_{k-1}Ar^{k-1} + \left(1 + \frac{\gamma_{k-1}\beta_{k-1}}{\gamma_{k-2}}\right)r^{k-1} - \frac{\gamma_{k-1}\beta_{k-1}}{\gamma_{k-2}}r^{k-2} + \Delta_r^k, \quad k \geq 2,$$
$$r^1 = r^0 - \gamma_0 Ar^0 + \Delta_r^0.$$

Introducing the CG–Lanczos vectors w^k determined from the iteratively computed CG residuals (in finite precision arithmetic w^k is not generally identical to the vector v^k computed via the finite precision Lanczos algorithm)

$$w^{k+1} = (-1)^k \frac{r^k}{\|r^k\|}, \quad k = 1, 2, \ldots,$$

we get the following theorem.

Theorem 5.1. The three-term recurrence for the CG–Lanczos vectors determined from the finite precision CG algorithm is

$$\eta_{k+1}w^{k+1} = Aw^k - \alpha_k w^k - \eta_k w^{k-1} + \Delta_w^k, \quad k = 2, 3, \ldots, \quad (5.2)$$

where

$$\eta_{k+1} = \frac{\sqrt{\beta_k}}{\gamma_{k-1}}, \qquad \alpha_k = \frac{1}{\gamma_{k-1}} + \frac{\beta_{k-1}}{\gamma_{k-2}}, \qquad \alpha_1 = \frac{1}{\gamma_0}$$

with

$$\gamma_k = \frac{\|r^k\|^2}{(Ap^k, p^k)}, \qquad \beta_k = \frac{\|r^k\|^2}{\|r^{k-1}\|^2}$$

and the initial vectors given by

$$w^1 = r^0/\|r^0\| + \Delta_w^0, \qquad \eta_2 w^2 = Aw^1 - \alpha_1 w^1 - \Delta_w^1.$$

Here the recurrence is based on the coefficients γ_k and β_k determined from the computed r^k, r^{k-1} and p^k *exactly.* If we want to refer to the computed coefficients $\overline{\gamma}_k$ and $\overline{\beta}_k$, we still have the same kind of relationship but with slightly different perturbation terms. We leave the bound on the perturbation terms Δ_w^k to Meurant (2006).

5.2. Results of the backward-like analysis of CG

Based on Theorem 5.1, the analysis of Section 4.3 will also apply to the finite precision CG algorithm: see Greenbaum (1989, p. 24). The tridiagonal matrix T_k, with the entries defined by the finite precision CG algorithm as described in Theorem 5.1, is equal to that generated by the *exact* CG algorithm for a matrix whose eigenvalues lie within small intervals of the original eigenvalues. This relationship implies that the Euclidean norms of the residuals in the finite precision CG algorithm are the same as those in the correspondingly constructed exact CG recurrence. With the A-norm of the error, which is minimized at each step of the ideal CG algorithm, the situation is technically more complicated, as the reader can find in Greenbaum (1989, Theorem 3, pp. 26–29), since the definition of the norm depends on the matrix. It can still be concluded, however, that the A-norm of the error in the finite precision CG algorithm is reduced at approximately the same rate as the corresponding energy-norm of the error in the constructed exact CG recurrence. This has been further discussed and illustrated numerically in Greenbaum and Strakoš (1992).

Here it is assumed that the maximal attainable accuracy, which is limited because of rounding errors, is far away. We are solely interested in the delay of convergence. In principle, the delay at step k is given by the rank-deficiency of a basis of the computed Krylov subspace. This is in fact determined from the numerical rank (for some appropriate threshold criterion) of the computed matrix $W_k = (w^1, \ldots, w^k)$, where the w^j are the CG–Lanczos vectors: cf. Paige and Strakoš (1999).

The results can be quantified in various ways using the polynomial formulation of the CG algorithm. Instead of working with orthogonal polynomials corresponding to the distribution function with n points of increase $\lambda_1 < \lambda_2 < \cdots < \lambda_n$ (we again assume, for simplicity of notation, that the eigenvalues of A are distinct), one must, however, consider orthogonal polynomials with respect to distribution functions having possibly many points of increase close to some or each λ_j.

In constructing the bounds one must consider the minimax polynomials on the union of tiny intervals containing the eigenvalues λ_j: see Greenbaum (1989), Greenbaum and Strakoš (1992), Greenbaum (1994). This seemingly small difference generally has a dramatic impact. We notice this from the fact that rounding errors can make a dramatic difference to the behaviour of the CG errors and residuals: see, e.g., the example presented in Figure 5.1 above. The last fact is obvious, but in terms of polynomials it is not always correctly understood. This sometimes leads to misleading statements relating convergence behaviour of finite precision CG to incorrectly interpreted and simplified approximations to the minimal polynomial of A.

An example of a rigorous and instructive extension of the results from Greenbaum (1989) and Greenbaum and Strakoš (1992) can be found in Notay (1993), where the author presents bounds for the delay of convergence of the finite precision CG algorithm in the presence of isolated outlying eigenvalues.

5.3. Estimates of the error norms

As we have seen in Section 3.3, the initial error $\epsilon^0 = x - x^0$ and the kth error $\epsilon^k = x - x^k$, measured in the A-norm, are in exact precision CG ideally related by the identity

$$\frac{\|\epsilon^0\|_A^2}{\|r^0\|^2} = k\text{th Gauss quadrature} + \frac{\|\epsilon^k\|_A^2}{\|r^0\|^2},$$

where ϵ^0 and ϵ^k are unknowns and the kth Gauss quadrature can be determined by

$$(e^1)^T T_k^{-1} e^1 = \sum_{l=1}^{k} \gamma_{l-1} \|r^{l-1}\|^2.$$

In order to get an estimate for $\|\epsilon^k\|_A^2$, we have to eliminate $\|\epsilon^0\|_A^2$: see Golub and Strakoš (1994, pp. 262–263). Subtracting the identities for k and $k+d$,

$$\frac{\|\epsilon^k\|_A^2}{\|r^0\|^2} = (k+d)\text{th Gauss quadrature} - k\text{th Gauss quadrature} + \frac{\|\epsilon^{k+d}\|_A^2}{\|r^0\|^2}.$$

Since the last term on the right-hand side is always nonnegative (and strictly smaller than the term on the left-hand side), the difference between the Gauss quadratures determines in exact arithmetic the square of the lower bound for $\|\epsilon^k\|_A/\|r^0\|$.

Based on the analysis of the Gauss quadrature, Golub and Strakoš (1994) proved that this bound also works in finite precision CG computations until $\|\epsilon^k\|_A/\|r^0\|$ drops below the level $\sqrt{\varepsilon_M}$. An appropriate numerically stable implementation of this estimate was proposed by Golub and Meurant (1997). Experimental evidence shows that estimates obtained with this implementation are not significantly affected by rounding errors until the finite precision CG algorithm reaches its maximal attainable accuracy level. The proof from Golub and Strakoš (1994) cannot, however, be extended in order to justify that.

As mentioned in Section 3.3, using some simple algebraic manipulations and a lengthy rounding error analysis Strakoš and Tichý (2002) proved that in the finite precision CG algorithm the A-norm of the error satisfies

$$\|\epsilon^k\|_A^2 - \|\epsilon^{k+1}\|_A^2 = \gamma_k \|r^k\|^2 + \delta_\epsilon^k, \tag{5.3}$$

where δ_ϵ^k depends on the loss of orthogonality between r^{k+1} and p^k. Based on (5.3),

$$\nu_{k,k+d} = \sum_{l=k}^{k+d-1} \gamma_l \|r^l\|^2 \tag{5.4}$$

can be used as a lower bound for $\|\epsilon^k\|_A^2$, and this lower bound is not significantly affected by rounding errors until $\|\epsilon^k\|_A/\|\epsilon^0\|_A$ reaches a level proportional to the machine precision: see Strakoš and Tichý (2002, Section 10).

We wish to emphasize an important point. The numerical justification for (5.4) as the squared lower bound for $\|\epsilon^k\|_A$ is in no way based on the fact that in finite precision arithmetic this term is evaluated with negligible additional errors (here we do not even consider them). It is based on the nontrivial fact that (5.3) holds for the finite precision CG approximate solutions, and that δ_ϵ^k is small. We see an analogy with the rounding error analysis of the accuracy of Ritz values in the finite precision Lanczos algorithm given by Paige: see Section 4. Here again, the error estimate is also valid in finite precision computations, but we know this *only because of rigorous and nontrivial mathematical proofs.* It can be easily shown that ideally equivalent but numerically different formulas can lead to highly misleading results: see Strakoš and Tichý (2002, Figure 6.1, p. 69) and Strakoš and Liesen (2005, Figure 8, p. 319). Error estimates without appropriate rounding error analyses represent a highly hazardous pursuit.

In order to get a lower bound for the A-norm of the error at step k, we need to perform d extra steps. If the A-norm of the error reasonably drops at around step k, then d can be small. If on the other hand the A-norm of the error almost stagnates, then a small d will not ensure a close lower bound. Of course, the actual convergence behaviour is not known: it is to be estimated. Therefore the choice of d represents a difficult open problem. In any case, the proposed lower bound offers extra information which is computable at negligible additional cost, and which can with great benefit complement the commonly used measures of convergence: see Arioli (2004), Arioli, Noulard and Russo (2001), Strakoš and Liesen (2005), Strakoš and Tichý (2005) and Meurant (1999a). Moreover, if we agree to store one additional real number per iteration, we can easily update the previous estimates at each step. Together with the estimate for $\|\epsilon^k\|_A$ based on d, we can get (at step $k+d$) an estimate for $\|\epsilon^{k-1}\|_A$ based on $d+1$, an estimate for $\|\epsilon^{k-2}\|_A$ based on $d+2$, etc. In this way, the convergence of CG measured by the A-norm of the error can be 'reconstructed' using lower bounds: see Strakoš and Tichý (2005, Figure 5.4).

In linear systems arising from finite element discretizations of self-adjoint elliptic partial differential equations, it is natural to evaluate CG convergence via the relative A-norm of the error

$$\frac{\|\epsilon^k\|_A}{\|x\|_A} = \frac{\|x - x^k\|_A}{\|x\|_A}$$

(see Arioli (2004)). Subtracting the ideal identities

$$\|\epsilon^0\|_A^2 = \nu_{0,k+d} + \|\epsilon^{k+d}\|_A^2,$$
$$\|\epsilon^0\|_A^2 = \|x - x^0\|_A^2 = \|x\|_A^2 - b^T x^0 - (r^0)^T x^0$$

gives

$$\|x\|_A^2 = \nu_{0,k+d} + b^T x^0 + (r^0)^T x^0 + \|\epsilon^{k+d}\|_A^2,$$
$$\|x\|_A^2 \geq \mu_{k+d} \equiv \nu_{0,k+d} + b^T x^0 + (r^0)^T x^0.$$

We will assume that $\|x - x^0\|_A \leq \|x\|_A$. This represents a very natural assumption which should never be violated in practical computations. Indeed, it is meaningless to use a nonzero x^0 without justification that guarantees that a nonzero initial approximation is better than taking $x^0 = 0$. For CG, the A-norm of the error represents the proper measure of 'goodness'. If in doubt, it is always possible to scale an initial approximation such that $\|x - \alpha x^0\|_A$ is minimal, which gives

$$\alpha = \frac{b^T x^0}{(x^0)^T A x^0}$$

(see Strakoš and Tichý (2005)). If $\|x - x^0\|_A \leq \|x\|_A$, then it is easy to show that $\mu_{k+d} > 0$, and an algebraic manipulation ideally gives

$$\frac{\|\epsilon^k\|_A^2}{\|x\|_A^2} \geq \frac{\nu_{k,d}}{\mu_{k+d}} > 0,$$

i.e., in exact precision CG, $\nu_{k,d}/\mu_{k+d}$ is a lower bound for the squared relative A-norm of the error. Since numerically all considerations leading to this bound are based on *local orthogonality only*, this estimate is also well established (though not always a lower bound) for the finite precision CG algorithm. For further details we refer to Strakoš and Tichý (2005) and also to Strakoš and Tichý (2002), who also describe estimation of the Euclidean norm of the error and presents open problems. For the estimation of the Euclidean norm see also Meurant (2005), and for that norm in finite precision see Meurant (2006).

Various other options for computing the error bounds in the CG algorithm are summarized by Calvetti, Morigi, Reichel and Sgallari (2000). Based on quadrature considerations, the bounds are more complicated. They cannot be easily justified for finite precision CG computations. Still, they can prove useful in some particular applications. Interesting ideas concerning the upper bounds for the A-norm of the error can also be found in Greenbaum (1997a, p. 108) and in Golub and Meurant (1997).

5.4. Maximal attainable accuracy

Rounding errors generally do not allow the finite precision CG algorithm to produce approximate solutions to arbitrary accuracy. It is therefore important to find out the maximal attainable accuracy which can be reached for a given A and b. The importance of this question is, however, more in the impact which the corresponding analysis has on understanding the CG algorithm and its implementations, than in practical applications of the results. In most applications, perhaps with the exception of some inner CG iterations used in nonlinear optimization, or difficult problems with $\|A\|$ large, the computation is stopped much before the maximal attainable accuracy is reached.

Here we will assume, as above, that A is symmetric positive definite and not close to singular, and we will concentrate on limitations on the maximal attainable accuracy caused by the possible amplification of elementary round-off throughout the recurrences. We leave other effects, which can be observed in indefinite systems or near-singular systems, to the literature: see, e.g., Sleijpen, van der Vorst and Modersitzki (2001). Work on maximal attainable accuracy has focused on the residual as the easiest and most common measure of convergence. Based on the residual, bounds for the maximal attainable accuracy measured by the Euclidean or the A-norm of the error can easily be obtained using the obvious relationships, together with the characterization of conditioning of the matrix A.

In the CG algorithm, the residual vector is recursively computed at each step as a part of the recurrence. In finite precision arithmetic, this recursively computed residual r^k (see (5.1)) can differ from the directly computed quantity $b - Ax^k$, which is generally called the true residual. Convergence of the recursive residuals was analysed by Wozniakowski (1978, 1980) and Bollen (1984), for example. Although some assumptions used there cannot in general be satisfied by the CG recurrence (5.1), the results proved useful in a further analysis: see Greenbaum (1994, 1997b). For a survey of the early developments see Higham (2002).

In Theorem 2 of Greenbaum (1989), the question of the difference between the true residual and the recursively computed residual was analysed for the first time, to our knowledge. It was shown that this difference at step k can be bounded by a simple sum of the elementary perturbation terms at steps 0 (which means computation of the initial residual) to k, i.e.,

$$\|r^k - (b - Ax^k)\| \leq \|\delta_r^0\| + \sum_{l=1}^{k}(\|\delta_r^l\| + \|A\delta_x^l\|).$$

Sleijpen, van der Vorst and Fokkema (1994), Greenbaum (1994), and slightly later Greenbaum (1997b) studied this problem further, resulting in the bound

$$\frac{\|r^k - (b - Ax^k)\|}{\|A\| \, \|x\|} \leq \mathcal{O}(k)\varepsilon_M \left(1 + \max_{l \leq k} \frac{\|x^l\|}{\|x\|}\right). \tag{5.5}$$

If $\|r^k\|$ becomes of the order of the machine precision, which is often observed numerically for large k but which has not yet been completely proved in the given literature, then (5.5) gives a bound for the maximal attainable accuracy measured by the true residual norm divided by $\|A\| \, \|x\|$.

This result offers the following insight into the behaviour of the finite precision CG algorithm. One can expect a high maximal attainable accuracy with the finite precision CG algorithm if the norms of the iterates do not significantly exceed the norm of the true solution. Since ideally the Euclidean norm of the error is strictly decreasing, $\|x - x^k\| < \|x - x^0\|$ implies

$$\|x^k\| \leq 2\|x\| + \|x^0\|.$$

Using the backward-like error analysis of Greenbaum described above, this upper bound holds true, to within a small error, in the finite precision CG algorithm. With a reasonable choice of $\|x^0\|$, the finite precision CG algorithm can therefore be expected to achieve a high maximal attainable accuracy if $\|A\|$ is not too large.

The situation is dramatically different in CG-like algorithms applied to nonsymmetric systems, to many of which the above analysis can also be applied. For detailed discussions see Greenbaum (1997b) and Greenbaum (1997a, Section 7.3).

When the CG algorithm is implemented via the mathematically equivalent three-term recurrence (for examples see Rutishauser (1959) and Hageman and Young (1981)), the maximal attainable accuracy is much more vulnerable to local errors. As shown by Gutknecht and Strakoš (2000), the difference $r^k - (b - Ax^k)$ is then equal to a sum of local error terms (different from those in the analysis of the two term recurrences above) plus multiples of the same terms by factors which can become large if the norm of the iteratively computed residual oscillates, $i.e.$, if

$$\max_{0 \leq l < j \leq k} \frac{\|r^j\|^2}{\|r^l\|^2} \quad \text{is large.}$$

Consequently a large increase in the norm of the computed iterative residuals can damage the maximal attainable accuracy. Moreover, damage caused at an early stage of the computation cannot in general be compensated for in the subsequent iterations. The technique used in Gutknecht and Strakoš (2000) is based on writing k steps of the second-order nonhomogeneous difference equation for the gap $r^k - (b - Ax^k)$ as a superposition of the $k + 1$ homogeneous difference equations, which resembles the technique used in a different context by Grcar (1981). For further details we refer to Gutknecht and Strakoš (2000). As pointed out in the concluding part of the last paper, the same result can also be attained by using matrix approach analogous to that of Paige (1980). The matrix approach allows easier further generalizations. In some applications the matrix A is not explicitly available, and the matrix–vector multiplication is performed by solving an auxiliary problem. It might therefore be convenient to relax the accuracy of this operation. That can, however, affect convergence behaviour and the maximal attainable accuracy. Analysis of this problem goes far beyond the investigation of numerical stability. Several authors have recently presented interesting results focused mostly on maximal attainable accuracy: see, $e.g.$, Bouras and Frayssé (2005), and the surveys in Simoncini and Szyld (2005, Section 11) and van den Eshof (2003, Chapter 5).

5.5. Recent developments

In this section we summarize some recent results about CG convergence in finite precision arithmetic: see Meurant (2006).

For the recurrence of w^k given in (5.2) we can directly apply the results we have reviewed for general three-term recurrences: see Theorem 4.16. Let us denote by $\bar{w}_i^k$ the component of w^k in the direction of the ith eigenvector of A, $i = 1, 2, \dots, n$.

Theorem 5.2. Let

$$\bar{\delta}^k \equiv (\bar{\delta}_1^k, \dots, \bar{\delta}_n^k)^T = Q^T \Delta_w^k,$$

let j be given and let $p_{j,k}$ be the polynomial determined by

$$p_{j,j-1}(\lambda) = 0, \qquad p_{j,j}(\lambda) = 1,$$

$$\eta_{k+1} p_{j,k+1}(\lambda) = (\lambda - \alpha_k) p_{j,k}(\lambda) - \eta_k p_{j,k-1}(\lambda), \quad k = j, j+1, \dots.$$

The solution of the perturbed recurrence

$$\eta_{k+1} \bar{w}_i^{k+1} = (\lambda_i - \alpha_k) \bar{w}_i^k - \eta_k \bar{w}_i^{k-1} + \bar{\delta}_i^k$$

starting from $w_i^0 = 0$ and w_i^1 is given by

$$\bar{w}_i^{k+1} = p_{1,k+1}(\lambda_i) \bar{w}_i^1 + \sum_{l=1}^{k} p_{l+1,k+1}(\lambda_i) \frac{\bar{\delta}_i^l}{\eta_{l+1}}, \quad i = 1, \dots, n.$$

This immediately leads to an expression for w^{k+1}:

$$w^{k+1} = p_{1,k+1}(A) w^1 + \sum_{l=1}^{k} p_{l+1,k+1}(A) \frac{\Delta_w^l}{\eta_{l+1}}.$$

Then, using the correspondence between w^{k+1} and r^k, we can express the recursively determined CG residual vector computed in finite precision arithmetic in the following form.

Theorem 5.3. Using the notation of Theorem 5.2,

$$r^k = (-1)^k \frac{\|r^k\|}{\|r^0\|} p_{1,k+1}(A) r^0 + (-1)^k \|r^k\| \sum_{l=1}^{k} p_{l+1,k+1}(A) \frac{\Delta_w^l}{\eta_{l+1}}.$$

In exact arithmetic, after a Ritz value has converged, the corresponding projections of the residual and of the error on the corresponding eigenvector vanish. This is not the case in finite precision arithmetic. After decreasing for a while, the projection of the residual on the subspace generated by the corresponding eigenvector of A rises back to contribute to the norm of the residual, because of the amplification of the local round-off. Once a new Ritz copy is formed, the component again decreases, *etc.* This can

534 G. MEURANT AND Z. STRAKOŠ

delay convergence and lead to oscillations of the residual components in the directions of the individual eigenvectors of A, and, as a consequence, to oscillations of the residual norm. In comparison with the expression for the finite precision Lanczos–CG vector w^{k+1}, the perturbation term in Theorem 5.3 is multiplied by $\|r^k\|$. Therefore, for small $\|r^k\|$ the possible oscillations caused by possible amplification of the error terms are typically much less pronounced in r^k than in w^{k+1}.

When considering CG convergence, we have to be careful on how to link the error to the computed quantities. Ideally, the error is $\epsilon^k = x - x^k$ where x is the exact solution and it is related to the residual by $A\epsilon^k = r^k$. But this is only true if the residual is $b - Ax^k$. We have seen in Section 5.4 that the computed iterative residual can be different from $b - Ax^k$. Hence there are more alternatives. Considering the ultimate stagnation of $\|b - Ax^k\|$, it seems reasonable to work (besides the true error linked to the true residual) with $(A^{-1}r^k, r^k) = \|A^{-\frac{1}{2}}r^k\|^2$, where r^k is the (recursively) computed iterative residual, as another useful measure. We denote it $\varepsilon^k \equiv A^{-1}r^k$. Then, we have the following result whose proof is based on a lengthy analysis of local orthogonality: see Meurant (2006) and Strakoš and Tichý (2002, (10.1)).

Proposition 5.4.

$$\|\varepsilon^{k+1}\|_A^2 = \|\varepsilon^k\|_A^2 - \gamma_k\|r^k\|^2 + \varepsilon_M C_1^k\|r^k\|^2 + \varepsilon_M^2 C_2^k\|r^k\|^2,$$

where $|C_1^k|$ and $|C_2^k|$ are bounded by quantities involving $\|r^k\|$ and $\|p^k\|$.

This proposition leads to a result about strict decrease of the error norm under a restriction on the condition number of A.

Theorem 5.5. If

$$\kappa(A) < \frac{1}{\varepsilon_M \lambda_1 |C_1^k|} + \mathcal{O}(\varepsilon_M), \quad \text{for all } k,$$

then

$$\|\varepsilon^{k+1}\|_A < \|\varepsilon^k\|_A.$$

Hence, if the condition number is not too large, $\|\varepsilon^k\|_A$ is, as in exact arithmetic, strictly decreasing. However, having a limitation on $\kappa(A)$ is not satisfactory, since in numerical computations we hardly observe an increase or oscillation of $\|\varepsilon^k\|_A$.

This result complements those of Anne Greenbaum (1989) who obtained a decrease of the A-norm of the error without an explicit restriction on the condition number but with additional small terms. A proof of the strict decrease of $\|\varepsilon^k\|_A$ and a proof that the computed iterative residual must ultimately vanish, i.e., $\|r^k\| \to 0$, without any restriction on the condition number, remains open.

6. Conclusions

The Lanczos and conjugate gradient algorithms are considered effective numerical tools for computing eigenvalues, approximating matrix functions and quadratic forms, and for solving (linear) algebraic equations. As we have seen, they also represent interesting mathematical objects with very deep links reaching far beyond the borders of numerical linear algebra, numerical mathematics or algebraic structures. This is perhaps why the investigation into their behaviour in exact and in finite precision arithmetic is leading to results which, piece by piece, are being assembled into a rigorous, consistent, rich and beautiful mathematical theory. In this way the Lanczos and conjugate gradient algorithms represent another example along the lines drawn by Baxter and Iserles (2003). The rigour and beauty of their mathematical structure, including the effects of rounding errors, reveals once again the presence of such attributes in the field called computational mathematics.

Acknowledgements

We sincerely thank Iveta Hnětynková, Jörg Liesen, Jurjen Duintjer Tebbens and Petr Tichý for comments and corrections, Chris Paige for a careful reading of earlier versions of the manuscript, suggesting valuable improvements and correcting our wording in many places, and Gene Golub for suggesting the writing of this paper.

REFERENCES

M. Arioli (2004), A stopping criterion for the conjugate gradient algorithms in a finite element method framework, *Numer. Math.* **97**, 1–24.

M. Arioli, E. Noulard and A. Russo (2001), Stopping criteria for iterative methods: applications to PDE's, *Calcolo* **38**, 97–112.

W. E. Arnoldi (1951), The principle of minimized iterations in the solution of the matrix eigenvalue problem, *Quart. Appl. Math.* **9**, 17–29.

O. Axelsson and I. Kaporin (2001), Error norm estimations and stopping criteria in preconditioned conjugate gradient iterations, *Numer. Linear Algebra Appl.* **8**, 265–286.

O. Axelsson and G. Linskog (1986), On the rate of convergence of the preconditioned conjugate gradient methods, *Numer. Math.* **51**, 209–227.

B. J. C. Baxter and A. Iserles (2003), On the foundations of computational mathematics, in Vol. XI of *Handbook of Numerical Analysis*, North-Holland, Amsterdam, pp. 3–34.

B. Beckermann and A. Kuijlaars (2002), Superlinear CG convergence for special right-hand sides, *Electron. Trans. Numer. Anal.* **14**, 1–19.

Å. Björck, T. Elfving and Z. Strakoš (1998), Stability of conjugate gradients and Lanczos methods for linear least squares problems, *SIAM J. Matrix Anal. Appl.* **19**, 720–736.

J. A. M. Bollen (1980), Round-off error analysis of descent methods for solving linear equations, PhD thesis, Technische Hogeschool Eindhoven, the Netherlands.

J. A. M. Bollen (1984), Numerical stability of descent methods for solving linear equations, *Numer. Math.* **43**, 361–377.

A. Bouras and V. Frayssé (2005), Inexact matrix–vector products in Krylov methods for solving linear systems: A relaxation strategy, *SIAM J. Matrix Anal. Appl.* **26**, 660–678.

D. Calvetti, S. Morigi, L. Reichel and F. Sgallari (2000), Computable error bounds and estimates for the conjugate gradient, *Numer. Algorithms* **25**, 79–88.

E. B. Christoffel (1877), Sur une classe particulière de fonctions entières et de fractions continues, *Ann. Mat. Pura Appl.* **8**, 1–10.

J. K. Cullum and R. A. Willoughby (1985), *Lanczos Algorithms for Large Symmetric Eigenvalue Computations*, Vol. I, Theory, Vol. II, Programs, Birkhäuser. Reprinted by SIAM in the series *Classics in Applied Mathematics*.

G. Dahlquist, S. C. Eisenstat and G. H. Golub (1972), Bounds for the error of linear systems of equations using the theory of moments, *J. Math. Anal. Appl.* **37**, 151–166.

G. Dahlquist, G. H. Golub and S. G. Nash (1978), Bounds for the error in linear systems, in *Proc. Workshop on Semi-Infinite Programming* (R. Hettich, ed.), Springer, pp. 154–172.

J. W. Daniel (1967), The conjugate gradient method for linear and nonlinear operator equations, *SIAM J. Numer. Anal.* **4**, 10–26.

G. Darboux (1878), Mémoire sur l'approximation des fonctions de très grand nombres et sur une classe étendue de développements en série, *J. Mat. Pures Appl.* **4**, 5–56, 377–416.

P. Davis and P. Rabinowitz (1984), *Methods of Numerical Integration*, second edition, Academic Press.

P. Deuflhard (1994), Cascadic conjugate gradient methods for elliptic partial differential equations: Algorithm and numerical results, in *Domain Decomposition Methods in Scientific and Engineering Computing* (University Park PA, 1993), Vol. 180 of *Contemporary Mathematics*, AMS, Providence, RI, pp. 29–42.

I. Dhillon and B. N. Parlett (2003), Orthogonal eigenvectors and relative gaps, *SIAM J. Matrix Anal. Appl.* **25**, 858–899.

I. Dhillon and B. N. Parlett (2004), Multiple representations to compute orthogonal eigenvectors of symmetric tridiagonal matrices, *Linear Algebra Appl.* **387**, 1–28.

V. Druskin, A. Greenbaum and L. Knizhnerman (1998), Using nonorthogonal Lanczos vectors in the computation of matrix functions, *SIAM J. Sci. Comput.* **19**, 38–54.

V. Druskin and L. Knizhnerman (1991), Error bounds in the simple Lanczos procedure for computing functions of symmetric matrices and eigenvalues, *Comput. Math. Math. Phys.* **31**, 20–30.

V. Druskin, L. Borcea and L. Knizhnerman (2005), On the sensitivity of Lanczos recursions to the spectrum, *Linear Algebra Appl.* **396**, 103–125.

S. Elhay, G. M. L. Gladwell, G. H. Golub and Y. M. Ram (1999), On some eigenvector-eigenvalue relations, *SIAM J. Matrix Anal. Appl.* **20**, 563–574.

J. van den Eshof (2003), Nested iteration methods for nonlinear matrix problems. PhD thesis, University of Utrecht.

B. Fischer (1996), *Polynomial Based Iteration Methods for Symmetric Linear Systems*, Wiley, Chichester.

R. Fletcher (1976), Conjugate gradient methods for indefinite systems, in *Numerical Analysis: Proc. 6th Biennial Dundee Conf., Univ. Dundee* (Dundee, 1975), Vol. 506 of *Lecture Notes in Mathematics*, Springer, Berlin, pp. 73–89.

V. M. Fridman (1963), The method of minimum iterations with minimum errors for a system of linear algebraic equations with a symmetrical matrix, *USSR Comput. Math. Math. Phys.* **2**, 362–363.

F. R. Gantmacher (1959), *The Theory of Matrices*, Vol. 1 and 2, Chelsea Publishing Co., New York.

W. Gautschi (1968), Construction of Gauss–Christoffel quadrature formulas, *Math. Comp.* **22**, 251–270.

W. Gautschi (1981), A survey of Gauss–Christoffel quadrature formulae, in *E. B. Christoffel: The Influence of His Work on Mathematics and the Physical Sciences* (P. L. Bultzer and F. Fehér, eds), Birkhauser, Boston, pp. 73–157.

W. Gautschi (1982), On generating orthogonal polynomials, *SIAM J. Sci. Statist. Comput.* **3**, 289–317.

W. Gautschi (1985), Orthogonal polynomials: Constructive theory and applications, *J. Comput. Appl. Math.* **12** & **13**, 61–76.

W. Gautschi (2002), The interplay between classical analysis and (numerical) linear algebra: A tribute to Gene H. Golub, *Electron. Trans. Numer. Anal.* **13**, 119–147.

W. Gautschi (2004), *Orthogonal Polynomials, Computation and Approximation*, Oxford University Press, Oxford.

G. H. Golub (1973), Some uses of the Lanczos algorithm in numerical linear algebra, in *Topics in Numerical Analysis* (J. H. H. Miller, ed.), Springer, Heidelberg/New York, pp. 23–31.

G. H. Golub and U. von Matt (1991), Quadratically constrained least squares and quadratic problem, *Numer. Math.* **59**, 561–580.

G. H. Golub and G. Meurant (1994), Matrices, moments and quadrature, in *Numerical Analysis 1993* (D. F. Griffiths and G. A. Watson, eds), *Pitman Research Notes in Mathematics*, pp. 105–156.

G. H. Golub and G. Meurant (1997), Matrices, moments and quadrature II: How to compute the norm of the error in iterative methods, *BIT* **37**, 687–705.

G. H. Golub and D. P. O'Leary (1989), Some history of the conjugate gradient and Lanczos algorithms: 1948–1976, *SIAM Rev.* **31**, 50–102.

G. H. Golub and Z. Strakoš (1994), Estimates in quadratic formulas, *Numer. Algorithms* **8**, 241–268.

G. H. Golub and J. H. Welsch (1969), Calculation of Gauss quadrature rules, *Math. Comp.* **23**, 221–230.

W. B. Gragg and W. J. Harrod (1984), The numerically stable reconstruction of Jacobi matrices from spectral data, *Numer. Math.* **44**, 317–335.

J. Grcar (1981), Analysis of the Lanczos algorithm and of the approximation problem in Richardson's method, PhD thesis, University of Illinois at Urbana–Champaign.

A. Greenbaum (1979), Comparison of splittings used with the conjugate gradient algorithm, *Numer. Math.* **33**, 181–194.

A. Greenbaum (1981), Convergence properties of the conjugate gradient algorithm in exact and finite precision arithmetic, PhD thesis, University of California, Berkeley.

A. Greenbaum (1989), Behavior of slightly perturbed Lanczos and conjugate gradient recurrences, *Linear Algebra Appl.* **113**, 7–63.

A. Greenbaum (1994), The Lanczos and conjugate gradient algorithms in finite precision arithmetic, in *Proc. Cornelius Lanczos International Centenary Conference, 1993* (J. D. Brown, M. T. Chu, D. C. Ellison and R. J. Plemmons, eds), SIAM, pp. 49–60.

A. Greenbaum (1997a), *Iterative Methods for Solving Linear Systems*, SIAM.

A. Greenbaum (1997b), Estimating the attainable accuracy of recursively computed residual methods, *SIAM J. Matrix Anal. Appl.* **18**, 535–551.

A. Greenbaum and Z. Strakoš (1992), Predicting the behavior of finite precision Lanczos and conjugate gradient computations, *SIAM J. Matrix Anal. Appl.* **13**, 121–137.

M. Gutknecht and Z. Strakoš (2000), Accuracy of two three-term and three two-term recurrences for Krylov space solvers, *SIAM J. Matrix Anal. Appl.* **22**, 213–229.

W. Hackbusch (1994), *Iterative Solution of Large Sparse Systems of Equations*, Vol. 95 of *Applied Mathematical Sciences*, Springer, New York. Translated and revised from the 1991 German original.

L. Hageman and D. Young (1981), *Applied Iterative Methods*, Academic Press, Orlando.

M. R. Hestenes and E. Stiefel (1952), Methods of conjugate gradients for solving linear systems, *J. Nat. Bur. Standards* **49**, 409–436.

M. R. Hestenes and J. Todd (1991), *Mathematicians Learning to Use Computers, National Institute of Standards and Technology Special Publication* **730**, US department of Commerce, National Institute of Standards and Technology, Washington, DC.

N. J. Higham (2002), *Accuracy and Stability of Numerical Algorithms*, second edition, SIAM.

A. S. Householder (1975), *The Theory of Matrices in Numerical Analysis*, Dover, New York. Reprint of 1964 edition.

S. Kaniel (1966), Estimates of some computational techniques in linear algebra, *Math. Comp.* **20**, 369–378.

J. Kautsky and G. H. Golub, (1983), On the calculation of Jacobi matrices, *Linear Algebra Appl.* **53/53**, 439–455.

L. Knizhnerman (1995a), The quality of approximations to an isolated eigenvalue and the distribution of 'Ritz numbers' in the simple Lanczos procedure, *Comput. Math. Math. Phys.* **35**, 1175–1187.

L. Knizhnerman (1995b), On adaptation of the Lanczos method to the spectrum. Report EMG-001-95-12, Schlumberger–Doll–Research.

L. Knizhnerman (1996), The simple Lanczos procedure: estimates of the error of the Gauss quadrature formula and their applications, *Comput. Math. Math. Phys.* **36**, 1481–1492.

A. N. Krylov (1931), O Čislemnon rešenii uravnenija, kotorym v techničeskih voprasah opredeljajutsja častoy malyh kolebanii material'nyh., *Izv. Adad. Nauk SSSR old. Mat. Estet.*, pp. 491–539.

A. B. J. Kuijlaars (2006), Convergence analysis of Krylov subspace iterations with methods from potential theory, *SIAM Review* **48**, 3–40.

C. Lanczos (1950), An iteration method for the solution of the eigenvalue problem of linear differential and integral operators, *J. Res. Nat. Bur. Standards* **45**, 255–282.

C. Lanczos (1952), Solution of systems of linear equations by minimized iterations, *J. Res. Nat. Bur. Standards* **49**, 33–53.

D. P. Laurie (1999), Accurate recovery of recursion coefficients from Gaussian quadrature formulas, *J. Comput. Appl. Math.* **112**, 165–180.

D. P. Laurie (2001), Computation of Gauss-type quadrature formulas, *J. Comput. Appl. Math.* **127**, 201–217.

J. G. Lewis (1977), Algorithms for sparse matrix eigenvalue problems. PhD thesis, Report STAN-CS-77-595, Computer Science Department, Stanford University, Stanford, CA.

Ren-Cang Li (2005), On Meinardus' examples for the conjugate gradient method. Technical Report 2005-06, Department of Mathematics, University of Kentucky.

J. Liesen and P. Tichý (2005), On the worst case convergence of MR and CG for symmetric positive definite tridiagonal Toeplitz matrices, *Electron. Trans. Numer. Anal.* **20**, 180–197.

D. G. Luenberger (1969), Hyperbolic pairs in the method of conjugate gradients, *SIAM J. Appl. Math.* **17**, 1263–1267.

D. G. Luenberger (1970), The conjugate residual method for constrained minimization problems, *SIAM J. Numer. Anal.* **7**, 390–398.

G. Meinardus (1963), Über eine Verallgemeinerung einer Ungleichung von L. V. Kantorowitsch, *Numer. Math.* **5**, 14–23.

G. Meurant (1997), The computation of bounds for the norm of the error in the conjugate gradient algorithm, *Numer. Algorithms* **16**, 77–87.

G. Meurant (1999*a*), Numerical experiments in computing bounds for the norm of the error in the preconditioned conjugate gradient algorithm, *Numer. Algorithms* **22**, 353–365.

G. Meurant (1999*b*), *Computer Solution of Large Linear Systems*, North-Holland.

G. Meurant (2005), Estimates of the l_2 norm of the error in the conjugate gradient algorithm, *Numer. Algorithms* **40**, 157–169.

G. Meurant (2006), *The Lanczos and Conjugate Gradient Algorithms: From Theory to Finite Precision Computations*, SIAM, book to appear.

Y. Notay (1993), On the convergence rate of the conjugate gradients in the presence of rounding errors, *Numer. Math.* **65**, 301–317.

D. P. O'Leary and Z. Strakoš (2004), On sensitivity of Gauss–Christoffel quadrature estimates. Computer Science Department Report CS-TR-4622, Institute for Advanced Computer Studies Report UMIACS-2004-64, University of Maryland.

M. M. Overton (2001), *Numerical Computing with IEEE Floating Point Arithmetic*, SIAM.

C. C. Paige (1969a), Error analysis of the generalized Hessenberg processes. Technical Note ICSI 179, London University Institute of Computer Science.

C. C. Paige (1969b), Eigenvalues of perturbed Hermitian matrices. Technical Note ICSI 179, London University Institute of Computer Science.

C. C. Paige (1970a), Error analysis of the symmetric Lanczos process for the eigenproblem. Technical Note ICSI 248, London University Institute of Computer Science.

C. C. Paige (1970b), Practical use of the symmetric Lanczos process with reorthogonalization, *BIT* **10**, 183–195.

C. C. Paige (1971), The computation of eigenvalues and eigenvectors of very large sparse matrices, PhD thesis, University of London.

C. C. Paige (1972), Computational variants of the Lanczos method for the eigenproblem, *J. Inst. Math. Appl.* **10**, 373–381.

C. C. Paige (1976), Error analysis of the Lanczos algorithm for tridiagonalizing a symmetric matrix, *J. Inst. Math. Appl.* **18**, 341–349.

C. C. Paige (1980), Accuracy and effectiveness of the Lanczos algorithm for the symmetric eigenproblem, *Linear Algebra Appl.* **34**, 235–258.

C. C. Paige and M. Saunders (1975), Solution of sparse indefinite systems of linear equations, *SIAM J. Numer. Anal.* **12**, 617–629.

C. C. Paige and Z. Strakoš (1999), Correspondence between exact arithmetic and finite precision behavior of Krylov space methods, in *XIV Householder Symposium* (J. Varah, ed.), University of British Columbia, pp. 250–253.

C. C. Paige, B. N. Parlett and H. van der Vorst (1995), Approximate solutions and eigenvalue bounds from Krylov subspaces, *Numer. Linear Algebra Appl.* **2**, 115–133.

B. N. Parlett (1980), *The Symmetric Eigenvalue Problem*, Prentice-Hall.

B. N. Parlett (1990), The contribution of J. H. Wilkinson to numerical analysis, in *A History of Scientific Computing* (Princeton, NJ, 1987), ACM Press Hist. Ser., ACM, New York, pp. 17–30.

B. N. Parlett (1992), The rewards for maintaining semi-orthogonality among Lanczos vectors, *Numer. Linear Algebra Appl.* **1**, 234–267.

B. N. Parlett (1994), Do we fully understand the symmetric Lanczos algorithms yet?, in *Proc. Cornelius Lanczos International Centenary Conference, 1993* (J. D. Brown, M. T. Chu, D. C. Ellison and R. J. Plemmons, eds), SIAM, pp. 93–108.

B. N. Parlett (1996), Invariant subspaces for tightly clustered eigenvalues of tridiagonals, *BIT* **36**, 542–562.

B. N. Parlett and I. Dhillon (2000), Relatively robust representations of symmetric tridiagonals, *Linear Algebra Appl.* **309**, 121–151.

B. N. Parlett and D. S. Scott (1979), The Lanczos algorithm with selective orthogonalization, *Math. Comp.* **33**, 217–238.

H. Rutishauser (1959), Theory of gradient methods, in *Refined Iterative Mehods for Computation of the Solution and the Eigenvalues of Self-Adjoint Boundary Value Problems*, Mitt. Inst. Angew. Math. ETH Zürich, Birkhäuser, Basel, pp. 24–49.

Y. Saad (1980), On the rates of convergence of the Lanczos and the block Lanczos methods, *SIAM J. Numer. Anal.* **17**, 687–706.

Y. Saad (1992), *Numerical Methods for Large Eigenvalue Problems*, Wiley.

D. S. Scott (1978), Analysis of the symmetric Lanczos algorithm, PhD thesis, University of California, Berkeley.

D. S. Scott (1979), How to make the Lanczos algorithm converge slowly, *Math. Comp.* **33**, 239–247.

D. S. Scott (1981), The Lanczos algorithm, in *Sparse Matrices and Their Use* (I. S. Duff, ed.), Academic Press, pp. 139–159.

H. D. Simon (1982), The Lanczos algorithm for solving symmetric linear systems, PhD thesis, University of California, Berkeley.

H. D. Simon (1984a), The Lanczos algorithm with partial reorthogonalization, *Math. Comp.* **42**, 115–142.

H. D. Simon (1984b), Analysis of the symmetric Lanczos algorithm with reorthogonalization methods, *Linear Algebra Appl.* **61**, 101–131.

V. Simoncini and D. Szyld (2005), Recent developments in Krylov subspace methods for linear systems. Research Report 05-9-25. Department of Mathematics, Temple University.

G. L. G. Sleijpen, H. A. van der Vorst and D. R. Fokkema (1994), BiCGstab(l) and other hybrid Bi-CG methods, *Numer. Algorithms* **7**, 75–109.

G. L. G. Sleijpen, H. A. van der Vorst and J. Modersitzki (2001), Difference in the effects of rounding errors in Krylov solvers for symmetric indefinite systems, *SIAM J. Matrix Anal. Appl.* **22**, 726–751.

A. van der Sluis and H. A. van der Vorst (1986), The rate of convergence of conjugate gradients, *Numer. Math.* **48**, 543–560.

A. van der Sluis and H. A. van der Vorst (1987), The convergence behavior of Ritz values in the presence of close eigenvalues, *Linear Algebra Appl.* **88**, 651–694.

G. W. Stewart (2001), *Matrix Algorithms*, Vol. II: *Eigensystems*, SIAM.

T. J. Stieltjes (1884), Quelques recherches sur la théorie des quadratures mécaniques, *Ann. Sci. Ecole Norm. Paris* **1**, 409–426. [Oeuvres I, 377–396.]

J. Stoer (1983), Solution of large linear systems of equations by conjugate gradient type methods, in *Mathematical Programming: The State of the Art* (A. Bachem, M. Grötschel and B. Korte, eds), Springer, pp. 540–565.

J. Stoer and R. Bulirsch (1983), *Introduction to Numerical Analysis*, second edition, Springer.

J. Stoer and R. W. Freund (1982), On the solution of large linear systems of equations by conjugate gradient type methods, in *Computer Methods in Applied Science and Engineering V* (R. Glowinski and J. L. Lions, eds), North-Holland, pp. 35–53.

Z. Strakoš (1991), On the real convergence rate of the conjugate gradient method, *Linear Algebra Appl.* **154–156**, 535–549.

Z. Strakoš (1998), Convergence and numerical behavior of the Krylov space methods, *NATO ASI Institute Algorithms for Large Sparse Linear Algebraic Systems: The State of the Art and Applications in Science and Engineering* (G. Winter Althaus and E. Spedicato, eds), Kluwer Academic, pp. 175–197.

Z. Strakoš and A. Greenbaum (1992), Open questions in the convergence analysis of the Lanczos process for the real symmetric eigenvalue problem, *IMA Preprint Series* **934**, University of Minnesota.

Z. Strakoš and J. Liesen (2005), On numerical stability in large scale linear algebraic computations, *ZAMM Z. Angew. Math. Mech.* **85**, No. 5, 307–325.

Z. Strakoš and P. Tichý (2002), On error estimation in the conjugate gradient method and why it works in finite precision computation, *Electron. Trans. Numer. Anal.* **13**, 56–80.

Z. Strakoš and P. Tichý (2005), Error estimation in preconditioned conjugate gradients, *BIT Numerical Mathematics* **45**, 789–817.

R. C. Thompson and P. McEnteggert (1968), Principal submatrices II: The upper and lower quadratic inequalities, *Linear Algebra Appl.* **1**, 211–243.

H. A. van der Vorst (1982), Preconditioning by incomplete decompositions, PhD thesis, Academic Computer Centrum Utrecht.

H. A. van der Vorst (2003), *Iterative Krylov Methods for Large Linear Systems*, Cambridge University Press.

J. H. Wilkinson (1965), *The Algebraic Eigenvalue Problem*, Oxford University Press.

H. Wozniakowski (1978), Round-off error analysis of iterations for large linear systems, *Numer. Math.* **30**, 301–314.

H. Wozniakowski (1980), Round-off error analysis of a new class of conjugate gradient algorithms, *Linear Algebra Appl.* **29**, 507–529.

W. Wülling (2005), The stabilization of weights in the Lanczos and conjugate gradient methods, *BIT Numerical Mathematics* **45**, 395–414.

W. Wülling (2006), On stabilization and convergence of clustered Ritz values in the Lanczos method, *SIAM J. Matrix Anal. Appl.* **27**, 891–908.

Q. Ye (1995), On close eigenvalues of tridiagonal matrices, *Numer. Math.* **70**, 507–514.

J.-P. M. Zemke (2003), Krylov subspace methods in finite precision: a unified approach, PhD thesis, Technical University of Hamburg–Harburg.

Acta Numerica (2006), pp. 543–639
doi: 10.1017/S0962492906270016

Kernel techniques: From machine learning to meshless methods

Robert Schaback and Holger Wendland
Institut für Numerische und Angewandte Mathematik,
Universität Göttingen, Lotzestraße 16–18,
D–37083 Göttingen, Germany
E-mail: {schaback}{wendland}@math.uni-goettingen.de
http://www.num.math.uni-goettingen.de/schaback
http://www.num.math.uni-goettingen.de/wendland

Kernels are valuable tools in various fields of numerical analysis, including approximation, interpolation, meshless methods for solving partial differential equations, neural networks, and machine learning. This contribution explains why and how kernels are applied in these disciplines. It uncovers the links between them, in so far as they are related to kernel techniques. It addresses non-expert readers and focuses on practical guidelines for using kernels in applications.

CONTENTS

1. Introduction

This article can be seen as an extension of Martin Buhmann's presentation of *radial basis functions* (Buhmann 2000) in this series. But we shall take a somewhat wider view and deal with *kernels* in general, focusing on their recent applications in areas such as *machine learning* and *meshless methods* for solving partial differential equations.

In their simplest form, kernels may be viewed as bell-shaped functions like Gaussians. They can be shifted around, dilated, and superimposed with weights in order to form very flexible spaces of multivariate functions having useful properties. The literature presents them under various names in contexts of different numerical techniques, for instance as *radial basis functions*, *generalized finite elements*, *shape functions* or even *particles*. They are useful both as *test functions* and *trial functions* in certain *meshless methods* for solving partial differential equations, and they arise naturally as *covariance kernels* in probabilistic models. In the case of learning methods, sigmoidal functions within neural networks were successfully superseded by radial basis functions, but now they have both been replaced by *kernel machines*[1] to implement the most successful algorithms for *machine learning* (Schölkopf and Smola 2002, Shawe-Taylor and Cristianini 2004). Even the term *kernel engineering* has been coined recently, because efficient learning algorithms require specially tailored application-dependent kernels.

With this slightly chaotic background in mind, we survey the major application areas while focusing on a few central issues that lead to guidelines for practical work with kernels. Section 2 starts with a general definition of kernels and provides a short account of their properties. The main reasons for using kernels at all will be described in Section 3, starting with their ability to recover functions optimally from given unstructured data. At this point, the connections between kernel methods for interpolation, approximation, learning, pattern recognition, and PDE solving become apparent. The probabilistic aspects of kernel techniques follow in Section 4, while practical guidelines for constructing new kernels follow in Section 5. Special application-oriented kernels are postponed to Section 6 to avoid too much detail at the beginning.

Since one of the major features of kernels is to generate spaces of trial functions with excellent approximation properties, we devote Section 7 to a short account of the current results concerning such questions. Together with strategies to handle large and ill-conditioned systems (Section 8), these results are of importance to the applications that follow later.

After a short interlude on kernels on spheres in Section 9 we start our survey of applications in Section 10 by looking at interpolation problems

---

[1] http://www.kernel-machines.org

first. These take advantage of the abilities of kernels to handle unstructured Birkhoff-type data while producing solutions of arbitrary smoothness and high accuracy. Then we review kernels in modern learning algorithms, but we can keep this section short because there are good books on the subject.

In Section 12 we survey meshless methods (Belytschko, Krongauz, Organ, Fleming and Krysl 1996b) for solving partial differential equations. We describe the different techniques currently sailing under this flag, and point out where and how kernels occur. Owing to an existing survey (Babuška, Banerjee and Osborn 2003) in this series, we keep the generalized finite element method short here, but we incorporate meshless local Petrov–Galerkin techniques (Atluri and Shen 2002).

The final two sections then focus on purely kernel-based meshless methods. We treat applications of symmetric and unsymmetric collocation, of kernels providing fundamental and particular solutions, and provide the state of the art of their mathematical foundation.

Altogether, we want to keep this survey digestible for the non-expert and casual reader who wants to know roughly what has happened so far in the area of application-oriented kernel techniques. This is why we omit most of the technical details and focus on the basic principles. Consequently, we have to refer as much as possible to background reading for proofs and extensions. Fortunately, there are two recent books, Buhmann (2004) and Wendland (2005b), which contain the core of the underlying general mathematics for kernels and radial basis functions. For kernels in learning theory, we have already cited two other books, Schölkopf and Smola (2002) and Shawe-Taylor and Cristianini (2004), providing further reading. If we omit pointers to proofs, these books will contain what is needed.

Current books and survey articles in the area of meshless methods are rather specialized, because they focus either on certain classes of methods or on applications. We cite them as needed, placing them into a more general context. Clearly, the list of references cannot contain all available papers on all possible kernel applications. This forces us to select a very small subset, and our main selection criterion is how a certain reference fits into the current line of argument at a certain place of this survey. Incorporation or omission of a certain publication does not express our opinion on its importance in general.

2. Kernels

Definition 2.1. A *kernel* is a function

$$K : \Omega \times \Omega \to \mathbb{R}$$

where Ω can be an arbitrary nonempty set.

Some readers may consider this to be far too general. However, in the context of learning algorithms, the set Ω defines the possible *learning inputs*. Thus Ω should be sufficiently general to allow Shakespeare texts or X-ray images, *i.e.*, Ω should preferably have no predefined structure at all. Thus the kernels occurring in machine learning are extremely general, but they still take a special form which can be tailored to meet the demands of applications. We shall now explain the recipes for their definition and usage.

2.1. Feature maps

In certain situations, a kernel is given *a priori*, *e.g.*, the *Gaussian*

$$K(x, y) := \exp(-\|x - y\|_2^2) \quad \text{for all } x, y \in \Omega := \mathbb{R}^d. \tag{2.1}$$

Each specific choice of a kernel has a number of important and possibly unexpected consequences which we shall describe later.

If no predefined kernel is available for a certain set Ω, an application-dependent *feature map* $\Phi : \Omega \to \mathcal{F}$ with values in a Hilbert '*feature*' space $\mathcal{F}$ is defined. It should provide for each $x \in \Omega$ a large collection $\Phi(x)$ of *features* of x which are characteristic for x and which live in the Hilbert space $\mathcal{F}$ of high or even infinite dimension. Note that $\mathcal{F}$ has plenty of useful structure, while Ω does not.

Guideline 2.2. Feature maps $\Omega \to \mathcal{F}$ allow us to apply linear techniques in their range $\mathcal{F}$, while their domain Ω is an unstructured set. They should be chosen carefully in an application-dependent way, capturing the essentials of elements of Ω.

With a feature map Φ at hand, there is a *kernel*

$$K(x, y) := \big(\Phi(x), \Phi(y)\big)_{\mathcal{F}} \quad \text{for all } x, y \in \Omega. \tag{2.2}$$

In another important class of cases, the set Ω consists of random variables. Then the *covariance* between two random variables x and y from Ω is a standard choice of a kernel. These and other kernels arising in nondeterministic settings will be the topic of Section 4. The connection to learning is obvious: two learning inputs x and y from Ω should be very similar, if they are closely 'correlated', if they have very similar features, or if (2.2) takes large positive values. These examples suggest the following definition.

Definition 2.3. A kernel K is *symmetric* if $K(x, y) = K(y, x)$ holds for all $x, y \in \Omega$.

2.2. Spaces of trial functions

A kernel K on Ω defines a function $K(\cdot, y)$ for all fixed $y \in \Omega$. This allows us to generate and manipulate spaces

$$\mathcal{K}_0 := \operatorname{span}\{K(\cdot, y) : y \in \Omega\} \tag{2.3}$$

of functions on Ω. In learning theory, the function $K(\cdot, y) = (\Phi(\cdot), \Phi(y))_{\mathcal{F}}$ relates each other input object to a fixed object y via its essential features. But in general $\mathcal{K}_0$ just provides a handy linear space of *trial* functions on Ω which is extremely useful for most applications of kernels, *e.g.*, when Ω consists of texts or images. For example, in meshless methods for solving partial differential equations, certain finite-dimensional subspaces of $\mathcal{K}_0$ are used as *trial* spaces to furnish good approximations to the solutions.

2.3. Convolution kernels

In certain other cases, the set Ω carries a measure μ, and then, under reasonable assumptions like $f, K(y, \cdot) \in L^2(\Omega, \mu)$, the generalized *convolution*

$$K *_\Omega f := \int_\Omega f(x) K(\cdot, x) \, d\mu(x) \tag{2.4}$$

defines an integral transform $f \mapsto K *_\Omega f$ which can be very useful. Note that Fourier or Hankel transforms arise this way, and recall the role of the Dirichlet kernel in the Fourier analysis of univariate periodic functions. The above approach to kernels via convolution works on locally compact topological groups using Haar measure, but we do not want to pursue this detour into abstract harmonic analysis too far. For space reasons, we also have to exclude complex-valued kernels and all transform-type applications of kernels here, but it should be pointed out that wavelets are special kernels of the above form, defining the *continuous wavelet transform* this way.

Note that discretization of the integral in the convolution transform leads to functions in the space $\mathcal{K}_0$ from (2.3). Using kernels as trial functions can be viewed as a discretized convolution. This is a very useful fact in the theoretical analysis of kernel-based techniques.

Guideline 2.4. Kernels have three major application fields: they generate convolutions, trial spaces, and covariances. The first two are related by discretization.

2.4. Scaling

Another important aspect in all kernel-based techniques is the *scaling problem*. If the kernel K in the convolution equation (2.4) is a sharp nonnegative spike with integral one, the convolution will reproduce f approximately, and the distributional 'delta kernel' will reproduce f exactly. This is theoretically nice, but discretization will need a very fine spatial resolution. On the other hand, convolution with a nonnegative smooth kernel of wide or infinite support acts as a *smoothing operator* which will not have good reproduction quality. To control this trade-off between approximation and smoothing, many kernel applications involve a free scaling parameter, and it is a serious problem to derive good strategies for its determination.

Guideline 2.5. Success and failure of kernel usage may crucially depend on proper scaling.

The scaling problem will come up at various places in this article.

2.5. Positive definiteness

For many applications, the space $\mathcal{K}_0$ needs more structure. In fact, it can be turned into a Hilbert space via the following construction.

Definition 2.6. A symmetric kernel K is *positive (semi-) definite* if, for all finite subsets $X := \{x_1, \ldots, x_N\}$ of distinct points of Ω, the symmetric *kernel matrices* $A_{K,X}$ with entries $K(x_j, x_k)$, $1 \leq j, k \leq N$ are positive (semi-) definite.

We delay the definition of *conditionally* positive definite kernels to Section 6. For a symmetric positive definite kernel K on Ω, the definition

$$(K(x, \cdot), K(y, \cdot))_{\mathcal{K}} = (K(\cdot, x), K(\cdot, y))_{\mathcal{K}} := K(x, y) \quad \text{for all } x, y \in \Omega \quad (2.5)$$

of an inner product of two generators of $\mathcal{K}_0$ easily generalizes to an inner product on all of $\mathcal{K}_0$ such that

$$\left\| \sum_{j=1}^{N} \alpha_j K(\cdot, x_j) \right\|_{\mathcal{K}}^2 := \sum_{j,k=1}^{N} \alpha_j \alpha_k K(x_j, x_k) = \alpha^T A_{K,X} \alpha \quad (2.6)$$

defines a *numerically accessible* norm on $\mathcal{K}_0$ which allows us to construct a *native* Hilbert space

$$\mathcal{K} := \operatorname{clos} \mathcal{K}_0 \quad (2.7)$$

as the completion, or Hilbert space closure, of $\mathcal{K}_0$ under the above norm. In most cases, the space $\mathcal{K}$ is much richer than $\mathcal{K}_0$ and does not seem to have any explicit connection to the kernel from which it is generated. For instance, Sobolev spaces $\mathcal{K} = W_2^k(\mathbb{R}^d)$ with $k > d/2$ result from the kernel

$$K(x, y) = \|x - y\|_2^{k-d/2} K_{k-d/2}(\|x - y\|_2) \quad (2.8)$$

where K_ν is the Bessel function of third kind. Starting from (2.8) it is not at all clear that the closure (2.7) of the span (2.3) of all translates of K generates the Sobolev space $W_2^k(\mathbb{R}^d)$. But it should be clear that the native Hilbert space for a kernel has important consequences for any kind of numerical work with the trial space $\mathcal{K}_0$ of (2.3).

Guideline 2.7. User of kernel techniques should always be aware of the specific native Hilbert space associated to the kernel.

Under certain additional assumptions, there is a one-to-one correspondence between symmetric positive definite kernels and Hilbert spaces of functions, so that such kernels cannot be separated from their native Hilbert space.

However, note that in general the Hilbert spaces $\mathcal{F}$ from (2.2) and $\mathcal{K}$ from (2.5) are different. The space $\mathcal{K}$ is always a Hilbert space of functions on Ω, while the 'feature space' $\mathcal{F}$ in general is not. However, the two notions coincide if we start with a given kernel, not with a feature map.

Theorem 2.8. Every symmetric positive definite kernel can be generated via a suitable feature map.

Proof. Given a symmetric positive definite kernel K, define $\Phi(x) := K(x, \cdot)$ and $\mathcal{F} := \mathcal{K}$ using (2.5) to get (2.2). □

2.6. Reproduction

By construction, the spaces $\mathcal{K}$ and $\mathcal{K}_0$ have a nice structure now, and there is a *reproduction property*

$$f(x) := (f, K(\cdot, x))_{\mathcal{K}} \quad \text{for all } f \in \mathcal{K}, \ x \in \Omega \qquad (2.9)$$

for all functions in $\mathcal{K}$. At this point, we are on the classical ground of *reproducing kernel Hilbert spaces* (RKHS) with a long history (Aronszajn 1950, Meschkowski 1962, Atteia 1992).

Guideline 2.9. Positive definite kernels *reproduce* all functions from their associated *native* Hilbert space. On the trial space (2.3) of translated positive definite kernels, the Hilbert space norm can be *numerically calculated* by plain kernel evaluations, without integration or derivatives. This is particularly useful if the Hilbert space norm theoretically involves integration and derivatives, *e.g.*, in the case of Sobolev spaces.

2.7. Invariance

Guideline 2.10. If the set Ω has some additional geometric structure, kernels may take a simplified form, making them *invariant* under geometric transformations on Ω.

For instance, kernels of the form

$$
\begin{array}{ll}
K(x - y) & \text{are \textit{translation-invariant} on abelian groups} \\
K(x^T y) & \text{are \textit{zonal} on multivariate spheres} \\
K(\|x - y\|_2) & \text{are \textit{radial} on } \mathbb{R}^d
\end{array}
$$

with a slight abuse of notation. Radial kernels are also called *radial basis functions*, and they are widely used because of their invariance under Euclidean (rigid-body-) transformations in $\mathbb{R}^d$. The most important example is the *Gaussian* kernel of (2.1), which is symmetric positive definite on $\mathbb{R}^d$, for any space dimension d. It naturally arises as a convolution kernel, a covariance, a perfectly smooth trial function, and a multivariate probability density, illustrating the various uses of kernels. Less obvious is the fact that it has a native Hilbert space of analytic functions on $\mathbb{R}^d$.

2.8. Metric structure

In many applications, for instance in machine learning, the kernel value $K(x, y)$ increases with the 'similarity' of x and y, like a correlation or a covariance, and is *bell-shaped* like the Gaussian. More precisely, any symmetric positive definite kernel K generates a *distance metric* $d : \Omega \times \Omega \to [0, \infty)$ via

$$d^2(x, y) := K(x, x) - 2K(x, y) + K(y, y) \quad \text{for all } x, y \in \Omega \qquad (2.10)$$

on a general set (Schoenberg 1937, Stewart 1976). Looking back at feature maps, we see that a well-chosen feature map defines a kernel that introduces a metric structure on the set Ω for which 'close' elements have 'similar' features.

Guideline 2.11. Symmetric positive definite kernels on Ω introduce a 'geometry' on the set Ω which can be tailored to meet the demands of applications.

The art of *kernel engineering* is to do this in a best possible way, depending on the application in question.

3. Optimal recovery

One of the key advantages of kernels is as follows.

Guideline 3.1. Kernel-based methods can make optimal use of the given information.

Results like this come up at various places in theory and applications, and they have a common background linking them to the interesting fields of *information-based complexity*[2] (Traub and Werschulz 1998) and *optimal recovery* (Micchelli, Rivlin and Winograd 1976, Micchelli and Rivlin 1977) which we have to ignore here. In a probabilistic context, Guideline 3.1 can be forged into an exact statement using Bayesian arguments, but we want to keep things simple first and postpone details to Section 4.

3.1. Recovery from unstructured data

Assume that we want to model a black-box transfer mechanism like Figure 3.1 that replies to an input $x \in \Omega$ by an output $f(x) \in \mathbb{R}$. This can be the reaction $f(x)$ of a well-trained individual or machine to a given stimulus

---

[2] http://www.ibc-research.org

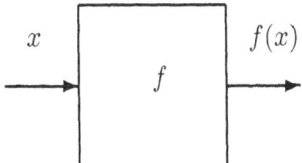

Figure 3.1. A black-box response mechanism.

x given to it. Finding a good response mechanism f can be called *learning* or *black-box modelling*. If the output should take only a finite number of possible values, this is *pattern recognition* or *classification*. We shall use the term 'recovery problem' (Micchelli *et al.* 1976, Micchelli and Rivlin 1977) to summarize all of these situations, which mathematically require the determination of a function. But we want to stick to an application-oriented view here.

At this point we do not have any further information on the model or the intended reactions to the stimuli. But usually we have some examples of 'good behaviour' that can be used. These take the form of a sequence $(x_1, y_1), \ldots, (x_N, y_N)$ of unstructured *training data*, pairing inputs $x_j \in \Omega$ with their expected responses $y_j \in \mathbb{R}$. The recovery task now is to find a function f such that

$$f(x_j) \approx y_j, \quad 1 \le j \le N, \tag{3.1}$$

and this is a standard interpolation or approximation problem, though posed on an unstructured set Ω using unstructured data.

If we slightly extend the meaning of the word 'data', we can try to find a smooth function f such that

$$(-\Delta f)(y_j) \approx \varphi(y_j), \quad 1 \le j \le M,$$
$$f(z_k) \approx \psi(z_k), \quad M+1 \le k \le N, \tag{3.2}$$

where $y_1, \ldots, y_M$ are points in a bounded domain Ω while $z_{M+1}, \ldots, z_N$ lie on the boundary. This would hopefully provide an approximate solution f to the Poisson problem

$$(-\Delta f)(y) = \varphi(y), \quad y \in \Omega,$$
$$f(z) = \psi(z), \quad z \in \partial\Omega,$$

for given functions φ on Ω and ψ on $\partial\Omega$. Note that this *collocation* technique is again a recovery problem for a function f from certain of its data, just replacing point evaluations in (3.1) by evaluations of certain derivatives. In general, one can replace (3.1) by

$$\lambda_j(f) \approx y_j, \quad 1 \le j \le N, \tag{3.3}$$

for a set of given linear *data functionals* $\lambda_1, \ldots, \lambda_N$ generalizing the point

evaluation functionals $\delta_{x_1}, \ldots, \delta_{x_N}$ of (3.1). Tradition in approximation theory would call this a recovery problem from *Hermite–Birkhoff* data, if the
data functionals are evaluations of derivatives at points. But there are much
more general functionals, *e.g.*, those defining *weak data* via

$$\lambda_j(f) = \int_\Omega \nabla f \cdot \nabla v_j$$

as in finite elements, using a *test function* v_j. This way, finite element
methods for solving linear partial differential equations can be written as
recovery problems (3.3).

For later sections of this article, the reader should keep in mind that
suitable generalizations (3.3) of the recovery problem (3.1) lead to methods
for solving partial differential equations. We shall stick to the simple form
of (3.1) for a while, but when reviewing large parts of numerical analysis,
e.g., finite element techniques, we have the following heuristic.

Guideline 3.2. Many applications can be rephrased as *recovery problems*
for functions from unstructured data.

3.2. Generalization

The resulting model function f should be such that it *generalizes* well,
i.e., it should give practically useful responses $f(x)$ to new inputs $x \in \Omega$.
Furthermore, it should be *stable* in the sense that small changes in the
training data do not change f too much. But these goals are in conflict
with good reproduction of the training data. A highly stable but useless
model would be $f = 1$, while *overfitting* occurs if there is too much emphasis
on data reproduction, leading to unstable models with bad generalization
properties.

Guideline 3.3. Recovery problems are subject to the *reproduction–generalization dilemma* and need a careful balance between generalization and
stability properties on one hand, and data reproduction quality on the other.

This is also called the *bias-variance dilemma* under certain probabilistic
hypotheses, but it also occurs in deterministic settings.

Given a recovery problem as in (3.1), there is not enough information to
come up with a useful solution of the recovery problem. In particular, we
have no idea how to define f or from which space of functions to pick it from.
From a theoretical point of view, we are facing an *ill-posed problem* with
plenty of indistinguishable approximate solutions. From a practical point of
view, all mathematical *a priori* assumptions on f are useless because they
do not take the application into account.

Instead, one should use additional application-dependent information con-
cerning the essentials of the inputs, *e.g.*, define a *feature map* $\Phi : \Omega \to \mathcal{F}$

as in (2.2), taking an object x to an object $\Phi(x)$ in $\mathcal{F}$ containing all essential features of x. With this additional information, we can define a kernel K using (2.2), and we get a space $\mathcal{K}$ of functions on Ω via (2.3) and (2.7). Since $\mathcal{K}$ usually turns out to be rather large (see the example in (2.8) for Sobolev spaces), this space serves as a natural reservoir from which to pick f, and if we have no other information, there is no other choice for a space defined on all of Ω. Of course, the choice of a feature map is just another way of adding hypotheses, but it is one that can be tailored perfectly to the application, using kernel engineering knowledge.

3.3. *Optimality*

We are now left with the problem to pick f somehow from the space $\mathcal{K}$, using our training set. If we insist on exact recovery, we get an instance of Guideline 3.1 from the following theorem.

Theorem 3.4. Let the kernel K be symmetric positive definite. Then a function of the form

$$f^* := \sum_{k=1}^{N} \alpha_k K(\cdot, x_k) \qquad (3.4)$$

is the unique minimizer of the Hilbert space norm in $\mathcal{K}$ amongst all functions $f \in \mathcal{K}$ with $f(x_j) = y_j$, $1 \leq j \leq N$. The coefficients α_k can be calculated from the linear system

$$\sum_{k=1}^{N} \alpha_k K(x_j, x_k) = y_j, \quad 1 \leq j \leq N. \qquad (3.5)$$

As Section 4 will show, the system (3.5) also arises for different nondeterministic recovery problems in exactly the same way, but with different semantics.

Clearly, symmetric positive definiteness of the kernel implies positive definiteness of the *kernel matrix* $A_{K,X}$ in (3.5) which we saw in Definition 2.6.

Guideline 3.5. Interpolation of unstructured data using a kernel is an optimal strategy for black-box modelling and learning from noiseless information.

The essential information on the application is built into the kernel. Once the kernel is there, things are simple, theoretically. The *generalization error* is optimal in the following sense.

Theorem 3.6. Consider all possible linear recovery schemes of the form

$$f_u(\cdot) := \sum_{j=1}^{N} u_j(\cdot) f(x_j)$$

which use the training data $(x_j, y_j) = (x_j, f(x_j))$, $1 \leq j \leq N$ for an un-known model $f \in \mathcal{K}$ and employ arbitrary functions u_j on Ω. Then the approximate solution f^* of Theorem 3.4 satisfies

$$\inf_{u} \sup_{\|f\|_{\mathcal{K}} \leq 1} |f(x) - f_u(x)| = \sup_{\|f\|_{\mathcal{K}} \leq 1} |f(x) - f^*(x)| \quad \text{for all } x \in \Omega \qquad (3.6)$$

and it has the form $f^* = f_{u^*}$ with Lagrange-type functions $u_1^*(x), \ldots, u_N^*(x)$ from $\text{span}\{K(\cdot, x_j) : 1 \leq j \leq N\}$ satisfying

$$\sum_{j=1}^{N} u_j^*(x) K(x_j, x_k) = K(x, x_k), \quad 1 \leq k \leq N, \quad \text{for all } x \in \Omega. \qquad (3.7)$$

Note that this is another instance of Guideline 3.1. The optimality results of the previous theorems are well-known properties of univariate splines.

Guideline 3.7. In the context of optimal recovery, kernel methods provide natural multivariate extensions of classical univariate spline techniques.

For later reference in Section 4, we should explain the connection between the linear systems (3.5) and (3.7) on one hand, and the representations (3.4) and (3.6) on the other. Theorem 3.4 works on the basis $K(\cdot, x_k)$ directly, while Theorem 3.6 produces a new basis of functions u_j^* which has the Lagrangian property $u_j^*(x_k) = \delta_{jk}$ but spans the same space. The optimal recovery solutions coincide, but have different basis representations.

This basis change, if executed only approximately, is important for applications. In fact, transition to a local Lagrange or 'cardinal' basis is one of the possible preconditioning strategies (Faul and Powell 1999, Ling and Kansa 2004, Brown, Ling, Kansa and Levesley 2005, Ling and Kansa 2005), and approximate Lagrangian bases yield *quasi-interpolants* (Buhmann 1988, Beatson and Light 1993, Buhmann 1993, Buhmann, Dyn and Levin 1995, Maz'ya and Schmidt 2001) which avoid solving linear systems because they provide approximate inverses. This is a promising research area.

3.4. Generalized recovery

If the recovery problem (3.1) is generalized to (3.3), there is a similar theory (Wu 1992, Luo and Levesley 1998) concerning optimal recovery, replacing the kernel matrix with entries $K(x_j, x_k)$ by a symmetric matrix with elements $\lambda_j^x \lambda_k^y K(x, y)$, where we used an upper index x at λ^x to indicate that the functional λ acts with respect to the variable x. The system (3.5) becomes

$$\sum_{k=1}^{N} \alpha_k \lambda_j^x \lambda_k^y K(x, y) = y_j, \quad 1 \leq j \leq N, \qquad (3.8)$$

while (3.7) will turn into

$$\sum_{j=1}^{N} u_j^*(x)\lambda_j^x \lambda_k^y K(x,y) = \lambda_k^y K(x,y), \quad 1 \le k \le N, \quad \text{for all } x \in \Omega.$$

Guideline 3.8. Kernel methods can recover a function f from very general unstructured data, if the kernel is sufficiently smooth and the 'data' of f are linear functionals acting on f.

This is used in applications described in Section 10. In the case of the recovery problem (3.2), we get a symmetric meshless *collocation* technique for solving Poisson's equation. This will be treated in more detail in Section 14.

3.5. Error, condition, and stability

Let us go back to the generalization error. We shall see in Section 7 that the generalization error of kernels on $\mathbb{R}^d$ dramatically improves with their smoothness while still maintaining applicability to recovery problems with unstructured data. This is one of the key features of kernel techniques.

Guideline 3.9. Methods based on fixed smooth positive definite kernels can provide recovery techniques with very small errors, using rather small amounts of data.

But the small generalization error comes at a high price, because there are serious practical problems with systems of the form (3.5). This is in sharp contrast to the encouraging optimality properties stated so far.

Guideline 3.10. The linear systems (3.5) and (3.8) can be very large, non-sparse and severely ill-conditioned.

However, the latter is no surprise because the method solves an ill-posed problem approximately. Thus the bad condition of the system (3.5) must be expected somehow. There is an apparent link between condition and scaling, since kernels with small supports will lead to approximately diagonal kernel matrices, while kernels with wide scales produce matrices with very similar rows and columns.

Guideline 3.11. Positive definite kernels with small scales lead to better matrix condition than kernels with wide scales.

Since we know that kernel systems (3.5) or (3.8) are solvable for symmetric positive definite kernels and linearly independent data functionals, we have the following guideline.

Guideline 3.12. Methods based on positive definite kernels have a built-in *regularization*.

In fact, they solve the ill-posed problem (3.1) by providing an approximate solution minimizing the Hilbert space norm in $\mathcal{K}$ under all conceivable exact recovery schemes, as if they were using a regularizing penalty term of the form $\|f\|_{\mathcal{K}}^2$, which can be a Sobolev space norm for certain kernels. This regularization property will come up later when we use kernels in collocation techniques for solving partial differential equations. If (3.5) is viewed as an approximate solution of the integral equation

$$\int_{\Omega} \alpha(x) K(y, x) \, dx = f(y) \quad \text{for all } y \in \Omega$$

via a quadrature formula, we have another aspect telling us that (3.5) solves an ill-posed problem approximately via some regularization in the background. Note the connection to convolution (2.4).

The generalization error $f(x) - f^*(x)$ and the condition of the system (3.5) have an unexpected connection. Theoretical results (Schaback 1995a) and simple numerical experiments with various kernels show the following.

Guideline 3.13. Increasing smoothness of kernels on $\mathbb{R}^d$ decreases the recovery error but increases the condition of the system (3.5). There are no kernels that provide small errors and good condition simultaneously.

Guideline 3.14. Increasing the scale of a kernel on $\mathbb{R}^d$ decreases the recovery error but increases the condition of the system (3.5).

Note that this limits the use of systems like (3.5) in their original form, but techniques like preconditioning (Faul and Powell 1999, Ling and Kansa 2004, Brown *et al.* 2005, Ling and Kansa 2005) or domain decomposition (see Section 8) should be applied.

Guideline 3.15. Programming of kernel methods should always use the kernel scaling as an adjustable parameter. Experiments with different scales often show that systems without preconditioning give best results when the scaling is as wide as numerically feasible. Following Guideline 3.17 below, one should use pivoting or SVD techniques, and this can work well beyond the standard condition limit of 10^{15}.

Reasons for this will be given at the end of this section. The limit behaviour of recovery problems for analytic kernels with increasing width is related to multivariate polynomials (Driscoll and Fornberg 2002). Unexpectedly, function recovery problems using wide-scaled Gaussians tend to the polynomial interpolation method of de Boor and Ron, if the scales get infinitely wide (Schaback 2005a). This will hopefully lead to a better understanding of preconditioning techniques in the future.

3.6. Relaxation and complexity

If N is huge, the exact solution (3.4) of a system (3.5) is much too complex to be useful. This is where another general rule comes up.

Guideline 3.16. Within kernel methods, relaxation of requirements can lead to reduction of complexity.

Under certain probabilistic hypotheses, this is another aspect of the *bias-variance dilemma* related to *overfitting*. As we mentioned at the beginning, insisting on exact reproduction of huge amounts of data increases the complexity of the model and makes it very sensible to changes in the training data, thus less reliable as a model. Conversely, relaxing the reproduction quality will allow a simpler model. Before we turn to specific relaxation methods used in kernel-based learning, we should look back at Guideline 3.9 to see that badly conditioned large systems of the form (3.5) using smooth kernels will often have subsystems that provide good approximate solutions to the full system. This occurs if the generalization error is small when going over from the training data of a subset to the full training data. Thus Guideline 3.16 can be satisfied by simply taking a small suitable subset of the data, relying on Guideline 3.9. As we shall see, this has serious implications for kernel-based techniques for solving partial differential equations or machine learning. For simple cases, the following suffices.

Guideline 3.17. Within kernel methods, large and ill-conditioned systems often have small and better conditioned subsystems furnishing good approximate solutions to the full system. Handling numerical rank loss by intelligent pivoting is useful.

However, large problems need special treatment, and we shall deal with such cases in Section 8.

The relaxation of (3.5) towards (3.1) can be done in several ways, and learning theory uses *loss functions* to quantify the admissible error in (3.1). We present this in Section 11 in more detail. Let us look at a simple special case. We allow a uniform tolerance ϵ on the reproduction of the training data, *i.e.*, we impose the linear constraints

$$-\epsilon \leq f(x_j) - y_j \leq \epsilon, \quad 1 \leq j \leq N. \tag{3.9}$$

We then minimize $\|f\|_{\mathcal{K}}$ while keeping ϵ fixed, or we minimize the weighted objective function $\frac{1}{2}\|f\|_{\mathcal{K}}^2 + C\epsilon$ when ϵ is varying and C is fixed. Optimization theory then tells us that the solution f^* is again of the form (3.4), but the Kuhn–Tucker conditions imply that the sum only contains terms where the constraints in (3.9) are *active*, *i.e.*, $\alpha_k \neq 0$ holds only for those k with $|f(x_k) - y_k| = \epsilon$. In view of principle (3.9) these *support vectors* will often be a rather small subset of the full data, and they provide an instance of complexity reduction via relaxation according to Guideline 3.16. This roughly describes the principles behind *support vector machines* for the implementation of learning algorithms. These principles are consequences of optimization, not of statistical learning theory, and they arise in other applications as well. We explain this in some more detail in Section 11 and apply it to adaptive collocation solvers for partial differential equations in Section 14.

Furthermore, we see via this optimization argument that the exact solution of a large system (3.5) can be replaced by an approximate solution of a smaller subsystem. This supports Guideline 3.17 again. It is in sharpest possible contrast to the large linear systems arising in finite element theory.

Guideline 3.18. Systems arising in kernel-based recovery problems should be solved approximately by adaptive or optimization algorithms, finding suitable subproblems.

At this point, the idea of *online learning* is helpful. It means that the training sample is viewed as a possibly infinite input sequence $(x_j, y_j) \approx (x_j, f(x_j))$, $j = 1, 2, \ldots$ which is used to update the current model function f_k if necessary. The connection to adaptive recovery algorithms is clear, since a new training data pair (x_{N+1}, y_{N+1}) will be discarded if the current model function f_k works well on it, *i.e.*, if $f_k(x_{N+1}) - y_{N+1}$ is small. Otherwise, the model function is carefully and efficiently updated to make optimal use of the new data (Schaback and Werner 2006). Along these lines, one can devise adaptive methods for the approximate solution of partial differential equations which 'learn' the solution in the sense of online learning, if they are given infinitely many data of the form (3.2).

Within approximation theory, the concept of adaptivity is closely related to the use of *dictionaries* and *frames*. In both cases, the user does not work with a finite and small set of trial functions to perform a recovery. Instead, a selection from a large reservoir of possible trial functions is made, *e.g.*, by *greedy* adaptive methods or by choosing frame representations with many vanishing coefficients via certain projections. This will be a promising research area in the coming years.

The final sections of this article will review several application areas of kernel techniques. However, we shall follow the principles stated above, and we shall work out the connections between recovery, learning, and equation solving at various places. This will have to start with a look at nondeterministic recovery problems.

4. Kernels in probabilistic models

There are several different ways in which kernels arise in probability theory and statistics. We shall describe the most important ones very briefly, ignoring the standard occurrence of certain kernels like the Gaussian as *densities* of probability distributions. Since *Acta Numerica* is aiming at readers in numerical analysis, we want to assume as little stochastic background as possible.

4.1. Nondeterministic recovery problems

If we go back to the recovery problem of Section 3 and rewrite it in a natural probabilistic setting, we get another instance of Guideline 3.1, because kernel-based techniques again turn out to have important optimality properties. Like in Section 3 we assume that we want to find the response $f(x)$ of an unknown model function f at a new point x of a set Ω, provided that we have a sample of input-response pairs $(x_j, y_j) = (x_j, f(x_j))$ given by observation or experiment. But now we assume that the whole setting is nondeterministic, *i.e.*, the response y_j at x_j is not a fixed function of x_j but rather a realization of a real-valued random variable $Z(x_j)$. Thus we assume that for each $x \in \Omega$ there is a real-valued random variable $Z(x)$ with expectation $E(Z(x))$ and bounded positive variance $E((Z(x) - E(Z(x))^2)$. The goal is to get information about the function $E(Z(x))$ which replaces our f in the deterministic setting. For two elements $x, y \in \Omega$ the random variables $Z(x)$ and $Z(y)$ will not be uncorrelated, because if x is close to y the random experiments described by $Z(x)$ and $Z(y)$ will often show similar behaviour. This is described by a *covariance kernel*

$$\text{cov}(x, y) := E(Z(x) \cdot Z(y)) \quad \text{for all } x, y \in \Omega. \tag{4.1}$$

Such a kernel exists and is positive semidefinite under weak additional assumptions. If there are no exact linear dependencies in the random variables $Z(x_i)$, a kernel matrix with entries $\text{cov}(x_j, x_k)$ will be positive definite. A special case is a *Gaussian process* on Ω, where for every subset $X = \{x_1, \ldots, x_N\} \subset \Omega$ the vectors $Z_X := (Z(x_1), \ldots, Z(x_N))$ have a multivariate Gaussian distribution with mean $E(Z_X) \in \mathbb{R}^N$ and a covariance yielding a matrix $A \in \mathbb{R}^{N \times N}$ which has entries $\text{cov}(x_j, x_k)$ in the above sense. Note that this takes us back to the *kernel matrix* of Definition 2.6 and the system (3.5).

Now there are several equivalent approaches to produce a good estimate for $Z(x)$ once we know data pairs (x_j, y_j) where the y_j are noiseless realizations of $Z(x_j)$. The case of additional noise will be treated later.

First, *Bayesian* thinking asks for the expectation of $Z(x)$ given the information $Z(x_1) = y_1, \ldots, Z(x_N) = y_N$ and write this as the expectation of a conditional probability

$$\tilde{Z}(x) := E(Z(x)|Z(x_1) = y_1, \ldots, Z(x_N) = y_N).$$

This is a function of x and all data pairs, and it serves as an approximation of $E(Z(x))$.

Second, *estimation theory* looks at all linear estimators of the form

$$\tilde{Z}(x) := \sum_{j=1}^{N} u_j(x) y_j$$

using the known data to predict $Z(x)$ optimally. It minimizes the *risk* defined as

$$E\left(\left(Z(x) - \sum_{j=1}^{N} u_j(x) Z(x_j)\right)^2\right)$$

by choosing appropriate coefficients $u_j(x)$.

Both approaches give the same result, repeating Theorems 3.4 and 3.6 with a new probabilistic interpretation. Furthermore, the result is computationally identical to the solution of the deterministic case using the kernel $K(x,y) = \text{cov}(x,y)$ right away, ignoring the probabilistic background completely. The system (3.5) has to be solved for the coefficients α_k, and the result can be written via either (3.4) or Theorem 3.6. The proof of this theorem is roughly the same as that for the estimation theory case in the probabilistic setting.

Guideline 4.1. Positive definite kernels allow a unified treatment of deterministic and probabilistic methods for recovery of functions from data.

Guideline 4.2. Applications using kernel-based trial spaces in non-deterministic settings should keep in mind that what they do is equivalent to an estimation process for spatial random variables with a covariance described by the chosen kernel.

This means that compactly supported or quickly decaying kernels lead to uncoupled spatial variables at larger distances. Furthermore, it explains why wide scales usually allow us to get along with fewer data (see Guideline 3.14). If there is a strong interdependence of local data, it suffices to use few data to explain the phenomena.

If the covariance kernel is positive definite, the general theory of Section 2 applies. It turns the space spanned by functions $\text{cov}(\cdot, y)$ on Ω into a reproducing kernel Hilbert space such that the inner product of two such functions is expressible via (2.5) by the covariance kernel itself. This is not directly apparent from where we started. In view of learning theory, the map $x \mapsto \text{cov}(x, y)$ is a special kind of feature map which assigns to each other input x a number indicating how closely related it is to y.

4.2. Noisy data

If we add a noise variable $\epsilon(x)$ at each point $x \in \Omega$ with mean zero and variance σ^2 such that the noise at different points is independent and also independent of Z, the covariance kernel with noise is

$$E((Z(x) + \epsilon(x)) \cdot (Z(y) + \epsilon(y))) = \mathrm{cov}(x, y) + \sigma^2 \delta_{xy}.$$

Thus, in the presence of noise we have to add a diagonal matrix with entries σ^2 to the kernel matrix in (3.5). This addition of noise makes the kernel matrices positive definite even if the covariance kernel is only positive semidefinite. In a deterministic setting, this reappears as *relaxed interpolation* and will be treated in Section 7.

If there is no *a priori* information on the covariance kernel and the noise variance σ, one can try to estimate these from a sufficiently large data sample. For details we refer to the vast statistical literature concerning noise estimation and techniques like cross-validation. Choosing the relaxation parameter in the deterministic case will be treated in some detail in Section 7, with references given there.

4.3. Random functions

In the above situation we had a random variable $Z(x)$ at each point $x \in \Omega$. But one can also consider random choices of functions f from a set or space $\mathcal{F}$ of real-valued functions on Ω. This requires a probability measure ρ on $\mathcal{F}$, and one can define another kind of *covariance kernel* via

$$\mathrm{cov}(x, y) := E(f(x) \cdot f(y))$$

$$= \int_{\mathcal{F}} f(x)f(y) \, d\rho(f) \qquad \text{for all } x, y \in \Omega \qquad (4.2)$$

$$= \int_{\mathcal{F}} \delta_x(f)\delta_y(f) \, d\rho(f) \quad \text{for all } x, y \in \Omega.$$

This is a completely different situation, both mathematically and 'experimentally', because the random events and probability spaces are different.

But now the connection to Hilbert spaces and feature maps is much clearer right from the start, since the final form of the covariance kernel can be seen as a bilinear form $\mathrm{cov}(x, y) = (\delta_x, \delta_y)$ in a suitable space. For this, we define a feature map

$$\Phi(x) := \delta_x \ : \ f \mapsto f(x) \quad \text{for all } f \in \mathcal{F} \qquad (4.3)$$

as a linear functional on $\mathcal{F}$. To a fixed input item x it assigns all possible 'attributes' or 'features' $f(x)$ where f varies over all random functions in $\mathcal{F}$. If we further assume that the range of the feature map is a pre-Hilbert subspace of the dual $\mathcal{F}^*$ of $\mathcal{F}$ under the inner product

$$(\lambda, \mu)_{\mathcal{F}^*} := E(\lambda(f) \cdot \mu(f)) = \int_{\mathcal{F}} \lambda(f)\mu(f) \, d\rho(f),$$

we are back to (2.2) in the form

$$\text{cov}(x, y) = (\Phi(x), \Phi(y))_{\mathcal{H}} \quad \text{for all } x, y \in \Omega \tag{4.4}$$

once we take $\mathcal{H}$ as the Hilbert space completion.

If we have training data pairs (x_i, y_i), $i = 1, \ldots, N$ as before, the y_i are simultaneous evaluations $y_i = f(x_i)$ of a random function $f \in \mathcal{F}$. A Bayesian recovery problem without noise would take the expected $f \in \mathcal{F}$ under the known information $y_i = f(x_i)$ for $i = 1, \ldots, N$. Another approach is to find functions u_j on Ω such that the expectation

$$E\left(\left(f(x) - \sum_{j=1}^{N} u_j(x) f(x_j)\right)^2\right)$$

is minimized. Again, these two recovery problems coincide and are computationally equivalent to those treated in Section 2 in the deterministic case, once the covariance kernel is specified.

The two different definitions for a covariance kernel cannot lead to serious confusion, because they are very closely related. If we start with random functions and (4.2), there are pointwise random variables $Z(x) := \{f(x)\}_{f \in \mathcal{F}}$ leading to the same covariance kernel via (4.1). Conversely, starting from random variables $Z(x)$ and (4.1) such that the covariance kernel is positive definite, a suitable function class $\mathcal{F}$ can be defined via the span of all $\text{cov}(\cdot, y)$, and point evaluations on this function class carry an inner product which allows us to define a Hilbert space $\mathcal{H}$ such that (4.3) and (4.4) hold.

From here on, statistical learning theory (Schölkopf and Smola 2002, Shawe-Taylor and Cristianini 2004) takes over, and we refer to the two cited books.

4.4. Density estimation by kernels

This is again a different story, because the standard approach does not solve a linear system. The problem is to recover the density f of a multivariate distribution over a domain Ω from a large sample $x_1, \ldots, x_N \in \Omega$ including repetitions. The true density function must take large values in regions where the density of sampling points is high. A primitive density estimate is possible via counting the samples in each cell of a grid, and to plot the resulting histogram. This yields a piecewise constant density estimate, but we can do better by using a nonnegative symmetric translation-invariant kernel K with total integral one, and defining

$$\tilde{f}(x) := \frac{1}{N} \sum_{j=1}^{N} K\left(\frac{x - x_i}{h}\right)$$

as a smooth estimator. If the *bandwidth h* is taken too small, the result just shows sharp peaks at the x_i. If h is too large, the result is smoothed too much to be useful. We have another instance of the *scaling problem* here. Statisticians have quite some literature on picking the 'right' bandwidth and kernel experimentally, using as much observational or *a priori* information as possible, but we cannot deal with these here.

5. Kernel construction

Before we delve into applications, we have to prepare by taking a closer and more application-oriented view at kernels. We want to give a short but comprehensive account of kernel construction techniques, making the reader able to assess features of given kernels or to construct new ones with prescribed properties.

If the domain Ω has no structure at all, we already know that the most important strategy to get a useful kernel is to construct a feature map $\Phi : \Omega \to \mathcal{F}$ with values in some Hilbert space $\mathcal{F}$ first, and then to use (2.2) for definition of a kernel. The resulting kernel is always positive semidefinite, but it will be hard to check for positive definiteness *a priori*, because this amounts to proving that, for arbitrary different $x_j \in \Omega$, the feature vectors $\Phi(x_j) \in \mathcal{F}$ are linearly independent. However, linearly dependent $\Phi(x_j)$ lead to linearly dependent functions $K(\cdot, x_j)$, and these are useless in the representation (3.4) and can be blended out by pivoting or a suitable optimization.

Guideline 5.1. If pivoting, adaptivity, or optimization is used according to Guidelines 3.17 and 3.18, one can safely work with positive *semi*definite kernels in practice.

5.1. Mercer kernels

A very common special case of a feature map occurs if there is a finite or countable set $\{\varphi_i\}_{i \in I}$ of functions on Ω. In applications, this arises if $\varphi_i(x)$ is the value of feature number i on an element $x \in \Omega$. The feature map Φ then takes an element x into the set $\Phi(x) := \{\varphi_i(x)\}_{i \in I} \in \mathbb{R}^{|I|}$. For a set $\{w_i\}_{i \in I}$ of positive weights one can define a weighted ℓ_2 space by

$$\ell_{2,w}(I) := \left\{ \{c_i\}_{i \in I} : \sum_{i \in I} w_i c_i^2 < \infty \right\}$$

and then assume that these weights and the functions φ_i satisfy

$$\sum_{i \in I} w_i \varphi_i^2(x) < \infty$$

on all of Ω. This means that the scaling of the functions φ_i together with the weights w_i must be properly chosen such that the above series converges. Then we define $\mathcal{F} := \ell_{2,w}(I)$ and (2.2) yields the kernel

$$K(x,y) := \sum_{i \in I} w_i \varphi_i(x) \varphi_i(y) \quad \text{for all } x, y \in \Omega \tag{5.1}$$

dating back to early work of Hilbert and Schmidt. Such kernels are called *Mercer* kernels in the context of learning algorithms because of their connection to the Mercer theorem on positive integral operators. But note that the latter theory is much more restrictive, decomposing a given positive integral operator with kernel K into orthogonal eigenfunctions φ_i corresponding to eigenvalues w_i. For our purposes, such assumptions are not necessary.

Even outside machine learning, many useful recovery algorithms use kernels of the above form. For instance, on spheres one can take spherical harmonics, and on tori one can take *sin* and *cos* functions as the φ_i. This is the standard way of handling kernels in these situations, and there is a huge literature on such methods, including applications to geophysics. We describe the case of the sphere in Section 9 and provide references there.

The reader may figure out that finite partial sums of (5.1) are well-known ingredients of calculus. For instance, classical Fourier analysis on $[0, 2\pi)$ or the unit circle in the complex plane using standard trigonometric functions and fixed weights leads to the well-known *Dirichlet* kernel this way. If the functions φ_i are orthogonal univariate polynomials, the corresponding kernel is provided by the *Christoffel–Darboux* formula.

Guideline 5.2. If expansion-type kernels (5.1) are used, kernel methods provide natural multivariate extensions not only of splines (see Guideline 3.7), but also of classical univariate techniques based on orthogonality.

A highly interesting new class of kernels arises when the functions φ_i are scaled shifts of compactly supported refinable functions in the sense of wavelet theory. The resulting *multiscale kernels* (Opfer 2006) have a built-in multiresolution structure relating them to wavelets and frames. Implementing these new kernels into known kernel techniques yields interesting multiscale algorithms which are currently investigated.

5.2. Convolution kernels

Of course, one can generalize (5.1) to a convolution-type formula, *i.e.*,

$$K(x,y) := \int_T \varphi(x,t) \varphi(y,t) w(t) \, dt \quad \text{for all } x, y \in \Omega \tag{5.2}$$

under certain integrability conditions and with a positive weight function w on an integration domain T. This always yields a positive semidefinite kernel, and positive definiteness follows if functions $\varphi(x, \cdot)$ are linearly independent on T for different $x \in \Omega$ (Levesley, Xu, Light and Cheney 1996). Together with (5.1), this technique is able to generate compactly supported kernels, but there are no useful special cases known which were constructed along these lines.

Guideline 5.3. Kernels obtained by weighted positive summation or by convolution of products of other functions are positive semidefinite.

5.3. Kernels and harmonic analysis

However, the most important case arises when the underlying set Ω has more structure, in particular if it is a locally compact abelian group and allows *transforms* of some sort.

Guideline 5.4. Invariant kernels with positive transforms are positive semidefinite.

We do not want to underpin this in full generality, *e.g.*, for Riemannian manifolds (Dyn, Narcowich and Ward 1999) or for topological groups (Gutzmer 1996, Levesley and Kushpel 1999). Instead, we focus on translation-invariant kernels on $\mathbb{R}^d$ and use Fourier transforms there, where the above result is well known and easy to prove. In fact, positivity of the Fourier transform almost everywhere is sufficient for positive definiteness of a kernel. This argument proves positive definiteness of the Gaussian and the Sobolev kernel in (2.8), because their Fourier transforms are well known (another Gaussian and the function $(1 + \|\cdot\|_2^2)^{-k}$, respectively, up to certain constants). By inverse argumentation, the *inverse multiquadric* kernels of the form

$$K(\|x - y\|_2) := (1 + \|x - y\|_2^2)^{-k}, \quad x, y \in \mathbb{R}^d, \ k > d/2 \qquad (5.3)$$

are also positive definite.

5.4. Compactly supported kernels

But note that all of these kernels have infinite support, and the kernel matrices arising in (3.5) will not be sparse. To generate sparse kernel matrices, we need explicitly known compactly supported kernels with positive Fourier transforms. This was quite a challenge for some years, but now there are classes of such kernels explicitly available via efficient representations (Wu 1995, Wendland 1995, Buhmann 1998). If they are dilated to have support on the unit ball, they have the simple radial form

$$K(x, y) = \phi(\|x - y\|_2) = \begin{cases} p(\|x - y\|_2), & \text{if } \|x - y\|_2 \leq 1, \\ 0, & \text{else,} \end{cases} \qquad (5.4)$$

Table 5.1. Wendland's functions $\phi_{d,k}$.

Space dimension	Function	Smoothness
$d = 1$	$\phi_{1,0}(r) = (1-r)_+$	C^0
	$\phi_{1,1}(r) \doteq (1-r)_+^3 (3r+1)$	C^2
	$\phi_{1,2}(r) \doteq (1-r)_+^5 (8r^2 + 5r + 1)$	C^4
$d \leq 3$	$\phi_{3,0}(r) = (1-r)_+^2$	C^0
	$\phi_{3,1}(r) \doteq (1-r)_+^4 (4r+1)$	C^2
	$\phi_{3,2}(r) \doteq (1-r)_+^6 (35r^2 + 18r + 3)$	C^4
	$\phi_{3,3}(r) \doteq (1-r)_+^8 (32r^3 + 25r^2 + 8r + 1)$	C^6
$d \leq 5$	$\phi_{5,0}(r) = (1-r)_+^3$	C^0
	$\phi_{5,1}(r) \doteq (1-r)_+^5 (5r+1)$	C^2
	$\phi_{5,2}(r) \doteq (1-r)_+^7 (16r^2 + 7r + 1)$	C^4

where p is a univariate polynomial in the case of Wu's and Wendland's functions. For Buhmann's functions, p contains an additional log-factor. In particular, Wendland's functions are well studied for various reasons. First of all, given a space dimension d and degree of smoothness $2k$, the polynomial $p = \phi_{d,k}$ in (5.4) has minimal degree among all positive definite functions of the form (5.4). Furthermore, their 'native' reproducing kernel Hilbert spaces are norm-equivalent to Sobolev spaces $H^\tau(\mathbb{R}^d)$ of order $\tau = d/2 + k + 1/2$. Finally, the simple structure allows a fast evaluation. Examples of these functions are given in Table 5.1.

5.5. Further cases

There are a few other construction techniques that allow us to generate new kernels out of known ones.

Theorem 5.5. Kernels obtained by weighted positive summation of positive (semi-) definite kernels on the same domain Ω are positive (semi-) definite.

For handling data involving differential operators, we need the following.

Guideline 5.6. If a nontrivial linear operator L is applied to both arguments of a positive semidefinite kernel, then it will in most cases be possible to construct another positive semidefinite kernel.

This can be carried out in detail by using the representations (5.1) or (5.2), if they are available. In general, one can work with (2.5) and assume that L can be applied inside the inner product.

There is a final construction technique we only mention here briefly. It is covered well in the literature, dating back to Hausdorff, Bernstein and Widder, and it was connected to completely monotone univariate functions by Schoenberg and Micchelli (Micchelli 1986). It is of minor importance for constructing application-oriented kernels, because it is restricted to radial kernels which are positive definite on $\mathbb{R}^d$ for all dimensions, and it cannot generate kernels with compact support. However, it provides useful theoretical tools for analysing the kernels which follow next.

6. Special kernels

So far, we have already presented the Gaussian kernel (2.1), the inverse multiquadric (5.3), and the Sobolev kernel (2.8). These have in common that they are *radial basis functions* which are globally positive and have positive Fourier transforms. Another important class of radial kernels is compactly supported and of local polynomial form, *i.e.*, the Wendland functions (5.4). But this is not the end of all possibilities.

6.1. Kernels as fundamental solutions

Guideline 6.1. There are other and somewhat more special kernels which are related to important partial differential equations.

The most prominent case is the *thin-plate spline* (Duchon 1976, 1979)

$$K(x, y) = \|x - y\|_2^2 \log \|x - y\|_2 \quad \text{for all } x, y \in \mathbb{R}^d, \qquad (6.1)$$

which models a thin elastic sheet suspended at y as a function of x and solves the biharmonic equation $\Delta^2 u = 0$ in two dimensions, everywhere except at y. More generally, there are *polyharmonic splines* defined as fundamental solutions of iterated Laplacians. They deserve a closer look, because they have special scaling properties, are of central importance for the meshless *method of fundamental solutions* in Section 13, and lead naturally to the notion of *conditionally positive definite functions* below.

The *fundamental solution* for a differential operator L at some point $x \in \mathbb{R}^d$ is defined as a kernel $K(x, \cdot)$ which satisfies $LK(x, \cdot) = \delta_x$ in the distributional sense. For the iterated Laplacian $L_m := (-\Delta)^m$ we get radial kernels

$$r^{2m-d} \quad \text{for } d \text{ odd},$$

$$r^{2m-d} \log r \quad \text{for } d \text{ even},$$

as functions of $r = \|x - y\|_2$ up to multiplicative constants and for $2m > d$. This contains the thin-plate splines of (6.1) for $m = d = 2$ and generalizes to positive real exponents as

$$r^\beta \quad \text{for } \beta \notin 2\mathbb{Z},$$
$$r^\beta \log r \quad \text{for } \beta \in 2\mathbb{Z}, \tag{6.2}$$

where now the space dimension no longer appears.

Unfortunately, these functions increase with r, and so they are neither bell-shaped nor globally integrable. Their Fourier transforms cannot be calculated in the classical sense, and thus there are no standard Fourier transform techniques to prove positive definiteness. The same holds for *multiquadrics*

$$(1 + r^2)^{\beta/2} \quad \text{for } \beta \notin 2\mathbb{Z}, \ \beta > 0,$$

which can be seen as a regularization of the polyharmonic spline r^β at zero, and which extends the inverse multiquadrics of (5.3) to positive exponents, the most widely used case being $\beta = 1$. Fortunately, these functions can be included within kernel theory by a simple generalization.

6.2. Conditionally positive definite kernels

Definition 6.2. A symmetric kernel $K : \Omega \times \Omega \to \mathbb{R}$ is *conditionally positive (semi-) definite* of order m on $\Omega \subseteq \mathbb{R}^d$, if for all finite subsets $X := \{x_1, \ldots, x_N\}$ of distinct points in Ω the symmetric matrices $A_{K,X}$ with entries $K(x_j, x_k)$, $1 \leq j, k \leq N$ define a positive (semi-) definite quadratic form on the subspace

$$V_{m,X} := \left\{ \alpha \in \mathbb{R}^N : \sum_{j=1}^N \alpha_j p(x_j) = 0 \quad \text{for all } p \in \pi_{m-1}(\mathbb{R}^d) \right\} \tag{6.3}$$

of coefficient vectors satisfying certain 'discrete vanishing moment' conditions with respect to the space $\pi_{m-1}(\mathbb{R}^d)$ of d-variate real polynomials of degree smaller than m.

Note that (unconditional) positive definiteness is identical to conditional positive definiteness of order zero, and that conditional positive definiteness of order m implies conditional positive definiteness of any larger order. Table 6.1 lists the appropriate orders of positive definiteness for special radial kernels.

Recovery problems using conditionally positive definite kernels of positive order m have to modify the trial space $\mathcal{K}_0$ to

$$\mathcal{K}_m := \pi_{m-1}(\mathbb{R}^d) + \mathcal{P}_m,$$

$$\mathcal{P}_m := \text{span}\left\{ \sum_{j=1}^N \alpha_j K(\cdot, x_j), \ \alpha \in V_{m,X}, \ X = \{x_1, \ldots, x_N\} \subset \Omega \right\}. \tag{6.4}$$

The norm (2.6) now works only on the space $\mathcal{P}_m$, and thus only clos $\mathcal{P}_m$ turns into a Hilbert space. The native space $\mathcal{K}$ for K is then

$$\mathcal{K} := \pi_{m-1}(\mathbb{R}^d) + \text{clos } \mathcal{P}_m,$$

but the reproduction (2.9) of functions via the kernel K needs a modification which we shall omit here.

If we have training data (x_k, y_k), $1 \leq k \leq N$ for a model $f(x_k) = y_k$, we now plug these equations into our new trial space, using a basis $p_1, \ldots, p_Q$ of $\pi_{m-1}(\mathbb{R}^d)$ and get a linear system

$$\sum_{j=1}^{N} \alpha_j K(x_k, x_j) + \sum_{i=1}^{Q} \beta_i p(x_k) = y_k, \quad 1 \leq k \leq N,$$

$$\sum_{j=1}^{N} \alpha_j p_\ell(x_j) + \quad 0 \quad = 0, \quad 1 \leq \ell \leq Q.$$

This system has $N + Q$ equations and unknowns, and it is uniquely solvable if there is no nonzero polynomial vanishing on the set $X = \{x_1, \ldots, x_N\}$. Since the order m of conditional positive definiteness is usually rather small ($m = 1$ for standard multiquadrics and $K(x, y) = \|x - y\|_2$, while $m = 2$ for thin-plate splines) this modification is not serious, and it can be made obsolete if the kernel is changed slightly (Light and Wayne 1998, Schaback 1999). However, many engineering applications use multiquadrics or thin-plate splines without adding constant or linear polynomials, and without caring for the moment conditions in (6.3). This often causes no visible problems, but violates restrictions imposed by conditional positive definiteness.

Note that trial spaces for polyharmonic functions are independent of scaling, if they are properly defined via (6.4). This eliminates many of the scaling problems arising in applications, but it comes at the price of limited smoothness of the kernels, thus reducing the attainable reproduction errors according to Guideline 3.13.

Table 6.1. Orders of conditional positive definiteness.

Kernel $\Phi(r)$, $r = \|x - y\|_2$	Order m	Conditions
$(-1)^{\lceil \beta/2 \rceil}(c^2 + r^2)^{\beta/2}$	$\lceil \beta/2 \rceil$	$\beta > 0$, $\beta \notin 2\mathbb{N}$
$(-1)^{\lceil \beta/2 \rceil} r^\beta$	$\lceil \beta/2 \rceil$	$\beta > 0$, $\beta \notin 2\mathbb{N}$
$(-1)^{k+1} r^{2k} \log r$	$k + 1$	$k \in \mathbb{N}$

6.3. Singular kernels

The condition $2m > d$ for the polyharmonic functions forbids useful cases like $m = 1$ in dimensions $d \geq 2$, and thus it excludes the fundamental solutions $\log r$ and r^{-1} of the Laplacian in dimensions 2 and 3. These kernels are radial, but they have singularities at zero. They still are useful reproducing kernels in Sobolev spaces $W_2^1(\mathbb{R}^d)$ for $d = 2, 3$, but the reproduction property now reads

$$\lambda(f) = (\lambda^x K(\cdot - x), f)_{W_2^1(\mathbb{R}^d)} \tag{6.5}$$

for all $f \in W_2^1(\mathbb{R}^d)$, $\lambda \in \left(W_2^1(\mathbb{R}^d)\right)^* = W_2^{-1}(\mathbb{R}^d)$. These kernels and their derivatives arise in integral equations as *single* or *double layer potentials*, and we shall encounter them again in Section 13 where they are used for the meshless *method of fundamental solutions*.

7. Approximation by kernels

This section serves to support Guideline 3.9 concerning the surprising quality of kernel-based approximations. We shall do this in a strictly deterministic setting, ignoring, for instance, the interesting results from statistical learning theory.

7.1. Convolution approximation

One of the oldest forms of kernel approximation is used for *series expansions* and *mollifiers*, and it takes the form of *convolution*. It is also at the core of *smoothed particle hydrodynamics*, a class of practically very useful meshless kernel-based methods we briefly describe in Section 12. Here, we use it as an introduction to the behaviour of kernel approximations in general.

Global convolution of a given function f with a kernel K is

$$K * f := \int_{\mathbb{R}^d} f(x) K(\cdot - x) \, dx,$$

where we have restricted ourselves to translation-invariant kernels on $\mathbb{R}^d$. Approximation of a function f by kernel convolution means

$$f \approx K * f \tag{7.1}$$

in some norm. Clearly, equality in (7.1) holds only if the kernel acts like a delta functional. Thus convolutions with kernels should achieve good reproduction if the kernels are approximations to the delta functional. This indicates that *scaling* is a crucial issue here again. If K is smoother than f, convolution allows us to construct smooth approximations to nonsmooth functions.

To deal with scaling properly, we observe Guideline 3.15 and introduce a positive scaling parameter δ to scale a fixed kernel K in $L_1(\mathbb{R}^d)$ by

$$K_\delta(\cdot) := \delta^{-d} K(\cdot/\delta)$$

to make the integral of K_δ on $\mathbb{R}^d$ independent of δ. Furthermore, the kernel convolution should approximate monomials $p_\alpha(x) := x^\alpha$ of order at most k well in a *pointwise* sense, i.e., for all $|\alpha| < k$, $\delta > 0$ we require

$$|K_\delta * p_\alpha - p_\alpha|(x) \le \delta^k A(x) \quad \text{for all } x \in \mathbb{R}^d \tag{7.2}$$

with a fixed function A on $\mathbb{R}^d$. For some positive integer k we finally assume that the kernel satisfies a *decay* condition

$$p_\alpha \cdot K \in L_1(\mathbb{R}^d) \quad \text{for all } |\alpha| < k. \tag{7.3}$$

Theorem 7.1. (Cheney, Light and Xu 1992, Cheney and Light 2000)
Under these assumptions, there is a constant c such that

$$\|K_\delta * f - f\|_{L_\infty(\mathbb{R}^d)} \le c\delta^k \max_{|\alpha| \le k} \|f^\alpha\|_{L_\infty(\mathbb{R}^d)} \tag{7.4}$$

holds for all functions $f \in C^k(\mathbb{R}^d)$.

Note that the convergence depends on the scale parameter δ going to zero, while the rate is dependent on the decay of K. Surprisingly, the reproduction condition (7.2) can always be achieved exactly (Cheney and Light 2000) by a suitable linear combination of scaled instances of the original kernel, provided that it satisfies the decay condition (7.3) and has integral one. However, this modification of the kernel will in general spoil positive definiteness. Similar kernel modifications arise in many application-oriented papers on meshless kernel methods.

7.2. Discretized convolution approximation

Discretization of the convolution integral leads to

$$(K_\delta * f)(x) \approx \sum_{i \in I_\delta} f(x_{i,\delta}) K_\delta(x - x_{i,\delta}) w_{i,\delta} \quad \text{for all } x \in \mathbb{R}^d$$

with integration weights $w_{i,\delta}$ at integration nodes $x_{i,\delta}$. This is a straightforward way to approximate f by a trial function of the form (3.4).

The error bound (7.4) now gets an additional term for the integration error. Near each $x \in \mathbb{R}^d$ there must be enough integration points to resolve the kernel at scale δ, and therefore the integration points must be closer than $\mathcal{O}(\delta)$. This approach will be called *stationary* below, and it needs more and more integration points for kernels of decreasing width.

But for reaching a prescribed accuracy, we can first choose a kernel scale δ such that (7.4) is sufficiently small. For this fixed δ we then perform a sufficiently good numerical integration to reproduce f sufficiently well by a finite linear combination of kernel translates. At this point, the smoothness of f and K determines the required density of integration points. This will be called a *nonstationary* setting below. This discussion will be resumed later, and it is crucial for the understanding of kernel approximations.

The *discretized convolution approach* leads to *quasi-interpolants* (Rabut 1989, Beatson and Light 1993, Buhmann *et al.* 1995, Maz'ya and Schmidt 2001, Ling 2005) of f which can be directly calculated from function values, without any linear system to solve. However, as is well known from the classical Bernstein or Schoenberg operators, there are better approximations using the same trial space. These will be dealt with later, but we note that quasi-interpolation works in quite a general fashion and is worth further investigation.

The theoretical consequence is that approximations from spaces spanned by translates of kernels result from an interaction between the scale of the kernel and the density of the translation points. This is a crucial issue for all kernel-based techniques, and it has consequences not only for the approximation, but also for its stability.

7.3. Fill distance and separation radius

In studying the approximation and stability properties of meshless methods, the following two geometric quantities are usually employed. Suppose we are confronted with a bounded set $\Omega \subseteq \mathbb{R}^d$ and a finite subset $X = \{x_1, \ldots, x_N\} \subseteq \Omega$ used for defining a trial space

$$\mathcal{K}_X := \mathrm{span}\{K(\cdot, x_j) \,:\, x_j \in X\} \subset \mathcal{K}_0 \subset \mathcal{K} \qquad (7.5)$$

in the terminology of Section 2. The *approximation power* of $\mathcal{K}_X$ is measured in terms of the *fill distance* of X in Ω, which is given by the radius of the largest data-site free ball in Ω, *i.e.*,

$$h_X := h_{X,\Omega} := \sup_{x \in \Omega} \min_{1 \le j \le N} \|x - x_j\|_2. \qquad (7.6)$$

The second geometric quantity is the *separation radius* of X, which is half the distance between the two closest data sites, *i.e.*,

$$q_X := \tfrac{1}{2} \min_{j \ne k} \|x_j - x_k\|_2, \qquad (7.7)$$

and does not depend on the domain. Obviously, the separation radius plays an important role in the stability analysis of the interpolation process, since a small q_X means that at least two points, and hence two rows in the system (3.5) are nearly the same. If the data in these two points are roughly the same or only differ by noise, it is reasonable to discard one of them. This is an instance of Guideline 3.17 and will be used by thinning algorithms within multiscale methods, as described in Section 10.

Finally, we will call a sequence of data sets $X = X_h$ *quasi-uniform* if there is a constant $c_q > 0$ independent of X such that

$$q_X \leq h_{X,\Omega} \leq c_q q_X. \tag{7.8}$$

The *mesh ratio* $\rho = \rho_{X,\Omega} := h_{X,\Omega}/q_X \geq 1$ provides a measure of how uniformly points in X are distributed in Ω. Remember that special results on convergence of univariate polynomial splines are valid only for cases with bounded mesh ratio; similar restrictions should be expected here as well.

7.4. Nonstationary versus stationary scales

There are two fundamentally different ways in which scales of kernel-based trial spaces are used in theory and practice. This often leads to misunderstandings of certain results, and therefore we have to be very explicit at this point.

In classical finite element and spline theory, the support of the nodal basis functions scales with the size of the mesh. For example, using classical hat functions to express a piecewise linear spline function over the node set $h\mathbb{Z}$ leads to a representation of the form

$$s_h(x) = \sum_{j \in \mathbb{Z}} \alpha_j B_1\left(\tfrac{x-jh}{h}\right) = \sum_{j \in \mathbb{Z}} \alpha_j B_1\left(\tfrac{x}{h} - j\right), \tag{7.9}$$

where B_1 is the standard hat function, which is zero outside $[0,2]$ and is defined to be $B_1(x) = x$ for $0 \leq x \leq 1$ and $B_1(x) = 2 - x$ for $1 \leq x \leq 2$.

From (7.9) it follows that each of the basis functions $B_1(\tfrac{\cdot}{h} - j)$ has support $[jh, (j+2)h]$, i.e., the support scales with the grid width. As a consequence, when setting up an interpolation system, each row in the interpolation matrix has the same number of nonzero entries (here actually only one); and this is independent of the current grid width. Hence, such a situation is usually referred to as a *stationary scheme*. Thus, for a stationary setting, the basis function scales linearly with the grid width.

In contrast to this, a *nonstationary scheme* keeps the basis function fixed for all fill distances h, i.e., the approximant now takes the form

$$s_h(x) = \sum_{j \in \mathbb{Z}} \alpha_j B_1(x - jh), \tag{7.10}$$

resulting in a denser and denser interpolation matrix if h tends to zero.

Note that for univariate polynomial spline spaces these two settings generate the same trial space. But this is not true for general kernels. In any case of kernel usage, one should follow Guideline 3.15 and introduce a scaling parameter δ to form a *scaled* trial space $K_{\delta,X}$ as in (7.5). Then a *stationary* setting scales δ proportional to the fill distance $h_{X,\Omega}$ of (7.6), while the *nonstationary setting* uses a fixed δ and varies $h_{X,\Omega}$ only.

7.5. Stationary scales

A stationary setting arises with a discretization of the convolution approximation (7.1) if using integration points whose fill distance h is proportional to the kernel width δ. It is also the standard choice for finite element methods, including their generalized version (Babuška *et al.* 2003) and large classes of meshless methods with nodal bases (see Section 12).

The standard analysis tools for stationary situations are Strang–Fix conditions for the case of gridded data, while for general cases the Bramble–Hilbert lemma is applied, relying on reproduction of polynomials. These tools do not work in the nonstationary setting.

Stationary settings based on (fixed, but shifted and scaled) nodal functions with compact support will generate matrices with a *sparsity* which is independent of the scaling or fill distance. For finite element cases, the condition of the matrices grows like some negative power of h, but can be kept constant under certain conditions by modern preconditioning methods. But this does not work in general.

Guideline 7.2.   For kernel methods, stationary settings have to be used with caution.

Without modification, the interpolants from Section 3 using stationary kernel-based trial spaces on regular data will not converge for $h \to 0$ for absolutely integrable kernels (Buhmann 1988, 1990), including the Gaussian and Wendland's compactly supported functions. But if kernels have no compact support, stationary kernel matrices will not be sparse, giving away one of the major advantages of stationary settings. There are certain methods to overcome this problem, and we shall deal with them in Section 8.

However, the practical situation is not as bad. Nobody can work for extremely small h anyway, such that convergence for $h \to 0$ is a purely theoretical issue. We summarize the experimental behaviour (Schaback 1997) as follows.

Guideline 7.3.   The error of stationary interpolation by kernel methods decreases with $h \to 0$ to some small positive threshold value. This value can be made smaller by increasing the starting scale of the kernel, *i.e.*, by using a larger sparsity.

This effect is called *approximate approximation* (Maz'ya 1994, Lanzara, Maz'ya and Schmidt 2005) and deserves further study, including useful bounds of the threshold value. It is remarkable that it occurred first in the context of parabolic equations.

Practical work with kernels should follow Guideline 3.15 and adjust the kernel scale experimentally. Once it is fixed, the nonstationary setting applies, and this is how we argued in the case of discretized convolution above, if a prescribed accuracy is required. We summarize as follows.

Guideline 7.4. In meshless methods using positive definite kernels, approximation orders refer in general to a nonstationary setting. However, nonstationary schemes lead to ill-conditioned interpolation matrices. On the other hand, a fully stationary scheme generally provides no convergence but interpolation matrices with a condition number being independent of the fill distance.

Guideline 7.4 describes another general trade-off or uncertainty principle in meshless methods; see also Guideline 3.13. As a consequence, when working in practice with scaled versions of a single translation-invariant kernel, the scale factor needs special care. This brings us back to what we said about scaling in Sections 2 and 3, in particular Guideline 2.5.

However, from now on we shall focus on the nonstationary case.

7.6. Nonstationary interpolation

While in classical spline theory nonstationary approximants of the form (7.10) play no role at all, they are crucial in meshless methods for approximating and interpolating with positive definite kernels. Thus we now study approximation properties of interpolants of the form (3.4) with a *fixed* kernel but for various data sets X. To make the dependence on X and $f \in C(\Omega)$ explicit, we will use the notation

$$s_{f,X} = \sum_{j=1}^{N} \alpha_j K(\cdot, x_j),$$

where the coefficient vector is determined by the interpolation conditions $s_{f,X}(x_j) = f(x_j)$, $1 \leq j \leq N$ and the linear system (3.5) involving the kernel matrix $A_{K,X}$.

Early convergence results and error bounds were restricted to target functions $f \in \mathcal{K}$ from the native function space $\mathcal{K}$ of (2.7) associated to the employed kernel (Madych and Nelson 1988, 1990, 1992, Wu and Schaback 1993, Light and Wayne 1998). They are local pointwise estimates of the form

$$|f(x) - s_{f,X}(x)| \leq CF(h)\|f\|_{\mathcal{K}}, \tag{7.11}$$

where F is a function depending on the kernel. For kernels of limited smoothness it is of the form $F(h) = h^{\beta/2}$, where β relates to the smoothness of K in the sense of Table 6.1. For infinitely smooth kernels such as Gaussians or (inverse) multiquadrics it has the form $F(h) = \exp(-c/h)$. A more detailed listing of kernels and their associated functions F can be found in the literature (Schaback 1995b, Wendland 2005b).

Guideline 7.5. If the kernel K and the interpolated function f are smooth enough, the obtainable approximation rate for nonstationary interpolation increases with the smoothness, and can be exponential in the case of analytic kernels and functions.

This supports Guideline 3.9 and is in sharpest-possible contrast to the stationary situation and finite element methods. It can also be observed when nonstationary trial spaces are used for solving partial differential equations.

Guideline 7.6. If the kernel K and the interpolated function f have different smoothness, the obtainable approximation rate for nonstationary interpolation depends on the smaller smoothness of the two.

Recent research has concentrated on the *escape scenario*, in which the smoothness of the kernel K exceeds the smoothness of f, *i.e.*, error estimates have to be established for target functions from outside the native Hilbert space. This is a realistic situation in applications, where a fixed kernel has to be chosen without knowledge of the smoothness of f. Surprisingly, these investigations have led far beyond kernel-based trial spaces.

To make this more precise, let us state two recent results (Narcowich, Ward and Wendland 2005*b*, 2004). As usual we let $W_p^k(\Omega)$ denote the Sobolev space of measurable functions having weak derivatives up to order k in $L_p(\Omega)$. Furthermore, we will employ *fractional* order Sobolev spaces $W_p^\tau(\Omega)$, which can, for example, be introduced using interpolation theory.

Theorem 7.7. Let k be a positive integer,

$$0 \le s < 1, \quad \tau = k + s, \quad 1 \le p < \infty, \quad 1 \le q \le \infty,$$

and let $m \in \mathbb{N}_0$ satisfy $k > m + d/p$ or, for $p = 1, k \ge m + d$. Let $X \subset \Omega$ be a discrete set with mesh norm $h_{X,\Omega}$ where Ω is a compact set with Lipschitz boundary which satisfies an interior cone condition. If $u \in W_p^\tau(\Omega)$ satisfies $u|_X = 0$, then

$$|u|_{W_q^m(\Omega)} \le Ch_{X,\Omega}^{\tau - m - d(1/p - 1/q)_+} |u|_{W_p^\tau(\Omega)},$$

where C is a constant independent of u and $h_{X,\Omega}$, and $(x)_+ = \max\{x, 0\}$.

Theorem 7.7 bounds lower Sobolev seminorms of functions in terms of a higher Sobolev seminorm, provided the functions have lots of zeros. It is entirely independent of any reconstruction method or trial space, and it can be successfully applied to any interpolation method that keeps a discretization-independent upper bound on a high Sobolev seminorm.

In fact, if $s_{f,X} \in W_p^\tau(\Omega)$ is an arbitrary function which interpolates $f \in W_p^\tau(\Omega)$ exactly in X, we have

$$|f - s_{f,X}|_{W_q^m(\Omega)} \le Ch_{X,\Omega}^{\tau - m - d(1/p - 1/q)_+} (|f|_{W_p^\tau(\Omega)} + |s_{f,X}|_{W_p^\tau(\Omega)}),$$

and if the interpolation manages to keep $|s_{f,X}|_{W_p^\tau(\Omega)}$ bounded independent of X, this is an error bound and an optimal order convergence result. This opens a new way to deal with all interpolation methods that are regularized properly.

For interpolation by kernels we can use Theorem 3.4 to provide the bound $|s_{f,X}|_\mathcal{K} \leq |f|_\mathcal{K}$ if the kernel's native Hilbert space $\mathcal{K}$ is continuously embedded in Sobolev space $W_2^\tau(\Omega)$ and contains f. By embedding, we also have $|s_{f,X}|_{W_p^\tau(\Omega)} \leq C|f|_\mathcal{K}$ and Theorem 7.7 yields error estimates of the form

$$|f - s_{f,X}|_{W_2^m(\Omega)} \leq Ch_{X,\Omega}^{\tau-m}\|f\|_\mathcal{K},$$
$$|f - s_{f,X}|_{W_\infty^m(\Omega)} \leq Ch_{X,\Omega}^{\tau-m-d/2}\|f\|_\mathcal{K}.$$

This still covers only the situation of target functions from the native Hilbert space, but it illustrates the regularization effect provided by Theorem 3.4 and described in Guideline 3.12.

The next result is concerned with the situation that the kernel's native space is norm-equivalent to a smooth Sobolev space $W_2^\tau(\Omega)$ while the target function comes from a rougher Sobolev space $W_2^\beta(\Omega)$. It employs the mesh ratio $\rho_{X,\Omega} = h_{X,\Omega}/q_X$.

Theorem 7.8. If $\tau \geq \beta$, $\beta = k + s$ with $0 < s \leq 1$ and $k > d/2$, and if $f \in W_2^\beta(\Omega)$, then

$$\|f - s_{f,X}\|_{W_2^\mu(\Omega)} \leq Ch_{X,\Omega}^{\beta-\mu}\rho_{X,\Omega}^{\tau-\mu}\|f\|_{W_2^\beta(\Omega)}, \quad 0 \leq \mu \leq \beta.$$

In particular, if X is quasi-uniform, this yields

$$\|f - s_{f,X}\|_{W_2^\mu(\Omega)} \leq Ch_{X,\Omega}^{\beta-\mu}\|f\|_{W_2^\beta(\Omega)}, \quad 0 \leq \mu \leq \beta. \tag{7.12}$$

Note that these error bounds are of optimal order. Furthermore, since they can be applied locally, they automatically require fewer data or boost the approximation order at places where the function is smooth.

Guideline 7.9. Nonstationary kernel approximations based on sufficiently smooth kernels have both h- and p-adaptivity.

7.7. Condition

But this superb approximation behaviour comes at the price of ill-conditioned matrices, if no precautions like preconditioning (Faul and Powell 1999, Ling and Kansa 2004, Brown *et al.* 2005, Ling and Kansa 2005) are taken. This is due to the fact that rows and columns of the kernel matrix $A_{K,X}$ with entries $K(x_i, x_j)$ relating to two close points x_i and x_j will be very similar. Thus the condition will increase when the separation radius q_X of (7.7) decreases, even if the fill distance $h_{X,\Omega}$ is kept constant, *i.e.*, when adding data points close to existing ones.

A thorough analysis shows that the condition number of the kernel matrix $A_{K,X} = (K(x_i, x_j))$ depends mainly on the smallest eigenvalue of $A_{K,X}$, while the largest usually does not increase more rapidly than the number N of data points. For the smallest eigenvalue it is known (Narcowich and Ward 1991, Ball 1992, Ball, Sivakumar and Ward 1992, Binev and Jetter 1992, Narcowich and Ward 1992, 1994b, Schaback 1995b, Wendland 2005b) that it can be bounded from below by

$$\lambda_{\min}(A_{K,X}) \geq cG(q_X).$$

Unfortunately, in many cases this inequality is sharp and the function G is related to the function F arising in (7.11) by $G(q) = \Theta(F(q^2))$ for $q \to 0$ (Schaback 1995b). This is the theoretical background of Guideline 3.13 relating error and condition.

7.8. Approximation via relaxed interpolation

The above discussion suggests the following heuristic.

Guideline 7.10. The best approximation error with the most stable system is achieved by using quasi-uniform data sets.

Sorting out nearly coalescing points by thinning (Floater and Iske 1998) and going over to suitable subproblems by adaptive methods (Schaback and Wendland 2000a, Bozzini, Lenarduzzi and Schaback 2002, Hon, Schaback and Zhou 2003, Ling and Schaback 2004, de Marchi, Schaback and Wendland 2005) are useful to ensure quasi-uniformity. However, these methods are dangerous in the presence of noise.

One possible remedy to both problems, coalescing points and noisy data, is to relax the interpolation condition and to solve instead the following smoothing problem:

$$\min\left\{\sum_{j=1}^{N}[f(x_j) - s(x_j)]^2 + \lambda\|s\|_{\mathcal{K}}^2, : s \in \mathcal{K}\right\}, \qquad (7.13)$$

where $\lambda > 0$ is a certain smoothing parameter balancing the resulting approximant between interpolation and approximation. This problem occurred in a probabilistic setting in Section 4. It is also intensively studied in the context of kernel learning (Cristianini and Shawe-Taylor 2000, Cucker and Smale 2001, Schölkopf and Smola 2002, Shawe-Taylor and Cristianini 2004) and in the theory of regularization networks (Evgeniou, Pontil and Poggio 2000).

A standard result of central importance for all noisy recovery problems and including learning theory is the following generalization of Theorem 3.4.

Theorem 7.11. Suppose K is the reproducing kernel of the Hilbert space $\mathcal{K}$. Then the solution to (7.13) is given by a function of the form (3.4), where the coefficient vector $\alpha = \{\alpha_j\}$ now can be calculated by solving the linear system

$$(A_{K,X} + \lambda I)\alpha = f|X.$$

Guideline 7.12. Relaxed interpolation along the lines of (7.13) is computationally equivalent to recovery from noisy observations. The relaxation parameter λ is connected to the noise variance σ by $\lambda = \sigma^2$.

Theorem 7.11 shows that the ill-conditioning problem is simply addressed by moving the eigenvalues of the interpolation matrix away from zero by an offset given by the smoothing parameter $\lambda > 0$.

However, this immediately introduces the problem of how to choose the smoothing parameter. There have been thorough investigations mainly motivated by probabilistic approaches along the lines of Section 4 (Reinsch 1967, 1971, Wahba 1975, Craven and Wahba 1979, Ragozin 1983, Cox 1984, Wahba 1990, Wei, Hon and Wang 2005).

Instead of going into details here, we follow a recent deterministic approach (Wendland and Rieger 2005) which is based upon the following simple observation. The solution s_λ of (7.13) allows the following bounds:

$$|f(x_j) - s_\lambda(x_j)| \leq \sqrt{\lambda}\|f\|_{\mathcal{K}} \quad \text{for all } 1 \leq j \leq N,$$
$$\|s_\lambda\|_{\mathcal{K}} \leq \|f\|_{\mathcal{K}}.$$

Both can easily be verified, since $f \in \mathcal{K}$ is feasible in (7.13). Hence, if λ is considered to be small, the error function $u = f - s_\lambda$ is approximately zero on X and its $\mathcal{K}$-norm can be bounded by twice the $\mathcal{K}$-norm of f.

Theorem 7.13. Assume that all assumptions of Theorem 7.7 hold, except for $u|X = 0$. Then the following generalized estimate holds:

$$|u|_{W_q^m(\Omega)} \leq C\left(h_{X,\Omega}^{\tau-m-d(1/p-1/q)_+}|u|_{W_p^\tau(\Omega)} + h_{X,\Omega}^{-m}\|u|X\|_\infty\right). \tag{7.14}$$

This is a generalization of a Poincaré–Friedrichs inequality, and it turns out to be very useful for the analysis of unsymmetric kernel-based methods for solving partial differential equations (Schaback 2005a). Under the assumptions of Theorem 7.8 this yields the estimate

$$\|f - s_\lambda\|_{L_\infty(\Omega)} \leq C\left(h_{X,\Omega}^{\tau-d/2} + \sqrt{\lambda}\right)\|f\|_{\mathcal{K}}.$$

for our smoothing problem. Keeping in mind that in this particular situation $F(h) = h^{\tau-d/2}$ and $G(q) = q^{2\tau-d}$, we have the following guideline.

Guideline 7.14. If the smoothing parameter $\lambda > 0$ is chosen as $\lambda = Ch^{2\tau-d}$, the relaxed technique still has an optimal order of approximation, while the smallest eigenvalue now behaves as in the case of quasi-uniformity.

In contrast to adaptive methods working on smaller subproblems, this relaxed approximation will still have a full set of coefficients in (3.4). It is a challenging open problem to prove deterministic results concerning the complexity reduction obtainable by a more general relaxation like (3.9).

7.9. Moving least squares

While our analysis in the previous subsections dealt with nonstationary approximation schemes based on kernel methods, we will now discuss a particular *stationary* scheme. Approximation by moving least squares has a long history (Shepard 1968, McLain 1974, 1976, Lancaster and Salkauskas 1981, Farwig 1986, 1987, 1991). It has become popular again in approximation theory (Levin 1999, Wendland 2001), in computer graphics (Mederos, Velho and de Figueiredo 2003, Fleishman, Cohen-Or and Silva 2005), and in meshless methods for solving partial differential equations (Belytschko *et al.* 1996*b*).

The idea of moving least squares approximation is to solve for every point x a locally weighted least squares problem, where a kernel is used as a weight function. This appears to be quite expensive at first sight, but actually it is a very efficient method, because it can come at constant cost per evaluation, independent of the number and complexity of the data. Moving least squares also arise in meshless methods, where they are used for a 'nodal basis' to generate data in nearby locations, *e.g.*, for performing the integrations to calculate entries of a stiffness matrix. Moreover, in many applications we are interested in only a few evaluations. For such cases, moving least squares techniques are even more attractive, because it is not necessary to set up and solve a large system.

The influence of the data points is governed by a weight function $w :$ $\Omega \times \Omega \to \mathbb{R}$, which becomes smaller the farther its arguments are away from each other. Ideally, w vanishes for arguments $x, y \in \Omega$ with $\|x - y\|_2$ greater than a certain threshold. Such behaviour can be modelled by using a translation-invariant nonnegative kernel of compact support, with no need for positive definiteness. As in any other kernel-based method, Guideline 2.5 makes *scaling* a serious issue, and Guideline 3.15 implies that w should be of the form $w(x, y) = \Phi_\delta(x - y)$ with a controllable scaled version $\Phi_\delta = \Phi(\cdot/\delta)$ of a compactly supported kernel $\Phi : \mathbb{R}^d \to \mathbb{R}$.

Definition 7.15. For $x \in \Omega$ the value $s_{f,X}(x)$ of the moving least squares approximant is given by $s_{f,X}(x) = p^*(x)$ where p^* is the solution of

$$\min \left\{ \sum_{i=1}^N (f(x_i) - p(x_i))^2 w(x, x_i) \ : \ p \in \pi_m(\mathbb{R}^d) \right\}. \tag{7.15}$$

Here, $\pi_m(\mathbb{R}^d)$ denotes the space of all d-variate polynomials of degree at most m. But it is not at all necessary to restrict oneself to polynomials.

It is, for example, even possible to incorporate singular functions into the finite-dimensional function space. This is a common trick for applications in meshless methods dealing with shocks and cracks in mechanics (Belytschko, Krongauz, Fleming, Organ and Liu 1996a).

The minimization problem (7.15) can be seen as a discretized version of the continuous problem

$$\min\left\{\int_{\mathbb{R}^d} |f(y) - p(y)|^2 w(x,y)\,\mathrm{d}y : p \in \pi_m(\mathbb{R}^d)\right\},$$

where the integral is supposed to be restricted by the support of the weight function to a region around the point x.

The simplest case of (7.15) is given by choosing only constant polynomials, i.e., $m = 0$. In this situation, the solution of (7.15) can easily be computed to the explicit form

$$s_{f,X}(x) = \sum_{j=1}^{N} f(x_j) \frac{w(x,x_j)}{\sum_{k=1}^{N} w(x,x_k)}, \qquad (7.16)$$

which is also called *Shepard approximant* (Shepard 1968). From the explicit form (7.16), one can already read off some specific properties, which also hold more generally for moving least squares. First of all, since the weight function $w(x,y)$ is supposed to be nonnegative, so are the 'basis' functions

$$u_j(x) = \frac{w(x,x_j)}{\sum_{k=1}^{N} w(x,x_k)}, \qquad 1 \le j \le N,$$

which also occur under the name 'shape functions' or 'particle functions' in meshless methods (see Section 12). Moreover, these functions form a *partition of unity*, i.e., they satisfy

$$\sum_{j=1}^{N} u_j(x) = 1. \qquad (7.17)$$

Note that partitions of unity arise again in Section 8, and they play an important role in computer-aided design because of their invariance under affine transformations.

Nonnegativity and partition of unity already guarantee linear convergence, if the weight functions are of the form $w(x,y) = \Phi((x-y)/h)$, since we have for all $p \in \pi_0(\mathbb{R}^d)$

$$|f(x) - s_{f,X}(x)| \le |f(x) - p(x)| + |p(x) - s_{f,X}(x)|$$

$$\le |f(x) - p(x)| + \sum_{j=1}^{N} u_j(x)|p(x) - f(x_j)|$$

$$\le 2\|f - p\|_{L_\infty(B(x,h))}$$

and the last term can be bounded by $C_f h$ if p is the local Taylor polynomial to f of degree zero.

To derive a similar result for the general moving least squares approximation scheme, it is important to rewrite the approximant in form of a quasi-interpolant

$$s_{f,X}(x) = \sum_{j=1}^{N} u_j^m(x) f(x_j),$$

as in Theorem 3.6. This is always possible under mild assumptions on the data sites. Though the basis functions $u_j^m(x)$ are in general not nonnegative, they satisfy a constrained minimization problem, which leads to a uniform bound of the ℓ_1-norm of $\{u_j^m(x)\}_{j=1}^{N}$. From this, convergence orders can be derived. The following result (Wendland 2001) summarizes this discussion.

Theorem 7.16. Suppose the data set $X \subseteq \Omega$ is quasi-uniform and $\pi_m(\mathbb{R}^d)$-unisolvent. If the support radius δ of the compactly supported, nonnegative weight function $w(x,y) = \Phi((x-y)/\delta)$ is chosen proportional to the fill distance $h_{X,\Omega}$ and if $f \in C^{m+1}(\mathbb{R}^d)$ is the target function, then the error can be bounded by

$$\|f - s_{f,X}\|_{L_\infty(\Omega)} \leq C_f h_{X,\Omega}^{m+1}.$$

It is remarkable that this result actually is local, *i.e.*, in regions where the target function is less smooth, the associated approximation order is automatically achieved. As in Guideline 7.9 we have the following.

Guideline 7.17. Moving least-squares methods have both h- and p-adaptivity, if the order m of the local polynomial space is large enough and if sufficiently many local data points are included.

Moreover, the assumption on the quasi-uniformity of the data set can be dropped if the support radius is continuously adapted to the local fill distance.

Finally, if for a point $x \in \Omega$ the positions of a bounded number of surrounding data sites in the ball of radius $\delta = Ch_{X,\Omega}$ are known, the minimization problem can be solved and hence the moving least squares approximation can be computed in constant time. Locating the relevant data sites can be done by employing an 'intelligent' data structure in at most $\mathcal{O}(\log N)$ time, if an additional $\mathcal{O}(N \log N)$ time is allowed to build the data structure. This, of course, is only relevant if a substantial number of evaluations is necessary. For only a few evaluations, all relevant data sites can be found by brute force methods in linear time.

8. Large and ill-conditioned kernel systems

Section 7 indicated that approximation by nonstationary scales of kernel-based trial spaces may lead to large, non-sparse systems which are often highly ill-conditioned. This will become important for applications in Section 10. Hence, it is now time to discuss efficient methods for solving large and dense systems arising from kernel approximations or interpolations. Note that these systems are qualitatively different from those arising in finite element methods (see Guidelines 3.17 and 3.18), and thus they call for different numerical techniques.

There are five major approaches in this area:

- multipole expansions, often coupled with
- domain decomposition methods,
- partition of unity methods,
- multilevel techniques using compactly supported kernels,
- preconditioning.

Each of these methods has its strengths and drawbacks and it depends on the users to decide which one suits their application best.

8.1. Multipole expansions

We start with the discussion of multipole expansions. They are, in the first place, only a tool to approximately evaluate sums of the form

$$s(x) = \sum_{j=1}^{N} \alpha_j K(x, x_j) \tag{8.1}$$

from (3.4) in a fast way. As a matter of fact, they have been developed in the context of the N-body problem, which appears in various scientific fields (Barnes and Hut 1986, Appel 1985, Greengard and Rokhlin 1987).

Large systems of the form (3.5) cannot be solved by any direct method. Instead, iterative methods have to be employed. No matter which iterative method is used, the main operation is a matrix by vector multiplication, which is nothing but the evaluation of N sums of the form (8.1).

Hence, not only for a fast evaluation of the interpolant or approximant but also for solving the linear equations (3.5) it is crucial to know how to calculate the above sums efficiently.

To derive a sufficiently fast evaluation of (8.1), for every evaluation point x the sum is split into the form

$$s(x) = \sum_{j \in I_1} \alpha_j K(x, x_j) + \sum_{j \in I_2} \alpha_j K(x, x_j), \tag{8.2}$$

where I_1 contains the indices of those points x_j that are close to x, while

I_2 contains the indices of those points x_j that are far away from x. Both sums can now be replaced by approximations to them. In the first case, since $\|x - x_j\|_2$ is small for $x_j \in I_1$, the associated sum can, for example, be approximated by a Taylor polynomial. This is sometimes called a *near-field expansion*. More important is a proper approximation to the second sum, which is done by a *unipole* or *far-field* expansion.

The main idea of such an expansion is based upon a kernel expansion of the form (5.1). Incorporating the weights w_i into the function φ_i and also allowing different functions for the two arguments, this can more generally be written as

$$K(x, t) = \sum_{i=1}^{\infty} \varphi_i(x)\psi_i(t), \qquad (8.3)$$

and we usually refer to t in $\Phi(x, t)$ as a *source point*, while x is called an *evaluation point*.

Now suppose that the source points x_j, $j \in I_2$, are located in a local *panel* with centre t_0, which is sufficiently far away from the evaluation point, *i.e.*, panel and evaluation points are *well separated*. Suppose further, (8.3) can be split into

$$K(x, t) = \sum_{k=1}^{p} \phi_k(x)\psi_k(t) + R_p(x, t) \qquad (8.4)$$

with a remainder R_p that tends to zero for $\|x - t_0\|_2 \to \infty$ or for $p \to \infty$ if $\|x - t_0\|_2$ is sufficiently large. Then, (8.4) allows us to evaluate the second sum s_2 in (8.2) by

$$s_2(x) := \sum_{j \in I_2} \alpha_j K(x, x_j)$$

$$= \sum_{j \in I_2} \alpha_j \sum_{k=1}^{p} \phi_k(x)\psi_k(x_j) + \sum_{j \in I_2} \alpha_j R(x, x_j)$$

$$= \sum_{k=1}^{p} \phi_k(x) \sum_{j \in I_2} \alpha_j \psi_k(x_j) + \sum_{j \in I_2} \alpha_j R(x, x_j)$$

$$=: \sum_{k=1}^{p} \beta_k \phi_k(x) + \sum_{j \in I_2} \alpha_j R(x, x_j).$$

Hence, if we use the approximation $\tilde{s}_2(x) = \sum_{k=1}^{p} \beta_k \phi_k(x)$ we have an error bound

$$|s_2(x) - \tilde{s}_2(x)| \leq \|\alpha\|_1 \max_{j \in I_2} |R(x, x_j)|,$$

which is small if x is sufficiently far away from the sources x_j, $j \in I_2$.

Moreover, each coefficient β_k can be computed in advance in linear time. Thus, if p is much smaller than N, we can consider it as constant and we need only constant time for each evaluation of $\tilde{s}_2$.

So far, we have developed an efficient method for evaluating a sum of the form (8.1) for one evaluation point or, more generally, for evaluation points from the same panel, which is well separated from the panel containing the source points. To derive a fast summation formula for arbitrary evaluation points $x \in \Omega$, the idea has to be refined. To this end, the underlying region of interest Ω is subdivided into cells or panels. To each panel a far field and a near field expansion is assigned. For evaluation, all panels are visited and, depending on whether or not the panel is well separated from the panel which contains the evaluation point, the near field or far field expansion is used.

The decomposition of Ω into panels can be done either uniformly or adaptively, dependent on the data. A uniform decomposition makes a near field expansion indispensable since the cardinality of I_1 cannot be controlled. However, its simple structure makes it easy to implement and hence it has been and is still often used. An adaptive decomposition is usually based on a tree-like data structure where the panels are derived by recursive subdivision of space. More details can be found in the literature (Greengard and Strain 1991, Beatson and Newsam 1992, Beatson, Goodsell and Powell 1996, Beatson and Greengard 1997, Beatson and Light 1997, Beatson and Newsam 1998, Roussos 1999, Beatson and Chacko 2000, Beatson, Cherrie and Ragozin 2000a, 2001).

In any case, since we now have to implement a unipole expansion for every panel, the resulting technique is called *multipole expansion*.

Unfortunately, the multipole expansion has to be precomputed for each kernel separately. However, for translation-invariant kernels $K(x, y) = K(x - y)$, it suffices to know the far field expansion around zero. Because this gives the far field expansion around any t_0 simply by

$$K(x - t) = K((x - t_0) - (t - t_0))$$

$$= \sum_{k=1}^{p} \phi_k(x - t_0)\psi_k(t - t_0) + R(x - t_0, t - t_0).$$

The far field expansion around zero can often be calculated using Laurent expansions of the translation-invariant kernel. Details can be found in the literature cited above.

8.2. Domain decomposition

Having a fast evaluation procedure for functions of the form (8.1) at hand, different iterative methods for solving the linear system (3.5) can be applied. However, the reader should be aware of the fact that the far field expansion may now lead to a nonsymmetric situation (Beatson, Cherrie and Mouat 1999).

Here, we want to describe a domain decomposition method (Beatson, Light and Billings 2000b), which can be extended to generalized interpolation problems (Wendland 2004). Domain decomposition is a standard technique in finite elements, involving interface conditions and related to Schwarz iteration. In kernel techniques, it is much simpler, has a fundamentally different flavour and already quite some history in the context of meshless methods for partial differential equations (Dubal 1994, Hon and Wu 2000a, Zhou, Hon and Li 2003, Ingber, Chen and Tanski 2004, Li and Hon 2004, Ling and Kansa 2004). However, the name is rather misleading here, since the domain or the analytic problem are not decomposed, but rather the approximation or the trial space. The technique itself is an iterative projection method.

To decompose the trial space it suffices to decompose the set of centres X, or generally the set of data functionals in the sense of (3.8). To be more precise, let us decompose X into subsets $X_1, \ldots, X_k$. These subsets need not be disjoint but their union must be X. Then the algorithm starts to interpolate on the first set X_1, forms the residual, interpolates this on X_2 and so on. After k steps one cycle of the algorithm is complete and it starts over again. A more formal description is

(1) Set $f_0 := f$, $s_0 := 0$.

(2) For $n = 0, 1, 2, \ldots$

      For $r = 1, \ldots, k$

$$f_{nk+r} := f_{nk+r-1} - s_{f_{nk+r-1}, X_r}$$
$$s_{nk+r} := s_{nk+r-1} + s_{f_{nk+r-1}X_r},$$

    If $\|f_{(n+1)k}\|_{L_\infty(X)} < \epsilon$ stop.

This algorithm approximates the interpolant $s_{f,X} = f^*$ from (3.4) up to the specified accuracy. The convergence result is based upon the fact that the interpolant $s_{f,X}$ is also the *best approximant* to f from the subspace K_X of (7.5) in the native Hilbert space norm. This optimality is another instance of Guideline 3.1 which we suppressed in Section 3 for brevity.

Convergence is achieved under very mild assumptions on the decomposition. The data sets X_j have to be *weakly disjoint* meaning that $X_j \neq Y_j$ and $Y_{j+1} \neq Y_j$ for each $1 \leq j \leq k - 1$, where $Y_j = \cup_{i=j}^k X_i$, $1 \leq j \leq k$. This is, for example, satisfied, if each X_j contains at least one data site, which is not contained in any other X_i.

Theorem 8.1. Let $f \in K$ be given. Suppose $X_1, \ldots, X_k$ are weakly distinct subsets of $\Omega \subseteq \mathbb{R}^d$. Set $Y_j = \cup_{i=j}^k X_i$, $1 \leq j \leq k$. Denote with $s^{(j)}$

the approximant after j completed cycles. Then there exists a constant $c \in (0,1)$ so that

$$\|f^* - s^{(n)}\|_\mathcal{K} \leq c^n \|f\|_\mathcal{K}.$$

For a proof of this theorem and for a more thorough discussion on how the subsets X_j have to be chosen we refer the reader to the literature (Beatson *et al.* 2000*b*, Wendland 2005*b*).

For an efficient implementation within multipole codes we need not only the far field or multipole expansion of the kernel. Since the coefficients of the sum (8.1) are now changing with every iteration, we also need intelligent update formulas. Finally, the decomposition of X into $X_1, \ldots, X_k$ has to be done in such a way that the local interpolants and the (global) residuals can be computed efficiently.

Theorem 8.1 suggests a hidden connection to preconditioning, if the local problems are solved by approximate inverses of the local submatrices. But this is an open research question.

8.3. Partitions of unity

Any iterative method for solving the system (3.5) leads to a full $\mathcal{O}(N)$-term solution of the form (3.4). Unless the inverse of the kernel matrix is sparse, every data site x_k has some influence on every evaluation point x, even if compactly supported kernels are used. To improve *locality* in the sense of letting only nearby data locations x_j influence the solution at x, multipole methods are a possible choice. Moving least squares approximants have this property by definition, but they need recalculation at each new evaluation point, because they calculate values, not functions. *Partitions of unity* are a compromise, because they allow us to patch local approximating functions together into a global approximating function, allowing a cheap local function evaluation.

While the 'domain decomposition' methods above decompose the data set rather than the domain, we now actually decompose the domain $\Omega \subseteq \cup_{j=1}^{M} \Omega_j$ in an overlapping manner into simple small subdomains Ω_j which may, for instance, be Euclidean balls. Associated to this covering $\{\Omega_j\}$ we choose a partition of unity, *i.e.*, a family of weight functions $w_j : \Omega_j \to \mathbb{R}$, which are nonnegative, supported in Ω_j, and satisfy

$$\sum_{j=1}^{M} w_j(x) = 1, \quad x \in \Omega.$$

These weight functions are conveniently chosen as translates of *kernels* which are smooth and compactly supported, but not necessarily positive definite. If balls are used, and if the problem is isotropic, the kernels should be compactly supported *radial basis functions*.

Finally, we associate to each cell Ω_j an approximation space V_j and an approximation process which maps a function $f : \Omega_j \to \mathbb{R}$ to an approximation $s_j : \Omega_j \to \mathbb{R}$. This approximation process can, for example, be given by local interpolants using only the data sites $X_j = X \cap \Omega_j$. However, the whole procedure works for arbitrary approximation processes, e.g., for approximations by augmented finite elements bases, thus leading to the *generalized finite element method* (Melenk and Babuška 1996, Babuška and Melenk 1997, Babuška, Banerjee and Osborn 2002, Babuška *et al.* 2003)

In the end, the global approximant is formed from the local approximants by weighting:

$$s(x) = \sum_{j=1}^{M} w_j(x) s_j(x), \quad x \in \Omega.$$

From the partition of unity property, we can immediately see that

$$|f(x) - s(x)| = \left| \sum_{j=1}^{M} [f(x) - s_j(x)] w_j(x) \right|$$

$$\leq \sum_{j=1}^{M} |f(x) - s_j(x)| w_j(x)$$

$$\leq \max_{1 \leq j \leq M} \|f - s_j\|_{L_\infty(\Omega_j)}$$

implies the following rule.

Guideline 8.2. The partition of unity approximant is at least as good as its worst local approximant.

More sophisticated error estimates can be found in the literature (Babuška and Melenk 1997, Wendland 2005b), also including bounds on the derivatives (simultaneous approximation). In the latter case, additional assumptions on the partitions and the weight functions have to be made. However, for an efficient implementation of the partition of unity method, these are automatically satisfied in general.

To control the complexity of *evaluating* the partition of unity approximant, the cells must not overlap too much, *i.e.*, every $x \in \Omega$ has to be contained in only a small number of cells and these cells must be easily determinable. Moreover, each local approximant has to be evaluated efficiently. Keeping Guideline 8.2 in mind, this often goes hand in hand with the fact that the regions are truly local, meaning that their diameter is of the size of the fill distance or the separation distance. For example, if the local approximation process employs polynomials, a diameter $\mathcal{O}(h_{X,\Omega})$ of local domains guarantees good approximation properties of the local approximants by a Taylor polynomial argument. If interpolation by kernels is employed, it is more important that the number of centres in each cell can be considered constant when compared to the global number of centres. In each case we have to assume that the number of cells is roughly proportional to the number of data sites. In this situation, all local interpolants can be computed in *linear* time provided that the local centres are known. Hence, everything depends upon a good data structure for both the centres and the cells, which can be provided by tree-like constructions again. We summarize as follows.

Guideline 8.3. Localization strategies within kernel methods should try to use a fixed or at least globally bounded number of data in each local domain. This applies to panels in multipole expansions, to domain decomposition methods, to partitions of unity, to preconditioning by local cardinal bases, and to all stationary methods.

The proper choice of scalings of kernels or influence regions is a major research area in theory, while proper programming and experimentation gives good practical results. Note that partitions of unity provide a localization strategy which helps with the scaling dilemma and mimics a stationary situation.

Finally, the easiest way to construct the partition of unity weight functions w_j is by employing moving least-squares in its simplest form, namely Shepard approximants (see Section 7).

8.4. Multilevel and compactly supported kernels

We now turn to a method tailored in particular to compactly supported kernels. We know from Section 7 that interpolation in the *stationary* setting will not lead to convergence. Moreover, to guarantee solvability, the same support radius for all basis functions has to be used. In a certain way, this contradicts a well-known rule from signal analysis as follows.

Guideline 8.4. Resolve coarse features by using a large support radius, and finer features with a smaller support radius.

To obey Guideline 8.4, the following multilevel scheme is useful. We first split our set X into a nested sequence

$$X_1 \subseteq X_2 \subseteq \cdots \subseteq X_k = X. \tag{8.5}$$

If X is quasi-uniform, meaning that the separation radius q_X of (7.7) has size comparable to the fill distance $h_{X,\Omega}$ of (7.6), then the subsets X_j should also be quasi-uniform. Moreover, they should satisfy $q_{X_{j+1}} \approx c q_{X_j}$ and $h_{X_{j+1},\Omega} \approx c h_{X_j,\Omega}$ with a fixed constant c.

Now the *multilevel method* (Floater and Iske 1996, Schaback 1997) is simply one cycle of the domain decomposition method. But this time we

use compactly supported basis functions with a different support radius at
each level. We could even use different basis functions at different levels.
Hence, a general formulation proceeds as follows. For every $1 \leq j \leq k$ we
choose a kernel K_j and form the interpolant

$$s_{f,X_j,K_j} = \sum_{x_j \in X_j} c_{x_j}(f) K_j(\cdot, x_j)$$

by using a kernel K_j on level j. We have in mind to take $K_j(x,y)$ as
$K((x-y)/\delta_j)$ with a compactly supported basis function K and a scaling
parameter δ_j proportional to $h_{X_j,\Omega}$. The idea behind this algorithm is that
one starts with a very thin, widely spread set of points and uses a smooth
basis function to recover the global behaviour of the function f. In the next
level a finer set of points is used and a less smooth function possibly with a
smaller support is employed to resolve more details, and so on.

As we said before, the algorithm performs one cycle of the domain decomposition algorithm:

set $f_0 = f$ and $s_0 = 0$.
for $1 \leq j \leq k$

$$s_j = s_{j-1} + s_{f_{j-1},X_j,K_j}$$
$$f_j = f_{j-1} - s_{f_{j-1},X_j,K_j}$$

The method shows linear convergence between levels (Schaback 1997), but
a thorough theoretical analysis is a hard research problem with only partial
results known (Narcowich, Schaback and Ward 1999, Hales and Levesley
2002).

8.5. Preconditioning

The localization techniques used above can be modified to enable specific
preconditioning methods. Any good preconditioning technique must somehow implement an approximate inverse to the linear system to be solved.
This can be done classically by partial LU or Cholesky factorization, but
it can also be done by approximately inverting the matrices of the subproblems introduced by localization. This approximate inversion of local
kernel matrices is a transition from the basis $K(\cdot, x_j)$ to local cardinal or
Lagrangian functions, as (3.7) and (3.5) show.

Such methods are around for a while (Faul and Powell 1999, Mouat 2001,
Schaback and Wendland 2000b) and have also been demonstrated to be
quite effective within meshless kernel-based methods for solving partial differential equations (Ling and Kansa 2004, Brown et al. 2005, Ling and
Kansa 2005). We have to leave details to the cited literature, but here is
again a promising research field. In particular, considering the limit for
wide-scaled analytic kernels reveals unexpected connections to polynomial

interpolation (Schaback 2005a) and allows us to handle cases beyond all condition limits (Driscoll and Fornberg 2002, Larsson and Fornberg 2005).

9. Kernels on spheres

Expansions of the form (5.1) play a less important role for kernels defined on $\mathbb{R}^d$. There, continuous Fourier or Laplace transform techniques dominate the theory of characterizing and analysing such kernels.

The situation changes, if kernels on tori and spheres, or more generally, on compact (homogenous) Riemannian manifolds (Narcowich 1995, Dyn *et al.* 1999) are considered. There, expansions of the form (5.1) are natural. Also, the summation of feature functions in learning theory, as described in Section 5 and leading to Mercer kernels, is a standard application area for kernels defined by summation of products.

As a placeholder for more general situations, we will shortly outline the theory of approximation by kernels on the sphere

$$S^{d-1} := \{x \in \mathbb{R}^d : \|x\|_2 = 1\} \subseteq \mathbb{R}^d.$$

However, since there are some nice overview articles and books on approximation on the sphere including results on positive definite kernels (Freeden, Schreiner and Franke 1997, Freeden, Gervens and Schreiner 1998, Fasshauer and Schumaker 1998), and since this topic has also been covered in surveys and books on radial basis functions (Buhmann 2000, Wendland 2005b), we will restrict ourselves only to a basic introduction and some very recent results.

9.1. Spherical harmonics

On the sphere, the basis functions φ_i in (5.1) are given by *spherical harmonics* (Müller 1966). Here, we use the fact that a spherical harmonic is the restriction of a homogenous harmonic polynomial to the sphere. We will denote a basis for the set of all homogenous harmonic polynomials of degree ℓ by

$$\{Y_{\ell,k} : 1 \le k \le N(d,\ell)\},$$

where $N(d,\ell)$ denotes the dimension of this space. Moreover, the space of polynomials of degree at most L, restricted to the sphere $\pi_L(S^{d-1}) = \pi_L(\mathbb{R}^d)|S^{d-1}$ possesses the orthonormal basis

$$\{Y_{\ell,k} : 1 \le k \le N(d,\ell), 0 \le \ell \le L\}$$

and any $L_2(S^{d-1})$ function f can be expanded into a Fourier series

$$f = \sum_{\ell=0}^{\infty} \sum_{k=1}^{N(d,\ell)} \widehat{f}_{\ell,k} Y_{\ell,k} \quad \text{with } \widehat{f}_{\ell,k} = (f, Y_{\ell,k})_{L_2(S^{d-1})},$$

where $(\cdot, \cdot)_{L_2(S^{d-1})}$ is the usual $L_2(S^{d-1})$ inner product.

To understand and investigate *zonal* basis functions, which are the analogue of radial basis functions on the sphere (see Section 2), we need the well-known addition theorem for spherical harmonics. Between the spherical harmonics of order ℓ and the generalized Legendre polynomial $P_\ell(d; \cdot)$ of degree ℓ there exists the relation

$$\sum_{k=1}^{N(d,\ell)} Y_{\ell,k}(x) Y_{\ell,k}(y) = \frac{N(d,\ell)}{\omega_{d-1}} P_\ell(d; x^T y), \quad x, y \in S^{d-1}. \tag{9.1}$$

Here ω_{d-1} denotes the surface area of the sphere in $\mathbb{R}^d$.

9.2. Positive definite functions on spheres

After introducing spherical harmonics, we can write down the analogue of the kernel expansion (5.1) as

$$K(x,y) = \sum_{\ell=0}^{\infty} \sum_{k=1}^{N(d,\ell)} a_{\ell,k} Y_{\ell,k}(x) Y_{\ell,k}(y), \quad x, y \in S^{d-1}. \tag{9.2}$$

Such a kernel is obviously positive definite if all coefficients $a_{\ell,k}$ are positive, following Guideline 5.4 concerning positive transforms. Here and in the rest of this section we will assume that the coefficients decay sufficiently fast, such that all series are absolutely convergent and lead to continuous kernels.

However, as in the $\mathbb{R}^d$ case, such general kernels are hardly used. Instead, radial or zonal kernels are employed, following Guideline 2.10.

Definition 9.1. A kernel $K : S^{d-1} \times S^{d-1}$ is called *radial* or *zonal* if $K(x,y) = \varphi(\mathrm{dist}(x,y)) = \psi(x^T y)$ with univariate functions φ, ψ and the geodesic distance $\mathrm{dist}(x,y) = \arccos(x^T y)$. The function ψ is sometimes called the *shape function* of the kernel K.

Suppose that $a_{\ell,k} = a_\ell$, $1 \leq k \leq N(d,\ell)$. Then, by the addition theorem (9.1), we have

$$K(x,y) = \sum_{\ell=0}^{\infty} \frac{a_\ell N(d,\ell)}{\omega_{d-1}} P_\ell(d; x^T y) =: \sum_{\ell=0}^{\infty} b_\ell P_\ell(d; x^T y),$$

which shows that K is radial or zonal. Conversely, if K is radial we can expand the shape function ψ using the orthogonal basis $P_\ell(d; \cdot)$ for $L_2[-1, 1]$ to get

$$K(x,y) = \sum_{\ell=0}^{\infty} b_\ell P_\ell(d; x^T y).$$

The addition theorem and the uniqueness of the Fourier series suffice to prove the following result.

Theorem 9.2. A kernel K of the form (9.2) with sufficiently fast decaying coefficients is zonal if and only if $a_{\ell,k} = a_\ell$, $1 \le k \le N(d, \ell)$.

Obviously, a zonal kernel is positive definite if all coefficients b_ℓ are positive. Moreover, it is also necessary that all coefficients are nonnegative (Schoenberg 1942). However, it is not necessary that all coefficients are strictly positive (Chen, Menegatto and Sun 2003b). The authors derived the following characterization.

Theorem 9.3. In order that a zonal kernel Φ is positive definite on S^{d-1} with $d \ge 3$ it is necessary and sufficient that the set $K = \{k \in \mathbb{N}_0 : b_k > 0\}$ contains infinitely many odd and even numbers.

The condition in this theorem is no longer sufficient on the unit circle (Pinkus 2004).

A first and most intuitive example of zonal functions comes from the $\mathbb{R}^d$ case. Suppose $K = \phi(\| \cdot \|_2) : \mathbb{R}^d \to \mathbb{R}$ is a positive definite and radial function on $\mathbb{R}^d$. Since we have $\|x - y\|_2^2 = 2 - 2x^T y$ for $x, y \in S^{d-1}$, we see that the restriction of K to S^{d-1} has the representation $K(x - y) = \phi(\|x - y\|_2) = \phi(\sqrt{2 - 2x^T y})$, so that it is indeed a zonal function with shape function $\psi = \phi(\sqrt{2 - 2\cdot})$. This immediately gives access to a huge class of zonal kernels on the sphere which are explicitly known and avoid calculation of any series.

This also raises the question if there is a connection between the (radial) Fourier transform $\widehat{K}$ of the positive definite function on $\mathbb{R}^d$ and the Fourier coefficients a_ℓ of the zonal function

$$\psi(x^T y) = \sum_{\ell=0}^{\infty} a_\ell \frac{N(d,\ell)}{\omega_{d-1}} P_\ell(d; x^T y).$$

Interestingly, there is a direct connection, which also shows that almost all positive definite and radial kernels on $\mathbb{R}^d$ define positive definite zonal kernels on S^{d-1} and the Fourier coefficients of the latter are all positive.

Theorem 9.4. (Narcowich and Ward 2004, zu Castell and Filbir 2005) Let K be a positive definite radial function having a nonnegative Fourier transform $\widehat{K} \in L^1(\mathbb{R}^d)$, and let $\psi(x^T y) := K(x - y)|_{x,y \in S^{d-1}}$. For $\ell \ge 0$, we have that

$$a_\ell = \int_0^{\infty} t\widehat{K}(t) J_\nu^2(t) \, dt, \quad \nu := \ell + \tfrac{n-1}{2}, \tag{9.3}$$

where J_ν is the order ν Bessel function of the first kind. Moreover, if $\widehat{K}$ is nontrivial, i.e., positive on a set of nonzero measure, then $a_\ell > 0$ for all ℓ.

This result was later generalized to conditionally positive definite basis functions (Narcowich, Sun and Ward 2006).

9.3. Error analysis

As in the case of (conditionally) positive definite kernels on $\mathbb{R}^d$, error estimates were first derived in the context of the associated reproducing kernel
Hilbert space (Jetter, Stöckler and Ward 1999, Golitschek and Light 2001,
Morton and Neamtu 2002, Hubbert and Morton 2004). Later, results on
escaping the native Hilbert space, *i.e.*, for target functions from a rougher
function space came up (Narcowich and Ward 2004). However, the involved
rougher function spaces were not given by standard Sobolev spaces.

Here, we want to mimic the situation of Theorem 7.8. To this end we
have to introduce Sobolev spaces on the sphere, which can be written as

$$W_2^\tau(S^{d-1}) = \left\{ f \in L_2(S^{d-1}) : \sum_{\ell,m}(1+\ell^2)^\tau|\widehat{f}_{\ell,m}| < \infty \right\}.$$

Naturally, to provide error estimates, the fill distance and the separation
radius have to be redefined using *geodesic* distance now. If this is done,
then it is possible to show (Narcowich, Sun, Ward and Wendland 2005*a*)
the following analogue to Theorem 7.8.

Theorem 9.5. Assume $\tau \geq \beta > (d-1)/2$ and let ψ generate $W_2^\tau(S^{d-1})$ as
its reproducing kernel Hilbert space. Given a target function $f \in W_2^\beta(S^{d-1})$
and a set of discrete points $X \subseteq S^{d-1}$ with mesh norm h_X, separation radius
q_X and mesh ratio $\rho_X = h_X/q_X$, the error between f and its interpolant
$s_{f,X}$ can be bounded by

$$\|f - s_{f,X}\|_{W_2^\mu(S^{d-1})} \leq C\rho_X^{\beta-\mu}h_X^{\beta-\mu}\|f\|_{W_2^\beta(S^{d-1})} \tag{9.4}$$

for all $0 \leq \mu \leq \beta$.

Note that (9.4) reduces to the expected error estimates when the approximation order is dictated by the rougher Sobolev space and if quasi-uniform
data sets are considered. Finally, we should remark that a zonal function ψ
generates $W_2^\tau(S^{d-1})$ if its coefficients a_ℓ in Theorem 9.2 decay like $\ell^{-2\tau}$.

10. Applications of kernel interpolation

Here, we review some practical application areas for kernel techniques which
fit neither into Section 11 on machine learning nor into the final sections on
solving partial differential equations. These techniques perform generalized
interpolation of smooth functions using unstructured data. The background
was described in Section 3 on optimal recovery, with conditional positive
definiteness added from Section 6. Finally, special techniques for handling
large-scale problems from Section 8 will occur at certain places. We group
the applications by certain features that are sufficiently general to enable

the reader to insert new applications into the right context. Unfortunately, our references cannot cover the application areas properly.

10.1. Modelling nonlinear transformations

Recovery problems like in (3.5) can of course be made vector-valued, and then they provide nonlinear multivariate mappings $F : \mathbb{R}^d \to \mathbb{R}^n$ with specified features expressible as linear conditions. Typical examples are *warping* and *morphing*. Warping is done by a fixed map taking an object of $\mathbb{R}^n$ to another object in $\mathbb{R}^n$, while morphing requires a parametrized scale of warping maps that describe all intermediate transformations. For these transformations, some input and output points have fixed prescribed locations, *e.g.*, keeping eyeballs fixed when morphing two faces, and these conditions take the form (3.1) or (3.3). Since kernel-based interpolation allows any kind of unstructured data, it is very easy to generate a warping or morphing map F with such conditions in any space dimension (Noh, Fidaleo and Neumann 2000, Glaunés, Vaillant and Miller 2004). However, the most popular applications (Gomes, Darsa, Costa and Velho 1998) avoid solving a linear system and prefer simple local techniques. Here is an open research field.

10.2. Exotic data functionals

This application area uses the fact that kernel techniques can recover functions from very general kinds of 'data' which need not be structured in any way. Any linear functional λ_j acting on multivariate functions is allowed in (3.3), provided that the kernel K is chosen to be sufficiently smooth to make $\lambda_j^x \lambda_j^y K(x,y)$ meaningful.

Guideline 10.1. Kernel methods can handle generalized recovery problems when the data are given by rather exotic linear functionals.

A typical example (Iske and Sonar 1996, Sonar 1996, Cecil, Qian and Osher 2004, Wendland 2005a) concerns postprocessing the output of *finite volume methods*. These calculate a set of values f_j of an unknown function f which are not evaluations of f at certain nodes x_j, but rather integrals of f over a certain small 'volume' V_j. Thus the functionals in (3.3) are

$$\lambda_j(f) := \int_{V_j} f(t)\,dt, \quad 1 \le j \le N.$$

Usually, the domains V_j form a non-overlapping decomposition of a domain Ω. Then any recovery $\tilde{f}$ of f along the lines of Sections 3 and 6 will have the same local integrals as f, and also the global integral of f is reproduced. Thus postprocessing a finite-volume calculation produces a smooth function with correct local 'finite volumes'. These functions can then be used for further postprocessing, *e.g.*, calculation of gradients, pressure, or contours.

This technique can be used in quite a general fashion. In fact, one can always add interpolation conditions of the above type to any other recovery problem, and the result will have the required conservation property.

Guideline 10.2. Within kernel-based reconstruction methods, it is possible to maintain conservation laws.

In some sense, morphing also maintains some kind of conservation.

Another similar case occurs when a certain algorithm produces an output function which does not have enough smoothness to be the input of a subsequent algorithm. An intermediate kernel-based interpolation will help.

Guideline 10.3. Using kernel-based techniques, we can replace a non-smooth function by a smooth one, preserving any finite number of data which are expressible via linear functionals.

We stated this in the context of conservation here, but it will occur again later with a different focus.

A somewhat more exotic case is the recovery of functions f from *orbital derivatives* along trajectories $X(t) \in \mathbb{R}^d$ of a dynamical system (Giesl 2005). The data at x_j are not $f(x_j)$ but the derivative of $t \mapsto f(X(t))$ at t_j of the trajectory passing through $x_j = X(t_j)$. This information, when plugged into a suitable recovery problem, can be used to prove stability of solutions of dynamical systems numerically, by constructing Lyapunov functions as solutions to recovery problems from unstructured orbital derivative data.

10.3. Recovery from many scattered values

The recovery of a multivariate function f from large samples of unstructured data $(x_j, f(x_j))$, $1 \le j \le N$ on a domain $\Omega \subseteq \mathbb{R}^d$ theoretically follows the outlines given in Sections 2 and 3. However, for large N there are specific problems that need special numerical techniques of Section 8. We do not repeat these here. Instead, we focus on *terrain modelling* as a typical application.

As long as terrains are modelled as elevations $z = f(x)$ described by a bivariate function f and using gridded data $(x_j, z_j) = (x_j, f(x_j)) \in \mathbb{R}^3$ as in current geographic databases (*e.g.*, the US Geological Survey), there are no serious problems. But the raw elevation data often come in an irregular distribution, because they are sampled along routes of ships, aeroplanes, or satellites. This means that the fill distance (7.6) will be much larger than the separation radius (7.7). The latter is given by the sampling rate along each route, while the first depends on how well the routes cover the domain. The problem data live on two different scales: a smaller one along the sampling trajectories and a larger one 'between' the trajectories. Section 7 tells us that the recovery error is dominated by the fill distance, while the condition is determined by the separation radius. This calls for *multiscale* methods, which are also necessary in many other applications in geometric modelling.

Multiscale techniques, as described in Section 8, use Guideline 8.4, but they have to split the given large dataset X into a nested sequence (8.5). Each subset X_j of the data should be quasi-uniform in the sense of (7.8). This can be done by sophisticated *thinning algorithms* (Floater and Iske 1998) and using kernels of different scales at different levels. For details, we refer the reader to recent books covering this subject (Iske 2004, Dodgson, Floater and Sabin 2004). A promising new approach via multiscale kernels (Opfer 2006) directly resolves such problems on several scales, but work is still in progress.

10.4. Recovery of implicit surfaces

This is different from the previous case, because the resulting surface should not be in *explicit* form $z = f(x)$ with $x \in \Omega \subset \mathbb{R}^2$. Instead, the goal is to find an *implicit* description of a surface as the level set $\{x \in \mathbb{R}^3 : g(x) = 0\}$ of a scalar function $g : \mathbb{R}^3 \to \mathbb{R}$. The given data consist of a large set of unstructured points $x_j \in \mathbb{R}^3$, $1 \le j \le N$ expected to lie on the surface, *i.e.*, to satisfy $g(x_j) = 0$ for all j in question. This is an important problem of *reverse engineering*, if the data come from a laser scan of a 3D object. The final goal is to produce an explicit piecewise CAD-compatible representation of the object from the implicit representation.

The basic trick for handling such problems is to view them as a plain interpolation problem for g with values 0 at the x_j. To avoid the trivial function a number of points 'outside' the object has to be added with values less than zero and points 'inside' with values larger than zero.

To this end, it is assumed that the surface indeed divides $\mathbb{R}^3$ into an inner and outer part, meaning that the surface is closed and orientable and has a well-defined outer normal vector at each point. With these additional assumptions at hand, the first task is to find outer normal vectors for each point. This can be done by using additional information, such as the position of the laser scanner, or by trying to fit in each point a local tangent plane to the surface. In the latter case, the so-calculated normals have to be oriented consistently, which is, unfortunately, an NP-hard problem. However, there exist good algorithms producing in most cases a satisfactory orientation (Hoppe, DeRose, Duchamp, McDonald and Stuetzle 1992, Hoppe 1994).

With these normals at hand, the additional points can be inserted along the normals. A function value which is proportional to the signed distance to the surface is assigned to each new point, making the interpolation problem nontrivial.

However, this procedure might triple the often already huge number of data sites, such that efficient algorithms, like those described in Section 8, are required. For example, this has been successfully demonstrated in various papers (Carr, Beatson, Cherrie, Mitchell, Fright, McCallum and Evans 2001, Turk and O'Brien 2002, Ohtake, Belyaev, Alexa, Turk and Seidel 2003a, Ohtake, Belyaev and Seidel 2003b) and is already well established in industry.[3]

10.5. Transition between different representations

Consider two different black-box numerical programs which have to be linked together, in the sense that the first produces multivariate discrete output data describing a function f while the second program needs different data of f as its input. This occurs if two FEM programs with different meshes and elements are used, or if results of a program need some post-processing.

Everything is fine if the two programs use function representations based on the same discrete data. Otherwise, an intermediate kernel-based recovery will be useful. The output data of the first program is taken as input of a kernel-based recovery process to find a function $\tilde{f}$ close to f. Then the input data for the second program is derived from $\tilde{f}$.

A typical field of this application is aeroelasticity. Here, the interaction between the flow field around an elastic aircraft during flight and the aircraft itself is studied. A deforming aircraft leads to more realistic lift and drag and, particularly in the design of large aircrafts, has to be taken into account.

The black-box solvers involved are the aerodynamic solver for the computation of the flow field and a structural solver for the computation of the deformation of the aircraft. While the flow field is often discretized using high-resolution finite volume methods in Eulerian coordinates, the structure of the aircraft is generally described by a coarse finite element discretization in Lagrangian coordinates. The exchange of information is limited to transfer forces from the aerodynamic program to the structural mesh and displacements from the structural mesh to the aerodynamical one. In particular the latter can be modelled as a scattered data interpolation problem. This has been done successfully, for example, in a series of papers (Farhat and Lesoinne 1998, Beckert 2000, Beckert and Wendland 2001, Ahrem, Beckert and Wendland 2005) and is already on its way to become an industry standard. The exchange of forces is in general differently achieved such that the sum of all forces and the virtual work are conserved between both models.

---

[3] http://www.farfieldtechnology.com/
    http://aranz.com/research/

Interestingly, an early application (Harder and Desmarais 1972) in aircraft engineering is the first paper in which thin-plate or surface splines were used in a scattered data interpolation problem, while the theory arrived four years later (Duchon 1976).

11. Kernels in machine learning

The older literature on radial basis functions was dominated by applications in *neural networks*, in which sigmoid response functions were gradually replaced by radial basis functions over the years. Many papers of this kind call a function (3.4) with a radial kernel a *radial basis function network*. We do not want to explain this machinery in detail here, because *kernels* provide a much more general and flexible technique replacing classical neural networks in learning algorithms. There are close connections of machine learning to pattern recognition and data mining, but we have to be brief here and prefer to focus on learning, leaving details to standard books on machine learning with kernels (Schölkopf and Smola 2002, Shawe-Taylor and Cristianini 2004).

11.1. Problems in machine learning

We start with an introduction to the notions of machine learning, based on the recovery problems in Section 3. These are subsumed under *supervised learning*, because the expected response y_j to an input $x_j \in \Omega$ is provided by the unknown 'supervisor' function $f : \Omega \to \mathbb{R}$. If the target data y_j can take non-discrete real values, the supervised recovery problem is called *regression*, while the case of discrete values is called *classification*. In the latter case the input set Ω is divided into the equivalence classes defined by the different target values. After learning, the resulting function $\tilde{f} \approx f$ should be able to classify arbitrary inputs $x \in \Omega$ by assigning one of the finitely many possible target values. For instance, a classification between 'good' and 'bad' inputs x_j^+ and x_j^- can be done by finding a hyperplane in feature space which separates the features $\Phi(x_j^+)$ and $\Phi(x_j^-)$ of 'good' and 'bad' inputs in a best possible way. This can be done by linear algebra or linear optimization, and is an instance of Guideline 2.2.

In many applications, classification is reduced to regression by:

(1) embedding the discrete target values into the real numbers,
(2) solving the resulting regression problem by some function $\tilde{f}$,
(3) classifying new inputs x by assigning the discrete target value closest to $\tilde{f}(x)$.

Thus we shall focus on regression problems later, ignoring special techniques for classification.

Unsupervised learning has inputs $x_j \in \Omega$ but no given target responses y_j associated to them. The goal for learning is given semantically instead. A frequent case is *clustering*, which is classification with just a few target values whose calculation is part of the problem. Another unsupervised technique is the determination of anomalies, outliers, or novelties. This can be seen as a classification where only the 'normal' inputs are known beforehand, while future 'abnormal' inputs have to be detected. A more general topic closely related to unsupervised learning is *data mining*, which attempts to discover unknown relations between given data (Hastie, Tibshirani and Friedman 2001), but we cannot go into details.

11.2. Linear algebra methods in feature space

Many pattern recognition or learning techniques apply a linear algebra technique in feature space, and thus they use Guideline 2.2. Since the kernel matrix contains all geometric information on the learning sample, the algorithms are based on the kernel matrix or on information derived from it. A simple novelty detection could, for instance, just check how far a new feature vector $\Phi(x)$ is away from the mean of the 'normal' feature vectors $\Phi(x_j)$ and declare it 'abnormal' if it is 'too far away'. Of course, there are statistical background arguments to support certain decision rules.

Primitive binary classification can take the means μ^+ and μ^- of the feature vectors $\Phi(x_j^+)$ and $\Phi(x_j^-)$ of the 'good' samples x_j^+ and 'bad' samples x_k^-, and then classify a new input x by checking whether $\Phi(x)$ is closer to μ^+ or μ^-. Of course, there are more sophisticated methods with statistical foundations, but the upshot is that a kernel defined via a feature map is all that is needed to start a linear algebra machinery, ending up with certain statistical decision rules.

A very important background technique for many pattern recognition and learning algorithms is to attempt a *complexity reduction* of the input data first. If this is possible, anomalies can be detected if they do not fit properly into the reduction pattern for the 'normal' data. The most widely used method for complexity reduction proceeds via *principal component analysis*, which in the case of kernel-based methods boils down to a singular-value decomposition of the kernel matrix followed by projection onto the eigenspaces associated to large singular values.

11.3. Optimization methods in feature space

But the most important numerical methods in machine learning are *optimizations*, not linear algebra techniques. For illustration, we take a closer look at unsupervised learning in the regression case, which in Section 3 was called a recovery problem. The *reproduction–generalization* dilemma stated in Guideline 3.3 is observed in machine learning by minimizing both a *loss function* penalizing the reproduction error and a *regularization* term penalizing instability and assuring generalization. These two penalty terms arise in various forms and under various assumptions, deterministic and nondeterministic, and they can be balanced by taking a weighted sum as an objective function for joint minimization. A typical deterministic example is (7.13) summing a least-squares loss function and a native space norm penalty term. Another case is the sup-norm loss function

$$\epsilon := \max_j |y_j - f(x_j)|$$

arising indirectly in (3.9) and added to the native space norm to define the objective function $\frac{1}{2}\|f\|_{\mathcal{K}}^2 + C\epsilon$ to be minimized.

Both cases, like many others in machine learning, boil down to quadratic optimization, because (2.6) allows explicit and efficient calculation of the native space norm on the trial space (2.3) via the kernel matrix defined for the training data. This applies to all techniques using the quadratic penalty

$$\alpha \in \mathbb{R}^N \mapsto \alpha^T A_{K,X}\alpha = \|f\|_{\mathcal{K}}^2 \tag{11.1}$$

to guarantee stability and generalization. For large training samples, the resulting quadratic programming problems have to cope with huge positive definite kernel matrices in their objective function, calling for various additional numerical techniques like principal component analysis to keep the complexity under control. Of course one can also get away with linear optimization if the quadratic term is replaced by minimization of terms like $\|A_{K,X}\alpha\|_\infty$ or $\|\sqrt{A_{K,X}}\alpha\|_\infty$ with a similar penalty effect. Again, the kernel matrix is the essential ingredient.

But this technique is not limited to learning algorithms. One can use it for regularizing many other methods, because one has a cheap grip on high derivatives.

Guideline 11.1. Quadratic penalty terms (11.1) using the square of the native space norm of a kernel-based trial function are convenient for regularizing ill-posed problems.

Since this only requires the trial space to consist of translates of a single positive definite kernel, and since such trial spaces have good approximation properties, kernel-based methods are good candidates for solving ill-posed and inverse problems (Lewitt, Matej and Herman 1997, Hon and Wu 2000*b*, Cheng, Hon, Wei and Yamamoto 2001*b*, Cheng, Hon and Yamamoto 2001*a*, Green 2002, Hon and Wei 2002, 2003, 2005). Solving ill-posed and inverse problems by kernel techniques has a promising future.

11.4. Loss functions

After looking at the penalty for instability, we have to focus on the *loss* function, while we assume an at least quadratic optimization using (11.1) as part of the objective function. There are various ways to define loss, but they have seriously different consequences, not only from a statistical, but also from a numerical viewpoint. We ignore the vast literature on statistical learning theory here and focus on computationally relevant questions with implications for other kernel-based techniques.

The quadratic least-squares loss in (7.13) has the consequence to add a constant diagonal to the kernel matrix. This is the old Levenberg–Marquardt regularization of least-squares problems, but it has the disadvantage that the solution will not have a reduced complexity. The resulting coefficient vector $\alpha \in \mathbb{R}^N$ for N training samples will not necessarily have many zeros, so that the kernel-based model (3.4) has full $\mathcal{O}(N)$ complexity.

On the other hand, Guidelines 3.16 and 3.17 tell us that a complexity reduction should be possible, using only $n \ll N$ terms in the solution (3.4). This is achieved by using *linear* loss constraints like (3.9) instead of quadratic ones. Then the Kuhn–Tucker theory restricts the optimal solution via the *active* constraints. In the literature on machine learning, this is the *support vector machine* philosophy, because the feature vectors $\Phi(x_j)$ for the 'active' indices j with $|f(x_j) - y_j| = \epsilon$ are called 'support vectors' for some reason or other.

Guideline 11.2. Complexity reduction via linear loss constraints is useful for most recovery situations, deterministic or non-deterministic.

Since many numerical methods can be reformulated as recovery problems, this has an unexpectedly wide range of possible applications. We use it for adaptive meshless collocation methods in Section 14. There are good chances that future methods for PDE solving will take the form of adaptive optimization routines with linear loss constraints leading to complexity reduction.

11.5. Kernels in learning theory

Theoretical research on learning has close connections to approximation theory, and it is naturally focusing on kernels (Smola and Schölkopf 1998, Cucker and Smale 2001, Schölkopf and Smola 2002, Zhou 2002, Smale and Zhou 2003, Poggio and Smale 2003). Most of this is based on statistical learning theory. Since we want to stay on the numerical analysis side, we only present the most important connection to approximation by kernels.

A central question in supervised learning is to have bounds for the necessary number N of training data (x_j, y_j) to guarantee the availability of a trained model $\tilde{f}$, based on these data, which has a small generalization error $\|f - \tilde{f}\|_\Omega \leq \epsilon$ in some norm $\| \cdot \|_\Omega$ over the input domain Ω. This problem can be handled using Theorem 7.8. In particular, if the true model f lies in some Sobolev space $W_2^\beta(\Omega)$ containing the native space for our kernel, and if X is a quasi-uniform sample set of N points in Ω with fill distance $h_{X,\Omega}$, we can find an exact data reproduction $s_{f,X}$ based on X such that Theorem 7.8 provides an error bound of order $h_{X,\Omega}^{\beta-\mu}\|f\|_{W_2^\beta(\Omega)}$ for the generalization error in the Sobolev norm $\| \cdot \|_{W_2^\mu}(\Omega)$. Thus the necessary number $N \approx h_{X,\Omega}^{-1/d}$ of training samples to handle all nonzero unknown functions $f \in W_2^\beta(\Omega)$ to an error $\|f - \tilde{f}\|_{W_2^\mu(\Omega)} \leq \epsilon$ behaves like

$$N \geq C \cdot \left(\frac{\epsilon}{\|f\|_{W_2^\beta(\Omega)}} \right)^{\frac{-d}{\beta-\mu}}$$

for $0 \leq \mu < \beta$. Guideline 3.13 arises here again, because smoothness of the kernel and the model pays off. There are similar bounds in other norms, but we do not go into details. Unfortunately, there are no deterministic results yet which support Guideline 3.16 in a quantitative way, reducing N if the reproduction quality is relaxed.

12. Meshless methods

Here, we start considering applications of kernels within methods solving partial differential equations. These are published in abundance, mainly in journals focusing on computational techniques in engineering and sciences, and this paper should help the user to sort them out properly. To this end, we derive some guidelines for using kernels in numerical methods, but this will need some general considerations first. To set the stage properly, we recall the fundamental dichotomies between

- strong and weak problem formulations
- test and trial functions
- stationary and nonstationary scales of trial spaces
- implicit or explicit shape functions
- symmetric and unsymmetric methods

and consider

- regularity of solutions
- consistency, *i.e.*, reproduction of polynomials
- adaptivity
- necessity of global spatial discretizations
- numerical integration.

These issues are intimately related, as we shall see.

12.1. Strong and weak problems

Strong problems define solutions as functions satisfying a partial differential equation and certain boundary conditions *pointwise*, employing evaluations of functions and classical derivatives. *Weak problems* replace point evaluations by local integrations against *test functions* or (weak) derivatives thereof, introducing numerical integrations. Both apply 'tests' to check whether a 'trial' function is a solution. Their difference is not on the 'trial' side, but on the 'test' side. We shall come back to this later.

Strong methods can be called 'integration-free', and this is often more important than the notion of 'mesh-free' or 'meshless'. As far as point evaluations are concerned, there is no big difference between weak and strong methods, since the weak methods also use strong function values for their integration routines. The crucial point of weak formulations, however, is to apply integration by parts to the integrals of derivatives against test functions, thus reducing the necessary order of differentiability and allowing Hilbert space methods like Dirichlet's principle.

Strong formulations imply stronger regularity assumptions, *i.e.*, classical differentiability with Hölder continuity of the highest derivatives occurring in the differential equations. Weak formulations can get away with lower regularity and lower-order derivatives, but the derivatives are not classical ones. While this argument is independent of numerical methods, regularity is closely connected to them, since convergence orders usually increase with regularity.

Guideline 12.1. If the PDE problem has a rather regular solution, the user should apply techniques that make use of this regularity, and can choose between weak and strong problem formulations. If the solution will definitely have low regularity, the user should first try to convert the problem to another with more regularity, *e.g.*, by giving expected singularities or discontinuities a special treatment. If the final problem still leads to a solution with low regularity, the user is forced to pose a weak problem, but must expect poor numerical performance of any numerical method.

If there is enough regularity to have a choice between weak and strong problems, the connection of the problem formulation to numerical integration becomes important. Weak formulations introduce additional numerical integrations which are not necessary for strong formulations. These numerical integrations increase the algorithmic complexity and introduce a possibly avoidable source of numerical errors.

Guideline 12.2. Strong problem formulations avoid certain numerical integrations, but they have to assume higher regularity than weak formulations.

The integration error can be quite serious (Ciarlet 1991, Babuška *et al.* 2002) and needs a careful selection of integration techniques. In particular, if regularity is high to allow high-order methods like h-p finite elements in a weak formulation, the integration quality must be increased properly to adjust to the convergence order, so that the final error is not dominated by the one induced by numerical integration. This makes it questionable to go for a weak problem formulation in the case of high regularity, because strong formulations without integrations become an option in that case.

12.2. Trial functions

If we rule out purely discrete techniques like plain finite differences, the approximate solutions of partial differential equations are usually represented as linear combinations of *trial functions*. These come in a great variety, *e.g.*, as polynomials, piecewise polynomials (splines, box splines, or finite elements), shape functions, particle functions, generalized finite elements, wavelets, or kernel translates. Furthermore, they do not come singly, but usually as a whole *scale* of spaces, and then the question of *stationary* or *nonstationary* scaling comes up as in Section 7. Let us have a closer look at trial spaces in general in order to see where kernels are useful.

Guideline 12.3. Trial functions should

- provide a good approximation to the solution,
- be effectively evaluable,
- easy to modify, and
- easy to integrate numerically, in the case of weak problems.

They should only in the latter situation be dependent on the test side. We shall now look at these properties one by one, starting with approximation properties.

In many cases, *e.g.*, for finite elements, scales of trial spaces attain their approximation power via a *geometric domain discretization* of the underlying domain up to some granularity h describing something like the maximum diameter of a local polyhedral support of a trial function. Certain methods using shape functions or translated kernels do not split the domain geometrically, but use a cloud of points that 'fills' the domain so that h is a *fill distance* such as (7.6), which measures the radius of the largest ball with centre in the domain but without one of the data points. In both cases, there is a *domain discretization* involved.

But as far as approximation power is concerned, it is by no means mandatory that a scale of trial spaces requires a geometric global domain discretization of any kind.

Guideline 12.4. If the expected solution of a problem has a good approximation from a low-dimensional space of global functions, the trial space should be selected accordingly, without discretizing the domain at all. If singularities of known form and place are to be expected, they should be included into the trial space, no matter what the actual numerical method is.

Note that a possibly missing space discretization for the trial space is just one aspect when looking at 'meshless methods'. There may be integration nodes in certain cases, and there may be a space discretization for the test side which we have not yet looked at. Currently, most meshless methods are still using global space discretizations, but allow us to add adaptive local refinement when necessary. However, the user should keep in mind that spectral methods (Fornberg and Sloan 1994) or general trial spaces without space discretization are to be considered as alternatives when the expected properties of the solution allow them.

Guideline 12.5. High approximation *orders* are not related to domain discretization, but to smoothness. They are achievable if the solution of the problem is sufficiently smooth. This is independent of the trial space. But they also require a trial space that can make use of that smoothness.

Such spaces must have higher smoothness themselves, as in the p-version of the finite element method. A trial space with good approximation properties should thus have p-adaptivity in the sense that it guarantees the highest possible approximation order attainable for the (unknown) smoothness of the solution. By Section 7 and Guideline 7.9 we know that nonstationary scales of kernel-based trial spaces have both a p- and h-adaptivity, but theory still requires a space discretization with a small fill distance h, because it focuses on a worst-case scenario. It is a future challenge to provide a sound mathematical basis for data-dependent h-type adaptivity such as the support vector technology within machine learning. Future adaptive optimization strategies for PDE solving should use Guideline 3.18 and select spatial resolutions locally where needed, and automatically yield optimal local approximation orders depending on the local smoothness of the solution.

Some applications require good approximations of higher derivatives of the solution, *e.g.*, if pressure or stress is to be evaluated from displacements ion mechanics. This calls for smooth trial functions.

Guideline 12.6. Because the node connectivity problems of piecewise polynomials increase dramatically with smoothness requirements and space dimension, it is much easier for meshless kernel-based methods than for finite elements to generate smooth trial spaces, in particular for higher space dimensions.

Standard results concerning numerical methods for solving ODEs and time-dependent PDEs suggest that good convergence orders are obtained by high-*consistency* orders, provided that *stability* is satisfied. This is not directly related to the approximation power of trial spaces. Unfortunately, *consistency* occurs in quite a number of application papers on meshless methods in a nonstandard way, and we shall later describe its questionable use there.

We now leave approximation quality and focus on evaluation efficiency of trial functions. Though not standard in the literature, we distinguish between *explicit* and *implicit* evaluation of trial functions. For *explicit* evaluation, there is a simple formula, *e.g.*, $\exp(-0.3 * \|x - x_j\|_2^2)$, for each trial function, and there is no need to look up a number of other nodes or to evaluate geometric data. This is the standard technique for kernel-based trial spaces. *Implicit* evaluation means that each trial function value is the result of a subroutine call to a function that depends on multiple data in a somewhat complex and geometry-dependent way. This applies to finite elements and all 'shape functions' which are the result of pointwise local optimizations like *moving least squares* of Section 7.

Guideline 12.7. If applications need to evaluate the solution on extremely many points, implicit trial spaces may not be the best choice.

It often happens that the calculation of the parameters of the representation of a solution is faster than the generation of all values needed for postprocessing, *e.g.*, for visualization. Then *evaluation* becomes more important than *solving*. *A posteriori* display of a scattered-data interpolant to the actual solution along the lines of Section 7 is always possible, of course, but it is a problem of its own and induces additional errors.

Another efficiency argument arises when the dimension of the trial space is large. This should be avoided following Guideline 12.4, but it always occurs if the trial space is using a space discretization with fine granularity. Even if there is not too much connectivity between geometric information, *i.e.*, if the method is meshless, we need to have a fast method for range queries retrieving neighbours of nodes. Similar problems always come up when trial spaces need some *localization*. There are various ways to cope with it, *e.g.*, wavelets, multipole, partition-of-unity, and multilevel methods, but they all seem to be closely connected to the choice of a useful basis, either *a priori* or adaptively. This brings us to the next issue: the adaptivity properties of trial spaces.

The really serious situations for the choice of the trial space occur when singularities will arise, but at places not known beforehand. This is the case for certain fluid dynamics, advection-diffusion, or crack propagation problems. However, it does not make sense to use a fine global space discretization when there will be just a local effect that calls for a finer local resolution. This is observed by plenty of *adaptive* methods. They sometimes just re-mesh a global space discretization locally where necessary, or they add new and more flexible elements into the fixed basic triangulation, but both of these tasks are not easy. Particle- or kernel-based methods using clouds of scattered points can adapt by adding or deleting points where necessary, but they usually do not need to update geometric contingency information that arises with meshes or triangulations. This is the punchline when *meshless methods* are characterized (Belytschko *et al.* 1996*b*) as *constructing the approximation entirely in terms of nodes*. The cited article considers meshless approximations based on

- moving least squares
- kernels
- partitions of unity

and states that these *three methods are in most cases identical except for the important fact that partitions of unity enable p-adaptivity to be achieved.* Furthermore, kernels occur in all three, and this is another reason why kernels are a central tool in meshless methods. Some authors even talk of *truly meshless methods* when they want to stress that they do not need numerical integration, but we suggest stating precisely to which extent spatial discretizations need to be maintained, and whether the trial functions can be accessed explicitly or implicitly.

12.3. Kernel-based trial spaces

At this point, we should show how 'representability in terms of nodes' is understood in meshless methods and how it is related to kernel-based trial functions, establishing a very close connection of nearly all meshless methods to kernels. The idea of 'nodes' is roughly the same as the 'centres' for standard kernel approximations as in Section 7. In the simplest case, the trial space should be spanned by multivariate functions $\varphi_i(x, x_i)$, $i \in I$, which are functions of $x \in \mathbb{R}^d$ depending on a single 'node' or 'centre' or 'particle position' $x_i \in \mathbb{R}^d$. This function can be seen as a 'smoothed particle' as in *smoothed particle hydrodynamics*, SPH (Monaghan 2005), and it is called *shape function* or *particle function* in the literature. For a meshless method, there should be no complicated geometric connection between nodes like a triangulation of the convex hull of the nodes with the nodes as vertices (this could then be called a 'mesh'). It should be easy to extend the trial space by adding some new nodes and associated trial functions (this is called '*h*-adaptivity' in FEM terms) without updating the connectivity information. In this sense, meshless methods can be seen as an alternative to adaptive finite element methods.

For many reasons, the functions $\varphi_i(x, x_i)$ in meshless methods should be

- translation-invariant and
- compactly supported around the node x_i.

This implies that they should necessarily have the form

$$\varphi_i(x, x_i) := K(x - x_i), \quad i \in I \qquad (12.1)$$

with a compactly supported translation-invariant kernel K of small support, provided that they depend on no other neighbouring node.

Guideline 12.8. Translation-invariant trial functions for meshless methods are always kernel-based, if they are dependent on a single node.

This implies that trial spaces spanned by functions of the form (3.4) occur canonically in meshless methods, and the previous sections have accumulated much information on those spaces.

But the literature on meshless methods also uses 'shape functions' defined *implicitly* via local processes such as moving least squares. Then the resulting trial functions depend on more than one node, though this is often ignored in the notation. In fact, for each node x_i there is a trial function φ_i depending on x_i and some of its neighbours, if they fall within the support of the weight function associated to the node x_i. Kernels occur here only via the weight functions used, and they need not be positive definite. For scattered nodes, the resulting trial functions will not be translation-invariant.

12.4. Reproduction of polynomials

For MLS-based shape functions, we know from Section 7 that polynomial reproduction

$$\sum_{i \in I} p(x_i)\varphi_i(x) = p(x) \quad \text{for all } x \in \mathbb{R}^d, \ p \in \pi_m(\mathbb{R}^d) \qquad (12.2)$$

can be achieved under mild additional assumptions, where $\pi_m(\mathbb{R}^d)$ stands for the space of d-variate polynomials of degree at most m. Note that the partition of unity property (7.17) coincides with polynomial reproduction of degree zero.

In application papers, polynomial reproduction properties are often called *consistency conditions* (Belytschko *et al.* 1996*b*), and very many papers seem to understand *reproducing kernels* via the above reproduction property, not via (2.9) in Hilbert spaces. Some also seem to assume that convergence follows as soon as there is a consistency condition of some nonnegative order in the above sense, but this argument has no solid foundation, since the usual Lax-type theory understands consistency differently and is modelled for discretizations of time-dependent problems. Mathematicians will find plenty of open questions concerning convergence and error bounds of meshless methods, while many engineers seem to believe themselves to be on solid ground once they have what they call consistency.

Sometimes the notion of *completeness* is used in the sense of *convergent approximate polynomial reproduction*, i.e., (12.2) holding for $h \to 0$. In particular, *linear completeness* often means convergent approximate reproduction of linear functions (Belytschko *et al.* 1996*b*). This is different from the usual notion of completeness in mathematics, and it must be used with extreme care, in particular when assuming that it implies convergent approximate reproduction of *piecewise* linear functions.

Guideline 12.9. Within meshless methods, the notions of *consistency* and *completeness* should be used with caution.

Anyway, the polynomial reproduction property (12.2) appears in many meshless methods. In fact, recent surveys (Li and Liu 2002, 2004, Fries and Matthies 2004) of meshless methods focus entirely on methods with exact polynomial reproduction. However, it must be stated clearly that exact polynomial reproduction is not necessary for convergence, as is shown, for example, by the rigorous convergence analysis of the generalized finite element method (Babuška *et al.* 2003), and the symmetric (Franke and Schaback 1998*b*) and unsymmetric (Schaback 2005*b*) meshless collocation methods. Polynomial reproduction appears to be popular because it is necessary in convergence arguments for *stationary* scales of trial functions, using Strang–Fix conditions or the Bramble–Hilbert lemma. But it is not mandatory to use these tools. By Theorem 7.8, optimal approximation orders in Sobolev spaces are attained without it in very general situations, not only for interpolation from nonstationary scales of kernel-based trial spaces.

In view of these remarks, future work should remove exact polynomial reproduction from the assumptions of many meshless methods. Instead, care must be taken to conserve physical properties like mass and momentum in applications. This is only loosely related to polynomial reproduction.

12.5. Particle methods

After this detour into polynomial reproduction we still have to look at a class of methods that arrives at meshless trial spaces via a slightly different approach. *Smoothed particle hydrodynamics* (SPH) use spatial kernel approximations that we called *discretized kernel convolutions* in Section 7. This means that a suitably scaled and normalized kernel K is chosen such that (7.1) holds, and a discretization of the convolution integral implies

$$(K * f)(x) \approx \sum_{i \in I} w_i K(x, x_i) f(x_i) \quad \text{for all } x \in \mathbb{R}^d$$

with integration weights w_i at integration nodes x_i. The linear unknowns here are $f(x_i)$, while the points x_i are interpreted as *particle positions* and can be considered as nonlinear parameters whose number and value can change. The name of the technique is derived from the fact that the right-hand side writes a function or vector field as a sum over the local kernel-controlled influences of discrete particles at the points x_i. Thus the logic of SPH does not directly aim at trial spaces, but rather parametrizes fields describing flows in the form (3.4) we had in the beginning, by using the right-hand side of the above approximation. All other operations, *e.g.*, setting up momentum equations, are performed using the parametrized flow. Since the background problems are time-dependent, the above spatial discretizations lead to large systems of ordinary differential equations, where time discretization is another issue we do not address here.

To achieve a good approximation in the continuous convolution error (7.1), Theorem 7.1 tells us that the kernel should reproduce low-order polynomials well, but not necessarily exactly. If the integration scheme is exact for low-order polynomials, and if the kernel convolution reproduces low-order polynomials exactly, this implies the partition-of-unity property for the trial functions $w_i K(\cdot, x_i)$, but there will be no exact reproduction of higher-order polynomials. This problem can be removed by dropping the philosophy of discretizing a convolution integral, and going radically over to functions (12.1) with exact or approximate polynomial reproduction. This is called the *reproducing kernel particle method* (RKPM), when the rest of the SPH is maintained, *i.e.*, when discretized systems are derived from parametrized kernel-based field representations some way or other. We refer the reader to a recent survey article (Li and Liu 2002) and a book (Li and Liu 2004) on SPH and RKPM techniques, containing long lists of references, and describing many variations induced by additional physical constraints. But remember that meshless methods based on stationary moving least squares, reproducing kernels, or partitions of unity are *in most cases identical* (Belytschko *et al.* 1996b), so that all variants have to be looked at carefully.

12.6. Residuals, test functionals and functions

After considering the trial side, we should now focus on the test side. If we assume that the trial side has somehow produced some trial function which is a candidate for an approximate solution of the partial differential equation and the boundary conditions, we want to conclude that this trial function is close to the real solution. This is the job of the *test* side. In contrast to Guideline 12.4, we have the following rule, since the test side has to consider *security*.

Guideline 12.10. Space discretization is much more important on the test side than on the trial side.

But we postpone discretization on the test side for a while, noting that the above guideline calls for unsymmetric methods we deal with later.

If we rewrite the differential equation and the boundary conditions as differences $L(u) - f = 0$ which should be zero for the exact solution u, an approximate solution $\tilde{u}$ should make the *residuals* $L(\tilde{u}) - f$ small everywhere. Usually, to conclude that the error $u - \tilde{u}$ is small, it suffices to make sure that the residuals are small, because the solution of any well-posed linear problem will be continuously dependent on the data, implying

$$\tilde{u} - u \text{ small, if all residuals } L(\tilde{u}) - L(u) = L(\tilde{u}) - f \text{ are small,}$$

where 'all' means residuals of differential equation(s) and boundary condition(s) altogether, as many as are present in the problem. This means that 'testing' should usually make sure that the residuals are zero or at least small *globally*. Numerical techniques aiming at globally small residuals are often called *methods of weighted residuals*.

Guideline 12.11.   Globally small residuals imply small errors for well-posed linear problems, *i.e.*, if the solution is continuously dependent on the data. But one must make sure that the notions of 'well-posedness' and 'globally small' are consistently defined.

In fact, if we pack differential equations and boundary conditions into one single linear operator $L : U \to F$, continuous dependence requires fixing spaces U and F for the solution u and the data f of the problem $L(u) = f$ such that

$$\|u\|_U \leq C\|L(u)\|_F \tag{12.3}$$

holds, *i.e.*, L has a continuous inverse taking the data into a solution having these data. Then one must ensure that 'globally small' residuals for an approximate solution $\tilde{u}$ implies that the corresponding non-discrete norm $\|L(u) - L(\tilde{u})\|_F$ is also small, and *vice versa*. Thus, even when discretization of residuals is not an issue, the choice of a *residual norm* is important.

This is closely connected to the distinction between strong and weak problems. For strong problems, the residual norm is usually something like the $\|\cdot\|_\infty$ norm, while weak problems will use 'weaker' norms such as $\|\cdot\|_2$. But in most cases small residuals in the $\|\cdot\|_\infty$ norm will also be small in the $\|\cdot\|_2$ norm, so that, even if the $\|\cdot\|_2$ norm is the correct one for continuous dependence, users are safe if they minimize $\|\cdot\|_\infty$ instead, *i.e.*, solving a strong instead of a weak problem. This requires the trial space and the data f to have enough smoothness for $\|L(\tilde{u}) - f\|_\infty$ to be well defined, but this is usually not a big problem in many applications.

Guideline 12.12.   If trial functions and data are smooth enough, users can often use a strong formulation even if a corresponding weak formulation is known to be well posed.

12.7. Global residual minimization

There is a natural class of numerical methods related to weighted residuals, *i.e.*, methods that globally optimize residuals in the correct residual norm. These will always lead to an optimization problem instead of a linear system, reminding us of Guideline 3.18 and the complexity-reducing optimization problems in Section 11 on machine learning. In the case of L_2 residual minimization, this is the well-known *method of (continuous) least squares*, and there the optimization problem is quadratic, boiling down again to a linear system of equations. With weak problems it shares the disadvantage of requiring integration, while it has the additional disadvantage of working with higher-order derivatives than weak techniques. It also requires additional regularity in excess of L_2 to conclude that numerical integration of residuals has a controllable error.

For L_∞ residual minimization for problems in strong form, we get a semi-infinite linear programming problem. Application-oriented users should know that there are good numerical techniques for solving such problems. Furthermore, Kuhn–Tucker conditions will help to reduce complexity, as for learning algorithms via support vector machines, while adaptivity on the test side is built in automatically. Thus there is some hope that linear programming codes will be very helpful in the future when it comes to calculate low-complexity solutions of partial differential equations by adaptive methods.

For both cases of residual minimization, there is a trial function with small residuals, if the true solution u has a good approximation $\hat{u}_r$ from the trial space U_r. We avoid h here and prefer r, because trial spaces should not be automatically connected to space discretizations with fill distance h as in finite elements. The existence of $\hat{u}_r$ is a problem of approximation theory which is dependent on the solution u, the trial space U_r, and the norm $\|\cdot\|_U$ in the solution space U only, but not on any partial differential equation. Thus the user should keep Guidelines 3.9, 12.4, and 12.5 in mind without looking at the partial differential equation. Then the numerical method for solving a PDE problem, in weak or strong form, just has to make sure not to discard the existing unknown good approximation $\hat{u}_r$, while it produces another approximation $\tilde{u}_r \in U_r$ based on PDE data which is not too much worse. For residual minimization algorithms, this means that there exists an admissible trial function yielding small residuals $\|L(u) - L(\hat{u}_r)\|_F$, such that the final optimal solution cannot have worse residuals. Error bounds and convergence results will then follow the simple estimates

$$\|u - \tilde{u}_r\|_U \leq C\|L(u) - L(\tilde{u}_r)\|_F$$

$$= C \inf_{v \in U_r} \|L(u) - L(v)\|_F$$

$$\leq C\|L(u) - L(\hat{u}_r)\|_F,$$

which is a well-known line of argument, known in finite elements as Cea's lemma.

Guideline 12.13. Residual minimization works if the problem is well posed and if the trial space contains a good approximation to the solution. This allows plenty of freedom to design useful residual minimization algorithms.

12.8. Discrete residual minimization

Because all residual-based techniques have to evaluate norms on the test side, they have problems when dealing with global L_2, L_∞, or Sobolev norms there. Therefore we now look at *discretization* on the test side. It means that only finitely many 'tests' are performed. Discretization of a strong problem means taking a finite subset of points where the differential equation or boundary conditions are satisfied. This is the standard technique of *collocation*. For weak problems, discrete testing means taking inner products of the residuals with finitely many test functions, and then the residuals are not zero or small, but orthogonal to the *test space* spanned by *test functions*. In both cases we have to make sure that small results of discrete testing lead to small results in (theoretical) infinite testing.

Guideline 12.14. Coping with only finitely many conditions on the test side is the most serious part of any error or convergence analysis for numerical methods solving partial differential equations.

Such an analysis usually requires a *stability* condition relating the test and the trial space, and making sure that a small discrete residual on the trial space implies a small full residual on the trial space. We shall see examples later, but we can already state at this point that there should be no nonzero trial function $\hat{u}_r$ with vanishing test residuals, if we want to have error bounds, because all functions $\tilde{u}_r + \alpha \cdot \hat{u}_r$ for arbitrary $\alpha \in \mathbb{R}$ would have the same discrete test residuals and spoil the error bound.

Guideline 12.15. The discretized residual norm on the test side should at least work like a norm on the trial space.

This is the core of recent work (Schaback 2005*b*) on convergence analysis of unsymmetric methods on which we will now focus.

12.9. Symmetric and unsymmetric methods

Following Guidelines 12.10 and 12.13, the test side will need more attention than the trial side, and this leads us to the distinction between *symmetric* and *unsymmetric* methods. Symmetric methods use discretizations with

- the same degree of freedom on the trial and test side,
- closely related test functionals and trial functions,
- square and possibly positive definite matrices.

For weak problems, this means that trial and test *functions* coincide, and usually the standard Galerkin method is employed, yielding a positive definite square matrix. This applies to finite elements and several generalizations, *e.g.*, the GFEM (Babuška *et al.* 2003), described in detail in this series. The GFEM is a meshless method which enlarges the admissible trial spaces far beyond classical piecewise polynomial finite elements, but still uses the basic symmetric Hilbert space formulation of the finite element method. In its actual form, the GFEM uses stationary scales of trial spaces spanned by a partition of unity. Since it is a symmetric Galerkin technique, the trial functions and the test functions coincide. Compactly supported kernels occur naturally in the partition of unity, but they need not be positive definite. Since the current theory uses stationary approximations (see Section 7) in its scales of local trial spaces, the only kernels providing useful approximation orders are conditionally positive definite with infinite support, like multiquadrics or thin-plate splines. When local trial spaces are generated by moving least squares (see Section 7), weight kernels occur again. But most applications just augment finite element spaces by useful additional trial functions, *e.g.*, for treating singularities. However, the overall axiomatic structure of the GFEM theory (Babuška *et al.* 2003) suggests that it should be possible to extend the theory of the GFEM to allow nonstationary scales of kernel-based trial spaces with high-approximation orders.

For strong problems, the test side contains point evaluation *functionals* and there are no test *functions*. But there is also a symmetric method taking the trial functions as results when these functionals are applied to one argument of a positive definite kernel. This establishes a close relation

$$\lambda \leftrightarrow v_\lambda := \lambda^x K(x, \cdot)$$

between test functionals λ and trial functions v_λ which is only possible because kernels are involved. We call this *symmetric collocation* and deal with it in Section 14. It follows the lines of *general recovery* in Section 3, leading to symmetric positive definite systems of the form (3.8).

Both kinds of symmetric methods can be rewritten as an approximation or optimization problem in Hilbert space, and their theoretical foundation strongly relies on this fact. This comes close to Guideline 3.18, because the problem itself is a quadratic optimization problem solved via a linear system.

Let us now look at unsymmetric methods. In the strong case, collocation (Kansa 1986) using nonstationary scales of trial spaces of radial basis functions, in particular multiquadrics occurs in many applications we cannot list here. Theoretical support was only given recently (Schaback 2005b), proving high convergence rates depending on the regularity assumptions. We provide more details in Section 14.

Unsymmetric methods for weak problems usually take the form of Petrov–Galerkin schemes, where trial and test functions differ. Their basic theory (Douglas, Dupont, Rachford and Wheeler 1977) was established for trial spaces spanned by multivariate polynomial splines and for elliptic problems, making use of coercivity. More modern applications (Bialecki and Fairweather 2001, Bialecki, Ganesh and Mustapha 2004) have the same theoretical basis, but also do not apply kernel techniques.

A more radical approach to solving weak problems by an unsymmetric Petrov–Galerkin technique is the *meshless local Petrov–Galerkin* (MLPG) technique developed by S. N. Atluri and his collaborators (Atluri and Zhu 1998) with a short and recent survey (Atluri and Shen 2005) and two books (Atluri and Shen 2002, Atluri 2005) reporting many successful applications. It can use a variety of test and trial functions, and owing to its general form it can claim to include formally many other methods, *e.g.*, Kansa's unsymmetric collocation and various forms of symmetric methods, meshless or not.

However, there is currently no general convergence proof or error estimate available unless the method is restricted to well-known special cases. The main obstacle for its analysis is the fact that it uses a practically very valuable *local weak form* which, as opposed to weak forms arising in standard or generalized finite element methods, cannot be written as a necessary condition for a minimizing trial function in some Hilbert space of functions. But the MLPG can be viewed as an unsymmetric technique which tries to minimize residuals, and thus there are good chances to use Guideline 12.13 for underpinning it, extending techniques (Schaback 2005b) which currently only handle the special case of strong testing.

12.10. Numerical integration

Let us finally return to numerical integration questions, and let us look at weak problems first. The integrals for stiffness matrix entries within weak problems usually contain products of test and trial functions or derivatives thereof. To make integration easy and precise, test and trial functions have to be chosen carefully and should be closely related. The standard choice of piecewise polynomial trial and test functions in the finite element method achieves this, since the integrals can be done exactly in the case of polyhedral domains, though one has to keep track of the polyhedra carefully.

The integration of test functions against arbitrary functions is required for the inhomogeneities, but this is an issue of the test side, not of the trial side. Anyway, integrating piecewise polynomials on polyhedra needs some domain triangulation first (the primary *mesh*), and then a careful choice of interpolation nodes (or transformation to standard elements) for the integration (the *integration mesh*). Even 'meshless' methods, if they require integration, may sometimes need an integration mesh and are subject to influences of integration error, if they are applied to weak problems.

Using translates of radial kernels on both the trial and test side of weak problems can be equally efficient as finite elements are, if the integration domains do not interfere with boundaries, because the integrals are univariate radial functions which are either analytically known or can be pretabulated. Certain variations of the MLPG method could take advantage of this. Integration of 'test' kernels against given functions may be simplified by first representing the function in terms of translates of a 'trial' kernel, followed by integrations of kernels against kernels, which again is easy if no boundaries are in the way. In the presence of nontrivial boundaries, all trial and test functions cause problems, unless the real boundary is replaced piecewise by boundaries of supports of trial and test functions.

For problems in strong form, this discussion is not necessary. The trial functions can be chosen freely to satisfy the first three properties of Guideline 12.3. These properties are independent of PDE solving. We shall take a closer look at them, but from a more general point of view.

12.11. Classification of meshless methods

Summarizing, the universe of time-independent meshless methods can be roughly split into four parts by the dichotomies between strong/weak and symmetric/unsymmetric problems.

Strong problems imply collocation techniques as numerical methods, and then there are the symmetric and unsymmetric meshless collocation methods we describe in Section 14. They have in common the use of nonstationary scales of trial functions based on explicit kernels.

Weak problems in unsymmetric form are handled by Petrov–Galerkin techniques or the more general MLPG method. Everything else falls into the category of symmetric techniques solving weak problems. These come in a big variety and mostly differ on the trial side, while one of their common features is to rely on minimization in Hilbert space.

We have to leave out time-dependent meshless methods for space reasons, but we want to point out that there are strange gaps in the above scenario. First, for strong problems there are no investigations of methods using stationary scales. Second, for weak problems there are no investigations of methods using nonstationary scales, though this should be possible using the partition-of-unity framework behind the generalized finite element method. There is plenty of leeway for future research.

13. Special meshless kernel techniques

Following Guideline 6.1 and applying techniques of Section 6, kernel engineering can provide kernels which are closely connected to standard differential equations. This is used by certain numerical methods to be described in this section.

13.1. Dual reciprocity method

This misleading name stands for a technique coming from boundary element methods (Nardini and Brebbia 1982, Partridge, Brebbia and Wrobel 1992) and proliferating by use of kernel techniques (Chen, Golberg and Schaback 2003a). The basic idea is to split the problem into an inhomogeneous and a homogeneous subproblem with respect to the differential equation. A problem $L(u) = f$ with a linear differential operator L and linear boundary conditions $B(u) = g$ is treated first by constructing a *particular* solution u_P with $L(u_P) = f$ without regard of boundary values. Then the *homogeneous* problem $L(u) = 0$ is solved by some function u_H under the boundary conditions $B(u) = g - B(u_P)$ to get the final solution as $u := u_P + u_H$.

The first problem uses trial spaces of known particular solutions. These are easy to construct for kernel-based trial functions. The second problem makes use of *a priori* information on homogeneous solutions either via *integral equations* or *fundamental solutions*, providing trial spaces of homogeneous solutions via a special kernel called *the* fundamental solution of the differential operator L. Because of this close connection to kernels, we have to treat this technique in some detail.

Guideline 13.1. The dual reciprocity method can be applied to well-posed linear problems with well-known fundamental and particular solutions which have good approximation properties.

13.2. Method of particular solutions

To find a *particular* solution u_P with $L(u_P) = f$ without regard to boundary values, one can use trial functions u_i whose images $f_i := L(u_i)$ under L are well known and numerically available. Then the right-hand side f of the differential equation is approximated by a linear combination

$$\tilde{f} := \sum_i \alpha_i f_i$$

of the f_i to some small error $\|f - \tilde{f}\|_F$ in some suitable function space F, and the approximation

$$\tilde{u}_P := \sum_i \alpha_i u_i$$

is the canonical approximation to a particular solution u_P. Note that this part of the algorithm is an approximation problem which is completely independent of partial differential equations. After construction of $\tilde{f}$ we know that the *residual* $L(u_P - \tilde{u}_P) = f - \tilde{f}$ is small, but we have to postpone a thorough error analysis based on residual minimization and Guideline 12.13 until we have looked at the homogeneous problem and boundary conditions.

Of course, there are plenty of ways to produce good approximations $\tilde{f}$ to f, provided that the approximation properties of the functions f_i are well known. But it is a problem to find functions f_i which are particular solutions *and* have good approximation properties. Starting from well-approximating multivariate functions f_i such as finite elements, it is often hard or impossible to find the functions u_i with $L(u_i) = f_i$. On the other hand, starting with nice functions u_i will only rarely lead to functions $f_i = L(u_i)$ with good approximation properties.

But things can be easy if kernels are used. The simplest way is to take a smooth symmetric translation-invariant positive definite kernel K and define

$$u_i := K(\cdot - x_i) \quad \text{and} \quad f_i := LK(\cdot - x_i)$$

for trial centres x_i. If the operator L is elliptic with constant coefficients, the resulting kernel LK for the f_i will be positive definite again, as inspection of Fourier transforms shows. Now all techniques of Section 7 can be applied to reconstruct f approximately using the trial functions f_i.

If the operator is not elliptic, the kernel LK will not be positive definite. In such cases, the reverse strategy can be helpful, starting with $f_i := K(\cdot - x_i)$ using a positive definite kernel K and finding another kernel K_L such that $L(K_L) = K$. This new kernel need not be positive definite, but since it is not used for approximation, there is no problem here.

Guideline 13.2. A natural kernel-based strategy for the method of particular solutions is to have pairs u_i, f_i with $f_i = L(u_i) = K(\cdot - x_i)$ such that one can perform approximation of f by the standard translates of the kernel K.

The literature contains many such pairs, and we can only cite a selection: Chen and Rashed (1998), Chen, Muleshkov and Golberg (1999b), Cheng (2000), Ramachandran and Balakrishnan (2000) and Golberg, Muleshkov, Chen and Cheng (2003).

13.3. Method of fundamental solutions

Once the problem $L(u) = f$ with boundary data $B(u) = g$ is transformed into homogeneous form $L(u) = 0$, $B(u) = g - B(\tilde{u}_P) =: g_H$ by the method of *particular solutions*, the method of *fundamental solutions* (Mathon and Johnston 1977, Fairweather and Karageorghis 1998) takes over. It uses a special kernel F called *the* fundamental solution of $L(u) = 0$ such that $LF(\cdot, x) = \delta_x$ in the distributional sense. These kernels are well known for a number of linear operators, and we presented those for the iterated Laplacian in Section 6, *i.e.*, the *thin-plate spline* of (6.1) and the *polyharmonic splines* of (6.2). This can be generalized to linear elliptic differential operators with constant coefficients, but we do not want to go into details and refer the reader to the literature on Fourier methods in partial differential equations (Hörmander 2003) and on special fundamental solutions (Kythe 1996, Golberg and Chen 1999, Chen, Marcozzi and Choi 1999a, Balakrishnan and Ramachandran 2000, Alves, Chen and Šarler 2002, Poullikkas, Karageorghis and Georgiou 2002, Hon and Wei 2004), where we again picked out just a few cases from different application areas.

However, as we pointed out at the end of Section 6, the kernel F providing the fundamental solution will have a singularity 'on the diagonal', *i.e.*, for $F(x, x)$ or derivatives thereof. For second-order equations in dimension 2 or more, F itself is already singular, while for higher order we get singularities in the derivatives of F. Singular kernels are not directly covered by the standard theory of positive definite kernels, but they work fine in the generalized sense of (6.5), avoiding point evaluation functionals.

Once a fundamental solution F is at hand, there are various ways to generate trial functions solving the homogeneous differential equation. Before we describe these techniques, we want to look at the error and convergence analysis. The trial functions are used for approximating the prescribed boundary values g_H on the boundary. If a numerical scheme comes up with a trial function $\tilde{u}_H$ satisfying $L(\tilde{u}_H) = 0$ and with a small residual $B(\tilde{u}_H) - g_H = B(\tilde{u}_H) - B(u) + B(\tilde{u}_P)$, we use $\tilde{u} := \tilde{u}_H + \tilde{u}_P$ for our full solution and residuals

$$\|L(\tilde{u}) - f\|_F = \|L(\tilde{u}_H + \tilde{u}_P) - f\|_F$$
$$= \|L(\tilde{u}_P) - f\|_F$$
$$= \|\tilde{f} - f\|_F,$$
$$\|B(\tilde{u}) - g\|_G = \|B(\tilde{u}_H + \tilde{u}_P) - g\|_G$$
$$= \|B(\tilde{u}_H) - g_H\|_G,$$

for a suitable norm on a space G where the boundary values live. If the problem is continuously dependent on the data in the sense that an *a priori* inequality

$$\|u\|_U \leq C(\|L(u)\|_F + \|B(u)\|_G) \tag{13.1}$$

holds, and if the exact solution u exists and lies in U, then there is an error bound

$$\begin{aligned}
\|\tilde{u} - u\|_U &\leq C(\|L(\tilde{u} - u)\|_F + \|B(\tilde{u} - u)\|_G) \\
&= C(\|L(\tilde{u}) - f\|_F + \|B(\tilde{u}) - g\|_G) \\
&= C(\|\tilde{f} - f\|_F + \|B(\tilde{u}_H) - g_H\|_G),
\end{aligned}$$

reducing the overall error to the error of the residuals.

Guideline 13.3. The dual reciprocity method has a solid mathematical foundation once continuous dependence holds and the residuals are small in the correct norms.

This confirms Guidelines 13.2 and 12.13. For elliptic operators satisfying a maximum principle, these error bounds can be improved, provided that the spaces F and G are chosen appropriately.

However, it remains to prove that certain approximation schemes in the methods of particular and fundamental solutions lead to small residuals in the correct spaces needed for continuous dependence. If methods of Section 7 based on positive definite kernels are applied within the method of particular solutions, there are no serious problems, because there are good error estimates like (7.12) in Sobolev spaces on bounded domains. We are thus left with the analysis of the approximation power of the method of fundamental solutions.

A particularly simple way to generate trial functions satisfying the homogeneous problem $L(u) = 0$ is to proceed as in Section 7 by taking linear combinations of translates $F(\cdot, x_i)$ of the fundamental solution. This is an approximation problem

$$g_H(t) \approx \sum_j \alpha_j F(t, x_j), \quad t \in \Gamma \tag{13.2}$$

to be posed on the boundary Γ of the domain. But in order to avoid singularities of trial functions inside the domain or on the boundary, the trial centres x_i should be placed outside the domain. But then the theory of Section 7 does not apply, because the approximation domain does not contain the centres and there is no notion like a fill distance as in (7.6) making sense. However, the references cited above support that this method performs very well in practice if the outside centres are placed with care. For very special domains and smooth boundary data the method can be proven to have spectral convergence (Li 2005), but a general theory is still missing.

A well-known and much older approach is to place infinitely many trial centres right on the boundary and to take a weighted sum over all such translates of the fundamental solution. This leads to the singular *single-layer potential* integral equation

$$g_H(t) = \int_\Gamma \alpha(x)F(t,x)\,\mathrm{d}x, \quad t \in \Gamma.$$

Note that this is a non-discrete form of (13.2). Owing to the singularities of F, this equation cannot be solved strongly, but it can be solved weakly, *e.g.*, via finite elements on the boundary. Such techniques are called *boundary element methods* and have a rich literature including various books.

Variations of this approach are possible by replacing $F(\cdot, x)$ by certain linear functionals acting on $F(\cdot, x)$ with respect to the second argument x. These new kernels, like the normal derivative $\frac{\partial F(\cdot, x)}{\partial n}$ will usually preserve the property that action of L on the first argument results in zero. The standard case is the integral equation of the *double-layer potential*, but there are plenty of other possibilities that are yet unexploited, *e.g.*, replacing $F(t, x)$ by local integrals around x of $F(t, s)$ with respect to s in order to remove the singularities. A special case of this is the recent *boundary knot method* (Chen and Tanaka 2002, Chen 2002, Chen and Hon 2003).

13.4. Divergence-free kernels

In analogy to the methods of fundamental and particular solutions, there is a trick (Narcowich and Ward 1994a, Lowitzsch 2005) to generate kernel-based divergence-free trial spaces from smooth kernels, and curl-free trial spaces are also possible. The general idea behind this is to employ *matrix-valued* kernels, which allow us to incorporate these additional features into the rows and/or columns of the matrix. From these matrix-valued kernels, *vector-valued* interpolants can be built, which satisfy the additional constraints analytically. Applications of such divergence-free kernels to Stokes, Navier–Stokes, Euler and Maxwell equations are currently under investigation. We see this as a further case of *kernel engineering* in the direction of partial differential equations.

Unfortunately, we cannot describe more general applications of kernels to transport problems, advection, and fluid dynamics here, but this is a very promising research area (Mai-Duy and Tran-Cong 2001, Behrens and Iske 2002, Iske 2003, Barba, Leonard and Allen 2005, Shu, Ding and Yeo 2005).

14. Meshless collocation

Within the classification of meshless methods in Section 12, the techniques of this section solve partial differential equations in *strong* form, using *collocation* on the test side and avoiding *numerical integration* completely. On the trial side, they use *nonstationary* scales of *explicit* kernel-based trial functions. They come in a *symmetric* and an *unsymmetric* form.

In both cases, a given partial differential equation $L(u) = f$ and various boundary conditions of the form $B(u) = g$ are discretized by point evaluations of both sides in certain *collocation* nodes. For instance, a Poisson problem on a domain Ω with Dirichlet conditions $u = g_D$ on $\Gamma_D \subseteq \Gamma := \partial\Omega$ and Neumann conditions $\frac{\partial u}{\partial n} = g_N$ on $\Gamma_N \subset \Gamma$ can be discretized by a set $\Lambda := \{\lambda_1, \ldots, \lambda_N\}$ of *test functionals* consisting of three parts:

$$\Lambda = \Lambda_1 \cup \Lambda_2 \cup \Lambda_3$$
$$\Lambda_1 := \{\lambda_1, \ldots, \lambda_{N_1}\}$$
$$\lambda_j(u) := -\Delta u(x_j), \qquad x_j \in \overline{\Omega}, \ 1 \leq j \leq N_1,$$
$$\Lambda_2 := \{\lambda_{1+N_1}, \ldots, \lambda_{N_2}\}$$
$$\lambda_j(u) := u(x_j), \qquad x_j \in \Gamma_D, \ 1 + N_1 \leq j \leq N_2,$$
$$\Lambda_3 := \{\lambda_{1+N_2}, \ldots, \lambda_{N_3}\}$$
$$\lambda_j(u) := \frac{\partial u}{\partial n}(x_j), \qquad x_j \in \Gamma_N, \ 1 + N_2 \leq j \leq N_3 =: N.$$

If the evaluation points within the three sets Λ_1, Λ_2, Λ_3 of functionals are different, all linear test functionals in $\Lambda = \Lambda_1 \cup \Lambda_2 \cup \Lambda_3$ are linearly independent.

This specifies the *test* part of the problem for both the symmetric and unsymmetric methods. In general, there may be several differential operators and several boundary conditions in any kind of mixture, provided that everything is linear in u and the test functionals are linearly independent. There is no numerical integration, no test functions, and up to now there are no kernels. But for practical reasons, we mention the following guideline.

Guideline 14.1. To give certain test functionals special importance, one should apply constant factors.

For example, boundary test functionals in two-dimensional Poisson problems should get a factor of about 1000 over the differential equation test functionals. Exact rules for this are not known, but the background is provided by continuous dependence inequalities such as (13.1) where the parts of the right-hand side should carry different weights.

14.1. Symmetric meshless collocation

The difference between symmetric and unsymmetric meshless collocation shows up when looking at the trial side, provided that they use the same testing strategy. For *unsymmetric* collocation, a standard nonstationary scale of kernel-based trial spaces is used, where the translates $K(\cdot, y_k)$ are taken with *trial* nodes y_k that are independent of the *test functionals*. This method goes back to Kansa (1986) and will be analysed later.

In the symmetric case, there must be a strong connection between trial functions and test functionals. This is done by taking the trial functions $\lambda_j^x K(\cdot, x)$, $1 \leq j \leq N$ for a sufficiently smooth kernel K guaranteeing that all test functionals $\lambda_1, \ldots, \lambda_N$ lie in the dual of its native space. This is a special case of general Hermite–Birkhoff interpolation as described in Section 3. Under mild additional assumptions, this leads to a symmetric nonsingular linear system (3.8) and error bounds along the lines of Section 7. A detailed theoretical analysis of symmetric collocation can be found in the literature (Wu 1992, Iske 1995, Franke and Schaback 1998a, 1998b), while reports on applications are somewhat scattered (Power and Barraco 2002, Larsson and Fornberg 2003, Fasshauer 2004, Šarler 2005) and often limited to small problems with regular solutions. For such cases, the method gives quick and useful results, provided that the general guidelines on scaling in Section 3 are observed. Future work should apply special techniques of Section 8 for handling large-scale and ill-conditioned systems.

14.2. Unsymmetric meshless collocation

Unsymmetric meshless collocation is much more popular than the symmetric case, because it is easier to handle and shows similarly good experimental results (Cheng, Golberg, Kansa and Zammito 2003). The matrix entries $\lambda_i^x \lambda_j^y K(x, y)$ of the symmetric case apply all derivatives twice, while the unsymmetric case with trial functions $K(\cdot, y_k)$ involves only $\lambda_i^x K(x, y_k)$, which is simpler to program. There is a huge number of papers on practical applications of this technique which we cannot cite here, unfortunately, and which would require a survey of its own. Some application areas with recent sample papers are

- convection-diffusion problems (Li and Chen 2003, La Rocca, Hernandez Rosales and Power 2005),
- ill-posed problems (Cheng and Cabral 2005),
- thermal analysis (Pepper and Šarler 2005),
- fluid dynamics (Šarler 2005),
- flows in porous media (Šarler, Perko and Chen 2004),
- viscous vortex flows (Barba *et al.* 2005),
- boundary-layer problems (Ling and Trummer 2004),
- transport problems (Lorentz, Narcowich and Ward 2003),
- free boundary value problems (Kovačevič, Poredoš and Šarler 2003),
- fracture problems (Lee and Yoon 2004),
- nonlinear problems in smart materials (Liu, Liew, Hon and Zhang 2005),

but this list is far from complete. These papers, however, are recent enough to enable readers starting in these areas to find older results and the application-oriented background.

A thorough theoretical analysis was missing for about 20 years, because the unsymmetric systems can in general be singular (Hon and Schaback 2001). However, if the method is changed along some of the guidelines of this survey, error bounds and a convergence analysis can be supplied (Schaback 2005b). We summarize the relevant issues as follows.

Guideline 14.2. The mathematical foundation of unsymmetric collocation requires four ingredients:

(a) a linear and well-posed PDE problem,

(b) a nonstationary scale of meshless trial spaces with good approximation properties and spanned by sufficiently smooth kernel translates $K(\cdot, y_k)$,

(c) a scale of test discretizations via sets of collocation functionals λ_i which is fine enough to guarantee at least a full rank of the unsymmetric linear systems with entries $\lambda_i^x K(x, y_k)$,

(d) an approximate solution of this linear system with small discrete residuals.

Items (a)–(c) above are (in a more detailed and rigid form) sufficient to guarantee approximate solvability in the final step. It can be implemented by various techniques including linear or least-squares optimization or greedy adaptive methods (Hon *et al.* 2003, Ling and Schaback 2004) described below. Of course, guidelines of Section 7 concerning scaling must be observed at all times. If the sup norm of residuals is minimized, the method reduces to linear optimization, and it can be implemented via the revised simplex method. By the Kuhn–Tucker theory, the final result will then be based only on a small finite set of test functionals. This is a connection to *support vector machines*.

14.3. Adaptive collocation solvers

In finite elements, there is a vast recent literature on adaptivity controlled by efficient error estimation techniques. Meshless kernel-based collocation methods can implement this in a very simple way by inspecting residuals of the differential equation and the boundary conditions. Since evaluations of trial functions are explicitly possible and very cheap, one can always evaluate the residuals on a large set of background test points, using only a few of these to define the test functionals entering the calculations.

The general recipe is as follows.

1 Start with $\tilde{u}$ being the zero trial function and set $N := 0$.

2 Iteration:

Assume that there is a trial function $\tilde{u}$ which is a linear combination of N trial functions $u_1, \ldots, u_N$ such that the $N \times N$ system with entries $\lambda_j(u_k)$ for N test functionals $\lambda_1, \ldots, \lambda_N$ is non-singular and has $\tilde{u}$ as an approximate solution.

(a) Find a point in the domain or on the boundary where there is a large or maximal residual. Stop if none can be found.

(b) Use this point to define a new test functional λ_{N+1} for further calculations.

(c) Add a new trial function u_{N+1} such that the enlarged system still is nonsingular.

(d) Solve the new system approximately for a new trial function $\tilde{u}$.

If candidates for test functionals and trial functions are chosen from a large reservoir satisfying the background theory for unsymmetric calculations as described in Guideline 14.2, this is an adaptive bootstrapping technique that automatically selects useful subsets of trial functions and test functionals without ever forming a huge matrix defined by all possible trial functions and test functionals. Connections to the notions of *dictionaries* in approximation theory and to *greedy algorithms* are apparent.

This technique works fine for small problems (Hon *et al.* 2003, Ling and Schaback 2004) but needs further theoretical and numerical research if N gets large and the systems get ill-conditioned. In particular, step 2(c) of the algorithm can be implemented in various ways, and it is not clear how to assess the performance for small N. The symmetric case can also be handled adaptively by omitting step 2(c), taking the new test point and the corresponding functional to define a new trial function. There is a theoretical background for this symmetric *greedy* strategy in the interpolation case (Schaback and Wendland 2000a), but a thorough analysis for general collocation is an open problem, as is the incorporation of methods from Section 8 dealing with large ill-conditioned systems.

REFERENCES

R. Ahrem, A. Beckert and H. Wendland (2005), A new multivariate interpolation method for large-scale spatial coupling problems in aeroelasticity. DGLR-Bericht 2005-04; *Proc. IFASD*, Munich 2005.

C. J. S. Alves, C. S. Chen and B. Šarler (2002), The method of fundamental solutions for solving Poisson problems, in *Boundary Elements XXIV, Sintra 2002*, Vol. 13 of *Int. Ser. Adv. Bound. Elem.*, WIT Press, Southampton, pp. 67–76.

A. Appel (1985), 'An efficient program for many-body simulation', *SIAM J. Sci. Statist. Comput.* **8**, 85.

N. Aronszajn (1950), 'Theory of reproducing kernels', *Trans. Amer. Math. Soc.* **68**, 337–404.

S. N. Atluri (2005), *The Meshless Method (MLPG) for Domain and BIE Discretizations*, Tech Science Press, Encino, CA.

S. N. Atluri and S. Shen (2002), *The Meshless Local Petrov–Galerkin (MLPG) Method*, Tech Science Press, Encino, CA.

S. N. Atluri and S. Shen (2005), 'The basis of meshless domain discretization: the meshless local Petrov–Galerkin (MLPG) method', *Adv. Comput. Math.* **23**, 73–93.

S. N. Atluri and T. L. Zhu (1998), 'A new meshless local Petrov–Galerkin (MLPG) approach to nonlinear problems in computer modeling and simulation', *Computer Modeling and Simulation in Engineering* **3**, 187–196.

M. Atteia (1992), *Hilbertian Kernels and Spline Functions*, North-Holland, Amsterdam, Vol. 4 of *Studies in Computational Mathematics*.

I. Babuška and J. M. Melenk (1997), 'The partition of unity method', *Internat. J. Numer. Methods Engrg.* **40**, 727–758.

I. Babuška, U. Banerjee and J. E. Osborn (2002), Meshless and generalized finite element methods: A survey of some major results, in *Meshfree Methods for Partial Differential Equations*, Vol. 26 of *Lecture Notes in Computational Science and Engineering*, Springer, pp. 1–20.

I. Babuška, U. Banerjee and J. E. Osborn (2003), Survey of meshless and generalized finite element methods: A unified approach, in *Acta Numerica*, Vol. 12, Cambridge University Press, pp. 1–215.

K. Balakrishnan and P. A. Ramachandran (2000), 'The method of fundamental solutions for linear diffusion-reaction equations', *Math. Comput. Modelling* **31**, 221–237.

K. Ball (1992), 'Eigenvalues of Euclidean distance matrices', *J. Approx. Theory* **68**, 74–82.

K. Ball, N. Sivakumar and J. D. Ward (1992), 'On the sensitivity of radial basis interpolation to minimal data separation distance', *Constr. Approx.* **8**, 401–426.

L. A. Barba, A. Leonard and C. B. Allen (2005), 'Advances in viscous vortex methods: Meshless spatial adaption based on radial basis function interpolation', *Internat. J. Numer. Meth. Fluids* **47**, 387–421.

J. E. Barnes and P. Hut (1986), 'A hierarchical $\mathcal{O}(N \log N)$ force-calculation algorithm', *Nature* **324**, 446–449.

R. K. Beatson and E. Chacko (2000), Fast evaluation of radial basis functions: A multivariate momentary evaluation scheme, in *Curve and Surface Fitting: Saint-Malo 1999* (A. Cohen, C. Rabut and L. L. Schumaker, eds), Vanderbilt University Press, Nashville, pp. 37–46.

R. K. Beatson and L. Greengard (1997), A short course on fast multipole methods, in *Wavelets, Multilevel Methods and Elliptic PDEs; 7th EPSRC Numerical Analysis Summer School, University of Leicester, Leicester, GB, July 8–19, 1996* (M. Ainsworth, J. Levesley, W. Light and M. Marletta, eds), Clarendon Press, Oxford, pp. 1–37.

R. K. Beatson and W. A. Light (1993), 'Quasi-interpolation with thin plate splines on a square', *Constr. Approx.* **9**, 407–433.

R. K. Beatson and W. A. Light (1997), 'Fast evaluation of radial basis functions: Methods for two-dimensional polyharmonic splines', *IMA J. Numer. Anal.* **17**, 343–372.

R. K. Beatson and G. N. Newsam (1992), 'Fast evaluation of radial basis functions: I', *Comput. Math. Appl.* **24**, 7–19.

R. K. Beatson and G. N. Newsam (1998), 'Fast evaluation of radial basis functions: Moment-based methods', *SIAM J. Sci. Comput.* **19**, 1428–1449.

R. K. Beatson, J. B. Cherrie and C. T. Mouat (1999), 'Fast fitting of radial basis functions: Methods based on preconditioned GMRES iteration', *Adv. Comput. Math.* **11**, 253–270.

R. K. Beatson, J. B. Cherrie and D. L. Ragozin (2000a), Polyharmonic splines in $\mathbb{R}^d$: Tools for fast evaluation, in *Curve and Surface Fitting: Saint-Malo 1999* (A. Cohen, C. Rabut and L. L. Schumaker, eds), Vanderbilt University Press, Nashville, pp. 47–56.

R. K. Beatson, J. B. Cherrie and D. L. Ragozin (2001), 'Fast evaluation of radial basis functions: Methods for four-dimensional polyharmonic splines', *SIAM J. Math. Anal.* **32**, 1272–1310.

R. K. Beatson, G. Goodsell and M. J. D. Powell (1996), 'On multigrid techniques for thin plate spline interpolation in two dimensions', *Lect. Appl. Math.* **32**, 77–97.

R. K. Beatson, W. A. Light and S. Billings (2000b), 'Fast solution of the radial basis function interpolation equations: Domain decomposition methods', *SIAM J. Sci. Comput.* **22**, 1717–1740.

A. Beckert (2000), 'Coupling fluid (CFD) and structural (FE) models using finite interpolation elements', *Aerosp. Sci. Technol.* **1**, 13–22.

A. Beckert and H. Wendland (2001), 'Multivariate interpolation for fluid-structure-interaction problems using radial basis functions', *Aerosp. Sci. Technol.* **5**, 125–134.

J. Behrens and A. Iske (2002), 'Grid-free adaptive semi-Lagrangian advection using radial basis functions', *Comput. Math. Appl.* **43**, 319–327.

T. Belytschko, Y. Krongauz, M. Fleming, D. Organ and W. K. Liu (1996a), 'Smoothing and accelerated computations in the element free Galerkin method', *J. Comput. Appl. Math.* **74**, 111–126.

T. Belytschko, Y. Krongauz, D. Organ, M. Fleming and P. Krysl (1996b), 'Meshless methods: An overview and recent developments', *Comput. Methods Appl. Mech. Engrg.* **139**, 3–47.

B. Bialecki and G. Fairweather (2001), 'Orthogonal spline collocation methods for partial differential equations', *J. Comput. Appl. Math.* **128**, 55–82.

B. Bialecki, M. Ganesh and K. Mustapha (2004), 'A Petrov–Galerkin method with quadrature for elliptic boundary value problems', *IMA J. Numer. Anal.* **24**, 157–177.

P. Binev and K. Jetter (1992), Estimating the condition number for multivariate interpolation problems, in *Numerical Methods in Approximation Theory, Vol. 9:*

Proc. Conf. Oberwolfach, Germany, November 24–30, 1991 (D. Braess *et al.*, eds), Vol. 105 of *International Series of Numerical Mathematics*, Birkhäuser, Basel, pp. 41–52.

M. Bozzini, L. Lenarduzzi and R. Schaback (2002), 'Adaptive interpolation by scaled multiquadrics', *Adv. Comput. Math.* **16**, 375–387.

D. Brown, L. Ling, E. Kansa and J. Levesley (2005), 'On approximate cardinal preconditioning methods for solving PDEs with radial basis functions', *Engrg. Anal. Boundary Elements* **19**, 343–353.

M. D. Buhmann (1988), 'Convergence of univariate quasi-interpolation using multiquadrics', *IMA J. Numer. Anal.* **8**, 365–383.

M. D. Buhmann (1990), 'Multivariate cardinal interpolation with radial-basis functions', *Constr. Approx.* **6**, 225–255.

M. D. Buhmann (1993), 'On quasi-interpolation with radial basis functions', *J. Approx. Theory* **72**, 103–130.

M. D. Buhmann (1998), 'Radial functions on compact support', *Proc. Edinb. Math. Soc. II* **41**, 33–46.

M. D. Buhmann (2000), 'Radial basis functions', in *Acta Numerica*, Vol. 9, Cambridge University Press, pp. 1–38.

M. D. Buhmann (2004), *Radial Basis Functions*, Cambridge Monographs on Applied and Computational Mathematics, Cambridge University Press.

M. D. Buhmann, N. Dyn and D. Levin (1995), 'On quasi-interpolation by radial basis functions with scattered centers', *Constr. Approx.* **11**, 239–254.

J. C. Carr, R. K. Beatson, J. B. Cherrie, T. J. Mitchell, W. R. Fright, B. C. McCallum and T. R. Evans (2001), Reconstruction and representation of 3D objects with radial basis functions, in *SIGGRAPH '01: Proc. 28th Annual Conf. Computer Graphics and Interactive Techniques*, ACM Press, New York, NY, USA, pp. 67–76.

W. zu Castell and F. Filbir (2005), 'Radial basis functions and corresponding zonal series expansion on the sphere', *J. Approx. Theory* **134**, 65–79.

T. Cecil, J. Qian and S. Osher (2004), 'Numerical methods for high dimensional Hamilton–Jacobi equations using radial basis functions', *J. Comput. Phys.* **196**, 327–347.

C. S. Chen and Y. F. Rashed (1998), 'Evaluation of thin plate spline based particular solutions for Helmholtz-type operators for the DRM', *Mech. Res. Comm.* **25**, 195–201.

C. S. Chen, M. A. Golberg and R. Schaback (2003a), Recent developments in the dual reciprocity method using compactly supported radial basis functions, in *Transformation of Domain Effects to the Boundary* (Y. F. Rashed and C. A. Brebbia, eds), WIT Press, Southampton, Boston, pp. 138–225.

C. S. Chen, M. D. Marcozzi and S. Choi (1999a), The method of fundamental solutions and compactly supported radial basis functions: A meshless approach to 3D problems, in *Boundary Elements XXI: Oxford, 1999*, Vol. 6 of *Internat. Ser. Adv. Bound. Elem.*, WIT Press, Southampton, pp. 561–570.

C. S. Chen, A. S. Muleshkov and M. A. Golberg (1999b), The numerical evaluation of particular solution for Poisson's equation: A revisit, in *Boundary Elements XXI* (C. Brebbia and H. Power, eds), WIT Press, pp. 313–322.

D. Chen, V. A. Menegatto and X. Sun (2003b), 'A necessary and sufficient condition for strictly positive definite functions on spheres', *Proc. Amer. Math. Soc.* **131**, 2733–2740.

W. Chen (2002), 'Symmetric boundary knot method', *Engrg. Anal. Boundary Elements* **26**, 489–494.

W. Chen and Y. C. Hon (2003), 'Numerical convergence of boundary knot method in the analysis of Helmholtz, modified Helmholtz, and convection-diffusion problems', *Comput. Methods Appl. Mech. Engrg.* **192**, 1859–1875.

W. Chen and M. Tanaka (2002), 'A meshless, integration-free, and boundary-only RBF technique', *Comput. Math. Appl.* **43**, 379–391.

E. W. Cheney and W. A. Light (2000), *A Course in Approximation Theory*, Brooks/Cole Publishing Company, Pacific Grove.

E. W. Cheney, W. A. Light and Y. Xu (1992), 'On kernels and approximation orders', *Lect. Notes Pure Appl. Math.* **138**, 227–242.

A. H.-D. Cheng (2000), 'Particular solutions of Laplacian, Helmholtz-type, and polyharmonic operators involving higher order radial basis functions', *Engrg. Anal. Boundary Elements* **24**, 531–538.

A. H.-D. Cheng and J. J. S. P. Cabral (2005), Direct solution of certain ill-posed boundary value problems by collocation method, in *Boundary Elements XXVII* (A. Kassab, C. A. Brebbia, E. Divo and D. Poljak, eds), pp. 35–44.

A. H.-D. Cheng, M. A. Golberg, E. J. Kansa and G. Zammito (2003), 'Exponential convergence and h-c multiquadric collocation method for partial differential equations', *Numer. Methods Partial Differential Equations* **19**, 571–594.

J. Cheng, Y. C. Hon and M. Yamamoto (2001a), 'Conditional stability estimation for an inverse boundary problem with non-smooth boundary in $\mathbb{R}^3$', *Trans. Amer. Math. Soc.* **353**, 4123–4138.

J. Cheng, Y. C. Hon, T. Wei and M. Yamamoto (2001b), 'Numerical computation of a Cauchy problem for Laplace's equation', *ZAMM Z. Angew. Math. Mech.* **81**, 665–674.

P. G. Ciarlet (1991), Basic error estimates for elliptic problems, in *Handbook of Numerical Analysis*, Vol. II, North-Holland, Amsterdam, pp. 17–351.

D. D. Cox (1984), 'Multivariate smoothing spline functions', *SIAM J. Numer. Anal.* **21**, 789–813.

P. Craven and G. Wahba (1979), 'Smoothing noisy data with spline functions: estimating the correct degree of smoothing by the method of generalized cross-validation', *Numer. Math.* **31**, 377–403.

N. Cristianini and J. Shawe-Taylor (2000), *An Introduction to Support Vector Machines and Other Kernel-Based Learning Methods*, Cambridge University Press, Cambridge.

F. Cucker and S. Smale (2001), 'On the mathematical foundation of learning', *Bull. Amer. Math. Soc.* **39**, 1–49.

N. Dodgson, M. Floater and M. Sabin (2004), *Advances in Multiresolution for Geometric Modelling*, Springer, Berlin, Germany.

J. Douglas, Jr., T. Dupont, H. H. Rachford, Jr. and M. F. Wheeler (1977), 'Local H^{-1} Galerkin procedures for elliptic equations', *RAIRO Anal. Numér.* **11**, 3–12.

T. A. Driscoll and B. Fornberg (2002), 'Interpolation in the limit of increasingly flat radial basis functions', *Comput. Math. Appl.* **43**, 413–422.

M. R. Dubal (1994), 'Domain decomposition and local refinement for multiquadric approximations I: Second-order equations in one-dimension', *J. Appl. Sci. Comput.* **1**, 146–171.

J. Duchon (1976), 'Interpolation des fonctions de deux variables suivant le principe de la flexion des plaques minces', *Rev. Française Automat. Informat. Rech. Opér. Anal. Numer.* **10**, 5–12.

J. Duchon (1979), Splines minimizing rotation-invariate semi-norms in Sobolev spaces, in *Constructive Theory of Functions of Several Variables* (W. Schempp and K. Zeller, eds), Springer, Berlin/Heidelberg, pp. 85–100.

N. Dyn, F. J. Narcowich and J. D. Ward (1999), 'Variational principles and Sobolev-type estimates for generalized interpolation on a Riemannian manifold', *Constr. Approx.* **15**, 174–208.

T. Evgeniou, M. Pontil and T. Poggio (2000), 'Regularization networks and support vector machines', *Adv. Comput. Math.* **13**, 1–50.

G. Fairweather and A. Karageorghis (1998), 'The method of fundamental solution for elliptic boundary value problems', *Adv. Comput. Math.* **9**, 69–95.

C. Farhat and M. Lesoinne (1998), Higher-order staggered and subiteration free algorithms for coupled dynamic aeroelasticity problems, in *36th Aerospace Sciences Meeting and Exhibit, AIAA 98-0516, Reno/NV*.

R. Farwig (1986), 'Multivariate interpolation of arbitrarily spaced data by moving least squares methods', *J. Comput. Appl. Math.* **16**, 79–83.

R. Farwig (1987), Multivariate interpolation of scattered data by moving least squares methods, in *Algorithms for Approximation* (J. C. Mason and M. G. Cox, eds), Clarendon Press, Oxford, pp. 193–211.

R. Farwig (1991), Rate of convergence of moving least squares interpolation methods: The univariate case, in *Progress in Approximation Theory* (P. Nevai and A. Pinkus, eds), Academic Press, Boston, pp. 313–327.

C. Fasshauer (2004), RBF collocation methods and pseudospectral methods. Preprint, http://amadeus.csam.iit.edu/~fass/.

G. Fasshauer and L. L. Schumaker (1998), Scattered data fitting on the sphere, in *Mathematical Methods for Curves and Surfaces II* (M. Dæhlen, T. Lyche and L. L. Schumaker, eds), Vanderbilt University Press, Nashville, pp. 117–166.

A. C. Faul and M. J. D. Powell (1999), 'Proof of convergence of an iterative technique for thin plate spline interpolation in two dimensions', *Adv. Comput. Math.* **11**, 183–192.

S. Fleishman, D. Cohen-Or and C. T. Silva (2005), 'Robust moving least-squares fitting with sharp features', *ACM Trans. Graph.* **24**, 544–552.

M. S. Floater and A. Iske (1996), 'Multistep scattered data interpolation using compactly supported radial basis functions', *J. Comput. Appl. Math.* **73**, 65–78.

M. S. Floater and A. Iske (1998), 'Thinning algorithms for scattered data interpolation', *BIT* **38**, 705–720.

B. Fornberg and D. M. Sloan (1994), A review of pseudospectral methods for solving partial differential equations, in *Acta Numerica*, Vol. 3, Cambridge University Press, pp. 203–267.

C. Franke and R. Schaback (1998*a*), 'Convergence order estimates of meshless collocation methods using radial basis functions', *Adv. Comput. Math.* **4**, 381–399.

C. Franke and R. Schaback (1998*b*), 'Solving partial differential equations by collocation using radial basis functions', *Appl. Math. Comput.* **93**, 73–82.

W. Freeden, T. Gervens and M. Schreiner (1998), *Constructive Approximation on the Sphere*, Clarendon Press, Oxford.

W. Freeden, M. Schreiner and R. Franke (1997), 'A survey on spherical spline approximation', *Surv. Math. Ind.* **7**, 29–85.

T.-P. Fries and H.-G. Matthies (2004), Classification and overview of meshfree methods. Informatikbericht Nr. 2003-3, Scientific Computing, Universität Braunschweig: http://opus.tu-bs.de/opus/volltexte/2003/418/.

P. Giesl (2005), Construction of global Lyapunov functions using radial basis functions, Habilitationsschrift, Technische Universität München.

J. Glaunés, M. Vaillant and M. I. Miller (2004), 'Landmark matching via large deformation diffeomorphisms on the sphere', *J. Math. Imaging Vis.* **20**, 179–200.

M. A. Golberg and C. S. Chen (1999), The method of fundamental solutions for potential, Helmholtz and diffusion problems, in *Boundary Integral Methods: Numerical and Mathematical Aspects*, Vol. 1 of *Comput. Eng.*, WIT Press/Comput. Mech. Publ., Boston, MA, pp. 103–176.

M. A. Golberg, A. S. Muleshkov, C. S. Chen and A. H.-D. Cheng (2003), 'Polynomial particular solutions for certain kinds of partial differential operators', *Numer. Methods Partial Differential Equations* **19**, 112–133.

M. v. Golitschek and W. A. Light (2001), 'Interpolation by polynomials and radial basis functions on spheres', *Constr. Approx.* **17**, 1–18.

J. Gomes, L. Darsa, B. Costa and L. Velho (1998), *Warping and Morphing of Graphical Objects*, Morgan Kaufmann, San Francisco, CA, USA.

J. J. Green (2002), Approximation with the radial basis functions of Lewitt, in *Algorithms for Approximation IV* (J. Levesley, I. Anderson and J. C. Mason, eds), University of Huddersfield, pp. 212–219.

L. Greengard and V. Rokhlin (1987), 'A fast algorithm for particle simulations', *J. Comput. Phys.* **73**, 325–348.

L. Greengard and J. Strain (1991), 'The fast Gauss transform', *SIAM J. Sci. Statist. Comput.* **12**, 79–94.

T. Gutzmer (1996), 'Interpolation by positive definite functions on locally compact groups with application to SO(3)', *Result. Math.* **29**, 69–77.

S. J. Hales and J. Levesley (2002), 'Error estimates for multilevel approximation using polyharmonic splines', *Numer. Algorithms* **30**, 1–10.

R. L. Harder and R. N. Desmarais (1972), 'Interpolation using surface splines', *J. Aircraft* **9**, 189–197.

T. Hastie, R. Tibshirani and J. Friedman (2001), *The Elements of Statistical Learning*, Springer Series in Statistics, Springer, New York.

Y. C. Hon and R. Schaback (2001), 'On unsymmetric collocation by radial basis functions', *Appl. Math. Comput.* **119**, 177–186.

Y. C. Hon and T. Wei (2002), A meshless computational method for solving inverse heat conduction problem, in *Boundary Elements XXIV* (C. Brebbia, ed.), WIT Press, pp. 135–144.

Y. C. Hon and T. Wei (2003), A meshless scheme for solving inverse problems of Laplace equation, in *Recent Development in Theories and Numerics*, World Scientific, River Edge, NJ, pp. 291–300.

Y. C. Hon and T. Wei (2004), 'A fundamental solution method for inverse heat conduction problem', *Engrg. Anal. Boundary Elements* **28**, 489–495.

Y. C. Hon and T. Wei (2005), 'The method of fundamental solution for solving multidimensional inverse heat conduction problems', *CMES Comput. Model. Eng. Sci.* **7**(2), 119–132.

Y. C. Hon and Z. Wu (2000a), 'Additive Schwarz domain decomposition with a radial basis approximation', *Internat. J. Appl. Math.* **4**, 81–98.

Y. C. Hon and Z. Wu (2000b), 'A numerical computation for inverse boundary determination problems', *Engrg. Anal. Boundary Elements* **24**, 599–606.

Y. C. Hon, R. Schaback and X. Zhou (2003), 'An adaptive greedy algorithm for solving large RBF collocation problems', *Numer. Algorithms* **32**, 13–25.

H. Hoppe (1994), Surface reconstruction from unorganized points. PhD thesis, University of Washington.

H. Hoppe, T. DeRose, T. Duchamp, J. McDonald and W. Stuetzle (1992), 'Surface reconstruction from unorganized points', *Computer Graphics (Proc. SIG-GRAPH'92)* **26**, 71–78.

L. Hörmander (2003), *The Analysis of Linear Partial Differential Operators I: Distribution Theory and Fourier Analysis*, Classics in Mathematics, Springer.

S. Hubbert and T. M. Morton (2004), 'L_p-error estimates for radial basis function interpolation on the sphere', *J. Approx. Theory* **129**, 58–77.

M. S. Ingber, C. S. Chen and J. A. Tanski (2004), 'A mesh free approach using radial basis functions and parallel domain decomposition for solving three-dimensional diffusion equations', *Internat. J. Numer. Methods Engrg.* **60**, 2183–2201.

A. Iske (1995), Reconstruction of functions from generalized Hermite–Birkhoff data, in *Approximation Theory VIII*, Vol. 1 (C. Chui and L. Schumaker, eds), World Scientific, Singapore, pp. 257–264.

A. Iske (2003), 'Radial basis functions: basics, advanced topics and meshfree methods for transport problems', *Rend. Sem. Mat. Univ. Pol. Torino* **61**, 247–285.

A. Iske (2004), *Multiresolution Methods in Scattered Data Modelling*, Vol. 37 of *Lecture Notes in Computational Science and Engineering*, Springer, Berlin.

A. Iske and T. Sonar (1996), 'On the structure of function spaces in optimal recovery of point functionals for ENO-schemes by radial basis functions', *Numer. Math.* **74**, 177–201.

K. Jetter, J. Stöckler and J. Ward (1999), 'Error estimates for scattered data interpolation on spheres', *Math. Comput.* **68**, 733–747.

E. J. Kansa (1986), Application of Hardy's multiquadric interpolation to hydrodynamics, in *Proc. 1986 Simul. Conf.*, pp. 111–117.

I. Kovačevič, A. Poredoš and B. Šarler (2003), 'Solving the Stefan problem with the radial basis function collocation method', *Numerical Heat Transfer, Part B: Fundamentals* **44**, 575–599.

P. K. Kythe (1996), *Fundamental Solutions for Differential Operators and Applications*, Birkhäuser, Boston, MA.

A. La Rocca, A. Hernandez Rosales and H. Power (2005), 'Radial basis function Hermite collocation approach for the solution of time dependent convection-diffusion problems', *Engrg. Anal. Boundary Elements* **29**, 359–370.

P. Lancaster and K. Salkauskas (1981), 'Surfaces generated by moving least squares methods', *Math. Comput.* **37**, 141–158.

F. Lanzara, V. Maz'ya and G. Schmidt (2005), Approximate approximations from scattered data. WIAS Preprint No. 1058.

E. Larsson and B. Fornberg (2003), 'A numerical study of some radial basis function based solution methods for elliptic PDEs', *Comput. Math. Appl.* **46**, 891–902.

E. Larsson and B. Fornberg (2005), 'Theoretical and computational aspects of multivariate interpolation with increasingly flat radial basis functions', *Comput. Math. Appl.* **49**, 103–130.

S.-H. Lee and Y.-C. Yoon (2004), 'Meshfree point collocation method for elasticity and crack problems', *Internat. J. Numer. Methods Engrg.* **61**, 22–48.

J. Levesley and A. K. Kushpel (1999), 'Generalised sk-spline interpolation on compact abelian groups', *J. Approx. Theory* **97**, 311–333.

J. Levesley, Y. Xu, W. A. Light and W. E. Cheney (1996), 'Convolution operators for radial basis approximation', *SIAM J. Math. Anal.* **27**, 286–304.

D. Levin (1999), 'Stable integration rules with scattered integration points', *J. Comput. Appl. Math.* **112**, 181–187.

R. Lewitt, S. Matej and G. Herman (1997), Discretization and iterative solution of inverse problems in 3D computed tomography using bell-shaped radial basis functions having compact support. Technical report, Medical Image Processing Group, Dept. of Radiology, University of Pennsylvania, Philadelphia.

J. Li and C. S. Chen (2003), 'Some observations on unsymmetric radial basis function collocation methods for convection-diffusion problems', *Internat. J. Numer. Methods Engrg.* **57**, 1085–1094.

J. Li and Y. C. Hon (2004), 'Domain decomposition for radial basis meshless methods', *Numer. Methods Partial Differential Equations* **20**, 450–462.

S. Li and W. K. Liu (2002), 'Meshfree and particle methods and applications', *Appl. Mech. Rev.* **55**, 1–34.

S. Li and W. K. Liu (2004), *Meshfree Particle Methods*, Springer, Berlin.

X. Li (2005), 'On convergence of the method of fundamental solutions for solving the Dirichlet problem of Poisson's equation', *Adv. Comput. Math.* **23**, 265–277.

W. A. Light and H. Wayne (1998), 'On power functions and error estimates for radial basis function interpolation', *J. Approx. Theory* **92**, 245–266.

L. Ling (2005), 'Multidimensional quasi-interpolation formula with dimension-splitting multiquadric basis', *Appl. Math. Comput.* **161**, 195–209.

L. Ling and E. J. Kansa (2004), 'Preconditioning for radial basis functions with domain decomposition methods', *Math. Comput. Modelling* **40**, 1413–1427.

L. Ling and E. J. Kansa (2005), 'A least-squares preconditioner for radial basis functions collocation methods', *Adv. Comput. Math.* **23**, 31–54.

L. Ling and R. Schaback (2004), On adaptive unsymmetric meshless collocation, in *Proc. 2004 International Conference on Computational and Experimental*

Engineering and Sciences (S. N. Atluri and A. J. B. Tadeu, eds). Vol. CD-ROM, *Advances in Computational and Experimental Engineering and Sciences*, Tech Science Press, Forsyth, USA, paper # 270.

L. Ling and M. R. Trummer (2004), 'Multiquadric collocation method with integral formulation for boundary layer problems', *Comput. Math. Appl.* **48**, 927–941.

Y. Liu, K. M. Liew, Y. Hon and X. Zhang (2005), 'Numerical simulation and analysis of an electroactuated beam using a radial basis function', *Smart Materials and Structures* **14**, 1163–1171.

R. Lorentz, F. J. Narcowich and J. D. Ward (2003), 'Collocation discretization of the transport equation with radial basis functions', *Appl. Math. Comput.* **145**, 97–116.

S. Lowitzsch (2005), 'Matrix-valued radial basis functions: Stability estimates and applications.', *Adv. Comput. Math.* **23**, 299–315.

Z. Luo and J. Levesley (1998), 'Error estimates and convergence rates for variational Hermite interpolation', *J. Approx. Theory* **95**, 264–279.

D. H. McLain (1974), 'Drawing contours from arbitrary data points', *Comput. J.* **17**, 318–324.

D. H. McLain (1976), 'Two dimensional interpolation from random data', *Comput. J.* **19**, 178–181.

W. R. Madych and S. A. Nelson (1988), 'Multivariate interpolation and conditionally positive definite functions', *Approx. Theory Appl.* **4**, 77–89.

W. R. Madych and S. A. Nelson (1990), 'Multivariate interpolation and conditionally positive definite functions II', *Math. Comput.* **54**, 211–230.

W. R. Madych and S. A. Nelson (1992), 'Bounds on multivariate polynomials and exponential error estimates for multiquadric interpolation', *J. Approx. Theory* **70**, 94–114.

N. Mai-Duy and T. Tran-Cong (2001), 'Numerical solution of Navier–Stokes equations using radial basis function networks', *Internat. J. Numer. Meth. Fluids* **37**, 65–86.

S. de Marchi, R. Schaback and H. Wendland (2005), 'Near-optimal data-independent point locations for radial basis function interpolation', *Adv. Comput. Math.* **23**, 317–330.

R. Mathon and R. L. Johnston (1977), 'The approximate solution of elliptic boundary-value problems by fundamental solutions', *SIAM J. Numer. Anal.* **14**, 638–650.

V. Maz'ya (1994), Approximate approximations, in *The Mathematics of Finite Elements and Applications*, Wiley, Chichester, pp. 77–104.

V. Maz'ya and G. Schmidt (2001), 'On quasi-interpolation with non-uniformly distributed centers on domains and manifolds', *J. Approx. Theory* **110**, 125–145.

B. Mederos, L. Velho and L. H. de Figueiredo (2003), Moving least squares multiresolution surface approximation, in *XVI Brazilian Symposium on Computer Graphics and Image Processing (SIBGRAPI'03)*, p. 19.

J. Melenk and I. Babuška (1996), 'The partition of unity finite element method: Basic theory and applications', *Comput. Methods Appl. Mech. Engrg.* **139**, 289–314.

H. Meschkowski (1962), *Hilbertsche Räume mit Kernfunktion*, Springer, Berlin.

C. A. Micchelli (1986), 'Interpolation of scattered data: Distance matrices and conditionally positive definite functions', *Constr. Approx.* **2**, 11–22.

C. A. Micchelli and T. J. Rivlin (1977), A survey of optimal recovery, in *Optimal Estimation in Approximation Theory* (C. A. Micchelli and T. J. Rivlin, eds), Plenum Press, pp. 1–54.

C. A. Micchelli, T. J. Rivlin and S. Winograd (1976), 'Optimal recovery of smooth functions', *Numer. Math.* **26**, 191–200.

J. J. Monaghan (2005), 'Smoothed particle hydrodynamics', *Reports on Progress in Physics* **68**, 1703–1759.

T. M. Morton and M. Neamtu (2002), 'Error bounds for solving pseudodifferential equations on spheres by collocation with zonal kernels', *J. Approx. Theory* **114**, 242–268.

C. T. Mouat (2001), Fast algorithms and preconditioning techniques for fitting radial basis functions. PhD thesis, Mathematics Department, University of Canterbury, Christchurch, New Zealand.

C. Müller (1966), *Spherical Harmonics*, Springer, Berlin.

F. J. Narcowich (1995), 'Generalized Hermite interpolation and positive definite kernels on a Riemannian manifold', *J. Math. Anal. Appl.* **190**, 165–193.

F. J. Narcowich and J. D. Ward (1991), 'Norms of inverses and condition numbers for matrices associated with scattered data', *J. Approx. Theory* **64**, 69–94.

F. J. Narcowich and J. D. Ward (1992), 'Norm estimates for the inverse of a general class of scattered-data radial-function interpolation matrices', *J. Approx. Theory* **69**, 84–109.

F. J. Narcowich and J. D. Ward (1994a), 'Generalized Hermite interpolation via matrix-valued conditionally positive definite functions', *Math. Comput.* **63**, 661–687.

F. J. Narcowich and J. D. Ward (1994b), 'On condition numbers associated with radial-function interpolation', *J. Math. Anal. Appl.* **186**, 457–485.

F. J. Narcowich and J. D. Ward (2004), 'Scattered data interpolation on spheres: Error estimates and locally supported basis functions', *SIAM J. Math. Anal.* **33**, 1393–1410.

F. J. Narcowich, R. Schaback and J. D. Ward (1999), 'Multilevel interpolation and approximation', *Appl. Comput. Harmon. Anal.* **7**, 243–261.

F. J. Narcowich, X. Sun and J. D. Ward (2006), 'Approximation power of RBFs and their associated SBFs: A connection', to appear in *Adv. Comput. Math.*

F. J. Narcowich, X. Sun, J. D. Ward and H. Wendland (2005a), Direct and inverse Sobolev error estimates for scattered data interpolation via spherical basis functions. Preprint, College Station, TX.

F. J. Narcowich, J. D. Ward and H. Wendland (2004), Sobolev error estimates and a Bernstein inequality for scattered data interpolation via radial basis functions. Preprint, College Station, TX. To appear in *Constr. Approx.*

F. J. Narcowich, J. D. Ward and H. Wendland (2005b), 'Sobolev bounds on functions with scattered zeros, with applications to radial basis function surface fitting', *Math. Comput.* **74**, 643–763.

D. Nardini and C. A. Brebbia (1982), A new approach to free vibration analysis using boundary elements, in *Boundary Element Methods in Engineering* (C. A. Brebbia, ed.), Springer, New York, pp. 312–326.

J. Y. Noh, D. Fidaleo and U. Neumann (2000), Animated deformations with radial basis functions, in *VRST '00: Proc. ACM Symposium on Virtual Reality Software and Technology*, ACM Press, New York, USA, pp. 166–174.

Y. Ohtake, A. Belyaev, M. Alexa, G. Turk and H.-P. Seidel (2003a), 'Multi-level partition of unity implicits', *ACM Trans. Graphics* **22**, 463–470.

Y. Ohtake, A. Belyaev and H.-P. Seidel (2003b), A multi-scale approach to 3D scattered data interpolation with compactly supported basis functions, in *Shape Modeling International*, IEEE Computer Society Press, pp. 153–164.

R. Opfer (2006), 'Multiscale kernels', to appear in *Adv. Comput. Math.*

P. Partridge, C. Brebbia and L. Wrobel (1992), *The Dual Reciprocity Boundary Element Method*, CMP/Elsevier.

D. W. Pepper and B. Šarler (2005), 'Application of meshless methods in thermal analysis', *Mechanical Engineering J.* **51**, 476–483.

A. Pinkus (2004), 'Strictly Hermitian positive definite functions', *J. d'Analyse Math.* **94**, 293–318.

T. Poggio and S. Smale (2003), 'The mathematics of learning: dealing with data', *Notices Amer. Math. Soc.* **50**, 537–544.

A. Poullikkas, A. Karageorghis and G. Georgiou (2002), 'The method of fundamental solutions for three-dimensional elastostatics problems', *Comput. & Structures* **80**, 365–370.

H. Power and V. Barraco (2002), 'A comparison analysis between unsymmetric and symmetric radial basis function collocation methods for the numerical solution of partial differential equations', *Comput. Math. Appl.* **43**, 551–583.

C. Rabut (1989), Fast quasi-interpolation of surfaces with generalized B-splines on regular nets, in *Mathematics of Surfaces III* (D. C. Handscomb, ed.), Clarendon Press, Oxford, pp. 429–449.

D. L. Ragozin (1983), 'Error bounds for derivative estimates based on spline smoothing of exact or noisy data', *J. Approx. Theory* **37**, 335–355.

P. Ramachandran and K. Balakrishnan (2000), 'Radial basis functions as approximate particular solutions: Review of recent progress', *Engrg. Anal. Boundary Elements* **24**, 575–582.

C. H. Reinsch (1967), 'Smoothing by spline functions', *Numer. Math.* **10**, 177–183.

C. H. Reinsch (1971), 'Smoothing by spline functions II', *Numer. Math.* **16**, 451–454.

G. Roussos (1999), Computation with radial basis functions, PhD thesis, Imperial College of Science Technology and Medicine, University of London.

B. Šarler (2005), 'A radial basis function collocation approach in computational fluid dynamics', *Comput. Methods Engineering Sci.* **7**, 185–193.

B. Šarler, J. Perko and C. S. Chen (2004), 'Radial basis function collocation method solution of natural convection in porous media', *Internat. J. Numer. Methods Heat Fluid Flow* **14**, 187–212.

R. Schaback (1995a), Creating surfaces from scattered data using radial basis functions, in *Mathematical Methods for Curves and Surfaces* (T. L. M. Daehlen and L. Schumaker, eds), Vanderbilt University Press, Nashville, TN, pp. 477–496.

R. Schaback (1995b), 'Error estimates and condition number for radial basis function interpolation', *Adv. Comput. Math.* **3**, 251–264.

R. Schaback (1997), On the efficiency of interpolation by radial basis functions, in *Surface Fitting and Multiresolution Methods* (A. LeMéhauté, C. Rabut and L. Schumaker, eds), Vanderbilt University Press, Nashville, TN, pp. 309–318.

R. Schaback (1999), Native Hilbert spaces for radial basis functions I, in *New Developments in Approximation Theory* (M. D. Buhmann, D. H. Mache, M. Felten and M. W. Müller, eds), Vol. 132 of *International Series of Numerical Mathematics*, Birkhäuser, pp. 255–282.

R. Schaback (2005*a*), Convergence analysis of methods for solving general equations, in *Boundary Elements XXVII* (A. Kassab, C. Brebbia, E. Divo and D. Poljak, eds), WIT Press, Southampton, pp. 17–24.

R. Schaback (2005*b*), Convergence of unsymmetric kernel-based meshless collocation methods. Preprint, Universität Göttingen.

R. Schaback and H. Wendland (2000*a*), 'Adaptive greedy techniques for approximate solution of large RBF systems', *Numer. Algorithms* **24**(3), 239–254.

R. Schaback and H. Wendland (2000*b*), Numerical techniques based on radial basis functions, in *Curve and Surface Fitting: Saint-Malo 1999* (A. Cohen, C. Rabut and L. L. Schumaker, eds), Vanderbilt University Press, Nashville, TN, pp. 359–374.

R. Schaback and J. Werner (2006), 'Linearly constrained reconstruction of functions by kernels with applications to machine learning', to appear in *Adv. Comput. Math.*

I. J. Schoenberg (1937), 'On certain metric spaces arising from Euclidean spaces by a change of metric and their imbedding in Hilbert space', *Ann. of Math.* **38**, 787–793.

I. J. Schoenberg (1942), 'Positive definite functions on spheres', *Duke Math. J.* **9**, 96–108.

B. Schölkopf and A. J. Smola (2002), *Learning with Kernels*, MIT Press, Cambridge.

J. Shawe-Taylor and N. Cristianini (2004), *Kernel Methods for Pattern Analysis*, Cambridge University Press.

D. Shepard (1968), A two dimensional interpolation function for irregularly spaced data, in *Proc. ACM National Conference*, pp. 517–524.

C. Shu, H. Ding and K. S. Yeo (2005), 'Computation of incompressible Navier–Stokes equations by local RBF-based differential quadrature method', *Comput. Methods Engrg. Sci.* **7**, 195–206.

S. Smale and D.-X. Zhou (2003), 'Estimating the approximation error in learning theory', *Anal. Appl. (Singap.)* **1**, 17–41.

A. J. Smola and B. Schölkopf (1998), 'On a kernel-based method for pattern recognition, regression, approximation and operator inversion', *Algorithmica* **22**, 211–231.

T. Sonar (1996), 'Optimal recovery using thin plate splines in finite volume methods for the numerical solution of hyperbolic conservation laws', *IMA J. Numer. Anal.* **16**, 549–581.

J. Stewart (1976), 'Positive definite functions and generalizations: An historical survey', *Rocky Mountain J. Math.* **6**, 409–434.

J. F. Traub and A. G. Werschulz (1998), *Complexity and Information*, Oxford University Press, Oxford, UK.

G. Turk and J. F. O'Brien (2002), 'Modelling with implicit surfaces that interpolate', *ACM Trans. Graphics* **21**, 855 – 873.

G. Wahba (1975), 'Smoothing noisy data by spline functions', *Numer. Math.* **24**, 383–393.

G. Wahba (1990), *Spline Models for Observational Data*, CBMS-NSF, Regional Conference Series in Applied Mathematics, SIAM, Philadelphia.

T. Wei, Y. Hon and Y. B. Wang (2005), 'Reconstruction of numerical derivatives from scattered noisy data', *Inverse Problems* **21**, 657 – 672.

H. Wendland (1995), 'Piecewise polynomial, positive definite and compactly supported radial functions of minimal degree', *Adv. Comput. Math.* **4**, 389–396.

H. Wendland (2001), 'Local polynomial reproduction and moving least squares approximation', *IMA J. Numer. Anal.* **21**, 285–300.

H. Wendland (2004), Solving large generalized interpolation problems efficiently, in *Advances in Constructive Approximation* (M. Neamtu and E. B. Saff, eds), Nashboro Press, Brentwood, TN, pp. 509–518.

H. Wendland (2005*a*), 'On the convergence of a general class of finite volume methods', *SIAM J. Numer. Anal.* **43**, 987–1002.

H. Wendland (2005*b*), *Scattered Data Approximation*, Cambridge Monographs on Applied and Computational Mathematics, Cambridge University Press, Cambridge, UK.

H. Wendland and C. Rieger (2005), 'Approximate interpolation with applications to selecting smoothing parameters', *Numer. Math.* **101**, 643–662.

Z. Wu (1992), 'Hermite–Birkhoff interpolation of scattered data by radial basis functions', *Approx. Theory Appl.* **8**, 1–10.

Z. Wu (1995), 'Multivariate compactly supported positive definite radial functions', *Adv. Comput. Math.* **4**, 283–292.

Z. Wu and R. Schaback (1993), 'Local error estimates for radial basis function interpolation of scattered data', *IMA J. Numer. Anal.* **13**, 13–27.

D. X. Zhou (2002), 'The covering number in learning theory', *J. Complexity* **18**, 739–767.

X. Zhou, Y. C. Hon and J. Li (2003), 'Overlapping domain decomposition method by radial basis functions', *Appl. Numer. Math.* **44**, 241–255.